Miloslav Rechcigl, Jr.
Editor-in-Chief

SECTION OUTLINE

SECTION A: Nutrition and Food Science
Minimum of 1 volume projected.
Nomenclature, Nutrition Literature, Nutrition Societies, Foundations, and Historical Milestones in Nutrition.

SECTION B: The Living Organisms, Their Chemical Constitution, Feeding and Digestive System, and Ecological Aspects
Minimum of 3 volumes projected.
Taxonomy, Distribution of Organisms, Ecology – Symbiosis, Feeding and Digestive System, Chemical Constitution of Organisms, and Biological Productivity.

SECTION C: The Nutrients and Their Metabolism
Minimum of 1 volume projected.
Nutrients and Growth Regulators, Antinutrients and Antimetabolites, Naturally Occurring Food Toxicants, Regulatory Aspects of Nutrition, and Availability of Nutrients.

SECTION D: Nutritional Requirements
Minimum of 4 volumes projected.
Comparative Requirements of Organisms, Qualitative Requirements of Specific Organisms, Quantitative Requirements (Nutritional Standards), and Nutritional Requirements for Specific Processes and Functions: Animals, Microorganisms, Plants.

SECTION E: Nutritional Disorders
Minimum of 2 volumes projected.
Nutritional Disorders in Living Organisms, Effect of Specific Nutrient Deficiencies and Toxicities, and Nutritional Disorders in Specific Tissues.

SECTION F: Food Composition
Minimum of 1 volume projected.
Nutrient Content and Energy Value of Food, Factors Affecting Nutrient Composition of Food, and Utilization and Biological Value of Food.

SECTION G: Diets, Culture Media, and Food Supplements
Minimum of 4 volumes projected.
Diets, Culture Media, and Food and Feed Supplements.

SECTION H: The State of World Food and Nutrition
Minimum of 1 volume projected.
World Population, Natural and Food Resources, Food Production, Food Losses, World Food Usage and Consumption, Geographic Distribution of Nutritional Diseases, Nutritional Requirements – Current and Projected, Agricultural Inputs – Current and Projected, Food Aid, Food Marketing and Distribution, Socioeconomic, Cultural, and Psychological Factors Affecting Nutrition.

SECTION I: Food Safety, Food Processing, Food Preservation
Minimum of 1 volume projected.
Food Contamination, Food Spoilage, Food Wastes, Food Laws, and Nutrition Labeling.

SECTION J: Nutrition and Food Methodology
Minimum of 1 volume projected.
Assessment of Nutritional Status, and Measuring Nutritive Value of Food.

CRC Handbook Series in Nutrition and Food

Miloslav Rechcigl, Jr., Editor-in-Chief

Nutrition Advisor and Director
Interregional Research Staff
Agency for International Development
U.S. Department of State

Section G: Diets, Culture Media, Food Supplements

Volume I

Diets for Mammals

CRC PRESS, Inc.
18901 Cranwood Parkway · Cleveland, Ohio 44128

Library of Congress Cataloging in Publication Data

Main entry under title:

Diets, culture media, and food supplements.

(CRC handbook series in nutrition and food; section G)
Bibliography: p.
Includes index.
1. Animal nutrition – Collected works.
2. Plants – Nutrition – Collected works.
3. Animals, Food habits of – Collected works.
4. Culture media (Biology) – Collected works.
I. Rechcigl, Miloslav. II. Series.

QH519.D54 591'.13 77-6287
ISBN 0-8493-2700-8 (Set)
0-8493-2736-9 (Volume I)

International Standard Book Number 0-8493-2700-8 (Complete Set)
International Standard Book Number 0-8493-2736-9 (Volume I)

Library of Congress Card Number 77-6287
Printed in the United States

PUBLISHER'S PREFACE

In 1913, when the First Edition of the *Handbook of Chemistry and Physics* appeared, scientific progress, particularly in chemistry and physics, had produced an extensive literature but its utility was seriously handicapped because it was fragmented and unorganized. The simple but invaluable contribution of the *Handbook of Chemistry and Physics* was to provide a systematic compilation of the most useful and reliable scientific data within the covers of a single volume. Referred to as the "bible," the Handbook soon became a universal and essential reference source for the scientific community. The 57th Edition, published in the bicentennial year of 1976, represents more than 63 years of continuous service to millions of professional scientists and students throughout the world.

In the years following World War II, scientific information expanded at an explosive rate due to the tremendous growth of research facilities and sophisticated analytical instrumentation. The single-volume Handbook concept, although providing a high level of convenience, was not adequate for the reference requirements of many of the newer scientific disciplines. Due to the sheer quantity of useful and reliable data being generated, it was no longer feasible or desirable to select only that information which could be contained in a single volume and arbitrarily to reject the remainder. **Comprehensiveness** had become as essential as **convenience**.

By the late 1960's, it was apparent that the solution to the problem was the development of the multi-volume Handbook. This answer arose out of necessity during the editorial processing of the *Handbook of Environmental Control.* A hybrid discipline or, to be more precise, an interdisciplinary field such as Environmental Science could be logically structured into major subject areas. This permitted individual volumes to be developed for each major subject. The individual volumes, published either simultaneously or by some predetermined sequence, collectively became a multi-volume Handbook series.

The logic of this new approach was irrefutable and the concept was promptly accepted by both the scientist and science librarian. It became the format of a growing number of CRC Handbook Series in fields such as Materials Science, Laboratory Animal Science, and Marine Science.

Within a few years, however, it was clear that even the multi-volume Handbook concept was not sufficient. It was necessary to create an information structure more compatible with the dynamic character of scientific information, and flexible enough to accommodate continuous but unpredictable growth, regardless of quantity or direction. This became the objective of a "third generation" Handbook concept.

This latest concept utilizes each major subject within an information field as a "Section" rather than the equivalent of a single volume. Each Section, therefore, may include as many volumes as the quantity and quality of available information will justify. The structure achieves permanent flexibility because it can, in effect, expand "vertically" and "horizontally." Any section can continue to grow (vertically) in number of volumes, and new sections can be added (horizontally) as and when required by the information field itself. A key innovation which makes this massive and complex information base almost as convenient to use as a single-volume Handbook is the utilization of computer technology to produce up-dated, cumulative index volumes.

The *Handbook Series in Nutrition and Food* is a notable example of the "sectionalized, multi-volume Handbook series." Currently underway are additional information programs based on the same organizational design. These include information fields such as Energy and Agricultural Science which are of critical importance not only to scientific progress but to the advancement of the total quality of life.

We are confident that the "third generation" CRC Handbook comprises a worthy contribution to both information science and the scientific community. We are equally certain that it does not represent the ultimate reference source. We predict that the most dramatic progress in the management of scientific information remains to be achieved.

B. J. Starkoff
President
CRC Press, Inc.

PREFACE
CRC HANDBOOK SERIES IN NUTRITION AND FOOD

Nutrition means different things to different people, and no other field of endeavor crosses the boundaries of so many different disciplines and abounds with such diverse dimensions. The growth of the field of nutrition, particularly in the last two decades, has been phenomenal, the nutritional data being scattered literally in thousands and thousands of not always accessible periodicals and monographs, many of which, furthermore, are not normally identified with nutrition.

To remedy this situation, we have undertaken an ambitious and monumental task of assembling in one publication all the critical data relevant in the field of nutrition.

The *CRC Handbook Series in Nutrition and Food* is intended to serve as a ready reference source of current information on experimental and applied human, animal, microbial, and plant nutrition presented in concise tabular, graphical, or narrative form and indexed for ease of use. It is hoped that this projected open-ended multivolume set will become for the nutritionist what the *CRC Handbook of Chemistry and Physics* has become for the chemist and physicist.

Apart from supplying specific data, the comprehensive, interdisciplinary, and comparative nature of the *CRC Handbook Series in Nutrition and Food* will provide the user with an easy overview of the state of the art, pinpointing the gaps in nutritional knowledge and providing a basis for further research. In addition, the *Handbook* will enable the researcher to analyze the data in various living systems for commonality or basic differences. On the other hand, an applied scientist or technician will be afforded the opportunity of evaluating a given problem and its solutions from the broadest possible point of view, including the aspects of agronomy, crop science, animal husbandry, aquaculture and fisheries, veterinary medicine, clinical medicine, pathology, parasitology, toxicology, pharmacology, therapeutics, dietetics, food science and technology, physiology, zoology, botany, biochemistry, developmental and cell biology, microbiology, sanitation, pest control, economics, marketing, sociology, anthropology, natural resources, ecology, environmental science, population, law, politics, nutritional and food methodology, and others.

To make more facile use of the *Handbook,* the publication has been divided into sections of one or more volumes each. In this manner the particular sections of the *Handbook* can be continuously updated by publishing additional volumes of new data as they become available.

The Editor wishes to thank the numerous contributors, many of whom have undertaken their assignment in pioneering spirit, and the Advisory Board members for their continuous counsel and cooperation. Last but not least, he wishes to express his sincere appreciation to the members of the CRC editorial and production staffs, particularly President Bernard J. Starkoff, Mr. Gerald A. Becker, Mrs. Kathryn H. Harter, Mrs. Karen G. Ketchaver, and Mr. Paul R. Gottehrer, for their encouragement and support.

We invite comments and criticism regarding format and selection of subject matter, as well as specific suggestions for new data (and additional contributors) which might be included in subsequent editions. We should also appreciate it if the readers would bring to the attention of the Editor any errors or omissions that might appear in the publication.

Miloslav Rechcigl, Jr.
Editor-in-Chief
October 1976

PREFACE
SECTION G: DIETS, CULTURE MEDIA, FOOD SUPPLEMENTS

This section, projected into several volumes, furnishes information relating to food habits and dietary regimens of various types of animals including livestock and laboratory and zoo animals as well as animals living in the wild. Data are furnished on the natural and semipurified and purified diets, listing individual ingredients, their concentrations and their sources. In the case of natural ingredients or mixtures, chemical composition is also given, if known.

The first volume contains information on diets and food habits of mammals. The subsequent volume will cover comparable data on other vertebrates and the invertebrates. A special volume will be devoted to media used for cultivation of microorganisms and plants, while another subsection will be concerned with culture media for cells, tissues, organs, and embryos.

Miloslav Rechcigl, Jr.
Editor
January 1977

MILOSLAV RECHCIGL, JR., EDITOR

Miloslav Rechcigl, Jr. is Nutrition Advisor and Director of the Interregional Research Staff in the Agency for International Development, U.S. Department of State.

He has a B.S. in Biochemistry (1954), a Master of Nutritional Science degree (1955), and a Ph.D. in nutrition, biochemistry, and physiology (1958), all from Cornell University. He was formerly a Research Biochemist in the National Cancer Institute, National Institutes of Health and subsequently served as Special Assistant for Nutrition and Health in the Health Services and Mental Health Administration, U.S. Department of Health, Education, and Welfare.

Dr. Rechcigl is a member of some 30 scientific and professional societies, including being a Fellow of the American Association for the Advancement of Science, Fellow of the Washington Academy of Sciences, Fellow of the American Institute of Chemists, and Fellow of the International College of Applied Nutrition. He holds membership in the Cosmos Club, the Honorary Society of Phi Kappa Pi, and the Society of Sigma Xi, and is recipient of numerous honors, including an honorary membership certificate from the International Social Science Honor Society Delta Tau Kappa. In 1969, he was a delegate to the White House Conference on Food, Nutrition, and Health and in the last two years served as President of the District of Columbia Institute of Chemists and a Councilor of the American Institute of Chemists.

His bibliography extends over 100 publications, including contributions to books, articles in periodicals, and monographs in the fields of nutrition, biochemistry, physiology, pathology, enzymology, and molecular biology. Most recently he authored and edited *World Food Problem: A Selective Bibliography of Reviews* (CRC Press, 1975), *Man, Food, and Nutrition: Strategies and Technological Measures for Alleviating the World Food Problem* (CRC Press, 1973), *Food, Nutrition and Health: A Multidisciplinary Treatise Addressed to the Major Nutrition Problems from a World Wide Perspective* (Karger, 1973), following his earlier pioneering treatise on *Enzyme Synthesis and Degradation in Mammalian Systems* (Karger, 1971), and that on *Microbodies and Related Particles. Morphology, Biochemistry and Physiology* (Academic Press, 1969). Dr. Rechcigl also has initiated and edits a new series on Comparative Animal Nutrition and is Associated Editor of *Nutrition Reports International.*

ADVISORY BOARD MEMBERS
SECTION G: DIETS, CULTURE MEDIA, FOOD SUPPLEMENTS
VOLUME I

ADVISORY BOARD MEMBERS
SECTION G: DIETS, CULTURE MEDIA, FOOD SUPPLEMENTS
VOLUME I

CONTRIBUTORS
SECTION G: DIETS, CULTURE MEDIA, FOOD SUPPLEMENTS
VOLUME I

John T. Abrams
Department of Clinical Veterinary Medicine
University of Cambridge
Madingley Road
Cambridge, England

Michael Aleksiuk
The Boreal Institute for Northern Studies
University of Alberta
Edmonton, Alberta, Canada

M. H. Beaver
Lobund Laboratory
Department of Microbiology
University of Notre Dame
Notre Dame, Indiana

A. T. Bergerud
Department of Biology
University of Victoria
Victoria, British Columbia, Canada

Lytle H. Blankenship
Agricultural Research and Extension Center
Texas Agricultural Experiments Station
The Texas A & M University System
Uvalde, Texas

Diane L. Bray
Department of Nutritional Sciences
University of California, Berkeley
Berkeley, California

George M. Briggs
Department of Nutritional Sciences
University of California, Berkeley
Berkeley, California

S. T. Chou
Ralston Purina Company
Purina Research Farm
Gray Summit, Missouri

Larry R. Collins
National Zoological Park
Conservation and Research Center
Front Royal, Virginia

G. L. Dryden
Biology Department
Slippery Rock State College
Slippery Rock, Pennsylvania

N. Fairall
Provincial Administration of the Cape of Good Hope
Department of Nature Conservation
Cape Town, Republic of South Africa

Saul Fernandez-Baca
Facultad de Medicina, Veterinaria, y Zootecnia
U.N.A.M.
Mexico

David Ford
MRC Laboratory Animals Centre
Medical Research Council Laboratories
Woodmansterne Road, Carshalton
Surrey, England

James L. Gobble
Department of Dairy and Animal Science
College of Agriculture
The Pennsylvania State University
University Park, Pennsylvania

Eduardo González-Jiménez
Universidad Central de Venezuela
Facultad de Agronomia
Instituto Produccion Animal
Maracay, Venezuela

Martyn L. Gorman
Department of Zoology
University of Aberdeen
Aberdeen, Scotland

CONTRIBUTORS
SECTION G: DIETS, CULTURE MEDIA, FOOD SUPPLEMENTS
VOLUME I

D. G. Govaert
Lobund Laboratory
Department of Microbiology
University of Notre Dame
Notre Dame, Indiana

P. Haberey
Institut de Physiologie
Faculte de Medecine
Université Louis Pasteur de Strasbourg
Strasbourg Cedex, France

Anthony J. Hall-Martin
Mountain Zebra National Park
Cradock, Republic of South Africa

John Hammond, Jr.
School of Agriculture
University of Cambridge
Cambridge, England

J. S. Ichhponani
Department of Animal Science
Punjab Agricultural University
Ludhiana, India

A. Sydney Johnson
Institute of Natural Resources
University of Georgia
Athens, Georgia

R. M. Jordan
Department of Animal Science
University of Minnesota
St. Paul, Minnesota

George R. Kerr
Department of Nutrition
School of Public Health
Harvard University
Boston, Massachusetts

Thomas A. Leege
Idaho Fish and Game Department
Kamiah, Idaho

Lim Boo Liat
Medical Ecology Division
Institute for Medical Research
Kuala Lumpur, Malaysia

Jack R. Luick
Institute of Arctic Biology
University of Alaska
Fairbanks, Alaska

Rex E. Marsh
Agricultural Experiment Station
Wildlife and Fisheries Biology
College of Agricultural and Environmental Sciences
University of California, Davis
Davis, California

J. K. Matsushima
Department of Animal Sciences
Colorado State University
Fort Collins, Colorado

Dennis A. Meritt, Jr.
Lincoln Park Zoological Gardens
Chicago, Illinois

R. P. Millar
Provincial Administration of the Cape of Good Hope
Department of Nature Conservation
Cape Town, Republic of South Africa

Julius G. Nagy
Department of Fishery and Wildlife Biology
Colorado State University
Fort Collins, Colorado

Jack R. Nelson
Department of Forestry and Range Management
Washington State University
Pullman, Washington

CONTRIBUTORS
SECTION G: DIETS, CULTURE MEDIA, FOOD SUPPLEMENTS
VOLUME I

John J. Rasweiler IV
Department of Anatomy
College of Physicians and Surgeons of Columbia University
New York, New York

Ronald T. Reuther
Philadelphia Zoological Garden
Philadelphia, Pennsylvania

Pilar S. Reyes
Department of Nutritional Sciences
University of California, Berkeley
Berkeley, California

Sam H. Ridgway
Biomedical Group
Undersea Sciences Department
Naval Undersea Center
San Diego, California

Charles C. Schwartz
Natural Resource Ecology Laboratory
Colorado State University
Fort Collins, Colorado

Patricia P. Scott
Interdepartmental Division of Cell Studies
Royal Free Hospital School of Medicine
University of London
London, England

Terry A. Schulz
School of Veterinary Medicine
University of California, Davis
Davis, California

Ann U. Shump
The Museum
Michigan State University
East Lansing, Michigan

G. S. Sidhu
Commonwealth Scientific and Industrial Research Organization
Division of Animal Production – Ian Clunies Ross Animal Research Laboratory – Prospect
Blacktown, New South Wales, Australia

J. D. Skinner
Mammal Research Institute
Department of Zoology
University of Pretoria
Pretoria, Republic of South Africa

Lyle K. Sowls
Arizona Cooperative Wildlife Research Unit
University of Arizona
Tucson, Arizona

Richard R. Tenaza
Department of Biological Sciences
University of the Pacific
Stockton, California

Jack W. Thomas
U.S. Department of Agriculture
Forest Service
Pacific Northwest Forest and Range Experiment Station
Range and Wildlife Habitat Laboratory
LaGrande, Oregon

J. H. Topps
Division of Agricultural Chemistry and Biochemistry
School of Agriculture, Aberdeen
Aberdeen, Scotland

D. E. Ullrey
Department of Animal Husbandry
Michigan State University
East Lansing, Michigan

CONTRIBUTORS
SECTION G: DIETS, CULTURE MEDIA, FOOD SUPPLEMENTS
VOLUME I

Don Van Dyke
Biomedical Group
Undersea Sciences Department
Naval Undersea Center
San Diego, California

Elver H. Voth
Department of Biology
George Fox College
Newberg, Oregon

A. Lorin Ward
U.S. Department of Agriculture
Forest Service
Rocky Mountain Forest and Range Experiment Station
Laramie, Wyoming

Barbara J. Weir
Journal of Reproduction and Fertility
Editorial Office
Cambridge, England

William C. Weir
Department of Nutrition
Agricultural Experiment Station
College of Agricultural and Environmental Sciences
University of California, Davis
Davis, California

John O. Whitaker, Jr.
Department of Life Sciences
Indiana State University
Terre Haute, Indiana

Bernard S. Wostmann
Lobund Laboratory
Department of Microbiology
University of Notre Dame
Notre Dame, Indiana

E. Young
Division of Veterinary Services
Kruger National Park
Republic of South Africa

To my inspiring teachers at Cornell: Harold H. Williams, John K. Loosli, Richard H. Barnes, the late Clive M. McCay, and the late Leonard A. Maynard.

And to my supportive and beloved family: Eva, Jack, and Karen.

TABLE OF CONTENTS
SECTION G: DIETS, CULTURE MEDIA, FOOD SUPPLEMENTS
VOLUME I

Carnivora

THE NUTRITION OF THE DOG

J. T. Abrams

ENERGY

When their needs for oxygen and water have been met, the paramount requirement of all animals is for sufficient dietary energy.

Factors Influencing Body Heat Loss

Studies of mammalian species have shown that the rate of loss of heat from the body depends upon (a) external factors, chiefly the ambient temperature, and (b) characteristics of the animal itself, of which the most important are weight, nutritional status, amount of physical activity, psychological condition, and age. Less important quantitatively are breed or strain within a species, sex, and type of coat.

Ambient temperature – If the ambient temperature is low, the dog will shiver and thereby increase its heat output by muscular action; correspondingly, a temperature that is too high leads to another form of compensatory muscular activity, namely, panting. There is doubt about the specific ambient temperature range, the "comfort zone," within which neither response is quantitatively significant.[1,2]

Nutritional status – This is important both in the short-range and in the long-range sense. For several hours after the ingestion of a meal, heat output is increased above the preprandial level. Furthermore, the heat loss per day by a well-nourished dog is different from that of the same animal when it is thin, other factors remaining the same.

Weight (W) – Other factors being equal, the daily heat output is closely proportional to $W^{0.7}$, rather than to W. For a species in which the different breeds show an approximately 50-fold range of adult live weights, this is important (see below).

Physical activity – The amount of voluntary muscular activity performed by dogs ranges from very little in the case of many house pets to the severe exercise of sheep dogs in hilly country at lambing time. Thus, it is desirable to use the resting level of heat production as the base line.

Psychological condition – Many dogs are excitable, especially if they are not trained to obedience. This makes them difficult subjects for laboratory study. Attempts have been made to overcome this difficulty by regularly training the animals to lie quietly,[3] by carrying out tests early in the morning under conditions of quiet,[4] by sedation,[1] or by the selection of quiet dogs for use in repeated tests.[5] It is not easy to relate such laboratory specimens to "normal" dogs.

Age – Only one group of workers appears to have measured the heat outputs of dogs during growth *and* in the adult state;[3] often, no information is provided about the age of the animals tested.

Length and type of coat – In one extensive series of trials, bull terriers, whose type of coat may be assumed, were used;[3] in another, "male dogs of mixed breeds, of varied colors and furs" were tested, but no differential treatment of results is given.[1]

Method of Calorimetry Employed

Direct calorimetry,[6] respiration calorimetry with the animal in a chamber,[1,5] and respiration calorimetry with a mask applied to the dog's face[3] have been used to measure canine heat output, directly or indirectly. The periods of measurement have varied from as little as 6 to 15 min to as long as 24 hr.

Basal Metabolism

The daily heat output of a resting dog – well-nourished, but in the postabsorptive state, and exposed to an ambient temperature within the comfort zone – is its basal metabolic rate.

It is difficult to compare the different sets of observations, not only for the reasons given in the preceding paragraphs, but also because of the different ways in which the results are presented. For example, one set is given in the form of a mean for each animal, together with the "variation from the average normal," but with no indication of the number of determinations other than "197 calorimeter experiments have been performed on Dog XIX during five successive years."[5]

In another study it was reported that "each of the results represents the lowest value of two determinations made on different days, but not differing more than 12%" – in other words, the results were twice selected.[1] For 34 dogs whose BMR was measured under anesthetic, a single value is given. For a further group of six trained animals, four values were obtained by use of a type of respiration chamber and two were obtained with a face mask. Comparisons between the trained and anesthetized states were made for 25 dogs, eight using the chamber and the rest wearing a face mask.

In yet another study, the means and standard deviations together with the number of observations for each of 11 dogs are given.[3] Still others show the average daily heat output together with the maximum and minimum for each of 17 animals.[4]

Basal Metabolism of the Adult Dog

The disparate observations discussed above – the most recent of them 30 years old, and all of them carried out before the isolation of the later members of the vitamin B complex – must form the basis for any generalization. They include 1,070 measurements relating to 80 adult dogs, which have been subjected to analysis here.

No attempt has been made to relate daily heat output J to a poorly defined canine "surface area." Instead, a plot of log J against log W has shown a scatter of the data about a straight line; the constants a and b in the equation

$$\log J = a + b \log W$$

have been determined by regression analysis, W is the dog's weight in kg, and J is the corresponding heat output in kJ.

An effort has been made to disclose differences between observers, even though experimental techniques and animal characteristics are confounded in such differences. Similarly, differences of sexual origin have been sought.

The derived equation is

$$\log J = 0.6548 \log W + 2.6088$$

The slope, 0.6548, has a standard error of ± 0.0086 (approximately ±2%), and there is a residual error (scatter about the regression line) of −7.1 to +7.6%. The range of weights covered in the analysis is 3.4 to 31.3 kg; the mean log W, corresponding to a weight of 16.17 kg, has an equivalent J of 2,513 kJ.

When the data for dogs and bitches are adjusted to the common mean log W by covariance analysis, they suggest that the value of J is 2,534 for dogs and 2,470 for bitches, a small (2.6%) but significant difference. Correspondingly, estimated values for log J, adjusted to a common log W, indicate that different "observers" (using different breeds of dog, experimental techniques, etc.) have obtained means differing by as much as 13% from the overall mean recorded here.

Although de Beer and Hjort[3] found the interindividual variance for their dogs to be significantly different from the intraindividual variance, such possibilities have had to be ignored in this chapter.

The generally unsatisfactory bases for the calculations having already been indicated, estimated values of J for adult dogs of different weights are given in Table 1, which covers a weight range greater than that for which experimental data are available.

Basal Metabolism of the Growing Dog

The one detailed study of the growing dog indicates that the recently weaned puppy may have a basal metabolism per m^2 that is 25 to 30% greater than that exhibited in the adult state. If one takes the reasonable view that 1 m^2 of body surface is closely proportional to $W^{0.65}$ and that there is supporting evidence from other mammalian species, the above percentage change may be accepted for general guidance: there is a steady decline to the adult level, which is reached at about 9 months of age.

Maintenance Energy Requirements

Maintenance is understood to be virtual constancy of body composition over the period in question, under conditions that (a) permit limited voluntary muscular activity – but exclude "work" – together with the involuntary muscular movements associated with the ingestion and digestion of food, and (b) allow for some deviation of the ambient temperature outside the limits of the comfort zone. These conditions are not well-defined, nor are they readily checked, except in laboratories, but they apply to most adult dogs, i.e., to most house dogs. Brody[7] attempted to discover the ratio of the amount of digestible energy that would suffice for maintenance to that of basal metabolism in different mammalian species; he had some difficulty, but decided that the ratio was about 2.

Evidence for the dog is limited to a few observations of reasonable constancy of live weight in relation to energy intake. This is unsatisfactory for two reasons: first, constancy of body weight does not necessarily imply constancy of the composition of the body; secondly, the gross energy of the food is to be expected to be greater than the amount of digestible energy it affords. Ward[8] found that adult beagles, under laboratory conditions, could maintain their weights (average ~11 kg) on 335 J or 80 kcal/kg live weight/day; this rate is about twice that of basal metabolism (Table 1). Indeed, the concept of

Table 1
BASAL METABOLIC RATES AND ESTIMATED MAINTENANCE DIGESTIBLE ENERGY REQUIREMENTS OF ADULT DOGS

Adult weight, kg	Basal metabolism, J[a]		Digestible energy for maintenance, J[a]	
	per day	per kg weight	per day	per kg weight
1	406	406	810	810
3	792	264	1,580	528
5	1,166	233	2,330	466
10	1,835	184	3,670	367
20	2,890	145	5,780	290
30	3,770	126	7,540	251
40	4,550	114	9,100	228
50	5,260	105	10,500	210
60	5,930	99	11,900	198
70	6,560	94	13,100	187

[a] 1 J = 1 kJ; log basal metabolic rate = 0.655 log W (kg) + 2.6088

maintenance as applied to the average adult dog is so very ill-defined and so difficult to check rigorously that the best course seems to be to adopt Brody's suggestion, i.e., twice the basal metabolism, as a rough guide to the amount of digestible energy needed; the quantity of feed can then be modified to suit the condition of the animal: a little more can be given if the animal is losing weight, and less if it shows signs of becoming fat. This view is responsible for the last column of Table 1.

Adult Requirements for Work and Adverse Environmental Conditions

The amount of work done by dogs varies from the very limited exercise of the house dog, through that of the greyhound in training, to the severe load of the sheepdog in rough country at the appropriate season. The greyhound's needs for energy are increased by some 25% above maintenance,[9] whereas those of the sheepdog may rise 100%.[10] For the former class, systematic feeding for a regular work schedule is possible; for the latter, feeding to condition is more appropriate, with the above suggestion in mind.

If the severity of polar climates is added to a heavy work load, as for huskies weighing 35 to 45 kg, a supply of digestible energy equivalent to twice that of maintenance is unlikely to avert a weight loss of 10 to 15% during a long period of sledging, with temperatures in the neighborhood of −20°C.[11]

Energy Needs of the Pregnant or Lactating Bitch

From limited measurements of the energy intake of the normal pregnant bitch, and from numerous studies of other mammalian species, it is safe to assume that, for the first two thirds of the period of gestation, energy requirements are only slightly more than those of maintenance. For the remainder of the time, however, they increase by as much as 50 to 60%.[12]

The requirements of lactation depend on the size of the litter. At the peak of lactation, about the third week, the food intake of a bitch of 16 kg live weight, with a litter of four pups, was three times her maintenance intake. A smaller litter would require less; a larger litter would not result in an appreciable increase in the intake of food, but could result in small pups at weaning. Diets with good levels of both fat and protein are desirable in these circumstances.

Energy Requirements of Growing Pups

What constitutes good bodily conformation in young and mature dogs is not a matter of general agreement among dog breeders. The target is thus one of appearance or of weight, neither of which is readily expressed in terms of energy. Furthermore, the end product is achieved by the use of arbitrary diets, so that it is unclear how far the performance of the pups is affected by the palatability of the food, how much of the gross energy is available, i.e., digestible, or how far the low energy density (J/kg) of the diet may have limited food intake.

From comparative mammalian studies there is no doubt that the newly weaned animal can cope with a digestible energy intake equal to twice the maintenance needs of an adult animal of the same weight. In reviewing the older literature on the dog, Walker[13] comes to the same conclusion and adds data of his own relating to six breeds with adult live weights ranging from about 5 to 30 kg.

In conformity with other experience, Walker[13] finds that *with his diets* the pups reach 90% of their adult weights by 7 months of age, although growth has slackened appreciably a month or two before, whereas efficiency of food conversion (kg dry food/kg weight gain) rises sharply at 4 to 5 months. With the above reservations in mind, Table 2 is presented as a guide to the energy requirements of growing dogs.

Table 2
SUGGESTED DIGESTIBLE ENERGY ALLOWANCES FOR GROWING PUPS

Weight, percentage of adult weight for breed	30–60	60–80	80–95
F[a]	2.0	1.7	1.3

$^{a}\ F = \dfrac{\text{daily energy allowance of the pup}}{\text{daily allowance of adult of the same weight (Table 1)}}$

PROTEIN

In theory, the study of protein metabolism involves following the fate of ingested protein through the organism. In practice, this is so difficult that the pursuit of protein nitrogen has to be substituted for it. So, in particular circumstances, one may measure nitrogen retention or loss from the body (and from its products, including milk) over a given time, or one needs to carry out serial nitrogen analyses of groups of comparable animals in order to estimate the average daily change in bodily status.

The first method has been applied to a variety of mammalian species, including limited studies of dogs; the second has been applied only to conveniently sized farm mammals, namely, pigs and sheep, but not to dogs, which are regarded as inedible in the western world.

Weight changes, however convenient to measure, are no substitute for studies of nitrogen turnover. This is the great obstacle to a better appreciation of canine nutrition, where at present there is a surfeit of calculations in a desert of facts.

Maintenance Protein Requirements of Adult Dogs

The need of animals is for some 20 amino acids; about half of these (essential amino acids) the organism cannot synthesize from simple carbon and nitrogen compounds. In the formulation of diets designed for particular physiological functions, it is desirable to know the amount of each essential amino acid in the diet, since the organism requires them in specific proportions; correspondingly, the digestibility of each amino acid should be known.

When measured under appropriate conditions, the biological value (BV) of the dietary protein is a fair index of the amounts and proportions of essential amino acids the latter contains. Furthermore, it is common to assume that all of the amino acids in a given protein are equally well digested, although this is not necessarily true.

Clearly, the value of true digestibility (%) multiplied by BV provides information concerning the suitability of the dietary protein as a source of available essential amino acids; this product has been called the *net protein utilization* (NPU).[14] Equally clearly, the NPU value must be multiplied by the protein content of the diet to indicate the weight of essential amino acids present in a given quantity of the diet, i.e., in 100 units of diet; this multiplication product, in turn, has been named the *net protein value* (NPV) of the diet. Finally, since diets may vary considerably in composition, and thus in energy content, the most general method of dietary protein evaluation is that which expresses the energy value of the NPV as a percentage of the digestible energy content of the feed.

Although the average BV of proteins of animal origin is higher than that from the vegetable kingdom, there is an overlap in the ranges of the two classes. It is also true that the digestibility of the former tends to be higher than that of the latter. These differences having been granted, there is no basic reason why dogs should be limited to meat as a protein source. This has long been known. More recently, adult beagles, under good experimental conditions, have been maintained in nitrogen equilibrium on semisynthetic diets in which commercial isolated soy protein, the sole dietary protein, provided some 7% of the dietary energy. During the trials, the soy protein was found to have a true digestibility of 86%, and its BV at the 7% level of inclusion was about 85; its NPU was thus 74, and its NPV on an energy basis was 5.2%. At that level of inclusion, dogs could maintain live weight for periods of at least 1 month.[15]

Observations of man and of sheep support this figure as a level of protein adequate for maintenance. However, there are three objections to its use in the *practical* feeding of common dogs. First, it is not known what protein catabolism is involved when the animal has to cope with the normal stresses of life; these include bacterial invasion, parasitic infestation, trivial accidents, and adverse physical environments. Secondly, the natural proteins are the chief vehicles for the vitamins of the B complex, so that efforts to lower the protein intake of the feed may lead to depressed intake of those vitamins at a time when, in fact, they may be needed in increased amounts, because of the compensatory rise in carbohydrate and/or fat intake. Finally, most of the foods commonly available for dogs contain at least 10% protein, the exceptions being fats and refined carbohydrates. For these reasons, the *practical* levels of Table 3 are suggested.

Protein Requirements for Growth

Even though growth can be defined in the laboratory in terms of daily protein (N × 6.25) accretion, the information is necessarily of limited value when nothing is known about a particular animal other than its daily rate of weight gain. Much of the literature relates simply to weight gain on diets of known protein content, with neither the digestibility of the protein nor its BV being measured.

In an attempt to link the protein requirement of the growing pup with its concurrent needs for energy, Payne[16] assumed that live weight gain includes 17% of protein, but neither the limited carcass analyses available to Widdowson[17] nor the more voluminous information about mammals – including cattle, sheep, and pigs, but excluding dogs – presented by Mitchell[18] give one confidence in the validity of this figure, nor in its independence of the fat content of the weight gains, quite apart from uncertainties about the data for energy.

In any case, Payne's[16] method requires knowledge of the digestibility of the protein of the food and of its BV. Information on the former is accumulating slowly; it has been reviewed by Göcke,[19] who has added her own original contributions. The range is wide, varying from 53% for meat-and-bone meal protein to 97% for casein. Ward[8] found the true digestibility of whole-egg protein for beagles to be 92%, and that of commercial isolated soy protein to be 86%. It does not follow, however, that the digestibility of protein is quite independent of other dietary components (see below).

Most biological values for dietary proteins have been determined on diets of low fat content, with the diet limited to the protein in question at 10% by weight; at higher

Table 3
PRACTICAL PROTEIN ALLOWANCES FOR ADULT MAINTENANCE

Dietary fat, wt %	5	10	15	20
Dietary protein, wt %	11.0	11.8	12.5	13.3

Note: It is assumed that, on the dry matter basis, the feed contains about 95% of organic substances having an average digestibility of 90%; it is also assumed that the dietary protein has an apparent digestibility of 80 to 90% and a biological value of 75 to 85, and that 4% of fat rich in essential fatty acids is present; 3 to 4% of crude fiber should also be present.

percentages of the diet, the BV for most proteins falls as the proportion rises.[20] Data in the literature are thus quite suitable for maintenance, but are of doubtful value where growth is concerned, since weight gains arbitrarily thought to be "satisfactory" nevertheless require dietary protein levels higher than 10%. A further complication is that the biological values of proteins in a mixed diet are not necessarily additive.

There is, indeed, no substitute for practical studies for canine nutritional requirements, preferably by nitrogen balance methods, since public opinion would find the alternative serial carcass analyses very repugnant.

Turning to growth on diets of known protein content, one finds that weaned beagles and collies, with feeds of low fat percentage and varying proportions of protein, made better gains during 12 weeks at the 17.2% level than on 15.0% or 19.4%.[21] The lowest dietary protein content was 10.6%.

Walker[13] gave weaned beagles solid foods with either 21% or 26% protein (dry matter basis); the pups also received about 280 cc of milk daily. Protein at the higher level was less efficiently used in relation to weight gain. Other workers have also found 20% to be better than higher levels, except when the energy density of the feed had been raised by the inclusion of higher percentages of fat.[22] Walker[13] concluded that a ratio of protein (g) to dietary energy (J) of about 1 to 75 would give satisfactory growth in weaned pups if feeding were ad libitum, if the diet were palatable, and if the feed provided about 14,000 J/kg dry matter.

If the life expectancy of a dog is assumed to be about 10 years, to err on the mildly generous side of protein provision during some 4 or 5 months of the immediate postweaning period is a trivial cost. Most dogs are house pets, kept under individually specific, but usually unknown conditions. In the light of all these circumstances, it would appear that the weaned pup should receive 20% of its dietary energy in the form of protein of reasonable quality until it is about three fourths grown; the protein level can then be gradually reduced to the maintenance level, with the dog's condition being closely observed.

Protein Requirements for Pregnancy and Lactation

The protein requirements of pregnancy and lactation are difficult to place precisely, partly for the reasons given in the preceding section. Another reason is the variability in the number of pups produced by the bitch.

Studies of a variety of species, though again excluding the dog, show quite clearly that the growth of the products of conception is relatively slight during the first two thirds of the period of gestation, but rises much more rapidly closer to term.[23] If anything, protein deposition increases more sharply than total energy. This is taken into account in the feeding of farm mammals, but the time scale is much shorter for bitches. Nevertheless, studies of beagle and spaniel bitches showed average intakes of food of 1 kg/week during the first 3 to 4 weeks of gestation, rising by two thirds for the next 2 weeks and falling somewhat when approaching term, on diets containing some 22 to 30% of the gross energy as protein. The basic diet (lower level mentioned above) contained casein. Other diets had their protein contents increased by one fourth, either with casein or with autoclaved egg albumen. Still another group had a supplement of 10% of fresh liver.[12]

The increased levels of protein led to (a) birth weights 10 to 30% higher, (b) pups that were more vigorous at birth, and (c) reducing mortality in the pups by one half between birth and weaning at 35 days, with most of the deaths occurring during the first two days after birth; the effect on litter size was relatively slight. During lactation, the food intake of bitches with five or more pups rose to almost 3.5 times the maintenance level. During 5 weeks of lactation, a litter of six pups showed an average increase of 900 g/week. Unfortunately, there are no data on nitrogen absorption or retention by the dams, although it is said that there was no evidence of thriftlessness; coats and stool consistency were good. Also unfortunately, the mortality of pups in individual litters within treatment groups is not stated, so that the improvement in survival rate due to supplementation of the basal diet with protein – though "physiologically appealing" – is statistically not convincing. In the absence of much-needed confirmatory data, the suggestions of Table 4 are offered.

FAT

The term "fat" means, almost exclusively, triglycerides. If one believes, as a first working hypothesis, that the dog has been shaped by evolutionary experience to its present state, it is difficult to imagine that it is fitted to deal with a great preponderance of dietary carbohydrate, relatively inexpensive though this food component may be; neither its dentition nor the rest of the alimentary tract indicate a capacity to deal with much fibrous food. Maltase and sucrase occur in the canine small intestine,[24] but it is difficult to believe that the feral animal ingests large amounts of starch and sucrose. Furthermore, the well-nourished but fasting dog, with water freely available, can work for many hours, using mobilized body fat very efficiently.[25] Finally, the bitch's milk, like that of most mammals, contains 40% fat on the dry matter basis.[9]

Saturated fatty acids, which predominate in fats that are solid at about 15°C, have no apparent dietary function, except as a concentrated energy source; for this reason, they are useful food components for both the lactating bitch with a large litter to support and for the dog with a heavy work load in a cold climate. In the latter event, they need to constitute 40% or more of the dietary dry matter. Normally (see below) they are well digested (95% or more), although the harder tallows may only be 75% digestible.

Unsaturated fatty acids must also be present in the canine diet; their presence in glycerides lessens the softening point of the fat and improves digestibility. The dog requires a supply of the unsaturated, essential fatty acids, i.e., linoleic, linolenic, or arachidonic acid; the second and third of these can be synthesized in the animal body from the first. Linoleic is abundant in many vegetable oils;[26] it cannot be synthesized in the animal organism, and thus is an essential dietary component. From the essential fatty acids arise the prostaglandins,[27] which have a wide variety of physiological functions, still to be fully explored in the dog, but evidently needed to maintain the health of the skin and coat[28] and to keep the balance of the serum lipoproteins.[29] Limited evidence indicates that a minimum of 4% of the dietary energy should be in the form of linoleate; this would be provided by 6 to 8% of energy in the form of linoleate-rich oils (e.g., sunflower seed, soy, or maize), according to plant species and variety. Such types of fat call for the use of biologically acceptable antioxidants, including vitamin E, in the feed.

Table 4
SUGGESTED PRACTICAL FEEDING FOR PREGNANCY AND LACTATION

	Weeks of pregnancy			
	1–4	5–6	7–9	Lactation
Fat, wt %	6	9	12	12
Protein, wt %	13.2	13.6	14.0	14.0

See note following Table 3.

CARBOHYDRATE

With inorganic (mineral) elements providing about 5% of the dry matter of a canine diet, protein up to 25%, and fat up to 40%, only carbohydrate is available to fill the balance. Lactose, for the very young puppy, is the natural carbohydrate of milk; however, diets containing much more than 15 to 20% of it seem liable to cause diarrhea in dogs.

Cooked starch appears to be the most suitable form of carbohydrate. Doubts about the inclusion of large amounts of sucrose have already been expressed; we need long-term studies of this field, including metabolic and clinical appraisal.

Finally, the dog appears to need fibrous matter, at least with particular types of diet. On semipurified isocaloric diets of varying protein content, diarrhea was most marked at the lower levels; in the stools, moisture content varied directly, and fat content inversely, with dietary protein level.[8] The diarrhea was prevented by the inclusion of 4% paper floc in the feed. In another study,[30] a 5% content of dietary fiber (e.g., as sawdust) gave firmer feces and more rapid food transit times. There seems to be no reason why normal vegetable foods should not provide the suggested level of crude fiber, but there is a clear need for further work in this area.

INORGANIC (MINERAL) ELEMENTS

The needs of dogs for inorganic elements have not been as well studied as those of the rat or the chicken, but there are no good grounds for believing that they are unusual. Nevertheless, research on different mammalian species over the last two decades has shown not only the need for elements, hitherto not recognized as nutritionally ncesssary, but also the complex interactions between many of them; this section should be read in that light. Other nutrients or dietary components also need consideration, the best example being calcium and vitamin D.

Major Elements

Potassium

Potassium, universally a component of living cells, is present in sufficient quantity in virtually all natural foods, unless the foods have been extracted with or cooked in water, so that the whole amount is not available to the dog.

Approximately 40 mEq/day has been found adequate for dogs of 16 to 23 kg live weight;[31] 30 mEq (1.2 g)/kg dry diet was sufficient to replenish potassium-depleted dogs weighing 11 to 19 kg.[32]

When potassium-containing tissue is being laid down, as in growth, a dietary level of 0.3% seems adequate, but the field has not been well studied, since the range of weights of the dogs examined was rather small. It is not clear, for example, whether requirements are proportional to live weight ($W^{1.0}$) or to $W^{0.7}$

Sodium

The distribution of sodium in foods is unlike that of potassium; the levels in vegetable matter, including cereals and oilseeds, are usually low (0.1% or less), except for processed foods. The rich sources are meat, fish, and, of course, sodium chloride.

For many animal species a level of 0.5% NaCl in the diet seems adequate, and the few canine studies support this view. Thus, male mongrels of about 12 kg weight were quite normal when allowed 40 mEq of sodium daily.[33] If one assumes a dry matter intake of 120 to 150 g, this would require dietary concentrations of about 0.6 to 0.75% sodium, equivalent to approximately 1.5 to 1.9% common salt. This is probably too high for regular diets, but levels of 0.15 to 0.20% sodium (10 mEq/day) are much too low.[34] A level of 1% seems a generous compromise.

Chlorine

No studies of chlorine requirements, as distinct from sodium, where chlorine was provided as sodium chloride, have apparently been made. In the few investigations of other species, sodium rather than chlorine has been the limiting factor.

Calcium and Phosphorus

Much comparative study of nutrition indicates that animals fare best when, in the first place, the ratio of the amount of calcium provided by the diet is about 1.5 times that of phosphorus (as phosphate). In cereal grains and oilseed meals, calcium percentages are low (~0.05 to 0.4%) compared to phosphorus (0.3 to 1.3%). Furthermore, in flesh the same kind of imbalance (~0.01% calcium to 0.20% phosphorus in lean beef, for example) is seen. A further complication is that most of the phosphorus of cereal grains and oilseeds occurs as phytic acid or its salts. Diets based on such components, supplemented perhaps only by boneless meat, will be low in calcium content. Bone ash, which forms approximately 55% of the dry, fat-free matter of adult femoral bone, consists largely of calcium phosphate, with the ratio of the two elements present being more than 2:1. The skeleton accounts for 99% of the body calcium, but for only about 70% of the phosphorus.

The growth of weanling pups over 10-week periods, when fed ad libitum either a basal diet low in phosphorus and relatively low in calcium or the same diet with known supplements of calcium or inorganic phosphate, has been studied. Response was assessed on the basis of weight gain, femoral bone ash percentage, and plasma calcium and phosphate levels.

It was concluded that a level of 0.33% phosphorus was about adequate when the feed provided 0.6% calcium. With the "availability" of the phosphorus in the diet being 75%, the true dietary requirement was set at 0.25%. Here the term "availability" is equivalent to apparent digestibility;[35] in some other texts, true digestibility is used.[36]

With the phosphorus (P) level of the diet at 0.33%, the performance of the pups deteriorated markedly when the calcium (Ca) level was raised to 0.9% or to 1.2%; there was slight impairment when it was reduced to 0.3%. Trials with dietary P at 0.41% and Ca at 0.37, 0.47, 0.57, and 0.67% showed improved bone ash percentage as Ca content increased. With Ca at 1.2%, the pups grew well if the P content was raised to 1.0%.[37]

Taking into account the fall in requirements as the growth of bone and soft tissues progresses, and the likelihood that the availability of Ca, at least, will fall with age, dietary levels of 0.6% Ca and 0.5% P would seem to be sufficient for small and medium-sized dogs; further study of the larger breeds needs to be made. It is probable that these levels will suffice also for pregnancy and lactation.

A breakdown of 20 to 40% of phytate in the canine gut has been reported.[35] This is probably an uncertain figure to apply to all diets; provision of about half the total P in inorganic form is recommended, apart from the need of sufficient vitamin D.

The most common bone disease of nutritional origin clinically encountered is claimed to be secondary hyperparathyroidism,[38] due to regular feeding of Ca-deficient diets. In this respect, as indicated by research on several species, there is a hierarchy of the bones in retaining calcium under conditions of dietary deprivation. Tests on adult beagles receiving feed with 0.12% Ca and 1.2% P contents showed the greatest loss of mineral matter from the jaws, then from the vertebrae, and lastly from the long bones. There is greater loss of P than of Ca, reparable with diets of adequate calcium content.[39] As indicated, pups suffer stunted growth in addition to skeletal dysfunction.

Minor Elements

Magnesium

On semipurified diets of casein, sucrose, and fat, with vitamin and low-magnesium supplements, beagle and mongrel pups did not grow well on a magnesium supplement of 80 mg/kg diet, but did as well on 180 as on 1,300 mg/kg. Supplementary tests, including measurements of serum magnesium (Mg), Ca, and P concentrations and of Mg, Ca, and P in aortic ash, led to the view that the minimal dietary concentration should be 140 mg/kg.[40] Further examination of the heavier breeds of dog is again desirable, because of their smaller intake of food/kg live weight. Cereal grains as a group will provide such a minimum: oilseed meals contain about twice as much. Meat and fish are rather poor sources of magnesium, unless they contain fair amounts of bone.

Iron

From studies of mammalian species, it can generally be said that the absorption of iron appears to depend on the form in which it is presented (e.g., ferrous iron is better than ferric iron) and on the nature of anions in the diet (increased concentrations of phosphate or phytate, for example, depress absorption). Other factors being equal, the amount absorbed is greater the more iron-deficient the subject is, although the percentage of ingested iron absorbed by the gut is usually low – about 10%. The numerous variables concerned make it difficult to frame requirements based upon experiments made specifically with dogs.

Early studies of growing dogs that had milk as a basal diet indicated that the animals required about 0.8 mg/kg live weight/day (as pyrophosphate) to maintain satisfactory hemoglobin levels. In animals made anemic, the iron of bran had an availability of 50 to 70%, compared to 10 to 20% for spinach.[41] In dogs, the iron of hemoglobin is well absorbed (33% and 54% respectively in two animals, each of about 10 kg live weight, but of unrecorded history); it is suggested that this trait may be characteristic of carnivores.[42] Pending the results of more tests with dogs, an intake of 1.5 mg/kg live weight/day may be advocated. It would be interesting to know whether the myoglobin status of the *working* dog is influenced by dietary iron intake.

Copper

Hemoglobin regeneration in dogs made anemic by their being confined to milk diets was shown not to occur unless the daily allowance of 30 mg of iron was supplemented with 3 mg of copper.[43] This latter amount was equivalent to 200 to 300 μg/kg live weight and would, in turn, correspond to about 10 mg/kg dietary dry matter, a level that appears to be adequate for many common mammals, provided that other inorganic elements, including zinc, are not present in excess.

Zinc

The precise requirements of dogs for zinc have not been determined. A diet containing substantial amounts of soybean meal and a calculated 0.3% calcium caused lesions in dogs similar to those induced in pigs, for example, when its calcium content was increased by 0.8%,[44] although its zinc level was 33 mg/kg. On comparative grounds, one would expect such a concentration to be adequate in the absence of high levels of calcium and phytate, the former being unnecessary and the latter being associated with large proportions of cereals and/or oilseeds, in which case an additional 20 mg/kg could be needed.

Manganese

The manganese requirements of dogs have not been studied, but those of rats and of common farm animals have had much attention.[45] Using this information as a guide, one would expect the needs of dogs to be in the range from 20 to 40 mg/kg dietary dry matter, with the higher level perhaps necessary in the presence of unnecessarily high proportions of calcium and phosphate.

Iodine

Information about iodine requirements is also sparse; it consists chiefly of assertions that dog colonies can be maintained when the iodine concentration of the feed is at a certain level. Iodine is definitely an essential element. One group of workers[46] suggested that the minimal level for health in beagles is 25 μg/day, but added that dogs in their colony had been taking nearly 20 times as much for 5 years. A rough calculation suggests that the latter dogs had consumed 0.2 mg iodine/kg dietary dry matter; this is about twice the level deemed suitable for man.

Iodine is not well distributed in the common canine feedstuffs. If common salt is provided as 1% of dry matter, iodine at a level of 0.01 to 0.02% of the salt would seem to be adequate. The stable iodate is perhaps to be preferred to iodide, though there are no reports of its having been tried on dogs.

Cobalt

See *Cyanocobalamin* (vitamin B_{12}).

Selenium

Requirements for selenium are best considered together with the review of vitamin E.

Other Trace Elements

Trace elements, which now appear to be necessary for the health of the animal organism, but about which information is still meager, include molybdenum, fluorine, chromium, nickel, vanadium, and silicon. Highly purified diets had to be used to prove that laboratory animals need them. If this suggests that they may not be of much significance in practical diets, the suggestion could be invalid if ordinary foods are also liable to contain specific antagonists – for example, unduly high dietary levels of other elements. Almost no work has been done on canine requirements.

Molybdenum – Xanthine oxidase, a group of enzymes widely distributed in animal tissues and also occurring in milk, contains molybdenum. The evidence, none of it from tests with dogs, suggests that dietary requirements may be very low, perhaps much less than 1 mg/kg dry matter.[47]

Fluorine – Fluorine is needed by rats and mice to stimulate growth and to prevent infertility and anemia; the required minimum level is about 2.5 mg/kg dry matter.[48] Although mottling of tooth enamel in growing dogs has been studied, the dietary fluorine levels then were of the order of 20 to 200 mg/kg, much higher than may be physiologically necessary.[49]

Chromium – It was shown in 1957 that chromium-deficient rats have impaired glucose tolerance;[50] a similar condition has since been found in mice, squirrel monkeys, and man. In terms of dietary dry matter, requirements seem to be of the order of 0.3 mg/kg. Like a number of other trace metals, chromium is found with the bran when grain is milled. Rich sources of a chromium complex of high biological activity are said to include dried brewer's yeast and the bran and germ of cereals.[51] Refined sucrose is a poor source.

Nickel – Nickel is now said to be a necessary dietary factor for the growth of rats, guinea pigs, pigs, goats, and chicks; consequently, one has grounds for believing that dogs may also need to be included in the list. Nickel is necessary, moreover, for reproduction in pigs.[52] The element seems to be widely distributed in foods; those of vegetable origin tend to be much richer than milk and meat, but many more analyses are desirable. Dietary requirements are of the order of 5 to 10 mg/kg for the species listed above.[53]

Vanadium – Over the last few years, evidence has been obtained to show that vanadium is a dietary essential for the growth of chicks[54] and for growth and reproduction in the rats.[55] It has been claimed that minimal requirements of 0.1 mg/kg dry matter may not be much below the vanadium content of a number of foods. There are no canine data available, but the suggestion has been made that one sign of mild deficiency could be an elevation of serum lipids.

Silicon – Silicon is an essential trace element for rats and chicks, at least, and appears to be a structural component of bone and connective tissue.[56] Requirements are about 500 mg/kg dietary dry matter. Silicon is so widely distributed in nature that one would not expect practical diets to be deficient. On the other hand, it has been said that only 2% of the silicon of wheat may appear in flour, the rest going into bran and germ. In foods of animal origin, the element is concentrated in organs such as skin and in cartilaginous tissues.

The use of refined products – such as sugar, white flour (as in biscuit waste), fat, and extracted proteins – in canine feeds needs examination. Against the suggestions given here, the data for *one* type of canned dog food (dry matter basis) may be set as follows: Na, 0.24%; K, 0.09%; Cl, 0.16%; Ca, 1.92%; Mg, 0.16%; Fe, 705%; Cu, 27%; Zn, 60%; Mn, 257%; I, 2.4%; Co, 5%; Se, 0.26%; Cr, 32%; V, 1.6%; Sn, 16%.[57]

VITAMINS

Canine requirements for fat-soluble vitamins will be reviewed before those for the water-soluble group.

The fat-soluble vitamins – A, D, E, and K – have some common properties of nutritional significance: (a) they are very unequally distributed through the common foodstuffs; (b) they are photolabile; (c) they are easily oxidized, even by air, and especially in the presence of traces of catalysts such as copper; and (d) their alimentary absorption depends on the copresence of fatty acid in the gut.

Vitamin A [Retinol; Carotene (Provitamin A)]

Carotene, found largely in green plants, and to a lesser extent in the tissues of animals that have consumed much of it, is converted, with apparently variable efficiency, into retinol in the epithelium of the small intestine.

Some of the reasons why the dog's requirements for retinol are poorly defined must be examined before suggestions for canine feeding can be made.

The Evidence Relating to Dogs

There appear to have been no critical studies of the most common type of dog, the adult house dog, and very few of the growing pup.

General Criteria for the Adequacy of Retinol Supply

In the body, some of the retinol in circulation is oxidized to the corresponding aldehyde, which then combines with a specific protein to form the visual pigment of the retina. Light falling on the retina breaks down more or less of the pigment, depending on the strength of the stimulus, and thus initiates the sensation of vision. The pigment must be reformed before visual capacity can be restored to its initial level; consequently, the time of adaptation to a fresh stimulus depends on the rate at which retinol is transported to the eye, and thus on the retinol status of the subject. The time taken for visual adaptation to dim light, after previous exposure to a bright source, has therefore been used in attempts to determine requirements for retinol. With carotene most frequently employed as the source of vitamin A, the requirements of horses, cattle, sheep, and pigs were so assessed.[58]

However, although its precise biochemical pathways are not known, vitamin A has other physiological functions than that concerned with vision. It is certainly concerned with the integrity of epithelial tissues, and therefore with the health of the skin and of the respiratory, alimentary, and reproductive tracts. There is no reason to believe that retinol sufficient for vision will necessarily satisfy *all* functions; requirements based solely on dark-adaptation tests need careful scrutiny.

Vitamin A Supply and the Physiological Functions Studied

Much of the work on larger farm animals has been performed during their late growth period, with carotene as the vitamin source. Other workers have looked at the effects of depletion on appetite, the pressure of the cerebrospinal fluid, renal clearance, plasma retinol levels, the size of hepatic stores of the vitamin, and general health and longevity. It is doubtful whether retinol intakes just sufficient to restore visual capacity are large enough for other functions.[59]

Vitamin A Requirements in Relation to Live Weight

Despite the heterogeneous evidence available, it has been firmly assumed, from comparisons between species, that requirements for retinol vary with body weight ($W^{1.0}$).[58] Moore,[60] however, observed that the rat needs 10 μg/day if it is to be able to store a little of the vitamin, and for full longevity it requires about three times as much; he compared this to human adult needs – a minimum based on dark-adaptation tests of about 750 μg daily, or 1,200 μg with the now assumed margin for safety and storage.[61] As Moore[60] pointed out, such requirements are not in proportion to the average adult weight (70 kg) of man and that (0.2 kg) of the strain of rat concerned.

Adult Requirements as an Exponential Function of Weight

If one assumes that the daily retinol requirements of sedentary, unproductive adult animals are proportional to aW^k, where a and k are constants and W is live weight in kg, use of the above data gives $a = 83$ and $k = 0.63$, on the assumption that the higher levels are needed for the two species. In turn, the calculated requirements for adult animals of different live weights are those shown below:

Live weight, kg	1	5	10	20	50	100	400
Retinol, μg/day	83	228	352	545	971	1,500	3,600
Retinol, μg/kg live weight/day	83	46	35	27	19	15	9

The value of k, which recalls the exponent of W relating to basal metabolism, implies that a common dietary level of the vitamin will satisfy the maintenance requirements of adult dogs generally. A ratio of about 0.1 μg retinol/J digestible maintenance energy seems appropriate.

Requirements For Growing Animals

Growth rates may be expressed either in terms of daily weight increase or as a percentage of the initial daily live weight. The maximum possible rate of growth is genetically determined; the practical rate possible is usually governed by economics. In any event, the rate is not constant (save in absolute daily gain in the middle of the growth period in many instances), so that the period chosen for nutritional study is often quite arbitrary and not necessarily typical of the whole growth. Thus, the time it takes to deplete young animals of their reserves of retinol (in order to make them responsive in tests of dark adaptation) can be quite long if initial reserves are large. Variability of this kind and in the method of assessment used probably accounts for the different suggestions found in the literature.

Puppies put on a vitamin A-free diet at weaning, with graded levels of retinol added to the diet over periods varying from 14 to 22 weeks, achieved very little hepatic storage of the vitamin at intakes of less than 120 μg/kg/day;[62] an intake of 300 μg gave substantial

In studies of growing animals, where an insufficient amount of absorbed calcium leads to soft, easily deformed, and poorly calcified bone (rickets), it has been claimed that the pups of large breeds require much larger amounts of dietary vitamin D than those of smaller breeds;[64] however, the daily allowance of 4 to 6 μg D_3/kg live weight, which was still insufficient to inhibit "severe symptoms of rickets," carried with it the equivalent of 1,500 to 2,400 μg retinol/kg live weight, with about three fourths of the vitamin A being carotene used to "reinforce" cod-liver oil. Such levels of vitamin A are considerably in excess of those necessary (see Table 7). In speaking of the intractable "rickets," the statement is made that "before there is any deformity of the legs or before there is any X-ray evidence of rickets, there is a peculiar arching of the neck." The dogs do not move around much; they "hold the head in that peculiar position, but with some effort they are able to straighten up." These statements, put in juxtaposition to the following – "The lesion of greatest significance, and that responsible for the clinical condition, is the massive bossing and deformation of the *cervical** spine" – which refers to chronic hypervitaminosis A in growing cats regularly fed too large amounts of fresh liver,[65] leave one convinced that the canine condition was largely, if not wholly, chronic hypervitaminosis A, a conclusion supported by the resolution of the condition in another dog "on the same vitamin intake," which "in 10 days was given 31,000 USP units" (approximately 0.8 mg) "of vitamin D and was freed of these symptoms."

The large breeds of dog may still present problems. They need to have sufficient calcium in their diet, and in early growth they may well need dietary calcium/phosphorus ratios within the narrow range of 1.2:1 to 1.4:1. It is unfortunate that much of the early – and only – work in this field seems open to question on various grounds; some basal diets have been of the semipurified type, whereas others have been cereal-based, with clear differences between the forms of phosphorus they provide. Subject to the above provisions concerning calcium and phosphorus, it may be suggested that the weight of vitamin D – and preferably D_3 until more is known of canine metabolism of D_2 – per kg live weight should be 1/50 of the appropriate weight of retinol. This suggestion implies a linkage of vitamin D intake to $W^{0.63}$. Some samples of cod-liver oil have as narrow a retinol/calciferol ratio as the above; others would have to be fortified with calciferol.

Hypervitaminosis D has been exhibited in dogs that had received ergocalciferol at the rate of 0.25 mg/kg live weight daily for several months.[66] Depression of growth, diarrhea, and other signs of disturbance, including calcification of soft tissues, result from such overdosage. Severe skeletal damage may follow attempts to remedy bone dystrophy, arising from *calcium* insufficiency, by giving dogs more vitamin D.[67]

Neither retinol nor calciferol should be given orally, or by any other route, in amounts equivalent to more than 10 to 20 days' requirements. If the animal is sufficiently sick to warrant such treatment, it is sick enough to need a visit to check its progress. The vogue for huge labor-saving doses has gone; the risks of overdosage are now better understood.

Vitamin E (the Tocopherols and Related Compounds) and Selenium

α-Tocopherol is the best-studied member of a closely related group of compounds; indeed, in the literature vitamin E and this particular compound seem to be regarded as synonymous. Although its biochemical function in the body is still a matter for debate, vitamin E has long been recognized as a powerful antioxidant in vitro.

The absence of vitamin E from the diet results in a variety of pathological conditions, depending on the animal species concerned, on the physiological function – e.g., reproduction – under consideration, and on the nature of the diet, especially its contents of polyunsaturated fatty acids (PUFA) and, in the light of more recent work, of selenium.

On diets relatively high in PUFA and low in vitamin E, body fat becomes brown, the color at post-mortem examination deepening on exposure of the fatty tissue to air. This

*Present writer's italics.

storage in the liver. The evidence is indeed meager, but the lower value above suggests that a dietary level sufficient for adult animals could well be enough for pups, because of the latters' higher food requirement per kg live weight.

Renal Losses of Retinol

Dogs appear to be more liable than many other mammalian species to suffer from chronic nephritis. The early estimates of the amounts of retinol that dogs may excrete in urine would now tend to be "indicative of a pathological condition,"[63] but losses of 1 to 15 μg/100 cc of urine are regarded as normal.

Hypervitaminosis A

It seems unlikely that retinol intakes 2 to 3 times as high as those suggested here would cause untoward effects, but large intakes – either by misguided use of concentrated sources or by regular ingestion of large amounts of liver rich in retinol – are toxic; even intakes that are very high, though apparently not harmful, are wasteful.

Vitamin D (Ergocalciferol, D_2; Cholecalciferol, D_3)

In many mammals, irradiation of the coat or skin by light of suitable wavelength – the near-ultraviolet component of sunlight, for example – converts the inactive precursor to the active vitamin D_3; corresponding irradiation of the plant sterol, ergosterol, leads to the production of D_2 in appropriate foods.

Like other mammals, dogs may thus obtain their supply of vitamin D either by the action of sunlight of sufficient potency or from the diet, the richest natural sources undoubtedly being certain fish liver oils (D_3). In view of early statements that dogs reared indoors require more dietary vitamin D than those reared "under ordinary light conditions,"[64] it is surprising that subsequent workers were unable to find the precursor of cholecalciferol in the lipids of dog skin;[65] consequently, dogs were presumed to derive their supply of the vitamin solely from their diet. This observation needs confirmation and extension.

Apparently because of the specific nature of their "carrier proteins," chicks and certain monkeys use D_2 much less efficiently than D_3. The situation in dogs does not seem to have been established; meanwhile, it is judicious to include D_3 only in the diet of the species.

In several species that have been studied, cholecalciferol is transported either from gut or skin to the liver, where it is converted to the 25-hydroxy derivative. The latter is then converted in the kidney to the 1,25-dihydroxy compound, the active or hormonal form of the vitamin, which is concerned with the transport of calcium across membranes, including the small intestine.[66] Critical phases for calcium transport include those of growth, when calcium needs to be absorbed and then laid down in bone, and of lactation, when the suckling must be provided with calcium, some of which may arise from the diet and some, at the height of lactation, may have to be mobilized from bone.

The dietary variables of importance include not only the supply of vitamin D, but also the amounts of calcium and phosphorus, and probably the ratio of inorganic to phytic acid phosphorus. Much of the investigation of the requirements of dogs for vitamin D has involved pups rather than adult animals, purified diets rather than commercial kinds, and D_2 rather than D_3. How far unsuspected chronic canine nephritis, with consequent inefficient conversion of the 25-hydroxy compound to the 1,25-dihydroxy metabolite, has influenced our views on canine requirements for vitamin D is impossible to say. Another cause for thought is the use by early investigators of cod-liver oil as a source of vitamin D; at the time this may have been almost inevitable, but it meant that the effects of two vitamins were being confounded.

has been seen in terriers given about 15 cc of maize or cottonseed oil two or three times weekly without a supplement of vitamin E. Breeding in the kennel was poor; the dogs exhibited muscular weakness and an abnormal gait, together with low plasma tocopherol levels (22 to 61 μg/100 cc). On diets known to produce normal performance, the plasma vitamin E concentrations of the dogs rose to levels ranging from 436 to 669 μg.[70]

The brown lipid described above contains peroxides presumed to arise from the oxidation of PUFA in vivo,[71] so that vitamin E has been accredited with antioxidant powers within the organism. Other tissues in affected animals may not contain peroxides, and it has been debated whether vitamin E, like retinol, has more than one metabolic function. Over an experimental period of 15 weeks from weaning, beagle pups given diets providing 1, 5, 10, or 15% of stripped safflower oil as a tocopherol-free source of PUFA showed a requirement for the vitamin that varied with the amount of PUFA in the feed, judged by terminal creatine phosphokinase levels in blood and by pigmentation both of intestinal muscularis and of smooth muscle. Pups that received no supplement of tocopherol had an average initial plasma level of 260 μg/100 cc; and a final concentration of 51 μg/100 cc. Pups that received 11 mg tocopherol/kg live weight daily, starting with the same initial plasma tocopherol level, had an average final concentration of 1,413 μg/100 cc.[72]

Much earlier studies of canine reproduction, using a basal diet of aerated milk with supplements of manganese, copper, and cod-liver oil, indicated that normal birth and development could be achieved on ~1 mg tocopherol/kg live weight/day.[73] One would suspect that copper would accelerate the oxidation of the PUFA of the liver oil in the intestine at least.

On this rather meager evidence alone it may be assumed that 1 mg of tocopherol/kg live weight/day will suffice for diets low in PUFA but adequate in other respects (see below). From a consideration of data relating to several mammalian species, but excluding dogs, it was suggested that the ratio of *d*-α-tocopherol to PUFA (mg/g) in the diet should not fall below 0.6.[74] Thus, diets including 6 to 8% oils, such as sunflower, ought to contain 20 to 30 mg vitamin E/kg dry matter, though doubt has been expressed whether this is quite sufficient.[72] Whether synthetic antioxidants should replace some of the tocopherol in diets relatively rich in PUFA is probably a matter for further experiment.

Some 20 years ago, selenium was first recognized as an essential micronutrient,[75] capable, in some circumstances, of sparing vitamin E and of preventing or alleviating some of the pathological conditions observed in animals on diets low in the vitamin.[76] Only quite recently, however, has it been possible to locate and ascribe a function to a selenium-containing metabolite that fits into the antioxidant scheme. Selenium is a component of an enzyme, a glutathione peroxidase,[77] with the ability to reduce the peroxides of the unsaturated fatty acids to the less toxic hydroxy derivatives.

On the dry matter basis, the necessary concentration of selenium in the diet appears to be about 0.5 mg/kg, judging by the limited evidence; slightly higher levels than this may be toxic, however. Moreover, different chemical forms of selenium, e.g., organic and inorganic, have different physiological potencies. In pups, muscular dystrophy has been found, although the proportions of selenium in diet, blood, and tissues were normal;[78] no evidence of tocopherol levels was given.

The situation of dogs is thus once more uncertain, especially of animals on feeds containing substantial amounts of vegetable matter (cereals and oilseeds), the selenium contents of which depend on where they are grown. The canned food mentioned earlier contained 0.26 mg/kg.

Vitamin K

In healthy dogs, sufficient vitamin K seems to be produced by microbial action in the gut; consequently, there is no dependence on dietary intake. However, diseases affecting the gut, or drugs that influence the alimentary microflora, require veterinary investigation.

The Vitamin B Complex

The different members of the vitamin B complex have common features of importance in canine nutrition: (1) their concentrations in *natural* foods is roughly proportional to the corresponding protein contents; (2) as components of various enzyme systems they have well-established roles in the metabolism of proteins, fats, carbohydrates and nucleic acids; (3) they are much more freely soluble in water than in nonionizing solvents; (4) according to circumstances, the gut microflora may add to or diminish the supply of the B complex in the ingested food; and (5) the storage capacity of the body for this group, in contrast to the fat-soluble vitamins, is equivalent to a few days' requirements.* These points will now be briefly elaborated.

Foods that have been extracted with water, either commercially, as with yeast, or in the course of canine husbandry, as in the cooking of meat, lose much of these vitamins to the extract. In the latter instance, if the volume of liquid is large, much of the B complex may be thrown away; a little liquid, well mixed with the diet, is desirable.

Proteins are relatively rare and expensive, and the trend must be toward canine diets containing no more protein than is biologically necessary; such diets must, therefore, provide fat and carbohydrate in place of protein. However, reduction of the protein content also leads to a reduction of intake of the B complex; therefore, supplements rich in the complex – e.g., dried, unextracted yeast – must be added.

Whether the protein replacement is fat or carbohydrate affects dietary requirements for particular members of the vitamin B group (see below).

Bacterial synthesis of the B-vitamins is not extensive enough to make dogs independent of dietary supplies, though the case of *fat*-soluble vitamin K has already been mentioned. There are microorganisms, however, that will destroy vitamins under appropriate conditions; the most widely reported group of these may be found in some offal from raw fish. If such material is mixed with the rest of the feed, which is then left for a few hours at moderate temperatures, the thiamine may be totally broken down. Good food hygiene and the cooking of all food of doubtful origin are thus necessary. Some food preservatives – e.g., sulfite – can also cause problems.

Finally, much canine research, never repeated, was carried out before the last member of the vitamin B complex – namely, cyanocobalamin – was isolated. This introduces some uncertainty into the story.

Thiamine (Vitamin B_1)

Thiamine pyrophosphate is a necessary component of the carboxylase system, which leads to the expulsion of CO_2 from α-keto acids, especially pyruvic and α-ketoglutaric acids. It is, thus, concerned with the catabolism of some amino acids and of glucose, and with the energy-yielding citric acid cycle. On diets containing little fat, one might expect thiamine requirements to vary with energy production, and thus, at least in the adult animal, with $W^{0.65}$. Early studies of mice, rats, and dogs supported this view, the acceptance of which requires a constant ratio of thiamine to energy in the diet.

Fat metabolism does not require thiamine in the above way; indeed, it has been shown that the nervous signs observed in dogs that have been made marginally deficient in thiamine can be relieved by an isocaloric replacement of dietary sucrose by fat.[79]

Trials with beagles, beginning when the pups were between 3 and 3.5 months of age, were carried out over a period of 6 months; different levels of supplementation of

*Sometimes "conserved" by the animal's going off feed.

thiamine to a basal diet were used. On the basis of a variety of diagnostic criteria, it was concluded that 22 μg/kg live weight daily was not sufficient.[80] At the beginning of the test period, the dogs were quite well-grown, with average weights of about 7.5 kg; at the end of the test period, their weights averaged 10 kg. One may estimate that, toward the end of the trial, the above level of supplementation would be provided by a diet containing about 0.75 mg thiamine/kg dry matter.

Since an earlier, very limited study had led to the suggestion that the dry feed should contain 0.8 mg thiamine/kg,[81] a dietary level of 1 mg/kg may be offered as adequate for diets in which the bulk of the energy is provided by carbohydrate and protein.

Riboflavine (Vitamin B_2)

Also in an appropriate enzyme system, riboflavine is concerned with metabolic dehydrogenation, more particularly the removal of a pair of hydrogen atoms from the adjacent carbon atoms of substrates, such as succinic acid or a fatty acid of the saturated type.

Studies with semipurified diets, incompletely defined by modern standards, led to the suggestion that the probable daily requirement of adult dogs weighing about 8 kg is 25 μg/kg live weight. In equivalent dietary (dry matter) terms, this is about 1 mg/kg.[82] Later work suggests that about 600 μg/kg diet is sufficient for adult animals of 10 kg live weight.[83] As might be expected, comparative studies point to riboflavine requirements as being exponentially related to weight, and thus to energy intake. However, as late growth was thought to need twice as much of the vitamin/kg live weight, the figure of 1 mg riboflavine/kg food is to be preferred.

Nicotinic Acid (Nicotinamide)

Nicotinic acid, as an integral component of specific enzyme systems, functions in dehydrogenations, as of the group –CHOH– to –CO–, and is thus also involved in energy release in the citric acid cycle. A minor complication in the assessment of canine requirements for the vitamin is that it can be produced in the organism, though not very efficiently, from the amino acid tryptophan. Requirements for nicotinic acid are thus affected by the protein levels, and especially by the tryptophan levels, of the diet.

On semipurified diets to which thiamine, riboflavine, pyridoxine, pantothenate, and choline were added, dogs were depleted of nicotinic acid in 2 to 3 weeks. Different levels of the vitamin were employed to assess daily requirements for growth and recovery; they were put at 200 to 225 μg/kg live weight for adult animals, and at 250 to 365 μg/kg for growing dogs.[84] Making the same assumptions as for thiamine and riboflavine, this may be translated into approximately 10 mg/kg dietary dry matter.

Pyridoxine (Vitamin B_6)

Pyridoxine is a component of many enzyme systems whose main functions relate to amino acid metabolism. One might thus expect requirements to vary with the protein content of the diet.

Specific information about the requirements of dogs is meager and unsatisfactory. When diets containing maize starch, salts, hydrogenated cottonseed oil, and casein at either 18% or 45.8%, and sucrose at 20% or 47.8%, were supplemented (daily doses) with thiamine, riboflavine, nicotinamide, pantothenate, *p*-aminobenzoic acid, choline, inositol, and, as a source of biotin and folic acid, a rice bran preparation, and then fed to 3-kg weanling spaniels, severe signs of deficiency were seen on the high-protein diets in 79 to 123 days; at the lower protein level, the depletion period was longer (169 to 190 days). An allowance of 0.1 mg/kg live weight/day was sufficient to prevent pyridoxine deficiency signs at both levels of protein.[85] A dietary level of 1 mg/kg dry matter has been suggested.[86] Surveys of the requirements of pigs and of mammals smaller than dogs all point to the above dietary level as being sufficient for growth on diets of appropriate protein content, which again suggests that requirements vary with $W^{0.7}$ (approximate).

Pantothenic Acid

Pantothenic acid is combined in coenzyme A, which – in relevant enzyme systems – is concerned with the movement of the acetyl group CH_3CO-, and is thus involved in the metabolism of carbohydrates, fats, and proteins, including energy release in the citric acid cycle. One might expect requirements to be proportional to metabolism (i.e., varying with $W^{0.7}$) rather than to weight. Evidence of canine needs is inadequate, but data collected by Mitchell,[86,87] relating to a variety of mammalian species, suggest that the necessary dietary level is about 8 mg/kg;[87] the species considered did not include dogs, but the meager evidence relating to the latter does not seem sufficient to cause any modification.

Biotin

Biotin is a component of enzyme systems that are involved in the metabolism of carbonate; the fixation of the latter can lead, for example, to the production of succinate from propionate. Requirements for biotin are small, compared to those of most members of the B complex, and the technical difficulties of determining them are considerable (see folic acid, below).

Folic Acid

In enzymic form, folic acid is involved in the transfer of single carbon groups (formyl, or closely related groups, e.g., –CH:NH); it is thus concerned with the metabolisms of amino acids, especially methionine, and of nucleic acids.

The use of semipurified diets indicates that dogs may require 15 μg/kg live weight/day.[88] Comparison with human requirements again suggests a constant value for requirements/kg "metabolic weight" ($W^{0.7}$).

There is evidence of alimentary microbial synthesis of both biotin and folic acid, apparently in amounts comparable at least to the host's requirements. Disturbance of the gut population – especially by drugs of the sulfonamide class with structural similarities to part of the folate molecule – often seems to be necessary to make the dog responsive to the amounts of these vitamins in its feed.[89]

Cyanocobalamin (Vitamin B_{12}; Cobalamin)

The cobalt-containing vitamin B_{12} is a coenzyme involved in fatty acid, protein, and nucleic acid metabolism; synthesis of methyl groups is one of its functions. The vitamin is synthesized by microorganisms, but not by higher plants or by animals, so that bacterial activity in the gut may be presumed to influence the need for dietary provision. There seem to have been few canine experiments other than one trial showing that the growth of pups was stimulated by the use of B_{12} at the rate of 50 μg/kg feed (DM basis).[90] This is comparable to the median value of rather disparate estimates obtained with other species.

Choline

Choline (trimethylamino ethanol) is sometimes excluded from the list of vitamins because, under appropriate conditions, it may be synthesized in the animal organism. It acts as a source of methyl groups in the synthesis of other metabolites, and also assists in the mobilization of fat from the liver. Primarily because of the first function, less choline is required at higher dietary levels of normal proteins than at lower ones; because of the second function, more choline is needed as the fat content of the diet is increased. The importance of choline in canine nutrition thus seems likely to increase as the cost and scarcity of proteins force down the proportions of protein in dog foods toward physiological levels.

For obvious reasons, it is not possible to give specific requirements for choline, regardless of the nature of the rest of the diet, but on diets of 15% protein content, 100 mg/kg live weight daily sufficed;[91] this is roughly equivalent to 4 g choline/kg dietary dry matter.

Ascorbic Acid (Vitamin C)

Although ascorbic acid is a physiological requirement of dogs, endogenous synthesis provides sufficient amounts in healthy animals; studies of such animals have failed to demonstrate a dietary need.

There is, however, rather unsatisfactory literature relating to the supposed needs of sick animals, whose plasma levels may fall during disease, and whose urinary excretion also declines. Such changes may be the result of inanition, or of fever, or of bacterial action. Evaluation of the particular situation needs to be made by the veterinary surgeon concerned.

WATER

Although water has been left until last in the list of nutritional requirements, it is, in fact, "first among equals" in importance. The ideal situation is that of a clean continuous supply; next best is access to a free supply at least four or five times daily. Growing and lactating dogs should never be put at risk, nor should the working dog, whose metabolism will function satisfactorily for many hours in the absence of food, but whose efficiency diminishes when water also is withheld. It must not be forgotten that, apart from its function as a fluid (e.g., in milk secretion and in the excretion of waste products in urine), water intake influences dry matter consumption.

REQUIREMENTS AND RECOMMENDED ALLOWANCES

Suggested levels and sources for major and minor inorganic elements are listed in Tables 5 and 6 respectively. Similar data for vitamins are presented in Table 7.

The requirement of an animal for a particular purpose – e.g., the amount of calcium to be deposited in a bone of known composition – can often be set with a high degree of precision. The recommended allowance, i.e., the amount that must be provided in the diet in order that such a quantity may be so deposited, is affected, for example, by the physical form in which the calcium compound is presented and by the relative proportions of phosphate, vitamin D, lactose, and other dietary components. Too little experimental work has been carried out to enable such distinctions to be made clearly for dogs. Thus, many of the data presented here derive from tests made with semipurified diets; their relevance to the kind of food fed to the dog population may therefore be uncertain, especially in view of the fact that many of the experimental diets would now be regarded as incomplete because of subsequent discoveries in nutrition. Finally, from the practical aspect there is little to be said in favor of setting requirements at levels significantly lower than can be found in those feedstuffs that present economic circumstances compel us to use. Tables 5, 6, and 7 should be viewed in that light.

Table 5
MAJOR INORGANIC ELEMENTS

Element	Suggested level,[a] g/kg	Comments
Potassium	5.0	Most whole cereal grains approach this level; in oilseed meals, the general level is twice as high; fresh muscle (~24% dry matter) contains about 3.7%.
Sodium (as common salt)	10.0	Few cereal grains or oilseed meals approach even one fourth of this value; fresh muscle varies from about 1.2 to 2.6% (as NaCl) in content.
Calcium	9.0	The level could be reduced to 6.0 to 7.0 g/kg for diets containing most of their phosphorus as phosphate at about 5 g/kg and also containing proper levels of cholecalciferol.
Phosphorus	7.0	Whole cereal grains contain about 0.4%, and oilseed meals contain an average of around 1.0% (within a fairly wide range), mostly as phytate phosphorus; fresh muscle contains ~2%, chiefly as phosphate. This is the most difficult element to fit in when significant amounts of vegetable concentrates appear in the feed; it is suggested that at least 3 g/kg should be phosphate from sources such as bone meal and flesh.

[a] Proportion of dietary dry matter.

Table 6
MINOR INORGANIC ELEMENTS

Element	Suggested level,[a] mg/kg	Comments
Magnesium	400	Most cereal grains contain about 0.14%; oilseed meals contain about twice as much; fresh muscle contains 0.3 g/kg; bone meal contains ~1%.
Iron	60	This level is approached only by foods of animal origin.
Copper	10	Content in whole cereal grains is often below this level.
Zinc	30	Possibly 40 to 50 mg/kg is needed for diets rich in phytate and with the higher levels of calcium.
Iodine	2	Probably 1 mg/kg is sufficient if stable iodate is used.
Manganese	10	Suggestions based on requirements of other mammals.
Selenium	0.3	Suggestions based on requirements of other mammals.
Molybdenum	1	Suggestions based on requirements of other mammals.
Fluorine	2	Suggestions based on requirements of other mammals.
Chromium	0.3	Suggestions based on requirements of other mammals.
Nickel	5	Suggestions based on requirements of other mammals.
Vanadium	0.1	Suggestions based on requirements of other mammals.
Silicon	250	Suggestions based on requirements of other mammals.

[a] Proportion of dietary dry matter.

Table 7
VITAMINS

Vitamin	Suggested level[a]	Comments
Retinol	1.5 mg/kg	Absent from cereal grains, oilseed meals, and lean meat.
Cholecalciferol	30 μg/kg	Absent from cereal grains, oilseed meals, and lean meat.
Vitamin E	30 mg/kg	This level is unattained by most feedstuffs.
Vitamin K	–	See text.
Thiamine	1 mg/kg	This level is attained by most whole cereal grains, and oilseed meals.
Riboflavine	1–2 mg/kg	This level is attained by most whole cereal grains and oilseed meals.
Nicotinamide	10 mg/kg	This level is attained by most whole cereal grains and oilseed meals, except maize, see text.
Pyridoxine	1 mg/kg	As for thiamine.
Pantothenate	8 mg/kg	As for thiamine; based on the responses of other mammals.
Biotin	200 μg/kg	Based on the responses of other mammals; probably sufficient microbial synthesis in the healthy gut and in the absence of drugs.
Folic acid	100 μg/kg	As for biotin.
Cyanocobalamin	20 μg/kg	As for biotin.
Choline	1–3 g/kg	Lower level on higher-protein, low-fat diets; see text.

[a] Proportion of dietary dry matter.

REFERENCES

1. **Galvao, P. E.,** *Am. J. Physiol.,* 148, 478–489, 1947.
2. **Benedict, F. G.,** *Vital Energetics,* Carnegie Institute, Washington, D.C., 1938, pp. 77–80.
3. **deBeer, E. J. and Hjort, A. M.,** *Am. J. Physiol.,* 124, 517–523, 1938.
4. **Kunde, M. M. and Steinhaus, A. H.,** *Am. J. Physiol.,* 78, 127–135, 1926.
5. **Lusk, G. and Du Bois, E. F.,** *J. Physiol.,* 59, 213–216, 1924.
6. **Rubner, M.,** in *The Science of Nutrition,* 4th ed., W. B. Saunders Co., Philadelphia, 1924, p. 42.
7. **Brody, S.,** *Bioenergetics and Growth,* Reinhold Publ. Co., New York, 1945, p. 475.
8. **Ward, J.,** Investigation of the Amino Acid Requirements of the Adult Dog, Thesis, Cambridge University, 1975, p. 278.
9. **Abrams, J. T.,** *Animal Nutrition and Veterinary Dietetics,* 4th ed., Williams & Wilkins, Baltimore, 1961, pp. 749 and 441.
10. **Greig, J. R.,** *Br. Vet. J.,* 109, 14–32, 1953.
11. **Wyatt, H. T.,** *Br. J. Nutr.,* 17, 273–279, 1963.
12. **Ontko, J. A. and Phillips, P. H.,** *J. Nutr.,* 65, 211–219, 1958.
13. **Walker, A. D.,** Nutritional Studies on the Domestic Dog and Cat, Including Investigations into Patterns of Food Acceptance, Thesis, London University, 1971, p. 181.
14. **Block, R. J. and Mitchell, H. H.,** *Nutr. Abstr. Rev.,* 16, 249–278, 1946–1947.
15. **Ward, J. and Abrams, J. T.,** unpublished observations, 1971.
16. **Payne, P. R.,** in *Canine and Feline Nutritional Requirements,* Graham-Jones, O., Ed., Pergamon Press, Oxford, 1965, pp. 19–31.
17. **Widdowson, E. M.,** in *Canine and Feline Nutritional Requirements,* Graham-Jones, O., Ed., Pergamon Press, Oxford, 1965, pp. 9–17.
18. **Mitchell, H. H.,** *Comparative Nutrition of Man and Animals,* Vol. 1, Academic Press, New York, 1962, p. 357 *et seq.*
19. **Göcke, A.,** Über die Zusammensetzung und Verdaulichkeit von Hundefertigfuttermitteln, Thesis, Tierärztliche Hochschule, Hannover, 1970.
20. **Allison, J. B.,** in *Protein and Amino Acid Nutrition,* Albanese, A. A., Ed., Academic Press, New York, 1959, p. 112.
21. **Gessert, C. F. and Phillips, P. H.,** *J. Nutr.,* 58, 415–421, 1956.
22. **Ontko, J. A., Wuthier, R. E., and Phillips, P. H.,** *J. Nutr.,* 62, 163–168, 1957.
23. **Abrams, J. T.,** in *Comparative Nutrition of Wild Animals* (21st Symposium of the Zoological Society of London), Crawford, M. A., Ed., Academic Press, New York, 1968, p. 49.
24. **Hove, H. P. and Messer, M.,** *Comp. Biochem. Physiol.,* 24, 717–725, 1968.
25. **Young, D. R. and Price, R.,** *J. Appl. Physiol.,* 16, 351–354, 1961.
26. **Hilditch, T. P. and Williams, P. N.,** in *The Chemical Constitution of Natural Fats,* 4th ed., Chapman and Hall, London, 1964, pp. 218–239.
27. **Gurr, M. I. and James, A. T.,** in *Lipid Biochemistry,* 2nd ed., Chapman and Hall, London, 1975, pp. 52–60.
28. **Wiese, H. F., Yamanaka, W., Coon, E., and Barber, S.,** *J. Nutr.,* 89, 113–122, 1966.
29. **Lindall, A. W., Grande, F., and Schultz, A.,** *Proc. Soc. Exp. Biol. Med.,* 136, 1032–1037, 1971.
30. **Riklin, M.,** Untersuchungen über den Einfluss von Strukturelementen in Futter auf Verdauung, Peristaltik und Kotkonsistenz beim Hund, Thesis, Tierärztliche Hochschule, Hannover, 1973.
31. **Abbrecht, P. H.,** *Am. J. Physiol.,* 223, 555–560, 1972.
32. **Burnell, J. M. and Dewborn, J. K.,** *Am. J. Physiol.,* 218, 1583–1589, 1970.
33. **Brubacher, E. S. and Vander, A. J.,** *Am. J. Physiol.,* 214, 15–21, 1968.
34. **Ganong, W. F. and Boryczka, A. T.,** *Proc. Soc. Exp. Biol. Med.,* 124, 1230–1231, 1967.
35. **Jenkins, K. J. and Phillips, P. H.,** *J. Nutr.,* 70, 235–240, 1960.
36. *The Nutritional Requirements of Farm Livestock, No. 2. Ruminants,* Agricultural Research Council, London, 1965, p. 26.
37. **Jenkins, K. J. and Phillips, P. H.,** *J. Nutr.,* 70, 241–246, 1960.
38. **Bennett, D.,** *Vet. Rec.,* 98, 313–321, 1976.
39. **Krook, L., Lutwak, L., Henrikson, P., Kallfelz, F., Hirsch, C., Romanus, B., Bélanger, L. F., Marier, J., and Sheffy, B. E.,** *J. Nutr.,* 101, 233–247, 1971.
40. **Bunce, G. E., Jenkins, K. J., and Phillips, P. H.,** *J. Nutr.,* 76, 17–22, 1962.
41. **Ruegamer, W. R., Michaud, L., Hart, E. B., and Elvehjem, C. A.,** *J. Nutr.,* 32, 101–111, 1946.
42. **Bannerman, R. M.,** *J. Lab. Clin. Med.,* 65, 944–950, 1965.
43. **Frost, D. V., Potter, V. R., Elvehjem, C. A., and Hart, E. B.,** *J. Nutr.,* 19, 207–211, 1940.
44. **Robertson, B. T. and Burns, M. S.,** *Am. J. Vet. Res.,* 24, 997–1002, 1963.
45. **Underwood, E. J.,** *Trace Elements in Human and Animal Nutrition,* 3rd ed., Academic Press, New York, 1971, pp. 195–198.

46. **Norris, W. P., Fritz, T. E., and Taylor, J. A.,** *Am. J. Vet. Res.*, 31, 21–33, 1970.
47. **Underwood, E. J.,** *Trace Elements in Human and Animal Nutrition*, 3rd ed., Academic Press, New York, 1971, p. 125.
48. **Schwarz, K.,** in *Trace Element Metabolism in Animals – 2*, Hoekstra, W. G., Ed., University Park Press, Baltimore, 1974, p. 364.
49. **Biester, H. E., Greenwood, D. A., and Nelson, V. E.,** *North Am. Vet.*, 17, 38–43, 1937.
50. **Schwarz, K. and Mertz, W.,** *Arch. Biochem. Biophys.*, 72, 515–518, 1959.
51. **Toepfer, E. W., Mertz, W., Roginski, E. E., and Polansky, M. M.,** *J. Agric. Food Chem.*, 21, 69–73, 1973.
52. **Anke, M., Grün, M., Dittrich, G., Groppel, B., and Hennig, A.,** in *Trace Element Metabolism in Animals – 2*, Hoekstra, W. G., Ed., University Park Press, Baltimore, 1974, p. 715.
53. **Nielsen, F. H.,** in *Trace Element Metabolism in Animals – 2*, Hoekstra, W. G., Ed., University Park Press, Baltimore, 1974, p. 381.
54. **Hopkins, L. H., Jr.,** in *Trace Element Metabolism in Animals – 2*, Hoekstra, W. G., Ed., University Park Press, Baltimore, 1974, p. 397.
55. **Schwarz, K.,** in *Trace Element Metabolism in Animals – 2*, Hoekstra, W. G., Ed., University Park Press, Baltimore, 1974, p. 361.
56. **Carlisle, E. M.,** in *Trace Element Metabolism in Animals – 2*, Hoekstra, W. G., Ed., University Park Press, Baltimore, 1974, p. 407.
57. **Furr, A. K., Kosikowski, F. V., Bache, C. A., and Lisk, D. J.,** *J. Food Sci.*, 39, 887–891, 1974.
58. **Mitchell, H. H.,** in *Comparative Nutrition of Man and Domestic Animals*, Vol. 2, Academic Press, New York, 1964, p. 20.
59. **Richards, J. I., Bridge, P. S., Spratling, F. R., Abrams, J. T., and Sharman, I. M.,** *Int. Z. Vitaminforsch.*, 40, 567–574, 1970.
60. **Moore, T.,** in *Vitamin A*, Elsevier, Amsterdam, 1957, p. 228.
61. **Sauberlich, H. E., Hodges, R. E., Wallace, D. L., Kolder, H., Canham, J. E., Hood, J., Raica, N., Jr., and Lowry, L. K.,** *Vitam. Horm. N.Y.*, 32, 251–275, 1974.
62. **Bradfield, D. and Smith, M. C.,** *Am. J. Physiol.*, 124, 168–173, 1938.
63. **Worden, A. N.,** *Vet. Rec.*, 72, 2–5, 1960.
64. **Morgan, A. F.,** *North Am. Vet.*, 21, 476–486, 1940.
65. **Wheatley, V. R. and Sher, D. W.,** *J. Invest. Dermatol.*, 36, 169–173, 1961.
66. **Edelstein, S.,** in *The Metabolism and Function of Vitamin D*, Fraser, D. R., Ed., Biochemical Society, London, 1974, p. 43.
67. **Seawright, A. A. and English, P. B.,** *J. Pathol. Bacteriol.*, 88, 503–509, 1964.
68. **Hendricks, J. B., Morgan, A. F., and Freytag, R. M.,** *Am. J. Physiol.*, 149, 319–332, 1947.
69. **Dam, H. and Granados, H.,** *Acta Physiol. Scand.*, 10, 162–171, 1945.
70. **Cordes, D. O. and Mosher, A. H.,** *J. Pathol. Bacteriol.*, 92, 197–206, 1966.
71. **Glavind, J.,** Vitamin E in animal nutrition, *Acta Agric. Scand. Suppl.*, 19, 105, 1973.
72. **Hayes, K. C., Nielsen, S. W., and Rousseau, J. E., Jr.,** *J. Nutr.*, 99, 196–209, 1969.
73. **Elvehjem, C. A., Gonce, J. E., and Newell, G. W.,** *J. Pediatr.*, 24, 436–441, 1944.
74. **Harris, P. L. and Embree, N. D.,** *Am. J. Clin. Nutr.*, 13, 385–392, 1963.
75. **Schwarz, K. and Foltz, C. M.,** *J. Am. Chem. Soc.*, 79, 3292–3293, 1957.
76. **Sharman, G. A. M., Blaxter, K. L., and Wilson, R. S.,** *Vet. Rec.*, 71, 536, 1953.
77. **Omaye, S. T. and Tappel, A. L.,** *J. Nutr.*, 104, 747–753, 1974.
78. **Mason, R. W.,** *N.Z. Vet. J.*, 22, 170–172, 1974.
79. **Stern, F. E., Arnold, A., and Elvehjem, C. A.,** *J. Nutr.*, 17, 485–495, 1939.
80. **Noel, P. R. B., Chesterman, H., Jolly, D. W., and Partington, H.,** *Vet. Rec.*, 89, 260–266, 1971.
81. **Arnold, A. and Elvehjem, C. A.,** *J. Am. Vet. Med. Assoc.*, 95, 187–194, 1939.
82. **Street, H. R. and Cowgill, G. R.,** *Am. J. Physiol.*, 125, 323–334, 1939.
83. **Spector, H., Maass, A. R., Michaud, L., Elvehjem, C. A., and Hart, E. B.,** *J. Biol. Chem.*, 150, 75–87, 1943.
84. **Schaefer, A. E., McKibbin, J. M., and Elvehjem, C. A.,** *J. Biol. Chem.*, 144, 679–685, 1942.
85. **Morgan, A. F., Groody, M., and Axelrod, H. E.,** *Am. J. Physiol.*, 146, 723–738, 1946.
86. **Mitchell, H. H.,** *Comparative Nutrition of Man and Animals*, Vol. 1, Academic Press, New York, 1962, p. 126.
87. **Mitchell, H. H.,** *Comparative Nutrition of Man and Animals*, Vol. 1, Academic Press, New York, 1962, p. 132.
88. **Afonsky, D.,** *Science*, 120, 803–805, 1954.
89. **Jukes, T. H.,** *Am. J. Clin. Nutr.*, 3, 56–63, 1955.
90. **Arnich, L., Lewis, E. M., and Morgan, A. F.,** *Proc. Soc. Exp. Biol. Med.*, 80, 401–404, 1952.
91. **Fouts, P. J.,** *J. Nutr.*, 25, 217–228, 1943.

DIETS (NATURAL AND SYNTHETIC): CATS AND OTHER FELIDAE

P. P. Scott

Domestic and wild cats are highly adapted to a carnivorous diet, so the composition of the mixed diet they are fed under domestic and zoo conditions ought to be related to the average body composition of small mammals, i.e., about 70% water, 14% protein, 10% fat, 5% carbohydrate, and 1% ash, with a calorie value of around 150 kcal/100 g wet diet. Provided that adequate levels of vitamins and minerals (as set out in Tables 1 and 2) are included, there is little doubt that such a diet will be adequate for growth and reproduction. Details of known nutritional requirements are included in recent publications.[1-5] Daily food requirements for domestic cats of various ages are given in Table 3.

To compound and manufacture diets suitable for domestic and zoo cats, it is desirable to use raw materials high in protein and fat, e.g., muscle meats; however, these are expensive and often in short supply. Attempts to replace protein of animal origin by soya or novel proteins derived from microorganisms have not been very successful, since it has proved to be difficult to persuade cats to maintain an adequate food intake when fed formulations containing them. The cat will starve rather than eat food it does not like. Other useful sources of protein for cats include good quality fish and fish meals, infertile hen's eggs, and chicken trimmings; care must be taken to avoid accidentally introducing estrogens, which are very toxic to cats.

On mixed and semipurified diets, the daily protein requirements were found to be equivalent to 1.7 g of N_2/kg body weight for growth and 0.5 g of N_2/kg body weight[6] for maintenance. This is equivalent to 3 g reference protein per kilogram[6] body weight per day for maintenance. The minimum caloric percent of net dietary protein for the adult cat is 8.3 g,[7] which is equivalent to 18% of the total calories as protein. On a practical mixed diet containing fish and meat, the minimum protein to maintain nitrogen balance in adults was found to be 21% dry weight of diet, but for growth in kittens it was necessary to increase this to a minimum of 33%.[8] Manufactured diets often contain as much as 40 to 50% protein, which allows for a reduction in biological value due to the use of less expensive ingredients of poorer biological quality and for processing damage to protein or imbalance of amino acids. However, these high levels of dietary protein may be wasteful, since the excess is used to provide energy. Diets low in protein are not acceptable (it is almost impossible to get a cat to eat a protein-free diet voluntarily); inadequate protein intake results in loss of weight and a fall in plasma protein.[9] Since cats are intermittent feeders in the wild, storing reserve protein in muscle and excess calories in depot fat subcutaneously and around the viscera, these reserves permit a well-nourished adult cat to survive without food for up to 6 weeks.

Cats digest animal fats very efficiently; in spite of suggestions to the contrary, we have successfully fed up to 60% of the dry weight of the diet as fat.[10] Recent reports suggest that a source of animal lipids is required, rather than linoleic and linolenic acids (the desaturation of which are rate limiting in the cat).[11] Polyunsaturated fatty acids, as in cod liver oil, can be toxic, especially in the absence of vitamin E.[1,3]

Carbohydrates are not essential in diets for cats, but to reduce the cost and improve manufacturing processes they are normally used to provide up to 50 or 60% of the calorie intake. Lactose is not well tolerated by some adult cats, due to lack of intestinal lactase. Sucrose is tolerated, but dextrins and cooked starches are better utilized.[1]

Some examples of formulations for semipurified diets based on casein and used in investigating the requirements of domestic cats are given in Tables 4 through 6.

Examples of diets used for cats are given below, showing the varied water content. Care must be taken to give plenty of fluid to drink, especially when feeding dry foods.

	Approximate percentage of dry solids
Fresh meat	30
Fresh fish	20
Processed canned foods	25
Soft-moist manufactured foods	75
Processed dried foods	90

Some suggestions for practical feeding of domestic cats[5] follow:

Morning snack (optional) – ½ teacup of cow's (or goat's) milk, with a teaspoonful of cod liver oil stirred in twice a week. Dried or diluted, unsweetened, condensed milk can also be used.

Evening meal (to be given at a regular time) – Portion of canned or fresh food (containing 75% moisture); 6 to 9 oz (170 to 250 g) is equivalent to 2 to 3 level teacups. Adjust the quantity according to the size, age, and activity of the cat. A varied diet is important and the following alternatives can be offered: small tin of meat-based canned cat food; portion of fresh white fish, herring, mackerel, or fish pieces cooked lightly and served with bones and juices; 4 to 6 oz (96 to 170 g) liver, preferably raw, served with biscuit meal or dry cat food; small tin of fish-based cat food; minced or chopped fresh or deep-frozen meat (thaw before serving); chicken pieces (with bones) or small tin of chicken-based food; portion of rabbit cooked lightly and served with bones; ½ teacup of dry cat food with beaten egg poured over; meat or fish plate scraps or cooking trimmings when available in sufficient quantity (must be fresh).

In 1968, the Veterinary Officer and the Curator of Mammals of the Zoological Society, London, with the help of the author and Dr. Alan Walker of Spillers, Ltd., formulated a tinned diet known as ZF6 for carnivores (Table 7). Other vitamins are added to the formula before processing (see Table 8) but have not been analyzed in the finished product. When the food is fed at the rate of approximately 70 g/kg body weight, adequate minerals and vitamins are provided; it should constitute at least 50% of the animal's total diet. Vitamin A has been fixed at a high level to fulfil the requirements of the cat family.

Since November 1968, nearly all the small mammals at the London Zoo, which normally eat food mainly of animal origin, have obtained the greater proportion of their nutrients from ZF6. Many of the larger mammals within this category also received part of their diet in this form. As had been anticipated, animals that had grown up accustomed to eating raw meat (with a supplement) proved resistant to change, but young animals accepted the tinned diet fairly readily. Routine clinical and radiological examinations of animals fed on ZF6 carried out over a period of 2 years revealed no evidence of the osteodystrophies encountered in the past in young animals which had been fed a diet of raw meat deficient in calcium.

Relatively little work has been performed regarding mineral requirements, but calcium and iodine deficiency occur when both domestic and zoo cats are fed unsupplemented muscle meat diets.[4] Excess magnesium presents a problem in urinary excretion to the domestic tom cat, and may be a contributory factor in the development of urolithiasis.[16] Possibly because of their carnivorous habit, cats have lost the ability to convert β-carotene to vitamin A,[4] so both domestic and wild cats are dependent on supplies of fully formed vitamin A, the principal source of which is the liver of the prey. Alternatively, stabilized vitamin A must be added to manufactured diets. Attention should also be paid to the levels of thiamine (B_1) in manufactured foods; owing to destruction in processing and loss in storage, thiamine deficiency occurs quite regularly in cats fed manufactured diets.[4] Requirements for vitamin D in food are known to be very small indeed, and those for vitamins C and K are negligible.

Milk mixtures for hand rearing kittens should simulate cat's milk as far as possible. Litters of more than five kittens may require supplementation, and the daily weight gain of kittens should be 10 g or more. There should never be a loss. The milk output of the queen depends on her food conversion efficiency; she may consume two to three times as much food as in the nonpregnant state. Even then she may lose weight. According to Widdowson,[17] cat's milk contains 9.5 g protein, 6.8 g fat, 10 g lactose (by difference), 35 mg calcium, and 70 mg phosphorus. It provides 142 cal/100 g milk. This explains why kittens do not thrive when hand reared on fresh cow's milk, but a mixture of dried cow's milk made up at twice the normal strength is satisfactory. A suitable mixture consists of 20 g skim milk powder dissolved in 90 ml of warm water, to which is added 10 ml of corn oil or other tocopherol-fortified fat. The oil should be fortified with 80 μg vitamin A (retinol) per milliliter for the first 2 days, then 50 μg/ml subsequently.[18] It may be replaced by yolk of hen's egg. The entire mixture is warmed to 37°C and fed at this temperature. The feeding regime is shown in Table 9.

Cats require drinking water in addition to the considerable quantity of water present in natural diets. Lack of water, especially in dry diets, is thought to be another of the causes contributing to urolithiasis.[16] Cats have an excretory problem in dealing with waste products containing nitrogen and sulfur; they form a unique amino acid, felinine,[19] in the kidney, apparently to assist in the excretion of these waste products of metabolism. The kidney is also exceptionally rich in vitamin A, present in the fat that is found in the proximal convoluted tubules.[20]

Benzoic acid, a preservative permitted in human foods, has been found to be extremely toxic to cats,[21] but it can safely be replaced in cat food by sorbic acid.

Table 1
RECOMMENDED VITAMIN ALLOWANCES FOR THE CAT

Vitamin	Daily dietary allowance	Comment
A (alcohol or ester)	500–700 μg (1500–2100 IU)	Cannot utilize carotene
D (cholecalciferol)	50–100 IU	May synthesize in skin
Essential fatty acids	1% of total fatty acids	Tolerates high saturated fatty acid intake
K (menaquinone)	Negligible	Intestinal synthesis (probable)
E (α-tocopherol)	0.4–4.0 mg	Proportional to polyunsaturated fatty acid content
B_1 (thiamine)	0.2–1.0 mg (or 0.1 mg/50 kcal diet)	Increase in lactation or fever
B_2 (riboflavin)	0.15–0.2 mg	Increase in lactation or fever and on high fat diet
Niacin (nicotinic acid)	2.6–4.0 mg	Increase in lactation or fever; cannot synthesize
B_6 (pyridoxine)	0.2–0.3 mg	Increase in lactation or fever
Pantothenic acid	0.25–1.0 mg	
Biotin	0.1 mg	
Choline	100 mg	
Inositol	10 mg	Essential
B_{12} (cyanocobalamin)	Unknown	Intestinal synthesis (Co present)
Folic acid (folacin)	Unknown	Must be present in food
C (ascorbic acid)	Negligible	Metabolic synthesis

From Scott, P. P., in *Gaines Basic Guide to Canine Nutrition,* 3rd ed., Gaines Dog Research Center, White Plains, N.Y., 1974, 81. With permission.

Table 2
RECOMMENDED DAILY MINERAL ALLOWANCES FOR THE CAT

	Range	Comment
Major Mineral Elements		
Na	20–10 mg	This is minimum intake
NaCl	1000–1500 mg	Common salt requirement
K	80–200 mg	Adequate in meat and fish
Ca	200–400 mg	In growth and lactation give 400 mg.
P	150–400 mg	Ca/P ratio between 0.9 and 1.0
Minor and Trace Elements		
Mg	8–11 mg	Usually present in larger amounts
Fe	5 mg	Iron from hemoglobin is available.
I	100–400 μg	Can be deficient in meat
Mn	200 μg	Normally adequate in a mixed diet
Zn	250–300 μg	Normally adequate in a mixed diet
Co	100–200 μg	Normally adequate in a mixed diet

From Scott, P. P., in *Gaines Basic Guide to Canine Nutrition,* 3rd ed., Gaines Dog Research Center, White Plains, N.Y., 1974, 85. With permission.

Table 3
DAILY FOOD REQUIREMENTS OF CATS ACCORDING TO AGE

	Expected weight		Daily cal per body weight		Daily ration	
Age	kg	lb	kcal/kg	kcal/lb	g	oz
Newborn	0.12	0.25	380	190	30	1
5 weeks	0.5	1.0	250	125	85	3
10 weeks	1.0	2.0	200	100	140	5
20 weeks	2.0	4.5	130	65	175	6
30 weeks	3.0	6.5	100	50	200	7
Adult ♂	4.5	10.0	80	40	240	8.5
Adult ♀ (pregnant)	3.5	7.5	100	50	240	8.5
Adult ♀ (lactating)	2.5	5.5	250	125	415	14
Neuter ♂	4.0	9.0	80	40	200	7
Neuter ♀	2.5	5.5	80	40	140	5

From Scott, P. P., in *Gaines Basic Guide to Canine Nutrition,* 3rd ed., Gaines Dog Research Center, White Plains, N.Y., 1974, 79. With permission.

Table 4
SEMIPURIFIED DIETS USED FOR CATS

Ingredient (g/kg dry diet)	Gershoff and Hegsted[12]	Miller and Allison[6]	Scott and Greaves[22]
Casein	350	250	400
Sucrose	337	352	287[a]
Lard	125	241	200
Arachris oil	–	–	50
Corn oil	125	50	–
Cod liver oil	–	20	–
Vitamin mix	–[b]	–[b]	15[b]
Liver powder	–	30	6
Choline chloride	3	2	3
Salt mix	50[c]	40[d]	20[e]
Gelatin or agar	10	66	–
Sugar beet residue	–	–	50

[a] Subsequently replaced by dextrin.
[b] Vitamins shown in Table 5.
[c] Salt mixture of Hegsted et al. (Table 6).
[d] Salt mixture of Wesson (Table 6).
[e] Salt mixture of Hubbell et al. (Table 6).

Table 5
VITAMIN SUPPLEMENTS, RECALCULATED PER KILOGRAM OF DIET, USED WITH THE DIETS SHOWN IN TABLE 4

	Gershoff and Hegsted[12]	Miller and Allison[6]	Scott and Greaves[22]
Thiamine (mg)	4	20	11.7–13.5
Riboflavin (mg)	8	20	22.5–24.0
Nicotinic acid (mg)	40	80	75
Pyridoxine (mg)	4	20	6–7.5
Ca pantothenate (mg)	20	80	30–36
Cyanocobalamin (μg)	–	–	30
Folic acid (mg)	1.0	1.0	1.0[b]
PABA (mg)	–	80	–
Biotin (mg)	0.2	1.0	1.0[b]
Inositol (mg)	–	200	100
Vitamin A (μg)	8,300	13,000[a]	5,000–10,000
Vitamin D (IU)	1,000	4,000	1,000–2,000
α-Tocopherol acetate (mg)	–	40	7.5–15.0
Vitamin K (mg)	1.0	20	–

[a] From cod liver oil.
[b] Supplied in liver powder.

Table 6
SALT MIXTURES (RECALCULATED) USED IN SEMIPURIFIED DIETS SHOWN IN TABLE 4

Ingredients (g/kg salt mix)	Hegsted et al. (salts IV)[13]	Wesson[14]	After Hubbell et al.[15]
$CaCO_3$	300	210	542
$Ca_3(PO_4)_2$	–	149	–
$Ca_2H(PO_4)_2 \cdot 4H_2O$	75	–	–
KH_2PO_4	–	310	212
K_2HPO_4	322.5	–	–
KCl	–	120	112
NaCl	167.5	105	69
$MgCO_3$	–	–	25
$MgSO_4$	–	90	16
$MgSO_4 \cdot 7H_2O$	102	–	–
$FePO_4 \cdot 4H_2O$	–	14.7	20.5
$Fe(C_6H_5O_7)_2 \cdot 6H_2O$	27.5	–	–
$CuSO_4 \cdot 5H_2O$	0.30	0.39	0.90
$ZnCl_2$	0.25	–	0.60
$ZnSO_4$ (anhydrous)	–	0.20	–
$MnSO_4 \cdot H_2O$	–	–	0.35
$MnSO_4 \cdot 4H_2O$	5.0	–	–
$CoCl_2 \cdot 6H_2O$	–	–	0.65
KI	0.8	0.05	0.08
$Al_2(SO_4)_3 \cdot K_2SO_4 \cdot 24H_2O$	–	–	0.17
NaF	–	–	0.75

Table 7
COMPOSITION OF DIET ZF6

Analysis	Mean	Range
Moisture (%)	75.0	73.0–76.0
Crude protein (N × 6.25) (%)	10.3	10.0–11.0
Crude fat (by hydrolysis) (%)	4.5	4.3–4.7
Ash (%)	1.8	1.6–2.0
Carbohydrate (by difference) (%)	8.4	8.0–9.0
	100.0	
Calcium (%)	0.5	0.4–0.6
Phosphorus (%)	0.3	0.27–0.35
Ca/P ratio	1.6	1.4–1.8
Sodium chloride (%)	0.2	0.15–0.25
Vitamin A (IU/g)	20	15–25
Vitamin E (as α-tocopherol) (μg/g)	40	35–45
Vitamin B_1 (μg/g)	2.5	2.3–2.8
Calories per 100 g	115	110–120

Table 8
VITAMINS ADDED TO DIET ZF6 BEFORE PROCESSING

Vitamin D (IU/g)	0.44
Vitamin B_2 (μg/g)	5.5
Vitamin B_6 (μg/g)	2.2
Vitamin B_{12} (μg/g)	0.022
Choline (mg/g)	0.88
Folic acid (μg/g)	0.022
Nicotinic acid (μg/g)	8.8
Pantothenic acid (μg/g)	5.5

Table 9
FEEDING ROUTINE FOR HAND-REARED KITTENS

Age (weeks post-partum)	Feed	Number of meals in 24 hr	Volume of milk per meal		Expected body weight	
			Teaspoon	ml	oz	g
1	Milk mixture in bottle	12–9	½–1½	2–7	3½–7	100–200
2	Milk mixture in bottle	9	1½–2	7–9	7–10½	200–300
3	Milk mixture in bottle	9	2	10	10½–13	300–360
4	Milk mixture in bottle; introduce solids	7	2	10	12½–15	350–420
5	Reduce bottle; increase solids	7			14½–18	400–500
6	Milk in bowl; solids	6			16–21½	450–600
7	Weaning completed	3			19½–25	550–700

From Raleigh, I., Scott, P. P., Jackson, E., and Jackson, O., *Practical Guide to Cats,* Hamlyn, London, 1976, 88. With permission.

REFERENCES

1. **Scott, P. P.**, in *Gaines Basic Guide to Canine Nutrition,* 3rd ed., Gaines Dog Research Center, White Plains, N.Y., 1974, 76–89.
2. **Scott, P. P.**, Nutrition and disease, in *Feline Medicine and Surgery,* 2nd ed., Catcott, E. J., Ed., American Veterinary Publications, Santa Barbara, Calif., 1975, 131–144.
3. National Research Council, *Nutrient Requirements of Domestic Animals, No. 10, Nutrient Requirements of Laboratory Animals,* National Academy of Sciences, Washington, D.C., 1972.
4. **Scott, P. P.,** *Symp. Zool. Soc. London,* 21, 21–36, 1968.
5. **Raleigh, I., Scott, P. P., Jackson, E., and Jackson, O.,** *Practical Guide to Cats,* Hamlyn, London, 1976, 83–88.
6. **Miller, S. A. and Allison, J. B.,** *J. Nutr.,* 64, 493–501, 1958.
7. **Greaves, J. P.,** in *Canine and Feline Nutritional Requirements,* Graham-Jones, O., Ed., Pergamon, New York, 1965, 33–45.
8. **Greaves, J. P. and Scott, P. P.,** *Br. J. Nutr.,* 14, 361–369, 1960.
9. **Cole, B. L.,** *Br. J. Nutr.,* 14, 419–424, 1960.
10. **Humphreys, E. R. and Scott, P. P.,** *Proc. Nutr. Soc.,* 21, xviii, 1962.
11. **Rivers, J. P. W., Sinclair, A. J., Moore, D. P., and Crawford, M. A.,** *Proc. Nutr. Soc.,* 35, 68–70A, 1975.
12. **Gershoff, S. N. and Hegsted, D. M.,** *Lab. Invest.,* 6, 227–240, 1957.
13. **Hegsted, D. M., Mills, R. C., Elvehjem, C. A., and Hart, H. B.,** *J. Biol. Chem.,* 138, 459–466, 1941.
14. **Wesson, L. G.,** *Science,* 75, 339–340, 1932.
15. **Hubbell, R. B., Mendel, L. B., and Wakeman, A. J.,** *J. Nutr.,* 14, 273–285, 1937.
16. **Finco, D. S., Kneller, S. K., and Crowell, W. A.,** Diseases of the urinary system, in *Feline Medicine and Surgery,* 2nd ed., Catcott, E. J., Ed., American Veterinary Publications, Santa Barbara, Calif., 1975, 284–295.
17. **Widdowson, E. M.,** in *Canine and Feline Nutritional Requirements,* Graham-Jones, O., Ed., Pergamon, New York, 1965, 9–17.
18. **Scott, P. P. and Thompson, S. Y.,** *Proc. Nutr. Soc.,* 28, 8A, 1969.
19. **Arizonis, P. V. and Wriston, J. C., Jr.,** *Biochim. Biophys. Acta,* 34, 279–281, 1959.
20. **Moore, T., Sharman, I. M., and Scott, P. P.,** *Res. Vet. Sci.,* 4, 397, 1937.
21. **Bedford, P. G. C. and Clarke, E. G. C.,** *Vet. Rec.,* pp. 53–58, January 1972.
22. **Scott, P. P. and Greaves, J. P.,** unpublished.

DIETS FOR MUSTELIDAE

A. U. Shump

The members of the family Mustelidae, although carnivores, are quite diverse in their diets. Weasels, polecats, minks, and fishers are almost exclusively flesh eaters, wolverines prefer carrion, otters exist mainly on aquatic life, and the rest eat both plant and animal matter.

Weasels are very quick and kill their prey very efficiently. Their prey varies from mouse-sized to squirrel- or rabbit-sized animals, depending on the size of the weasel.

Martens and fishers are somewhat more specialized in their diets. Martens are quite arboreal and prey mainly on squirrels in many areas; fishers seem to prefer porcupines.

Although wolverines prefer carrion, they feed on a variety of other items, depending on the season. Eggs of ground-nesting birds are often eaten in early summer, the larvae of wasps in late summer, and berries in the fall. In the winter, living animals are preyed upon more than at other times of the year.

Ratels are particularly fond of insects, but also eat lizards, tortoises, and many snakes, including cobras, as well as small mammals. They have a particular fondness for honey, which accounts for their other name, honey badger.

Badgers are probably more omnivorous than any of the other mustelids. Besides eating a variety of small mammals, they feed on insects, slugs, worms, berries, and seeds. Large amounts of beech mast and acorns are eaten in the autumn. Suckling females apparently feed entirely on worms.

Otters feed primarily on aquatic animals, but birds and small land mammals are also consumed. *Aonyx* feed mainly on crabs and other crustaceans, *Lutra* more on fish, and *Paraonyx* primarily on relatively soft matter, such as frogs, eggs, and small land vertebrates; therefore, there is little direct competition where these three genera occur together. The sea otter feeds on sea urchins, mollusks, crabs, and fish; it floats on its back with a rock on its chest, and breaks open shells of mussels and clams by hitting them on the rock.

Table 1 shows how variable the natural diets of the animals within this group are, as well as the specific preferences of some of the species.

Table 1
NATURAL DIETS

Species (common name)	Food of animal origin	Fruits	Seeds and crops	Vegetative plant parts	Other foods	Reference
Mustela vison (Mink)	Small mammals, birds, fish, aquatic invertebrates					1
Mustela nivalis (Least weasel) *Mustela erminea* (Short-tailed weasel) *Mustela frenata* (Long-tailed weasel)	Mice, other small mammals, birds, cold-blooded vertebrates, invertebrates (earthworms)					1, 2
Mustela nigripes (Black-footed ferret)	Prairie dogs					2
Mephitis (Striped skunk, hooded skunk)	Small mammals, birds, eggs, reptiles, insects	Various				1
Conepatus leuconotus (Hog-nosed skunk)	Insects, grubs, small vertebrates (snakes)	Various				2
Spilogale putorius (Spotted skunk)	Rodents, other small animals, insects			Some in summer		2
Meles meles (Eurasian badger)	Beetles, larvae of bees and wasps, worms,[a] slugs, young rabbits, carrion, small mammals	Berries, grapes	Acorns, seeds, corn, oats	Various tubers and roots	Combs of bees and wasps	1, 2

[a] Suckling females feed entirely on worms.

Table 1 (continued) NATURAL DIETS

Species (common name)	Food of animal origin	Fruits	Seeds and crops	Vegetative plant parts	Other foods	Reference
Gulo gulo (Wolverine)	Carrion, eggs of ground-nesting birds, wasp larvae, some small mammals	Berries				2
Grison (Grison)	Mice, birds, cold-blooded vertebrates, invertebrates	Some fruit				2
Eira (Tayra)	Birds, occasionally chickens, small mammals – cricetid rodents, agoutis, rabbits, tree squirrels, small deer, marmosets	Bananas and other fruit			Honey	2, 3
Martes americana (American marten)	Squirrels	Fruits	Nuts			1, 2
Martes pennanti (Fisher)	Porcupines, sometimes deer	Fruits	Nuts			1, 2
Enhydra (Sea otter)	Sea urchins, crabs, fish, mussels, clams, sea snails					2

Table 1 (continued)
NATURAL DIETS

Species (common name)	Food of animal origin	Fruits	Seeds and crops	Vegetative plant parts	Other foods	Reference
Lutra (Otter)	Fish, crayfish, frogs, turtles, aquatic invertebrates, birds, land mammals – rodents and rabbits, insects					2
Paraonyx (African small-clawed otter)	Frogs, eggs, small land vertebrates					2
Amblonyx (Oriental small-clawed otter)	Mussels, snails, crabs, fish					2
Pteronura (Giant otter)	Fish, eggs, birds, aquatic mammals					2
Aonyx	Mainly crabs, also mollusks, turtles, fish, frogs, lizards, aquatic birds, small mammals					2
Mydaus (Malay badger)	Worms, insects, entrails of chickens[b]					2
Poecilogale (White-naped weasel)	Small mammals, birds, snakes, insects, larvae					2

[b] Captives.

Table 1 (continued)
NATURAL DIETS

Species (common name)	Food of animal origin	Fruits	Seeds and crops	Vegetative plant parts	Other foods	Reference
Ictonyx (Zorilla)	Mice, large insects, eggs, snakes					2
Poecilictis (Libyan striped weasel)	Mice, large insects, eggs, snakes					2
Vormela peregusna (Marbled polecat)	Rodents, birds, reptiles, sometimes poultry					2
Melogale (Ferret badger)	Small animals, insects, earthworms	Fruits				2
Arctonyx (Hog badger)	Flesh, fish, earthworms	Fruits				1
Mellivora (Ratel)	Insects, small mammals, lizards, tortoises, snakes (cobras), carrion; attack sheep			Plant matter	Honey	2

Table 2
CHEMICALLY DEFINED DIETS

Species (common name)	Ingredients	% Composition				Reference
Mustela vison[a] (Mink)		Diet 1[b]	Diet 2[c]	Diet 3[d]	Diet 4[e]	4
	Horsemeat	10	10			
	Rabbit	10	10	10	20	
	Kidneys	5	5			
	Cottage cheese	5	5			
	Liver	10	10	5	5	
	Poultry by-products (heads and entrails)			35	25	
	Tripe	20	20	10	10	
	Lung	5	5	5	5	
	Fish	20	20	15	15	
	Cereal mixture, fortified	15	15	20	20	
	Fat content of diet (% of dry matter)	20	22–24	22–24	20–22	
Mustela vison[f] (Mink)	Cereal	20				5, 6
	Tripe	15				
	Liver (beef, inedible)	5				
	Poultry ("spent chickens")[g]	20				
	Fish (mixture of cod, haddock and flounder)	30				
	Lung (beef)	10				

[a] Various diets are used on different mink farms. Those listed here are just examples.
[b] Reproduction, March to May.
[c] Lactation, May to July.
[d] Growth, July to September.
[e] Fur production, September to December.
[f] This diet is used by the Michigan State University mink farm and has also been used successfully on the following species: *Mustela putorius, M. nivalis, M. frenata, M. erminea, Conepatus, Spilogale putorius*, and *S. pygmaea.*
[g] Birds 18 to 20 months old, which are no longer profitable for egg production and are not in demand as food for human consumption.[7]

Table 3
PARTIALLY PURIFIED DIETS

Species (common name)	Ingredients	% Composition[a]	Reference
Mustela vison (Mink)	Growth diet:		4, 8
	Sucrose	66	
	Casein	19	
	Cottonseed oil	8	
	Salts	4	
	Cod-liver oil	3	
	Choline chloride	100 mg/100 g	
	p-Aminobenzoic acid	50 mg/100 g	
	Inositol	25 mg/100 g	
	Nicotinic acid	4 mg/100 g	
	α-Tocopherol	2 mg/100 g	
	Calcium pantothenate	1.5 mg/100 g	
	2-Methyl-1,4-naphthoquinone	0.5 mg/100 g	
	Riboflavine	0.4 mg/100 g	
	Pyridoxine hydrochloride	0.2 mg/100 g	
	Thiamine chloride	0.2 mg/100 g	
	Folic acid	0.1 mg/100 g	
	Biotin	0.025 mg/100 g	

[a] Unless otherwise specified.

REFERENCES

1. **Crandell, L. S.,** *Management of Wild Mammals in Captivity,* University of Chicago Press, Chicago, 1964.
2. **Walker, E. P.,** *Mammals of the World,* Johns Hopkins Press, Baltimore, 1968.
3. **Dawson, G. A.,** personal communication, 1975.
4. **Leoschke, W. L.,** Mink Nutrition Research at the University of Wisconsin, University of Wisconsin Research Bulletin No. 222, 1960.
5. **Aulerich, R. J.,** personal communication, 1975.
6. **Shump, A. U.,** Maternal Behavior of the Ferret, *Mustela putorius,* Ph.D. thesis, Michigan State University, 1975.
7. **Aulerich, R. J. and Schaible, P. J.,** The use of "spent" chickens for mink feeding, *Mich. Agric. Exp. Stn. Q. Bull.,* 47(3), 451–458, 1965.
8. **Aitken, F. C.,** Feeding of Fur-bearing Animals, Technical Communication No. 23, Commonwealth Bureau of Animal Nutrition, Rowett Research Institute, Bucksburn, Aberdeen, Scotland, 1963.

REFERENCES

1. **Chao, J. and Ball, G. H.**, in *Proc. 4th Int. Colloq. on Invertebrate Tissue Culture,* Maramorosch, K., Ed., Academic Press, New York, in press.
2. **Hirumi, H.**, in *Invetebrate Tissue Culture: Applications to Fundamental Research,* Maramorosch, K., Ed., Academic Press, New York, 1976, 233–268.
3. **Green, A. E., Charney, J., Nichols, W. W., and Coriel, L. L.,** *In Vitro,* 7, 313–322, 1972.
4. **Goodwin, R. H.,** *In Vitro,* 11, 369–378, 1975.
5. **Schneider, I.**, in *Tissue Culture Methods and Applications,* Kruse, P. F., Jr. and Patterson, M. K., Jr., Eds., Academic Press, New York, 1973, 150–152.
6. **Hink, W. F., Richardson, B. L., Schenk, D. K., and Ellis, B. J.**, in *Proc. 3rd. Int. Colloq. on Invertebrate Tissue Culture,* Řeháček, J., Blaškovič, D., and Hink, W. F., Eds., Slovak Academy of Science, Bratislava, 1973, 195–208.
7. **Federoff, S.,** *J. Natl. Cancer Inst.,* 38, 607–611, 1967.
8. **Grace, T. D. C.,** *Nature,* 195, 788–789, 1962.
9. **Yunker, C. E., Vaughn, J. L., and Cory, J.,** *Science,* 155, 1565–1566, 1967.
10. **Hink, W. F. and Ellis, B. J.,** *Curr. Top. Microbiol. Immunol.,* 55, 19–28, 1971.
11. **Hsu, S. H., Mao, W. H., and Cross, J. H.,** *J. Med. Entomol.,* 7, 703–707, 1970.
12. **Chao, J. and Ball, G. H.,** *In Vitro,* 8, 406, 1973.
13. **Chao, J. and Ball, G. H.**, in *Proc. 4th Int. Colloq. on Invertebrate Tissue Culture,* Maramorosch, K., Ed., Academic Press, New York, in press.
14. **Mitsuhashi, J.,** *Appl. Entomol. Zool.,* 8, 64–72, 1973.
15. **Landureau, J. C.,** *Exp. Cell Res.,* 50, 323–337, 1968.
16. **Louloudes, S. J., Vaughn, J. L., and Dougherty, K. A.,** *In Vitro,* 8, 473–479, 1973.
17. **Hink, W. F.,** *Nature,* 226, 466–467, 1970.
18. **Řeháček, J. and Brzostowski, H. W.,** *J. Insect Physiol.,* 15, 1683–1686, 1969.
19. **Clements, A. N. and Grace, T. D. C.,** *J. Insect Physiol.,* 13, 1327–1332, 1967.
20. **Grace, T. D. C. and Brzostowski, H. W.,** *J. Insect Physiol.,* 12, 625–633, 1966.
21. **Gilby, A. R. and McKellar, J. W.,** *J. Insect Physiol.,* 20, 2219–2224, 1974.
22. **Mitsuhashi, J. and Grace, T. D. C.,** *Appl. Entomol. Zool.,* 5, 182–188, 1970.
23. **Vaughn, J. L., Louloudes, S. J., and Dougherty, K.,** *Curr. Top. Microbial. Immunol.,* 55, 92–97, 1971.
24. **Larsen, W. P.,** *J. Insect Physiol.,* 13, 613–619, 1967.
25. **Gallagher, B. M. and Hartig, W. J.,** *In Vitro,* 12, 165–172, 1976.
26. **Parker, R., Edgar, R., and Yunker, C. E.,** in *Proc. Conf. Insect and Mite Nutrition,* Rodrigues, J. G., Ed., North Holland, Amsterdam, 1972, 397–406.
27. **Saska, J., Grzelakowska-Sztabert, B., and Zielinska, Z. M.,** *J. Insect Physiol.,* 18, 1733–1737, 1972.
28. **Marks, E. P., Reinecke, J. P., and Caldwell, J. M.,** *In Vitro,* 3, 85–92, 1968.
29. **Mitsuhashi, J.,** *J. Insect Physiol.,* 22, 397–402, 1976.
30. **Landureau, J. C. and Jollés, P.,** *Exp. Cell Res.,* 54, 391–398, 1969.
31. **Landureau, J. C.,** *Exp. Cell Res.,* 54, 399–402, 1969.
32. **Landureau, J. C.,** *C. R. Acad. Sci. Paris,* 270, 3288–3291, 1970.
33. **Sanborn, R. C. and Haskell, J. A.,** in *Proc. 11th Int. Congr. Entomol.,* B3, Strouhal, H. and Beir, M., Eds., Organisationskomittee der XI Internationalen Kongresses fuer Entomologie, Vienna, 1960, 237–243.
34. **Stockdale, H. and Gardiner, G. R.,** in *Proc. 4th Int. Colloq. on Invertebrate Tissue Culture,* Maramorosch, K., Ed., Academic Press, New York, in press.
35. **Kurtti, T. J., Chaudhary, S. P. S., and Brooks, M. A.,** *In Vitro,* 11, 274–285, 1975.
36. **Eide, P. E.,** *J. Insect Physiol.,* 21, 1431–1438, 1975.
37. **Eschalier, G. and Ohanessian, A.,** *In Vitro,* 6, 162–172, 1970.
38. **Goodwin, R. H.,** *In Vitro,* 11, 369–378, 1975.

DIETS (NATURAL AND SYNTHETIC): VIVERRIDAE

M. L. Gorman

NATURAL DIETS

As a group, the viverrids are not highly specialized predators; rather, they eat what they can get: small vertebrates, carrion, eggs, a wide range of invertebrates, and plant material. The exact composition of their diet thus depends to a large extent on the local availability of prey.

To describe fully the natural diet of an animal, one must have accurate, quantitative data on what it eats in a variety of habitats, throughout its geographical range, and at different times of the year. Complete information of this kind is available for no species of viverrid; indeed, for most species information is very scanty. Many of the African species are fairly well known qualitatively, but the Asiatic and Madagascan ones are almost totally unstudied.

Qualitative Studies

Most studies on viverrid diets have been qualitative, based on an examination of what kinds of remains are found in a restricted number of stomachs or scats (feces). Table 1 gives examples of such studies, including species from all the subfamilies. Most of the information is based on very small samples, and the actual range of items taken is probably greater than that indicated, in almost all cases.

Quantitative Studies

Detailed quantitative studies have been made on very few species indeed. The most detailed are those made on introduced populations of *Herpestes auropunctatus* in the West Indies, Hawaii, and Fiji.

Several methods of recording the data have been used. A study of 493 scats from Hawaii[19] noted in what percentage of the scats a particular prey item occurred: 72% of scats from cane fields and 24% from pastures contained rodent remains, while 80% of pasture scats and 41% of cane field scats contained insects. Plant remains were present in almost all scats, while birds appeared in about 5%. This type of analysis is unsatisfactory in that equal importance is given to large and small prey items; clearly, they are not of equal nutritional value.

In a more laborious method, all the remains in the scat are separated and identified. The weight (or volume) of each type of remains is then expressed as a percentage of the weight (or volume) of the total remains. In such an analysis of 86 Hawaiian scats,[18] mammals accounted for 39.6% of the volume of remains, birds 4.1%, insects 27%, and plants 29.3%. The disadvantage of this method stems from the fact that different prey types contain differing proportions of undigestible material. Consequently, the weights of undigested material in scats do not give a clear indication of the real importance of each prey type in the diet. This problem can be overcome by deriving correction factors[29] which relate the weight of undigested material in the scat to the actual weight of prey eaten. These correction factors are derived from feeding trials with captive animals and are calculated by dividing the wet weight of prey eaten by the dry weight of undigested material of that prey type remaining in the scats.

Correction factors have been calculated for *H. auropunctatus* using a variety of prey types,[20] and are shown in Table 2. These correction factors have been applied to prey remains in scats from a variety of Fijian habitats;[20] the results are shown in Table 3.

SYNTHETIC DIETS

There are currently no detailed studies on the nutritional requirements of captive viverrids. All species thrive on a mixture of fruit and meat.

The Zoological Society of London feed their viverrids 55% by weight fruit and vegetables and 45% Spillers ZF 6* (ZF 6 is a canned food for carnivores based on the nutritional requirements of the domestic cat. Basic ingredients consist of mammalian and fish flesh, cereal, and a standard vitamin and mineral concentrate).

The Royal Zoological Society of Scotland uses the following diets:

Species	Diet (by weight)
Suricata suricatta and mongooses	ZF 6 or meat and egg
Paguma larvata	2/3 ZF 6 + 1/3 mixed fruit, or 1/2 meat + 1/2 fruit
Genetta sp.	2/3 ZF 6 + 1/3 mixed fruit, or 1/2 meat + 1/2 fruit
Civettictis civetta	ZF 6 or 3/4 meat + 1/4 fruit
Crossarchus obscurus	3/4 ZF 6 + 1/4 fruit or 1/2 meat + 1/2 fruit
Arctictis binturong	1/2 ZF 6 + 1/2 fruit

All given mice, day-old chicks, and rabbit when available.

The New York Zoological Park feeds the following diets:[30]

Species	Diet (by weight)
Nandinia binotatata	30% banana, 20% apple, 20% Diet M,[a] 20% orange, 10% meat
Arctictis binturong	50% banana, 20% horsemeat, 20% orange, 5% apple, 5% yam
Herpestes edwardsi	50% Diet M, 15% apple, 15% banana, 15% chopped meat, 5% kale
Mungos mungo	60% Diet M, 15% chopped meat, 10% apple, 10% banana, 5% kale
Suricata suricatta	60% Diet M, 20% whole meat, 10% apple, 10% banana
Crossarchus obscurus	60% Diet M, 15% chopped meat, 10% apple, 10% banana, 5% kale

[a] Diet M: 30 lb chopped meat, 20 cups meat, 1/2 cup ground limestone, 1/2 cup Paltone®.

* Spillers Food Ltd., Old Change House, 4-6 Cannon Street, London, EC4M 6XB, England, U.K.

Table 1
CATEGORIES OF FOODS FOR VIVERRIDS

This table summarizes the range of food items in the diets of various viverrids, based on an analysis of stomach contents and scat remains. An x implies that a broad category of food is taken. Where a more precise identification (e.g., frogs rather than amphibia) is available it is indicated in the appropriate column.

			Foods											
Subfamily	Genus and species	Comment	Earth-worms	Molluscs	Crustaceans	Insects	Fish	Amphibia	Reptiles	Birds	Mammals	Carrion	Plants	Ref.
Cryptoproctinae: fossas	*Cryptoprocta ferox*	Forest dweller; highly predaceous								Poultry	Lemurs	x		1, 2
Viverrinae: civets and genets	*Genetta tigrina*	Predaceous								x	Bats, rodents			3, 4
	Genetta rubiginosa					x	x	Frogs		x			Seeds	5
	Poiana richardsoni	Omnivorous				x				x			Kola nuts	6
	Osbornictis piscivora	Semiaquatic			x		x	x						6, 7
	Civettictis civetta	Omnivorous				x		Frogs	Lizards	x eggs	x	x	x	8–10
	Viverra spp.				Crabs	x	x	Frogs	Snakes, lizards	x eggs	x			2
Paradoxurinae: palm civets	*Nandinia binotata*	Arboreal, frugivorous								x	Rodents, pottos		Fruit	2, 11
	Paguma larvata	Arboreal, omnivorous				x	← Small vertebrates →	→	→	→	→	→	Fruit	2, 12
	Arctictis binturong	Arboreal, frugivorous										x	Fruit	2, 6
Hemigalinae: banded and otter civets	*Fossa fossa*					x			Lizards					13
	Hemigalus derbyanus		x			Ants								2
	Eupleres spp.	Dentition implies invertebrate diet	x			x		Frogs	Lizards					2, 6

Table 1 (continued)
CATEGORIES OF FOODS FOR VIVERRIDS

Subfamily	Genus and species	Comment	Foods: Earthworms	Molluscs	Crustaceans	Insects	Fish	Amphibia	Reptiles	Birds	Mammals	Carrion	Plants	Ref.
Galidiinae: Madagascar mongooses	*Galidia elegans*					x			Lizards	x	x			13, 14
Herpestinae: mongooses	*Herpestes sanguineus*	Predaceous				x			Lizards	x	x			3, 11, 15, 16
	Herpestes auropunctatus	Omnivorous		x	Crabs, prawns	x	x	Toads, frogs	Snakes, lizards	x eggs	Rodents	x	x	17–20
	Herpestes urva				x			Frogs						6
	Cynictis penicillata	Omnivorous				x			x	x	Rodents			2, 21, 22
	Mungos mungo	Largely insectivorous				x								
	Crossarchus obscurus		x		Crabs	x		Toads	Snakes, lizards	Nestlings, eggs	x		Fruit	7, 23–25
	Ichneumia albicauda	Largely insectivorous				x			Lizards, eggs		Rodents			26, 27
	Rhynchogale melleri	Dentition implies nonpredaceous diet				Termites and centipedes							Fruit	6
	Suricata suricatta	Largely insectivorous				x		Frogs	Snakes, lizards	Nestlings, eggs	x			22
	Atilax paludinosus	Semiaquatic		Snails	x	x	x	Frogs	Snakes	x	Rodents			6, 28

Table 2
CORRECTION FACTORS FOR DIFFERENT CATEGORIES OF MONGOOSE (*HERPESTES AUROPUNCTATUS*) PREY

Each factor is the weight of prey eaten divided by the dry weight of undigested material of that prey type remaining in the scats.

	Correction factors								
	Mongoose number								
Category of prey	1	2	3	4	5	6	7	8	Mean
Fresh-water prawns	3	4	–	–	–	–	–	–	3.5
Littoral crabs	1.8	3.0	2.7	2.8	–	–	–	–	2.6
Brachiopoda	6.3	6.8	7.2	6.4	6.9	6.1	–	–	6.6
Diptera	20	21	28	20	18	20	21	29	22
Cockroaches	17	17	17	18	23	16	17	19	18
Dragonflies	21	20	19	21	23	18	19	21	20
Adult beetles	9	4	9	9	8	13	9	8	9
Adult lepidoptera	14	17	19	18	14	12	18	23	17
Larval lepidoptera	16	21	17	18	16	23	16	18	18
Grasshoppers and crickets	6	10	13	11	10	9	8	6	9
Stick insects and mantids	12	10	12	13	12	16	12	11	12
Frogs	14	–	–	–	–	–	–	–	14
Toad *Bufo marinus*	15	17	19	18	18	19	21	14	18
Birds	34	30	32	38	29	30	32	33	32
Snakes	27	–	–	–	–	–	–	–	27
Lizards	19	12	22	18	17	19	21	16	18
Mus musculus	20	19	29	19	26	28	25	23	24
Rattus spp.	36	30	37	41	30	29	42	34	35
Plant remains	101	100	71	218	78	82	118	176	118

From Gorman, M. L., *J. Zool.*, 175, 273–278, 1975. With permission.

Table 3
DIET OF *HERPESTES AUROPUNCTATUS* IN DIFFERENT HABITATS ON THE ISLAND OF VITI LEVU, FIJI

The weight of each prey category taken at each location is expressed as a percentage of the total weight of prey taken at that location. The weight figures were calculated by applying the correction factors in Table 2 to the dry weights of prey remains in scats. The number of scats analyzed for each location is shown in the table.

Number of scats analyzed	Agricultural land 383	Urban 708	Coconut plantation 109	Sugar cane 831	Dry grassland 396	High rain forest 208	Lowland rain forest 1001	Mangrove forest 768
Fresh-water prawns	0	0	0	0	0	0	4	0
Littoral crabs	0	0	0	0	0	0	0	50
Brachiopoda	0	0	0	0	0	0	0	25
Diptera	1	3	0	2	6	5	1	0
Cockroaches	0	8	0	0	4	0	5	0
Dragonflies	14	12	9	4	8	12	11	5
Adult beetles	4	2	3	2	9	11	5	4
Adult lepidoptera	2	3	2	1	3	9	3	0
Larval lepidoptera	3	4	4	4	4	15	10	0
Grasshoppers and crickets	4	2	3	0	8	4	9	0
Stick insects	6	0	5	0	4	2	0	0
Frogs	0	0	0	0	0	11	4	0
Toad *Bufo marinus*	19	14	14	6	4	0	15	10
Birds	12	12	0	0	4	10	5	0
Snakes	0	0	0	0	0	0	5	0
Lizards	8	9	8	9	13	8	7	5
Mus musculus	10	14	0	10	6	0	0	0
Rattus spp.	14	15	50	61	24	12	15	0
Plants	3	2	2	1	3	1	1	1

From Gorman, M. L., *J. Zool.*, 175, 273–278, 1975. With permission.

REFERENCES

1. **Grassé, P. P.,** *Traité de Zoologie,* Vol. 17, Masson et Cie, Paris, 1955.
2. **Walker, E. P.,** *Mammals of the World,* Vol. 2, Johns Hopkins Press, Baltimore, 1968.
3. **Verschuren, J.,** *Ecologie et Biologie des Grandes Mammifères. Exploration du Parc National de Garamba,* Vol. 9, Institut Parcs National Congo Belge, Brussels, 1958.
4. **Azzaroli, L. and Simonetta, A. M.,** *Monit. Zool. Ital. Suppl.,* 74, 102–195, 1966.
5. **Ingles, J. M.,** *Puku Occas. Pap. Dep. Wildl. Fish. Natl. Parks,* 3, 75–86, 1965.
6. **Ewer, R. F.,** *The Carnivores,* Cornell University Press, New York, 1973.
7. **Verheyen, W.,** *Rev. Zool. Bot. Afr.,* 65, 121–128, 1962.
8. **Bothma, J. du P.,** *Fauna Flora,* 16, 16–22, 1965.
9. **Astley-Maberley, C. T.,** *Afr. Wildl.,* 9, 55–58, 1955.
10. **Leakey, L. S. B.,** *Animals of East Africa,* National Geographic Society, Washington, D.C., 1969.
11. **Bates, G. L.,** *Proc. Zool. Soc. London,* pp. 65–85, January–April 1905.
12. **Martin, S. J.,** *J. Bombay Nat. Hist. Soc.,* 33, 703, 1929.
13. **Rand, A. L.,** *J. Mammal.,* 16, 89–104, 1935.
14. **Eisenberg, J. F. and Gauld, E.,** *Smithson. Contrib. Zool.,* 27, 1–138, 1970.
15. **Ansell, W. F. H.,** *Puku Occas. Pap. Dep. Wildl. Fish. Natl. Parks,* 2, 14–52, 1970.
16. **Ansell, W. F. H.,** *Puku Occas. Pap. Dep. Wildl. Fish. Natl. Parks,* 3, 1–14, 1965.
17. **Williams, C. B.,** *Bull. Dep. Agric. Trinidad Tobago,* 17, 167–186, 1918.
18. **Baldwin, P. H., Schwartz, C. W., and Schwartz, E. R.,** *J. Mammal.,* 33, 335–356, 1952.
19. **Kami, H. T.** *Zoonoses Res.,* 3, 165–170, 1964.
20. **Gorman, M. L.,** *J. Zool.,* 175, 273–278, 1975.
21. **Zumpt, I. F.,** *J. S. Afr. Vet. Med. Assoc.,* 39, 105–108, 1968.
22. **Zumpt, I. F.,** *J. S. Afr. Vet. Med. Assoc.,* 39, 89–91, 1968.
23. **Simpson, C. D.,** *Anim. Kingdom,* 69, 52–57, 1966.
24. **Neal, E.,** *Animals London,* 13, 29–31, 1970.
25. **Neal, E.,** *East Afr. Wildl. J.,* 8, 63–71, 1970.
26. **Sanderson, I. T.,** *Trans. Zool. Soc. London,* 24, 623–725, 1940.
27. **Booth, A. H.,** *Small Mammals of West Africa,* Longmans, Green & Co., London, 1960.
28. **Booth, A. H.,** *J. West. Afr. Sci. Assoc.,* 5, 26–36, 1959.
29. **Lockie, J. D.,** *J. Wildl. Manage.,* 23, 224–227, 1959.
30. **Altman, P. L. and Dittmer, D. S., Eds.,** *Metabolism,* Federation of American Societies for Experimental Biology, Bethesda, Md., 1968.

DIETS (NATURAL AND SYNTHETIC): FERRETS

J. Hammond, Jr.

The ferret (*Mustela putorius furo.* L.) is generally held to be a domesticated form of the European polecat, or possibly a hybrid. As such, it can be said to have no natural diet; however, it has established itself as a feral species in New Zealand.

In New Zealand the principal prey is the rabbit, and the diet includes other small mammals, birds, birds' eggs, and a variety of other vertebrates and invertebrates, the contribution of these varying seasonally.[1] Carrion is a common article of diet.[2] The polecat eats part of a fresh kill and returns for later meals; a "high" carcass is noted to be far more effective bait for a trap than is fresh meat,[3] and fruit is said to be sometimes eaten.

The domestic ferret (kept for rabbiting, rat catching, or as a pet) is usually fed either on bread and milk or slaughterhouse waste.* On the former diet the droppings are rather wet and unpleasant. As a laboratory animal – though a possible alternative to the cat in pharmacological work – the ferret has been rather neglected, and used mainly for work on virology and photoperiodism. The best estimate of its nutritional requirements is probably to be obtained by reference to those of the mink,[4] a related species of similar size.

The ferret's large intestine is short, and there is no caecum; it seems evident that there can be little digestion of fiber. The practice of the caged animal of taking meat, or milk-soaked bread, from its food dish and storing it in a "larder" corner of its cage – and of the wild animal eating carrion – is perhaps in some way a substitute for bacterial fermentation in the hind gut.

A pelleted diet is undoubtedly of great practical convenience.[5] Over a period of 20 years I have myself reared 1700 ferrets on a commercial pelleted diet (mainly Amvilac No. 2, Glaxo) intended for early weaned piglets; they have also been maintained on one (Spratt's laboratory diet No. 1) designed particularly for S.P.F. rodents.[6] The stated crude compositions of these diets are

	Protein	Fat	Fiber
Amvilac No. 2	23%	7.5%	1.5%
Spiller's No. 1	21.3%	3.4%	2.2%

Both are based largely on mixed cereals, with some fish meal, and with added fat and mineral and vitamin supplements. Amvilac pellets contain about 25% powdered milk products, plus an antibiotic; the Spratt's diet is cooked and expanded in the process of pelleting.

I have found the Amvilac pellets generally satisfactory as a diet, provided that the pellets are supplied also as a wet mash from about the 20th day of lactation until weaning at 7 weeks old; even so, it is difficult to avoid some check in the growth of unweaned young. They seem to lack the strength of jaw, and probably also the saliva, to deal with dry pellets. Provision of the mash appears to result also in greater maternal intake. On a meat diet, food is taken to the young in the nest from about Day 20; young venture from the nest to get food 10 days or so before their eyes open at Day 34. Amvilac pellets are not taken into the nest. Access to a bowl of water is recommended[6] for lactating females fed the Spratt's diet so that they can wet the cubes.

* Which of course should include some bone.

REFERENCES

1. **Gibb, J. A. and Flux, J. E. C.,** in *The Natural History of New Zealand,* Williams, G. R., Ed., Reed, Wellington, N.Z., 1973, 360–361.
2. **Marshall, W. H.,** *The Ecology of Mustelids in New Zealand,* Dept. of Science and Industrial Research, New Zealand, 1963, 20.
3. **Lagoutte, A.,** *Le Furet; son Origin, ses Moeurs, son Élevage, son Utilisation, sa Pathologie,* Bosc Frères, Lyon, 1935, 19–35.
4. *Nutrient Requirements for Domestic Animals, No. VII.,* Pub. No. 1676, National Academy of Sciences, Washington, D.C., 1968 (see also "Diets for Mustelidae" by A. U. Shump elsewhere in this Handbook).
5. **Hammond, J., Jr. and Chesterman, F. C.,** in *The UFAW Handbook on the Care and Management of Laboratory Animals,* 4th ed., Churchill Livingstone, Edinburgh, 1972, 357.
6. **Frape, D. L.,** Spiller's Nutritional Centre, Kennett, Newmarket, Engl., personal communications, 1975.

DIETS FOR THE RACCOON AND OTHER PROCYONIDAE

A. S. Johnson

All members of the Procyonidae are adaptable omnivores, and their natural diet includes a wide variety of plant and animal material.* The various species utilize similar foods, but the proportions of items in the diet differ substantially.

NATURAL DIETS

The raccoon (*Procyon lotor*) is the only procyonid that has received much study under natural conditions or in the laboratory. Its food habits have been studied in many different habitats and geographic regions. Typically, the annual diet includes about 75% plant material (mostly fruits and nuts) and about 25% animal matter (mostly invertebrates). In some habitats, however, there may be considerable deviation from this typical diet. In coastal marshes, for example, crabs and other invertebrates compose the bulk of the diet.

The diet varies seasonally with physiological needs and with the availability of certain types of foods. There is a corresponding variation in intake of various food fractions and an associated seasonal weight cycle. A diet high in carbohydrates and fats generally is taken during summer and fall, and considerable fat is deposited during this time. The diet then consists mostly of fruits and nuts, such as blackberries (*Rubus* spp.), plums and cherries (*Prunus* spp.), corn (*Zea mays*), persimmons (*Diospyros* spp.), grapes (*Vitis* spp.), and acorns (*Quercus* spp.). During the winter and spring, when these foods are no longer available, protein content of the diet increases greatly. It is taken primarily in the form of invertebrate animals, especially insects, earthworms, mollusks, and crustaceans. Vertebrates generally compose only a small part of the diet. Although activity is reduced greatly during winter, there generally is considerable weight loss at this time. Animals may lose one third to one half of their total weight, and in some areas there is considerable mortality associated with malnutrition.[1,2]

The ring-tailed cat (*Bassariscus* spp.) consumes proportionately more vertebrates than any procyonid. About 80% of its diet is animal matter. In Texas, the only place where its food habits have been studied, small mammals and birds compose a major part of the diet.[3,4]

Food habits of other procyonids have received little study. Olingos (*Bassaricyon* spp.) seem to be primarily frugivorous, but insects and vertebrates are an important part of the diet.[5] Coatimundis (*Nasua* spp.) feed heavily on invertebrates found in litter on the forest floor, but they also feed on a variety of fruits.[6] Kinkajous (*Potos flavus*) feed mainly on fruits and insects. In captivity they eat meat reluctantly and refuse dead animals and lizards.[7]

NUTRITIONAL REQUIREMENTS

Studies of the nutritional requirements of procyonids are limited to Neseni's work with raccoons.[8-10] Neseni conducted feeding trials with two adult raccoons and found the average maintenance diet to be about 40 kcal/kg body weight/day on a mixed ration.[8] He concluded that the diet should include 2.8 to 3.0 g protein/kg body weight/day. Efficiency of utilization increased as meat content of the ration increased.

* Pandas, formerly included in the Procyonidae as the subfamily Ailurinae, are not included in this section. Their mainly vegetarian diet is much more limited.

Digestibility of cereals was improved by cooking. For efficient utilization, the quantity of food should not fall below 100 g/day, and generally should be at least 150 g/day.

From later studies, Neseni reported that mature raccoons depositing fat reserves for winter needed 80 to 90 kcal/kg body weight, including 3.5 to 4.5 g crude protein,[10] and growing raccoons (about 9 months old) needed 100 to 120 kcal/kg body weight, including 6 to 8 g crude protein.[9]

In similar studies with two adult male raccoons fed canned dog food (Ken-L Ration®), Kinard[11] reported an average of 86 kcal/kg/day needed for maintenance, and Golley[12] reported that an adult raccoon was fed 110 kcal/kg/day, with 96% assimilation of the energy consumed, but diet and weight changes were not specified.

Almost nothing is known of vitamin and mineral requirements of raccoons. Schultz[13] reported depleted vitamin A reserves in the livers of two emaciated raccoons found dead in box traps and suggested that vitamin A deficiency may have been a factor in the deaths. Holmes et al.[14] found that livers of 12 raccoons from a fur farm in Vermont contained no vitamin A reserves, whereas the liver of a wild raccoon contained a small amount of vitamin A. The captive animals had been fed a commercial fox ration supplemented with apples.

NURSING YOUNG

Montgomery,[15] in studies of weaning raccoons, found that milk consumption steadily increased from birth to a maximum (55 g/1-hr feeding after 24 hr of deprivation) at 8 to 9 weeks of age, at which time the young began feeding on solid food (Purina Laboratory Chow). The young were no longer dependent on mother's milk after 10 weeks, and by 15 weeks little or no milk was consumed.

The composition of raccoon milk has been reported as follows:[16]

Water	88.0%
Fat	3.9%
Protein	4.0%
Carbohydrate	4.7%
Ash	0.75%

DIETS FOR CAPTIVE RACCOONS

Scott and Scott[17] reviewed the general requirements of carnivores and reported that very little was known about the precise requirements of raccoons for particular substances. They indicated that raccoons kept in laboratories usually are fed in much the same way as cats, but are maintained successfully in captivity with a variety of diets. Captive raccoons usually are fed any of a number of commercially available synthetic diets, supplemented by natural foods.

Raccoons on commercial fur ranches are fed a variety of foods, including dried bread and milk, mushes of cereal and milk, cooked meat with cereals, sweet fruits, and mushes of cereal meals mixed with meat, broth, and milk.[18-21]

Crandall[22] reported that raccoons in the New York Zoological Park were fed a mixture of chopped raw horse meat, fine bone meal, cod-liver oil, and a small allowance of fish, fruits, and vegetables.

Wackernagel[23] contended that the dietary needs of most mammals can be satisfied by a few types of basic mixtures and that it is not necessary to attempt to copy nature. He reported basic diets prepared for herbivores, omnivores, carnivores, and fish-eating animals.[23,24] The omnivore diet (Table 1) was used with good results for bears and for pigs, and is probably suitable for raccoons and other procyonids. Similar omnivore diets are available commercially (Table 2). They should be supplemented with fresh fruits and vegetables.

Raccoons maintained for research purposes at the University of Georgia have been fed various commercial dog meals and feeds for laboratory animals with good results. These rations usually are supplemented with scraps of fruits, vegetables, and meat.

Generally, raccoons in captivity are fed once daily. Lactating females and young animals should be fed twice daily. Overfeeding is possible; deaths of captive raccoons in Georgia have been attributed to obesity and pancreatitis caused by overfeeding (necropsy report of Southeastern Cooperative Wildlife Disease Study, University of Georgia, Athens).

In hand-rearing baby raccoons, various formulas have been used with success. Usually these are comprised of condensed milk, water, cod-liver oil, eggs, and honey or syrup. Young raccoons perform poorly on cow's milk, but Borden's Esbilac® (Table 3) has been used with good results.[25,26] The formula is given ad libitum every 2 to 4 hr, with the amount per feeding and the interval between feedings gradually increasing. At 7 weeks or earlier the formula should be supplemented with increasing amounts of baby food and gradual addition of small amounts of solid food.

DIETS FOR OTHER PROCYONIDS

Published information on feeding other procyonids in captivity is scarce.[5,7,22] With slight modifications, diets suitable for raccoons are probably acceptable for other species. All procyonids are normally fed a mixed diet of fruits, vegetables, meat, and dog food, the proportions varying with the species; ring-tailed cats, for example, usually receive more meat, and kinkajous less. Crandall[22] reported providing similar diets for all procyonids in the New York Zoological Park. These included fruits (especially apples, bananas, and grapes), raw vegetables, meat (cooked or raw), dog meal, bone meal, and cod-liver oil.

Table 1
EXAMPLES OF OMNIVORE FEEDS SUITABLE FOR RACCOON DIETS[a]

Cakes

Basic Mixture (%)

Ground plate maize	16
Ground wheat	12
Ground barley	10
Rolled oats	10
Groundnut oil meal	10
Soybean oil meal	8
Wheat germ	5
Dried yeast	10
Skimmed-milk powder	10
Fat	5
Bone meal	2
Salt	0.9
Trace element mixture	0.1
Vitamin mixture	1

Trace Elements Added (mg/kg feed)

Fe	20
Cu	2
Mn	50
I	2
Zn	10
Co	1

Vitamins Added (amount/kg feed)

Vitamin A	40,000 IU
Vitamin D_3	6,000 IU
Vitamin B_2	40 mg

Composition (%)

Crude protein	~24
Crude fat	7.5
Crude fiber	2.5
Calcium	0.9
Phosphorus	0.8

Method of Preparation

Mix 9 parts basic mixture with 1 part minced cooked meat to a stiff mash with meat broth or water. Press the mash into a shallow pan. The mash hardens quickly, and the resulting cake can then be easily cut or broken into pieces for feeding. Keep refrigerated. The composition (in %) of the finished cake does not differ much from that of the basic mixture:

Crude protein	~25
Crude fat	7.5
Crude fiber	2.25
Calcium	0.8
Phosphorus	0.7

Note: The cake is not baked – "dough" would be a more appropriate designation.

Pellets

Basic Mixture (%)

Biscuit meal	25
Rolled oats	15
Rolled barley	5
Potato flakes	8
Carob bean meal	6
Dried sugar beet pulp	10
Skimmed-milk powder	5
Fish meal	5
Meat meal	8
Dried yeast	4
Wheat germ	5
Wheat bran	2
Bone meal	1
Salt	0.5
Trace element mixture	0.1
Vitamin mixture	0.4

Trace Elements Added (amount/kg feed)

Fe	40 mg
Mn	20 mg
Cu	5 mg
I	1 mg
Co	0.1 %
Zn	75 mg

Vitamins Added (amount/kg feed)

Vitamin A	8,000 IU
Vitamin D_3	1,200 IU
Vitamin B_1	1.8 mg
Vitamin B_2	8 mg
Vitamin B_6	0.7 mg
Pantothenic acid	7 mg
Nicotinic acid	20 mg
Choline	18 mg
Folic acid	0.07 mg
Vitamin B_{12}	20 μg

Composition (%)

Crude protein	20.5
Crude fat	3.3
Crude fiber	3.3
Calcium	1.2
Phosphorus	0.9

[a] Used by the zoo at Basel, Switzerland.

From Wackernagel, H., *Int. Zoo Yearb.*, 2, 95– 102, 1960. With permission.

Table 2
EXAMPLE OF A COMMERCIAL OMNIVORE FEED[a]

Average Analysis

Moisture	7.4%	Fiber	4.4%
Dry matter	92.6%	Calcium	1.5%
Protein	21.1%	Phosphorus	1.1%
Fat	5.4%	Sodium	0.4%
NFE	54.6%	Potassium	0.8%
Ash	7.1%	Energy	1,680 kcal/lb

Calculated Vitamin and Mineral Content
amount/lb feed

Magnesium	0.16%	Vitamin K	0.2 mg
Iron	71.28 mg	Thiamine	5.2 mg
Copper	4.34 mg	Riboflavine	4.6 mg
Cobalt	0.05 mg	Pyridoxine	2.0 mg
Zinc	7.0 mg	Niacin	24.0 mg
Iodine	0.26 mg	Pantothenic acid	8.0 mg
Manganese	10.8 mg	Biotine	0.07 mg
Vitamin A	1,650 IU	B_{12}	1.0 mg
Vitamin D	175 IU	Folic acid	0.4 mg
Vitamin E	6 mg	Choline	576 mg

Ingredients

Ground corn, meat and bone meal, soy grits, Brewer's rice, animal fat, oats, wheat germ meal, dehydrated alfalfa meal, iodized salt, Brewer's dried yeast, potassium sorbate, magnesium sulfate, iron carbonate, zinc chloride, manganese sulfate, copper oxide, vitamin A palmitate, D-activated animal sterol, niacin, calcium pantothenate, riboflavine, thiamine, vitamin B_{12} supplement

[a] ZU/PREEM omnivore diet. Hill's Division, Riviana Foods, Inc., P.O. Box 148, Topeka, KS 66601.

Table 3
COMMERCIAL MILK SUBSTITUTE (BORDEN'S ESBILAC®) SUCCESSFULLY USED IN HAND-REARING RACCOONS[a]

Composition (%)

Crude protein (minimum)	33.00
Crude fat (minimum)	40.00
Crude fiber (minimum)	None
Moisture (maximum)	5.00
NFE (minimum)	14.00
Ash (maximum)	6.00

Vitamins Added (amount/lb Esbilac)

Vitamin A	50,000 USP units
Vitamin D_2	3,500 USP units
Thiamine	1.4 mg
D-Pantothenic acid	4.1 mg
Niacin	15.4 mg
Choline chloride	1,300.0 mg
Riboflavine	3.2 mg

Ingredients

Hydrogenated soybean oil and cottonseed oil, dried skimmed milk, alkaline caseinate, butter, lactalbumin, dried egg yolk, vitamin A palmitate, D-activated plant sterol (source of vitamin D_2), riboflavine, thiamine hydrochloride, niacin, DL-calcium pantothenate, choline chloride, and salts as follows: ferric ammonium citrate, 0.01%; potassium phosphate, monobasic, 0.40%; potassium phosphate, dibasic, 0.10%; calcium carbonate, 1.30%; magnesium carbonate, 0.60%; magnesium sulfate, 0.04%; sodium chloride, 0.20%; potassium chloride, 0.30%; potassium iodide, 0.002%; manganese sulfate, 0.001%; aluminum potassium sulfate, 0.001%; cupric sulfate, 0.002%; sodium fluoride, 0.002%.

[a] Formulation of Borden, Inc., Smith-Douglass Division of Borden Chemical, Norfolk, VA 23501. Reproduced with permission.

REFERENCES

1. **Mech, L. D., Barnes, D. M., and Tester, J. R.,** *J. Mammal.,* 49, 63–73, 1968.
2. **Johnson, A. S.,** *Auburn Univ. Agric. Exp. Stn. Bull.,* 402, 1970.
3. **Taylor, W. P.,** *J. Mammal.,* 35, 55–63, 1954.
4. **Wood, J. E.,** *J. Mammal.,* 35, 406–414, 1954.
5. **Poglayen-Neuwall, I. and Poglayen-Neuwall, I.,** *Z. Saeugetierkd.,* 30, 321–366, 1965.
6. **Kaufmann, J. H.,** *Univ. Calif. Berkeley Publ. Zool.,* 60, 95–222, 1962.
7. **Poglayen-Neuwall, I.,** *Z. Saeugetierkd.,* 27, 1–44, 1962.
8. **Neseni, R.,** *Z. Saeugetierkd.,* 13, 77–113, 1938.
9. **Neseni, R.,** *Z. Tierernaehrung Futtermittelkd.,* 3, 125–146, 1940.
10. **Neseni, R.,** *Z. Tierernaehrung Futtermittelkd.,* 6, 107–118, 1943.
11. **Kinard, F. W.,** Food Habits of the Eastern Raccoon, *Procyon lotor* (L), in West Central South Carolina, M. S. thesis, University of Georgia, Athens, 1964.
12. **Golley, F. B.,** *Mammals of Georgia,* University of Georgia Press, Athens, 1962, p. 183.
13. **Schultz, V.,** *J. Wildl. Manage.,* 12, 332–333, 1948.
14. **Holmes, A. D., Tripp, F., and Satterfield, G. H.,** *Am. J. Physiol.,* 123, 693, 1938.
15. **Montgomery, G. G.,** *J. Wildl. Manage.,* 33, 154–159, 1969.
16. **Ben Shaul, D. M.,** *Int. Zoo Yearb.,* 4, 333–342, 1962.
17. **Scott, P. P. and Scott, M. G.,** in *Husbandry of Laboratory Animals,* Conalty, M. L., Ed., Academic Press, New York, 1967, p. 166.
18. **Ashbrook, F. C.,** *Fur Farming for Profit,* Orange Judd Publ. Co., New York, 1948, pp. 262–263.
19. **Aitken, F. C.,** *Feeding of Fur-Bearing Animals,* Commonwealth Agricultural Bureaux, Farnham Royal, Bucks, England, 1963, pp. 99–103.
20. **Stuber, J. W.,** *Trans. Am. Game Conf.,* 19, 328–333, 1932.
21. **Whitney, L. F. and Underwood, A. B.,** *The Raccoon,* Practical Sciences Publ. Co., Orange, Conn., 1952, pp. 148–155.
22. **Crandall, L. S.,** *The Management of Wild Animals in Captivity,* University of Chicago Press, Chicago, 1964, pp. 309–323.
23. **Wackernagel, H.,** in *Comparative Nutrition of Wild Animals,* Crawford, M. A., Ed., Academic Press, New York, 1968, pp. 1–12.
24. **Wackernagel, H.,** *Int. Zoo Yearb.,* 2, 95–102, 1960.
25. **McInnis, F. G.,** *Int. Zoo Yearb.,* 4, 312, 1962.
26. **Maberry, M.,** *Int. Zoo Yearb.,* 4, 301, 1962.

Rodentia

SCIURID DIETS

R. E. Marsh

INTRODUCTION

The family Sciuridae in the order Rodentia is separated into approximately 50 genera and is distributed throughout most temperate areas of the world except Australia, Madagascar, and southern South America.[1] Although the animals' habitats vary greatly, they are absent in certain desert regions of Arabia and Egypt and in areas approaching the polar regions.

This discussion is mostly restricted to North American Sciuridae in the United States and Canada since they are fairly typical for the world. The "squirrel family," as Sciuridae are commonly referred to, is a relatively diversified group, with the North American representatives including chipmunks, marmots, antelope squirrels, ground squirrels, prairie dogs, tree squirrels, and the flying squirrels. Table 1* is a complete list of the species; brevity does not permit listing of the numerous subspecies. As a group, squirrels are probably our best-known wild rodents because they are conspicuous and easily recognized.

This section on sciurid diets is divided between natural diets and the diets used to maintain certain squirrel species in captivity.

NATURAL DIETS

The diets and nutritional needs of wild animals such as squirrels (sciurids) have been determined by studying the animals' food habits. The natural diets of sciurids as a group have been studied much more than have the diets of many other rodent families. However, much more is known of the qualitative and quantitative nutritional needs of the family Muridae than of any other group of rodents because rats and mice are preferred laboratory animals in many scientific investigations.

Although diet or food habits have been studied for a number of sciurids, only a few of those studies include the nutritional or caloric value of the foods consumed. Studies of the chemical composition of natural diets have been largely lacking.

As generally assumed, Sciurids are herbivorous, though not to the exclusion of insects and other animals, including birds and their eggs. North American sciurids would have to be categorized as omnivorous rather than herbivorous, although exceptions may be cited in certain habitats. The foods and feeding habits of squirrels have been most often studied by three methods: analysis of stomach contents, field observations, and the feeding of captive animals.

General food categories taken from a composite of studies provide the overall picture for a genus or species. No sciurid species is considered to have developed specialized dietary needs, although in a few instances some species have come to be relatively dependent on a small group of plants.[3,4] Table 2 lists the general plant diet of a number of sciurid species (including the common names of some of the plants). It should be remembered that the plants or parts fed upon by the various species depend greatly on the habitat.

Members of some genera cause agricultural and forest damage or have public health importance as disease carriers. Squirrels and chipmunks have been kept as pets and have become part of the pet industry, and still other species are game animals and are used for food. Many sciurids, ranking high in aesthetic value, are popular subjects of natural

* Tables found at the end of the text.

history studies. All of these factors have contributed to numerous food-habit studies. In the analysis of stomach contents, the food present in the greatest quantity (i.e., volume) seems to be of greatest significance in the natural history of the species, though not necessarily of the greatest significance to the nutritional needs of the animal. A better indication of dietary needs may be the frequency of occurrence of a particular food item.

Stomach analyses (expressed quantitatively by either volume or frequency of occurrence) do not clearly reflect either food preference or availability, only some varying degree of compromise between the two.[6]

Food preferences appear to be related most closely to food availability, shifting as the supply is exhausted or as seasons change. Figure 1, illustrating the makeup of the diets of a number of species in relation to the four seasons of the year, shows that the changes are significant.

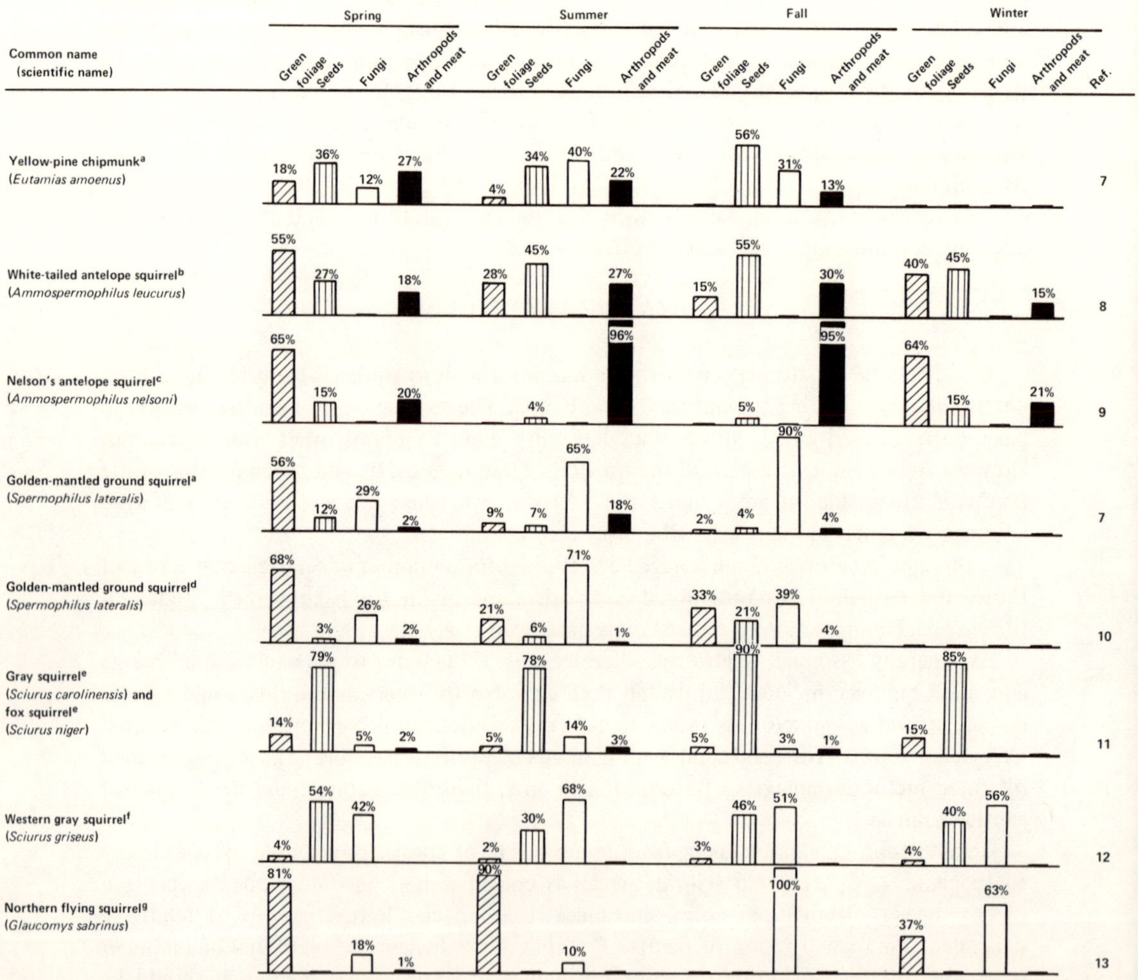

FIGURE 1. Graphic illustration of the make-up of the diets of a number of sciurid species in relation to the four seasons of the year. The information is derived from stomach-content analysis and calculated on percent by volume. Habitats are indicated as follows: [a]Predominantly pine, with some fir, Plumas County, California; [b]Desert shrub and pinyon-juniper woodlands, southern Nevada; [c]Dry plains and uplands of San Joaquin Valley, California; [d]Ponderosa pine forests, northern California; [e]Oak-hickory hardwood forest, Ohio; [f]Pine-fir forests, Trinity County, California; [g]Pine-fir forests, Lassen County, California. In the food categories, *green foliage* also includes bulbs and lichens, *seeds* include fruits and flowers also, while *fungi* and *arthropods and meat* are self-explanatory. There are no winter (December, January, February) foods indicated for the hibernating species.

Since the foods selected by a species may differ with the particular habitat occupied, food-habit studies frequently give varied results. To provide a good indicator of a species as a whole, a food-habit study must be drawn from a wide area and from a range of occupied habitats.

Specific physiological needs may influence preferences for certain food items; for example, increased amount of succulent foods appear in the diet to maintain water balance when free water is limited or unavailable. Snow cover, floods, drought, fire, and other natural events may temporarily deplete a food supply, causing the animal to switch to foods that are neither highly preferred nor very nutritious but are the best available.

The various parts of plants differ in food value. For example, the most nutritious part of a leaf is the growing tissue, which is at the tips of leaves of forbs and at the base of leaves of grasses. Plant seeds are generally more nutritious than the foliage. Also influencing nutrition are the season and stage of growth. Immature fruits and seeds become more nutritious as they mature. Rhizomes, stolons, roots, and other storage tissues of plants usually have the highest nutritive value, while the leafy portions are the least nutritious. When alternatives are available, it is believed that sciurids generally select plant food high in nutrients.[14]

Carnivorous Behavior

Although sciurids are not generally considered to be carnivorous, squirrels have been seen taking other vertebrates. The California ground squirrel, *Spermophilus beecheyi*, for example, has been seen a number of times feeding on animals killed by motor vehicles. As a rule, birds and small mammals comprise a rather small part of sciurid diets. Squirrels may be simply opportunistic, but the evidence suggests that certain species have a definite liking for meat. A specific dietary deficiency may motivate this behavior. It is believed by some that the carnivorous behavior of sciurids may be related to their need for water (i.e., food with high water content).[15] Examples of the carnivorous behavior of sciurids are given in Table 3. The table omits the insects and other invertebrates that stomach-content analysis and direct observations indicate are represented to various degrees in the diets of many species of sciurids.

Robbing of eggs from birds' nests is also commonly reported. Glading[27] reported that, where abundant, California ground squirrels, *S. beecheyi*, reduce the number of quail eggs by 31.25%. Most reports of reduced bird production are attributed to egg consumption, although the killing of young birds cannot be dismissed. Reports of egg consumption by certain sciurids is common.[23,28–32]

Cannibalism

Certain squirrels are cannibalistic to a minor degree, feeding on very young, dead, sick, or handicapped individuals of their own kind as the opportunity arises. Accounts are rare of squirrels actually pursuing and killing healthy animals of their own kind. Scavenging on road kills seems the most prevalent manifestation of cannibalism observed in sciurids. This subject is discussed below in "Diets for Captive Sciurids."

Bark Stripping

It has been suggested that bark stripping by squirrels may partly be a craving for some needed nutrients.[33] Bark stripping is done not only by tree squirrels but by a number of other sciurids species, including ground squirrels.[34] It occurs in different types of habitats and generally on a very irregular basis at a particular period in one year, sometimes not recurring to any degree for several years. In a particular locality, bark stripping may be limited to a single tree species or to several. This could suggest a nutritional need, but with ground squirrels it may be simply a need for water, which may be most readily available in the bark and cambium layer in the drier season in certain habitats, particularly in the arid west. Others believe that bark stripping by tree squirrels is primarily for nest-building materials.[35] This behavior and its significance to nutrition obviously need further study.

Food Rejection

Tree squirrels appear to have the ability to reject wormy nuts without opening them, and this is believed to be a learned behavioral trait.[36] The ability of tree squirrels to distinguish between sound nuts and insect-infested or empty nuts has been studied by other investigators, who also concluded it to be a learned discrimination rather than innate behavior.

Certain (though not all) native plants known to be toxic to other animals are rejected by squirrels even when other food may be scarce.

DIETS FOR CAPTIVE SCIURIDS

Animals in captivity must generally be offered substitutes for their natural foods, and the diet is most often far less varied than nature's. It is also difficult to make correlations between the animals' natural diet and nutritional needs for growth, reproduction, development of hibernation fat, etc. The varied diet of sciurids in nature changes from season to season, presumably because of changes in food availability. This complicates any attempt to duplicate a diet obtainable in the wild.

Table 4 lists diets used to maintain sciurids in captivity. These diets have evolved mostly from available knowledge of laboratory rats, through simple trial and error, and through attempts to match to some degree the natural diets of sciurids.

Diets used for sciurids mostly represent natural food products (complete natural diets). Some commercial laboratory diets occasionally include some semipurified ingredients in what would otherwise be a natural diet. No specific synthetic diets have been developed for sciurids. If we can make a broad assumption that the dietary needs of sciurids do not differ substantially from those of other rodents such as rats and mice, then we are relatively safe in providing captive sciurids with a basic diet of a commercial pelletized rodent laboratory chow, supplemented with natural food items known to be in the animal's diet during specific seasons of the year.

Experience has shown that commercial laboratory diets are not highly preferred by sciurids if offered in the presence of many other food items, but that should not interfere with their use as the staple diet. With few exceptions, a balanced commercial laboratory rat diet should be offered ad lib.

Diets prepared for laboratory rodents that can be used for sciurids may be either in meal or pelleted form. Each has advantages. The pelleted diet provides a gnawable surface (which may help keep teeth in condition), is suitable for self-feeding hoppers, and may prevent food waste. Pelleted chow, however, will be cached in great amounts by some species if they have free access to loose pellets. The meal form may be easier for the very young to feed upon and, unlike pellets, will rarely be cached.

It may be erroneous to equate short-term food preferences in captivity with the nutritional needs of the squirrel. Squirrels and other sciurids will eat great quantities of highly preferred foods offered in captivity, yet these food items by themselves are in all likelihood nutritionally deficient. Table 4 also gives examples of supplemental foods for the various species.

Ground squirrels do not breed in close confinement with anywhere near the regularity they do in the wild.[55] Diet may be a factor, though perhaps only one of several. If, as many believe, it is basically a nutritional problem, it remains to be resolved.

Sciurids are not normally considered to be carnivorous despite the number of observations of squirrels feeding on other vertebrates. As a rule, birds and small mammals killed or scavenged comprise a small (though possibly important) part of the total sciurid diet. Squirrels may be simply opportunistic, although the evidence suggests that certain species have a definite liking for meat.

Ground squirrels penned in groups may exhibit some cannibalistic behavior and may occasionally feed on the very young, injured, or dead. This seems most likely when ground squirrels are placed together in confinement for the first time and shortly after capture in the wild. Some groups of squirrels show no evidence of cannibalism, whereas another population of the same species will exhibit this behavior. Although cannibalism is known to exist in the wild, its prevalence in captivity has not been determined. Stress may play an undetermined role. Table 5 lists references to cannibalistic tendencies.

Invertebrates, rather than the higher animals, make up an even greater portion of the diet of some sciurids during certain periods of the year. No references were found of insects and other invertebrates being fed on a regular basis to captive sciurids, undoubtedly reflecting the expense of feeding insects or worms. It is relatively common for squirrels to feed on bird eggs when available, as discussed above in the section on natural diets. Either raw or hardboiled eggs may be incorporated into the diets of captive sciurids.

Present knowledge suggests that some fresh or canned meat in the diet of many species of sciurids would be highly desirable, if not essential. Squirrels often relish ground horse meat, some canned and dry dog food, and other foods containing meat by-products relatively high in protein. Meat may be fed on a schedule of once or twice a week.

There seems to be no evidence that squirrels have an unusual need or craving for salt, as do porcupines, *Erethizon dorsatum.* Salt blocks commonly used for domestic rabbits seem to be ignored by both California and Belding ground squirrels under field and laboratory conditions. Similar observations have been made of the Nelson antelope ground squirrel.[9]

Species that occupy arid regions can survive very well without free water if their diets contain enough moisture to maintain their water balance. Because of the difficulty in determining exact water needs in captivity, free water should be provided ad lib.

Although concise information is generally lacking on the nutritional needs of the various sciurid species, we can learn much from the experiences of others. Many of the diets in Table 4 mention Purina Laboratory Chow as the staple diet, proven adequate for sustaining the animals in apparent good health. Table 6 (a list of the approximate nutrients of Purina Laboratory Chow) may be a valuable guide for maintaining sciurids in captivity.

Crandall's book *The Management of Wild Mammals in Captivity*[59] is a good source of sciurid diets and is most enlightening for those wishing to keep sciurids as pets, in zoological gardens, or in the laboratory for research purposes. Crandall suggests the following staple diet for maintaining tree squirrels in zoological gardens: pecans, filberts, walnuts, and other hardshelled nuts; whole corn, oats, and canary grass seed; raw vegetables and greens; and fruits and berries. He also states that tropical sciurid species from both the New and Old World showed a preference for fruit such as mealy apples and bananas. Crandall believes that the Stanger's squirrel, *Protexerus stangeri,* from equatorial Africa is at least partly carnivorous and hence should receive some animal protein. Hardboiled eggs, dog biscuits, and raw or cooked ground meat can be provided in its diet. Finely ground bone meal and cod liver oil may be added to the ground meat. These same animal protein recommendations would be suitable as a portion of most diets of North American sciurids. Old dry bones placed in the cages will provide needed calcium.

The Cape ground squirrel, *Xerus inauris,* a South African squirrel quite similar to our genus *Spermophilus,* should be fed much the same diet as tree squirrels, including animal proteins.[59]

Young marmots have been reared on whole cow's milk and gradually converted to solid food. Mature marmots are more inclined to take greens such as lettuce and cabbage, raw vegetables (including carrots and potatoes), and a variety of grains, which may include whole-wheat or rye bread. They are somewhat less interested in nuts and fruits than are tree squirrels.[59]

The diet suggested by Crandall for prairie dogs consists of raw vegetables, greens, crushed oats, and canned dog food or dog biscuits. Cuttings of fresh clover or alfalfa are provided when available.[59]

A suitable diet for ground squirrels (*Spermophilus* spp.) consists of grain and other seeds, especially sunflower seed.[59] Crandall suggests greens and raw vegetables in as great a variety as may be available. In addition, a small portion of the daily diet consists of any dry or canned dog food, or a mixture of chopped raw meat to which bone meal and cod liver oil are added.

Diets used to maintain chipmunks and flying squirrels closely approximate those of tree squirrels and include canary grass, sunflower seed, cut carrots, greens, fruit, nuts, and a small amount of either dry or canned dog food or raw ground meat with bone meal and cod liver oil added.[59]

SUPPLEMENTAL FEEDING OF SCIURIDS

Supplementing the diets of free-living sciurids requires that the foods proffered be selected for a high degree of preference, that which entices tree squirrels, chipmunks, and other squirrels to feeders or feeding areas. A food less attractive than that already abundantly available will be largely ignored. Even so, for the best nutrition of wild animals, some portion of the proffered diet should consist of a commercially available balanced diet meal or pellet. The feeding of tree squirrels has been a popular hobby or pastime and offers a form of entertainment which, correctly or not, is generally believed beneficial to the animal. Tree squirrels in parks that feed heavily on peanuts may become sickly, appearing thin and having ruffled fur, whereas the same animals with a diet of walnuts, pecans, and other nuts become fat and sleek looking.[33] With few exceptions, supplemental or artificial feeding is not needed for squirrels to survive and might merely stave off an inevitable starvation or population decline from poor nutrition.

Supplemental feeding should be done sparingly, providing relatively small amounts at regular intervals. Since the purpose is generally to bring visual pleasure to the provider through watching the animal's feeding activities, these activities can be prolonged without overfeeding. Offering free squirrels nuts in the shell or only partially cracked will prolong the feeding process. Securing the food items to the feeding site will force the animal to feed on the spot. Self-feeders that permit only small amounts of food to be reached at a time are preferred.

Supplemental feeding of tree squirrels as a tool of wildlife management is a rare practice but has been considered. Davison[60] suggested that "chicken scratch feed," a blend of cracked corn, milo, and wheat, would probably be satisfactory for supplemental food for tree squirrels during winter when acorns and other mast crops are known to be scarce. In situations where food may be the limiting factor in maintaining some rarer species, the merits of supplemental feeding may be explored.

Table 1
LIST OF THE NORTH AMERICAN GENERA AND SPECIES OF THE FAMILY SCIURIDAE OF THE ORDER RODENTIA[2]

Scientific name	Common name
Tamias striatus	Eastern chipmunk
Eutamias alpinus	Alpine chipmunk
Eutamias minimus	Least chipmunk
Eutamias amoenus	Yellow-pine chipmunk
Eutamias townsendii	Townsend's chipmunk
Eutamias ochrogenys	Yellow-cheeked chipmunk
Eutamias senex	Allen's chipmunk
Eutamias siskiyou	Siskiyou chipmunk
Eutamias sonomae	Sonoma chipmunk
Eutamias merriami	Merriam's chipmunk
Eutamias dorsalis	Cliff chipmunk
Eutamias quadrivittatus	Colorado chipmunk
Eutamias ruficaudus	Red-tailed chipmunk
Eutamias cinereicollis	Gray-collared chipmunk
Eutamias canipes	Gray-footed chipmunk
Eutamias quadrimaculatus	Long-eared chipmunk
Eutamias speciosus	Lodgepole chipmunk
Eutamias panamintinus	Panamint chipmunk
Eutamias umbrinus	Uinta chipmunk
Eutamias palmeri	Palmer's chipmunk
Marmota monax	Woodchuck
Marmota flaviventris	Yellow-bellied marmot
Marmota broweri	Alaska marmot
Marmota caligata	Hoary marmot
Marmota olympus	Olympic marmot
Marmota vancouverensis	Vancouver marmot
Ammospermophilus harrisii	Harris' antelope squirrel
Ammospermophilus leucurus	White-tailed antelope squirrel
Ammospermophilus interpres	Texas antelope squirrel
Ammospermophilus nelsoni	Nelson's antelope squirrel
Spermophilus townsendii	Townsend's ground squirrel
Spermophilus washingtoni	Washington ground squirrel
Spermophilus brunneus	Idaho ground squirrel
Spermophilus richardsonii	Richardson's ground squirrel
Spermophilus armatus	Uinta ground squirrel
Spermophilus beldingi	Belding's ground squirrel
Spermophilus columbianus	Columbian ground squirrel
Spermophilus parryii	Arctic ground squirrel
Spermophilus tridecemlineatus	Thirteen-lined ground squirrel
Spermophilus mexicanus	Mexican ground squirrel
Spermophilus spilosoma	Spotted ground squirrel
Spermophilus franklinii	Franklin's ground squirrel
Spermophilus variegatus	Rock squirrel
Spermophilus beecheyi	California ground squirrel
Spermophilus mohavensis	Mohave ground squirrel
Spermophilus tereticaudus	Round-tailed ground squirrel
Spermophilus lateralis	Golden-mantled ground squirrel
Spermophilus saturatus	Cascade golden-mantled ground squirrel
Cynomys ludovicianus	Black-tailed prairie dog
Cynomys leucurus	White-tailed prairie dog
Cynomys parvidens	Utah prairie dog
Cynomys gunnisoni	Gunnison's prairie dog
Sciurus carolinensis	Gray squirrel
Sciurus griseus	Western gray squirrel
Sciurus aberti	Abert's squirrel
Sciurus kaibabensis	Kaibab squirrel

Table 1 (continued)
LIST OF THE NORTH AMERICAN GENERA AND SPECIES OF THE FAMILY SCIURIDAE OF THE ORDER RODENTIA[2]

Scientific name	Common name
Sciurus niger	Fox squirrel
Sciurus nayaritensis	Nayarit squirrel
Sciurus arizonensis	Arizona gray squirrel
Tamiasciurus hudsonicus	Red squirrel
Tamiasciurus douglasii	Douglas' squirrel
Glaucomys volans	Southern flying squirrel
Glaucomys sabrinus	Northern flying squirrel

Table 2
SOME REPRESENTATIVE SCIURIDS, THEIR GENERAL DIETS, AND COMMON NAMES OF CERTAIN PLANTS USED AS FOOD[5]

Currently Accepted Common and Scientific Names are Substituted Where Necessary for Those Previously Used

Common name	Scientific name	General diet	Common names of plants utilized
Eastern chipmunk	*Tamias striatus*	Seed and nuts of woody plants	Maple, hickory, oak, hazelnut, basswood, beech
Least chipmunk	*Eutamias minimum*	Seeds and fruits	Blackberry, blueberry, bitterbrush, wheat
Yellow-pine chipmunk	*Eutamias amoenus*	Seed of herbaceous and woody plants	Pine, fireweed, serviceberry, tarweed, brome grass
Harris' antelope squirrel	*Ammospermophilus harrisii*	Seed of herbaceous and woody plants	Mesquite, gooseberry, paloverde, grasses
White-tailed antelope squirrel	*Ammospermophilus leucurus*	Seed and fruit of herbaceous and woody plants	Prickly pear, pine, salt grass
Nelson's antelope squirrel	*Ammospermophilus nelsoni*	Seed of grasses and herbaceous plants	Turkeymullein, filaree, barley
Woodchuck	*Marmota monax*	Foliage of herbaceous and grassy plants	Clover, grasses, vegetables, soybeans, alfalfa
Yellow-bellied marmot	*Marmota flaviventris*	Foliage of grasses and herbaceous and woody plants	Grasses, locoweed, serviceberry, blackberry
Richardson's ground squirrel	*Spermophilus richardsonii*	Seed and/or foliage of a variety of plants	Goosefoot, wheat, bluegrass, dandelion
Columbian ground squirrel	*Spermophilus columbianus*	Seed and/or foliage of a variety of plants	Tarweed, turkeymullein, oak, filaree, wyethia
Thirteen-lined ground squirrel	*Spermophilus tridecemlineatus*	Seed of a variety of wild and cultivated plants	Ragweed, knotweed, bristlegrass, sunflower, clover, wheat
Franklin's ground squirrel	*Spermophilus franklinii*	Seed and/or foliage of a variety of plants	Cocklebur, needlegrass
California ground squirrel	*Spermophilus beecheyi*	Seed and/or foliage of a variety of plants	Oak, filaree, tarweed, lupine, oats

Table 2 (continued)
SOME REPRESENTATIVE SCIURIDS, THEIR GENERAL DIETS, AND COMMON NAMES OF CERTAIN PLANTS USED AS FOOD[5]

Currently Accepted Common and Scientific Names are Substituted Where Necessary for Those Previously Used

Common name	Scientific name	General diet	Common names of plants utilized
Golden-mantled ground squirrel	*Spermophilus lateralis*	Seeds, nuts, and fleshy fruits of herbaceous and woody plants	Penstemon, knotweed, gooseberry bitterbrush, pine
Black-tailed prairie dog	*Cynomys ludovicianus*	Foliage of a variety of plants	Wheatgrass, Russian thistle, brome grass, fescue, bluegrass, grama grass
White-tailed prairie dog	*Cynomys leucurus*	Foliage of a variety of plants including woody plants	Saltbush, Russian thistle, wheatgrass, sagebrush
Gunnison's prairie dog	*Cynomys gunnisoni*	Foliage of a variety of plants	Dandelion, Russian thistle, wheatgrass, nightshade, grama grass
Gray squirrel	*Sciurus carolinensis*	Nuts and fruits of trees and woody plants	Oak, hickory, beech, maple, walnut
Western gray squirrel	*Scuirus griseus*	Nuts and fruits of trees and woody plants	Pine, oak
Fox squirrel	*Sciurus niger*	Nuts and fruits of trees and woody plants	Oak, hickory, elm, beech, maple, walnut
Red squirrel	*Tamiasciurus hudsonicus*	Seeds, nuts, and fruits of trees	Hickory, serviceberry, oak, maple, pine, Douglas fir, spruce
Douglas' squirrel	*Tamiasciurus douglasii*	Seeds, nuts, and fruits of trees and woody plants	Pine, fir, hemlock, salol, hazelnut, ash
Southern flying squirrel	*Glaucomys volans*	Nuts and fruits of trees and woody plants	Beech, oak, hackberry, maple

Table 3
REPRESENTATIVE EXAMPLES OF CARNIVOROUS FEEDING IN SCIURIDS

Common name	Scientific name	Prey or food item	Ref.
Eastern chipmumk	*Tamias striatus*	Meadow mouse	16
	Tamias striatus lysters	Red-bellied snake	17
Least chipmunk	*Eutamias minimus jacksoni*	Passerine birds	18
Colorado chipmunk	*Eutamias quadrimaculatus*	Snake as carrion	7
White-tailed antelope squirrel	*Ammospermophilus leucurus*	Lizards, birds, other rodents, carrion	8
Arctic ground squirrel	*Spermophilus parryii barrowensis*	Meat scraps under natural conditions	19
Thirteen-lined ground squirrel	*Spermophilus tridecemlineatus*	Birds, shrews	20
	Spermophilus tridecemlineatus	Cottontail rabbits, snakes	21
Mexican ground squirrel	*Spermophilus mexicanus*	Cottontail rabbit	22
Franklin's ground squirrel	*Spermophilus franklinii*	Mice, hares, toads, carrion, birds, young ducks	23
Golden-mantled ground squirrel	*Spermophilus lateralis*	Chipmunks, deer mice	24
	Spermophilus lateralis	Carrion, chipmunks	10
	Spermophilus lateralis	Lizards, birds, eggs	7
White-tailed prairie dog	*Cynomys leucurus*	Young chickens	25
Southern flying squirrel	*Glaucomys volans*	Yellow-bellied sapsucker	26

Table 4
DIETS FOR CAPTIVE SCIURIDS

The following diets have been used to maintain certain sciurids in captivity. Whether they are optimal for good reproduction or other physiological events (e.g., hibernation) is mostly unknown. For greater details, review the cited references. The currently accepted common and scientific names are substituted (where necessary) for those used by the authors.

Common name	Scientific name	Diet	Ref.
Eastern chipmunk	*Tamias striatus*	Purina® laboratory chow	37
Nelson's antelope ground squirrel	*Ammospermophilus nelsoni*	Rolled oats, rolled barley, chicken mash, rabbit pellets, occasionally dog food. Meat readily eaten, bones sometimes gnawed.	9
Richardson's ground squirrel	*Spermophilus richardsonii*	Dandelion greens supplemented with oats, sunflower seed, apple, carrot, and rat pellets	38
	Spermophilus richardsonii	Oats, sunflower seeds, dog food cubes, lettuce, dandelion, and sow thistle greens	39
Belding's ground squirrel	*Spermophilus beldingi*	Rat or rabbit chow, sunflower seeds, grass, fruits, and vegetables	40
	Spermophilus beldingi	Purina laboratory chow	41
Arctic ground squirrel	*Spermophilus parryii*	Purina laboratory chow, carrots, cabbage	19
Thirteen-lined ground squirrel	*Spermophilus tridecemlineatus*	Purina laboratory chow supplemented by some green vegetation	21
	Spermophilus tridecemlineatus	A dry-meal type of laboratory diet proven to be adequate for rats and mice; also, sprouted oats or other green feed	42
	Spermophilus tridecemlineatus	Live grasshoppers, corn, oats, wheat, dried peas[a]	43

Table 4 (continued)
DIETS FOR CAPTIVE SCIURIDS

Common name	Scientific name	Diet	Ref.
Thirteen-lined ground squirrel (continued)	*Spermophilus tridecemlineatus*	Purina chow, whole-wheat bread, apples, lettuce, carrots, sunflower seeds, hamburger, eggnog	44
Franklin ground squirrel	*Spermophilus franklinii*	Fond of carrots, fresh peas, string beans, tomatoes, and potatoes; rejected rhubarb, onions, peppers[a]	23
	Spermophilus franklinii	Purina laboratory chow	45
California ground squirrel	*Spermophilus beecheyi beecheyi*	Purina laboratory chow, green vegetation (lettuce, alfalfa, sprouted oats, barley, and wheat)	46
	Spermophilus beecheyi	Oranges, dry bread, cabbage, sunflower seeds, scratch feed, cantaloupes, walnuts	47
	Spermophilus beecheyi douglasii	Dry bread, carrots, apples, potatoes, green grass, occasionally assorted grains (scratch feed)[b]	48
Golden-mantled ground squirrel	*Spermophilus lateralis*	Purina Commercial Rabbit Checkers (rabbit pellets)	49
Black-tailed prairie dog	*Cynomys ludovicianus*	Oats, corn, raw carrots, cabbage, lettuce, clover, orange pulp, salted peanuts	50
	Cynomys ludovicianus	Dog chow; lettuce and cabbage were provided for water	51
Gray squirrel	*Sciurus carolinensis carolinensis*	Sweet potatoes, raw peanuts, black walnuts, raw rabbit meat; old bones were gnawed on.	52
Fox squirrel	*Sciurus niger*	Preferred foods: white oak acorns, walnuts, corn[a]	53
Northern flying squirrel	*Glauconys sabrinus bungsi*	Wheat, corn, rolled oats, apples, carrots, lettuce, walnuts, Purina dog chow	54

[a] Partial diet; complete diet not given.
[b] Young offspring cannabilized in captivity; undetermined if diet contributed to this.

Table 5
EXAMPLES OF REPORTED INSTANCES OF CANNIBALISM IN VARIOUS SPECIES OF GROUND SQUIRRELS

Common name	Scientific name	Ref.
Belding's ground squirrel	*Spermophilus beldingi*	41
Arctic ground squirrel	*Spermophilus parryii*	19, 56
Thirteen-lined ground squirrel	*Spermophilus tridecemlineatus*	44
Franklin's ground squirrel	*Spermophilus franklinii*	23, 57
California ground squirrel	*Spermophilus beecheyi*	41, 58

Table 6
THE CHEMICAL COMPOSITION OF A RAT LABORATORY DIET[a] FOUND APPROPRIATE FOR MAINTAINING MOST SCIURIDS IN CAPTIVITY

Nutrient	Amount	Nutrient	Amount
Protein (%)[b]	23.93	Chlorine (%)	0.51
Arginine (%)	1.38	Fluorine (ppm)	35.0
Glycine (%)	1.26	Iron (ppm)	319.97
Lysine (%)	1.34	Zinc (ppm)	59.08
Methionine (%)	0.43	Manganese (ppm)	50.49
Tryptophan (%)	0.28	Copper (ppm)	17.96
Cystine (%)	0.327	Cobalt (ppm)	0.37
Histidine (%)	0.54	Iodine (ppm)	1.73
Leucine (%)	1.80	Carotene (ppm)	6.50
Isoleucine (%)	1.22	Vitamin A (IU/g)	12.00
Phenylalanine (%)	1.03	Vitamin D (IU/g)	5.31
Threonine (%)	0.94	α-Tocopherol (IU/lb)	22.57
Valine (%)	1.24	Thiamine (ppm)	17.73
Fat (%)	4.40	Riboflavin (ppm)	8.63
Fiber (%)	5.35	Niacin (ppm)	104.47
Ash (%)	7.27	Pantothenic acid (ppm)	24.80
Calcium (%)	1.30	Choline (ppm)	1968.89
Phosphorus (%)	0.94	Folic acid (ppm)	6.15
Potassium (%)	0.82	Pyridoxine (ppm)	4.30
Magnesium (%)	0.26	Biotin (ppm)	0.17
Sodium (%)	0.49	Vitamin B_{12} (μg/lb)	15.30

[a] Purina Laboratory Chow; information supplied by Ralston Purina Company, St. Louis, Missouri.

[b] Amino acids expressed as percent of ration.

REFERENCES

1. **Walker, E. P., Ed.,** *Mammals of the World,* Vol. 2, Johns Hopkins Press, Baltimore, 1975, 669–727.
2. **Jones, J. K., Jr., Carter, D. C., and Genoways, H. H.,** Revised checklist of North American mammals, *Occas. Pap. Mus. Tex. Tech. Univ.,* No. 28, 1–14, 1975.
3. **Keith, J. O.,** The Abert squirrel and its dependence on ponderosa pine, *Ecology,* 46, 150–161, 1965.
4. **Patton, D. R.,** Abert Squirrel Cover Requirements in Southwestern Ponderosa Pine, Res. Paper RM-145, U.S. Dep. of Agriculture Forest Service, Rocky Mountain Experiment Station, 1975, 1–12.
5. **Martin, A. C., Zim, H. S., and Nelson, A. L.,** *American Wildlife and Plants,* Dover, New York, 1961, 1–500.
6. **Davison, V. E. and Sullivan, E. G.,** Mourning doves' selection of foods, *J. Wildl. Manage.,* 27(3), 373–383, 1963.
7. **Tevis, L., Jr.,** Stomach contents of chipmunks and mantled squirrels in northeastern California, *J. Mammal.,* 34(3), 316–324, 1953.
8. **Bradley, W. G.,** Food habits of the antelope ground squirrel in southern Nevada, *J. Mammal.,* 49(1), 14–21, 1968.
9. **Hawbecker, A. C.,** Food and moisture requirements of the Nelson antelope ground squirrel, *J. Mammal.,* 28(2), 115–125, 1947.
10. **McKeever, S.,** The biology of the golden-mantled ground squirrel, *Ecol. Monogr.,* 34(4), 383–401, 1964.
11. **Nixon, C. M., Worley, D. M., and McClain, M. W.,** Food habits of squirrels in southeast Ohio, *J. Wildl. Manage.,* 32(1), 294–305, 1968.
12. **Stienecker, W. and Browning, B. M.,** Food habits of the western gray squirrel, *Calif. Fish Game,* 56(1), 36–48, 1970.
13. **McKeever, S.,** Food of the northern flying squirrel in northeastern California, *J. Mammal.,* 41(2), 270–271, 1960.
14. **Koford, C. B.,** Prairie dogs, whitefaces, and blue grama, *Wildl. Monogr.,* No. 3, 1–78, 1958.
15. **Bartholomew, G. A. and Hudson, J. W.,** Effects of sodium chloride on weight and drinking in the antelope ground squirrel, *J. Mammal.,* 40(3), 354–360, 1959.
16. **Torres, K. J.,** A chipmunk captures a mouse, *J. Mammal.,* 18(1), 100, 1937.
17. **Harriot, S. C.,** Chipmunk eating a red-bellied snake, *J. Mammal.,* 21, 92, 1940.
18. **Aldous, S. E.,** Food habits of chipmunks, *J. Mammal.,* 22, 18–24, 1941.
19. **Musacchia, X. J.,** Cannibalism and other observations of captive ground squirrels, *J. Mammal.,* 35(3), 445–447, 1954.
20. **Whitaker, J. O., Jr.,** Food and external parasites of *Spermophilus tridecemlineatus* in Vigo County, Indiana, *J. Mammal.,* 53(3), 643–648, 1972.
21. **Bridgwater, D. D. and Penney, D. F.,** Predation by *Citellus tridecemlineatus* on other vertebrates, *J. Mammal.,* 47(2), 345–346, 1966.
22. **Packard, R. L.,** Carnivorous behavior in the Mexican ground squirrel, *J. Mammal.,* 39(1), 154, 1958.
23. **Sowls, L. K.,** The Franklin ground squirrel, *Citellus franklinii* (Sabine), and its relationship to nesting ducks, *J. Mammal.,* 29(2), 113–137, 1948.
24. **Cameron, D. M., Jr.,** Carnivorous behavior of the golden-mantled ground squirrel, *Murrelet,* 48(1), 13, 1967.
25. **Kelso, L. H.,** Food habits of prairie dogs, U.S. Dep. Agric. Biol. Survey, Circ. 529, 1939, 1–15.
26. **Stoddard, H. L.,** The flying squirrel as a bird killer, *J. Mammal.,* 1(1), 95, 1919.
27. **Glading, B.,** Studies on the nesting cycle of the California valley quail in 1937, *Calif. Fish Game,* 24(4), 318–340, 1938.
28. **Emlen, J. T., Jr. and Glading, B.,** California ground squirrel robs nest of valley quail, *Condor,* 39, 41–42, 1938.
29. **Errington, P. L. and Hamestrom, F. N., Jr.,** The evaluation of nesting losses and juvenile mortality of the ring-necked pheasant, *J. Wildl. Manage.,* 1, 3–20, 1937.
30. **Errington, P. L.,** Experimental evidence on the eating of bob-white eggs by small Sciuridae, *J. Mammal.,* 19, 108, 1938.
31. **Horn, E. T.,** Some relationships of quail and ground squirrels in California, *J. Wildl. Manage.,* 2, 58–60, 1938.
32. **Stanton, F. W.,** Douglas ground squirrel as a predator on nests of upland game birds in Oregon, *J. Wildl. Manage.,* 8, 153–161, 1944.

33. **Madson, J.,** *Gray and Fox Squirrels,* Olin Mathieson Chemical Corp., East Alton, Ill., 1964, 1–112.
34. **Marsh, R. E.,** unpublished data.
35. **Davidson, A. M.,** British squirrels – between ourselves, *Q. J. For.,* 69(4), 225–232, 1975.
36. **Dennis, W.,** Rejection of wormy nuts by squirrels, *J. Mammal.,* 11, 195–209, 1930.
37. **Panuska, J. A.,** Weight patterns and hibernation in *Tamias striatus, J. Mammal.,* 40(4), 554–566, 1959.
38. **Sheppard, D. H. and Yoshida, S. M.,** Social behavior in captive Richardson's ground squirrels, *J. Mammal.,* 52(4), 793–799, 1971.
39. **Michener, G. R. and Sheppard, D. H.,** Social behavior between adult female Richardson's ground squirrels (*Spermophilus richardsonii*) and their own and alien young, *Can. J. Zool.,* 50(11), 1343–1349, 1972.
40. **Mochardt, S. S. and Gates, D. M.,** Energy-exchange analysis of the belding ground squirrel and its habitat, *Ecol. Monogr.,* 44(1), 17–44, 1974.
41. **Marsh, R. E. and Howard, W. E.,** unpublished data.
42. **Johnson, G. E. and Wade, N. J.,** Laboratory reproduction studies on the ground squirrel, *Citellus tridecemlineatus pallidus,* Allen, *Biol. Bull.,* 61, 101–114, 1931.
43. **Hisaw, F. L. and Emery, F. E.,** Food selection of ground squirrels, *Citellus tridecemlineatus, J. Mammal.,* 8, 41–44, 1927.
44. **Zimny, M. L.,** Thirteen-lined ground squirrels born in captivity, *J. Mammal.,* 46(3), 521–522, 1965.
45. **Haberman, C. G. and Fleharty, E. D.,** Energy flow in *Spermophilus franklinii, J. Mammal.,* 52(4), 710–716, 1971.
46. **Marsh, R. E. and Howard, W. E.,** Care of ground squirrels (*Spermophilus* spp.) in captivity, *Lab. Anim. Sci.,* 21(1), 367–371, 1971.
47. **Selle, R.,** Golden-mantled ground squirrel raised in captivity, *J. Mammal.,* 20(1), 106–107, 1939.
48. **Edge, E.,** Seasonal activity and growth of the Douglas ground squirrel, *J. Mammal.,* 12, 194–199, 1931.
49. **Jameson, E. W., Jr.,** Food consumption of hibernating and nonhibernating *Cittellus lateralis, J. Mammal.,* 46(4), 634–640, 1965.
50. **Young, S. P.,** Longevity and other data on a male and female prairie dogs kept as pets, *J. Mammal.,* 25(3), 317–319, 1944.
51. **Anthony, A.,** Behavior patterns in a laboratory colony of prairie dogs, *Cynomys ludovicianus, J. Mammal.,* 36(1), 69–78, 1955.
52. **Kilham, L.,** Grey squirrels born and raised in captivity, *J. Mammal.,* 34(4), 509–510, 1953.
53. **Baumgras, P.,** Experimental feeding of captive fox squirrels, *J. Wildl. Manage.,* 8(4), 296–300, 1944.
54. **Booth, E. S.,** Notes on the life history of the flying squirrel, *J. Mammal.,* 27(1), 28–30, 1946.
55. **Marsh, R. E. and Howard, W. E.,** Breeding ground squirrels, *Spermophilus beecheyi,* in captivity, *J. Mammal.,* 49(4), 781–783, 1968.
56. **Cade, T.,** Carnivorous ground squirrels on St. Lawrence Island, *J. Mammal.,* 32(3), 358–360, 1951.
57. **Bailey, V.,** A biological survey of North Dakota, *North Am. Fauna,* 49, 55–58, 1926.
58. **Dana, R. H.,** personal communication.
59. **Crandall, L. S.,** *The Management of Wild Mammals in Captivity,* University of Chicago Press, Chicago, 1964, 1–769.
60. **Davison, V. E.,** Selection of foods by gray squirrels, *J. Wildl. Manage.,* 28(2), 346–352, 1964.

DIETS (NATURAL AND SYNTHETIC): GEOMIDAE*

A. L. Ward

INTRODUCTION

Pocket gophers are considered to be strictly herbivorous (the traces of insect material occasionally found in their stomachs probably result from accidental ingestion along with plants). They are very adaptable rodents, found from sea level to mountain tops in many ecological habitats with many plants available for food. Because they eat too wide a variety of plant species to treat briefly, this section is confined to the food habits of one species, the northern pocket gopher (*Thomomys talpoides*), in the Rocky Mountain region.

Early food habit investigations in Colorado, Montana, and Utah were made by recording the kinds of plants found in underground food caches or clipped above the ground near mounds.[1,14,115] More recent studies have used detailed microscopic examination of stomach contents to determine plants eaten. Four areas and three range types have been studied in Colorado: shortgrass prairie at an elevation of 4900 ft on the Central Plains Experimental Range near Nunn, Colorado;[122] hayfields (mixed alfalfa and grasses) at an elevation of 5500 ft on the east slope of the Rocky Mountains near Livermore, Colorado;[125] and high-altitude native grass-forb ranges at an elevation of about 10,000 ft on Black Mesa Experimental Forest and Range and on Grand Mesa, Colorado.[60,125,129] These four studies are the principal sources of the following summary of food habits information. Stomachs were collected throughout the year on the shortgrass prairie and alfalfa hayfields and during the snow-free months on the two high-altitude ranges. On Grand Mesa work was also done on some herbicide-treated areas, but those results are not included here. Unless otherwise noted, comparisons of various foods in the following discussion are based on examination of stomach contents and are expressed as percent by volume (or in the case of Vaughan[122] as approximate percent by volume).

PLANT PARTS EATEN

Because of their burrowing habits and year-round activity, pocket gophers are able to utilize all portions of plants. They find roots as they extend their tunnel systems, and they can also make openings from their tunnels and forage for food and nest material above ground. In some cases they have been observed pulling entire plants down through the soil into their burrows.

Pocket gophers tend to change their diet seasonally, apparently in response to habitat conditions and availability and nutritional quality of plants. During the growing season, when soils are warm and moist and plants are green and succulent, leaves and stems make up the major portion of the diet. During the winter and periods of drought, when fresh vegetation is scarce and dry or frozen soil makes above-ground foraging difficult, roots provide most of the food.

On the shortgrass prairie (Central Plains Range) the diet was 76% leaves and stems in spring and summer and 61% in winter, with roots making up the remainder. In the mixed alfalfa hayfields near Livermore, leaves and stems made up more than 50% of the diet in May, June, and July (98% in June); annually, however, above-ground plant parts made up only 28% of the food consumed. On high-altitude native grass-forb ranges (Black Mesa

* From Ward, A. L., *Colo. State Univ. Exp. Sta. Bull.*, 5545, 43–90, 1973. With permission.

and Grand Mesa) over 70% of the diet was leaves and stems in summer, but this decreased to about 40 to 50% by October.

USE OF FORBS

Forbs were the most important food plants of northern pocket gophers in all four study areas. On the shortgrass prairie, forbs and shrubs made up about 67% of the yearly diet. Plains prickly pear (*Opuntia polyacantha*) was the most important, averaging 50% throughout the year, but scarlet globemallow (*Sphaeralcea coccineax*), fourwing saltbush (*Atriplex canescens*), and slimflower scurfpea (*Psoralea tenuiflora*) also made up at least 1% each.

In the mixed grass and alfalfa hayfields, alfalfa (essentially the only forb present) made up about 94% of the yearly diet; alfalfa leaves and stems accounted for 23% and roots for 71%.

On Black Mesa, the summer and fall diet consisted of 93% forbs. Aspen fleabane (*Erigeron macranthus*), lupine (*Lupinus* spp.), hairy goldaster (*Chrysopsis villosa*), slenderleaf gilia (*Gilia linearis*), aspen peavine (*Lathyrus leucanthus*), common dandelion (*Taraxacum officinale*), Fremont geranium (*Geranium fremonti*), beauty cinquefoil (*Potentilla pulcherrima*), western yarrow (*Achillea lanuloso*), and agoseris (*Agoseris* spp.) were most commonly eaten.

On Grand Mesa forbs made up about 82% of the summer and fall diet. Common dandelion, agoseris, western yarrow, Rydberg penstemon (*Penstemon rydbergi*), peavine (*Lathyrus* spp.), lupine, prairiesmoke sieversia (*Sieversia ciliata*), erigonum (*Eriogonum* spp.), and knotweed (*Polygonum* spp.) were most commonly eaten.

USE OF GRASSES

Grasses (here including grass-like plants) were eaten throughout the year in all four study areas. The amount varied with the season, but they were never a major component of the diet.

On the shortgrass prairie the yearly diet consisted of about 50% grasses. Needle-and-thread (*Stipa comata*) and bluestem wheatgrass (*Agropyron smithii*) made up 12 and 10%, respectively. Blue grama (*Bouteloua gracilis*), red three-awn (*Aristida longiseta*), and sand dropseed (*Sporobolus cryptandrus*) were also important. Needle-and-thread was eaten in all seasons and was an important winter food; other grasses were eaten mostly in summer.

In the alfalfa hayfields, grasses made up only 6% of the yearly diet. Grasses (mostly leaves and stems) were eaten principally in April through July, peaking at 15% in June. Grass roots never exceeded 5% of the diet (Figure 1).

On Black Mesa, where grasses account for about 50% of the total herbage, grass was found in 240 of 397 stomachs (62%) collected in summer but made up only 6% of the volume. Fescue (*Festuca* spp.), needlegrass (*Stipa* spp.), bromes (*Bromus* spp.), and prairie Junegrass (*Koeleria cristata*) were the most commonly eaten.

On Grand Mesa, where grasses account for about 20% of the total herbage, they made up 29% of the diet in June, 19% in July, 8% in August, and 36% in October. Species most commonly eaten were needlegrass, slender wheatgrass (*Agropyron pauciflorum*), spike trisetum (*Trisetum spicatum*), prairie Junegrass, sedges (*Carex* spp.), and bromes.

USE OF SHRUBS AND TREES

There were very few shrubs and no trees in the four study areas, but northern pocket gophers will eat both when available. On Black Mesa, Parry rabbitbrush (*Chrysothamnus*

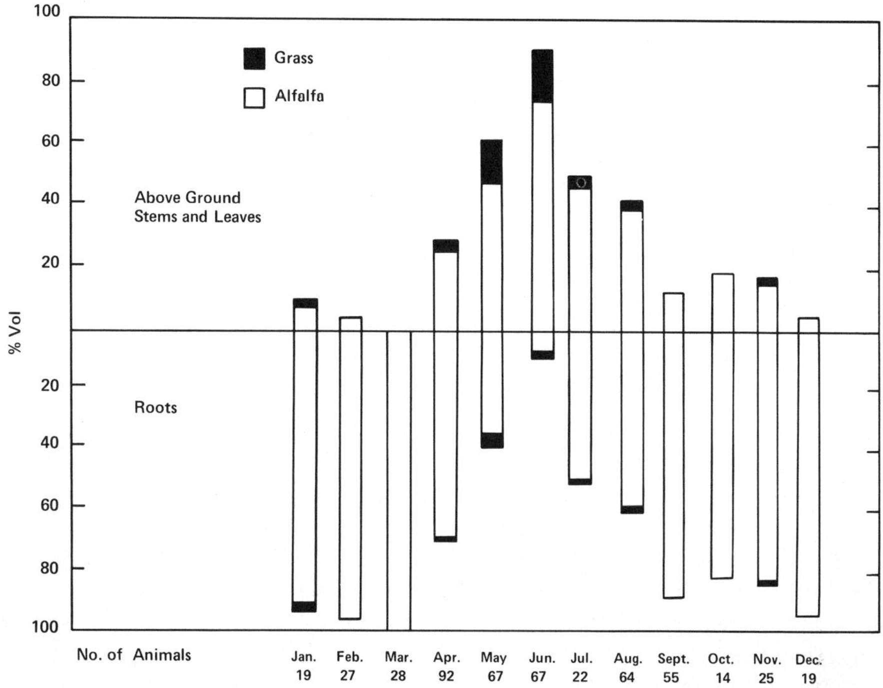

FIGURE 1. Mountain pocket gophers at Livermore, Colorado, ate mostly alfalfa in hay fields containing a mixture of range grasses and alfalfa. Stems and leaves are taken in greater amounts than roots during the growing season.

parryi) made up 3% of the summer and fall diet (20% in October). Reid and Turner[94] mentioned the use of Parry rabbitbrush during the winter on this same area.

Ronco[98] reported pocket gopher damage to seedlings of Engelmann spruce (*Picea engelmanni*) and lodgepole pine (*Pinus contorta*) in subalpine forests in western Colorado. Nearly all seedlings on the study area were clipped just above ground level, principally during winter.

FOOD CACHES

Northern pocket gophers store food in shallow underground caches, usually located 3 to 4 in. below the ground in lateral chambers loosely plugged off from the main burrow system. Large amounts of plant materials are often stored; one cache found on Black Mesa contained 577 g (1.2 lb) fresh weight. Root fragments about 1½ in. long generally predominate in the caches, and the three food habits studies based on cache contents[1,14,125] tend to indicate a higher use of roots than those studies based on stomach analysis.

In these three studies of cache contents and in observations at Livermore, Black Mesa, and Grand Mesa, the most common root materials identified in caches were: common dandelion, agoseris, western yarrow, orange sneezeweed (*Helenium hoopesi*), rockjasmine (*Androsace septentrionalis*), blueleaf strawberry (*Fragaria glauca*), Menzies larkspur (*Delphinium menziesi*), bluebell (*Campanula rotundifolia*), alfalfa, and showy oniongrass (*Melica spectabilis*). At Black Mesa common dandelion was usually the dominant species; one cache contained 752 pieces, weighting 452 g. A cache collected on March 30 beneath a 56-in. snowpack and 3 in. below ground contained about 400 g of green basal leaves and roots of rockjasmine.

The habit of storing food in underground caches may account for the harvest of considerably more vegetation than is actually eaten. In some instances full caches have been found apparently unused in pocket gopher burrows.

LABORATORY FEEDING TESTS

Experience in holding caged northern pocket gophers indicates that oat groats and soft milo maize are the most readily accepted grains and fresh carrots are the preferred vegetable. A diet of alfalfa pellets and fresh carrots without free water is sufficient to keep them healthy under laboratory conditions. Vaughan[122] showed that they could also survive on only dry alfalfa pellets and plains prickly pear.

Tietjen et al.[113] reported on extensive feeding tests to determine the reaction of northern pocket gophers to five forbs sprayed with 2,4-D. No toxic effects were observed. Individually caged animals, given only water and a single plant species, consumed a daily average of 85 g (air-dry weight) of hairy goldaster, 74 g of Fremont geranium, 53 g of aspen fleabane, 52 g of common dandelion, and 47 g of aspen peavine. However, only animals on dandelion and fleabane gained weight; those on geranium and goldaster lost weight and some died, despite the higher consumption, suggesting that these species were low in food value. Nine grasses were offered in the same manner. Consumption varied considerably, depending on season and species. For example, daily consumption of Thurber fescue (*Festuca thurberi*) averaged 98 g in June but only 20 g in early August. Animals lost weight on most grass diets, and only five – smooth brome (*Bromus inermis*), mountain brome (*B. carinatus*), timothy (*Phleum pratense*), Kentucky Bluegrass (*Poa pratensis*), and bluestem wheatgrass – did not result in mortality. These results indicated that only grasses with high moisture content or food storage structures yield a marginal maintenance diet.

If we accept 80 g fresh weight as an average daily consumption and 22 pocket gophers per acre as an average population, intake would amount to 4 lb of fresh vegetation per acre consumed each day. This amounts to 1460 lb of fresh weight, or 365 lb dry weight, per acre per year, not including vegetation stored uneaten in food caches or used to make nests.

COMPETITION WITH LIVESTOCK

There is no doubt that some competition for food exsists between northern pocket gophers and other animals, but the degree of competition is difficult to evaluate because of lack of knowledge about the specific plant use by each animal on a particular range. To further complicate matters, the native rangelands where pocket gophers are found have had a history of generally unexplained high and low cycles in both herbage production and wildlife populations. Grazing by livestock has introduced another variable that is both biologically and economically important to the use of these rangelands. Until the last 10 years, it was generally assumed that cattle on these areas ate grass and sheep ate forbs. However, recent studies have shown that cattle eat considerable amounts of some forbs and that sheep eat some grass.[27,41,69,89,109,118]

Paulsen[89] measured herbage production and cattle utilization of selected plant species on the grasslands of Black Mesa in 1961 to 1963. Idaho fescue (*Festuca idahoensis*) was found to be the primary forage species for cattle. Other important species were Letterman needlegrass (*Stipa lettermani*), Thurber fescue, nodding brome (*Bromus anomalus*), slender wheatgrass, and Kentucky bluegrass. However, forbs such as aspen fleabane, aspen peavine, agoseris, false sunflower (*Helianthella quinquenervis*), goldenrod (*Solidago ciliosa*), common dandelion, sharpleaf valerian (*Valeriana acutiloba*), and many other species made up a considerable percentage of the cattle's diet during the summer.

Many of these were also found to be major food plants of northern pocket gophers on this same range.[129]

The three forbs that furnished appreciable amounts of forage for cattle (and are also important for pocket gophers), aspen peavine, agoseris, and aspen fleabane, were collected regularly and analyzed for nutritional content.[89] They proved to be nutritionally superior to grasses through the first half of the grazing period, and even after becoming dry and decumbent, they maintained levels above the minimum nutrient requirements for normal growth of 600-lb steers and heifers.

These food studies of cattle and pocket gophers on Black Mesa show why it is important in any study of competiton to determine the food habits of the animals and the nutritional quality of the vegetation that they eat as well as the availability and productivity of the plant species. We know that elk (*Cervus canadensis*), mule deer (*Odocoileus hemionus*), pronghorn antelope (*Antilocapra americana*), and bighorn sheep (*Ovis canadensis*) are also present in many of the areas where northern pocket gophers are found and use many of the same plant species for food. The degree of competition, however, cannot be determined until the results of the various studies now planned or in progress are completed. The management of native rangelands cannot be really effective until comparative data for all competing species and their food plants are available.

REFERENCES

1. **Aldous, C. M.,** The feeding habits of pocket gophers (*Thomomys talpoides moorei*) in the high mountain ranges of central Utah, *J. Mammal.*, 32(1), 84–87, 1951.
2. **Aldous, C. M.,** Fluctuations in pocket gopher populations, *J. Mammal.*, 38(2), 266–267, 1957.
3. **Arkley, R. J. and Brown, H. C.,** The origin of Mima mound (hogwallow) microrelief in the far western states, *Proc. Soil Sci. Soc. Am.*, 18, 195–199, 1954.
4. **Barnes, V. G., Jr., Martin, P., and Tietjen, H. P.,** Pocket gopher control on Oregon ponderosa pine plantations, *J. For.*, 68(7), 433–435, 1970.
5. **Branson, F. A. and Payne, G. F.,** Effects of sheep and gophers on meadows of the Bridger Mountains of Montana, *J. Range Manage.*, 11(4), 165–169, 1958.
6. **Branson, F. A., Miller, R. F., and McQueen, I. S.,** Plant communities and soil moisture relationships near Denver, Colorado, *Ecology*, 46(3), 311–319, 1965.
7. **Buechner, H. K.,** Interrelationships between the pocket gopher and land use, *J. Mammal.*, 23(3), 346–348, 1942.
8. **Burnett, W. L.,** Pocket Gophers with Suggestions for Control, Office of State Entomology, Fort Collins, Colo., Circ. 23, 1917.
9. **Canutt, P. R.,** New Developments in Gopher Control, Proc. 1968 Annu. Meet. Western Reforest. Coord. Comm., San Francisco, 1968, 24–27.
10. **Canutt, P. R.,** Pocket Gopher Problems and Control Practices on National Forest Lands in the Pacific Northwest Region, Proc. 4th Vert. Pest Conf., University of California, Davis, 1970, 120–125.
11. Colorado Cooperative Pocket Gopher Project, Pocket gophers in Colorado, *Colo. State Univ. Agric. Exp. Sta. Bull.*, 508–S, 1960.
12. **Craighead, J. J. and Craighead, F. C., Jr.,** Hawks, Owls, and Wildlife, The Stackpole Co., Harrisburg, and Wildlife Management Institute, Washington, D.C., 1956.
13. **Crouch, W. E.,** Pocket Gopher Control, U.S. Dept. Int. Conserv. Bull. 23, 1942.
14. **Cummings, M. W.,** unpublished.
15. **Davis, W. B.,** Distribution and variation of pocket gophers (Genus *Geomys*) in the southwestern United States, *Tex. Agric. Exp. Sta. Bull.*, 590, 1–38, 1940.
16. **Davis, W. B., Ramsey, R. R., and Arendale, J. M., Jr.,** Distribution of pocket gophers (*Geomys breviceps*) in relation to soils, *J. Mammal.*, 19 (4), 412–418, 1938.
17. **Dingle, R. W.,** Pocket gophers as a cause of mortality in eastern Washington pine plantations, *J. For.*, 54(12), 832–835, 1956.
18. **Ellison, L.,** The pocket gopher in relation to soil erosion on mountain range, *Ecology*, 27(2), 101–114, 1946.
19. **Ellison, L. and Aldous, C. M.,** Influence of pocket gophers on vegetation of subalpine grassland in central Utah, *Ecology*, 33(2), 177–186, 1952.
20. **Evans, F. C. and Emlen, J. T., Jr.,** Ecological notes on the prey selected by a barn owl, *Condor*, 49(1), 3–9, 1947.

21. **Evans, R. A. and Love, R. M.**, The step-point method of sampling – a practical tool in range research, *J. Range Manage.*, 10(5), 208–212, 1957.
22. **Fichter, E., Schildman, G., and Sather, J. H.**, Some feeding patterns of coyotes in Nebraska, *Ecol. Monogr.*, 25, 1–37, 1955.
23. **Fitch, H. S.**, Predation by owls in the sierran foothills of California, *Condor,* 49(4), 137–151, 1947.
24. **Fitch, H. S., Swenson, F., and Tillotson, D. F.**, Behavior and food habits of the red-tailed hawk, *Condor,* 48(5), 205–237, 1946.
25. **Fitch, H. S. and Bentley, J. R.**, Use of California annual-plant forage by range rodents, *Ecology,* 30(3), 306–321, 1949.
26. Forest Service, U.S. Dept. Agric., An observation of environmental rodent control, *Range Impr. Notes,* 16(3), 12, 1971.
27. **Galt, H. D., Theurer, B., Ehrenreich, J. H., Hale, W. H., and Martin, S. C.**, Botanical composition of the diet of steers grazing a desert grassland range, *Proc. West. Sect. Am. Soc. Anim. Sci.*, 17, 397–401, 1966.
28. **Garrison, G. A. and Moore, A. W.**, Relation of the Dalles pocket gopher to establishment and maintenance of range grass plantings, *J. Range Manage.*, 9(4), 181–184, 1956.
29. **Grinnel, J.**, The burrowing rodents of California as agents in soil formation, *J. Mammal.*, 4(3), 137–149, 1923.
30. **Hall, E. R. and Kelson, K. R.**, *The Mammals of North America,* Vol. 1, Ronald Press, New York, 1959.
31. **Hansen, R. M.**, New Dispenser Aids Gopher Control, Colorado A & M College Agric. Exp. Sta. Pamphlet 1-S, Fort Collins, 1956.
32. **Hansen, R. M.**, Age and reproductive characteristics of mountain pocket gophers in Colorado, *J. Mammal.*, 41(3), 323–335, 1960.
33. **Hansen, R. M.**, Movements and survival of *Thomomys talpoides* in a mima-mound habitat, *Ecology,* 43(1), 151–154, 1962.
34. **Hansen, R. M.**, Pocket gopher density in an enclosure of native habitat, *J. Mammal.*, 46(3), 508–509, 1965.
35. **Hansen, R. M. and Miller, R. S.**, Observations on the plural occupancy of pocket gopher burrow systems, *J. Mammal.*, 40(4), 577–584, 1959.
36. **Hansen, R. M. and Remmenga, E. E.**, Nearest-neighbor concept applied to pocket gopher populations, *Ecology,* 42(4), 812–814, 1961.
37. **Hansen, R. M. and Bear, G. D.**, Comparison of pocket gophers from alpine, subalpine and shrub-grassland habitats, *J. Mammal.*, 45(4), 638–640, 1964.
38. **Hansen, R. M. and Ward, A. L.**, Some relations of pocket gophers to rangelands on Grand Mesa, Colorado, *Colo. State Univ. Agric. Exp. Sta. Tech. Bull.*, 88, 1–22, 1966.
39. **Hansen, R. M. and Beck, R. F.**, Habitat of pocket gophers in Cochetopa Creek drainage, Colorado, *Am. Midl. Nat.*, 79(1), 103–117, 1968.
40. **Hansen, R. M. and Morris, M. J.**, Movement of rocks by northern pocket gophers, *J. Mammal.*, 49(3), 391–399, 1968.
41. **Heady, H. F. and Torell, D. T.**, Forage preference exhibited by sheep with esophageal fistulas, *J. Range Manage.*, 12(1), 28–34, 1959.
42. **Hill, J. E.**, Morphology of the pocket gopher mammalian genus *Thomomys, Univ. Calif. Publ. Zool.*, 42(2), 81–171, 1937.
43. **Hisaw, F. L.**, The absorption of the public symphysis of the pocket gopher, *Geomys bursarius* (Shaw), *Am. Nat.*, 58, 93–96, 1924.
44. **Hisaw, F. L. and Gloyd, H. K.**, The bullsnake as a natural enemy of injurious rodents, *J. Mammal.*, 7(3), 200–205, 1926.
45. **Holliger, C. D.**, Anatomical adaptations in the thoracic limb of the California pocket gopher and other rodents, *Univ. Calif. Publ. Zool.*, 13(12), 447–494, 1916.
46. **Horn, E. E. and Fitch, H. S.**, Interrelations of rodents and other wildlife of the range, in The San Joaquin Experimental Range, University of California Agric. Exp. Sta. Bull. 663, 1942, 96–129.
47. **Howard, W. E.**, A pocket gopher population crash, *J. Mammal.*, 42(2), 258–260, 1961.
48. **Howard, W. E. and Childs, H. E., Jr.**, Ecology of pocket gophers with emphasis on *Thomomys bottae mewa, Hilgardia,* 29(7), 277–358, 1959.
49. **Hull, A. C., Jr.**, Effect of spraying with 2,4-D upon abundance of pocket gophers in Franklin Basin, Idaho, *J. Range Manage.*, 24(3), 230–232, 1971.
50. **Imler, R. H.**, Bullsnakes and their control on a Nebraska wildlife refuge, *J. Wildl. Manage.*, 9(4), 265–273, 1945.
51. **Ingles, L. G.**, Field observations on the growth of young mountain pocket gophers, *Murrelet,* 30(2), 35–36, 1949.

52. **Ingles, L. G.**, Ground water and snow as factors affecting the seasonal distribution of pocket gophers, *Thomomys monticola, J. Mammal.*, 30(4), 343–350, 1949.
53. **Ingles, L. G.**, The ecology of the mountain pocket gopher, *Thomomys monticola, Ecology*, 33(1), 87–95, 1952.
54. **Ingles, L. G., Clothier, R., and Crawford, L. A.**, Methods of estimating pocket gopher populations, *J. Wildl. Manage.*, 13(3), 311–312, 1949.
55. **Jellison, W. L.**, *Cuterebra thomomuris* sp. nov., a warble from the pocket gopher, *Thomomys talpoides* (Rodentia: Geomyidae), *J. Parasitol.*, 35(5), 482–486, 1949.
56. **Johnson, D. R.**, Effects of range treatment with 2,4-D on food habits of rodents, *Ecology*, 45(2), 241–249, 1964.
57. **Julander, O., Low, J. B., and Morris, O. W.**, Influence of pocket gophers on seeded mountain range in Utah, *J. Range Manage.*, 12(5), 219–224, 1959.
58. **Julander, O., Low, J. B., and Morris, O. W.**, Pocket gophers on seeded Utah mountain range, *J. Range Manage.*, 22(5), 325–329, 1969.
59. **Keith, J. O.**, An efficient and economical pocket gopher exclosure, *J. Range Manage.*, 14(6), 332–334, 1961.
60. **Keith, J. O., Hansen, R. M., and Ward, A. L.**, Effect of 2,4-D on abundance and foods of pocket gophers, *J. Wildl. Manage.*, 23(2), 137–145, 1959.
61. **Kelsey, H. P. and Dayton, W. A.**, *Standardized Plant Names*, 2nd ed., J. Horace McFarland, Harrisburg, 1942.
62. **Kennerly, T. E., Jr.**, Microenvironmental conditions of the pocket gopher burrow, *Tex. J. Sci.*, 16(4), 395–441, 1964.
63. **Kimbal, J., Poulson, T. A., and Savage, W. F.**, An observation of environmental rodent control, *U.S. For. Serv. Range Impr. Notes*, 15(2), 8–9, 1970.
64. **Lantz, D. E.**, Pocket Gophers as Enemies of Trees, U.S. Department of Agriculture Yearbook for 1909, 209–218.
65. **Larrison, E. J.**, Pocket gophers and ecological succession in the Wenas region of Washington, *Murrelet*, 23(2), 34–41, 1942.
66. **Laycock, W. A.**, An ecological study of the pocket gopher (*Thomomys talpoides tenellus* Goldman) in the Jackson Hole region of Wyoming, *J. Range Manage.*, 6(6), 461–462, 1953.
67. **Laycock, W. A.**, Seasonal periods of surface inactivity of the pocket gopher, *J. Mammal.*, 38(1), 132–133, 1957.
68. **Laycock, W. A.**, The initial pattern of revegetation of pocket gopher mounds, *Ecology*, 39(2), 346–351, 1958.
69. **Malechek, J. C.**, Cattle Diets on Native and Seeded Ranges in the Ponderosa Pine Zone of Colorado, U.S. Forest Service Res. Note RM-77, 1966.
70. **Marston, R. B. and Julander, O.**, Plant cover reductions by pocket gophers following experimental removal of aspen from a watershed area in Utah, *J. For.*, 59(2), 100–102, 1961.
71. **Marti, C. D.**, Some comparisons of the feeding ecology of four owls in northcentral Colorado, *Southwest. Nat.*, 14(2), 163–170, 1969.
72. **McGinnies, W. J.**, Effect of Mima-type microrelief on herbage production of five seeded grasses in western Colorado, *J. Range Manage.*, 13(5), 231–234, 1960.
73. **McNab, B. K.**, The metabolism of fossorial rodents: A study of convergence, *Ecology*, 47(5), 712–733, 1966.
74. **Miller, M. A.**, Reproductive rates and cycles in the pocket gopher, *J. Mammal.*, 27(4), 335–358, 1946.
75. **Miller, M. A.**, Seasonal trends in burrowing of pocket gophers (*Thomomys*), *J. Mammal.*, 29(1), 38–44, 1948.
76. **Miller, M. A.**, Experimental studies on poisoning pocket gophers, *Hilgardia*, 22(4), 131–166, 1953.
77. **Miller, M. A.**, Burrows of the Sacramento Valley pocket gopher in flood-irrigated alfalfa fields, *Hilgardia*, 26(8), 431–452, 1957.
78. **Miller, M. A. and Howard, W. E.**, Size of bait for pocket gopher control, *J. Wildl. Manage.*, 15(1), 62–68, 1951.
79. **Miller, R. S.**, Ecology and distribution of pocket gophers (*Geomyidae*) in Colorado, *Ecology*, 45(2), 256–272, 1964.
80. **Miller, R. S. and Bond, H. E.**, The summer burrowing activity of pocket gophers, *J. Mammal.*, 41(4), 469–475, 1960.
81. **Miller, R. S. and Ward, R. A.**, Ectoparasites of pocket gophers from Colorado, *Am. Midl. Nat.*, 64(2), 382–391, 1960.
82. **Mohr, C. O. and Mohr, W. P.**, Abundance and digging rate of pocket gophers, *Geomys bursarius*, *Ecology*, 17(2), 325–327, 1936.

83. **Moore, A. W. and Reid, E. H.**, The Dalles pocket gopher and its influence on forage production of Oregon mountain meadows, *U.S. Dept. Agric. Circ.*, 884, 1–36, 1951.
84. **Murie, A.**, Ecology of the Coyote in the Yellowstone, U.S. Department of the Interior, Fauna of the National Parks of the U.S. Bull. No. 4, 1940.
85. **Murie, O. J.**, Food habits of the coyote in Jackson Hole, Wyo., *U.S. Dept. Agric. Circ.*, 362, 1–24, 1935.
86. **Murray, D. F.**, Gravel mounds at Rocky Flats, Colorado, *Mt. Geol.*, 4(3), 99–107, 1967.
87. **Myers, G. T. and Vaughan, T. A.**, Food habits of the plains pocket gopher in eastern Colorado, *J. Mammal.*, 45(4), 588–598, 1964.
88. **Parker, K. W.**, A Method for Measuring Trend in Range Condition on National Forest Ranges, U.S. Forest Service Administration Studies, 1951.
89. **Paulsen, H. A., Jr.**, Forage values on a mountain grassland-aspen range in western Colorado, *J. Range Manage.*, 22(2), 102–107, 1969.
90. **Pechanec, J. F. and Pickford, G. D.**, A weight estimate method for the determination of range or pasture production, *J. Am. Soc. Agron.*, 29(10), 894–904, 1937.
91. **Petrides, G. A.**, Viewpoints on the analysis of open season sex and age ratios, *Trans. North Am. Wildl. Conf.*, 14, 391–410, 1949.
92. **Phillips, P.**, The distribution of rodents in overgrazed and normal grasslands of central Oklahoma, *Ecology*, 17(4), 673–679, 1936.
93. **Reid, V. H.**, Population Dynamics of Northern Pocket Gophers on Rangelands in Western Colorado, 1971.
94. **Reid, V. H. and Turner, G. T.**, personal communication.
95. **Reid, V. H., Hansen, R. M., and Ward, A. L.**, Counting mounds and earth plugs to census mountain pocket gophers, *J. Wildl. Manage.*, 30(2), 327–334, 1966.
96. **Richens, V. B.**, An evaluation of control on the Wasatch pocket gopher, *J. Wildl. Manage.*, 29(3), 413–425, 1965.
97. **Richens, V. B.**, Notes on the digging activity of a northern pocket gopher, *J. Mammal.*, 47(3), 531–533, 1966.
98. **Ronco, F.**, Lessons from Artificial Regeneration Studies in a Cutover Beetle-killed Spruce Stand in Western Colorado, U.S. Forest Service Res. Note RM-90, 1967.
99. **Ross, B. A., Tester, J. R., and Breckenridge, W. J.**, Ecology of Mima-type mounds in northwestern Minnesota, *Ecology*, 49(1), 172–177, 1968.
100. **Scheffer, T. H.**, Habits and economic status of the pocket gophers, *U.S. Dept. Agric. Tech. Bull.*, 224, 1–27, 1931.
101. **Scheffer, T. H.**, Breeding records of Pacific coast pocket gophers, *J. Mammal.*, 19(2), 220–224, 1938.
102. **Scheffer, T. H.**, Excavation of a runway of the pocket gopher (*Geomys bursarius*), *Trans. Kans. Acad. Sci.*, 43, 473–478, 1940.
103. **Scheffer, T. H. and Garlough, F. E.**, Rodents and moles as pests in bulb plantings, *U.S. Dept. Agric. Circ.*, 381, 1–16, 1936.
104. **Scheffer, V. B.**, Do fossorial rodents originate Mima-type microrelief?, *Am. Midl. Nat.*, 59(2), 505–510, 1958.
105. **Schramm, P.**, Copulation and gestation in the pocket gopher, *J. Mammal.*, 42(2), 167–170, 1961.
106. **Smith, C. F.**, A burrow of the pocket gopher (*Geomys bursarius*) in eastern Kansas, *Trans. Kans. Acad. Sci.*, 51, 313–315, 1948.
107. **Sperry, C. C.**, Food Habits of the Coyote, U.S. Department of the Interior Fish and Wildlife Service Wildlife Res. Bull. 4, 1941.
108. **Storer, T. I.**, Controlling field rodents in California, *Calif. Agric. Exp. Sta. Ext. Serv. Circ.*, 434, 1–47, 1953.
109. **Strasia, C. A., Thorn, M., Rice, R. W., and Smith, D. R.**, Grazing habits, diet, and performance of sheep on alpine ranges, *J. Range Manage.*, 23(3), 201–208, 1970.
110. **Talbot, M. W. and Biswell, H. H.**, The forage crop and its management, in The San Joaquin Experimental Range, University of California Agric. Exp. Sta. Bull. 663, 1942, 13–19.
111. **Taylor, W. P.**, Some animal relations to soils, *Ecology*, 16(2), 127–136, 1935.
112. **Tester, J. R. and Marshall, W. H.**, A Study of Certain Plant and Animal Interrelations on a Native Prairie in Northwestern Minnesota, Paper No. 8, Minnesota Museum of Natural History, University of Minnesota Occas, 1961.
113. **Tietjen, H. P., Halvorson, C. H., Hegdal, P. L., and Johnson, A. M.**, 2,4-D herbicide, vegetation, and pocket gopher relationships, Black Mesa, Colorado, *Ecology*, 48(4), 634–643, 1967.
114. **Tryon, C. A., Jr.**, The great gray owl as a predator of pocket gophers, *Wilson Bull.*, 55(2), 130–131, 1943.

115. **Tryon, C. A., Jr.**, The Biology of the Pocket Gopher (*Thomomys talpoides*) in Montana, Mont. State Coll. Agric. Exp. Sta. Tech. Bull. 448, 1947.
116. **Tryon, C. A. and Cunningham, H. N.**, Characteristics of pocket gophers along an altitudinal transect, *J. Mammal.*, 49(4), 699–705, 1968.
117. **Turner, G. T.**, Responses of mountain grassland vegetation to gopher control, reduced grazing, and herbicide, *J. Range Manage.*, 22(6), 377–383, 1969.
118. **Van Dyne, G. M. and Heady, H. F.**, Botanical composition of sheep and cattle diets on a mature annual range, *Hilgardia*, 36(13), 465–492, 1965.
119. **Vaughan, T. A.**, Vertebrates inhabiting pocket gopher burrows in Colorado, *J. Mammal.*, 42(2), 171–174, 1961.
120. **Vaughan, T. A.**, Movements made by two species of pocket gophers, *Am. Midl. Nat.*, 69(2), 367–372, 1963.
121. **Vaughan, T. A.**, Two parapatric species of pocket gophers, *Evolution*, 21(1), 148–158, 1967.
122. **Vaughan, T. A.**, Food habits of the northern pocket gopher on shortgrass prairie, *Am. Midl. Nat.*, 77(1), 176–189, 1967.
123. **Vaughan, T. A. and Hansen, R. M.**, Activity rhythm of plains pocket gopher, *J. Mammal.*, 42(4), 541–543, 1961.
124. **Vaughan, T. A. and Hansen, R. M.**, Experiments on interspecific competition between two species of pocket gophers, *Am. Midl. Nat.*, 72(2), 444–452, 1964.
125. **Ward, A. L.**, Mountain pocket gopher food habits in Colorado, *J. Wildl. Manage.*, 24(1), 89–92, 1960.
126. **Ward, A. L. and Hansen, R. M.**, The Burrow-builder and its Use for Control of Pocket Gophers, U.S. Department of the Interior Fish and Wildlife Service Special Report on Wildlife No. 47, 1960.
127. **Ward, A. L. and Hansen, R. M.**, Pocket gopher control with the burrow-builder in forest nurseries and plantations, *J. For.*, 60(1), 42–44, 1962.
128. **Ward, A. L., Hegdal, P. L., Richens, V. B., and Tietjen, H. P.**, Gophacide, a new pocket gopher control agent, *J. Wildl. Manage.*, 31(2), 332–338, 1967.
129. **Ward, A. L. and Keith, J. O.**, Feeding habits of pocket gophers in mountain grasslands, Black Mesa, Colorado, *Ecology*, 43(4), 744–749, 1962.
130. **Wight, H. M.**, Breeding habits and economic relations of the Dalles pocket gopher, *J. Mammal.*, 11(1), 40–48, 1930.
131. **Wilks, B. J.**, Some aspects of the ecology and population dynamics of the pocket gopher (*Geomys bursarius*) in southern Texas, *Tex. J. Sci.*, 15(3), 241–283, 1963.
132. **Young, S. P. and Jackson, H. H. T.**, *The Clever Coyote*, The Stackpole Co., Harrisburg and Wildlife Management Institute, Washington, D.C., 1951.

DIETS (NATURAL AND SYNTHETIC): CASTORIDAE

M. Aleksiuk

The family Castoridae contains two species: the Canadian beaver (*Castor canadensis*), which occurs throughout most of North America,[1] and the European beaver (*Castor fiber*), which occurs in Eurasia.[2] The diets of these two species are similar; they will be discussed together with reference to the family Castoridae rather than to either species. However, because the information published in English pertains mostly to *Castor canadensis,* most of the data herein relate primarily to North America.

NATURAL DIETS

Although the beaver is noted for its use of poplar (*Populus* spp.) and willow (*Salix* spp.) bark, it is a highly versatile feeder and consumes a broad range of plant species (Tables 1 and 2). The beaver has a clear preference for poplar and willow, but can in fact be classified as a generalized herbivore; it consumes almost any plant species available, including species too numerous to list in Tables 1 and 2.

The diet of the beaver varies considerably on a seasonal basis.[3] During the summer, the beaver consumes a great variety of plants, including the bark, leaves, and growing tips of woody plants (Table 1) as well as almost any nonwoody species available (Table 2). Geographical variation in species used as food simply reflects variation in the species available, although there is some latitudinal variation in species and portions of a plant preferred.[3] In late summer and early autumn, the beaver stores a large cache of tree stems below the surface of the water in front of the lodge for use during the winter. In most cases, sufficient food is stored in the cache to last throughout the period of ice cover.[3,4]

SYNTHETIC DIETS

Because the beaver is a versatile feeder, the diet of the animal is easily satisfied in captivity. Any pelleted ration containing all the nutritional requirements of rodents will suffice. A pelleted ration that has been used with a high degree of success is presented in Table 3. Supplements to this ration are not necessary, although some researchers prefer to make natural food items available in addition to the ration. The beaver's energy requirements generally are lowest during the winter.[13] As a result, food intake in captivity is noticeably lower in winter.

The beaver's incisors grow continuously and may become too long if the animal is fed a soft diet over long periods of time. The pelleted ration used by Aleksiuk and Cowan[12] was sufficiently hard to cause wearing of the incisors, and thus prevented them from becoming too long. This potential problem can be avoided by providing beavers with short sections of hard, woody plant stems.

Table 1
MAJOR SPECIES OF WOODY PLANTS USED AS FOOD BY BEAVERS

Plant species	Locality	Reference
Balsam poplar (*Populus balsamifera*)	Mackenzie Delta, Canada	3
	Wood Buffalo National Park, Canada	4
Quaking aspen (*Populus tremuloides*)	Northern California, U.S.A.	5
	Wood Buffalo National Park, Canada	4
	Pennsylvania, U.S.A.	6
Willow (*Salix* spp.)	Mackenzie Delta, Canada	3
	Wood Buffalo National Park, Canada	4
	Northern California, U.S.A.	5
	Louisiana, U.S.A.	7
	Manitoba, Canada	8
Alder (*Alnus crispa*)	Mackenzie Delta, Canada	3
Alder (*Alnus incana*)	Maine, U.S.A.	9
Red osier (*Cornus stolonifera*)	Wood Buffalo National Park, Canada	4
Paper birch (*Betula papyrifera*)	Maine, U.S.A.	9
Red oak (*Quercus rubra*)	Maine, U.S.A.	9
White fir (*Abies concolor*)	Northern California, U.S.A.	5
Loblolly pine (*Pinus taeda*)	Louisiana, U.S.A.	7
Sweetgum (*Liquidambar styraciflera*)	Louisiana, U.S.A.	7

Table 2
MAJOR SPECIES OF NONWOODY PLANTS USED AS FOOD BY BEAVERS

Plant species	Locality	Reference
Water lily (*Nuphar variegatum*)	Mackenzie Delta, Canada	3
Sedges (*Carex* spp.)	Archangel Province, U.S.S.R.	10
	Wandering River, Alberta, Canada	11
Cattail (*Typha* spp.)	Archangel Province, U.S.S.R.	10
Coon tail (*Cerataphyllum* spp.)	Maine, U.S.A.	9
Bulrushes (*Scirpus* spp.)	Maine, U.S.A.	9
Rose (*Rosa bourgeauiana*)	Mackenzie Delta, Canada	11
Pondweed (*Potamogeton* spp.)	Mackenzie Delta, Canada	11
Arrowhead (*Sagittaria* spp.)	Mackenzie Delta, Canada	11

Table 3
COMPOSITION OF A SYNTHETIC PELLETED RATION (UNIVERSITY OF BRITISH COLUMBIA RATION 42-57) THAT MEETS THE NUTRITIONAL REQUIREMENTS OF BEAVERS IN CAPTIVITY[12]

Constituent	Parts/2,000
Alfalfa	475
Vitagrass	475
Cornmeal	375
Ground wheat	375
Bran	100
Molasses	100
Stabilized fat	50
Fish oil	10
Bone meal	20
Iodized salt	20

REFERENCES

1. **Hall, E. R. and Kelson, K. R.,** *The Mammals of North America,* Ronald Press, New York, 1959, pp. 1–1083.
2. **Djoshkin, W. W. and Safonow, W. G.,** *Die Biber der Alten und Neuen Welt,* A. Ziemsen Verlag, Wittenberg, Lutherstadt, 1972, pp. 1–168.
3. **Aleksiuk, M.,** *Ecology,* 51, 264–270, 1970.
4. **Novakowski, N. S.,** *Can. J. Zool.,* 45, 1107–1118, 1967.
5. **Hall, J. G.,** *Ecology,* 41, 484–494, 1960.
6. **Brenner, F. J.,** *J. Wildl. Manage.,* 26, 104–107, 1962.
7. **Chabreck, R. H.,** *J. Wildl. Manage.,* 27, 179–183, 1958.
8. **Nash, J. B.,** *An Investigation of Some Problems of Ecology of the Beaver In Northern Manitoba,* Department Mines and Natural Resources, Government of Manitoba, Canada, 1951, pp. 1–64.
9. **Hodgdon, K. W. and Hunt, J. H.,** *Beaver Management in Maine,* Department of Inland Fisheries and Game, Government of Maine, U.S.A., 1966, pp. 1–102.
10. **Semyonoff, B. T.,** *The River Beaver in Archangel Province, Biology of Furbearers, II, 1951,* Translation of Russian Game Reports, Canadian Wildlife Service, 1957, pp. 5–45.
11. **Aleksiuk, M.,** unpublished results.
12. **Aleksiuk, M. and Cowan, I. McT.,** *Can. J. Zool.,* 47, 965–979, 1969.
13. **Aleksiuk, M. and Cowan, I. McT.,** *Can. J. Zool.,* 47, 471–481, 1969.

DIETS (NATURAL AND SYNTHETIC): MOUNTAIN BEAVER

E. H. Voth

Order Rodentia
Suborder Sciuromorpha
Family Aplodontidae
Species *Aplodontia rufa;* mountain beaver

Resident in the Pacific Northwest rain forest, the mountain beaver is a nonhibernating semifossorial herbivore, monogastric with functional cecum, and weighing up to 1,000 g or more.

Laboratory research interest in the mountain beaver has recognized its primitive rodent model configurations. Investigations to date have focused on renal (limited urine concentrating capacity),[1] digestive (extent of synthesis of thiamine), and nervous[3] systems. The animal has been used in an attempt to standardize histologic research techniques with respect to feeding habits.[6] It is not tractable and has not met requirements for the usual laboratory model. It has not been bred successfully in laboratory confinement. Animals at the Forest Research Laboratory, Oregon State University, Corvallis, have survived in captivity for nearly 3 years when captured soon after weaning. Outdoor colonies have been maintained indefinitely in fenced enclosures by Weyerhaeuser Company, Centralia, Washington. Squeeze cages or anesthesia is required for handling adults.

Research of food habits in the field has indicated that the animal prefers a diet of primitive vascular plants, i.e., ferns.[5] Proximate analysis of two fern species, either of which can apparently supply all nutritional requirements of adults when fed fresh, is presented in Table 1.

Animals have been maintained for weeks on dry alfalfa hay and water. The digestive system resembles that of the guinea pig, and under most conditions guinea pig ration[4] may be suitable.* Protein content of diets for growth of the young probably cannot be less than 20%. Incisors must be clipped every 2 months on animals without access to earth balls.

The animal will eat almost any plant in its habitat in the wild. Literature naming such plants has been reviewed, and a cumulative list has been prepared by Godin.[3]

Table 1
PROXIMATE ANALYSIS[a] OF FRONDS OF TWO FERN SPECIES, AUGUST 1966

	Oven-dry percent					
Species	Dry matter	Crude protein	Crude fat	Crude fiber	Ash	N-free extract
Polystichum munitum	35.2	8.2	1.8	40.6	5.0	44.4
Pteridium aquilinum	28.1	14.6	3.2	24.1	6.2	51.9

[a] Dr. Paul H. Weswig, Forage Analysis Laboratory, Oregon State University, Corvallis (see Reference 5).

* Also see Bray, D. L. et al., Natural and semipurified diets for guinea pigs, elsewhere in this volume.

REFERENCES

1. **Dolph, C. I., Braun, H. A., and Pfeiffer, E. W.,** The effect of vasopressin upon urine concentration in *Aplodontia rufa* (Sewellel) and the rabbit, *Physiol. Zool.,* 35(3), 263–269, 1962.
2. **Godin, A. J.,** A Review of the Literature on the Mountain Beaver (Special Scientific Report, Wildlife No. 78), U.S. Department of the Interior, Fish and Wildlife Service, Bureau of Sport Fisheries and Wildlife, Washington, D.C., 1964.
3. **Merzenich, M. M., Kitzes, L., and Aitkin, L.,** Anatomical and physiological evidence for auditory specialization in the mountain beaver, *Brain Res.,* 58(2), 331–344, 1973.
4. **Reid, M. E.,** The Guinea Pig in Research: Biology, Nutrition, Physiology, Human Factors Research Bureau, Washington, D.C., 1958.
5. **Voth, E. H.,** Food Habits of the Pacific Mountain Beaver, *Aplodontia rufa pacifica* Merriam, Ph.D. dissertation, Oregon State University, Corvallis, 1968.
6. **Voth, E. H. and Black, H. C.,** A histologic technique for determining feeding habits of small herbivores, *J. Wildl. Manage.,* 37(2), 223–231, 1973.

DIETS FOR CRICETIDAE

A. U. Shump

Most members of the family Cricetidae feed on plant material and invertebrates. Some, however, also feed on carrion, and some have become essentially carnivorous. For example, the grasshopper mouse, *Onychomys,* eats insects, larvae, and worms, and such small rodents as it can overpower, with grasshoppers and scorpions being their favorite foods. Some of the more aquatic rats, such as *Ichthyomys,* feed primarily on fish. Still others, such as *Microtus,* are entirely herbivorous. On the whole, the species within this family show great variability in their natural diets, as is shown in Table 1.

Table 1
NATURAL DIETS

Species (Common name)	Food of animal origin	Fruits	Seeds and crops	Vegetative plant parts	Reference
Alticola (Mountain vole)				Stems and leaves of herbaceous plants and shrubs	1
Andinomys (Andean mouse)				Green plants	1
Anotomys (Ecuador fish-eating rat)	Fish				1
Arvicola (Water vole)				Reeds and other aquatic plants	1
Baiomys (Pygmy mouse)	Some insects		Some seeds	Green vegetation	1
Calomys (Vesper mouse)	Some insects			Plant material	1
Calomyscus (Mouse-like hamster)	Some animal food		Mainly seeds	Flowers and leaves	1
Chinchillula (Chinchilla mouse)				Plants	1
Clethrionomys (Red-backed vole)	Snails, insects, young birds	Berries	Nuts, seeds	Tender vegetation, roots, bark, lichens, fungi	1, 2
Cricetulus (Eurasian hamster)			Millet seeds and others, soybeans, peas	Young shoots	1
Cricetus (Common hamster)	Insect larvae, frogs		Lentils, grains, beans, cereal seeds, peas, potatoes	Roots, green parts of plants	1

Table 1 (continued)
NATURAL DIETS

Species (Common name)	Food of animal origin	Fruits	Seeds and crops	Vegetative plant parts	Reference
Desmodillus (Cape short-tailed gerbil)	Insects – locusts and grasshoppers		Seeds, grain		1
Dicrostonyx (Collared lemming)	Some insects and meat			Mainly plant materials	1
Dolomys (Snow vole)				Grasses	1
Eligmodontin (Highland desert mouse)	Mainly insects				1
Ellobius (Mole lemming)				Roots, bulbs, rhizomes of herbaceous plants	1
Gerbillus (Smaller gerbil)	Insects		Seeds, nuts	Roots and greenstuff, grass	1
Holochilus (Web-footed marsh rat)	Molluscs			Marsh plants	1
Hypogeomys (Malagasy rat)		Fallen fruit			1
Ichthyomys (Fish-eating rat)	Fish				1
Lagurus (Steppe lemming)				Mainly green foods	1
Lemmus (Lemming)		Berries		Sedges, grasses, bark, leaves, lichens, roots	1

Table 1 (continued)
NATURAL DIETS

Species (Common name)	Food of animal origin	Fruits	Seeds and crops	Vegetative plant parts	Reference
Lophiomys (Maned rat)				Leaves and tender shoots	1
Macrotarsomys (Malagasy rat)		Berries, fruits	Seeds	Plant stems, roots	1
Melanomys (Rice rat)			Some seeds	Mainly green vegetation	1
Mesocricetus (Golden hamster)	Meat	Fruits	Seeds	Green vegetation	1
Microtus (Meadow vole)			Seeds	Roots, bark, leaves	1, 3
Myopus (Wood lemming)				Mosses, stems of red wortleberry, juniper bark	1
Myospalax (Mole mouse)			Grain	Roots	1
Mystromys (White-tailed rat)			Seeds	Vegetable matter	1
Nelsonia (Diminutive wood rat)				Needles of Douglas fir and juniper	1
Neofiber (Round-tailed musk-rat			Sugar cane	Rushes, sedges, sawgrass, mangrove bark	1
Neotoma (Wood rat)	Some invertebrates		Seeds	Roots, stems, leaves, fleshy stems of cacti	1

Table 1 (continued)
NATURAL DIETS

Species (Common name)	Food of animal origin	Fruits	Seeds and crops	Vegetative plant parts	Reference
Nyctomys (Vesper rat)		Wild figs, avocados, bananas			1
Ochrotomys (Golden mouse)			Mainly seeds		1
Ondatra (Muskrat)	Mussels, crayfish	Apples (when available)	Corn and other crops	Roots and stems of cattails and lilies	1, 3, 4
Onychomys (Grasshopper mouse)	Insects – grasshoppers, larvae, worms, small rodents, scorpions, lizards		Some seeds		1, 5
Oryzomys (Rice rat)	Fish, invertebrates	Fruits	Seeds	Grasses, sedges	1
Peromyscus (Deer mouse)	Insects and other small invertebrates, carrion	Berries, fruits	Seeds, nuts		1, 3
Phenacomys (Heather vole, tree mouse)		Berries		Terminal twigs of coniferous trees, needles, willow, bark	1
Phodopus (Asian hamster)			Seeds	Any available plant material	1
Phyllotis (Long-eared mouse)			Seeds	Lichens, green plant material	1

Table 1 (continued)
NATURAL DIETS

Species (Common name)	Food of animal origin	Fruits	Seeds and crops	Vegetative plant parts	Reference
Pitymys (Pine vole)		Fruits	Seeds, nuts	Bark from roots and lower trunk, roots and tubers, leaves	1
Prometheomys (Long-clawed mole vole)				Roots	1
Psammomys (Sand rat)			Barley	Plant parts	1
Punomys (Puna mouse)				Twigs, *Senecio adenophylloides, Werneria digitata*	1
Reithrodon (Pampas gerbil)				Plant material	1
Reithrodontomys (Harvest mouse)	Some insects		Seeds	Green shoots	1
Rhombomys (Great gerbil)				Desert plants	1
Scapteromys (South American water rat)	Earthworms, insect larvae				1
Sigmodon (Cotton rat)	Insects, eggs		Grass seeds, cultivated crops – sugar cane and sweet potatoes, sunflower seeds, oatmeal	Stems, foliage	6

Table 1 (continued)
NATURAL DIETS

Species (Common name)	Food of animal origin	Fruits	Seeds and crops	Vegetative plant parts	Reference
Synaptomys (Bog lemming)	Slugs, snails, other invertebrates	Huckleberries, blueberries		Green plant tissues, mainly low-growing, sphagnum moss	1, 7
Tatera (Larger gerbil)	Insects, young birds, eggs		Seeds	Bulbs, roots, green plant material	1
Zygodontomys (American cane rat)		Fruits	Seeds, corn, rice	Grasses	1

Table 2
CHEMICALLY DEFINED DIETS

Species (Common name)	Ingredients	% Composition	Reference
Sigmodon (Cotton rat)	Breeding diet:		8
	Wholemeal flour	50	
	Dried full-cream milk	27	
	Dried yeast	12	
	Meat and bone meal	6	
	Cod-liver oil	3	
	Mineral mixture[a]	2	
	Additional greens		
	Growth diet:		8
	Wholemeal flour	35	
	Rolled oats	30	
	Dried skim milk	15	
	Meat and bone meal	10	
	Dried yeast	6	
	Cod-liver oil	2	
	Calcium carbonate	1	
	Sodium chloride	1	
	Additional water		
	Additional sprouted oats, spinach, carrots		
Cricetulus (Eurasian hamster)	Fine thirds (wheatings)	19.2	9
	Ground wheat	19.2	
	Sussex ground oats	19.2	
	Ground barley	9.5	
	Ground maize	9.5	
	Meat and bone meal (50% protein)	9.5	
	Dried skim milk	7.0	
	White fish meal	4.8	
	Dried yeast	1.3	
	Cod-liver oil	0.44	
	Sodium chloride	0.4	
(Gerbil)	Wholemeal flour	46	10, 11
	Sussex ground oats	40	
	Best white fish meal	8	
	Dried skim milk	3	
	Cod-liver oil	1	
	Dried yeast	1	
	Sodium chloride	1	
Mesocricetus (Hamster)	Sucrose	28	12
	Corn starch	28	
	Crude casein	14	
	Whole-wheat flour	8	
	Powdered yellow corn	8	
	Dried brewer's yeast	8	
	Dried hog's liver, powdered	3	
	Salt mixture[b]	3	
	Alfalfa meal	1	

[a] Calcium carbonate, 45.4%; sodium chloride, 45.4%; ferric citrate, 9%; potassium iodide, 0.2%.

[b] McCollums salt mixture 185, with supplement of 13.5 mg potassium iodide, 139 mg copper sulfate, and and 556 mg manganous sulfate/100 g.

Table 2 (continued)
CHEMICALLY DEFINED DIETS

Species (Common name)	Ingredients	% Composition	Reference
Onychomys (Grasshopper mouse)	Rolled oats or barley	60	5
	Alfalfa hay leaves	20	
	Pigeon mix (corn, wheat, milo, vetch, kafir, peas)	10	
	Sunflower seeds	10	
	Additional dog food		
Oryzomys (Rice rat)	Wholemeal flour	46	10, 13
	Sussex ground oats	40	
	Best white fish meal	8	
	Dried skim milk	3	
	Cod-liver oil	1	
	Dried yeast	1	
	Sodium chloride	1	
Peromyscus (Deer mouse)	Rolled oats	62	14
	Hemp seed	10	
	Millet seed	9	
	Dry meat scraps	8	
	Wheat seed	5	
	Wheat germ	4	
	Cod-liver oil	1.5	
	Calcium carbonate, powdered	0.5	

Table 3
PARTIALLY PURIFIED DIETS

Species (Common name)	Ingredients	% Composition[a]	Reference
Mesocricetus (Golden hamster)	Sucrose	62.4	15
	Casein, purified	24	
	Corn oil	5	
	Salt mixture[b]	4	
	Cellulose	3	
	Liver extract	1	
	Cod-liver oil	0.3	
	Cystine	0.3	
	Choline chloride	100 mg/100 g	
	Inositol	100 mg/100 g	
	p-Aminobenzoic acid	30 mg/100 g	
	Calcium pantothenate	4 mg/100 g	
	α-Tocopherol	2.5 mg/100 g	
	Niacin	2 mg/100 g	
	Menadione	0.6 mg/100 g	
	Pyridoxine hydrochloride	0.6 mg/100 g	
	Riboflavine	0.6 mg/100 g	
	Thiamine hydrochloride	0.6 mg/100 g	
	Folic acid	0.2 mg/100 g	
	Biotin	0.01 mg/100 g	
	Vitamin B_{12}	0.005 mg/100 g	
Microtus[c] (Meadow vole)	α-Cellulose	45–0	16
	Carbohydrate mix[d]	32–59	
	Casein	14–26	
	Corn oil	4.4–8.0	
	Vitamin mix, NBC[e]	V[f]	
	Minerals[g]	V[f]	
Sigmodon (Cotton rat)	Carbohydrate mix[d]	70 (63; 56)	17
	Casein	7 (14; 21)	
	Minerals[g]	15	
	Corn oil	3	
	Vitamin mix, NBC[e]	2	
	Sucrose	67	18
	Casein, purified	24	

[a] Unless otherwise specified.

[b] Hegsted's Salts IV (9), composed of the following (parts/weight): calcium carbonate, 600; potassium phosphate, dibasic, 645; calcium phosphate, dibasic, 150; magnesium sulfate (epsom salts), 204; sodium chloride, 335; ferric citrate·$6H_2O$, 55; potassium iodide, 1.6; manganous sulfate, 10.0; zinc chloride, 0.5; copper sulfate, 0.6.

[c] Apparently the most efficient of several experimental diets.

[d] Consists of 2 parts cornstarch, 1 part dextrin, and 1 part sucrose.

[e] Contains the following per kg of mix: vitamin A concentrate, 4.5 g (200,000 IU/g); vitamin D concentrate, 0.25 g (400,000 IU/g); α-tocopherol, 5.0 g; ascorbic acid, 45.0 g; inositol, 5.0 g; choline chloride, 75.0 g; menadione, 2.25 g; *p*-aminobenzoic acid, 5.0 g; niacin, 4.5 g; riboflavine, 1.0 g; pyridoxine hydrochloride, 1.0 g; thiamine hydrochloride, 1.0 g; calcium pantothenate, 3.0 g; biotin, 20 mg; folic acid, 90 mg; vitamin B_{12}, 1.35 mg.

[f] Variable; at 0% α-cellulose the diet contained 4% mineral and 3% vitamin mixtures; at 45% α-cellulose the diet contained 2.2% minerals and 1.7% vitamins.

[g] Salt mixture W-NBC.

Table 3 (continued)
PARTIALLY PURIFIED DIETS

Species (Common name)	Ingredients	% Composition[a]	Reference
Sigmodon (continued)	Corn oil (Mazola)	4.7	
	Salt mixture	4	
	Cod-liver oil, fortified	0.3	
	Choline chloride	100 mg/100 g	
	Inositol	100 mg/100 g	
	p-Aminobenzoic acid	30 mg/100 g	
	Niacin	25 mg/100 g	
	Calcium pantothenate	2.0 mg/100 g	
	Riboflavine	0.30 mg/100 g	
	Pyridoxine hydrochloride	0.25 mg/100 g	
	Thiamine hydrochloride	0.25 mg/100 g	
	Pteroylglutamic acid	0.2 mg/100 g	
	Biotin	0.01 mg/100 g	

REFERENCES

1. **Walker, E. P.,** *Mammals of the World,* Johns Hopkins Press, Baltimore, 1968.
2. **Steven, D. M.,** in *UFAW Handbook on the Care and Management of Laboratory Animals,* 2nd ed., Worden, A. N. and Land-Petter, W., Eds., Courier Printing and Publishing Co., Kent, England, 1957.
3. **Crandell, L. S.,** *Management of Wild Mammals in Captivity,* University of Chicago Press, Chicago, 1964.
4. **Svihla, A. and Svihla, R. D.,** The Louisiana muskrat, *J. Mammal.,* 12(1), 12–28, 1931.
5. **Egoscue, H. J.,** Laboratory and field studies of the northern grasshopper mouse, *J. Mammal.,* 41(1), 99–110, 1960.
6. **Meyer, B. J. and Meyer, R. K.,** Growth and reproduction of the cotton rat, *Sigmodon hispidus hispidus,* under laboratory conditions, *J. Mammal.,* 25(2), 107–129, 1944.
7. **Connor, P. F.,** The bog lemming, *Synaptomys cooperi,* in southern New Jersey, *Biol. Ser. Mich. State Univ.,* 1(5), 165–248, 1959.
8. **Rowlands, I. W.,** Tyburn Report, Wellcome Foundation, London, 1945.
9. **Schwentker, V.,** in *UFAW Handbook on the Care and Management of Laboratory Animals,* 2nd ed., Worden, A. N. and Land-Petter, W., Eds., Courier Printing and Publishing Co., Kent, England, 1957.
10. **Bruce, H. M.,** Feeding and breeding of laboratory animals. XI, *J. Hyg.,* 48, 171–183, 1950.
11. **Newsome, J.,** in *UFAW Handbook on the Care and Management of Laboratory Animals,* 2nd ed., Worden, A. N. and Land-Petter, W., Eds., Courier Printing and Publishing Co., Kent, England, 1957.
12. **Granados, H.,** Nutritional studies on growth and reproduction of the golden hamster *(Mesocricetus auratus auratus), Acta Physiol., Scand. Suppl.,* 87, 1–138, 1951.
13. **Steward, J. S.,** in *UFAW Handbook on the Care and Management of Laboratory Animals,* 2nd ed., Worden, A. N. and Land-Petter, W., Eds., Courier Printing and Publishing Co., Kent, England, 1957.
14. **Dice, L. R.,** in *UFAW Handbook on the Care and Management of Laboratory Animals,* 2nd ed., Worden, A. N. and Land-Petter, W., Eds., Courier Printing and Publishing Co., Kent, England, 1957.
15. UFAW, *Handbook,* reprinted from Nutrient Requirements of Laboratory Animals, Publ. 990, Committee on Animal Nutrition, National Academy of Sciences, National Research Council, Washington, D. C., 1962.
16. **Shenk, J. S., Elliott, F. C., and Thomas, J. W.,** Meadow vole nutrition studies with semisynthetic diets, *J. Nutr.,* 100, 1437–1446, 1970.
17. **Underhill, A. C.,** Use of neotropical rodents in protein efficiency studies, *Lab. Anim. Sci.,* 23(4), 499–503, 1973.
18. **Schweigert, B. S.,** Nutritional requirements of the cotton rat and hamster, in *Advances in Research and Applications,* Harris, R. S. and Thimann, K. V., Eds., Academic Press, New York, 1948.

DIETS (NATURAL AND SYNTHETIC): RATS AND MICE

D. Ford

INTRODUCTION

Rats and mice are the experimental animals most widespread in use, and consequently are the two species for which the majority of diets have been devised.

The tables that follow are not exhaustive but merely serve to give selected examples of diets designed for specific purposes. The criteria used for choosing the diets were (1) proof that the diets carried out the functions for which they were intended and (2) comprehensive description of their composition.

Up to the present, little evidence has been offered to show any marked differences between the nutrient requirements of rats and mice, so in this chapter it has been assumed that a diet designed to carry out a function in one species is adequate for the same job in another. Similarly, although there is some evidence suggesting differences in requirements for individual nutrients between inbred strains of mice,[1,2] this is not substantial or widespread enough to justify specific diets for individual inbred strains or groups of strains.

For many purposes, diets must conform to microbiological standards, which may involve complete sterility in the case of those for germ-free animals or absence of specific organisms in the case of those for specified pathogen-free animals. These diets usually undergo some form of sterilization or a "pasteurization" process to eliminate or reduce microbiological activity; in the case of solid diets, this involves treatment by autoclaving,[3] gamma irradiation,[4] or gaseous fumigation,[5] while in liquid diets it is often carried out by filtration.[6] Although this latter method has little effect on the dietary value, the former methods, especially autoclaving, may reduce the level of certain nutrients such as amino acids and vitamins.[3,7-12] Furthermore, animals kept in conditions with restricted flora may not be able to benefit from vitamins produced by the gut microflora, and so require additional dietary sources. For these reasons, diets for germ- and other defined flora-free animals requiring sterilization may include elevated protein and vitamin levels. Examples of diets suitable for germ-free rats and mice after sterilization by autoclaving are included in the tables, and Diet PRD sterilized by gamma radiation has been used for many years as a suitable diet for breeding SPF (MRC Category 4 star)[13] rats and mice.[14]

NATURAL DIETS

These diets (Table 1) are compounded from unpurified raw materials which inevitably vary between batches. Maintenance of constant nutrient content can only be achieved by selection of raw materials based on comprehensive quality control.

The examples given are all suitable for breeding and growing rats and mice, but these diets may be unsuitable for stock or experimental animals. Such animals have lower nutrient requirements than breeding stock,[15] so diets containing lower nutrient levels would probably result in increased longevity and reduced incidence of age-associated degenerative diseases.[16] Diets of this sort would be particularly valuable in long-term assays such as chronic toxicity experiments.

Natural diets are used in circumstances where nutrient constancy over a long period is not an overriding consideration, but where economy often is. Therefore, they are used in the breeding of most experimental rats and mice and in the maintenance of animals for certain experimental purposes.

PURIFIED DIETS

These diets (Table 2) are compounded from raw materials which have been extracted and purified, and consequently are better defined and more reproducible between batches than diets using natural ingredients. This characteristic is important in studies where nutrient concentrations may influence the results obtained, a possibility which is often disregarded or not considered by experimental workers.[20,21]

Purified diets also allow simple manipulation of nutrients to give desired concentrations or mixtures, making them suitable for study of nutrient requirements and the interrelationship between nutrients and other experimental conditions. In many cases a basic diet such as that of Drepper and Weik[18] with modifications to obtain the required nutrient balances is satisfactory. At other times, however, specific ingredients are necessary to produce the desired result, e.g., a low tryptophan diet to produce niacin deficiency or one containing raw egg white to produce a biotin deficiency.

The purified diets included in Table 2 are quoted as they were reported in the original references. Where obvious omissions of vitamins were made in the original papers, these have been rectified and are so signified. The mineral supplements have been unaltered and the formulations taken from the original references where necessary. This has been done despite the fact that commercial companies have been quoted in some papers as the supplier; several discrepancies have been found between catalogues of these companies and the original papers to which they refer.[20] Therefore, compared with published requirements,[20] mineral supplements may contain omissions or imbalances. Salt mixes that have been recommended as supplying all required minerals in the correct proportions[20] are those of Bernhart and Tomarelli[22] (see Table 2.F), Greenfield et al.,[23] Rogers and Harper[24] (see Table 4.A), the 19th edition of the *United States Pharmacopeia*,[25] and Cohen et al.[26] The formulations of those not presented elsewhere are included in Table 3.

It should be borne in mind that deficiency of the B-group vitamins and vitamin K in animals with a conventional gut flora is more quickly achieved by prevention of coprophagy. This may be carried out by means of a tail cup or wire mesh tube which prevents the animal consuming feces from the anus.

CHEMICALLY DEFINED DIETS

The diets in Table 4 consist of defined ingredients in which amino acids are supplied as pure substances. They are homogenous, may be easily supplied in a soluble form, and can be produced with a low antigen level, making them suitable for immunological work.

Table 1
BREEDING DIETS

A. PRD Labsure Animal Foods[a]

	g/kg diet
Barley (ground)	56.2
Wheat (ground)	200.0
Maize (ground)	100.0
Oats (ground)	181.2
Wheatfeed	200.0
Soya (extracted)	100.0
Roller dried skim milk	75.0
Unextracted dried yeast	25.0
English white fish meal	50.0
Vitamin supplement	5.0
Mineral supplement	7.6

Vitamin Supplement Supplies per Kilogram of Diet:

Vitamin A (IU)	8,000
Cholecalciferol (IU)	1,000
Vitamin E (mg)	25
Vitamin K (mg)	10
Thiamin (mg)	2
Riboflavin (mg)	8
Nicotinic acid (mg)	20
Pantothenic acid (mg)	4
Cyanocobalamin (μg)	12
Folic acid (mg)	6
Choline chloride (mg)	200

Mineral Mix Supplies per Kilogram of Diet:

$CaCO_3$ (g)	1.65
$CaHPO_4$ (g)	2.23
NaCl (g)	2.50
MgO (mg)	167
$FeSO_4 \cdot 7H_2O$ (mg)	147
$MnSO_4 \cdot 4H_2O$ (mg)	77
$CuSO_4 \cdot 5H_2O$ (mg)	27
$CoSO_4 \cdot 7H_2O$ (mg)	1.8
$CaIO_3$ (mg)	2.2
Fine maize meal (carrier) (mg)	695

Note: This diet is a development of Diet PRM.[4]

[a] Christopher Hill Ltd., Poole, Dorset, England.

B. Diet of Knapka et al.[17]

	g/kg diet
Ground hard winter wheat	230
Wheat middlings	100
Ground maize	245
Maize gluten meal[a]	30
Dried alfalfa meal[b]	40
Soya bean meal[c]	120
Fish meal[d]	100
Dried skim milk	50
Dried brewer's yeast	20
Dry molasses	15
Soya bean oil	25
Minerals and vitamins	25

Vitamin Mix Supplies per Kilogram of Diet:

Vitamin A (IU)	6,050
Cholecalciferol (IU)	5,060
α-Tocopherol acetate (mg)	22
Menadione sodium bisulfite (mg)	3.1
Thiamin (mg)	11
Riboflavin (mg)	3.7
Niacin (mg)	33
Calcium pantothenate (mg)	19.8
Pyridoxine (mg)	1.9
Vitamin B_{12} (μg)	4.4
Folic acid (mg)	2.4
Choline chloride (mg)	616

Mineral Mix Supplies per Kilogram of Diet:

$CaHPO_4$ (g)	12.50
$CaCO_3$ (g)	5.00
NaCl (g)	5.00
$FeSO_4 \cdot 7H_2O$ (mg)	655
MnO (mg)	85
ZnO (mg)	22
$CuSO_4 \cdot 5H_2O$ (mg)	17
$CoCO_3$ (μg)	890
KIO_3 (mg)	2.5

[a] 60% protein.
[b] 17% protein.
[c] 49% protein.
[d] 60% protein.

Table 1 (continued)
BREEDING DIETS

C. Diet VM 1225[18]

	Amount per kilogram of diet
Ground wheat (g)	280
Ground maize (g)	300
Soya meal (extracted) (g)	130
Fish meal (g)	180
Sucrose (g)	20
Soya oil (g)	40
Vitamin mix (g)	20
Mineral mix (g)	30
DL-Methionine (mg)	1,200

Vitamin Supplement Supplies per Kilogram of Diet:

Vitamin A (IU)	15,000
Cholecalciferol (IU)	500
Vitamin E (mg)	150
Menadione (mg)	10
Thiamin HCl (mg)	20
Riboflavin (mg)	20
Nicotinamide (mg)	50
Calcium pantothenate (mg)	50
Pyridoxine HCl (mg)	15
Vitamin B_{12} (μg)	30
Folic acid (mg)	10
Choline chloride (mg)	1,000
Inositol (mg)	100
p-Aminobenzoic acid (mg)	100
Biotin (μg)	200
Ascorbic acid (mg)	20

Mineral Supplement Supplies per Kilogram of Diet:

$CaCO_3$ (g)	7.00
$Ca_3(PO_4)_2$ (g)	7.00
K_2HPO_4 (g)	5.00
NaCl (g)	4.00
Na_2HPO_4 (g)	3.50
$MgSO_4 \cdot 7H_2O$ (g)	2.50
Fe(II)gluconate$\cdot 2H_2O$ (mg)	740
$MnSO_4 \cdot 4H_2O$ (mg)	225
$ZnCO_3$ (mg)	20
$CuSO_4 \cdot 5H_2O$ (mg)	10
KI (μg)	250
$Na_2MoO_4 \cdot 2H_2O$ (μg)	250
NaF (mg)	5

D. Diet L-485 for Germ-free Rats and Mice[19]

	g/kg diet
Ground maize	589.62
Soya bean meal[a]	300.0
Alfalfa meal[b]	35.0
Maize oil[c]	30.0
Lysine	5.0
Methionine	5.0
BHT	0.125
Mineral mix	25.25
Vitamin mix	10.0

Vitamin Mix Supplies per Kilogram of Diet:

Vitamin A (IU)	26,400
Cholecalciferol (IU)	1,012
α-Tocopherol acetate (mg)	220
Menadione (mg)	88
Thiamin (mg)	66
Riboflavin (mg)	31
Pantothenic acid (mg)	286
Niacin (mg)	66
Pyridoxine HCl (mg)	22
Vitamin B_{12}[d] (μg)	4.4
Folic acid (mg)	11
Choline chloride (mg)	1,980
p-Aminobenzoic acid (mg)	51

Mineral Mix Supplies per Kilogram of Diet:

$CaCO_3$ (g)	5.00
$CaHPO_4$ (g)	10.00
NaCl (g)	10.00
Fe (mg)	22
Mn (mg)	66
Zn (mg)	15
Cu (mg)	2.2
Co (mg)	0.66
I (mg)	1.32

Note: This diet can be sterilized by autoclaving at 122°C for 25 min.

[a] 50% protein.
[b] 17% protein.
[c] Once refined.
[d] As 0.1% triturate.

Table 2
PURIFIED DIETS

A. Diet of Drepper and Weik[18]

	g/kg diet
Casein[a]	300
Rice starch	560
Soya oil	20
Powdered cellulose	40
Vitamin mix[b]	20
Mineral mix	60

Mineral Mix Supplies per Kilogram of Diet:

$CaCO_3$ (g)	14.00
$Ca_3(PO_4)_2$ (g)	14.00
K_2HPO_4 (g)	10.00
NaCl (g)	8.00
Na_2HPO_4 (g)	7.00
$MgSO_4 \cdot 7H_2O$ (g)	5.00
Fe(II)gluconate$\cdot 2H_2O$ (g)	1.48
$MnSO_4 \cdot 4H_2O$ (mg)	450
$ZnCO_3$ (mg)	40
$CuSO_4 \cdot 5H_2O$ (mg)	20
KI (μg)	500
$Na_2MoO_4 \cdot 2H_2O$ (μg)	500
NaF (mg)	10

[a] Vitamin free.
[b] Same as for Diet VM 1225 (Table 1.C).

B. Diet of Knudsen and Meyer[27]

Casein	200
Dextrin	235
Maize starch	168
Sucrose	135
Potato starch	135
Lactose	5.4
Cellulose	30
Soya oil	50
Methionine	4
Vitamin mix	2.5
Mineral mix	35.1

B. Diet of Knudsen and Meyer[27] (continued)

g/kg diet

Vitamin Mix Supplies per Kilogram of Diet:

Vitamin A (IU)	10,000
Ergocalciferol (IU)	1,000
dl-α-Tocopherol acetate (mg)	100
Menadione sodium bisulfite (mg)	50
Thiamin HCl (mg)	4
Riboflavin (mg)	4
Nicotinamide (mg)	50
Calcium pantothenate (mg)	40
Pyridoxal HCl (mg)	4
Cyanocobalamin (μg)	30
Choline chloride (g)	2
Inositol (mg)	400
p-Aminobenzoic acid (mg)	50
Biotin (μg)	400

Mineral Mix Supplies per Kilogram of Diet:

$CaCO_3$ (g)	2.13
$CaHPO_4$ (g)	18.90
K_2HPO_4 (g)	4.20
Na_2HPO_4 (mg)	600
NaCl (g)	3.81
$MgSO_4$ (70%) (g)	3.00
Fe(III)citrate$\cdot 5H_2O$ (mg)	150
$MnSO_4 \cdot H_2O$ (mg)	150
$ZnSO_4 \cdot 7H_2O$ (mg)	60
$CuSO_4 \cdot 5H_2O$ (mg)	30
$CoCl_2 \cdot 6H_2O$ (μg)	360
KI (μg)	300
$MoCl_2$ (μg)	150
$NaSeO_3 \cdot 5H_2O$ (μg)	150
SiO_2 (g)	2.00

Table 2 (continued)
PURIFIED DIETS

C. Diet of Nalder et al.[28]

	g/kg diet
Casein	200
Glucose	590
Maize oil	100
Cellulose	50
Vitamin mix	20
Mineral mix	40

Vitamin Mix Supplies per Kilogram of Diet:

Retinyl acetate (IU)	18,000
Calciferol (IU)	2,000
α-Tocopherol (mg)	100
Menadione (mg)	45
Thiamin HCl (mg)	20
Riboflavin (mg)	20
Niacin (mg)	90
Calcium pantothenate (mg)	60
Pyridoxine HCl (mg)	20
Cyanocobalamin (μg)	27
Folic acid (mg)	1.8
Choline HCl (mg)	1,500
Inositol (mg)	100
p-Aminobenzoic acid (mg)	100
Biotin (μg)	400
Ascorbic acid (mg)	900
Glucose to 20 g	

Mineral Mix Supplies per Kilogram of Diet:

$CaCO_3$ (g)	15.80
$NaHPO_4 \cdot H_2O$ (g)	14.40
KCl (g)	8.00
$MgCO_3$ (g)	1.50
$FeSO_4 \cdot 7H_2O$ (mg)	99.2
$MnSO_4 \cdot H_2O$ (mg)	151.2
$ZnSO_4$ (mg)	11.5
$CuSO_4$ (mg)	11.6
$CoCl_2 \cdot 6H_2O$ (μg)	904
KI (mg)	31.6
$Na_2MoO_4 \cdot 2H_2O$ (mg)	1.7

D. Liquid Diet[a] of Pilström et al.[29]

	g/l
Dry milk powder[b]	67.5
Sucrose	103.0
Maize oil	7.0
Vegetable oil	30.0
Cystine	0.5
Methionine	0.2
Vitamin mix	6.0
Mineral mix	5.0
Sodium carraginate	1.0
Distilled water to 1 liter	

Vitamin Mix[30] Supplies per Liter of Diet:

Retinol (IU)[c]	5,775
Ergocalciferol (IU)[c]	400
α-Tocopheryl acetate (mg)[c]	30
Menadione (mg)	0.25
Thiamin (μg)	725
Riboflavin (mg)	1.25
Calcium pantothenate (mg)	5.0
Nicotinamide (mg)	3.75
Pyridoxine HCl (μg)	720
Cyanocobalamin (μg)	25
Folic acid (μg)	250
Choline chloride (mg)	250
Inositol (mg)	25
p-Aminobenzoic acid	12.5
Biotin (μg)	25
Glucose to 5 g	

Mineral Mix[31] Supplies per Liter of Diet:

$CaCO_3$ (g)	1.50
$CaHPO_4 \cdot 2H_2O$ (mg)	375
K_2HPO_4 (g)	1.61
NaCl (mg)	837
$MgSO_4 \cdot 7H_2O$ (mg)	510
Fe(III)citrate·$5H_2O$ (mg)	137
$MnSO_4 \cdot 4H_2O$ (mg)	25
$ZnCl_2$ (mg)	1.2
$CuSO_4 \cdot 5H_2O$ (mg)	1.5
KI (mg)	4

[a] Ingredients mixed and homogenized, and diet stored at 4°C for not more than 3 days.
[b] Skim milk powder containing (g/kg) milk protein, 670; lactose, 200; fat, 10.
[c] Made up in 970 mg maize oil and added with maize oil.

Table 2 (continued)
PURIFIED DIETS

E. Diet L 474E12[a] for Germ-free Rats and Mice[32]

	g/kg diet
Casein[b]	240
Maize oil	30
Rice starch	604
Cellophane spangles	50
B vitamin mix	6
Fat-soluble vitamin mix	20
Mineral mix	47
DL-Methionine	3

B-vitamin Mix Supplies per Kilogram of Diet:

Thiamine (mg)	60
Riboflavin (mg)	30
Nicotinamide (mg)	50
Nicotinic acid (mg)	50
Calcium pantothenate (mg)	300
Pyridoxine HCl (mg)	20
Pyridoxamine dihydrochloride (mg)	4
Cyanocobalamin (μg)[c]	25
Folic acid (mg)	10
Choline chloride (mg)	2,000
Inositol (mg)	1,000
p-Aminobenzoic acid (mg)	50
Biotin (mg)	1
Rice starch carrier (mg)	2,175

[a] This diet may be formed into pellets 25 cm in diameter and sterilized by autoclaving at 122°C for 25 min.
[b] Vitamin free.
[c] As 0.1% triturate in mannitol.

Fat-soluble Vitamin Mix Supplies per Kilogram of Diet:

	g/kg diet
Retinol ester (IU)	16,000
Cholecalciferol (IU)	1,000
dl-α-Tocopherol acetate (mg)	100
Mixed tocopherols (mg)	375
Menadione (mg)	100
Corn oil to 20 g	

Mineral Mix Supplies per Kilogram of Diet:

$CaCO_3$ (g)	17.00
K_2HPO_4 (g)	10.00
Na_2HPO_4 (g)	10.00
NaCl (g)[d]	5.15
$MgSO_4$ (g)	4.00
Fe(II)citrate (mg)	600
$MnCO_3$ (mg)	200
ZnO (mg)	25
CuO (mg)	25
$CoCl_2 \cdot 6H_2O$ (μg)	500
MoO_3 (μg)	50
NaF (μg)	100
KBr (μg)	100
Na_2SeO_3 (μg)	100

[d] Iodized.

Table 2 (continued)
PURIFIED DIETS

F. Vitamin-A-deficient Diet[33]

	g/kg diet
Casein[a]	180
Glucose[b]	340
Cornstarch	340
Arachis oil	50
Cellulose	28
Vitamin mix	22
Mineral mix	40

Vitamin Mix Supplies per Kilogram of Diet:

Vitamin D (IU)	2,200
α-Tocopherol (mg)	110
Menadione (mg)	49.5
Thiamin HCl (mg)	22
Riboflavin (mg)	22
Niacin (mg)	100
Calcium pantothenate (mg)	66
Pyridoxine HCl (mg)	22
Cyanocobalamin (μg)	29.7
Folic acid (mg)	2
Choline chloride (mg)	1,650
Inositol (mg)	110
p-Aminobenzoic acid (mg)	110
Biotin (μg)	440
Ascorbic acid (mg)	1,000

[a] Vitamin-free casein was refluxed with two and one half volumes (w/v) of 1% HCl in ethanol for 8 hr. Extracted casein was then washed in ethanol, and dried first in a Buchner funnel, then in an oven at 50°C for 8 hr.
[b] Anhydrous.

Mineral Mix[22] Supplies per Kilogram of Diet:[c]

	g/kg diet
$CaCO_3$ (mg)	840
$CaHPO_4$ (g)	29.40
K_2HPO_4 (g)	3.24
Na_2HPO_4 (mg)	856
K_2SO_4 (g)	2.72
NaCl (g)	1.22
MgO (g)	1.00
Fe(III)citrate·$5H_2O$ (mg)[d]	223
$MnCO_3$ (mg)[e]	168
$ZnCO_3$ (mg)[f]	32
$CuCO_3$ (mg)[g]	13
KI (mg)	288
Citric acid to 40 g	

[c] Prepare by adding KI, Fe citrate, and some of the citric acid to 7 ml water; heat to dissolve. Mix $ZnCO_3$, $MnCO_3$, and $CuCO_3$, and add slowly to hot solution. Allow to stand overnight and dry at 100°C.
[d] 16.7% Fe.
[e] 44.4% Mn.
[f] 56% Zn.
[g] 55.5% Cu.

Table 2 (continued)
PURIFIED DIETS

G. Vitamin-D-deficient Diet[34]

A basic diet is given, the calcium and phosphorus ratios of which can be adjusted to suit the requirement.[a]

	Amount per kilogram of diet
Basal diet[b]	
Egg albumen (g)	205
Sucrose (g)	683
Groundnut oil (g)	51
Vitamin mix (ml)	20
Mineral mix (g)	41
Vitamin Mix Supplies[c] per Kilogram of Diet:	
Retinyl acetate (IU)[d]	15,000
α-Tocopheryl acetate (mg)	250
Menadione (mg)	20
Thiamin HCl (mg)	15
Riboflavin (mg)	15
Nicotinamide (mg)	25
Calcium pantothenate (mg)[e]	20
Pyridoxine (mg)	5
Choline chloride (mg)	1,500
Cyanocobalamin (μg)[e]	30
p-Aminobenzoic acid (mg)	500
Biotin (mg)	40

[a] $CaCO_3$, $CaHPO_4$, and KH_2PO_4 are added in different proportions to supply the correct levels of calcium and phosphorus, and the sucrose level is adjusted accordingly.[34,36,37] $CaCO_3$ supplies 400 mg Ca/g, $CaHPO_4$ supplies 294 mg Ca/g + 228 mg P/g, and KH_2PO_4 supplies 228 mg P/g.

[b] The basal diet contains 0.26 g Ca/kg and 0.6 g P/kg.

[c] The fat-soluble vitamins are included in the oil, and the B vitamins are dissolved in 20 ml water before careful addition to the diet.

[d] This level is included by the compiler, since the 66,000 IU specified by the original reference is much higher than necessary.

[e] These vitamins were omitted in the original paper, so suitable levels have been included by the compiler.

G. Vitamin-D-deficient Diet[34] (continued)

	Amount per kilogram of diet
Basic Mineral Mix[35] Supplies per Kilogram of Diet:	
NaCl (g)	20.54
KCl (g)	16.18
$MgSO_4$ (g)	1.87
Fe(III)citrate (g)	1.04
$MnSO_4 \cdot 4H_2O$ (mg)	174.3
$ZnCl_2$ (mg)	74.7
$CuSO_4 \cdot 5H_2O$ (mg)	47.3
$CoCl_2$ (mg)	5.0
KI (mg)	32.4

H. Vitamin-E-deficient Diet (G15F)[38]

	g/kg diet
Casein[a]	150
Sucrose	540
Glucose	204.6
Lard	50
Vitamin mix	4
Mineral mix	51.4

Vitamin Mix[39] Supplies per Kilogram of Diet:

Vitamin A (IU)	11,000
Cholecalciferol (IU)	1,500
Menadione (μg)	280
Thiamin (mg)	9
Riboflavin (mg)	19
Nicotinic acid (mg)	90
Calcium pantothenate (mg)	90
Pyridoxine (mg)	9
Cyanocobalamin (μg)	300
Folic acid (mg)	2
Choline dihydrogen tartrate (mg)	900
Inositol (mg)	90
p-Aminobenzoic acid (mg)	90

[a] Low vitamin.

Table 2 (continued)
PURIFIED DIETS

H. Vitamin-E-deficient Diet (G15F)[38] (continued)

Mineral Mix Supplies per Kilogram of Diet:	g/kg diet
$CaCO_3$ (g)	18.20
$NaH_2PO_4 \cdot 2H_2O$ (g)	24.00
Na_2CO_3 (g)	1.20
KCl (g)	3.50
$MgSO_4 \cdot 7H_2O$ (g)	4.00
Fe(III)citrate (mg)	150
$MnSO_4 \cdot 4H_2O$ (mg)	200
$ZnSO_4 \cdot 7H_2O$ (mg)	60
$CuSO_4 \cdot 5H_2O$ (mg)	20
$CoSO_4 \cdot 7H_2O$ (mg)	10
KI (μg)	300
$(NH_4)_6Mo_7O_{24} \cdot 4H_2O$ (mg)	2
NaF (μg)	250
$AlK(SO_4)_2 \cdot 12H_2O$ (μg)	700
Na_2SeO_3 (μg)	70

I. Vitamin-K-deficient[a] Diet[40]

	g/kg diet
Soya protein	200
Sucrose	669
Glucose[b]	50
Triolein	30
Methyl linoleate urea inclusion compound[c]	6
DL-Methionine	5
Mineral mix	40

[a] The original reference claimed that this diet will produce a deficiency when coprophagy is allowed, but faster results will be obtained if it is prevented.
[b] To which vitamin mix was added.
[c] 25% methyl linoleate.

I. Vitamin-K-deficient[a] Diet[40] (continued)

Vitamin Mix[41] Supplies per Kilogram of Diet:	g/kg diet
Vitamin A (IU)	30,000
Cholecalciferol (IU)	1,950
α-Tocopherol (mg)	120
Thiamin HCl (mg)	10
Riboflavin (mg)	10
Nicotinic acid (mg)	20
Calcium pantothenate (mg)	50
Pyridoxine HCl (mg)	5
Cyanocobalamin (μg)	100
Folic acid (mg)	1
Choline chloride (mg)	1,000
Biotin (μg)	100

Mineral Mix[42] Supplies per Kilogram of Diet:	
$CaCO_3$ (g)	6.54
$CaHPO_4 \cdot 2H_2O$ (g)	14.22
K_2HPO_4 (g)	3.09
NaCl (g)	4.32
K citrate·H_2O (g)	9.47
$MgCO_3$ (g)	1.63
Fe(III)citrate·$3H_2O$ (mg)	640
$MnSO_4$ (mg)	49.8
$ZnCO_3$ (mg)	1.8
$CuSO_4 \cdot 5H_2O$ (mg)	7.1
$CoCl_2 \cdot 6H_2O$ (mg)	3.6
KI (mg)	1.8
NaF (μg)	35.5
$KAl(SO_4)_2 \cdot 12H_2O$ (mg)	3.6

Table 2 (continued)
PURIFIED DIETS

J. Thiamin- or Riboflavin-deficient Diet[43]

	g/kg
Casein[a]	200
Sucrose	650
Arachis oil	100
Vitamin mix	10
Mineral mix	40

Vitamin Mix Supplies per Kilogram of Diet:

Vitamin A (IU)	11,000
Ergocalciferol (IU)	2,000
Tocopherol (mg)	100
Menadiol (μg)	500
Niacin (mg)	25
Calcium pantothenate (mg)	20
Pyridoxine (mg)	2.5
Cyanocobalamin (μg)	100
Folic acid (μg)	200
Choline chloride (μg)	1,000
Inositol (mg)	100
Biotin (μg)	100
Ascorbic acid (mg)	50
Sucrose to 10 g	
In thiamin-deficient diet also add 5 mg riboflavin; in riboflavin-deficient diet also add 5 mg thiamin HCl	

Mineral Mix[44] Supplies per Kilogram of Diet:

$CaCO_3$ (g)	21.7
KH_2PO_4 (g)	8.5
NaCl (g)	2.8
KCl (g)	4.5
$MgCO_3$ (g)	1.0
$MgSO_4$ (mg)	640
$FePO_4 \cdot 4H_2O$ (mg)	820
$MnSO_4$ (mg)	14
$CuSO_4$ (mg)	36
KI (mg)	3.2
NaF (mg)	4.0
$KAl(SO_4)_2$ (mg)	6.8

[a] Vitamin-free casein is produced by stirring 2.5 g of casein in 30 l tap water containing 600 g NaCl and 30 ml glacial acetic acid for 30 min. The supernatant is poured off after 1 hr and washing repeated six times. The casein is then pressed as dry as possible in Buchner funnel, suspended in 96% ethanol, filtered, and dried in open trays.

K. Nicotinic-acid-deficient[a] Diet[45]

	g/kg diet
Casein[b]	50
Gelatin	70
Sucrose	785
Maize oil	50
L-Cystine	2
Vitamin mix	3
Mineral mix	40

Vitamin Mix Supplies per Kilogram of Diet:

Vitamin A (IU)[c]	10,000
Vitamin D (IU)[c]	1,000
α-Tocopherol (mg)	100
Menadione (mg)	45
Thiamin (mg)	20
Riboflavin (mg)	20
Calcium pantothenate (mg)	60
Pyridoxine (mg)	20
Cyanocobalamin (μg)[d]	
Folic acid (mg)	1.8
Choline chloride (mg)	1,500
Inositol (mg)	100
p-Aminobenzoic acid (mg)	100
Biotin (μg)	390
Ascorbic acid (mg)	900

Mineral Mix[46] Supplies per Kilogram of Diet:

$CaCO_3$ (g)	10.75
$CaHPO_4 \cdot 2H_2O$ (g)	3.40
$K_2HPO_4 \cdot 3H_2O$ (g)	15.14
NaCl (g)	6.00
$MgSO_4 \cdot 7H_2O$ (g)	3.66
Fe(III)citrate$\cdot 6H_2O$ (mg)	984
$MnSO_4 \cdot 4H_2O$ (mg)	13
$ZnCl_2$ (mg)	9
$CuSO_4 \cdot 5H_2O$ (mg)	11
KI (mg)	29

[a] Because tryptophan spares nicotinic acid, production of a deficiency of the vitamin is facilitated by the use of a low-tryptophan diet.
[b] Vitamin free.
[c] These values were inserted by the compiler as actual levels of vitamins A and D, and were not included in the original paper.
[d] Supplied as a 0.1% triturate in mannitol.

Table 2 (continued)
PURIFIED DIETS

L. Pantothenic-acid-deficient Diet[47]

	g/kg diet
Casein[a]	180
Glucose	710
Cottonseed oil	50
Cellulose flour	20
L-Cystine	2
L-Cysteine HCl[b]	1
Vitamin mix[c]	
Mineral mix	37

Vitamin Mix Supplies per Kilogram of Diet:

Vitamin A (as carotene) (IU)	15,000
Cholecalciferol (IU)	1,000
α-Tocopherol acetate (mg)	50
Thiamin HCl (mg)	10
Riboflavin (mg)	20
Niacin (mg)	40
Pyridoxine HCl (mg)	10
Cyanocobalamin (μg)	12
Pteroylglutamic acid (mg)	2
Choline chloride (mg)	1,000
Inositol (mg)	200
p-Aminobenzoic acid (mg)	100
Biotin (μg)	200

Mineral Mix[48] Supplies per Kilogram of Diet:

$CaCO_3$ (g)[d]	7.80
$Ca_3(PO_4)_2$ (g)[d]	5.50
KH_2PO_4 (g)[d]	11.50
KCl (g)	4.30
NaCl (g)	3.80
$MgSO_4$ (g)	3.30
$FeSO_4$ (mg)	570
$MnSO_4$ (mg)	7.2
ZnO (mg)	12.0
CuO (mg)	4.0
KI (mg)	1.8
NaF (mg)	21.0
$KAl(SO_4)_2 \cdot 12H_2O$ (mg)	3.3

[a] Vitamin-free casein.
[b] As an antioxidant.
[c] B vitamin mix incorporated in cellulose flour; fat-soluble vitamins incorporated in cottonseed oil.
[d] The compiler has replaced the bone ash specified in the original reference by these minerals in the proportions originally recommended in the Wesson[55] salt mix, upon which this is based.

M. Pyridoxine-deficient[a] Diet[49]

	g/kg diet
Casein[b]	220
Sucrose	334
Maize starch	150
Glucose	120
Cellulose	20
Maize oil	20
Hydrogenated cottonseed oil	80
Cystine	2
Mineral mix	41
Vitamin mix	13

Vitamin Mix Supplies per Kilogram of Diet:

Retinyl acetate (IU)	20,000
Calciferol (IU)	2,000
α-Tocopherol (mg)	110
Menadione (mg)	49
Thiamin HCl (mg)	22
Riboflavin (mg)	22
Niacin (mg)	99
Calcium pantothenate (mg)	66
Cyanocobalamin (μg)	29
Folic acid (mg)	2
Choline (g)	10
Choline chloride (mg)	1,650
Inositol (mg)	110
p-Aminobenzoic acid (mg)	110
Biotin (μg)	440
L-Ascorbic acid (mg)	990

Mineral Mix Supplies per Kilogram of Diet:

$CaCO_3$ (g)	2.70
$Ca(H_2PO_4)_2 \cdot H_2O$ (g)	4.50
Ca citrate·$4H_2O$ (g)	12.30
KH_2PO_4 (g)	8.80
KCl (g)	5.00
NaCl (g)	3.80
$MgCO_3$ (g)	1.40
$MgSO_4$ (g)	1.50
NH_4Fe(III)citrate (mg)	61
$MnSO_4$ (μg)	800
KI (mg)	2
NaF (mg)	20
$NH_4Al(SO_4)_2$ (mg)	4

[a] Certain alterations have been made to this diet to correct typographical errors made in the original publication[49] after personal communication with Dr. Lupien.
[b] Vitamin free.

Table 2 (continued)
PURIFIED DIETS

N. Vitamin-B_{12}-deficient Diet[50]

	g/kg diet
Soya flour	600
Sucrose	205
Sugar-salt mix	179
Vitamins	11
Choline dihydrogen citrate	5

Vitamin Mix[51] Supplies per Kilogram of Diet:[a]

Retinyl palmitate (IU)[b]	8,000
Ergocalciferol (IU)[b]	1,000
α-Tocopherol (mg)[b]	300
Menadione (mg)	1
Thiamin HCl (mg)	10
Riboflavin (mg)	12
Nicotinic acid (mg)	10
Calcium pantothenate (mg)	12
Pyridoxine (mg)	5
Folic acid (mg)	1
Choline (mg)	2,000
Inositol (mg)	200
p-Aminobenzoic acid (mg)	10
Biotin (μg)	200

Mineral Mix[51] Supplies per Kilogram of Diet:

Ca lactate (g)	22.00
$CaHPO_4$ (g)	11.01
NaCl (g)	8.00
$(NH_4)_2SO_4 \cdot FeSO_4 \cdot 6H_2O$ (g)	1.62
$MnSO_4 \cdot 4H_2O$ (mg)	324
$CuSO_4 \cdot 5H_2O$ (mg)	82
$NaIO_3 \cdot H_2O$ (μg)	660
D-Glucose (g)	22
Lactose (g)	113.9

[a] Fat-soluble vitamins made up in 1 ml soya oil; water-soluble vitamins and menadione made up to 10 g with glucose.
[b] Values taken from table of calculated contribution of supplements.[51]

O. Folic-acid-deficient Diet[52]

	g/kg diet
Gelatin	80
Casein[a]	200
Glucose	447
Maize oil	20
Vegetable hydrogenated oil	188
Vitamin mix[b]	
Mineral mix	60
Succinyl sulfathiazole	5

Vitamin Mix Supplies per Kilogram of Diet:

Retinyl acetate (IU)	18,000
Cholecalciferol (IU)	200
Tocopherol (mg)	110
Menadione (mg)	45
Thiamin HCl (mg)	20
Riboflavin (mg)	20
Niacin (mg)	90
Calcium pantothenate (mg)	60
Pyridoxine HCl (mg)	20
Cyanocobalamin (μg)	27
Choline chloride (mg)	1,500
Inositol (mg)	100
p-Aminobenzoic acid (mg)	100
Biotin (μg)	400
Ascorbic acid (mg)	900

Mineral Mix[53] Supplies per Kilogram of Diet:

$CaCO_3$ (g)	10.00
$CaHPO_4$ (g)	28.40
Na_2HPO_4 (g)	7.00
NaCl (g)	4.00
KCl (g)	7.00
$MgSO_4$ (g)	3.00
Fe(III)citrate (mg)	200
$MnSO_4 \cdot H_2O$ (mg)	250
$ZnCO_3$ (mg)	130
$CuSO_4$ (mg)	10
KIO_3 (mg)	10

[a] Vitamin free.
[b] Incorporated in glucose.

Table 2 (continued)
PURIFIED DIETS

P. Choline-deficient[a] Diet[54]

	g/kg diet
Soya protein[b]	160
Sucrose	397
Dextrin	198
Partially hydrogenated shortening	100
Maize oil	40
Cellulose	40
L-Cystine	5
Vitamin mix	40
Mineral mix	20

Vitamin Mix Supplies per Kilogram of Diet:

Retinyl acetate (IU)	36,000
Cholecalciferol (IU)	4,000
α-Tocopherol (mg)	200
Menadione (mg)	90
Thiamin (mg)[c]	40
Riboflavin (mg)	40
Calcium pantothenate (mg)	120
Niacin (mg)	180
Pyridoxine HCl (mg)	40
Folic acid (mg)	3.6
Inositol (mg)	200
p-Aminobenzoic acid (mg)	200
Biotin (μg)	800
Ascorbic acid (mg)	1,800
Glucose to 40 g	

Mineral Mix[55] Supplies per Kilogram of Diet:

$CaCO_3$ (g)	4.20
$Ca_3(PO_4)_2$ (g)	3.00
KH_2PO_4 (g)	6.20
KCl (g)	2.40
NaCl (g)	2.10
$MgSO_4$ (g)	1.80
$FePO_4$ (mg)	294
$MnSO_4$ (mg)	4
$CuSO_4 \cdot 5H_2O$ (mg)	8
KI (mg)	1
NaF (mg)	11
$KAl(SO_4)_2 \cdot 12H_2O$ (mg)	2

[a] Vitamin B_{12} is also omitted from the diet because it plays a role in choline synthesis and because the incidence of renal necrosis is reduced.[56]
[b] Assay protein C-1.
[c] This vitamin was omitted from the original reference, so a suitable level has been included by the compiler.

Table 2 (continued)
PURIFIED DIETS

Q. Biotin-deficient Diet[57]

Symptoms of biotin deficiency can only be reliably produced by the inclusion of egg white in the diet. This contains avidin, which binds any small quantities of biotin in the diet or those quantities synthesized by the intestinal microflora.

	g/kg diet
Spray-dried egg white	200
Glucose	706.4
Maize oil	40
Vitamin mix[a]	13.6
Mineral mix	40

Vitamin Mix Supplies per Kilogram of Diet:

Retinyl palmitate (IU)	20,000
Ergocalciferol (IU)	2,000
Glucose to 2.5 g	
dl-α-Tocopherol acetate (mg)	100
Menadione (mg)	5
Thiamin HCl (mg)	100
Riboflavin (mg)	15
Niacin (mg)	100
Calcium pantothenate (mg)	40
Pyridoxine HCl (mg)	6
Cyanocobalamin (μg)[b]	20
Folic acid (mg)	4
Choline chloride (mg)[c]	1,000
Inositol (mg)	100
p-Aminobenzoic acid (mg)	2
Ascorbic acid (mg)	250
Glucose to 10 g	

[a] Water-dispersible forms of retinyl palmitate and ergocalciferol mixed together with glucose and added separately. α-Tocopherol is mixed in the maize oil.
[b] Added as 0.1% triturate with mannitol.
[c] Added separately.

Mineral Mix[58] Supplies per Kilogram of Diet:

	g/kg diet
$CaCO_3$ (g)	6.54
$CaHPO_4 \cdot 2H_2O$ (g)	14.22
K_2HPO_4 (g)	3.09
K citrate·H_2O (g)	9.46
NaCl (g)	4.32
$MgCO_3$ (g)	1.64
Fe(III)citrate·$3H_2O$ (mg)	640
$MnSO_4 \cdot H_2O$ (mg)	55.2
$ZnCO_3$ (mg)	17.6
$CuSO_4 \cdot 5H_2O$ (mg)	7.2
KI (mg)	1.6

Table 2 (continued)
PURIFIED DIETS

R. Low-potassium Synthetic Diet[59]

	g/kg diet
Casein	239.3
Glucose	622.8
Maize oil	79.8
Vitamin mix	2.7
Mineral mix	55.4

Vitamin Mix Supplies per Kilogram of Diet:

Vitamin A (IU)[a]	10,000
Vitamin D (IU)[a]	1,000
α-Tocopherol (mg)	40
Thiamin HCl (mg)	4
Riboflavin (mg)	8
Niacin (mg)	100
Calcium pantothenate (mg)	20
Pyridoxine HCl (mg)	4
Cyanocobalamin (μg)	8
Folic acid (mg)	3.8
Inositol (mg)	1,000
p-Aminobenzoic acid (mg)	300

Mineral Mix Supplies per Kilogram of Diet:

$CaCO_3$ (g)	7.58
$CaHPO_4$ (g)	17.55
NaH_2PO_4 (g)	2.79
NaCl (g)	16.16
$MgSO_4 \cdot 7H_2O$ (g)	10.20
Fe(II)citrate·$5H_2O$ (mg)	595
$MnSO_4 \cdot H_2O$ (mg)	226
$ZnCl_2$ (mg)	99.6
$CuSO_4 \cdot 5H_2O$ (mg)	249
NaI (mg)	1.5

[a] Actual levels of vitamins A and D used in this diet were not given, so a suitable amount has been included by the compiler.

S. Sodium-deficient Diet[60]

	g/kg diet
Casein	200
Sucrose	650
Maize oil	20
Salt-free butter	50
Cellulose	20
Vitamin mix	20
Mineral mix	40

Vitamin Mix Supplies per Kilogram of Diet:

Vitamin A (IU)	18,000
Vitamin D (IU)	2,000
α-Tocopherol (mg)	100
Menadione (mg)	45
Thiamin (mg)[a]	20
Riboflavin (mg)	20
Niacin (mg)	90
Calcium pantothenate (mg)	60
Pyridoxine HCl (mg)	20
Cyanocobalamin (μg)	27
Folic acid (μg)	1,800
Choline chloride (mg)	1,500
Inositol (mg)	100
p-Aminobenzoic acid (mg)	100
Biotin (μg)	400
Ascorbic acid (mg)	900
Glucose to 20 g	

Mineral Mix Supplies per Kilogram of Diet:

$CaCO_3$ (g)	9.04
$Ca_3(PO_4)_2$ (g)	6.64
K_2HPO_4 (g)	9.00
KCl (g)	10.60
$MgSO_4$ (g)	4.00
$FePO_4 \cdot 2H_2O$ (mg)	648
$MnSO_4$ (mg)	8.4
$CuSO_4 \cdot 5H_2O$ (mg)	30
$CoCl_2 \cdot 6H_2O$ (mg)	4
KI (mg)	2.4
KF (mg)	23.2

[a] The diet as published did not contain thiamin, so this level has been added by the compiler.

Table 2 (continued)
PURIFIED DIETS

T. Magnesium-deficient[a] Diet[61]

	g/kg diet
Casein[b]	230
Sucrose	595
Lard	100
Vitamin mixture	25
Salt mixture	50

Vitamin Mixture Supplies per Kilogram of Diet:

Retinyl acetate (IU)	6,600
Cholecalciferol (IU)	10,000
α-Tocopherol (mg)	300
Menadione (mg)	5
Thiamin (mg)	20
Riboflavin (mg)	20
Nicotinic acid (mg)	25
Calcium pantothenate (mg)	50
Pyridoxine (mg)	20
Cyanocobalamin (μg)	20
Pteroylmonoglutamic acid (mg)	2
Choline (g)	2
Inositol (mg)	850
Biotin (mg)	1
Agar to 25 g	

Mineral Mix Supplies per Kilogram of Diet:

$CaHPO_4$ (g)	14.50
K_2HPO_4 (g)	14.50
NaH_2PO_4 (g)	7.00
NaCl (g)	4.75
Fe(III) lactate (g)	1.25
$MnSO_4 \cdot H_2O$ (mg)	3.75
KI (mg)	42.5
NaF (mg)	1.5
$AlK(SO_4)_2 \cdot 12H_2O$ (mg)	5

[a] When analyzed by Andrieux-Domont and Van Hung,[61] this diet contained 30 mg magnesium/kg.
[b] Vitamin free.

U. Iron-deficient[a] Diet[62]

	g/kg diet
Casein[b]	220
Sucrose	297.6
Maize starch	297.6
Maize oil	100
Alpha cellulose[c]	20
Vitamin mix	10
Mineral mix	54.7

Vitamin Mix Supplies per Kilogram of Diet:

Retinyl palmitate (IU)	19,824
Ergocalciferol (IU)	2,203
Tocopheryl acetate (mg)	121
Menadione (mg)	50
Thiamin HCl (mg)	22
Riboflavin (mg)	22
Niacin (mg)	99
Calcium pantothenate (mg)	66
Pyridoxine HCl (mg)	22
Cyanocobalamin (μg)[d]	30
Folic acid (mg)	2
Choline dihydrogen citrate (mg)	3,497
Inositol (mg)	110
p-Aminobenzoic acid (mg)	110
Biotin (μg)	440
Ascorbic acid (mg)	1,017
Corn starch to 10 g	

Mineral Mix Supplies per Kilogram of Diet:

$CaCO_3$ (g)	16.80
K_2HPO_4 (g)	24.84
NaCl (g)	10.00
$MgSO_4 \cdot 7H_2O$ (g)	2.78
$MnSO_4 \cdot H_2O$ (mg)	191
$ZnCl_2$ (mg)	30
$CuSO_4 \cdot 5H_2O$ (mg)	98
$CoCl_2 \cdot 6H_2O$ (mg)	1
KI (μg)	220

[a] On analysis, this diet contained 5 mg iron/kg.
[b] Vitamin free.
[c] Iron extracted[63] by soaking cellulose in 0.2 *M* HCl for 24 hr, filtering, and repeating. Cellulose was then washed several times with distilled water and dried at 100°C.
[d] Added as a 0.1% trituration in mannitol.

Table 2 (continued)
PURIFIED DIETS

V. Manganese-deficient Diet[1]

	g/kg diet
Casein	300
Dextrose	545
Maize oil	80
Vitamin mix	15
Mineral mix	60

Vitamin Mix Supplies per Kilogram of Diet:

Retinyl palmitate (IU)	11,360
Cholecalciferol (IU)	1,121
dl-α-Tocopherol acetate (mg)[a]	40
Menadione (mg)	19
Thiamin HCl (mg)	23
Riboflavin (mg)	8
Nicotinic acid (mg)	23
Calcium pantothenate (mg)	38
Pyridoxine (mg)	23
Cyanocobalamin (μg)[b]	23
Folic acid (μg)	450
Choline chloride (mg)	750
Inositol (mg)	375
p-Aminobenzoic acid (mg)	7.5
Biotin (mg)	1.9
Ascorbic acid (mg)	75
Dextrose to 15 g	

[a] The concentration of this vitamin in the original paper was inaccurate, so a value considered suitable has been included.
[b] Triturated in mannitol.

Mineral Mix Supplies per Kilogram of Diet:

	g/kg diet
$CaCO_3$ (g)	18.00
$CaHPO_4$ (g)	3.60
K_2HPO_4 (g)	19.26
NaCl (g)	10.08
$MgSO_4 \cdot 7H_2O$ (g)	7.50
$FeSO_4 \cdot 7H_2O$ (g)	1.50
$ZnCO_3$ (mg)	15
$CuSO_4 \cdot 5H_2O$ (mg)	18
KI (mg)	48

Table 2 (continued)
PURIFIED DIETS

W. Zinc-deficient Diet [64]

	g/kg diet
Dried egg albumen	200
Sucrose	659.5
Arachis oil	100
Mineral mix[a]	38.5
Vitamin mix[b]	2

Vitamin Mix Supplies per Kilogram of Diet:

Retinyl acetate (IU)	13,900
Ergocalciferol (IU)	8,000
α-Tocopherol acetate (mg)	200
Thiamin (mg)	5
Riboflavin (mg)[c]	5
Nicotinic acid (mg)[c]	40
Calcium pantothenate (mg)	80
Pyridoxol (mg)	5
Cyanocobalamin (μg)	25
Folic acid (mg)	5
Choline HCl (mg)	1,000
Inositol (mg)	400
Biotin (mg)	10
Sucrose (mg)	455

[a] Prepared by method of Mills & Murray.[65] Solutions of 200 g NH_4CO_3 in 500 ml water and 440 g hydrated $CaCl_2$ in 275 ml water which have 0.5 g sodium diethyldithiocarbamate added are separately washed with CCl_4 until washing is colorless. Ammonia solution (0.880 SG) is added to $CaCl_2$ solution until point of $Ca(OH)_2$ precipitation is reached. Washed solutions are then mixed with gentle swirling in a closed 10 l polyethylene bottle. Precipitation of $CaCO_3$ takes place during the next 2 hr, during which period discretion must be exercised to maintain high CO_2 pressures in the bottle without bursting. Precipitated $CaCO_3$ is washed twice with alcohol and twice with water and then dried at 100°C.

[b] Water-soluble vitamins are mixed with part of the sucrose, and fat-soluble vitamins are added using the oil as a carrier.

[c] These vitamins have been included by the compiler but were not in the diet described by the original authors.

Mineral Mix[d] Supplies per Kilogram of Diet:

	g/kg diet
$CaCO_3$ (g)[a]	13.50
KH_2PO_4 (g)[e]	8.20
Na_2HPO_4 (g)[e]	8.60
KCl (g)[e]	6.00
$MgCl_2$ (g)[e]	1.59
$FeSO_4 \cdot 7H_2O$ (mg)	260
$MnSO_4 \cdot 4H_2O$ (mg)	210
$CuSO_4 \cdot 5H_2O$ (mg)	98
KI (mg)[f]	21

[d] Purified forms of minerals should be used.

[e] Purified by dissolving in distilled water and adding 10 ml of saturated 8-hydroxy-quinoline in ethanol. Extract six times with 70 to 80 ml chloroform. Add solution of 0.5 g sodium diethyldithiocarbamate in 5 ml of water and again wash with chloroform six times. Continue extraction with chloroform until no yellow reaction is observed on adding Cu^{2+} to 1 to 2 ml of aqueous phase. Salt solution is then concentrated by evaporation and dried under infrared lamps until nearly granular, at which point the drying is completed in an air oven at 85°C. All apparatus used is washed in dilute nitric acid and distilled water before use.

[f] Powdered separately with sucrose to prevent reaction with metal salts.

Table 2 (continued)
PURIFIED DIETS

X. Copper-deficient Diet[66]

	g/kg diet
Milk, dry, fat free	550
Starch	170
Sucrose	165
Maize oil	50
Cellulose	20
Vitamin mix	20
Mineral mix	25

Vitamin Mix Supplies per Kilogram of Diet:

Vitamin A (IU)	18,000
Vitamin D (IU)	2,000
α-Tocopherol (mg)	100
Menadione (mg)	45
Thiamin HCl (mg)	20
Riboflavin (mg)	20
Niacin (mg)	90
Calcium pantothenate (mg)	60
Pyridoxine HCl (mg)	20
Cyanocobalamin (μg)	30
Folic acid (mg)	1.8
Choline chloride (mg)	1,500
Inositol (mg)	100
p-Aminobenzoic acid (mg)	100
Biotin (μg)	400
Ascorbic acid (mg)	900

Mineral Mix Supplies per Kilogram of Diet:

	g/kg diet
KCl (g)	5.73
NaCl (g)	7.63
$MgCO_3$ (g)	1.75
Fe_2O_3 (mg)	77
$MnCO_3$ (mg)	83.7
$ZnCO_3$ (mg)	95.9
$CoCO_3$ (μg)	807
KI (μg)	261
NaF (μg)	22
$Na_2MoO_4 \cdot 2H_2O$ (mg)	1.3
Na_2SeO_3 (μg)	190

Table 3
MISCELLANEOUS MINERAL MIXES

A. Mineral Mix of Greenfield et al.[23]

	Supplies per kilogram of diet
$CaCO_3$ (g)	8.20
$CaHPO_4$ (g)	13.00
Na_2HPO_4 (g)	7.40
KCl (g)	8.20
$MgSO_4 \cdot H_2O$ (g)	2.80
Fe(III)citrate$\cdot 5H_2O$ (mg)	174
$MnSO_4 \cdot H_2O$ (mg)	180
$ZnCO_3$ (mg)	30
$CuSO_4$ (mg)	15
KIO_3 (mg)	1

B. Mineral Mix as Given in U.S. Pharmacopeia[25]

	Supplies per kilogram of diet
$CaCO_3$ (g)	15.26
KH_2PO_4 (g)	15.56
NaCl (g)	5.57
$MgSO_4$ (g)	2.29
$FeSO_4 \cdot 7H_2O$ (g)	1.08
$MnSO_4 \cdot H_2O$ (mg)	160
$ZnSO_4 \cdot 7H_2O$ (mg)	22
$CuSO_4 \cdot 5H_2O$ (mg)	19
$CoCl_2 \cdot 6H_2O$ (μg)	920
KI (mg)	32

C. Mineral Mix of Cohen et al.[26]

	Supplies per kilogram of diet
$CaCO_3$ (g)	7.25
$CaHPO_4$ (g)	11.30
Na_2HPO_4 (g)	6.51
KCl (g)	7.30
$MgSO_4$ (g)	2.30
Fe(III)citrate (mg)[a]	151
$MnSO_4 \cdot H_2O$ (mg)	154
$ZnCO_3$ (mg)	21
$CuSO_4$ (mg)	13
KIO_3 (mg)	1

[a] 16.7% Fe.

Table 4
CHEMICALLY DEFINED DIETS

A. Solid Diet[24]

	g/kg diet mix[a]
L-Arginine HCl	13.5
L-Histidine HCl·H_2O	4.5
L-Isoleucine	8.2
L-Leucine	11.1
L-Lysine HCl	18.0
L-Methionine	8.2
L-Phenylalanine	11.6
L-Threonine	8.2
L-Tryptophan	1.7
L-Valine	8.2
L-Alanine	3.5
L-Aspartic acid	3.5
L-Glutamic acid	35.0
Glycine	23.3
L-Proline	3.5
L-Cystine	3.5
L-Serine	3.5
L-Tyrosine	3.5
L-Asparagine	6.0
Glucose	443.0
Sucrose	221.5
Maize oil	100
Mineral mix	50
Vitamin mix	7

Vitamin Mix Supplies per Kilogram of Diet:

Retinyl acetate (IU)	4000
Ergocalciferol (IU)	400
α-Tocopherol acetate (mg)	100
Menadione (μg)	500
Thiamin HCl (mg)	5
Riboflavin (mg)	5
Nicotinamide (mg)	25
Calcium pantothenate (mg)	20
Pyridoxine HCl (mg)	5
Cyanocobalamin (μg)[b]	30
Folic acid (μg)	500
Choline chloride (g)	2
Biotin (μg)	200
Ascorbic acid (mg)	50

[a] The dry diet after blending is mixed with an equal quantity of boiling 3% agar solution before being poured into trays and allowed to cool. The diet was then refrigerated before use. The gel diet gave a better growth rate than the dry diet.

[b] Added as 0.1% triturate in mannitol.

Mineral Mix Supplies per Kilogram of Diet:

	g/kg diet mix
$CaCO_3$ (g)	14.64
$CaHPO_4 \cdot 2H_2O$ (mg)	215
KH_2PO_4 (g)	17.15
NaCl (g)	12.53
$MgSO_4 \cdot 7H_2O$ (g)	4.99
Fe(III)citrate·$5H_2O$ (mg)	311
$MnSO_4 \cdot H_2O$ (mg)	60
$ZnCl_2$ (mg)	10
$CuSO_4$ (mg)	78
KI (μg)	250
$(NH_4)_6Mo_7O_{24} \cdot 4H_2O$ (mg)	1.2
$Na_2SeO_3 \cdot 5H_2O$ (μg)	750

Table 4 (continued)
CHEMICALLY DEFINED DIETS

B. Liquid Diet[67]

	g/l diet
L-Isoleucine	2.77
L-Leucine	3.60
L-Lysine HCl	3.42
L-Phenylalanine	2.02
L-Threonine	1.77
DL-Methionine	2.05
L-Tryptophan	0.48
L-Valine	2.60
L-Tyrosine ethyl ester HCl	1.92
L-Arginine HCl	3.33
L-Histidine HCl·H_2O	0.85
L-Alanine	0.44
L-Aspartic acid	1.47
L-Glutamine	3.78
L-Proline	0.74
L-Serine	0.93
Maltodextrin	162.90
Dextrose	115.10
Safflower oil	1.20
Sorbic acid	0.20
Minerals	
Vitamins	

Vitamins Supply per Liter of Diet:

Vitamin A palmitate (IU)	3755
Ergocalciferol (IU)	300
DL-α-Tocopherol acetate (mg)	22.5
Menadione sodium bisulfate (mg)	1.6
Thiamin mononitrate (μg)	901
Riboflavin (mg)	1.2
D-Calcium pantothenate (mg)	7.5
Nicotinamide (mg)	12.8
Pyridoxine HCl (mg)	1.9
Cyanocobalamin (μg)	4
Folic acid (μg)	75
Choline bitartrate (mg)	160
Inositol (mg)	87.5
p-Aminobenzoic acid (mg)	266.5
D-Biotin (μg)	150
Ascorbic acid (mg)	52.6

Minerals Supply per Liter of Diet:

	g/l diet
Ca acetate·H_2O (mg)	15
Ca gluconate (g)	9.99
$NaHPO_4$ (g)	4.09
KCl (g)	2.34
Mg gluconate (g)	2.43
$Fe(NH_4)_2(SO_4)_2 \cdot 6H_2O$ (mg)	511
Mn gluconate·$3H_2O$ (mg)	19.5
Cu gluconate (mg)	10.5
$ZnSO_4 \cdot H_2O$ (μg)	729
KI (μg)	135

Table 4 (continued)
CHEMICALLY DEFINED DIETS

C. Diet L-489 E8 for Germ-free Rats[68]

	g/l diet
L-Leucine[a]	4.58
L-Phenylalanine[a]	1.68
L-Isoleucine[a]	2.75
L-Methionine[a]	2.60
L-Tryptophan[a]	0.84
L-Valine[a]	2.19
L-Asparagine[a]	2.37
L-Arginine HCl[a]	1.86
L-Threonine[a]	1.68
L-Lysine HCl[a]	4.04
L-Histidine HCl·H_2O[a]	1.68
Glycine	0.69
L-Proline	3.35
L-Serine	3.05
L-Alanine	1.35
L-Na glutamate	7.78
L-Tyrosine ethyl HCl	1.41
Glucose[b]	221.5
Maize oil[c]	15.0
Vitamins	
Minerals	
Water to one liter	

[a] These amino acids are first dissolved in 468 ml of glass-distilled water at 70°C. When cool, other reagents except glucose are added. (The solutions are sterilized by filtration through a Millipore® GS filter (pore size 0.22 μm) and stored separately. For feeding they are mixed 1:1.)

[b] Made up to 500 ml in glass-distilled water. (The solutions are sterilized by filtration through a Millipore® GS filter (pore size 0.22 μm) and stored separately. For feeding they are mixed 1:1.)

[c] Added separately as a proportion of daily intake of diet.

	g/l diet
Vitamins Contained per Liter of Diet:	
Retinyl palmitate (IU)[e]	9724
Cholecalciferol (IU)[e]	52
DL-α-Tocopherol (mg)[e]	146
DL-α-Tocopherol acetate (mg)[e]	293
Phylloquinone (mg)	3.2
Thiamin HCl (mg)	3.0
Riboflavin (mg)	4.5
Calcium pantothenate (mg)	29.9
Niacin (mg)	22.4
Pyridoxine (mg)	3.7
Cyanocobalamin (μg)	359
Folic acid (μg)	898
Choline chloride (mg)	689
i-Inositol (mg)	150
p-Aminobenzoic acid (mg)	180
Biotin (μg)	599
Minerals Contained per Liter of Diet:	
Ca glycerophosphate (g)	11.98
$CaCl_2 \cdot H_2O$ (mg)	419.2
NaCl (mg)[f]	196
K acetate (g)	4.25
$MgCl_2 \cdot 6H_2O$ (g)	3.29
Mg glycerophosphate (g)	3.29
Fe(II)gluconate (mg)	119.8
Mn acetate·$4H_2O$ (mg)	134.7
Cu acetate·H_2O (mg)	9.0
$ZnSO_4 \cdot H_2O$ (mg)	32.9
Co acetate·$4H_2O$ (μg)	270
KI (mg)[f]	1.6
NaF (mg)	5.0
Cr(III)acetate·H_2O (mg)	6.0
$(NH_4)_6Mo_7O_{24} \cdot 4H_2O$ (μg)	900
Na_2SeO_3 (μg)	54
$SnSO_4 \cdot 2H_2O$ (μg)	898
$NiCl_2 \cdot 3H_2O$ (μg)	898
Na_3VO_4 (μg)	269

Note: All practical details discussed in the footnotes to Table 4.C are described by Pleasants et al.[69]

[e] Added to maize oil.

[f] KI previously mixed with NaCl.

REFERENCES

1. **Hurley, H. S. and Bell, L. T.,** *J. Nutr.*, 104, 133–137, 1974.
2. **Coleman, D. L. and Schlesinger, K.,** *Proc. Soc. Exp. Biol. Med.*, 119, 264–266, 1965.
3. **Coates, M. E.,** in *Nutrition and Disease in Experimental Animals,* Tavernor, W. D., Ed., Baillère Tindall & Cassell, London, 1970, 38–48.
4. **Ley, F. J., Bleby, J., Coates, M. E., and Paterson, J. S.,** *Lab Anim.*, 3, 221–254, 1969.
5. **Charles, R. T., Stevenson, D. E., and Walker, A. I. T.,** *Lab. Anim. Care,* 15, 321–324, 1965.
6. **Reddy, B. S., Wostmann, B. S., and Pleasants, J. R.,** in *The Germ-Free Animal in Research,* Coates, M. E., Ed., Academic Press, New York, 1968, 92.
7. **Coates, M. E., Ford, J. E., Gregory, M. E., and Thompson, S. Y.,** *Lab Anim.*, 3, 39–49, 1969.
8. **Zimmerman, D. R. and Wostmann, B. S.,** *J. Nutr.*, 70, 318–322, 1963.
9. **Ford, D. J.,** *Br. J. Nutr.*, 35, 267–276, 1976.
10. **Eggum, B. O.,** *Tierphysiol. Tierernähr. Futtermittelk.*, 25, 204–210, 1969.
11. **Schoen, A. and Hiller, H. H.,** *Tierphysiol. Tierernähr. Futtermittelk.*, 27, 338–343, 1971.
12. **Udes, H., Hiller, H. H., and Juhr, N. C.,** *Z. Versuchstierkd.*, 13, 160–166, 1971.
13. Medical Research Council, Laboratory Animals Centre, *The Accreditation and Recognition Schemes for Supplies of Laboratory Animals,* 2nd ed., MRC, Carshalton, Surrey, Engl., 1974.
14. **Ford, D. J.,** unpublished observation, 1976.
15. National Research Council, *Nutrient Requirements of Laboratory Animals,* 2nd ed., National Academy of Sciences, Washington, D.C., 1972, 64.
16. **Nolen, G. A.,** *J. Nutr.*, 102, 1477–1493, 1972.
17. **Knapka, J. J., Smith, K. P., and Judge, F. J.,** *Lab. Anim. Sci.*, 24, 480–487, 1974.
18. **Drepper, K. and Weik, H.,** *Versuchstiernahrung,* Verlag Paul Parey, Berlin, 1972, 12–18.
19. **Kellogg, T. F. and Wostmann, B. S.,** *Lab. Anim. Care,* 19, 812–814, 1969.
20. **Greenfield, H. and Briggs, G. M.,** *Ann. Rev. Biochem.*, 40, 549–572, 1971.
21. **Newberne, P. M.,** *Fed. Proc. Fed. Am. Soc. Exp. Biol.*, 34, 209–215, 1975.
22. **Bernhart, F. W. and Tomarelli, R. M.,** *J. Nutr.*, 89, 495–500, 1966.
23. **Greenfield, H., Briggs, G. M., Watson, R. H., and Yudkin, J.,** *Proc. Nutr. Soc. Engl. Scot.*, 28, 43A, 1969.
24. **Rogers, Q. R. and Harper, A. E.,** *J. Nutr.*, 87, 267–273, 1965.
25. *United States Pharmacopeia,* 19th ed., Mack, Easton, Pa., 1975, 612.
26. **Cohen, N. L., Reyes, P., Typpo, J. T., and Briggs, G. M.,** *J. Nutr.*, 91, 482–488, 1967.
27. **Knudsen, I. and Meyer, O.,** *Toxicology,* 4, 203–206, 1975.
28. **Nalder, B. N., Mahoney, A. W., Ramarkrishnan, R., and Hendricks, D. G.,** *J. Nutr.*, 102, 535–542, 1972.
29. **Pilström, L., Fellenius, E., and Kiessling, K. H.,** *Br. J. Nutr.*, 29, 297–306, 1973.
30. **Leiber, C. S. and De Carli, L. M.,** *Am. J. Clin. Nutr.*, 28, 474–478, 1970.
31. **Hegsted, D. M., Mills, R. C., Elvehjem, C. A., and Hart, E. B.,** *J. Biol. Chem.*, 138, 459–466, 1941.
32. **Wostmann, B. S. and Kellogg, T. F.,** *Lab. Anim. Care,* 17, 589–593, 1967.
33. **Muto, Y., Smith, J. E., Milch, P. O., and Goodman, G. S.,** *J. Biol. Chem.*, 247, 2542–2550, 1972.
34. **Harrand, R. B. and Hartles, R. L.,** *Br. J. Nutr.*, 22, 45–56, 1968.
35. **Hartles, R. L. and Leaver, A. G.,** *Arch. Oral Biol.*, 5, 38–44, 1961.
36. **Harrand, R. B. and Hartles, R. L.,** *Br. J. Nutr.*, 23, 523–531, 1969.
37. **Harrand, R. B. and Hartles, R. L.,** *Br. J. Nutr.*, 24, 929–942, 1970.
38. **Cawthorne, M. A., Diplock, A. T., Muthy, I. R., Bunyan, J., Murrell, E., and Green, J.,** *Br. J. Nutr.*, 21, 671–679, 1967.
39. **Bunyan, J., Green, J., Diplock, A. T., and Robinson, D.,** *Br. J. Nutr.*, 21, 137–145, 1967.
40. **Johnson, B. C., Mameesh, M. S., Metta, V. C., and Rano, P. B.,** *Fed. Proc. Fed. Am. Soc. Exp. Biol.*, 19, 1038–1044, 1960.
41. **Mameesh, M. S. and Johnson, B. C.,** *Proc. Soc. Exp. Biol. Med.*, 101, 467–468, 1959.
42. **Mameesh, M. S. and Johnson, B. C.,** *J. Nutr.*, 65, 161–167, 1958.
43. **Pandit, V. I. and Chakrabarti, C. H.,** *Enzymologia,* 38, 111–119, 1970.
44. **Hubbell, R. B., Mendel, L. B., and Wakeman, A. J.,** *J. Nutr.*, 14, 273–285, 1937.
45. **Fratta, I. D.,** *Lab. Anim. Sci.*, 19, 727–732, 1969.
46. **Phillips, P. H. and Hart, E. B.,** *J. Biol. Chem.*, 109, 656–663, 1935.
47. **Seronde, J.,** *Proc. Soc. Exp. Biol. Med.*, 132, 722–723, 1969.
48. **Seronde, J., Zucker, T. F., and Zucker, L. M.,** *J. Nutr.*, 59, 287–298, 1956.
49. **Audet, A. and Lupien, P. J.,** *J. Nutr.*, 104, 91–100, 1974.

50. Williams, D. L., Spray, G. H., Newman, G. E., and O'Brien, J. R. P., *Br. J. Nutr.*, 23, 343–352, 1969.
51. McCall, M. G., Newman, G. E., O'Brien, J. R. P., Valberg, L. S., and Witts, L. J., *Br. J. Nutr.*, 16, 297–304, 1962.
52. Howard, L., Wagner, C., and Schenker, S., *J. Nutr.*, 104, 1024–1032, 1974.
53. Fox, M. R. S. and Briggs, G. M., *J. Nutr.*, 72, 243–250, 1960.
54. Monserrat, A. J., Porta, E. A., Ghoshul, A. K., and Hartman, S. B., *J. Nutr.*, 104, 1496–1502, 1974.
55. Wesson, L. G., *Science*, 75, 339–340, 1943.
56. Schaefer, A. E., Salmon, W. D., and Strength, D. R., *Proc. Soc. Exp. Biol. Med.*, 71, 193–196, 1949.
57. Patel, M. S. and Mistry, S. P., *J. Nutr.*, 96, 409–414, 1968.
58. Leveille, G. A. and O'Hea, E. K., *J. Nutr.*, 93, 541–545, 1967.
59. Sanslone, W. R. and Muntwyler, E., *Metab. Clin. Exp.*, 17, 1084–1093, 1968.
60. Hartcroft, P. M. and Eisenstein, A. B., *Endocrinology*, 60, 641–651, 1957.
61. Andrieux-Domont, C. and Van Hung, L., *Br. J. Nutr.*, 29, 203–210, 1973.
62. Guthrie, H. A., Froozani, M., Sherman, A. R., and Barron, G. P., *J. Nutr.*, 104, 1273–1278, 1974.
63. Houk, A. E. H., Thomas, A. W., and Sherman, H. C., *J. Nutr.*, 31, 609–620, 1946.
64. Williams, R. B. and Mills, C. F., *Br. J. Nutr.*, 24, 989–1003, 1970.
65. Mills, C. F. and Murray, G., *J. Sci. Food Agric.*, 11, 547–552, 1960.
66. Evans, J. L. and Brar, B. S., *J. Nutr.*, 104, 1285–1291, 1974.
67. Campbell, A. H., Sewell, W. R., Chudkowski, M., Willson, J. E., Lord, G. H., and Mohammed, K., *Toxicol. Appl. Pharmacol.*, 26, 63–71, 1973.
68. Pleasants, J. R., personal communication, 1975.
69. Pleasants, J. R., Wostmann, B. S., and Reddy, B. S., in *Germfree Research, Biological Effects of Gnotobiotic Environments,* Heneghan, J. B., Ed., Academic Press, New York, 1973, 245–250.

DIETS (NATURAL AND SYNTHETIC): DORMICE

P. Haberey

INTRODUCTION

The dormouse (*Glis glis* L.) is a rodent of about 100 g body weight. It belongs to the family Gliridae Thomas, 1897 (= Myoxidae Waterhouse, 1839; Muscardinidae Palmer, 1899).

THE DORMOUSE IN ITS NATURAL HABITAT

Methods

Direct observations of wild dormice in their natural environment and half-free dormice bred in zoological research stations[1-13] may be made, but these are difficult because dormice are active only when it is dark.

The stomach contents of individuals captured during the different seasons may be examined.[14] Table 1 gives the results of such an investigation.

Results

Dormice typically eat vegetable matter, mainly fruit, plus grains, leaves, roots, wood, bark, mushrooms, and moss. Thus, their choice of food depends on their environment. Dormice do not travel much, and the diameter of their habitat rarely exceeds 1200 m.

Dormice are intermittent carnivores, from time to time adding to their vegetal diet some animal matter such as insects, eggs, and even small birds or fledglings.

The food specialization of dormice is not strict. Thus, dormice live in woods with

Table 1
REVIEW OF FOOD FOUND IN THE STOMACHS OF DORMICE CAPTURED IN DIFFERENT MONTHS OF THE YEAR[a]

	June	August	October
Vegetable matter	4 (58)	12 (60)	12 (82)
Leaf	3	4	3
Stalk, rhizome, root	–	1	1
Bark and buds	3	4	2
Wood	–	3	1
Fruit	–	8	11
Mushroom	–	1	5
Moss	–	1	1
Bait	3 (57)	7 (51)	7 (59)
Animal matter	3	6 (21)	3 (2)
Remains of insects	3	–	1
Number of stomachs examined	4	12	14

[a] Numbers outside parentheses are numbers of stomachs examined; values within parentheses indicate average percentage volume of food.

From Holišová, V., *Zool. Listy,* 17, 109–114, 1968. With permission.

deciduous trees as well as in those with perennial ones. Being fond of fruit, they sometimes devastate an orchard. They do not disregard human food, especially when sweet, and sometimes steal from kitchens or larders.

Since the amount of food supplies found in the habitat vary during the year, dormice adapt through hibernation and, as do most hibernators, body fatness variations.[15-27] Since dormice display no hoarding behavior,[28] they may be considered obligatory hibernators.[15,16]

DORMICE IN THE LABORATORY

Dormice adapt to usual commercial laboratory foods prepared for hamsters or even for rats, e.g., chow pellets. They need water with such food. If they are given a choice between pellets and grain (for instance sunflower), they prefer the latter.[18]

When vegetables (e.g., carrots, beet roots) or fruit (e.g., apples) are given, the water intake through them is enough, and dormice do not require more water.[29]

ARTIFICIAL FOOD

Methods

Artificial food adapted for dormice was prepared in order to record the self-selection between the three major dietary components (carbohydrates, proteins, and lipids), each being presented in a distinct container.[25-27,30] These foods were gelatinized by mixing the following into warm agar solution:

Carbohydrates: starch and sucrose
Proteins: casein supplemented with methionine
Lipids: peanut oil.

Vitamins, mineral salts, water, and other components are shown in Table 2.

Such preparations are well accepted by dormice, and allow the animals to remain quite healthy over long periods (more than 18 months).

Results

Cycles of body weight lasting a few weeks occurred in dormice fed ad libitum throughout the year with such food (Figure 1). These findings confirm the observations[18] made with laboratory foods (i.e., pellets).

The patterns of the fluctuations suggest an oscillation in body weight between an upper and a lower limit.[24] In summer, body weight stays near the upper limit, the lower limit approaching the upper one.

Large variations in food intake[18,25-27] accompany the variations in body weight. Figure 2 shows a cycle observed in one dormouse: the body weight varies in a ratio of 1 to 1.5. The total food intake varies from 10 to 100 kcal/day, i.e., a ratio of 1 to 10. The maximum body weight occurs about 5 days after the maximum food intake.

The increase in total food intake is mainly due to an increase in glucid intake followed by an increase in lipid intake, which is not continuous. The protein intake remains small and intermittent. Its maximum is synchronous with that of the total intake.

INTERPRETATION

The great fluctuations in food intake allow dormice to rapidly accumulate large fat reserves (for instance, within the short period of fruit harvesting in their habitat). These fluctuations may be a phylogenic adaptation to the alimentary environment of animals that eat mainly fruit and which exhibit no hoarding behavior.[25,27,31]

Table 2
ARTIFICIAL FOOD ADAPTED FOR DORMICE

Composition

	Carbohydrates	Protein	Lipids
Vitamins (g)	0.10	0.08	0.10
Salts (g)	0.97	0.78	0.94
Agar (g)	1.22	1.16	1.88
Sucrose (g)	10.99	–	–
Starch (g)	17.10	–	–
Casein + methionine (g)	–	18.90	–
Oil (arachid oil) (g)	–	–	23.94
Water (g)	69.61	79.08	72.80
Flavor	Apple juice	Meat extract	0.35 g onion powder
Total	100.	100.	100.
Caloric value (kcal/g)	1.12	0.76	2.16

Vitamins (Per 100 g of Food)

Thiamine	0.66 mg
Riboflavin	0.66 mg
Nicotinamid	2.97 mg
Vitamin B_6 (hydrochloride)	0.66 mg
Biotin	13.20 μg
Calcium panthotenate	1.98 mg
Folic acid	59.40 μg
p-Aminobenzoic acid	3.30 mg
Inositol	3.30 mg
Choline chloride	49.50 mg
Vitamin B_{12}	0.90 μg
Vitamin A	594.60 IU
Vitamin D	66.00 IU
Vitamin E	3.30 IU
Menadione	1.49 mg
Ascorbic acid	1.00 mg

Mineral Salts

Calcium carbonate ($CaCO_3$)	458.00 mg
Potassium phosphate (KH_2PO_4)	467.00 mg
Sodium chloride (NaCl)	167.00 mg
Mangesium chloride ($MgCl_2$)	69.00 mg
Iron II sulfate ($FeSO_4 \cdot 7\ H_2O$)	32.00 mg
Potassium iodide (KI)	948.00 μg
Manganese II sulfate ($MnSO_4 \cdot 2\ H_2O$)	340.00 μg
Copper II sulfate ($CuSO_4 \cdot 5H_2O$)	576.00 μg
Zinc chloride ($ZnCl_2$)	311.00 μg
Cobalt II chloride ($CoCl_2 \cdot 6\ H_2O$)	24.00 μg

From Schaefer, A., Interactions Entre Prise Alimentaire et Poids Corporel Chez un Hibernant: le Loir (*Glis glis* L.), Thèse présentée pour le Doctorat de Troisième Cycle, Université Louis Pasteur, Strasbourg, France, 1976.

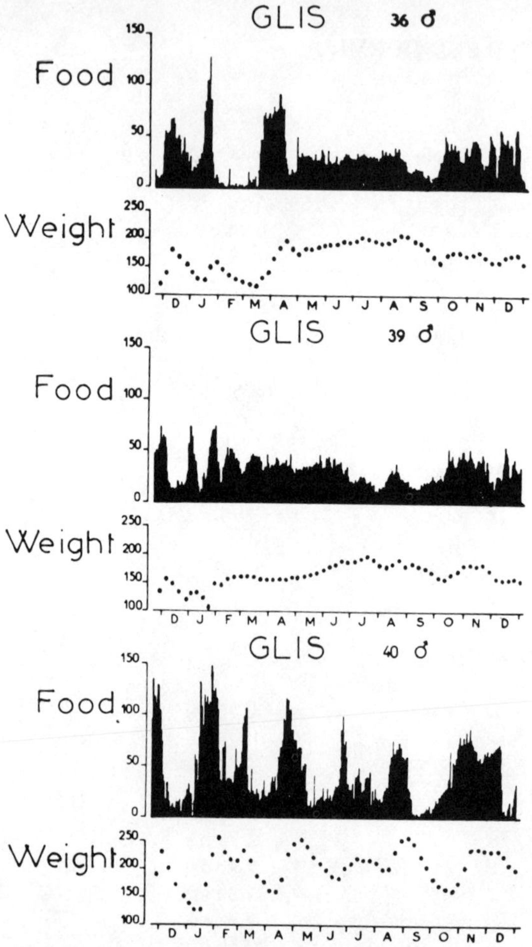

FIGURE 1. Body weight (g) and food intake (cal per day) recorded in three dormice during 1 year. (From Schaefer, A., Piquard, F., Kalinkov, D., and Haberey, P., *C. R. Soc. Biol.*, 168, 1428–1431, 1974. With permission.)

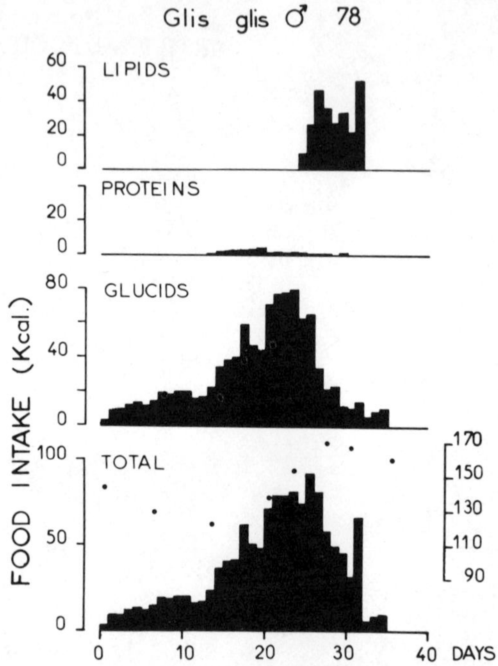

FIGURE 2. Changes in food self-selection and body weight (shown by the dots and the scale at right) in a dormouse. (From Schaefer, A., Piquard, F., and Haberey, P., *Comp. Biochem. Physiol.*, 55A, 115–118, 1976. With permission.)

REFERENCES

1. **Zimmermann, R.,** Das Vorkommen des Siebenschläfers und Beobachtungen über seine Lebensweise im Königreich Sachsen, *Zool. Gart.,* 46, 180–185, 1905.
2. **Hennings, C.,** *Die Saügetiere Deutschlands,* Quelle & Meyer, Leipzig, 1909.
3. **Zimmermann, R.,** Weiteres vom Siebenschläfers, *Zool Beobachtung.,* 51, 135–137, 1910.
4. **Pax, F.,** Die Verbreitung des Siebenschläfers in Schlesien, *Arch. Naturgesch.,* 84 Abt. A.S., 151–161, 1918.
5. **Zimmermann, R.,** Die Verbreitung und Lebenweise der Schlafmäuse in Sachsen, *Arch. Naturgesch.,* 87, 199–223, 1921.
6. **Kahmann, H.,** Uber das Klettervermögen des Siebenschläfers (*Glis glis* L.) nebst Bemerkungen über einige andere kletternde Säugetiere, *Zool. Jahrb.,* 54, 137–157, 1931.
7. **Kahmann, H.,** Beobachtungen an heimischen Nagetieren, *Zool. Säugetierkd.,* 6, 143–151, 1931.
8. **Mansfeld, K.,** Uber das Auftreten von Bilchen in Nistkästen, *Dtsch. Vogelwelt,* 67, 13–20, 1942.
9. **Ognew, S. I.,** *Die Säugetiere der UDSSR und benachbarter Länder,* Länder 5, Moskau–Leningrad, USSR, 1947.
10. **Kahmann, H.,** Die deutschen Schlafmäuse, *Kosmos,* 47, 491–496, 1951.
11. **Vietinghoff-Riesch, A.,** Beiträge zur Biologie des Siebenschläfers. *Bonn. Zool. Beitr.,* 3, 167–186, 1952.
12. **Koenig, L.,** Das Aktionssystem des Siebenschläfen (*Glis glis* L.), *Z. Tierpsychol.,* 17, 427–505, 1960.
13. **König, C.,** Einflüsse von Licht und Temperatur auf den Winterschlaf des Siebenschläfers *Glis G Glis* (Linnaeus 1766), *Z. Morphol. Oekol. Tiere,* 48, 545–575, 1960.
14. **Holišová, V.,** Notes on the food of dormice (Gliridae), *Zool. Listy,* 17, 109–114, 1968.
15. **Morrison, P.,** Some interrelations between weight and hibernation functions, *Bull. Mus. Comp. Zool. Harv. Univ.,* 124, 75–91, 1960.
16. **Kayser, C.,** *The Physiology of Natural Hibernation,* Pergamon, New York, 1961.
17. **Pengelley, E. T. and Fisher, K. C.,** The effect of temperature and photoperiod on the yearly hibernating behavior of captive golden mantled ground squirrels (*Citellus lateralis tescorum*), *Can. J. Zool.,* 41, 1103–1120, 1963.
18. **Mrosovsky, N.,** Acceleration of animal hibernating cycle to six weeks in captive dormice, *Can. J. Zool.,* 44, 903–910, 1966.
19. **Pengelley, E. T. and Asmundson, S. J.,** Free running periods of endogenous circannian rhythms in the golden mantled ground squirrel (*Citellus lateralis*), *Comp. Biochem. Physiol.,* 30, 177–183, 1969.
20. **Mrosovsky, N. and Fisher, K. C.,** Sliding set points for body weight in ground squirrels during the hibernation season, *Can. J. Zool.,* 48, 241-247, 1970.
21. **Mrosovsky, N.,** *Hibernation and the Hypothalamus,* Appleton-Century-Crofts, New York, 1971.
22. **Canguilhem, B. and Marx, Ch.,** Regulation of the body weight of the European hamster during the annual cycle, *Pflügers Arch.,* 338, 169–175, 1973.
23. **Barnes, D. S. and Mrosovsky, N.,** Body weight regulation in ground squirrels and hypothalamically lesioned rats: slow and sudden set point changes, *Physiol. Behav.,* 12, 251–258, 1974.
24. **Schaefer, A., Piquard, F., Kalinkov, D., and Haberey, P.,** Variations circannuelles de la régulation pondérale chez le loir, *C. R. Soc. Biol.,* 168, 1428–1431, 1974.
25. **Haberey, P., Schaefer, A., and Piquard, F.,** Comportement alimentaire et variations pondérales chez un hibernant: le loir (*Glis glis* L.), *Arch. Sci. Physiol.,* 28, 291–302, 1974.
26. **Schaefer, A., Piquard, F., and Haberey, P.,** Evolution saisonnière de la régulation du poids chez le loir (*Glis glis* L.), *J. Physiol.* (Paris), 70, 649–658, 1975.
27. **Schaefer, A., Piquard, F., and Haberey, P.,** Food self-selection during spontaneous body weight variations in the dormouse (*Glis glis* L.), *Comp. Biochem. Physiol.,* 55A, 115–118, 1976.
28. **Lanier, D. L., Estep, D. Q., and Dewsbury, D. A.,** Food hoarding in muroid rodents, *Behav. Biol.,* 11, 177–187, 1974.
29. **Meixner, A.,** Meine Erfahrungen mit dem Bilch, *Z. Säugetierkd.,* 15, 328–330, 1940.
30. **Haberey, P., Piquard, F., and Schaefer, A.,** Choix alimentaire spontané après le jeûne chez le rat, *C. R. Soc. Biol.,* 166, 1384–1387, 1972.
31. **Schaefer, A.,** Interactions entre Prise Alimentaire et Poids Corporel chez un Hibernant: le Loir (*Glis glis* L.), Thèse présentée pour le Doctorat de Troisième Cycle, Université Louis Pasteur, Strasbourg, France, 1976.

FOOD HABITS OF ZAPODIDAE

J. O. Whitaker, Jr.

INTRODUCTION

There are two genera and four North American species of mice currently recognized in the family Zapodidae: *Napaeozapus insignis, Zapus hudsonius, Z. princeps,* and *Z. trinotatus. Napaeozapus insignis* is northeastern, *Z. hudsonius* inhabits a large range in eastern and northern North America, *Z. trinotatus* is found along the west coast from British Columbia to California, and *Z. princeps* inhabits a large part of west central North America, extending north into Alaska. In addition, *Eozapus,* with one species, *E. setchuanus,* occurs in two provinces of China. This completes a resumé of the living members of the subfamily Zapodinae. The subfamily Sicistinae includes six species, all in the genus *Sicista,* and all occurring in Eurasia.

Unfortunately, of the 11 species, there is rather detailed published information on the food habits of only two, *Zapus hudsonius* from New York and Indiana[28,30] and *Napaeozapus insignis* from New York.[29] However, we have examined stomach contents of a number of individuals of *Zapus trinotatus* and *Z. princeps* from western North America,[14] so this information is also included here. To my knowledge, there is no published information on the food habits of *Eozapus setchuanus,* nor on four of the six species of *Sicista.*

Most of the information in this paper concerns the natural food of the various species. There is no published information on nutritional requirements, nor on laboratory diets of any of the species. I have kept a number of meadow jumping mice alive, and found them to do very well if kept supplied with rolled oats and fresh water, supplemented occasionally with miscellaneous other items such as corn, nuts, apple, or various types of seeds. In some individuals kept in the laboratory, a great increase in the amount of food eaten during a 2-week period in September or October during obvious preparation for hibernation is very noticeable. Information on the food habits of various species follows.

SUBFAMILY SICISTINAE

Sicista betulina Pall

Ivanter[13] examined the stomachs of 16 individuals of *Sicista betulina* from the Karelian Autonomous Soviet Socialist Republic. Insects, green plant material, and seeds predominated in that sample (Table 1).

Sicista subtilis

Hellwig[12] states, "*Sicista subtilis* can be considered a true insectivorous rodent, since Noctuidae and Curculionidae are consumed in large quantities."

I have no information concerning the food of *S. caucasica, S. caudata, S. concolor,* or *S. napaea.*

SUBFAMILY ZAPODINAE

Zapus hudsonius (Zimmermann)

Quimby[20] made various plant and animal materials available to caged meadow jumping mice. The starchy fruits of the Gramineae and the more or less fleshy fruits of various groups of plants were the foods most commonly eaten, but many animal foods

Table 1
FOODS EATEN BY 16 INDIVIDUALS OF THE BIRCHMOUSE, *SICISTA BETULINA*, FROM KARELIA (ASSR)

Food	Number of stomachs containing given food type	Percentage of stomachs containing food
Insect materials	12	75.0
Eggs	1	6.2
Larvae	3	18.7
Unidentified insects	4	25.0
Coleoptera	8	50.0
Blattidae	1	6.2
Lumbricidae	2	12.5
Molluscs	1	6.2
Plant remains	6	37.5
Green plant material	4	25.0
Seeds	3	18.7
Berries	1	6.2

From Ivanter, E. V., *Aquilo Ser. Zool.*, 13, 103–108, 1972. With permission.

(mostly insects) were also utilized. Hamilton[10] concluded that seeds are the main food of the species, but that berries, nuts, fruits, and insects are also eaten. Goodwin[9] found blackberries and cranberries in several stomachs. Stoner[26] found the food of the species to be seeds and nuts. Bailey[4] found only the pulp from "shelled, well masticated seeds" in many stomachs examined. Schmidt[24] found the remains of geometrid larvae in two stomachs. Richmond and Roslund[22] stated that the species feeds mostly on seeds, but they also saw *Zapus* feed on ripe strawberries and timothy. They twice observed the species climbing the stems of dock and feeding on the nearly ripe seeds. Whitaker[28] reported on the food habits of *Zapus hudsonius* from New York: When meadow jumping mice emerged from hibernation in the latter part of April and during May, they fed on a variety of foods. Animal materials constituted about half of the volume, and various seeds about 20% (Table 2). Few fungi were eaten. As the season progressed, more fungi and seeds but less animal materials were consumed. However, the seeds of various species of plants could be described as the basic or staple food of *Zapus hudsonius.* Small, fully ripe seeds with hard pericarps were most frequently eaten. The particular kinds of foods seemed to be determined by availability rather than by preference. Early in the season few seeds were available, so few were eaten, but as seeds began to ripen, those most common in the various areas were most apt to be eaten.

Monthly Information

May – The stomachs of 136 mice collected in May were examined. About half the food was animal, including mostly lepidopterous larvae and adult Coleoptera (particularly Curculionidae and Carabidae), although other animal material, including sowbugs and spiders, was present. The fungus *Endogone* was seldom found in stomachs in May, and then only towards the end of the month. Seeds, mostly of *Echinochloa* and *Asclepias,* together with unidentified seeds composed about a fifth of the volume. The seeds were mostly those which had been on the ground since the previous season, but seeds of *Ulmus* were eaten as they became available towards the end of the month.

June – The food of *Zapus* in early June was similar to that in May, but about the middle of June the amount of animal food decreased to about a tenth of the total, while the consumption of seeds increased. This change correlated with the ripening of vernal sweet grass, *Anthoxanthum odoratum,* and chickweed, *Cerastium vulgatum,* and with the

Table 2
FOOD OF *ZAPUS HUDSONIUS*

Ithaca, New York, 1958–1961

Food	April (8 stomachs)		May (136 stomachs)		June (109 stomachs)		July (159 stomachs)		August (138 stomachs)		September (191 stomachs)		October (55 stomachs)	
	% freq.	% vol.	% freq.	% vol.	% freq.	% vol.	% freq.	% vol.	% freq.	% vol.	% freq.	% vol.	% freq.	% vol.
Miscellaneous vegetation	100.0	61.3	41.5	14.7	27.6	9.6	22.6	4.8	31.9	10.9	32.5	5.7	14.5	4.3
Coleoptera	62.5	24.6	54.2	16.7	22.0	4.5	4.4	0.8	2.2	0.8	12.0	1.8	5.5	0.4
Lepidopterous larvae	25.0	9.8	65.3	30.2	56.9	20.6	23.3	5.6	23.9	6.5	26.7	8.4	38.2	8.3
Acer rubrum flowers	12.5	4.4	–	–	–	–	–	–	–	–	–	–	–	–
Unidentified seeds	–	–	11.9	10.7	23.9	12.9	19.5	12.1	34.8	18.2	29.3	14.2	30.9	13.0
Unidentified material	–	–	15.3	9.4	0.9	Trace	9.4	6.3	3.6	1.3	2.6	1.1	3.6	1.4
Echinochloa seeds	–	–	9.3	4.7	–	–	–	–	2.9	0.9	23.0	15.2	16.4	12.5
Unidentified animal material	–	–	11.9	3.7	11.9	4.5	5.7	1.0	2.2	1.3	12.6	1.0	3.6	0.9
Asclepias seeds	–	–	5.9	2.8	0.9	0.8	–	–	–	–	–	–	1.8	1.8
Ulmus seeds	–	–	2.5	1.7	8.3	5.7	1.9	0.8	–	–	–	–	–	–
Endogone	–	–	4.2	1.2	23.9	6.9	46.5	16.7	50.0	15.1	48.7	19.5	32.7	8.3
Impatiens seeds	–	–	1.7	1.2	–	–	–	–	8.0	0.8	8.9	4.5	29.1	19.3
Spiders	–	–	1.7	1.1	4.6	0.9	1.3	0.1	–	–	2.1	0.2	–	–
Anthoxanthum odoratum seeds	–	–	0.8	0.8	25.7	10.5	29.6	14.6	0.7	0.4	–	–	–	–
Sowbugs	–	–	0.8	0.6	–	–	–	–	–	–	–	–	–	–
Poa compressa seeds	–	–	0.8	0.1	0.9	0.9	3.1	1.8	4.3	4.1	12.6	6.1	9.1	2.8
Juncus effusus seeds	–	–	0.8	0.1	–	–	–	–	0.7	Trace	–	–	–	–
Earthworms	–	–	0.8	0.1	–	–	–	–	–	–	–	–	–	–
Cerastium seeds	–	–	–	–	34.9	15.5	6.9	1.3	0.7	Trace	1.6	0.1	1.8	0.1
Fragaria fruit	–	–	–	–	22.0	5.0	16.4	3.0	–	–	–	–	–	–
Hymenogaster	–	–	–	–	4.6	1.2	4.4	1.4	2.2	0.8	4.7	1.9	5.5	1.2
Rumex acetosello	–	–	–	–	0.9	0.2	15.0	5.8	9.4	3.7	2.6	1.3	–	–
Zapus hair	–	–	–	–	0.9	0.2	–	–	–	–	–	–	–	–
Dipterous larvae	–	–	–	–	0.9	Trace	–	–	–	–	3.1	0.2	18.2	1.4
Formicidae	–	–	–	–	0.9	Trace	–	–	–	–	–	–	–	–
Dactylis glomerata seeds	–	–	–	–	–	–	24.5	13.9	1.4	0.1	0.5	0.2	–	–
Phleum pratense seeds	–	–	–	–	–	–	6.3	3.7	24.6	18.6	7.8	6.1	1.8	1.1
Arrenatherum elatius seeds	–	–	–	–	–	–	6.9	2.4	5.1	2.3	–	–	–	–
Rubus fruit	–	–	–	–	–	–	3.8	1.6	10.1	4.8	1.0	0.1	–	–
Potentilla recta seeds	–	–	–	–	–	–	1.9	1.3	18.8	5.8	4.7	0.7	3.6	0.6
Oxalis europaea seeds	–	–	–	–	–	–	1.3	0.4	–	–	3.1	1.7	14.5	4.9
Adult Lepidoptera	–	–	–	–	–	–	0.6	0.3	–	–	0.5	0.1	–	–

Table 2 (continued)
FOOD OF *ZAPUS HUDSONIUS*

Ithaca, New York, 1958–1961

Food	April (8 stomachs)		May (136 stomachs)		June (109 stomachs)		July (159 stomachs)		August (138 stomachs)		September (191 stomachs)		October (55 stomachs)	
	% freq.	% vol.	% freq.	% vol.	% freq.	% vol.	% freq.	% vol.	% freq.	% vol.	% freq.	% vol.	% freq.	% vol.
Algae	–	–	–	–	–	–	0.6	0.1	–	–	–	–	–	–
Vaccinium fruit	–	–	–	–	–	–	0.6	Trace	6.5	2.8	–	–	–	–
Trifolium agaricus flowers	–	–	–	–	–	–	0.6	Trace	–	–	–	–	–	–
Mites	–	–	–	–	–	–	0.6	Trace	0.7	Trace	0.5	Trace	–	–
Agropyron repens seeds	–	–	–	–	–	–	–	–	0.7	0.3	–	–	–	–
Danthonia spicata seeds	–	–	–	–	–	–	–	–	0.7	0.2	–	–	–	–
Viburnum dentatum fruit	–	–	–	–	–	–	–	–	0.7	Trace	11.5	5.6	–	–
Culicidae	–	–	–	–	–	–	–	–	0.7	Trace	–	–	–	–
Coleopterous larvae	–	–	–	–	–	–	–	–	0.7	Trace	–	–	–	–
Thysanoptera	–	–	–	–	–	–	–	–	0.7	Trace	–	–	–	–
Leerzia oryzoides seeds	–	–	–	–	–	–	–	–	–	–	5.2	3.2	21.8	15.9
Scirpus seeds	–	–	–	–	–	–	–	–	–	–	1.0	0.8	–	–
Hymenopterous larvae	–	–	–	–	–	–	–	–	–	–	0.5	0.2	–	–
Slugs	–	–	–	–	–	–	–	–	–	–	0.5	0.2	1.8	0.7
Zapus toe	–	–	–	–	–	–	–	–	–	–	0.5	Trace	–	–
Setaria lutescens seeds	–	–	–	–	–	–	–	–	–	–	–	–	1.8	1.2
Chilopoda	–	–	–	–	–	–	–	–	–	–	–	–	1.8	Trace
Cicadellidae	–	–	–	–	–	–	–	–	–	–	–	–	1.8	Trace
Total		100.1		99.8		99.9		99.8		99.7		100.1		100.1

From Whitaker, J. O., Jr., *Ecol. Monogr.*, 33, 215–254. Copyright 1963 by the Ecological Society of America. With permission.

increased number of *Ulmus* seeds available. The fruits of strawberry, *Fragaria virginiana,* were eaten as they ripened in late June, and the fungi *Endogone* and *Hymenogaster* became important in the latter part of the month. Miscellaneous vegetation (particularly rootlets) was often in stomachs, some of which was probably ingested when mice were feeding on *Endogone.*

July – Seeds, especially of grasses, comprised over half of the food in July. *Anthoxanthum* was still available, and its seeds were eaten. By this time the seeds of orchard grass, *Dactylis glomerata,* had ripened and were heavily utilized, while the seeds of timothy, *Phleum pratense,* were just ripening and were slightly used. Seeds of sheep sorrel, *Rumex acetosella,* had ripened and become an important part of the diet. Other seeds eaten during July were those of *Ulmus, Poa, Oxalis, Potentilla, Arrhenatherum,* and *Cerastium.* Unidentified seeds accounted for about a tenth of the food of the mice examined. *Fragaria* and *Rubus* were eaten in slight amounts while *Endogone* and *Hymenogaster* made up about a fifth of the food.

August – In August, seeds again comprised over half the food. Important ones were those of *Phleum pratense, Potentilla recta, Poa compressa, Rumex acetosella,* and *Arrenatherum elatius.* Fruits of *Rubus* and *Vaccinium* were utilized slightly, and lepidopterous larvae were also eaten. The fungal foods comprised nearly a fifth of the food in the sample.

September – During September, principal seeds eaten were those of *Echinochloa, Poa,* and *Phleum,* while those of *Impatiens, Oxalis europaea, Rumex acetosella, Scirpus, Potentilla recta, Dactylis, Leerzia,* and *Cerastium vulgatum* were eaten to a lesser extent. *Viburnum dentatum* was occasionally eaten. Animal material, mostly lepidopterous larvae, formed slightly over a tenth of the food, while fungal foods comprised over a fifth of the volume.

October – In October, various seeds comprised nearly 70% of the food. Seeds of *Impatiens, Leerzia,* and *Echinochloa,* together with unidentified seeds, were the most important. Seeds of *Poa, Phleum, Setaria,* and *Oxalis* were less important; those of *Asclepias, Potentilla recta,* and *Cerastium vulgatum* were also eaten. Animal materials and fungal foods each comprised about 10% of the volume.

Bailey[4] stated that *Zapus* reached high on the culm of some grasses, bit through it, pulled it down, and then repeated the procedure until the head was reached. At one locality near Ithaca, New York, where nearly all the mice taken were feeding on *Phleum* seeds, many piles of glumes from this grass were found. The naked rachis was usually on top of the pile. Also, there often were piles of pieces of culm, each piece being about 5 to 10 cm long. These observations suggest that the method of feeding observed by Bailey also occurred at Ithaca. In other cases in the same area, the head was removed from the culm, which was left standing. It appeared that the mice had climbed the culms and bitten off the head. Captive *Zapus* had no trouble climbing the culms of grasses. In one area of about 2 m^2 where several jumping mice but no other vertebrates were taken, most of the heads were removed from the standing culms, and other culms were in match-length piles. It appeared that both types of feeding on grasses occurred here.

Seeds of orchard grass, *Dactylis glomerata,* were an important food when available. Since piles of culms or glumes of this grass (and most other grasses eaten) were never found, most seeds eaten were probably those that had fallen to the ground.

Seeds of most species of plants were finely chewed, but the seeds of some species that were small (*Poa* and *Cerastium,* for example) were occasionally unchewed and entered the intestines whole. In such cases, *Zapus* could easily act as a dispersal agent.

Two species of touch-me-not, *Impatiens pallida* and *I. biflora,* occur in the northeastern United States, and seeds of both were eaten by *Zapus.* The seeds could be easily identified by their turquoise endosperm and yellow seed coats. They were chewed into a blue pulpy mass with the yellowish, membranous coats interspersed throughout.

The blue color often could be seen through the stomach wall before the stomach was opened. This food was utilized when ripe in late summer and fall.

The food habits of juveniles did not differ appreciably from those of the adults (Table 3).

The stomachs of 131 *Zapus hudsonius* from Indiana were examined (Table 4).[30] Seeds of touch-me-not (*Impatiens*) were most important, followed by the fungus *Endogone*, unidentified grass seeds, seeds of foxtail grass (*Setaria*), and seeds of purple-top grass (*Triodia flava*).

Other Information

Water consumption – Caged mice kept by me drank water from containers. The same was found in studies by Sheldon[25] and Quimby.[20] It is not known whether mice in the natural state normally drink water or get it from green vegetation.

Food storage – Audubon and Bachman[1] fed captive meadow jumping mice wheat, maize, and buckwheat. The mice were said to store it in their holes before morning and to dig new burrows when needed. Kennicott[16] stated that *Zapus* had been observed to store nuts, grain, and seeds in its burrows. Preble[19] wrote, "Although they often lay up stores

Table 3
FOOD HABITS OF JUVENILE *ZAPUS* AS COMPARED TO THOSE OF ALL *ZAPUS* AND TO ADULT *ZAPUS*[a]

July to October, 1958–1961

Food	Percent volume		
	All *Zapus*	Juvenile *Zapus*	Older *Zapus*
Endogone	16.5	23.1	14.3
Unidentified seeds	14.5	15.8	14.1
Phleum pratense seeds	8.0	6.8	8.5
Lepidopterous larvae	7.1	4.1	8.0
Echinochloa seeds	6.8	4.2	7.7
Unidentified vegetation	6.6	8.1	6.2
Anthoxanthum odoratum seeds	4.4	6.3	3.8
Poa compressa seeds	4.0	4.6	3.8
Impatiens seeds	3.7	2.4	4.2
Rumex acetosella seeds	3.1	4.0	2.8
Leerzia oryzoides seeds	2.7	0.8	3.4
Potentilla recta seeds	2.2	2.9	1.9
Viburnum dentatum fruit	2.0	1.8	2.0
Hymenogaster	1.4	0.9	1.5
Oxalis europaea seeds	1.2	0.2	1.5
Coleoptera	1.1	1.3	1.0
Unidentified animal	1.1	0.6	1.2
Fragaria virginiana fruit	0.9	1.6	1.0
Vaccinium fruit	0.7	0.9	0.7
Cerastium seeds	0.4	0.8	0.3
Scirpus seeds	0.3	0.7	0.1
Setaria lutescens seeds	0.2	0.5	0.0
Dactylis glomerata seeds	0.1	2.3	0.7

[a] 130 juveniles, 413 adults (543 total).

From Whitaker, J. O., Jr., *Ecol. Monogr.*, 33, 215–254. Copyright 1963 by the Ecological Society of America. With permission.

Table 4
FOOD OF 131 MEADOW JUMPING MICE, *ZAPUS HUDSONIUS*, FROM INDIANA

Items	Percent volume	Percent frequency
Impatiens seeds	21.4	27.5
Endogone	15.6	50.4
Setaria seeds	10.0	10.7
Unidentified grass seeds	9.6	18.3
Triodea flava seeds	9.0	13.7
Lepidopterous larvae	5.7	13.7
Miscellaneous vegetation	4.6	15.3
Phleum pratense seeds	3.7	3.8
Green grass seeds	3.4	6.1
Hymenogaster	1.8	4.6
Unidentified seeds	1.7	4.6
Dactylus glomerata seeds	1.5	1.6
Slugs	1.3	1.6
Curculionidae	1.1	2.3
Green vegetation	1.0	1.6
Cerastium seeds	0.9	4.6
Rumex acetosella seeds	0.8	2.3
Bromus seeds	0.8	0.8
Panicum seeds	0.8	0.8
Elymus seeds	0.8	0.8
Echinochloa seeds	0.6	0.8
Rubus fruit	0.6	1.6
Unidentified animal	0.5	2.3
Oxalis seeds	0.5	0.8
Other items comprising 0.5% volume[a]	2.5	
Total	100.2	

[a] Other items, in order of decreasing importance: coleopterous larvae, *Poa* seeds, mast, *Chenopodium* seeds, Carabidae, *Prunus* seeds, Chrysomelidae, unidentified Coleoptera, *Festuca* seeds, *Leerzia* seeds, Reduviidae, dipterous larvae, and *Polygonum* seeds.

From Whitaker, J. O., Jr. and Mumford, R. E., *Proc. Indiana Acad. Sci.*, 80, 201–209, 1971. With permission.

of food in nests or burrows during summer, it is not known that they use this food during winter." The information given by Audubon and Bachman is the only record in which evidence interpreted as food storage is given, and statements by recent authors[3,17] apparently refer to this record.

During my studies, no indication was found that food storage was practiced. No food caches were discovered in the few instances when nesting burrows were dug up, and none of the mice in captivity showed any signs of storing food. Likewise, Sheldon[25] and Babcock[2] found no evidence of food storage. I believe the record of Audubon and Bachman to be a case of the animals incidentally burying their food supply in conjunction with moving their nests.

Method of feeding – Quimby[20] described a caged mouse feeding on grass seeds. It sat on its haunches and fed by means of the front feet. First the fruiting head of a grass was cut off, then all parts, beginning at one end, were stripped. Only the seeds were eaten, but many of these dropped to the ground. No effort was made to retrieve them. I have made similar observations.

***Zapus trinotatus* Rhoads and *Z. princeps* J. A. Allen**

There is essentially no detailed published information on the food habits of either *Zapus trinotatus* or *Z. princeps,* although there are some general statements in the literature that they feed on seeds of various sorts, especially of grasses. Some other foods mentioned are dock, blackberries, and skunk cabbage;[6] thimbleberry, salmonberry, evergreen huckleberry, and underground fungi;[18] and spruce seeds.[21]

Jones et al.[14] examined 93 stomachs of *Zapus trinotatus* and 63 of *Z. princeps* from various western North American states and provinces (Table 5). Specific foods varied, but the general types of foods eaten by *Z. princeps* and *Z. trinotatus* were similar to those eaten by *Z. hudsonius.* Foods consisted primarily of seeds (52.7% volume in *Z. trinotatus* and 44.2% volume in *Z. princeps*), with grass seeds predominating. Various fruits such as salmonberry (*Rubus spectabilis*), thimbleberry (*Rubus parviflorus*), and red huckleberry (*Vaccinium parviflorum*) were eaten. The subterranean fungus *Endogone* comprised 8.9 and 8.6% of the food volume in the *Z. trinotatus* and *Z. princeps* stomachs, respectively, and animal material accounted for 4.7 and 10.0%, respectively, of the total volume of food in these two species.

Table 5
FOOD FROM 93 STOMACHS OF *ZAPUS TRINOTATUS* AND 63 STOMACHS OF *ZAPUS PRINCEPS* FROM WESTERN NORTH AMERICA[14]

	Zapus trinotatus		*Zapus princeps*	
	% vol.	% freq.	% vol.	% freq.
Grass seeds	38.5	80.6	36.4	58.7
Unidentified seeds	12.5	38.7	5.9	25.4
Rubus spectabilis (salmonberry)	9.7	22.6	11.5	17.5
Endogone	8.9	45.2	8.6	31.7
Rubus parviflorus (thimbleberry)	8.9	17.2	–	–
Starchy material	5.7	7.5	1.5	1.6
Lepidopterous larvae	4.2	32.3	8.1	22.2
Miscellaneous vegetation	3.6	38.7	6.3	38.1
Vaccinium parviflorum (red huckleberry)	2.3	4.3	0.2	1.6
Cerastium seeds	1.6	9.7	1.9	6.3
Miscellaneous fungi	1.5	12.2	1.4	6.3
Lonicera involucrata (twinberry)	1.3	7.5	–	–
Gaultheria sp. (salal)	0.4	4.3	–	–
Unidentified animal	0.4	12.9	0.6	7.9
Rubus sp.	0.2	2.2	1.0	1.6
Moss	0.2	2.2	–	–
Formicidae	0.1	2.2	0.1	1.6
Rumex seeds	0.1	3.2	–	–
Aphididae	Trace	2.2	–	–
Mites	Trace	2.2	–	–
Coleoptera	Trace	1.1	0.5	1.6
Chalcidoidea	Trace	1.1	–	–
Unidentified fruit	–	–	15.3	17.5
Scarabaeidae	–	–	0.3	3.2
Diptera	–	–	0.2	1.6
Chrysomelidae	–	–	0.2	1.6
Total	100.1		100.0	

Napaeozapus insignis

Hamilton[11] examined 40 stomachs of *Napaeozapus* and found insect larvae, particularly of Lepidoptera and Diptera. Blueberries, raspberries, small seeds and nuts, green leaves, fronds of *Asplenium,* small soil worms, centipedes, and various small invertebrates were also consumed. Earlier, Hamilton[10] reported mitrewort (*Mitella diphylla*) and May apple (*Podophyllum peltatum*) in stomachs; three contained "several hundred tiny black seeds," while grubs, enchytrid worms, green vegetation, and parts of a crane fly were found in others. Connor[7] examined 14 stomachs and found that plant materials (seeds, starchy roots, and basal parts of plants) comprised 70% of the food. The remainder was animal material, including lepidopterous larvae. Saunders[23] believed that *Napaeozapus* had eaten alder fruits.

Collections from New York were made from April to October in 1960 and 1961.[29] The subterranean fungus *Endogone* was the single most important food (Table 6), occurring in 68.9% of the stomachs and forming 33.3% of the total volume of food in stomachs. The 103 mice examined included the 53 earlier reported upon by Whitaker.[27] Seeds, collectively, composed 24.2% of the total volume. Fruits (*Rubus, Fragaria,* and *Vaccinium*) together made up 11.2% of the total. The seeds of *Anthoxanthum, Cerastium, Dactylis,* and *Setaria,* primarily plants of the open fields at Ithaca, were in stomachs of mice from one trapline. Here the field plants grew in a narrow opening in the woods, adjacent to an open area. Hence, the occurrence of these foods in stomachs does not indicate the moving of *Napaeozapus* into the fields.

The stomachs of 24 woodland jumping mice collected by John G. New from Slide Mountain in New York and the stomachs of 18 individuals collected by the United States Biological Survey were also examined.[29] The stomachs from the Slide Mountain mice

Table 6
CONTENTS OF 103 STOMACHS OF *NAPAEOZAPUS INSIGNIS*

Ithaca, New York, Summers of 1960, 1961

Stomach contents	Percent frequency	Percent volume
Endogone	68.9	33.3
Unidentified seeds	33.0	14.1
Lepidopterous larvae	30.1	10.3
Rubus fruit	29.1	9.7
Miscellaneous vegetation	37.9	7.9
Coleoptera	40.8	7.5
Impatiens seeds	13.6	6.0
Hymenogaster	8.7	3.7
Anthoxanthum odoratum seeds	6.8	3.2
Unidentified animal	8.7	1.4
Fragaria virginiana fruit	3.9	1.0
Epifagus virginiana seeds	1.0	0.8
Vaccinium fruit	1.0	0.5
Lysimachia nummularia leaves	1.0	0.2
Cerastium vulgatum seeds	1.0	0.1
Dactylis glomerata seeds	1.0	Trace
Setaria lutescens seeds	1.0	Trace
Spiders	1.0	Trace
Diptera	1.0	Trace
Mites	1.0	Trace
Total		99.7

From Whitaker, J. O., Jr., *J. Mammal.,* 44, 316–321, 1963. With permission.

contained, by volume, 38.5% miscellaneous vegetation, 28.7% unidentified seeds, 27.6% *Endogone,* 2.8% *Rubus,* 2.0% Chilopoda, 0.9% unidentified animal materials, and 0.4% lepidopterous larvae. Of the mice collected by the Biological Survey, 16 were from New Hampshire, one was from Tennessee, and one was from North Carolina. The latter two were collected in June; one of the New Hampshire stomachs was collected in August, while the remainder were collected in July. Foods, by volume, were 33.9% miscellaneous vegetation, 33.5% *Endogone,* 11.1% seeds, 10.3% *Fragaria,* 5.3% unidentified animal materials, 5.2% Coleoptera, and 0.8% Lepidoptera.

DISCUSSION

Certain of the foods deserve further comment.

***Endogone* and other fungi** – Small subterranean fungi, particularly *Endogone,* are a staple food of *Zapus hudsonius* and *Napaeozapus insignis,* as well as of other small mammals.[27] They are of minor importance in the spring but become increasingly important in summer and fall. From the high incidence in stomachs, and from the fact that many stomachs contained from 50 to 100% of this material, it appears that this food is actively sought and not merely ingested along with other foods, as has been suggested.[5,8] Percent volume estimates of *Endogone* in 17 of the 131 Indiana stomachs examined were 100, 85, 80 (3 stomachs), 70, 60 (3), 55 (2), and 50 (6). Also, 50.4% of the mice had some *Endogone* in their stomachs, a greater percentage than was the case for any other food. Another factor supporting the conclusion that this food is actively sought is that neither *Zapus* nor *Napaeozapus* eats any food that seems a likely source for simultaneous ingestion of fungi. With the exception of Coleoptera, which made up only a minor portion of the diet, no subterranean foods are extensively eaten. That the mice actually do dig or scratch the earth was evident, since on many occasions mice were taken which had particles of dirt, sometimes including *Endogone* spores, clinging to hairs of the cervical and pectoral regions of their bodies. Also, teeth and nails often had dirt in the grooves and corners. This dirt must have been acquired while digging. Since these species do not burrow extensively, the dirt and spores seemed to indicate active digging for food, and the fungi seemed the likely food. Possibly the mice locate this food by olfaction.

It would be interesting to know the nutritive values obtained by jumping mice from all the major foods eaten by them, but such values for fungal foods seemed of particular interest, and a small amount of information on this was collected by Whitaker and Mumford.[30]

Potential energy value of *Endogone* collected from the stomachs and intestines of four meadow jumping mice from Indiana was determined by means of a Parr® Oxygen Bomb Calorimeter. *Endogone* samples from the stomach and intestines were each divided into two equal masses, weighed, and burned. The two masses in the stomach averaged 2735.50 cal/g (2610.31 and 2860.69), while the two masses in the intestine averaged 2266.44 cal/g (2169.42 and 2363.46), or significantly less than the material in the stomach ($t = 9.23$, 2 df). For comparison, some calorific measurements of seeds obtained by Kendeigh and West[15] are *Echinochloa crusgalli* (4819, hulled), *Elymus viginicus* (4695), *Setaria faberii* (4585, hulled), *Zea mays* (4317), *Polygonum pennsylvanicum* (4515), and *Fraxinus nigra* (5625). It would appear that *Z. hudsonius* gets energy from *Endogone,* but that the potential amount of energy of *Endogone* is much less than in many kinds of seeds. Additional study is needed.

Lepidopterous larvae – These were a very important animal food of jumping mice, being found in stomachs throughout the season, although their greatest utilization was in spring. The larvae were mainly noctuids such as cutworms (*Leucania*) which could be commonly found on the ground between clumps of grasses and sedges. A few geometrid and tortricid larvae were also identified in *Zapus* stomachs from New York.

In summary, the food of jumping mice seems to be governed in great part by availability rather than by preference, seeds and other foods becoming utilized as they ripen throughout the season.

REFERENCES

1. **Audubon, J. J. and Bachman, J.,** *The Viviparous Quadrupeds of North America,* V. G. Audubon, New York, 1851.
2. **Babcock, H. L.,** Notes on the meadow jumping mouse (*Zapus hudsonius*), especially regarding hibernation, *Am. Nat.,* 48, 485–490, 1914.
3. **Bailey, J. W.,** *The Mammals of Virginia,* Williams Printing Co., Richmond, Va., 1946.
4. **Bailey, V.,** A biological survey of North Dakota, *North Am. Fauna,* Vol. 49, U.S. Government Printing Office, Washington, D.C., 1926.
5. **Bakerspigel, A.,** The spores of *Endogone* and *Melanogaster* in the digestive tracts of rodents, *Mycologia,* 50, 440–442, 1958.
6. **Banfield, A. W. F.,** *The Mammals of Canada,* University of Toronto Press, Toronto, 1974.
7. **Connor, P. F.,** The small mammals of Otsego and Schoharie counties, New York, *N.Y. State Mus. Sci. Serv. Bull.,* 382, 1–84, 1960.
8. **Dowding, E. S.,** Ecology of *Endogone, Trans. Br. Mycol. Soc.,* 42, 449–457, 1959.
9. **Goodwin, G. G.,** The mammals of Connecticut, *Conn. State Geol. Nat. Hist. Surv. Bull.,* 53, 1–221, 1932.
10. **Hamilton, W. J., Jr.,** Habits of jumping mice, *Am. Midl. Nat.,* 16, 187–200, 1935.
11. **Hamilton, W. J., Jr.,** The food of small forest mammals in eastern United States, *J. Mammal.,* 22, 250–263, 1941.
12. **Hellwig, S.,** On the systematic position and biology of *Sicista subtilis nordmanni* Keys. et. Blas. (1840), *Proc. Isr. Soc. Zool.,* 15, 30, 1966 (1964).
13. **Ivanter, E. V.,** Contribution to the ecology of *Sicista betulina* Pall., *Aquilo Ser. Zool.,* 13, 103–108, 1972.
14. **Jones, G. S., Whitaker, J. O., Jr., and Maser, C.,** Food habits of *Zapus* of western North America, manuscript in preparation.
15. **Kendeigh, S. C. and West, G. C.,** Caloric values of plant seeds eaten by birds, *Ecology,* 46, 553–555, 1965.
16. **Kennicott, R.,** The Quadrupeds of Illinois, Report of the Commissioner of Patents for the Year of 1856, Illinois House of Representatives Ex. Doc. No. 65, Vol. 8, 1857, 52–110.
17. **Krutzsch, P. H.,** North American jumping mice (genus *Zapus*), *Univ. Kans. Publ.,* 7, 349–472, 1954.
18. **Maser, C. and Franklin, R.,** Checklist of Vertebrate Animals of the Cascade Head Experimental Forest, U. S. Department of Agriculture Forest Service Res. Bull. PNW-51, 1974.
19. **Preble, E. A.,** Revision of the jumping mice of the genus *Zapus, North Am. Fauna,* Vol. 15, U.S. Government Printing Office, Washington, D.C., 1899.
20. **Quimby, D. C.,** The life history and ecology of the jumping mouse, *Zapus hudsonius, Ecol. Monogr.,* 21, 61–95, 1951.
21. **Radvanizi, A.,** Seed Losses to Small Mammals and Birds, Direct Seeding Symp., Timmins, Ontario, 1973, 67–75.
22. **Richmond, N. D. and Roslund, H. R.,** Mammal Survey of Northwestern Pennsylvania, Pennsylvania Game Commission, Harrisburg, 1949.
23. **Saunders, W. E.,** Notes on *Napaeozapus, J. Mammal.,* 237–238, 1921.
24. **Schmidt, F. J. W.,** Mammals of western Clark County, Wisconsin, *J. Mammal.,* 12, 99–117, 1931.
25. **Sheldon, C.,** Studies on the life histories of *Zapus* and *Napaeozapus* in Nova Scotia, *J. Mammal.,* 15, 290–300, 1934.
26. **Stoner, D.,** The rodents of Iowa, *Iowa Geol. Surv. Bull.,* 5, 1–172, 1918.
27. **Whitaker, J. O., Jr.,** *Endogone, Hymenogaster,* and *Melanogaster* as small mammal foods, *Am. Midl. Nat.,* 67, 152–156, 1962.
28. **Whitaker, J. O., Jr.,** A study of the meadow jumping mouse, *Zapus hudsonius* (Zimmermann), in central New York, *Ecol. Monogr.,* 33, 215–254, 1963.
29. **Whitaker, J. O., Jr.,** Food, habitat and parasites of the woodland jumping mouse in central New York, *J. Mammal.,* 44, 316–321, 1963.
30. **Whitaker, J. O., Jr. and Mumford, R. E.,** Jumping mice (Zapodidae) in Indiana, *Proc. Indiana Acad. Sci.,* 80, 201–209, 1971.

DIETS (NATURAL AND SYNTHETIC): HYSTRICOMORPH AND CAVIOMORPH RODENTS (OTHER THAN THE GUINEA PIG)

B. J. Weir

Introduction

There are few species in the suborder Hystricomorpha[1] (about 170) compared with the other two rodent suborders, Sciuromorpha (ca. 366 species) and Myomorpha (ca. 1200 species). Most hystricomorph rodents are found in South America (45 genera, ca. 140 species) where they have evolved in isolation since the Eocene; the rest occur in Africa and Asia, except for the North American porcupine. There is much controversy about the relationship between the Old World and New World forms (see References 2 and 3); however, whether or not they are descended from a common ancestor, it is now customary to distinguish the New World group as the Caviomorpha[4] without necessarily giving the taxon subordinal status.

Apart from the guinea pig (dealt with elsewhere in this book), the chinchilla and the coypu (nutria) are the most commonly recognized members of this suborder outside South America because both are farmed commercially in many countries for their pelts. In recent years, however, many other species have been kept in captivity to extend knowledge of the group in order to determine suitability as new laboratory species[5] and to investigate further the many interesting physiological features.[6-15]

There is little detailed information about the diets of these rodents in the wild, but what is known is reported by Walker[16] and in Table 1. The rodents of this suborder are generally more vegetarian than some myomorphs (e.g., rats and mice) and have an almost totally vegetable diet, although the placenta and fetal membranes are normally ingested and cannibalism of sick animals has been observed. The readiness with which hystricomorphs and caviomorphs eat vegetable matter facilitates their feeding in captivity, but their continuous feeding habit increases their susceptibility to obesity when exercise is limited. A list of hystricomorphs and caviomorphs which, as far as I know, have been kept in some sort of laboratory environment is given in Table 1. The details of the diets given in the various papers cited inevitably vary according to place and circumstances. Therefore, the following account relates only to those species that we have kept in London and Cambridge (U.K.) and is intended as a guide for establishing colonies of these or other hystricomorphs or caviomorphs. The diets described are those that were found to be best for our animals after feeding selected groups differently and assessing growth and reproduction; they do not necessarily represent the minimum diet. A wide variety of foods is usually offered to a newly introduced laboratory species, and it is only when a colony is well established that feeding trials can be undertaken without endangering breeding potential. However, because the source of further animals is remote (over 11,000 km for many species of caviomorph) and our primary interest has been in reproductive[6,7,17] rather than nutritional physiology, detailed tests on dietary requirements have not been performed.

Chinchilla

These animals usually averaging 500 g in weight have been bred in captivity for about 50 years. Most chinchilla are kept commercially for their pelts, although live sales for breeding stock and pets are an important aspect of their rearing. Ranch chinchilla usually receive a commercial pelleted diet; the composition varies according to the manufacturer, but it usually includes ground grains and grain by-products, protein supplement of plant or animal origin, alfalfa fiber, minerals, vitamins, and antibiotics. Some ranchers may feed commercial rabbit diet, and chinchilla in my laboratory have survived and bred well when

Table 1
A REFERENCE LIST OF PAPERS ON THE HYSTRICOMORPH AND CAVIOMORPH RODENTS IN WHICH MENTION IS MADE OF DIETS IN THE WILD AND IN CAPTIVITY

Family and species	Common name	References: Diet in wild	References: Diet in captivity
	Hystricomorpha		
Hystricidae			
Hystrix cristata	African porcupine	44, 65	37
Atherurus africanus	Brush-tailed porcupine		45, 46
Thryonomyidae			
Thryonomys swinderianus	Cane rat, grasscutter	47, 48	49
Bathyergidae	Mole rats	50	51
Cryptomys hottentotus	Common mole rat	52	53, 54
Heterocephalus glaber	Naked mole rat		55
	Caviomorpha		
Erethizontidae			
Erethizon dorsatum	North American porcupine	56–59	60
Coendou prehensilis	Brazilian tree porcupine	61, 62	
Caviidae			
Cavia spp. (excluding *C. porcellus* but including hybrids)	Cavies	63, 66, 67	64, 65, 68–72
Galea musteloides	Cuis	63	65, 68
Microcavia australis	Desert cavy	63, 73	64
Dolichotis patagonum	Mara		74–77
Pediolagus salinicola	Salt desert cavy		77
Hydrochoeridae			
Hydrochoerus spp.	Capybara	78	79–81
Chinchillidae			
Chinchilla laniger	Chinchilla	82–84	19, 37, 68
Lagostomus maximus	Plains viscacha	28–31, 85–87	32, 65, 68
Lagidium peruanum	Mountain viscacha	34	
Abrocomidae			
Abrocoma cinerea	Chinchilla rat		77
Dasyproctidae			
Dasyprocta spp.	Agouti	35, 88	37, 89
Myoprocta spp.	Acouchi		37, 90–92
Cuniculus paca	Paca	62, 93, 117	77, 94
Dinomyidae			
Dinomys branickii	Pacarana		95, 96
Ctenomyidae			
Ctenomys spp.	Tuco-tuco	30, 31, 85, 97–100	11, 43, 101
Octodontidae			
Octodon degus	Degu	102	65, 77, 102, 104
Spalacopus cyanus	Coruro	105	77
Octodontomys gliroides	Chozchoz		77
Capromyidae			
Myocastor coypus	Coypu, nutria	20–22	68, 106–109
Capromys spp.	Hutia	110, 111	
Geocapromys spp.	Hutia	112, 113	114
Plagiodontia aedium	Hispaniolan hutia		115
Echimyidae		116, 117	
Proechimys spp.	Spiny rat, casiragua	118	119–121, 126
Hoplomys gymnurus	Spiny rat	122, 123	119
Diplomys darlingi	Spiny rat	124	

fed a diet formulated for guinea pigs (e.g., "Diet 18," Dixon & Sons, Ware, Herts, U.K.). Pellets are usually available to the animals at all times, even if this necessitates a special food hopper to prevent the animals from wasting the food through the mesh floor of the cage.[18] Some ranchers alter the diet or give additional food for pregnant and/or weanling animals, but this should not be necessary. The feeding of tidbits such as raisins, apples, sunflower seeds, carrots, and greens is a matter of choice; it is rarely possible under ranch conditions (because of expense and time) but is frequently done for pet chinchilla. Small amounts of these substances do not appear to be detrimental, but they could be if adequate roughage is not provided in the diet. Although some commercial breeders have successfully raised chinchilla without feeding them hay, the majority still do so and the chinchilla clearly enjoy it. It is not necessary, as it is for guinea pigs, to give chinchilla green feed.

If young chinchilla have to be hand-reared, they do well on goat's milk or evaporated cow's milk diluted with water and with added glucose. As the baby grows the milk can be given undiluted, but feeding[18,19] should be continued for 6 to 8 weeks even though solid food is eaten from about 4 weeks of age.

Coypu (Nutria)

To do well in captivity coypu need to have access to water for swimming, defecating, and washing their food.[68] Food normally consists of a mixture of root crops (such as carrots, potatoes, turnips, swedes), and cabbage and commercial pellets of the rat type (e.g., "Diet 86" or "Diet 41B," Dixon & Sons, Ware, Herts, U.K.). Bulk may be added to the diet in the form of bread, but we prefer to feed large dog biscuits which the coypu like to chew. Hay is not required with this sort of diet, although it probably would be if it were not possible to feed them vegetables for some time.

The diets of commercially farmed coypu vary according to the locality and conditions of husbandry. Where the animals are kept in seminatural conditions (e.g., in Louisiana,[20] North Carolina,[21] or the U.S.S.R.[22]), they eat the available vegetation such as reeds, submerged parts of water plants, grasses, and fungi. However, supplementary feeding is also given (its nature again depending on availability), and foods such as hay, maize, barley, sunflower seed cake, potato, beet fodder, and other root crops are provided, especially in winter. It has been estimated that coypu weighing 4 to 5 kg and living in nearly natural conditions and at a density of two animals per hectare will eat 0.7 to 2.0 kg of food per day.[22] Vitamin requirements are said to be 200 to 300 IU of vitamin A or 0.24 to 0.36 mg of carotene per 100 kcal/day[23] and 3 to 5 mg of pyridoxine per day.[24]

Cavies

In the wild, cavies are distributed throughout much of South America, from the pampas of Argentina to the altiplano of Peru and the campos of Brazil.[25-27] Their diet usually consists of grass supplemented with seeds, lichen, bark, fruits, and other plants when available.

In addition to the domestic guinea pig, we have bred two species of the subfamily Caviinae – the native guinea pig (*Cavia*) and the cuis (*Galea*). Animals of both species were given guinea pig compound diet pellets and hay daily and cabbage twice a week. Occasionally carrots and potatoes were also provided, but this was to add variety to the diet. There is no experimental evidence showing that other cavies resemble *Cavia porcellus* in being unable to synthesize vitamin C. Cuis will sometimes attack and eat dying young.

Plains Viscacha

The feeding habits of these large chinchillids (3 to 8 kg) on the pampas of Argentina and Uruguay have been described by a number of authors.[28-31] Animals have been kept alive for 6 months on a diet consisting only of oats and water,[32] but our animals are supplied with hay daily and rat pellets (e.g., Diet 41B) and large dog biscuits 3 (or 4) days a week; a vegetable mixture is given on alternate days (or twice a week). On 1 day each week (usually Sunday) the animals are given only hay, which is available from a hopper. This fast day has been found to be beneficial since viscacha tend to become obese in captivity. Viscacha kept as pets reveal widely different food choices, from ice cream to bread and butter or roast potatoes and melon. An adult viscacha eats about 100 g of the pelleted rat cubes per day, which is about 3% of its body weight (cf. a mouse which consumes about 20% of its body weight in food daily). Like rabbits, viscacha are cecotrophic although only one type of fecal pellet is produced.

Mountain Viscacha

In appearance these chinchillids resemble chinchilla more closely than they resemble plains viscacha.[33] Our animals preferred the smaller guinea pig-type pellet (e.g., Diet 18) over the larger rat type and were given a seed, vegetable, and fruit mixture in addition to hay. Mountain viscacha live at high altitude or high latitude; therefore, their natural diet is limited to grasses, lichens, and bark.[34]

Degu

These small octodontid (80 to 200 g) rat-like animals grow and breed when fed only a pelleted rat diet (e.g., Diet 41B). However, they are particularly prone to obesity on such a concentrated diet and, like the plains viscacha, benefit from fasting 1 day each week and eating a vegetable mixture instead of pellets for 2 or 3 days of the week. A small amount of hay is provided and a few sunflower seeds (about 3 to 4 per animal) are given with the pelleted food. Moribund young are sometimes attacked and eaten, and during lactation pellets are usually provided every day.

Some other octodontids (chozchoris) were kept for over 1 year. The animals were offered a seed and pellet mixture ad libitum and fruit and vegetables daily, but the incidence of impaction in the animals was high, suggesting that this diet was not appropriate.

Agouti and Acouchi

These dasyproctids are commonly kept in zoological gardens, where a variety of foods may be provided. Our animals thrived on a diet of compound rat pellets (e.g., Diet 41B) ad libitum and a fruit and vegetable mixture given daily. This is almost certainly not a minimum diet, but dasyproctids eat fruit in their native subtropical forests[35] and the seasonal changes in the fruits available have been suggested as a factor regulating breeding in these animals. Acouchis breed readily in captivity[36-38] but agoutis do not, at least not under restricted conditions.[37,39,40]

Casiragua

These echimyid rodents (often called "spiny rats" because of a very rat-like appearance) can be kept successfully under the same sort of laboratory conditions as rats. A diet of compound rat pellets (e.g., Diet 41B) and water only is adequate for growth and some reproduction. Atlhough the assessment of different diets was not made in the context of a formally designed experiment, reproduction of animals fed only a pelleted diet was less successful than that of animals also given hay daily and a vegetable mixture 2 or 3 times a week in that fewer females were involved in breeding. The hay is eaten by the animals and is also used for shelter, even if boxes are available. Sick and dying animals are often attacked, but only young animals are eaten.

Tuco-tuco

The members of at least one species of tuco-tuco (*Ctenomys talarum*) are of great interest nutritionally. In captivity, tuco-tucos of this species have been shown to develop symptoms resembling those of diabetes mellitus.[11,41–43] The condition (marked by high blood glucose levels, cataracts, sudden death, and infertility) appears to be due to the increased food intake and reduced exercise in captivity. When tuco-tucos were first caught they were offered a variety of food (for the reasons stated in the Introduction above). It was found that diabetic symptoms developed when animals were given a daily diet of 5 g of compound rat pellets (Diet 86), 150 g of cabbage and carrot, and 200 g of hay. The removal of pellets from the diet was sufficient to cause a fall in blood glucose levels; however, even on a pellet-free diet from the time of capture, it was impossible to prevent tuco-tucos from becoming diabetic.[125] In an activity wheel, our tuco-tucos "ran" about 6 km/day, but this type of exercise, or even that encouraged by housing in extensive cages, was not as effective as burrowing in a large tank of earth in preventing the onset of some symptoms of the disease, particularly infertility.

REFERENCES

1. **Simpson, G. G.,** *Bull. Am. Mus. Nat. Hist.,* 85, 1–35, 1945.
2. **Lavocat, R.,** *Symp. Zool. Soc. London,* 34, 7–20, 1974.
3. **Wood, A. E.,** *Symp. Zool. Soc. London,* 34, 21–60, 1974.
4. **Simpson, G. G.,** *Symp. Zool. Soc. London,* 34, 1–5, 1974.
5. **Rowlands, I. W.,** *WHO Chron.,* 24, 169–172, 1970.
6. **Weir, B. J.,** *Symp. Zool. Soc. London,* 34, 265–301, 1974.
7. **Weir, B . J. and Rowlands, I. W.,** *Symp. Zool. Soc. London,* 34, 303–332, 1974.
8. **Tam, W. H.,** *Symp. Zool. Soc. London,* 34, 363–384, 1974.
9. **Heap, R. B. and Illingworth, D. V.,** *Symp. Zool. Soc. London,* 34, 385–415, 1974.
10. **Neville, R. W. J., Weir, B. J., and Lazarus, N. R.,** *Symp. Zool. Soc. London,* 34, 417–435, 1974.
11. **Wise, P. H., Weir, B. J., Hime, J. M., and Forrest, E.,** *Nature,* 219, 1374–1376, 1968.
12. **Kleiman, D. G.,** *Symp. Zool. Soc. London,* 34, 171–209, 1974.
13. **Eisenberg, J. F.,** *Symp. Zool. Soc. London,* 34, 211–247, 1974.
14. **George, W. and Weir, B. J.,** *Symp. Zool. Soc. London,* 34, 79–108, 1974.
15. **Roberts, C. M. and Perry, J. S.,** *Symp. Zool. Soc. London,* 34, 333–360, 1974.
16. **Walker, E. P., Ed.,** *Mammals of the World,* 3rd ed., Johns Hopkins, Baltimore, 1974, 1004–1077.
17. **Weir, B. J. and Rowlands, I. W.,** *Annu. Rev. Ecol. Syst.,* 4, 139–163, 1973.
18. **Weir, B. J.,** in *UFAW Handbook on the Care and Management of Laboratory Animals,* 4th ed., Churchill Livingstone, Edinburgh, 1972, 269–277.
19. **Weir, B. J.,** in *Reproduction and Breeding Techniques for Laboratory Animals,* Hafez, E. S. E., Ed., Lea & Febiger, Philadelphia, 1970, 209–223.
20. **Adams, W. H.,** *Proc. La. Acad. Sci.,* 19, 28–41, 1956.
21. **Milne, R. C. and Quay, T. L.,** *Proc. 20th Annu. Conf. State Assoc. Game and Fish Commission,* 1966, 112–123.
22. **Vinograd, V. V.,** *Zool. Zh.,* 44, 1712–1721, 1965.
23. **Kladovschikov, V. F.,** *Krolikovod. Zverovod.,* 9, 18–19, 1964.
24. **Karkalickij, I. M., Karkalickaja, G. V., Asihmina, E. M., Kovrizhnyh, N. D., Tuzova, G. P., Plotnikova, G. F., and Bardnikov, M. P.,** *Vopr. Pitan.,* 29, 23–28, 1970.
25. **Tate, G. H. H.,** *Bull. Am. Mus. Nat. Hist.,* 68, 295–447, 1935.
26. **Cabrera, A.,** *Rev. Mus. Argent. Cienc. Nat. Bernadino Rivadavia Inst. Nac. Invest. Cienc. Nat.,* 4, 1–724, 1957–61.
27. **Cabrera, A. and Yepes, J.,** *Mamíferos Sud Americanos,* 2nd ed., Historia Natural Ediar, Buenos Aires, 1960.
28. **Hudson, W. H.,** *Proc. Zool. Soc. London,* 1872, 822–833, 1872.
29. **Llanos, A. C. and Crespo, J. A.,** *Rev. Invest. Agric.,* 6, 289–378, 1952.
30. **Barlow, J. C.,** *R. Ont. Mus. Life Sci. Contrib.,* 75, 1–59, 1969.
31. **Weir, B. J.,** *Symp. Zool. Soc. London,* 34, 113–130, 1974.
32. **Giusti, L. and Hug, E.,** *C. R. Soc. Biol.,* 87, 688–689, 1972.
33. **Rowlands, I. W.,** *Symp. Zool. Soc. London,* 34, 131–141, 1974.
34. **Pearson, O. P.,** *J. Mammal.,* 29, 345–375, 1948.
35. **Smythe, N.,** *Am. Nat.,* 104, 25–35, 1970.

36. Kleiman, D. G., *J. Reprod. Fertil.*, 23, 55–65, 1970.
37. Weir, B. J., *Lab. Anim.*, 1, 95–104, 1967.
38. Weir, B. J., *J. Reprod. Fertil.*, 24, 193–201, 1970.
39. Crandall, L. S., *Anim. Kingdom*, 47, 125, 1944.
40. Weir, B. J., *J. Reprod. Fertil.*, 24, 203–211, 1970.
41. Weir, B. J., Wise, P. H., Hime, J. M., and Forrest, E., *J. Endocrinol.*, 43, 7–8, 1969.
42. Wise, P. H., Weir, B. J., Hime, J. M., and Forrest, E., *Diabetologia*, 8, 165–172, 1972.
43. Weir, B. J., *Proc. Roy. Soc. Med.*, 67, 843–846, 1974.
44. Mohr, E., *Neue Brehm Buch.*, 350, 1–164, 1965.
45. Rahm, U., *Acta Trop.*, 13, 63–68, 1956.
46. Rahm, U., *Mammalia*, 26, 1–9, 1962.
47. Rosevear, D. R., *The Rodents of West Africa*, British Museum (Natural History), London, 1969.
48. Asibey, E. O. A., *Symp. Zool. Soc. London*, 34, 161–170, 1974.
49. Asibey, E. O. A., *Symp. Zool. Soc. London*, 34, 251–263, 1974.
50. Delaney, M. J., *Mammal Rev.*, 2, 1–42, 1972.
51. Jarvis, J. U. M. and Sale, J. B., *J. Zool.* (London), 163, 451–479, 1971.
52. Genelly, R. E., *J. Mammal.*, 46, 647–665, 1965.
53. Hill, W. C. O., Porter, A., Broom, R. T., Seago, J., and Southwick, M. D., *Proc. Zool. Soc. London*, 128, 455–514, 1955.
54. Bateman, J. A., *Afr. Wildl.*, 13, 65–71, 1959.
55. Davis, D. H. S., *S. Afr. J. Med. Sci.*, 28, 53–69, 1963.
56. Murie, O., *J. Mammal.*, 7, 109–113, 1926.
57. Taylor, W. P., *Univ. Ariz. Bull.*, 6, 1–177, 1935.
58. Shapiro, J., *J. Mammal.*, 30, 247–257, 1949.
59. Gabrielson, I. N., *J. Mammal.*, 9, 33–39, 1928.
60. Shadle, A. R., *J. Mammal.*, 31, 411–420, 1950.
61. Krieg, H., *Z. Morphol. Oekol. Tiere*, 15, 75–86, 1929.
62. Dalquest, W. W., *La. State Univ. Stud. Biol. Sci. Ser.*, 1, 1–229, 1953.
63. Rood, J. P., *Anim. Behav. Monogr.*, 5, 1–83, 1972.
64. Rood, J. P. and Weir, B. J., *J. Reprod. Fertil.*, 23, 393–409, 1970.
65. Weir, B. J., *Lab. Anim.*, 4, 83–97, 1970.
66. Barera, J. M. de la, *Rev. Inst. Bacteriol. Dep. Nac. Hig. Argent.*, 7, 439–506, 1936.
67. Pearson, O. P., *Bull. Mus. Comp. Zool. Harv. Univ.*, 106, 117–174, 1951.
68. Weir, B. J., in *UFAW Handbook on the Care and Management of Laboratory Animals*, 4th ed., Churchill Livingstone, Edinburgh, 1972, 278–286.
69. Detlefson, J. A., *Publ. Carnegie Inst.*, 205, 1–134, 1914.
70. Castle, W. E. and Wright, S., *Publ. Carnegie Inst.*, 241, 1–129, 1916.
71. Ubisch, G. and Mello, R. F., *J. Hered.*, 31, 389–398, 1940.
72. Pictet, A. and Ferero, A., *Genetica*, 25, 357–515, 1951.
73. Rood, J. P., *Am. Midl. Nat.*, 83, 415–454, 1970.
74. Kirchshofer, R., *Z. Saeugetierkd.*, 25, 112–127, 1960.
75. Mohr, E., *Zool. Gart.*, 16, 111–133, 1949.
76. Rosenthal, M. A., *Int. Zoo Yearb.*, 14, 214–215, 1974.
77. Kleiman, D. G., National Zoological Park, Smithsonian Institution, Washington, D.C. (see Reference 12).
78. Dubost, G., *Terre Vie*, 22, 3–28, 1968.
79. Trapido, H., *J. Mammal.*, 30, 433–436, 1949.
80. Ojasti, J. and Padilla, C. M., *Trans. North Am. Wildl. Nat. Resour. Conf.*, 37, 268–277, 1972.
81. Zara, J. L., *Int. Zoo Yearb.*, 13, 137–139, 1973.
82. Bidlingmaier, T. C., *J. Mammal.*, 18, 159–163, 1937.
83. Pavlov, M. P., Kornilov, B. E., Kiris, I. B., and Baranceva, E. D., *Krolikovod. Zverovod.*, 4(8), 16–18, 1961; 4(9), 12–15, 1961.
84. Thomasen, H. B., *Dtsch. Pelztier.*, 40, 190–192, 1966.
85. Eisentraut, M., *Z. Saeugetierkd.*, 8, 47–69, 1933.
86. Fernandez, M., *Bol. Acad. Nac. Cienc. Cordoba, Argent.*, 38, 347–378, 1949.
87. Gibson, E., *Proc. Nat. Hist. Soc. Glasgow*, 3, 136–140, 1877.
88. Enders, R. K., *Bull. Mus. Comp. Zool. Harv. Univ.*, 78, 383–502, 1935.
89. Roth-Kolar, H., *Z. Tierpsychol.*, 14, 363–375, 1957.
90. Wayre, P. and Wood, J. G. P., *Int. Zoo Yearb.*, 3, 25–27, 1961.
91. Morris, D., *Proc. Zool. Soc. London*, 139, 701–732, 1962.
92. Kleiman, D. G., *J. Reprod. Fertil.*, 23, 55–65, 1970.
93. Hingston, R. W. A., *A Naturalist in the Guiana Forest*, Edward Arnold, London, 1932.

94. **Kraus, C., Gihr, M., and Pilleri, G.,** *Rev. Suisse Zool.*, 77, 333–338, 1970.
95. **Mohr, E.,** *Zool. Garten,* 9, 204–209, 1937.
96. **Collins, L. R. and Eisenberg, J. F.,** *Int. Zoo Yearb.*, 12, 108–114, 1972.
97. **Darwin, C.,** *Journal of a Voyage Round the World,* 2nd ed., Nelson & Sons, London, 1890.
98. **Talice, R. V. and Momigliano, E.,** *Rev. Fac. Human. Cienc. Montevideo,* 13, 5–21, 1954.
99. **Pearson, O. P.,** *Mem. Mus. Hist. Nat. Javier Prado,* 9, 1–56, 1959.
100. **Pearson, O. P., Binzstein, N., Boiry, L., Busch, C., di Pace, M., Gallopin, G., Penchaszadeh, P., and Piantanida, M.,** *Invest. Zool. Chil.*, 13, 47–80, 1968.
101. **Talice, R. V. and Momigliano, E.,** *Arch. Soc. Biol. Montevideo,* 21, 179–185, 1954.
102. **Osgood, W. H.,** *Publ. Field Mus. Nat. Hist. Zool. Ser.*, 30, 1–268, 1943.
103. **Boraker, D. K.,** University of Vermont, College of Medicine, Burlington, Vt.
104. **Barros, C.,** Laboratorio de Embriologia, Instituto de Ciencias Biologicas, Universidad Catolica de Chile, Casilla 114–D, Santiago, Chile.
105. **Reig, O. A.,** *J. Mammal.*, 51, 592–601, 1970.
106. **Illman, O.,** *J. Anim. Tech. Assoc.*, 12, 8–10, 1961.
107. **Flack, M. B.,** *J. Anim. Tech. Assoc.*, 14, 145–156, 1964.
108. **Perry, J. S.,** Agricultural Research Council Institute of Animal Physiology, Babraham, Cambridge, Engl.
109. **Gosling, L. M.,** Coypu Laboratory, Government Buildings, Mile Cross Lane, Norwich, Engl.
110. **Allen, J. A.,** *Bull. Am. Mus. Nat. Hist.*, 3, 329–336, 1896.
111. **Bucher, G. C.,** *Mem. Soc. Cuban Hist. Nat.*, 11, 93–107, 1937.
112. **Clough, G. C.,** *J. Mammal.*, 53, 807–823, 1972.
113. **Mittermeier, R. A.,** *Oryx J. Fauna Preserv. Soc.*, 11, 258–262, 1972.
114. **Howe, R. and Clough, G. C.,** *Int. Zoo Yearb.*, 11, 89–93, 1971.
115. **Eisenberg, J. F.,** National Zoological Park, Smithsonian Institution, Washington, D.C. (see Reference 13).
116. **Davis, D. E.,** *Bol. Mus. Nac. Rio de Janiero Zool.*, 76, 1–8, 1947.
117. **Fleming, T. H.,** *J. Mammal.*, 51, 473–490, 1970.
118. **Moojen, J.,** *Kansas Publ. Mus. Nat. Hist.*, 1, 301–406, 1948.
119. **Tesh, R. B.,** *J. Mammal.*, 51, 199–202, 1970.
120. **Malliniak, E. and Eisenberg, J. F.,** *Int. Zoo Yearb.*, 11, 93–98, 1971.
121. **Weir, B. J.,** *Lab. Anim.*, 7, 125–134, 1973.
122. **Pine, R. H. and Carter, D. C.,** *J. Mammal.*, 51, 804, 1970.
123. **Buchanan, O. M. and Howell, T. R.,** *Ann. Mag. Nat. Hist.*, (13)8, 549–559, 1965.
124. **Tesh, R. B.,** *J. Mammal.*, 51, 197–199, 1970.
125. **Weir, B. J.,** unpublished work.
126. **Neville, R. W. J.,** Diabetes Research Unit, Burroughs Wellcome, Dartford, Kent, Engl.

NATURAL AND SEMIPURIFIED DIETS FOR GUINEA PIGS

D. L. Bray, P. S. Reyes and G. M. Briggs

INTRODUCTION

The guinea pig, *Cavia porcellus,* is the domesticated species of the genus *Cavia.* The newborn guinea pig weighs about 70 to 100 g, depending on the number in the litter, usually from one to four. Guinea pigs are so well developed at birth that they can be weaned within 2 to 4 days,[11] but the weanling guinea pigs used in most nutrition experiments are 2 to 3 weeks old and weigh about 200 to 250 g.

Metal cages with wire mesh floors are recommended housing for guinea pigs in studies using semipurified diets. The cages should be washed and sterilized at least once a week, and more often if diarrhea is present. Lice are often a problem, but can be controlled by monthly dusting with a pyrethrin-containing powder. Room temperatures between 70 and 80°F seem optimal. Fresh feed and clean water should be provided daily.

The guinea pig is one of the few animals requiring a dietary source of vitamin C; it is, therefore, often used in studies concerning this vitamin. Ground rabbit stock ration is a convenient scorbutic diet for guinea pigs; vitamin C can easily be added for a control diet. The exact requirement has not yet been established, but suggested dietary levels range from 0.02% to 0.2%.[7,11,12]

Weight gains may vary with the guinea pig strain, source of the animals, housing conditions, sex, initial weight of the animals, and diet composition. Weanlings weighing between 100 to 200 g and fed a natural-ingredient stock diet, such as Purina Guinea Pig Chow (see Table 1), usually gain 7 to 9 g/day, whereas those fed a semipurified diet (see Table 3) gain 4 to 6 g/day. Animals weighing 90 to 110 g show up to 30% increase in weight gain when 15% dehydrated alfalfa is added to the diet,[5] suggesting that the very young guinea pig may require an unidentified "plant factor."[1,5,7] Guinea pigs attain adult weights from 800 to 1,400 g in 4 to 6 months.[11]

Although the pelleted form in which a stock diet is provided seems to be preferred, guinea pigs will eat ground stock diet, which offers more appropriate control for powdery semipurified diets. Conversely, semipurified diets may be pelleted to increase acceptability, but nutritional losses must be monitored; a gel diet, which avoids this problem, has recently been formulated.[8]

Although not all the nutritional requirements for guinea pigs have as yet been determined, two typical semipurified diets that allow satisfactory growth are presented in Table 3. To insure homogeneous dispersion of all ingredients, vitamin and mineral premixes are often made. Previously weighed vitamins are ground together with a mortar and pestle, then thoroughly mixed with a carrier (we use the glucose hydrate Cerelose®). The mixtures may be stored in brown bottles under refrigeration up to 4 months. Minerals are ground with mortar and pestle or, if several kilograms are made at once, mixed in an electric mixer. Mineral mixes are stable indefinitely at room temperature. Mixing of a diet is most thorough if the fat is first blended with a large portion of the carbohydrate; the vitamins may then be mixed with the remaining carbohydrate, and the protein with the minerals. All ingredients are then mixed by hand or placed together in an electric mixer for 10 to 20 min, until homogeneous. Diets should be kept under refrigeration until each daily use, but not longer than 3 months.

For further information on guinea pig care and physiology, see References 6 and 11.

Table 1
A TYPICAL COMMERCIAL PELLETED DIET FOR GUINEA PIGS[a]

A. Ingredients

Ground wheat, ground yellow corn, dried skimmed milk, soybean meal, dehydrated alfalfa meal, wheat middlings, animal fat preserved with BHA, vitamin B_{12} supplement, riboflavin supplement, choline chloride, calcium pantothenate, niacin, folic acid, ascorbic acid, thiamine, vitamin A supplement, D-activated plant sterol, methionine hydroxy analog calcium, vitamin E supplement, dicalcium phosphate, calcium carbonate, iodized salt, magnesium oxide, manganous oxide, cobalt carbonate, zinc oxide.

B. Approximate Chemical Composition[b]

Nutrients[c]

Protein, %	18.5	Gross energy, kcal/g	4.2	Vitamins	
Arginine, %	1.04	Ash, %	8.5	Carotene, ppm	35.9
Cystine, %	0.27	Calcium, %	1.08	Menadione (added), ppm	–
Glycine, %	0.86	Phosphorus, %	0.76	Thiamine, ppm	3.5
Histidine, %	0.47	Potassium, %	1.36	Riboflavin, ppm	5.0
Isoleucine, %	0.95	Magnesium, %	0.35	Niacin, ppm	50.0
Leucine, %	1.56	Sodium, %	0.40	Pantothenic acid, ppm	19.0
Lysine, %	1.01	Chlorine, %	0.50	Choline, ppm × 100	18.5
Methionine, %	0.40	Fluorine, ppm	–	Folic acid, ppm	4.2
Phenylalanine, %	0.91	Iron, ppm	298.7	Pyridoxine, ppm	4.0
Threonine, %	0.74	Zinc, ppm	121.8	Biotin, ppm	0.30
Tryptophan, %	0.25	Manganese, ppm	120.7	B_{12}, µg/lb	6.0
Valine, %	0.98	Copper, ppm	20.7	Vitamin A, IU/g	30.0
Fat, %	4.2	Cobalt, ppm	2.8	Vitamin D, IU/g	3.4
Fiber, %	11.5	Iodine, ppm	1.6	α-Tocopherol, IU/lb	28.0
TDN, %	70			Ascorbic acid, mg/g	1.0
NFE (by difference), %	47.3				

[a] Purina Guinea Pig Chow information supplied by Ralston Purina Company, St. Louis, Missouri.

[b] Based on the latest ingredient analysis information. Since nutrient composition of natural ingredients varies, analysis will differ accordingly. Guinea Pig Chow contains a minimum of 18% protein and 4% fat, and a maximum of 16% fiber.

[c] Nutrients are expressed as percent of ration, except where otherwise indicated. Moisture content, though variable, is assumed to be 10.0% for the purposes of calculations.

Table 2
NATURAL-INGREDIENT DIET FOR GUINEA PIGS USED BY THE NATIONAL INSTITUTES OF HEALTH[12]

Ingredient	g/kg diet
Oats, whole, ground fine	177
Wheat, whole, ground fine	289
Alfalfa meal[a]	380
Soybean meal[b]	132.5
Vitamin D_2 premix (1,730,000 IU/kg)	1.0
Ascorbic acid	0.5
Sodium chloride, iodized	5.0
Limestone, ground	11.0
Dicalcium phosphate, minimum of 0.2% fluorine	2.5
Dela mix[c]	1.5
Total	1,000.0[d]

[a] Dehydrated, 17% crude protein.

[b] Dehulled, solvent-extracted, containing not less than 49% protein.

[c] A trace mineral mix produced by Limestone Products Company, Newton, N. J. Contains not less than 6% Mn, 2% Fe, 0.2% Cu, 0.12% I, 0.02% Co, and 26.5% Ca.

[d] Finished product at time of delivery should conform to the following calculated standards (not less than): 18% crude protein, 2.25% crude fat, 13.50% crude fiber, 48% nitrogen-free extract (NFE), 8.2% ash, 1.2% calcium, 0.4% phosphorus, 0.15% sodium, 1 ppm iodine, 80 ppm carotene, 40 ppm niacin, 4 ppm thiamine, 15 ppm pantothenic acid, 6 ppm riboflavin, 0.15 ppm biotin, 40 IU/g α-tocopherol, 2 IU/g vitamin D, and 450 ppm vitamin C.

From Reid, M. E. and Bieri, J., in *Nutrient Requirements of Laboratory Animals,* National Research Council, Agricultural Board, National Academy of Sciences, Washington, D.C., 1972. With permission.

Table 3
SATISFACTORY SEMIPURIFIED DIETS FOR GUINEA PIGS[a]

Ingredient	O'Dell et al.,[2] g/kg diet	Briggs GPB-50A,[b] g/kg diet
Casein	300	300
Corn starch	–	155
Sucrose	–[c]	131
Glucose hydrate (Cerelose®)	–	122
Fiber[d]	150	150
Vegetable oil[e]	40	30
Mineral mix[f]	70	75
Water-soluble vitamin mix[g]	–[g]	20
Fat-soluble vitamin mix[h]	–[h]	10
Ascorbic acid	2	2
Arginine HCl	–	3
Choline chloride	1	2
Chlortetracycline, mg/kg	2.5	–
Total	1,000.0	1,000.0

[a] Not recommended for guinea pigs 3 to 21 day old unless 15% dehydrated alfalfa is added.
[b] This diet is used for most guinea pig research in the Department of Nutritional Sciences at the University of California, Berkeley.
[c] Sucrose is added to make up to 1 kg.
[d] Solka Floc is recommended; may be obtained from Brown Company, 650 Main Street, Berlin, NH 03570.
[e] Briggs: corn oil; O'Dell: soybean oil.
[f] See Table 6.
[g] See Table 4.
[h] See Table 5.

Table 4
WATER-SOLUBLE VITAMINS IN GUINEA PIG DIETS[a]

Ingredient	1972 NRC,[b] mg/kg diet	O'Dell et al.,[2] mg/kg diet	Briggs GPWSV-50A[c] mg/kg mix	Briggs GPWSV-50A[c] mg/kg diet
Thiamine HCl	16	10	800	16
Riboflavin	16	10	800	16
Pyridoxine HCl	16	10	800	16
Ca pantothenate	(20)	30	2,000	40
Niacin	50 (20)	50	10,000	200
Folic acid	10 (6)	6	500	10
Biotin	–[d]	0.2	25	0.5
Vitamin B_{12}	–[d]	0.03	2.5[e]	0.05[f]
Inositol	–[d]	1,000	–	–
Cerelose®	–	–	to 1,000	–

[a] Ascorbic acid and choline are added separately; see Table 3.
[b] NRC states that the values in parentheses are tentative estimates of the minimum requirement and contain no margin of safety.
[c] Guinea pig water-soluble vitamins used in Briggs GPB-50A (see Table 3), 20 g mix/kg diet.
[d] Requirement not known.
[e] 2.5 g of 0.1% vitamin B_{12} triturated in mannitol.
[f] 50 mg of 0.1% vitamin B_{12} triturated in mannitol.

Table 5
FAT-SOLUBLE VITAMINS IN GUINEA PIG DIETS

Ingredient	1972 NRC, amt/kg diet	O'Dell et al.,[2] amt/kg diet	Briggs GPFSV-50A[a] amt/kg mix	Briggs GPFSV-50A[a] amt/kg diet
Vitamin A	20,000 IU[b]	20,000 IU[b]	6,960 mg vitamin A palmitate (250,000 IU/g) (1,740,000 IU)	69.6 mg (17,400 IU)
α-Tocopherol	60 mg[c]	20 mg[c]	12,000 mg *dl*-α-tocopherol (250 IU/g) (3,000 IU)	120 mg (30 IU)
Vitamin D_3	–[d]	2,800 IU	400 mg vitamin D_3 (400,000 IU/g) (160,000 IU)	4 mg (1,600 IU)
Cerelose	–	–	to 1,000	

[a] Guinea pig fat-soluble vitamin mix used in Briggs GPB-50A (see Table 3), 10 g/kg diet.
[b] Weights are not indicated.
[c] Potencies are not indicated.
[d] NRC states that vitamin D may be required in diets with unsatisfactory calcium-to-phosphorus ratio.

Table 6
MINERAL MIXTURES FOR GUINEA PIGS[a]

		O'Dell et al.,[2]		Briggs GPS-3[7]	
Ingredient	Unit	amt/kg mix	amt/kg diet[b]	amt/kg mix	amt/kg diet[c]
$CaHPO_4$	g	294	20.6	388	29.1
$CaCO_3$	g	130	9.10	66	4.95
Na_2HPO_4	g	93.1	6.52	–	–
NaCl	g	37.5	2.63	64	4.80
K acetate	g	357	25[d]	277	20.8
KCl	g	–	–	86	6.45
MgO	g	71.4	5.00	55	4.125
$MgSO_4$	g	–	–	51.1	3.83
$MnSO_4 \cdot H_2O$	g	9.1	0.64	4.1	0.3075
$ZnCO_3$	g	–	–	1.43	0.1073
$ZnSO_4$	g	2.1	0.145	–	–
$FeC_6H_5O_7 \cdot 5H_2O$	g	5.3	0.37	7.1	0.533
$CuSO_4$	mg	290	20	56	4.2
KIO_3	mg	230	16	170	12.8
$KCr(SO_4)_2 \cdot 12H_2O$	mg	–	–	32.0	2.4
$Na_2MoO_4 \cdot 2H_2O$	mg	–	–	1.67	0.125
$NiCl_2 \cdot 6H_2O$	mg	–	–	0.67	0.05
Na_2SeO_3	mg	–	–	0.67	0.05
Total	g	1,000.0	70.0	1,000.0	75.0

[a] HMW, Hegsted, and Chick Salts A may also be used for guinea pigs[1,3,4] if 2.5% K acetate and 0.5% MgO are added to the diet. If HMW is used, zinc must be added.
[b] 70 g mix/kg diet.
[c] 75 g mix/kg diet.
[d] Grace, N. D., and O'Dell, B. L., *Can. J. Biochem.*, 48, 21–26, 1970.

Table 7
MINERAL ELEMENTS SUPPLIED BY SALT MIXTURES IN GUINEA PIG DIETS

Element	Unit	Briggs GPS-3,[7] per kg diet	O'Dell et al.,[2] per kg diet	NRC 1972,[12] per kg diet
Ca	g	10.59	9.7	12
P	g	6.7	6.12	6
Na	g	1.9	3.13	–
Cl	g	6.0	1.58	–
K	g	11.64	10.0	14
Mg	g	3.3	3.015	3.5
Mn	g	0.0998	0.208	0.04
Fe	g	0.084	0.0618	0.033
Zn	g	0.060	0.0587	–
Cu	mg	1.70	7.96	0.006
I	mg	7.6	9.5	–
Cr	mg	0.24	–	–
Ni	mg	0.013	–	–
Mo	mg	0.05	–	–
Se	mg	0.023	–	–
F	mg	–	–	–
Al	mg	–	–	–

REFERENCES

1. **Briggs, G. M., Spivey, M. R., Keresztesy, J. C., and Silverman, M.,** Activity of citrovorum factor for the chick, *Proc. Soc. Exp. Biol. Med.,* 81, 113–115, 1952.
2. **Grace, N. D. and O'Dell, B. L.,** Potassium requirement of the weanling guinea pig, *J. Nutr.,* 94, 166–170, 1968.
3. **Hegsted, D. M., Mill, R. C., Elvehjem, C. A., and Hart, E. B.,** Choline in the nutrition of chicks, *J. Biol. Chem.,* 138, 459–466, 1941.
4. **Hubbell, R. B., Mendel, L. B., and Wakeman, A. J.,** A new salt mixture for use in experimental diets, *J. Nutr.,* 14, 273–285, 1937.
5. **Lakhanpal, R. K., Davis, J. R., Typpo, J. T., and Briggs, G. M.,** Evidence of an unidentified growth factor(s) from alfalfa and other plant sources for young guinea pigs, *J. Nutr.,* 89, 341–346, 1966.
6. **Lane-Petter, W. and Porter, G.,** Guinea pigs, in *Animals for Research: Principles of Breeding and Management,* Lane-Petter, W., Ed., Academic Press, New York, 1963, pp. 287–321.
7. **Lofgren, P. A., Reyes, P. S., McNutt, K. W., Briggs, G. M., and Kohler, G. O.,** Unidentified growth factor(s) present in alfalfa water-solubles for young guinea pigs, *Proc. Soc. Exp. Biol. Med.,* 147, 331–336, 1974.
8. **Navia, J. M. and Lopez, H.,** A purified gel diet for guinea pigs, *Lab. Anim. Sci.,* 23, 111–114, 1973.
9. **O'Dell, B. L., Morris, E. R., and Regan, W. O.,** Magnesium requirements of guinea pigs and rats. Effect of calcium and phosphorus and symptoms of magnesium deficiency, *J. Nutr.,* 70, 103–111, 1960.
10. **O'Dell, B. L., Moroni, R. I., and Regan, W. O.,** Interaction of dietary fluoride and magnesium in guinea pigs, *J. Nutr.,* 103, 841–850, 1973.
11. **Reid, M. E.,** in *The Guinea Pig in Research,* Human Factors Research Bureau, Inc., Washington, D.C., 1958, pp. 1–87.
12. **Reid, M. E. and Bieri, J.,** Nutrient requirements of the guinea pig, in *Nutrient Requirements of Laboratory Animals,* National Research Council, Agricultural Board, National Academy of Sciences, Washington, D.C., 1972, pp. 9–19.

DIGESTIVE PHYSIOLOGY AND FEEDING OF CAPYBARAS

E. González-Jiménez

The South American rodent capybara (*Hydrochoerus hydrochaeris*) is found from Argentina to Panama. Some of its outstanding features are that it is the largest living rodent (up to 70 kg mature body weight), it has a high reproductive potential (approximately five offspring per parturition in the wild), it is a terrestrial grazer with a digestive efficiency comparable to that of ruminants, and the possibility of domestication is real. These aspects make the capybara useful for the study of animal production in South America. Its meat and leather are used in many countries. Capybara meat has good taste qualities and properties such as a high emulsification score, allowing its competitive use with pork and other meats in the food industry. "Carpincho" leather (as it is known on the market) is one of the best hides for glove making, as its one-way directional elongation allows gloves to be fitted without lengthening. From the economic point of view, these two qualities justify the interest that this rodent has received in the last decade.

DIGESTIVE PHYSIOLOGY OF THE CAPYBARA

As are all rodents, the capybara is a simple-stomach animal; but, unlike most of the family, it is a "true" herbivore. The cecum of this animal shows an impressive development (Figure 1) and is the site of important digestive events. The capacities of the different compartments of the alimentary canal of the capybara are listed in Table 1. It is evident that the capybara is a nonruminant herbivore with an enlarged hindgut, fermentation occurring posterior to the site of enzymatic digestion. This arrangement may impart some advantage to this animal, as will be illustrated later. Comparison of the relative gut capacities of capybaras (nonruminant herbivores) and sheep (ruminant herbivores) indicates that the cecum has an equivalent holding capacity to that of the rumen (Table 2). The lengths of the different compartments of the capybara's digestive tract appear comparable to those of other herbivores. Garrod[9] gives similar values (Table 3). The pH of digesta in each gut compartment varies from 3.50 ± 0.29 at the stomach (indicating the acid digestion that follows in the compartments) to 6.86 ± 2.22 at the small intestine and 6.03 ± 0.29 in the cecum. The pH of the cecum is close enough to neutrality to suggest bacterial fermentation. The digestive capacity of the capybara appears to be similar to that of ruminants of similar size, with its cecum being the principal fermentation site.

When structural carbohydrates are the main components of the diet (as is the case with the capybara) they are degraded through the fermentation process. It remains to be explained how this nonruminant herbivore takes advantage of the endproducts of fermentation such as volatile fatty acids (VFA) and microbial protein. However, there is good evidence[1,16,28] that nonruminant herbivores efficiently absorb the VFA produced in the hindgut. The method of utilizing the microbial protein is not as clear. Two alternatives appear possible: (1) a well-established, highly efficient behavior adaptation known as coprophagy exists, allowing for part of the fecal material to be subjected to the gastric enzymatic digestion followed by absorption of the nutrients liberated from the feces, or (2) digestion and absorption of microbial protein occurs in the hindgut.

Comparison of the fecal excretion curves of capybaras, sheep (non coprophagic), and rabbits (coprophagic) following a high dose of chromium oxide (Figure 2) shows that capybaras do not practice coprophagy, as its fecal excretion curve is similar to that of sheep[11] and different from the staircase-like rabbit excretion curve. Thus, the question of what nutritional strategy capybara has adopted seems difficult to answer. Nevertheless, it is helpful to consider other aspects of its digestive physiology before attempting an explanation.

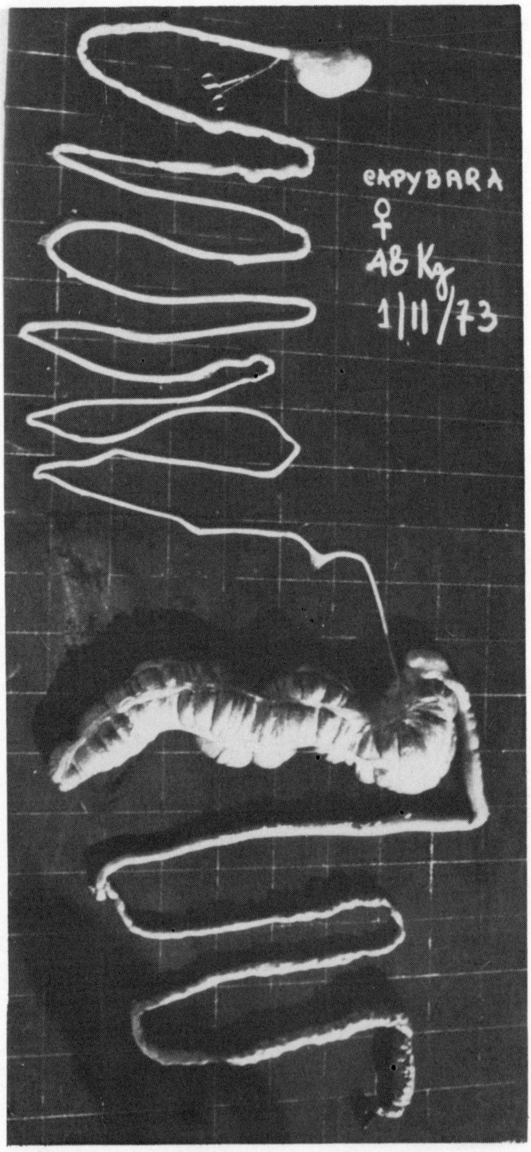

FIGURE 1. Digestive tract of the capybara. Each square = 10 cm.

Table 1
CHARACTERISTICS OF THE CAPYBARA DIGESTIVE TRACT

	Weights of tissues, free contents (g)	Weights of tissue organs (% body weight)	Dry contents of compart-ments	% of total
Stomach	223 ± 33	0.55	113 ± 58	10
Small intestine	650 ± 152	1.61	38 ± 16	3
Cecum	798 ± 132	1.97	869 ± 274	74
Colon	273 ± 97	0.68	78 ± 31	7
Rectum	175 ± 43	0.43	76 ± 59	6
Total	2188 ± 37	5.24	1175 ± 315	100

Table 2
LENGTHS IN METERS OF THE DIFFERENT COMPARTMENTS OF THE CAPYBARA DIGESTIVE TRACT

	González-Jiménez and Parra[11] [a]	Percent total length	Garrod[a]
Stomach	0.29 ± 0.08	3	
Small intestine	5.97 ± 0.91	59	6.3
Cecum	0.95 ± 0.28	10	0.6
Colon	1.98 ± 0.67	20	
Rectum	0.80 ± 0.29	8	2.2[a]
Total	9.99 ± 0.40	100	

[a] Gross intestine.

Table 3
RELATIVE CAPACITIES OF THE DIGESTIVE TRACTS OF CAPYBARAS AND SHEEP

	Capybara (%)	Sheep (%)
Rumen	–	54
Reticulum	–	5
Omasum	–	2
Abomasum or stomach	10	7
Small intestine	3	20
Cecum	74	2
Large intestine	13	10
Total	100	100

From González-Jiménez, E. and Parra, R., in *Proc. 3rd World Conf. on Animal Production,* Melbourne, May 1973, Sydney University Press, 1975, 80–86. With permission.

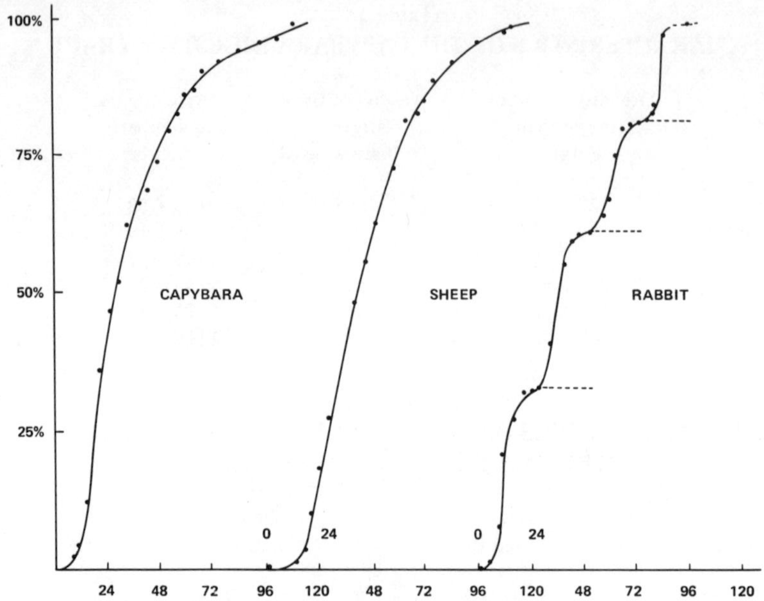

FIGURE 2. Fecal excretion curves of capybaras, sheep, and rabbits.

DRY MATTER CONSUMPTION AND DIGESTIVE EFFICIENCY

Dry matter intake of capybaras, sheep, and rabbits has been compared.[13] They were fed dried pelleted grass and pelleted concentrate in different proportions. The results (Table 4) showed minor differences when expressed as percent of body weight, but these differences were magnified when intake is expressed per kilogram of metabolic weight due to the body weight differences in the animals compared. The intakes were not high and, in the case of capybara data, there was a high variation. Comparison of dry matter digestibility of these pelleted dried grass and concentrate rations (Table 5) indicates that capybara and sheep do not differ significantly on forage digestibility; however, on the concentrate dry matter digestibility, capybara show an appreciable superiority (significant at 0.05), indicating the possibility of using high proportions of concentrate in its feeding with a high efficiency of utilization.

Nitrogen metabolism – In studying comparative protein utilization of capybaras, sheep, and rabbits it was found (Table 6) that capybaras are less efficient than sheep and rabbits for the apparent digestibility of forage protein, but as efficient as rabbits in the utilization of protein of concentrates. Bell[2] states that ruminants utilize protein more efficiently than nonruminants; Bonazzi,[3] in studies with horses, found exactly the reverse to be true. It is possible to say, as does Slade,[26] that ruminant and nonruminant herbivores are equally efficient in protein digestion. Rabbits are between sheep and capybaras in endogenous nitrogen losses, and capybara have higher losses than the others. The rabbit has a good mechanism for nitrogen utilization (its coprophagic strategy), but the capybara does not. It has also been found that sheep and capybara have similar requirements of protein per kilogram metabolic weight.[10] If, as was shown previously, capybara is not a coprophagic herbivore, then recycling nitrogen to the cecum appears to be a wasteful process if there is no other mechanism of using the nitrogen synthetized by bacteria and protozoa in the cecum. From preliminary observations (Figure 3) it appears that the capybara cecum has an enlarged, vascularized and thick-walled appendage at the entrance of the colon. This appendage may be the site for some digestion or autolysis of microorganisms produced in the cecum. Much more research is needed to prove this hypothesis.

Table 4
DRY MATTER CONSUMPTION EXPRESSED IN DIFFERENT WAYS

	Body weight (kg)	Dry matter/day (g)	% body weight	$DM/BW^{0.75}$ (g)
Capybaras	24.49 ± 3.23	444 ± 61	2.57 ± 0.59	60.40 ± 15.73
Rabbits	3.41 ± 0.15	94.9 ± 16.2	2.78 ± 0.50	37.93 ± 6.65
Sheep	46.88 ± 2.23	1259 ± 147	2.68 ± 0.29	70.20 ± 7.78

Table 5
PERCENT DRY MATTER DIGESTIBILITY (%) OF DIETS WITH DIFFERENT LEVELS OF CONCENTRATE[a]

% concentrate in diet (concentrate: forage ratio in %)	Capybara	Sheep
0:100	50.56	49.15
20:80	57.4	53.5
40:60	64.2	57.8
60:40	71.0	62.1
80:20	77.9	66.5
100:0	84.7	70.8

[a] Calculated from the equations $Y = 0.217\ X + 49.15$ for sheep and $Y = 0.342\ X + 50.56$ for capybara, where Y = dry matter digestibility and X = percent concentrate in ration.

Table 6
PROTEIN UTILIZATION OF CAPYBARAS, SHEEP, AND RABBITS[a]

	Percent Apparent Digestibility of Protein	
	Concentrate	Forage
Capybaras	76.01	16.51
Rabbits	78.80	37.60
Sheep	70.04	37.74

[a] From equations

Capybara $Y = 0.595X + 16.51$
r 0.88** Sy.x 8.18
Rabbits $Y = 0.412X + 37.60$
r 0.97** Sy.x 3.29
Sheep $Y = 0.323X + 37.74$
r 0.93** Sy.x 4.36

where

Y = digestibility of nitrogen
X = percent concentrate in ration
r = correlation coefficient
** = significant at 0.01%

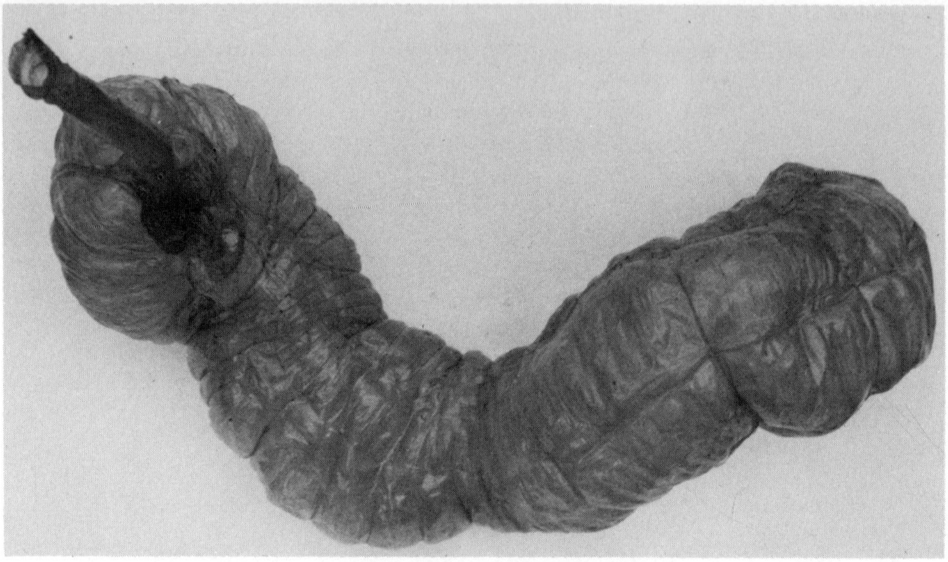

FIGURE 3. Cecum of a capybara.

Cell wall utilization – It is generally accepted that nonruminant herbivores have a lower digestive efficiency of the fiber components than do ruminants, but no satisfactory explanation is offered.[23] Figure 4 is a compilation of some of the available information on the effect of lignification on cell wall digestibility for ruminant and nonruminant herbivores. Parra et al.[24] found with tropical forages that lignin is the most important factor in reducing digestibility of fiber. This was shown previously by Van Soest[27] with temperate forages. This figure shows that small nonruminant herbivores are less efficient in fiber digestion than ruminants, but large nonruminant herbivores (horses, ponies, zebras, and capybaras) seem to have a capacity to digest fiber comparable to that of ruminants.[23] The values for capybara are equal to those for ruminants.

Rate of passage – Ojasti[20] found that 68% of a daily ration passed through the gut in 24 hr and the whole ration passed in 72 hr. Our studies[13] (Figure 2) showed that 5% of the indicator (chromium oxide) was recovered in the feces 12 hr after dosage and 85% was recovered 60 hr after dosage. Similar values for rabbit (allowed coprophagy) were 9 hr and 80 hr and for sheep 18 hr and 70 hr.

Particle size (as a component of rate of passage) – The capybara has a very efficient mastication process which is achieved by two incisors, one pair of premolars, and three pairs of molars in each jaw, the movements of which are forward and backward due to a special articulation.[11a] The incisors permit the ingestion of very short grass (e.g., *Paratheria prostata*) in natural grazings. The third molar, as shown in Figure 5, has a grinding surface as important as the other two together, masticating the forages to very fine particles between 0.001 and 0.3 mm^2.[20] This should contribute to the excellent digestive ability of the capybara when grazing on native pastures. Ruminants are very inefficient grinders of grasses despite the rumination process that helps shred it.

It is interesting to consider the quality of the digesta reaching the fermentation site, which should be affected in two different ways: (1) by dietary selection, and (2) by whatever action, if any, the acid digestion of the stomach and the enzymatic digestion of the small intestine might have on the more refractory components of the diet. This later assumption has been studied[10] by an in vitro method using different grass samples. It was found that when samples of native grass that had previously been incubated with pepsin in an acid medium were incubating with cecal liquor or ruminal liquor, an increment of dry matter digestibility of 17% and 11%, respectively, was found. This may be explained by the effect of the acid medium in the stomach and the effect of enzymes on certain components of cell wall such as hemicellulose or pentosans, permitting better conditions for the attack of microorganisms on the remainder of the cell wall. Using the method of Cavender and Hansen[5] to identify the cuticle remants of grass in herbivore feces, we[11] were able to show that capybaras choose the best forages available. Forages more commonly chosen by the capybara are those of higher protein content (Table 7). This ability may be explained by the capybara's semiaquatic adaptation and rodent incisors. In inundated savannas, grass quality generally follows water gradient, as shown by our studies on the adaptation of grasses to floods.[12]

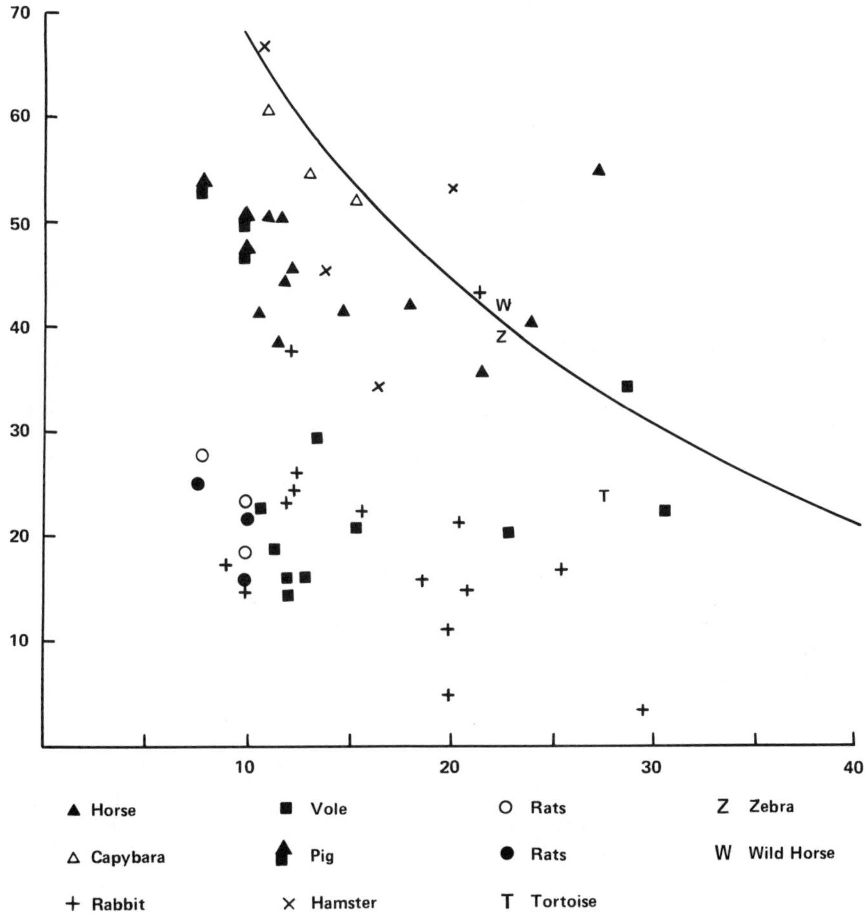

FIGURE 4. The effect of lignification on cell wall digestibility in ruminant and nonruminant herbivores. X = index of lignification (lignin × 100/ acid detergent fiber); Y = Percent digestibility coefficient of cell wall.

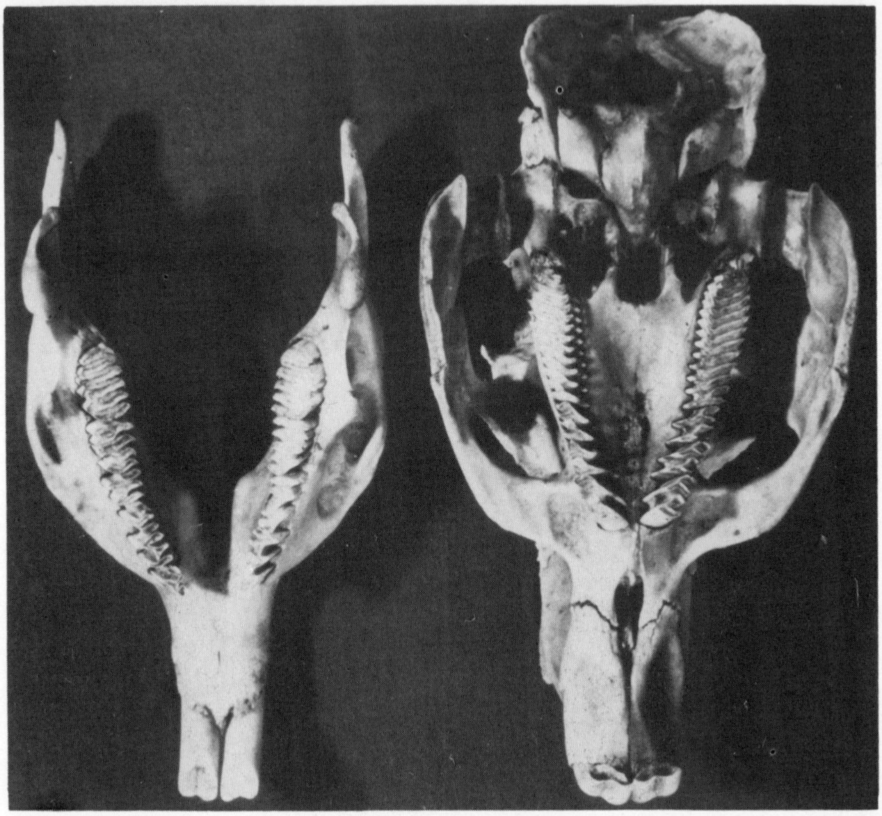

FIGURE 5. Capybara jawbones showing the third molars.

Table 7
COMPOSITION OF THE CAPYBARA'S DIET IN NATURAL CONDITIONS AND PERCENT PROTEIN OF THE MAIN GRASSES SELECTED

Plant species	Relative % in diet	% crude protein
Hymenachne amplexicaulis	34.96	10.44 ± 1.82
Leersia hexandra	29.16	9.43 ± 0.98
Panicum laxum	11.88	8.52 ± 0.93
Paspalum chaffanjonni	4.32	5.66 ± 0.87
Sporobolus indicus	2.72	4.10 ± 0.61
Eragrostis acutiflora	1.04	3.96 ± 0.28
Paspalum orbiculatum	0.94	4.00 ± 0.53
Total	85.02	

FEEDING IN NATURAL CONDITIONS

Patterns of forage consumption by the capybara were established using the same cuticle method; relative frequencies transformed give the consumption preferences of this animal. Measurements were performed three times during the year: at the end of the rainy season, at the end of the dry season, and during the rainy season. Table 8 shows the proportion of different species consumed by capybara. It is clear from this data that the preferred grass is *Hymenachne amplexicaulis*, but it is also noteworthy that 80% of the diet consists of only the five grasses *Hymenachne amplexicaulis, Leersia hexandra,*

Table 8
PROPORTION OF DIFFERENT SPECIES CONSUMED BY CAPYBARAS DURING THREE PERIODS OF THE YEAR

Percent of Dry Matter Ingested

	Rainy season	End rainy season	End dry season
Hymenachne amplexicaulis	43	25	20
Paratheria prostata	0	17	16
Leersia hexandra	35	15	8
Sporobolus indicus	3	9	17
Panicum dichotomiflorum	1	5	<1
Axonopus sp.	0	4	14
Luziola sp.	0	2	<1
Panicum laxum	12	<1	10
Panicum junceum	<1	<1	<1
Setaria geniculata	<1	<1	1
Oryza perennis	<1	<1	0
Panicum zizinioides	<1	<1	<1
Eragrostis acutiflora	<1	<1	<1
Paspalum chaffanjonii	4	<1	1
Paspalum orbiculatum	1	<1	<1
Ciperaceas	3	16	8
Eichhornia sp.	1	2	1
Others not identified	3	3	7

Panicum laxum, Paratheria prostata, and *Sporobolus indicus.* Cyperaceae and water hyacinth are consumed mainly in young stages throughout the year.

It has been reported that capybaras compete with cattle for the best pastures, but after his detailed study of capybara, Ojasti[20] did not reach a definitive answer. Escobar and González-Jiménez[7] measured food competition and diet overlapping of capybara and cattle using the disimilarity index of Hansen et al.[14] and found no competition between them at the end of the rainy season. This situation is explained by the presence of the grass *Paratheria prostata,* which is not available for cattle because of its short height and is consumed by capybaras. At the end of the dry season, strong competition between domesticated and wild animals is observed because of lower forage availability.

Out of its habitat capybara eats diverse botanical species, but in its niche it is always a strict herbivore. Nevertheless, some zoos feed capybara meat and fish.[29] Buffon[4] reported that some capybaras at Fontainebleau (France) eat fish. Osgood[21] reported that the digestive contents of capybaras shot in western Venezuela contained mostly algae from the bottoms of swamps. Mondolfi[18] and Humboldt[15] reported that capybaras eat the grass *Paspalum fasciculatum* (called capybara grass), which is abundant on river edges of the Orinoco basin. Nogueira-Neto[19] reported that capybaras harm sugar cane and rice plantations. In many parts of Brazil the capybara is considered an agricultural pest and is shot.

FEEDING IN CAPTIVITY

In captivity capybaras eat at any time, but preferentially during early morning and end of afternoon. It is easy to train capybaras to eat new food, but it is essential to offer fresh grass daily on feeding troughs, avoiding footing and soiling that contaminate its feed with parasites. Grasses most offered are Para grass (*Brachiaria mutia*), *Hyparrhenia rufa, Melinis minutiflora, Panicum maximum, Paspalum plicatulum, Tripsacum laxum, Axonopus scoparius,* and some crops such as sugar cane (*Saccharum officinarun*) and rice (*Oriza sativa*).[8] Similar grasses are mentioned by Mondolfi,[18] Otero de la Espriella,[22]

Cruz,[6] and Piccinini,[25] together with *Panicum purpurascens* and other grasses. In some zoos (such as the zoo in Evansville, Indiana), capybaras are offered spinach leaves, fruits such as apples, and roots such as carrots. In Brazil, Goeldi Zoo offers *Brassica oleracea* and *Lactuca sativa*, roots such as *Ipomea*, and fruits including *Cucurbita pepo* and *Musa* sp. Generally, feeding in zoos is very peculiar. For example, in Evansville they offer capybara primate biscuits, dog food, rodent pellets, raw meat, and fresh fish during the summer, as well as soft-billed bird mix (flamingo mixture); this seems unnecessary and wasteful. South American zoos give mostly grasses and vegetables and sometimes some starchy foods and roots and tubers such as *Ipomea batatas* and *Manihot utilisima*. Otero de la Espriella[22] reports feeding *Solanum tuberosum* and *Colocasia sculenta*, and also some milling byproducts such as rice bran and wheat bran and grains such as corn (*Zea mays*). Fuerbringer[8] reports mixing sugar cane molasses with powdered foods to avoid inhaling dust.

It is important to offer a variety of foods to capybaras kept in captivity, not only in quality but also in quantity as well as in protein content and vitamins. In our laboratory[23a] we have successfully raised capybaras on green chop (*Penissetum purpureum*) and pig meal mixture (16% protein). Many people believe that because capybaras thrive in poor conditions in the wild they do not need much attention when in captivity. However, in the wild, capybaras select their own diets and remain healthy primarily because they stay in their habitat in open areas with enough uncontaminated grazing regions.

Capybaras are easy to keep in captivity, providing a variety of foods such as forage, roots, fruits, and vegetables is available.

REFERENCES

1. **Barcroft, J., McAnally, R. A., and Phillipson, A. T.,** Absorption of volatile fatty acids from the alimentary tract of the sheep and other animals, *J. Exp. Biol.*, 20, 120–129, 1944.
2. **Bell, R. H. V.,** A grazing ecosystem in the Serengeti, *Sci. Am.*, 225, 86–93, 1971.
3. **Bonazzi, A.,** Estudios sobre las escretas de varias especies de animales domésticos, *Acta Cient. Venez.*, 2, 199–202, 1951.
4. **de Buffon, G. L. L.,** *Histoire Naturelle*, Vol. 4, Bureau de Publications Illustrées, Paris, 1844.
5. **Cavender, B. R. and Hansen, R. M.,** The Microscope Method Used for Herbivores Diets Estimates and Botanical Analysis of Litter and Mulch at the Pawnee Site, Tech. Rep. No. 18, U.S. International Biological Program Grassland Biome, Fort Collins, Colo., 1970.
6. **Cruz, C. A.,** Notas sobre comportamiento del chigüire (*Hydrochoerus hydrochaeris hydrochaeris*) en confinamiento. 1o Seminario Colombo-Venezolano de Chigüires y Babillas, Bogotá, August 1974.
7. **Escobar, A. and González-Jiménez, E.,** Variación estacional de la frecuencia relativa de las especies vegetales consumidas por los chigüires (*Hydrochoerus hydrochaeris*) en el llano inundable, *Acta Cient. Venez. Supl.*, Abstr., 25(Supl.1), 15, 1974.
8. **Fuerbringer, J.,** El Chigüire, su Cría y Explotación Racional, Temas de Orientación Agropecuaria No. 99, Bogotá, Colombia, 1974, 1–59.
9. **Garrod, A. H.,** On the *Caecum coli* of the capybara, *Proc. Zool. Soc. London*, 20–23, 1876.
10. **González-Jiménez, E. and Escobar, A.,** Digestibilidad comparada entre chigüires (*Hydrochoerus hydrochaeris*), conejos y ovinos, con raciones de diferentes proporciones de forrajes y concentrados, *Agron. Trop.* (Maracay, Venez.), 25, 283–290, 1975.
11. **González-Jiménez, E. and Escobar, A.,** Estudio de la competencia alimenticia de los herbívoros mayores del llano inundable, con referencia especial al chigüire (*Hydrochoerus hydrochaeris*). I. Salida de aguas, *Agron. Trop.* (Maracay, Venez.), in press.

11a. **González-Jiménez, E. and Parra, R.,** The capybara, a meat producing animal for the seasonally flooded areas of the tropics, in *Proc. 3rd World Conf. on Animal Production*, Melbourne, May 1973, Sidney University Press, 1975, 80–86.

12. **González-Jiménez, E. and Escobar, A.,** Food adaptation and productivity of savanna grasses, in *Proc. 13th Int. Grassland Congr.*, Leipzig, May 18–27, 1977, in press.

13. **González-Jiménez, E., Escobar, A., and Caires, O.,** Un método para detectar coprofagia. Resultados en Chigüires, Proc. 2o Seminario sobre Chigüires y Babas, Universidad Central de Venezuela, Maracay, 1976.

14. **Hansen, R. M., Peder, P. V., and Rice, R. W.,** Discerned fragments in feces indicate diet overlap, *J. Range Manage.*, 26(2), 103–105, 1971.

15. **von Humboldt, A.,** *Voyage aux Régions Equinoxiales du Nouveau Continent Fait en 1799, 1800, 1801, 1802, 1803 et 1804,* Chez N. Maze, Paris, 1820.

16. **McBee, R. H.,** Metabolic contribution of the cecal flora, *Am. J. Clin. Nutr.*, 23, 1514–1518, 1970.

17. **Moir, R. J.,** Ruminant digestion and evolution, in *Handbook of Physiology*, Sect. 6, Vol. 5, Code, C. F. and Heidel, W., Eds., Williams & Wilkins, Baltimore, 1968, 2673–2694.

18. **Mondolfi, E.,** Mamiferos de Venezuela: el chigüire, *El Farol,* 168, 38–40, 1957.

19. **Nogueira-Neto, P.,** *A Criacao de Animais Indigenras Vertebrados. Peixes-anfibios-repteis-aves-mamiferos,* Edicoes Tecnapis, Saõ Paulo, 1973, 1–327.

20. **Ojasti, J.,** *Estudio Biológico del Chigüire Capibara,* Fonaiap, P., Ed., Sucre, Caracas, 1973.

21. **Osgood, W. H.,** Mammals from western Venezuela and eastern Colombia, *Field Mus. Nat. Hist. Publ. Zool. Ser.*, 10(5), 29–66, 1912.

22. **Otero de la Espriella, R.,** El chigüire como recurso económico y alimenticio, *Rev. Vista Agric.* (Bogota), 19(1), 5–10, 1973.

23. **Parra, R.,** Comparative aspects of the digestive physiology of ruminants and non-ruminant herbivores, in Literature Review of Selected Topics in Comparative Gastroenterology, Vol. 2, Cornell University College of Veterinary Medicine, Ithaca, 1973.

23a. **Parra, R.,** personal communication, 1976.

24. **Parra, R., González-Jiménez, E., and Combellas, J.,** Composición y valor nutritivo de los forrajes producidos en el trópico. 2. Fracciones quimicas que afectan la disponibilidad de los componentes fibrosos, *Agron. Trop.* (Maracay, Venez.), 22, 219–230, 1972.

25. **Piccinini, R., Vale, W. G., and Gómez, F. W. R.,** Criadouros Artificiais de Animais Silvestres. I. Criadouro de Capivaras, Assessoria de Programaco e Coordenacao Divisao de Documentacao, Ministerio de Interior Superintendencia do Desenvolvimiento da Amazonia, Belém, Brazil, 1971, 1–31.

26. **Slade, L. M., Robinson, D. W., and Casey, K. E.,** Nitrogen metabolism in non-ruminant herbivores. 1. The influence of non-protein nitrogen and protein quality on the nitrogen retention of adult mares, *J. Anim. Sci.*, 30, 753–760, 1970.

27. **Van Soest, P. J.,** Development of a comprehensive system of feed analysis and its application to forages, *J. Anim. Sci.*, 26, 119–128, 1967.

28. **Yang, M. G., Manoharan, K., and Mickelsen, O.,** Nutritional contribution of volatile fatty acids from the cecum of rats, *J. Nutr.*, 100, 545–550, 1970.

29. **Zara, J. L.,** Breeding and husbandry of the Capybara, in *International Zoo Yearbook,* Vol. 13, International Pubns. Service, New York, 1973, 137–139.

Artiodactyla

DIETS (NATURAL AND SYNTHETIC): SWINE

J. L. Gobble

INTRODUCTION

Although great strides have been made in increasing the understanding of the nutrient requirements of various classes of swine, there is still a lack of complete information in this area. Therefore, the diets presented herein were chosen because they produced satisfactory results in experimental feeding trials or in field use.

These diets are grouped according to the functions being performed by the animals to which they were fed.

Consideration is given to the fact that swine are omnivorous monogastrics having digestive tracts somewhat limited in capacity. This is particularly true of the very young animal. Limited digestive tract capacity coupled with genetic potential for rapid growth indicates a necessity for high-nutrient-density diets to permit realization of growth potential.

With developing and mature breeding animals, it is necessary to restrict energy intake in order to prevent excessive fatness, with the possible exception of lactating females suckling large litters over a lactation period exceeding 3 weeks.

SOW MILK SUBSTITUTES

Table 1
SYNTHETIC SUBSTITUTE[a] FOR SOW'S MILK FOR PIGS WEANED AT 3 DAYS OF AGE[1]

Item	Amount, % of solids
Casein, Labco®	30
Glucose	54
Lard	10
Mineral mixture[b]	6
	Milligrams per kilogram solids
Thiamine	4.4
Riboflavin	3.1
Nicotinic acid	16.5
Pyridoxine	4.4
Calcium pantothenate	19.8
p-Aminobenzoic acid	17.2
Inositol	171.6
Choline	1,716.0
Biotin	0.07
Pteroylglutamic acid	0.35
Ascorbic acid	105.6
α-Tocopherol acetate (vitamin E)	6.6
2-Methyl-1:4-naphtaquinone (vitamin K)	1.8
	IU per kilogram solids
Vitamin A	13,000
Vitamin D	1,320

[a] The diet was made up with water to contain 15.1% solids.

[b] The mineral mixture in g per kg was NaCl, 80; K_2HPO_4, 175; $CaHPO_4$, 229; $MgSO_4$, 24; calcium lactate, 477; ferric citrate·$6H_2O$, 13.4; KI, 0.39; $MnSO_4$, 0.68; $ZnCl_2$, 0.11; $CuSO_4 \cdot 5H_2O$, 0.14; $CoCl_2$, 0.05; CaF_2, 0.23.

Table 2
SUBSTITUTE[a] FOR SOW'S MILK GIVEN TO ORPHAN PIGS[2]

A. Components of Solids

	%
Dried skim milk	87.2
Lard	6.5
Lecithin	3.5
Mineral, vitamin, and antibiotic supplements	2.8

B. Minerals, Vitamins, and Antibiotic Supplied by Supplements, per Kilogram Milk Substitute Solids

Minerals		Vitamins and antibiotic	
$CaCO_3$ (g)	7.5	Vitamin A (IU)	29,920
$CaHPO_4$ (g)	2.0	Vitamin D_2 (IU)	2,992
$MgSO_4$ (g)	1.5	Vitamin E (mg)	29.9
$FeSO_4$ (g)	1.8	Vitamin K (mg)	2.0
$CuSO_4$ (g)	0.2	Ascorbic acid (mg)	150.0
$MnSO_4$ (mg)	70.4	Thiamine (mg)	9.9
$CoSO_4$ (mg)	9.9	Riboflavin (mg)	9.9
CaF_2 (mg)	20.0	Nicotinic acid (mg)	49.9
KI (mg)	20.0	Calcium pantothenate (mg)	60.1
$ZnCl_2$ (mg)	9.9	Pyridoxine (mg)	9.9
		Choline chloride (mg)	499.4
		Vitamin B_{12} (μg)	39.6
		Chlortetracycline (mg)	99.9

[a] The diet was made up with water to contain about 14% dry matter.

Table 3
SEMISYNTHETIC SOW'S MILK SUBSTITUTE[3]

A. Composition of Solids[a]

	%
Dry skim milk	48.24
Casein	15.68
Lactose	8.85
Dicalcium phosphate	1.00
Trace mineralized salt	0.50
Corn oil	25.13
Premix	0.60

B. Premix Supplied the Following per 100 kg

Milk Subsitute Solids	
Furazolidone 10 (g)	454
Vitamin E (mg)	543
Vitamin K (mg)	35
Ascorbic acid (mg)	2,728
Vitamin A (IU)	544,000
Vitamin D (IU)	54,400
Ethoxyquin (g)	12.2
Methionine hydroxy analogue (g)	64
Niacin (mg)	1,375

[a] The sow's milk substitute solids were combined with water to provide a diet containing 20% solids. The diet was homogenized.

Table 4
FORTIFIED MILK DIET[4]

Prepare by adding the following to one liter fresh milk:

Dried whole milk (g)	168
Citric acid (g)	1.3
Niacin (mg)	4.4
Vitamin K (mg)	0.8
Vitamin D_2 (IU)	27
Vitamin E (mg)	2.8
$ZnSO_4 \cdot 7H_2O$ (mg)	85
$FeSO_4 \cdot 7H_2O$ (mg)	138
$MnCl_2$ (mg)	32
$CuSO_4 \cdot 4H_2O$ (mg)	11
$MgSO_4 \cdot 7H_2O$	530

Notes: For colostrum-free pigs taken by Caesarean section and grown in individual sterile incubators to 3 weeks of age. All ingredients were mixed together and homogenized at 1750 psi, pasteurized, and stored in containers until used.

PIG PRESTARTER DIETS (To be fed to suckling or early weaned pigs up to 4 or 5 weeks of age.)

Table 5
PIG PRESTARTER[5]

Ingredient	%
Ground yellow dent corn	19.2
Soybean meal, 49% protein	26.6
Feeding grade rolled oats	20.0
Thermolac (proprietary feedstuff, 20% protein, of milk origin, 10% fat)	20.0
Cane sugar	5.0
Corn sugar	5.0
Dicalcium phosphate	0.9
Ground limestone, 38% Ca	0.9
Trace mineralized salt, 1.0% Zn	0.5
Vitamin premix[a]	1.0
Selenium premix (adds 0.1 ppm Se in complete diet)	1.0

[a] Vitamin premix supplies the following per kilogram:

Vitamin A (IU)	330,000
Vitamin D_3 (IU)	110,000
Riboflavin (mg)	220
d-Calcium pantothenate (mg)	359
Niacin (g)	3.3
Choline chloride (g)	25.3
Vitamin B_{12} (mg)	1.0
Vitamin E (mg)	1,100
Ethoxyquin (g)	12.5

Table 6
DIET FOR PIGS WEANED AT 3 WEEKS OF AGE[6]

Ingredient	%
Finely ground yellow corn	66.55
Soybean meal, solvent, 48.5% protein	30.50
Ground limestone	1.10
Dicalcium phosphate	1.30
Iodized salt	0.50
Trace mineral mixture[a]	0.05
Vitamin-antibiotic mixture[b]	+

[a] Provided the following in the complete diet (ppm): Zn, 100; Fe, 50; Mn, 27.5; Cu, 5.5; I, 0.75.
[b] Provided the following per kilogram of diet:

Vitamin A (IU)	4400
Vitamin D_3 (IU)	440
Riboflavin (mg)	2.2
Niacin (mg)	11
d-Calcium pantothenate (mg)	6.6
Vitamin B_{12} (μg)	11
Vitamin K (μg)	440
Antibiotic-drug mixture (1:1 mixture of neomycin sulfate and oxytetracycline) (mg)	110

Table 7
DIET FOR PIGS FROM 3 TO 9 WEEKS OF AGE AND WEANED AT 5 WEEKS[7]

Ingredient	%
Finely ground yellow corn	57.5
Soybean meal, 44% protein	17.5
Sucrose	10.0
Dried skim milk	10.0
Fish meal	3.0
Dicalcium phosphate	0.7
Ground limestone	0.8
Salt[a]	0.5
Vitamin-antibiotic premix[b]	+

[a] Contained 94.3 to 96.8% NaCl, 0.008% Co, 0.064% Cu, 0.32% Fe, 0.64% Mn, 0.008% I, and 0.80% Zn. Provided 40 ppm supplemental Zn.

[b] Provided the following per kilogram of diet:

Vitamin A (IU)	4400
Vitamin D (IU)	440
Riboflavin (mg)	2.2
d-Calcium pantotheante (mg)	6.6
Niacin (mg)	6.6
Vitamin B_{12} (μg)	11
Antibiotic-drug mixture (2:2:1 mixture of chlortetracycline: sulfamethazine: procaine penicillin) (mg)	275

Table 8
DIET FOR PIGS WEANED AT 10 DAYS OF AGE[8]

Ingredient	%
Ground corn	30.7
Rolled wheat	12.0
Oat groats	10.0
Soybean meal, 50% protein	27.0
Dried skim milk	15.0
Corn oil	2.0
Iodized salt	0.5
Ground limestone (38% Ca)	0.3
Calcium phosphate (22% Ca, 20% P)	1.9
Vitamin and trace mineral supplement[a]	0.5

[a] Supplied the following levels of nutrients per kilogram of diet:

Vitamin A (IU)	8800
Vitamin D_3 (IU)	1650
Vitamin B_{12} (μg)	33
Riboflavin (mg)	7
d-Calcium pantothenate (mg)	17
Choline chloride (mg)	1210
Niacin (mg)	20
Folic acid (mg)	1
Medadione sodium bisulfite (mg)	2
Thiamine HCl (mg)	2
Pyridoxine (mg)	2
Neomycin (mg)	1.4
Copper (mg)	11
Zinc (mg)	110
D-Biotin (μg)	200
Santoquin®	63
Manganese (mg)	44
Iron (mg)	110

PIG GROWER DIETS

Table 9
DIETS FOR GROWING PIGS[9]

To Be Fed from 16–34 kg Live Weight

Ingredient	Diet 1 (%)	Diet 2 (%)
Ground shelled corn	60.5	46.9
Ground whole oats	–	20.0
Feeding grade rolled oats	15.0	12.0
Selenium premix[a]	1.0	1.0
Sow and pig supplement[b]	23.4	20.0
Chlortetracycline (100 g/ton)	0.1	0.1

[a] Adds 0.1 ppm Se. Carrier is finely ground shelled corn.
[b] Sow and pig supplement contains:

Ingredient	%
Dehydrated alfalfa meal, 17% protein	10.0
Soybean meal, 49% protein	61.0
Linseed meal, solvent process	5.9
Distiller's dried solubles	6.6
Dicalcium phosphate	7.5
Ground limestone	3.0
Trace mineralized salt, 1.0% Zn	3.0
Vitamin premix[c]	3.0

[c] Supplies the following per kilogram of premix:

Vitamin D_3 (IU)	95,337
Riboflavin (mg)	399
d-Calcium pantothenate (g)	2.7
Niacin (g)	4.0
Vitamin K (mg)	433
Choline chloride (g)	212.7
Vitamin B_{12} (mg)	3.67
Vitamin E (mg)	2,156
Ethoxyquin (mg)	4,160

Table 10
PIG GROWER DIET USING HIGH-MOISTURE ACID-TREATED SHELLED CORN[10]

Yellow shelled corn, 27% moisture, treated with 1.2% by weight at time of storage of propionic:actic acid (60:40) mixture. The diet contains corn and the following protein, mineral, and vitamin supplement self fed free choice.

Ingredient	%
Soybean meal, 49% protein	52.9
Dicalcium phosphate	1.5
Calcium carbonate	1.5
Salt, trace minerals added[a]	2.5
Vitamin premix[b]	0.8
Antibiotic[c]	0.8
Dehydrated alfalfa meal, 17% protein	20.0
Meat and bone meal, 50% protein	20.0

[a] Contained as a percent of the mix:

NaCl	97.0	Cu	0.048
Zn	0.800	I	0.011
Mn	0.400	Co	0.022
Fe	0.330		

[b] Contained per kilogram of premix:

Riboflavin (g)	1.1
Calcium pantothenate (g)	6.0
Nicotinic acid (g)	16.5
Choline chloride (g)	165.0
Vitamin B_{12} activity (mg)	18.0
Vitamin A (IU)	3,300,000
Vitamin D_2 (IU)	330,000
Vitamin E (D-α-tocopherol acetate) (IU)	22,000

[c] Contained 33.0 g streptomycin (as streptomycin acetate) and 11 g procaine penicillin per kilogram.

Table 11
PIG GROWER DIET[10]

Ingredient	%
Ground yellow shelled corn	78.1
Soybean meal, 49% protein	19.0
Dicalcium phosphate	1.0
Calcium carbonate	1.0
Salt, trace minerals added[a]	0.5
Vitamin premix[b]	0.2
Antibiotic[c]	0.2

[a] Contained as a percent of the mix:

NaCl	97.0	Zn	0.8000
Mn	0.400	Fe	0.330
Cu	0.048	I	0.011
Co	0.022		

[b] Contained per kilogram:

Riboflavin (g)	1.1
Calcium pantothenate (g)	6.0
Nicotinic acid (g)	16.5
Choline chloride (g)	165.0
Vitamin B_{12} activity (mg)	18.0
Vitamin A (IU)	3,300,000
Vitamin D (IU)	330,000
Vitamin E (IU)	22,000

[c] Contained 33.0 g streptomycin (as streptomycin acetate) and 11 g procaine penicillin per kilogram.

Table 12
PIG GROWER DIET[11]

Ingredient	%
Ground yellow shelled corn	55.2
Peanut meal, expeller process, 45% protein	17.5
Blood meal	4.3
Rice bran, solvent process, 14% protein	10.0
Cane molasses	10.0
Ground oyster shell, 33% Ca	1.0
Dicalcium phosphate	1.5
Vitamin – mineral premix[a]	+

[a] A Pfizer® product supplying the following per kilogram of diet:

Vitamins		Minerals	
A (IU)	9823	Manganese (mg)	341
D_3 (IU)	1965	Copper (mg)	344
B_{12} (μg)	10	Zinc (mg)	100
Riboflavin (mg)	41	Iodine (mg)	20
Niacin (mg)	246		
Pantothenic acid (mg)	98		
Folic acid (mg)	10		

Oxytetracycline hydrochloride was supplied at the level of 209 g/ton of diet.

Table 13
PIG GROWER DIET[12]

Ingredient	%
Ground yellow shelled corn	72.7
Soybean seed[a]	20.0
Alfalfa meal, dehydrated, 17% protein	2.5
Poultry offal, dry or wet rendered, dehydrated	2.5
Defluorinated phosphate	1.0
Ground limestone	0.5
Trace mineralized salt[b]	0.4
Trace minerals[c]	0.05
Vitamin-antibiotic premix[d]	0.35

[a] Whole soybeans were processed in an extruder-type cooker (Brady® 206 Crop Cooker, Koehring Farm Division, Des Moines, Iowa) for 22 sec at 143°C.
[b] Provided in addition to salt (in ppm): Zn, 100; Fe, 50; Mn, 27.5; Cu, 5.5; Co, 0.5; I, 0.75.
[c] Provided (in ppm): Zn, 10; Fe, 10; Mn, 30; Cu, 1.0; I, 0.75; Ca, 150.
[d] Provided per kilogram of diet:

Vitamin A (IU)	2500
Vitamin D (IU)	330
Riboflavin (mg)	4.5
d-Calcium pantothenate (mg)	9.0
Niacin (mg)	20
Vitamin B_{12} (μg)	13
Choline (mg)	22
Chlortetracycline (mg)	30

DIETS FOR FINISHING HOGS

Table 14
HOG FINISHING DIET[13]

Pelleted

Ingredient	%
Ground shelled corn	71.0
Soybean meal, 49% protein	13.9
Fish meal	2.0
Citrus pulp, dried	10.0
Prime stabilized tallow	1.8
Dicalcium phosphate	0.6
Ground limestone	0.2
Trace mineralized salt[a]	0.5
Vitamin-antibiotic premix[b]	0.3

[a] Mineral content (%):

Salt	97.500
Manganese sulfate	0.200
Ferrous oxide	0.200
Copper oxide	0.033
Cobalt oxide	0.010
Calcium iodate	0.007
Zinc oxide	0.007
Manganese sulfate	0.100

[b] Each kilogram of vitamin-antibiotic premix contains:

Vitamin A (IU)	880,000
Vitamin D (IU)	88,000
Riboflavin (g)	1.5
Pantothenic acid (g)	4.4
Bacitracin (g)	3.3
Niacin (g)	6.6
Choline chloride (g)	65.1
Vitamin B_{12} (mg)	8.8
Penicillin (g)	0.66

Table 15
HOG FINISHING DIETS[14]

Ingredient	Diet 1 (%)	Diet 2 (%)
Yellow corn meal	65.50	83.90
Soybean meal, 49%	11.40	12.90
Rice bran	20.00	–
Biofos®[a]	1.72	1.85
Ground limestone	0.63	0.60
Iodized salt	0.50	0.50
Trace minerals[b]	0.15	0.15
Vitamin premix[c]	0.10	0.10

[a] Product of International Mineral and Chemical Corporation, Skokie, Illinois.
[b] Supplied by Calcium Carbonate Company, Quincy, Illinois, Contained 20% zinc, 10% iron, 5.5% manganese, 1.1% copper, 0.15% iodine, 0.1% cobalt, and 2% calcium.
[c] Contained per pound premix:

Riboflavin (mg)	6,000
Niacin (mg)	20,000
Pantothenic acid (mg)	12,000
Choline chloride (mg)	80,000
Vitamin B_{12} (μg)	10,000
Vitamin A (IU)	2,500,000
Vitamin D_3 (IU)	400,000
Vitamin E (IU)	10,000

Table 16
DIETS FOR GROWING AND FINISHING HOGS[a;15]

Ingredient	Growing (%)	Finishing (%)
Ground shelled corn	68.79	82.84
Ground oats	10.00	–
Meat scraps, 50% protein	3.00	2.00
Alfalfa meal, dehydrated, 17% protein	2.50	2.50
Soybean meal	14.00	11.00
Dicalcium phosphate	1.00	1.00
Salt, trace mineralized	0.50	0.50
Vitamin-zinc-antioxidant premix[b]	0.13	0.13
Antibiotic-antibacterial additive[c]	0.03	0.03
Pro-Gen®[d]	0.05	

[a] Both diets may be self fed as a dry mash or as a paste (1 part dry diet to 1.2 to 1.5 parts water by weight).
[b] Provides the following per pound of diet:

Riboflavin (mg)	4.0	Vitamin B_{12} (μg)	10
Pantothenic acid (mg)	7.3	Vitamin D_2 (IU)	90
Niacin (mg)	18.0	Zinc (mg)	36.3
Choline (mg)	20.0	BHT (antioxidant) (mg)	57

[c] Provides 5.0 mg Aureomycin® (Aurofac 50), 3.75 mg zinc bacitracin, and 2.5 mg penicillin per pound of diet.
[d] Provides 45.4 mg arsanilic acid per pound of diet.

Table 17
DIETS FOR GROWING-FINISHING HOGS[16]

Ingredient	Diet 1 (%)	Diet 2 (%)
Ground shelled corn	77.25	–
Ground wheat	–	77.25
Soybean meal, 49% protein	18.75	18.75
Defluorinated phosphate	1.50	1.50
Ground limestone	0.40	0.40
Iodized salt	0.50	0.50
Vitamin mix[a]	0.50	0.50
Trace mineral mix[b]	0.10	0.10
Fat	1.00	1.00
Antibiotics[c]	+	+

[a] Provides per pound of complete feed:

Vitamin A (IU)	5,000
Vitamin D (IU)	2,500
Vitamin E (IU)	5
Vitamin B_{12} (μg)	20
Riboflavin (mg)	5
Pantothenic acid (mg)	12
Niacin (mg)	25
Choline chloride (mg)	500

[b] Provides (in ppm of diet): zinc, 200; iron, 100; manganese, 55; copper, 11; iodine, 1.5; and cobalt, 1.0.

[c] Aureo SP-250® fed at level of 5 lb per ton of diet to 125 lb body weight.

Table 18
DIETS FOR GROWING AND FINISHING HOGS[a;17]

Ingredient	Grower (%)	Finisher (%)	Grower (%)	Finisher (%)
Ground yellow corn	71.85	76.90	–	–
Sorghum grain[b]	–	–	72.55	77.60
Soybean meal, 49% protein	21.60	16.60	20.90	15.90
Stabilized animal fat	3.00	3.00	3.00	3.00
Biofos®[c]	1.60	1.80	1.60	1.80
Ground limestone	1.00	1.00	1.00	1.00
Iodized salt	0.50	0.50	0.50	0.50
Trace minerals[d]	0.10	0.10	0.10	0.10
Vitamin mix[e]	0.10	0.10	0.10	0.10
Antibiotic	0.25	–	0.25	–

[a] Grower diets were fed until pigs averaged 125 lb in each group; then the finisher diets were fed.
[b] Non-bird-resistant sorghum grain (McNair 652).
[c] International Mineral and Chemical Corporation, Skokie, Illinois.
[d] Contained 20% zinc, 10% iron, 5.5% manganese, 1.1% copper, 0.15% iodine, 0.1% cobalt, and 2.0% calcium.
[e] Contained the following per pound of premix:

Vitamin A (IU)	2,500,000	Riboflavin (mg)	6,000
Vitamin D_3 (ICU)	400,000	Niacin (mg)	20,000
Vitamin E (IU)	10,000	Pantothenic acid (mg)	12,000
Vitamin B_{12} (μg)	10,000	Choline chloride (mg)	80,000

Table 19
DIETS FOR GROWING AND FINISHING HOGS[18]

Ingredient	Growing (%)	Finishing (%)
Ground shelled corn	80.6	84.3
Selenium premix[a]	1.0	1.0
Hog supplement[b]	18.4	14.7

[a] Supplies 0.1 ppm of selenium.
[b] Hog supplement contains:

Ingredient	%
Alfalfa meal, dehydrated, 17% protein	12.6
Soybean meal, 49% protein	66.9
Corn distillers dried solubles	5.0
Dicalcium phosphate	6.5
Ground limestone	5.0
Trace mineralized salt	3.0
Vitamin premix[c]	1.0

[c] Supplies the following per pound premix:

Vitamin D_3 (IU)	204,000	Riboflavin (mg)	444
Vitamin B_{12} (μg)	4,000	*d*-Calcium pantothenate (mg)	2,914
Vitamin E (IU)	5,000	Niacin (mg)	4,198
Ethoxyquin (mg)	5,670		

Table 20
DIETS FOR GROWING AND FINISHING HOGS[19]

	Growing		Finishing	
Ingredient	%	%	%	%
Normal corn, 0.32% lysine	80.80	–	85.75	–
High lysine corn, 0.43% lysine[a]	–	85.05	–	90.10
Soybean meal, 49% protein	15.35	11.05	10.15	5.90
Dicalcium phosphate	1.85	2.00	2.00	2.15
Ground limestone	0.45	0.35	0.35	0.30
Salt	0.50	0.50	0.50	0.50
Trace mineral[b]	0.05	0.05	0.05	0.05
Vitamins A, B premix[c]	1.00	1.00	1.00	1.00

[a] Rations based on high lysine corn should be formulated on the basis of the *determined lysine and protein* contents of the corn.
[b] Swine 10% Zn, Calcium Carbonate Company, Quincy, Illinois.
[c] Contributed the following amounts per pound of complete diet:

Vitamin A (IU)	3,000	Pantothenic acid (mg)	6.0
Vitamin D (IU)	400	Choline chloride (mg)	100
Vitamin B_{12} (μg)	20	Menadione sodium bisulfite (mg)	1.26
Riboflavin (mg)	1.6	Aureomycin® (mg)	20
Niacin (mg)	16.0		

Table 21
DIETS FOR GROWING AND FINISHING HOGS CONTAINING CORN, WHEAT, MILO, OR BARLEY AND SOYBEAN MEAL OR ROASTED SOYBEANS[20]

Percent Composition of Diets

	Weight ranges fed			
	21–57 kg[a]		57–101 kg[b]	
Ingredient	**SBM**	**RSB[c]**	**SBM**	**RSB**
Grain[d]	78.3	71.3	83.6	78.6
Soybean meal, 49%	18.6	–	13.3	–
Roasted soybeans	–	25.6	–	18.3
Dicalcium phosphate	1.6	1.6	1.6	1.6
Ground limestone	0.5	0.5	0.5	0.5
Trace mineralized salt[e]	0.5	0.5	0.5	0.5
Vitamin mixture[f]	0.5	0.5	0.5	0.5

[a] With corn as the grain, each of these diets was calculated to contain 16% crude protein. Wheat, milo, or barley as substituted on an equal weight basis for corn to formulate the remaining diets.

[b] With corn as the grain, each of these diets was calculated to contain 14% crude protein. Wheat, milo, or barley as substituted on an equal weight basis for corn to formulate the remaining diets.

[c] Roasted whole soybeans, 38% crude protein, processed in an electric Roast-a-tron cooker (Mix Mill, Inc., Bluffton, Indiana) with soybeans having an exit temperature of 110°C. Roasted soybeans were ground before diets were mixed.

[d] Grains were either corn (8.7% crude protein), barley (11.5% crude protein), milo (9.6% crude protein), or wheat (11.6% crude protein).

[e] Contained 91 to 94% salt and provided to the complete diet 50 ppm Zn, 40 ppm Mn, 25 ppm Fe, 5 ppm Cu, 0.5 ppm I, and 0.5 ppm Co.

[f] Each kg of mixture contained:

Vitamin A (IU)	880,000	Riboflavin (mg)	880
Vitamin D (IU)	88,000	Pantothenic acid (g)	3.52
Vitamin E (IU)	2,200	Niacinamide (g)	4.4
Vitamin B_{12} (mg)	4.4	Choline (g)	22

Table 22
DIETS FOR FINISHING HOGS[a,21]

Ingredient	Diet 1 (%)	Diet 2 (%)	Diet 3 (%)
Ground yellow corn	76.6	64.3	52.1
Soybean meal, 49% protein	14.7	17.0	19.2
Linseed meal, solvent process	5.0	5.0	5.0
Dicalcium phosphate	1.3	1.4	1.4
Ground limestone	0.8	0.8	0.8
Trace mineralized salt, 1.0% Zn	0.5	0.5	0.5
Vitamin premix[b]	1.0	1.0	1.0
Ground sawdust	–	10.0	20.0

[a] All diets pelleted and self fed.
[b] Supplied the following per kilogram premix:

Vitamin A (IU)	220,000	Riboflavin (mg)	264
Vitamin D_3 (IU)	88,000	Niacin (mg)	2,200
Vitamin E (IU)	11,000	Choline chloride (g)	26.4
Vitamin K (menadione sodium bisulfite) (mg)	220	Vitamin B_{12} (mg)	2.0
d-Calcium pantothenate (mg)	1,100	Ethoxyquin (g)	12.5

DIETS FOR BREEDING STOCK

Table 23
DIET FOR ALL CLASSES OF BREEDING ANIMALS[22]

Energy intake is controlled by regulating daily feed intake.

Ingredient	%
Ground shelled corn	37.0
Ground oats	20.0
Wheat standard middlings	15.0
Alfalfa meal, dehydrated, 17% protein	10.0
Sow supplement[a]	17.0
Selenium premix (supplies 0.1 ppm Se)	1.0
Zinc oxide	5.7 g

[a] Sow supplement contains:

Ingredient	%
Alfalfa meal, dehydrated, 17% protein	10.0
Soybean meal, 49% protein	61.0
Linseed meal, solvent process	5.9
Distillers dried solubles	6.6
Dicalcium phosphate	7.5
Ground limestone	3.0
Trace mineralized salt, 1.0% Zn	3.0
Vitamin premix[b]	3.0

[b] Sow supplement vitamin premix supplies the following per kilogram of premix:

Vitamin D_3 (IU)	95,337
Riboflavin (mg)	339
d-Calcium pantothenate (g)	2.7
Niacin (g)	4.0
Vitamin K (mg)	433
Choline chloride (g)	212.7
Vitamin B_{12} (mg)	3.67
Vitamin E (mg)	2156
Ethoxyquin (mg)	4160

Table 24
GESTATION DIET FOR SWINE[23]

Ingredient	%
Ground yellow corn	80.055
Soybean meal, 48.5% protein	16.30
Calcium carbonate	1.30
Dicalcium phosphate	1.10
Salt	0.60
Vitamin premix[a]	0.50
Trace mineral premix[b]	0.05
Choline chloride, 434 g choline/kg	0.095

[a] Provided the following per sow per day during gestation:

Vitamin A (IU)	12,000
Vitamin D (IU)	1,100
Vitamin B_{12} (μg)	30
Vitamin E (mg)	100
Riboflavin (mg)	7
Niacin (mg)	24
Pantothenic acid (mg)	28

[b] Provided the following in milligrams per kilogram diet: Mn, 66; Fe, 44; Cu, 20; Co, 3.1; Zn, 69; I, 1.6.

Table 25
DIET FOR GESTATION AND LACTATION[24]

Ingredient	%
Ground yellow corn	79.3
Soybean meal, 49% protein	18.0
Ground limestone	1.0
Dicalcium phosphate	1.0
Vitamin mixture[a]	0.2
Trace mineralized salt[b]	0.5

[a] Contained per kilogram mix:

Vitamin A (IU)	3,300,000
Vitamin D (IU)	330,000
Vitamin B_{12} activity (mg)	18.0
Calcium pantothenate (g)	5.5
Riboflavin (g)	1.1
Nicotinic acid (g)	16.5
Choline chloride (g)	110.3

[b] Contained as a percent of the mix: salt, 97.0; zinc, 0.800; manganese, 0.400; iron, 0.300; copper, 0.048; iodine, 0.011; cobalt, 0.022.

Table 26
DIETS FOR LACTATION[a;25]

Ingredient	Diet 1 (%)	Diet 2 (%)	Diet 3 (%)	Diet 4 (%)
Ground yellow dent corn	79.1	75.2	74.8	71.0
Soybean meal, 49% protein	18.1	17.1	17.4	16.4
Alfalfa meal, dehydrated, 17%	–	5.0	–	5.0
Wheat bran, dry milled	–	–	5.0	5.0
Trace mineralized salt[b]	0.5	0.5	0.5	0.5
Ground limestone	1.3	1.1	1.5	1.3
Dicalcium phosphate	0.9	1.0	0.7	0.7
Vitamin mixture[c]	0.1	0.1	0.1	0.1

[a] Diets were fed *ad libitum* during lactation.
[b] Contributed the following per pound diet: NaCl, 2.2 g; Zn, 18.2 mg; Mn, 9.1 mg; Fe, 7.5 mg; Cu, 1.1 mg; I, 0.25 mg; Co, 0.5 mg.
[c] Contributed the following per kilogram diet:

Vitamin A (IU)	344	Riboflavin (mg)	0.45
Vitamin D_2 (IU)	34.5	Calcium pantothenate (mg)	2.3
α-Tocopheryl acetate (IU)	10	Niacin (mg)	7.3
Vitamin B_{12} (μg)	8.2	Choline chloride (mg)	50.0

Table 27
DIETS FOR GESTATION[a;26]

Ingredient	Diet 1 (%)	Diet 2 (%)	Diet 3 (%)	Diet 4 (%)	Diet 5 (%)	Diet 6 (%)
Ground yellow corn	12.00	40.00	40.00	12.00	12.00	–
Ground whole oats	65.00	17.50	17.50	32.50	–	–
Wheat bran, dry milled	–	2.50	2.50	–	–	–
Soybean meal, 44% protein	18.50	12.50	12.50	18.00	17.50	–
Alfalfa meal, dehydrated, 17% protein	–	25.00	–	–	–	–
Alfalfa hay, sun cured, midbloom	–	–	25.00	33.00	66.00	96.75
Iodized salt	0.50	0.50	0.50	0.50	0.50	0.50
Steamed bone meal	3.02	–	–	1.51	–	–
Sodium tripolyphosphate	0.40	1.50	1.50	1.30	1.90	2.40
Trace minerals[b]	0.10	0.10	0.10	0.10	0.10	0.10
Vitamin mixture	0.48[c]	0.40[d]	0.40[d]	1.09[d]	2.00[d]	0.25[d]

[a] All diets pelleted and fed once daily in individual stalls.
[b] Contributed the following per kilogram diet (in milligrams): Zn, 200; Fe, 100; Mn, 54; Cu, 11; Co, 1.0; I, 1.5.
[c] Contributed the following per kilogram diet:

Vitamin A (IU)	3435	Calcium pantothenate (mg)	4.0
Vitamin D_3 (IU)	248	Niacin (mg)	9.9
Vitamin B_{12} (μg)	11	Choline chloride (mg)	11.0
Riboflavin (mg)	2.2		

[d] Contributed the following per Kilogram diet:

Vitamin D_3 (IU)	248	Calcium pantothenate (mg)	4.0
Vitamin B_{12} (μg)	11	Niacin (mg)	9.9
Riboflavin (mg)	2.2	Choline chloride (mg)	11.0

REFERENCES

1. **Miller, E. R., Johnson, R. L., Hoefer, J. A., and Luecke, R. W.,** *J. Nutr.,* 52, 405–413, 1954.
2. **Lucas, I. A. M. and Lodge, G. A.,** *The Nutrition of the Young Pig. A Review,* Commonwealth Agricultural Bureaux, Farnham Royal, Slough, Bucks, Engl., 1961, 66.
3. **Gobble, J. L.,** Pennsylvania Agricultural Experiment Station, University Park, unpublished data.
4. **Coalson, J. A., Maxwell, C. V., Hillier, J. C., Nelson, E. C., Anderson, I. L., and Corley, L. D.,** *J. Anim. Sci.,* 36, 259–263, 1973.
5. **Gobble, J. L.,** This diet is currently being used as creep feed for suckling pigs at the Swine Research Center, Dept. of Dairy and Animal Science, Pennsylvania Agricultural Research Station, University Park.
6. **Rust, J. W., Meade, R. J., and Hanson, L. E.,** *J. Anim. Sci.,* 35, 963–966, 1972.
7. **Meade, R. J., Rust, J. W., Miller, K. P., Hanke, H. E., Grant, R. S., Vermedahl, L. D., Wass, D. F., and Hanson, L. E.,** *J. Anim. Sci.,* 29, 303–308, 1969.
8. **Bayley, H. S. and Carlson, W. E.,** *J. Anim. Sci.,* 30, 394–401, 1970.
9. **Gobble, J. L.,** These are the standard pig grower diets currently fed at the Swine Research Center, Dept. of Dairy and Animal Science, Pennsylvania Agricultural Experiment Station, University Park.
10. **Lynch, P. B., Hall, G. E., Hill, L. D., Hatfield, E. E., and Jensen, A. H.,** *J. Anim. Sci.,* 40, 1063–1069, 1975.
11. **Babatunde, G. M., Fetuga, B. L., and Oyenuga, V. A.,** *J. Anim. Sci.,* 40, 632–639, 1975.
12. **Seerley, R. W., Emberson, J. W., McCambell, H. C., Burdick, D., and Grimes, L. W.,** *J. Anim. Sci.,* 39, 1082–1091, 1974.
13. **Baird, D. M., Allison, J. R., and Heaton, E. K.,** *J. Anim. Sci.,* 38, 545–553, 1974.
14. **Compabadal, C. M., Wallace, H. D., and Combs, G. E.,** Res. Rep. No. AL-1975-9, Dept. of Animal Science, Florida Agricultural Experiment Station, Gainesville, 1975, 1–5.
15. **Grifo, A. P., Jr., Teague, H. G., and Roller, W. L.,** Res. Summary 78, Swine Research, Ohio Agricultural Research and Development Center, Wooster, 1974, 21–25.
16. **Hammell, D. L., Becker, H. N., and Combs, G. E.,** Animal Science Mimeo Series No. ARCLO 75-3, University of Florida Agricultural Research Center, Live Oak, 1975, 1–4.
17. **Wallace, H. D., Combs, G. E., and Honser, R. H.,** Res. Rep. No. AL-1974-11, Dept. of Animal Science, Florida Agricultural Experiment Station, Gainesville, 1975, 1–3.
18. **Gobble, J. L.,** Standard growing and finishing diets currently fed at the Swine Research Center, Dept. of Dairy and Animal Science, Pennsylvania Agricultural Experiment Station, University Park.
19. **Moser, R. D.,** Nebraska Swine Rep. E. C. 73–219, Dept. of Animal Science, Lincoln College of Agriculture, University of Nebraska, 1973, 11–13.
20. **McConnell, J. C., Skelley, G. C., Handlin, D. L., and Johnston, W. E.,** *J. Anim. Sci.,* 41, 1021–1030, 1975.
21. **Gobble, J. L., Hazlett, V. E., Sherritt, G. W., and Wagsness, P. J.,** Pennsylvania Livestock Day, Animal Science Res. Summary AS-SW-75-7, Dept. of Animal Science, Pennsylvania State University, University Park, 1975, 135–356.
22. **Gobble, J. L.,** Breeding stock diet currently fed at the Swine Research Center, Dept. of Dairy and Animal Science, Pennsylvania Agricultural Experiment Station, University Park.
23. **Stockland, W. L. and Blaylock, L. G.,** *J. Anim. Sci.,* 39, 1113–1116, 1974.
24. **Harmon, B. G., Liu, C. T., Cornelius, S. G., Pettigrew, J. E., Baker, D. H., and Jensen, A. H.,** *J. Animal. Sci.,* 39, 1117–1122, 1974.
25. **Baker, D. H., Harmon, B. G., and Jensen, A. H.,** *J. Anim. Sci.,* 39, 838–840, 1974.
26. **Danielson, D. M. and Noonan, J. J.,** *J. Anim. Sci.,* 41, 94–99, 1975.

DIETS (NATURAL AND SYNTHETIC): WILD PIGS, PECCARIES, AND HIPPOPOTAMUSES

L. K. Sowls

Information on the wild diets of Suidae (wild pigs), Tayassuidae (peccaries), and Hippopotamidae (hippopotamuses) is sketchy, except for some localized detailed studies.

Wild Boar (*Sus scrofa*)

The wild boar is primarily a vegetarian, but takes some meat. Because of its wide range in Europe and Asia, a list of plant species eaten by this species would be very long. Species of plants taken depends upon locality, but groups include a variety of green forbs, tubers, roots, acorns, and various other fruits.[2,21]

Crandall[1] reports that wild boars do well in captivity on almost any available greens, along with potatoes, carrots, apples mixed with meal, bread, chopped horse meat, and some alfalfa hay.

Warthog (*Phacochoerus aethiopicus*)

This common species is present in a wide variety of vegetative types in Africa. It is considered by many to be essentially a grazer, and also takes roots and tubers, leaves, and fruits of a wide variety of plants.[2,13,16,21] Smith and Ledger[17] found the wet stomach contents of 23 adult warthogs examined in Kenya to range from 127 to 1800 g, averaging 653 g. Crandall[1] says that the diet previously described for *Sus scrofa* is satisfactory for warthogs in captivity. Sowls[19] found commercial hog pellets to be an excellent diet for growing, immature warthogs.

Bushpig (*Potamochoerus porcus*)

The bushpig has a wide variety of habitats on the African continent, so a large number of plant species are taken. According to Fitzsimmons,[5] South African bushpigs take roots, bulbs, and fruits. Shortridge[15] agrees, but adds that they also devour reptiles, eggs, and birds, and tells of one instance where the carcass of a bushbuck was eaten by bushpigs. Roberts[13] says that the normal diet of the bushpig is roots and edible vegetable matter. Phillips[12] gives a rather extensive list of plants eaten by bushpigs in South Africa, including 7 species of ferns, 9 species of monocotyledons, 8 species of dicotyledons, and 26 species of forest tree fruits. Smithers[16] says bushpigs are predominantly vegetarian, feeding on roots, fallen fruits, and agricultural crops. Dorst and Dandelot[2] state that they also eat insects, bird eggs, carrion, and reptiles, but prefer roots and bulbs as their favorite diet. In agricultural areas surrounded by brushy timber, the bushpig makes domestic crops a large part of his diet. Maize, groundnuts, and field peas are taken by Rhodesian bushpigs in large quantities as the fruits become mature,[20] while Smith[17] found bushpigs feeding in garbage dumps in Uganda.

Sowls and Phelps[20] found that a diet of commercial hog pellets, along with some greens and vegetables, was an excellent diet for captive animals; young on this diet gained up to 2 lb per week.

Babirusa (*Babirussa babyrussa*)

Little has been written regarding this wild hog of the north Celebes Islands. Its diet consists of roots, berries, and other vegetation.[21] Though the babirusa is not a common species in American and European zoos, Crandall[1] states that it does well on the same diet as that given to the wild boar.

Tayassuidae

There are three living species of this family in the western hemisphere:

Collared peccary (*Dicotyles tajacu*) – The habitat of species ranges from the southwestern United States to the Rio Plata of Argentina; because of the variety of vegetative areas inhabited by it, a large number of plant species are eaten. Detailed studies have shown that in the southwestern United States it eats succulent plants (genera *Cacti* and *Agave*), roots and tubers, green forbs, grass, fruits, and nuts of a wide variety of plants. Numerous local studies of the foods eaten by this species have been made in Arizona and Texas.[3,4,6,7,10,11,14] Sowls[18] found that *D. tajacu* in captivity thrives and reproduces on a variety of fruits, vegetables, and prepared pelleted hog feeds, with added vitamins and minerals.

White-lipped peccary (*Tayassu pecari*) – Very little has been published concerning the diet of this tropical species. Leopold[9] says that they eat a wide variety of fruits, green vegetation, and roots. Crandall[1] found that it thrived well in captivity on the same foods given to the collared peccary.

***Gran Chaco peccary* (*Catagonus wagneri*)** – No information has been published regarding the foods taken by this peccary since it was discovered in 1972.[22]

Hippopotamus

The hippopotamus is a nocturnal feeder, emerging from pools during the night and feeding in an area that may extend several miles from the water; grasses are the principal plants eaten.[2,13,16,21] Dorst and Dandelot[2] list the following genera of grasses: *Therneda, Panicum,* and *Heteropogon*; they also include an aquatic plant, *Pistia stratiotes.* Hippopotamuses have also been found to eat the fruits of the sausage tree, *Kigelia pinnata.*

Although not a ruminant, the hippopotamus has a four-chambered stomach which allows for the holding of large quantities of food and the breakdown of cellulose by microbial action.[8] Smith and Ledger[17] have found that the wet contents of the stomach of nine adult hippopotamuses ranged from 134 to 293 kg, averaging 199 kg.

Crandall[1] lists the daily ration of an adult hippopotamus in the New York Zoological Park as composed of 80 to 100 lb of alfalfa hay, 12 lb of pelleted grain mixture with vitamin and mineral supplements, and about 8 qt of cut potatoes, apples, cabbage, and carrots, plus two loaves of stale bread. In large enclosures, grass for grazing is desirable, and rotating use of different paddocks provides fresh food.

REFERENCES

1. **Crandall, L. S.,** *The Management of Wild Mammals in Captivity,* University of Chicago Press, Chicago, 1964.
2. **Dorst, J. and Dandelot, P.,** *A Field Guide to the Larger Mammals of Africa,* Collins, London, 1970.
3. **Eddy, T. A.,** Foods and feeding patterns of the collared peccary in southern Arizona, *J. Wildl. Manage.,* 25, 248–257, 1961.
4. **Ellisor, J. E. and Harwell, W. F.,** Mobility and home range of collared peccary in southern Texas, *J. Wildl. Manage.,* 33, 425–427, 1969.
5. **Fitzsimmons, F. W.,** *The Natural History of South Africa,* Vol. 3, Longmann, Green & Co., London, 1920.
6. **Jennings, W. S. and Harris, J. T.,** The Collared Peccary in Texas, Ser. 12, Texas Parks and Wildlife Dept., Austin, 1953, 1–31.
7. **Knipe, T.,** The Javelina in Arizona, Wildlife Bull. No. 2, Arizona Game and Fish Dept., Tucson, 1957.
8. **Langer, P.,** Macroscopic anatomy of the stomach of the *Hippopotamidae,* Gray, 1821, *Zentralbl. Veterinaermed. Reihe C,* 4, 334–359, 1975.
9. **Leopold, A.,** *Wildlife of Mexico, the Game Birds and Mammals,* University of California Press, Berkeley, 1959.
10. **Low, W.,** The Influence of Aridity on Reproduction of the Collared Peccary (*Dicotyles tajacu* L.) in Texas, Ph.D. dissertation, University of British Columbia, Vancouver, 1970.
11. **Neal, B. J.,** A contribution on the life history of the collared peccary in Arizona, *Am. Midl. Nat.,* 61, 177–190, 1959.
12. **Phillips, J. F. V.,** "Wild pig" *Potamchoerus choeropotamus* at the Knysna: notes by a naturalist, *S. Afr. J. Sci.,* 23, 655–660, 1926.
13. **Roberts, A.,** *The Mammals of South Africa,* The Mammals of South Africa Book Fund, Central News Agency, Capetown, 1951.
14. **Schweinsburg, R.,** The home range, movements and herd integrity of the collared peccary in southern Arizona, *J. Wildl. Manage.,* 35, 455–460, 1971.
15. **Shortridge, G. C.,** *The Mammals of South West Africa,* William Heinemann, London, 1934.
16. **Smithers, R. H. N.,** *The Mammals of Rhodesia, Zambia and Malawi,* Collins, London, 1966.
17. **Smith, N. S. and Ledger, H. P.,** unpublished.
18. **Sowls, L. K.,** Reproduction in the collared peccary (*Pecari tajacu*), in *Comparative Biology of Reproduction in Mammals,* Academic Press, London, 1966, 155–172.
19. **Sowls, L. K.,** unpublished.
20. **Sowls, L. K. and Phelps, R. J.,** Observations on the African bushpig, *Potamocheorus porcus* L., in Rhodesia, *Zoologica,* 53, 75–84, 1968.
21. **Walker, E. P. et al.,** *Mammals of the World,* Vol. 2, Johns Hopkins Press, Baltimore, 1964.
22. **Wetzel, R. M., Dubos, R. E., Martin, R. L., and Myers, P.,** Catagonus, an "extinct" peccary, alive in Paraguay, *Science,* 189, 379–380, 1975.

DIETS (NATURAL AND SYNTHETIC): NEW WORLD CAMELIDAE

S. Fernandez-Baca

NEW WORLD CAMELIDAE

Four species constitute the group of New World Camelidae: the llama, *Lama glama* (Linné); the alpaca, *Lama pacos* (Linné); the guanaco, *Lama guanicoe* (Müller); and the vicuña, *Vicugna vicugna* (Molina). These four species are sometimes grouped within a single genus (*Lama*). However, it is generally accepted that the peculiar features of the vicuña, such as the rodent-like lower incisor teeth with enamel on only one face and an open root, justify its separate generic classification. These four species are often referred to as *Auchenidae* or *Auquenidos*. This designation originated with Illiger in 1811, who gave the name *Auchenia* to the genus without realizing that it had previously been named *Lama* by Frish in 1775. These four species and the one-humped (*Camelus dromedarius* L.) and two-humped camels (*C. bactrianus* L.), which constitute the group called Old World Camelidae, belong to the Camelidae family. Among the living artiodactyls, Camelidae is the only family in the suborder Tylopoda.

Fossil discoveries indicate that most of the evolution of Camelidae occurred in western North America. From there, the New World forms extended their range to South America while the Old World forms spread across the Bering Straits into Asia. Although the separation of these two groups occurred more than 1 million years ago and considerable morphological and physiological changes have taken place in response to their different environments, their karyotypes have undergone remarkably few changes. Similar karyotypes and a modal chromosome number of 74 have been reported for all of the six species.[1,2]

Crosses occur between members of the New World Camelidae, especially between the llama and the alpaca and the male alpaca and the female vicuña. The resulting hybrids are fertile. The similarity of karyotypes probably explains the reproductive ability of the hybrids.

The llama and alpaca exist only under domestication and are the most economically important Camelidae. They are raised primarily for fiber and meat. In addition, llamas are still used as beasts of burden in some Andean regions of South America. They are raised in an extensive area of the Andes, mostly situated over 4000 m above sea level, where the pasture grasses are woody and fibrous and of very poor quality.

The vicuña and guanaco exist only in the wild. The vicuña provides a fiber of extraordinary quality; this contributed to indiscriminate hunting which threatened the extinction of the species.

ANATOMIC AND PHYSIOLOGIC CHARACTERISTICS OF THE DIGESTIVE SYSTEM

Like sheep and cattle, the New World camelids have a capacious compartmentalized stomach and are ruminating animals. However, there are distinct structural differences between the stomachs of these two groups of animals. Although the existence of four stomachs in camelids has been recognized by some authors, recent work[3] shows the existence of only three well-differentiated compartments. The first compartment, which has frequently been compared to the rumen of the true ruminants, contains approximately 83% of the total gastric volume. It is ventrally divided into cranial and caudal sacs by a prominent transverse sulcus. The walls of both sacs contain oblique grooves with glandular pouches. The second compartment, a reniform, thick-walled organ, contains approximately 6% of the total gastric volume. The elongate third compartment contains about 11% of the total gastric volume. In the alpaca, the total capacity of the four compartments has been estimated to be 11.5 l.[4]

Much of the total gastric area of the camelid is occupied by a mucigenous glandular mucosa which covers large areas in all three compartments. Electron microscopic studies of this mucosal epithelium have been reported.[5] However, the role of these glandular structures in the process of digestion remains to be determined. A relatively high concentration of neutral proteases was found in the glandular-pouch epithelial tissue of the second compartment, but no cellulase or amylase activity was noted in the glandular epithelium of any of the three compartments.[6] Despite these structural differences, the camelids perform the basic processes of ruminant digestion, including microbial fermentation, regurgitation, and eructation.[7,8]

Camelids do not possess a gall bladder. Blood glucose levels in the alpaca are higher than in other ruminants, usually above 90 mg glucose per 100 ml blood.[4]

Motility studies indicate that the first and second compartments of the stomachs of llama and guanaco undergo cyclic activity similar to that of the rumen and reticulum in cattle and sheep, but with a greater frequency of contraction and rumination cycles. There appears to be a more rapid increase in the concentration of volatile fatty acids after feeding in the llama and guanaco than in sheep.[8] However, in comparative studies between sheep and alpaca at 3270 m above sea level under the same feeding and management conditions, a slower rate of increase in the concentration of volatile fatty acids after feeding was observed in the alpaca, but the levels remained high for a significantly longer period of time than in sheep.[9] It has been postulated that, because of more effective means of buffering, the short chain fatty acids produced might, in turn, tend to encourage volatile fatty acid production; thus, their level remains high for a longer period.[8]

DIGESTIBILITY OF NUTRIENTS

Digestion trials have shown that alpacas, llamas, and guanacos are more efficient in the digestion of forage than sheep, especially with respect to dry matter and crude fiber. Table 1 presents a summary of the digestibility coefficients of crude fiber reported for llamas, alpacas, and sheep, using different roughages at an altitude of over 3500 m above sea level.

In another trial[13] it was found that llama and guanaco are significantly more efficient in the digestion of dry matter – neutral detergent fiber, acid detergent fiber, and cellulose

Table 1
DIGESTIBILITY OF CRUDE FIBER CONSUMED BY SHEEP, ALPACA, AND LLAMA

	Digestibility coefficient (%)[a]			
Forage	Sheep	Alpaca	Llama	Ref.
Oat hay	49.7	77.5		10
Oat hay	51.6	58.0		11
Alfalfa hay	60.5		68.4	12
Alfalfa hay	46.5	52.8		11
Barley hay	48.0		65.0	12
Oat silage	54.6	71.2		10
Totora (*Scirpus* spp.)	66.4	79.1		10

[a] Digestibility coefficients were determined in all cases by total collection of feces.

– than sheep and ponies (Table 2). These results appear to indicate that New World camelids have a more efficient digestive system than sheep. It has been suggested that the greater efficiency of digestion of the camelids may be due to the greater frequency of their forestomach contractions and rumination cycles, which may provide more efficient maceration, mixing, and absorption.

The greater efficiency of digestion of the camelids may explain their capacity to utilize the fibrous, poor quality grasses that exist at the very high altitudes at which neither sheep nor cattle can be profitably raised.

DIETS COMMONLY USED FOR CAMELIDS

In their natural environment, camelids depend entirely on pastures for food. The quality and availability of these pastures depend on the altitude and season of the year. They are abundant during the rainy season, but scarce during the dry season, which is approximately 8 months of the year (April to November).

Grasses are the predominant forage species, while legumes are very scarce. This is probably a consequence of inadequate pasture management rather than lack of adaptation of legumes. For alpaca grazing, legume species such as *Trifolium repens* can be established in association with grasses such as *Lolium perenne* at altitudes as high as 4200 m above sea level.

Alpacas and llamas never consume any mineral supplement and there are no reports of specific nutritional deficiencies such as those caused by a lack of trace minerals or vitamins. Camelids are unable to lick; therefore, any mineral supplement, if needed, would have to be incorporated into the diet rather than provided as the mineral licks usually used for sheep and cattle.

When confined, New World camelids can eat any forage eaten by other ruminants. As

Table 2
DIGESTIBILITY OF TWO RATIONS BY NEW WORLD CAMELS, SHEEP, AND PONIES

	Dry matter		Neutral detergent fiber		Acid detergent fiber		Cellulose		Crude protein	
	Mean	S.E	Mean	S.E	Mean	S.E	Mean	S.E	Mean	S.E
				Digestibility of Alfalfa (%)						
Ponies	64.8	1.7	60.1	0.6	46.7	1.3	51.2	3.0	65.7	3.1
Sheep	63.8	1.1	58.5	1.1	50.5	0.2	64.8	4.1	69.7	0.2
Camels	71.5	1.7	69.4	2.4	61.0	4.0	77.6	0.9	74.7	2.7
				Digestibility of Pelleted Ration (%)						
Ponies	69.7	0.9	45.4	1.7	29.3	2.3	34.5	2.3	68.8	1.4
Sheep	71.3	1.7	44.7	1.3	34.0	0.4	39.4	1.4	63.1	1.0
Camels	78.0	1.1	58.6	3.0	38.7	4.5	47.2	5.2	69.2	2.4
				Ration Composition (%)						
Alfalfa	90.8		44.2		29.3		21.3		20.5	
Pelleted ration	91.2		35.2		13.7		10.6		12.7	

From Hintz, H. F., Schryver, H. F., and Halbert, M., *Anim. Prod.*, 16, 303, 1973. With permission.

with most animals, an adaptation period may be necessary before the camelids entirely accept a new feed. The length of the adaptation period will depend on the previous feeding and management practices. Forages such as alfalfa hay, oat hay, or barley are readily accepted. Concentrate mixtures such as those used for cattle or sheep are also readily eaten by confined camelids.

A summary of approximate live weights and daily dry matter intake is presented in Table 3. A daily intake of 2 kg dry matter has been estimated on the basis of the data reported in the digestion trials, i.e., the equivalent of approximately 65 g dry matter per kilogram of body weight$^{0.75}$. Weights of the animals may show great individual differences within each species.

DIET AND ITS RELATION TO OVERGROWTH OF INCISOR TEETH

When camelids are confined with no grazing, there is usually an overgrowth of the lower incisors. In some cases, this may cause considerable discomfort. Although there is no precise information, it appears that when animals have free access to grazing (which is usually done on very short grasses), the incisors wear off gradually, thus preventing their excessive growth. However, it has been observed that there is considerable individual variation in the extent of incisor overgrowth. This suggests that, rather than being a characteristic common to all animals, it is an inherited characteristic which manifests itself according to the feeding conditions to which the animals are subjected.

Table 3
ADULT WEIGHT AND APPROXIMATE FEED CONSUMPTION OF NEW WORLD CAMELIDS

Species	Live weight[a] (kg)	Feed consumption,[b] dry matter/day (kg)
Alpaca	75	1.5
Llama	120	2.4
Guanaco	100	2.0
Vicuña	40	0.8

[a] These weights are estimates based on published data and personal observations by the author. Adult weights may vary considerably with feeding and management conditions which, in their natural habitat, are markedly influenced by altitude.

[b] A daily intake of 2 kg dry matter per 100 kg body weight has been estimated on the basis of the scarce data available for alpaca and llama.

REFERENCES

1. **Benirschke, K.,** in *Comparative Aspects of Reproductive Failure,* Benirschke, K., Ed., Springer-Verlag, New York, 1967.
2. **Taylor, K. M., Hungerford, D. A., Snyder, R. L., and Ulmer, F. A., Jr.,** *Cytogenetics,* 7, 7, 1968.
3. **Vallenas, A., Cummings, J. F., and Munnell, J. F.,** *J. Morphol.,* 134, 399–423, 1971.
4. **Vallenas, A.,** in *Physiology of Digestion in the Ruminant,* Dougherty, R., Ed., Butterworths, London, 1965, 147.
5. **Cummings, J. F., Munnell, J. F., and Vallenas, A.,** *J. Morphol.,* 137, 71–110, 1972.
6. **Lucio, E. Z.,** Thesis, Cornell University, 1969.
7. **Vallenas, A. and Stevens, C. E.,** *Am. J. Physiol.,* 220, 275–282, 1971.
8. **Vallenas, A. and Stevens C. E.,** *Cornell Vet.,* 61, 239–252, 1971.
9. **Vallenas, A., Esquerre, J., Valenzuela, A., Candela, E., and Chauca, D.,** *Univ. Nac. San Marcos Rev. Inv. Pec.,* 2, 115–130, 1973.
10. **Fernandez-Baca, S.,** *Annu. Congr. Panam. Med. Vet. Zoot. Caracas,* 1, 353–358, 1966.
11. **Bardales, J. D.,** B.S. thesis, Veterinary Medicine Faculty, San Marcos University, Lima, 1969.
12. **Camargo, F. R. and Cardozo, A.,** III ALPA, Document R-18, Bogota, 1971, 46.
13. **Hintz, H. F., Schryver, H. F., and Halbert, M.,** *Anim. Prod.,* 16, 303–305, 1973.

DIETS (NATURAL AND SYNTHETIC): GIRAFFIDAE

A. J. Hall-Martin

INTRODUCTION

There are only two modern giraffids, the okapi *Okapia johnstoni* and the giraffe *Giraffa camelopardalis.* The recent natural range of both these species is confined to sub-Saharan Africa – in savanna and woodland areas in the case of the giraffe and the lowland evergreen forests of Zaire in the case of the okapi. The giraffids are ruminants and are predominantly browsers (i.e., subsist almost entirely on the leaves of woody plants) in the wild, but adapt well to artificial diets in captivity.

THE DIET IN THE WILD

Food selection by giraffe has been studied in different parts of Africa. Most workers[1-8] relied on visual observations of feeding animals, while stomach contents of a series of giraffes collected over a 1-year period provided quantitative data on the frequency of utilization of different plant species.[9] All these studies confirmed that giraffe are predominantly browsers, taking the leaves of trees, shrubs, creepers, and vines, but also eating flowers, fruit, buds, bark, twigs, dicot herbs, and more rarely, grass. The intensive studies[3,5,6,9] showed that there were major seasonal differences in the plant species selected, with deciduous plants predominant during the wet season and evergreen plants during the dry season. An example is the study of Hall-Martin,[9] which showed very clearly that during the wet season deciduous species such as *Combretum apiculatum, Acacia nigrescens,* and *Colophospermum mopane* were the major items in the diet, while during the dry season giraffes subsisted largely on evergreen species such as *Euclea undulata* and *Maytenus senegalensis.* The change in diet was apparently determined by the decreasing availability of the preferred deciduous species, resulting in an increasing dependence on the nonpreferred species that by virtue of their evergreen habit were constantly available. The dry season evergreen staples were seldom, if ever, taken during the wet season.

The nutritional requirements of the giraffe were satisfied by a relatively small number of plant species during the wet season. During the dry season, however, individual species contributed less to the diet, and a greater variety of species were taken.[1,3-7,9] These data confirm that the giraffe had to take a greater number of less-preferred species to satisfy their needs during the dry season, which was generally a lean time for them.

Osteophagia by giraffes has been reported as an occasional phenomenon, usually being seen during the dry season.[2,3,9-12] This behavior has been speculatively linked to nutritional deficiencies;[2] in fact, it has been shown that, at least in the Transvaal, South Africa, giraffe are subjected sometimes to severe malnutrition during the late dry season.[13] When mineral licks* are available, such as in cattle-ranching areas and in some parks, these are regularly utilized by giraffe.[2,5,6,10,12] Giraffe have also been recorded as having eaten the stomach contents from the remains of an eland (*Taurotragus oryx*) carcass as well as soil.[2] Giraffe have also been seen to pick up stones in their mouths,[2] and quantities of gravel have been found in giraffe stomachs.[9]

* The composition of mineral licks utilized in South Africa has been reported[10] as containing 8% phosphorus, 35% sodium chloride, and trace amounts of iron, copper, zinc, magnesium, manganese, and iodine.

Though giraffe can survive for long periods without access to free water,[14] they normally drink readily if water is available.[1,2,5] No quantitative information on the daily rate of intake of food in the wild is available. It is known, however, that the mean wet mass of stomach contents in adult males (6 years and older) in the Transvaal is 104.5 kg (230 lb) or 8.9% of mean total body mass (= live weight).[13] In the case of adult females, the mean stomach contents mass is 72.6 kg (160 lb) or 10.9% of the mean total body mass, but could be as much as 17.5% of total body mass in lactating females.[13] Once the passage time of food going through the digestive tract of the giraffe is established, it will be a relatively simple matter to calculate the daily intake of food of wild giraffe.

Though it has been observed that giraffe only occasionally eat grass in the wild,[4,8,9,15] certain game ranchers in the Transvaal have helped their wild giraffe through periods of critical food scarcity by hanging bales of hay in trees. These were readily eaten by the hungry animals.

Very little is known of the natural diet of the okapi, which to our knowledge has not been extensively studied in the wild. However, it is reported to be a pure browser, eating leaves, twigs, and fruits of trees and shrubs.[16,17] Earlier writers thought that okapi also grazed and made use of swamp-loving herbs.[18]

Chemical Composition of Giraffe's Diet

The rumen contents of a series of giraffe collected monthly over a period of 1 year in the Transvaal were subjected to proximal analyses.[19] The data so obtained are given in Table 1. The crude protein content of the rumen contents was above 13% for 10 months of the year, with a mean value of 17.76%. As could be expected, the lowest values were found at the end of the dry season (September and October). At Akira Ranch in Kenya the mean protein content of giraffe stomach contents was found to be 12.73%.[20] The variations in crude protein, ash, and gross energy content of the stomach contents of the Transvaal giraffe (Table 1) were thought to reflect changes in the composition of the diet through the seasons,[19] as described earlier (i.e., deciduous species during the wet season, evergreen species during the dry season), as well as to seasonal phenological changes

Table 1
MEAN COMPOSITION[a] OF COMBINED GRAB-SAMPLES OF GIRAFFE RUMEN CONTENTS ON A DRY MATTER BASIS, TRANSVAAL, SOUTH AFRICA, JULY 1971–JUNE 1972[19]

Month	Protein (%)	Ether extract (%)	Crude fiber (%)	Ash (%)	Heat of combustion (kJ/g)
July 1971	14.38	5.66	24.04	21.64	19.25
August	15.88	5.26	26.92	14.27	21.88
September	12.18	3.66	25.54	20.87	17.74
October	12.50	2.06	23.18	35.38	14.10
November	22.17	4.11	24.81	10.81	20.09
December	21.19	3.43	22.88	15.63	19.96
January 1972	19.19	2.64	25.18	12.58	19.39
February	20.34	4.01	26.58	12.52	19.94
March	21.55	3.25	23.38	14.90	18.64
April	16.64	3.13	25.07	14.40	19.14
May	18.62	3.91	24.40	15.23	18.90
June	18.50	4.70	25.50	12.55	19.98

[a] Crude protein (N × 6.25), ether extract (fat), crude fiber, and ash were determined on triplicate subsamples (±5 g each) by standard AOAC[25] procedures for animal feeds. The heat of combustion (gross energy content of the samples) was determined using an automatic adiabatic calorimeter.

within the food species. It has been shown, for example, that the first flush of leaves in the spring (early wet season) has a lower nutritive value than the same species later in the season.[21-23] There is no evident correlation for the crude fiber or ether extract values (Table 1) with the seasonal changes. The extent to which any data obtained from analyses of stomach contents might be influenced by fermentation processes before collection of the material is unknown. An exceptionally high ash content during October (Table 1) at the end of the dry season was attributed to the large amounts of woody material consumed by the giraffe; it was also the month when the largest quantity of gravel was found in the stomachs of the giraffe.[19] It was perplexing to find that there was not, because of the increased woody intake, a concomitant increase in crude fiber percentage in October; a similar state of affairs was described by Dagg.[24]

SYNTHETIC DIETS

Giraffe feeding in captivity is a relatively simple matter, as substitutes for a natural diet are readily taken,[26] and zoo animals generally are healthy and breed well. Of several zoological gardens polled, none reported any difficulties in feeding giraffe. The average daily diet of giraffes in the New York Zoological Gardens comprises, according to Crandall,[26] approximately 10 kg (20 lb) of alfalfa hay, 5 kg (10 lb) of commercial feeding pellets, and 2.5 kg (5 lb) of rolled oats mixed with cut apples, carrots, potatoes, cabbage, and bread. Bananas are also fed occasionally. Crandall reports that during the summer fresh leafy branches are hung around the enclosures at convenient heights and are readily taken. Several other institutions report similar diets for their giraffe,* such as King's Dominion animal park in Virginia,[35] where good quality alfalfa hay is available winter and summer in elevated racks. With the exception of the winter housing period, these animals have a limited amount of browse in addition to their hay and daily grain ration which is provided in the following proportions: 7 kg (15 lb) Purina® horse chow checkers, 7 kg (15 lb) Purina Omolene® checkers, and 2.5 kg (5 lb) Purina Omolene.

It is likely that some zoos use the diet described by Ratcliffe[27] for herbivorous mammals. However, it should be noted that such a diet predominantly consists of grain substances which are of no consequence in the diet of wild giraffe. Such diets can also lead to a low pH of the rumen contents, to possible defaunation, and to other undesirable consequences.[28] The stomach anatomy of the giraffe is specialized for a tree and shrub foliage diet, having a relatively small, simple rumen, adapted to a quick turnover of food and a high fermentation rate.[29]

The provision of freshly cut branches should be an essential part of any synthetic diet for giraffe. It provides a seminatural means of occupation, especially for the teeth and digestive tract, and so presumably gives greater contentment to the animals, as pointed out by Hediger.[30]

The diet fed by Wackernagel[31] contains 18.0% protein, 2.5% crude fat, and 7.0% crude fiber, among other constituents. A comparison of these data shows that the protein content is close to that found in the natural diet of giraffe; fat is slightly low, while the crude fiber is about one third of that found naturally (Table 1). If animals are fed on high protein diets, it is of course necessary that they should have access to suitable quantities of water, as it is reported[32] that if the dietary protein exceeds 15% the kidneys of some mammals cannot excrete waste nitrogen without an increased intake of water.

Feeding and hand rearing of giraffe calves has been thoroughly documented.[33] The calves were fed on cow's (*Bos taurus*) colostrum for the first 48 hr after birth and

* In South Africa, Epol Antelope Cubes are fed as a concentrate. (See the last part of the text for a listing of the chemical composition of the concentrates described here.)

thereafter on a diet of fortified and balanced canned milk food.* The dietary requirements of the young calf were estimated at 2000 cal per day per 45 kg (100 lb) body mass, this intake being increased as the calf grew older. Weaning was started at about 8 months of age and completed by 11 months, by which time the calf was taking concentrates† and hay. For the first 2 weeks of its life the calf was injected intramuscularly with 2 cc of vitamins A, D, and E, 600,000 units of penicillin, and 0.75 g of streptomycin daily. The importance of calcium and vitamin D in giraffe calves to prevent rickets has also been mentioned by Crandall.[26]

Information on the diet of the okapi in captivity is scant, though it may be assumed that if fed the same diet as giraffes (in lesser quantities) there should be no undue problems. A male okapi in New York was fed approximately 3.5 kg (7 to 8 lb) of alfalfa hay, 2.5 kg (5 lb) of concentrates, and 0.5 kg (1 lb) of rolled oats daily.[26] This okapi also received greens, vegetables, and fruits, particularly bananas, and during the summer an abundance of fresh foliage. Salt blocks impregnated with phenothiazine were always available.[26] An okapi kept in London in 1935 was fed on elm leaves, tares, carrots, and biscuits, but this might not have been an adequate diet, as the animal succumbed to a heavy internal parasite load within 6 months.[34]

Chemical Composition of Concentrates Mentioned

Epol Antelope Cubes: A concentrate based on yellow maize meal with a guaranteed analysis of minimum protein 16%, minimum fat 2.5%, maximum fiber 8.0%, maximum calcium 1.6%, and minimum phosphorus 0.7%; vitamins and minerals were also added.

Purina Horse Chow Checkers: Guaranteed analysis of crude protein, not less than 12.5%; crude fat, not less than 1.5%; and crude fiber, not more than 25.0%. Ingredients: soybean meal and/or linseed meal, ground yellow corn and/or ground grain sorghums, ground barley and/or ground oats, alfalfa meal, cut alfalfa, wheat middlings, cane molasses, vitamin A supplement, dicalcium phosphate, iodized salt, iron oxide, manganous oxide, copper oxide, cobalt carbonate, zinc oxide.

Purina Omolene Checkers: Guaranteed analysis of crude protein, not less than 13.0%; crude fat, not less than 2.0%; and crude fiber, not more than 9.0%. Ingredients: soybean meal and/or linseed meal, ground yellow corn and/or ground grain sorghums, ground barley and/or ground oats, alfalfa meal, wheat middlings, cane molasses, vitamin A supplement, vitamin E supplement, D activated animal sterol, vitamin B_{12} supplement, niacin riboflavin supplement, calcium pantothenate, choline chloride, defluorinated phosphate, calcium carbonate, iodized salt, copper sulfate, manganous oxide, cobalt carbonate, zinc oxide.

Purina Omolene: Guaranteed analysis of crude protein, not less than 10%; crude fat, not less than 2%; and crude fiber, not more than 9%. Ingredients: soybean meal and/or linseed meal, cracked corn, rolled barley and/or crimped oats, alfalfa meal, cane molasses, vitamin A supplement, calcium carbonate, dicalcium phosphate, iodized salt, iron oxide, manganese sulfate, manganous oxide, copper oxide, cobalt carbonate, zinc oxide.

Olac: A sterile canned milk food, balanced and fortified with minerals and vitamins, containing 40 cal per fluid ounce.

* Olac (Mead-Johnson).

† Purina D & F Chow.

Purina D & F Chow: Guaranteed analysis of crude protein, not less than 12.5%; crude fat, not less than 2.5%; and crude fiber, not more than 11.0%. Ingredients: soybean meal, linseed meal, crimped oats, cracked yellow corn, alfalfa meal, cane molasses, wheat bran, wheat middlings, dried beet pulp, vitamin A supplement, D activated animal sterol, 0.5% low fluorine rock phosphate, 2% calcium carbonate, 0.9% iodized salt, and traces of iron oxide, manganous oxide, copper oxide, cobalt carbonate, zinc oxide.

REFERENCES

1. **Innis, A. C.,** *Proc. Zool. Soc. London,* 131(2), 245–278, 1958.
2. **Nesbit Evans, E. M.,** *East Afr. Wildl. J.,* 8, 53–62, 1970.
3. **Leuthold, B. M. and Leuthold, W.,** *East Afr. Wildl. J.,* 10(2), 129–142, 1972.
4. **Oates, L. G.,** *J. S. Afr. Wildl. Manage. Assoc.,* 2(2), 21–23, 1972.
5. **van Aarde, R. J. and Skinner, J. D.,** *Publ. Univ. Pret.,* 97, 59–68, 1975.
6. **Sauer, J. J. C.,** An Ecological Study of Giraffe Feeding in the Koos Meintjies Nature Reserve, B.Sc. Hon. Report, University of Pretoria, 1975.
7. **Stephens, J. M.,** *Arnoldia Rhodesia,* 7(28), 1–7, 1975.
8. **Lamprey, H. F.,** *East Afr. Wildl. J.,* 1, 63–92, 1963.
9. **Hall-Martin, A. J.,** *J. S. Afr. Wildl. Manage. Assoc.,* 4(3), 191–202, 1974.
10. **Mason, D. R.,** Range Use and Population Structure of Ungulates in the Jack Scott Nature Reserve, M.Sc. thesis, University of Pretoria, 1973.
11. **Western, D.,** *East Afr. Wildl. J.,* 9, 156–157, 1971.
12. **Wyatt, J. R.,** *East Afr. Wildl. J.,* 9, 157, 1971.
13. **Hall-Martin, A. J.,** Studies on the Biology and Productivity of the giraffe *Giraffa camelopardalis,* D.Sc. thesis, University of Pretoria, 1975.
14. **Foster, J. B. and Dagg, A. I.,** *East Afr. Wildl. J.,* 10(1), 1–16, 1972.
15. **Smithers, R. H. N.,** *The Mammals of Botswana,* Museum Memoir No. 4, National Museums of Rhodesia, Salisbury, 1971, 210–212.
16. **Dorst, J. and Dandelot, P.,** *A Field Guide to the Larger Mammals of Africa,* Collins, London, 1970, 184–186.
17. **Ionides, C. J. P.,** *Afr. Wildl.,* 9(1), 63–65, 1955.
18. **Lydekker, R. and Dollman, J. G.,** *The Game Animals of Africa,* Rowland Ward, London, 1926, 367–374.
19. **Hall-Martin, A. J. and Basson, W. D.,** *J. S. Afr. Wildl. Manage. Assoc.,* 5(1), 19–21, 1975.
20. **Field, C. R. and Blankenship, L. H.,** *J. Reprod. Fertil. Suppl.,* 19, 287–302, 1973.
21. **Myre, M.,** *Reconhecimento Pasćicola ao Vale do Save,* Comm. 75, Instituto Investigacao Agronomica de Moçambique, Lourenco Marques, 1972, 1–171.
22. **Groenewald, J. W., Joubert, D. M., and Tölken, H.,** *Proc. S. Afr. Soc. Anim. Prod.,* 6, 117–128, 1967.
23. **Joubert, E. and Eloff, F. C.,** *Madoqua Ser. 1,* 3, 5–54, 1971.
24. **Dagg, A. I.,** *Proc. Zool. Soc. London,* 135, 640–642, 1960.
25. **Horwitz, W., Ed.,** *Official Methods of Analysis of the Association of Official Agricultural Chemists,* 9th ed., Benjamin Franklin Station, Washington, D.C., 1970, 1–832.
26. **Crandall, L. S.,** *The Management of Wild Mammals in Captivity,* University of Chicago Press, Chicago, 1964.
27. **Ratcliffe, H. L.,** *Int. Zoo Yearb.,* 6, 4–23, 1966.
28. **Abrams, J. T.,** *Symp. Zool. Soc. London,* 21, 41–62, 1968.
29. **Hofman, R. R. and Stewart, D. R. M.,** *Mammalia,* 36(2), 226–240, 1972.
30. **Hediger, H.,** *Int. Zoo Yearb.,* 6, 37–58, 1966.
31. **Wackernagel, H.,** *Int. Zoo Yearb.,* 6, 23–37, 1966.
32. **Fiennes, R.,** *Int. Zoo Yearb.,* 6, 58–67, 1966.
33. **Savoy, J. C.,** *Int. Zoo Yearb.,* 6, 202–204, 1966.
34. **Seth-Smith, D.,** *J. Soc. Preserv. Fauna Emp. New Ser.,* 27, 14–15, 1936.
35. **Barton, R. A.,** personal communication.

DIETS (NATURAL AND SYNTHETIC): DEER

Table 1
DIET FOR ORPHAN FAWNS (*ODOCOILEUS* SPP.)[a,1]

Age (days)	Feeding schedule (time, hr)	Amount per feeding (ml)	Formula[b]
0–3	0400, 0800, 1200, 1600, 2000, 2400	40–60	Deer or bovine colostrum
4–10	0800, 1200, 1600, 2000, 2400	60–100	Evaporated bovine milk (undiluted)
11–21	0800, 1200, 1600, 2000, 2400	100–200	Evaporated bovine milk (undiluted)
22–42	0800, 1200, 1600, 2000	200–300	Evaporated bovine milk (undiluted) (plus pelleted diet, Table 2)
43–84	0800, 1200, 1600	300–0	Evaporated bovine milk (undiluted) (plus pelleted diet, Table 2)
85	Weaned to pelleted diet		

[a] Fawns should preferably be kept in a penned grassy area with shrubs, trees, or an artificial shelter. Fawns will begin to browse or graze within a week or two and will frequently eat dirt (which may provide certain nutrients and a rumen inoculum). If confined inside or on paved floors, it may be desirable to inject intramuscularly 500 mg of iron as iron-dextran (two 2-1/2 ml injections of Armidexan® 100, Bradley Products Inc., Chicago) at the start of the rearing period. An intramuscular injection of 1 mg of selenium and 50 mg D-α-tocopheryl acetate (1 ml of BO-SE®, H.C. Burns Co., Inc., Oakland, Calif.) may also help prevent white muscle disease.

It is important that normal patterns of urination and defecation be established by gently rubbing the perineal region with a warm, moist cloth at the time of feeding. It may be necessary to perform this task for 2 to 3 weeks until it is obviously no longer required.

[b] A soft-drink bottle equipped with a calf nipple is a suitable feeding device. The bottle and nipple should be thoroughly cleaned between feedings. Since evaporated milk is sterilized, it may be stored and fed at room temperature. Water should be available ad libitum. Buckland et al.[2] have described a similar rearing system using raw bovine colostrum and milk, plus a commercial calf feed, hay, and water. They also described a diarrhea medication containing 400 ml Kaopectate® (The Upjohn Co.), 10 g Biosol® (The Upjohn Co.), 10 g Vetisulid® (Ciba Pharmaceutical Co.), and 10 g Furacin® water mix (Eaton Laboratories). Usually, 5 to 8 ml of the mixture was administered in the milk twice daily for no more than two successive days before the diarrhea was terminated.

Compiled by D. E. Ullrey.

REFERENCES

1. **Verme, L. J. and Ullrey, D. E.,** Feeding and nutrition of deer, in *Digestive Physiology and Nutrition of Ruminants,* Vol. 3, Church, D. C., Ed., Oregon State University Book Stores, Corvallis, 1972, 275–291.
2. **Buckland, D. E., Abler, W. A., Kirkpatrick, R. L., and Whelan, J. B.,** *J. Wildl. Manage.,* 39, 211–214, 1975.

Table 2
DIET FOR WEANED DEER (*ODOCOILEUS* SPP.)[a,1]

Ingredient	Percent
Corn cob product[b]	35.0
Corn grain	18.7
Soybean meal (49% crude protein)	23.95
Alfalfa meal, dehydrated (17% crude protein)	5.0
Cane molasses	5.0
Wheat	10.0
Soybean oil	1.0
Trace mineral salt (0.8% zinc)	0.5
Limestone, ground (38% calcium)	0.4
Vitamin ADE and selenium premix[c]	0.25
Calcium propionate	0.2
	100.00
Composition	
Crude protein	16
Ether extract	2
Cell wall constituents	36
Soluble carbohydrates	31
Ash	4
Calcium	0.4
Phosphorus	0.3
Digestible energy	2.8 mcal/kg

[a] Formed into pellets, 5 mm diameter. Suitable as the sole diet for weaned deer of any age or productive state. May also be used as creepfeed for nursing fawns and to supplement milk feeding of orphan fawns.

[b] Consists of bracts and pith (soft parenchyma without vascular bundles). Cell wall constituents comprise 81.2%, acid-detergent fiber 37.5%, and lignin 6.5% by weight.

[c] Supplies per kilogram of diet: 3300 IU vitamin A, 220 IU vitamin D, 44 IU vitamin E, and 0.2 mg selenium from sodium selenite.

Compiled by D. E. Ullrey.

REFERENCE

1. **Ullrey, D. E.,** *Proc. Am. Assoc. Zoo Vet.,* Atlanta, 1974, American Association of Zoo Veterinarians, Washington, D.C., 1974, 170–177.

FOOD HABITS AND DIET QUALITY OF NORTH AMERICAN ELK

T. A. Leege, J. R. Nelson, and J. W. Thomas

INTRODUCTION

Two major and interrelated premises followed in elk management have been that, provided adequate cover, water, and space, (1) the physical well-being of the herd is directly related to the quantity and quality of its diet and (2) a herd maintained on a high nutritional plane will generally be more productive and subject to lower losses from winter-kill, parasites and diseases, and predation.

Elk food habits have been studied reasonably thoroughly throughout most of the elk's range of distribution. However, compared to whitetailed and mule deer, relatively little is known about elk diet quality and nutritional requirements. With such limited knowledge about the elk's basic nutritional needs, managers have had to assume that elk are similar to their smaller relatives in nutritional requirements, and habitat management has proceeded accordingly. Where basic knowledge of deer physiology is lacking, domestic livestock parameters, principally those of cattle, have been applied.[61]

FOOD HABITS

An understanding of food preferences is required before attempts can be made to improve the forage supply in an animal's habitat. Numerous studies of elk food habits have been completed in recent years. These studies have usually involved one or more of the following study methods: directly observing the animals, observing the plants where the animals are known to have been feeding, examining stomach contents, and microscopically analyzing fecal samples.

Kufeld[53] summarized the information available on Rocky Mountain elk (*Cervus elaphus nelsoni*) food habits, and his techniques for listing and evaluating food preferences are followed in this section. The principal differences between this review and the one by Kufeld are that we have updated his information on Rocky Mountain elk by the inclusion of several recent publications and we have conducted a similar review of food habits literature on the Roosevelt (*Cervus elaphus roosevelti*) and Tule elk (*Cervus elaphus nannodes*).

When reviewing the literature, several conditions had to be met before we included a reference in the summary tables for Rocky Mountain and Roosevelt elk. Diet had to be listed by species and in such terms that relative value of the species could be estimated; season of use had to be indicated; species preference had to be for elk only, not for deer and elk combined or for big game in general; and sample size had to be large enough that the findings were reliable.

The procedure for calculating relative species value involved assigning a numerical rating of 1 (light), 2 (moderate), or 3 (heavy) to each plant species, depending on their use by elk. Plants were not assigned a numerical rating unless they constituted at least 1% of the diet or were eaten to a greater extent than their availability as indicated by the author. An average numerical value for each species was then computed using all references. Highly valuable plants were considered to be those having average scores of from 2.25 to 3.0, valuable plants were in the 1.50 to 2.24 range, and least valuable plants had values of 1.00 to 1.49. However, even a low value of 1.00 indicates that elk made some use of that plant species and may suggest a greater preference for that species than for one not mentioned at all. The numerical value does not reflect nutrient quality, but does indicate both palatability and quantity consumed.

Elk food habits were divided into four seasons of use: winter (December, January, February); spring (March, April, May); summer (June, July, August); and fall (September, October, November).

Rocky Mountain Elk

The food habits of Rocky mountain elk are extremely variable as these animals are found in many different vegetation regions in the United States. (Tables 1, 2, and 3).* Some evidence indicates that they prefer to be a grazing animal, yet some of the largest herds are found in areas where browse constitutes a large portion of their diet. Although elk are found in many areas and eat a variety of plant species, they do exhibit certain food preferences on each of their ranges.

Winter

The winter diet is strongly influenced by forage availability as affected by snow conditions. Elk move to ranges with the smallest snow depths, and they must exist on what is available there. On predominantly grass ranges in Montana, elk eat predominantly grass. On the shrub winter ranges in northern Idaho, the diet is primarily woody. Where both grass and shrubs are available, both are found in the diet, although a preference is usually shown for the grasses until snow conditions make feeding on shrubs easier.[19,81] When snow depths limit the availability of shrubs, elk increase their consumption of conifers.[43] When elk winter under mature forests, lichens are also present in the diet.[9,43]

Spring

Almost all studies agree that grass is the most important forage for elk during the spring "green-up" months; it usually constitutes as much as 85% or more of the diet. An exception is when deep snow conditions extend into March and limit the use of grass.

Summer

As grasses begin to dry and mature in summer, elk use shifts to forbs and the deciduous leaves and succulent new twig growth of woody plants. Some studies have reported that forbs constituted almost the entire summer diet.[50,73,81] Others have reported that shrubs were most important,[9,43,92] whereas in some cases, high use of grass continued during the summer months.[11,66]

Fall

On many Rocky Mountain elk ranges, precipitation increases during the fall months and a "green-up" occurs before winter weather arrives. At this time, elk again turn to grasses as a primary food when available. When herbaceous species are cured by cold temperatures, shrubs sometimes become more important in the diet.

Roosevelt Elk

Like the Rocky Mountain elk, seasonal forage preferences of the Roosevelt elk are strongly influenced by availability and phenology (Tables 4, 5, and 6). In northern California, grasses constitute more than one half of the diet during all seasons and are especially important during the winter months, when they constitute over three fourths of the elk's forage intake. Forbs are utilized during summer and fall, but are unimportant in winter and spring. Some browse is taken during all seasons.[41]

On Afognak Island, Alaska, Batchelor[6] found that grasses, grasslike plants, and perennial forbs are important in the diet during spring and summer. When these plants

* Tables are found at the end of the text.

become dry in the fall, browse becomes more important in the diet. During the winter months, Afognak elk feed almost exclusively on browse.

In western Oregon, browse comprises approximately 70% of the diet during spring and summer, but drops to approximately 50% during fall and winter. Forbs are most important in the fall and winter, when they constitute approximately 30% of the diet. This use drops to approximately 15% during the spring and summer. Grass consumption ranges from 12 to 15% during all seasons except winter, when it rises to 21%.[40]

The diet of elk on the Olympic Peninsula in Washington is similar to that of elk on Afognak Island, Alaska. Grasses and forbs constitute 75 to 90% of the spring diet. As the season progresses, browse, which constitutes less than one half of the summer and fall diet, becomes more important. During winter, the Olympic elk feed almost exclusively on browse, presumably because of the effect of deep snow on forage availability.[76]

Tule Elk

The distribution of the Tule elk is limited to three small areas in California. The most complete information on their year-round food habits was reported by McCullough.[59] He found that annual forbs are extremely important in spring and early summer. Genera of greatest significance in the diet include *Eriogonum, Layia, Anisocoma, Mentzelia, Cryptanthus, Gilia,* and *Erodium.* As they begin to dry up in early summer, the later curing annual forbs, especially *Bassia hyssopifolia, Melilotus albus,* and *Heliantus annuus,* are eagerly consumed. Perennial forbs are important as forage in summer and fall. Important species include *Glycyrrhiza lepidota, Sida hederacea,* and domestic alfalfa.

Browse is especially important in the winter, although some is consumed throughout the year. The most sought-after species include *Purshia tridentata, Artemesia tridentata, Eriogonum fasciculatum, Eurotia lanata, Sarcobatus vermiculatus, Atriplex lentiformis,* and *Salix* spp.

Annual and perennial grasses are eaten by Tule elk throughout the year. Some of the preferred species are *Stipa speciosa, Sitanion hystrix, Oryzopsis hymenoides, Poa scabrella, Hilaria jamesii, Hordeum jubatum,* and *Paspalum distichum.* Sedges (*Carex* spp.) are also eaten. The rumens of 100 elk from Owens Valley, California contained almost 50% grass in October, indicating the importance of that type of forage during the fall.[26-28]

Artificial Feed for Elk

Wild elk are normally fed artificial rations only during the winter months when natural foods are limited by snow conditions or on those traditional elk winter ranges where man has taken over.[22] Almost all of the states with large elk populations occasionally feed elk artificially, but usually because of severe winter weather. The National Elk Range near Jackson, Wyoming is probably the most famous area in which a large elk population is fed each winter and thus maintained at a much larger number than would naturally survive. Experience has shown that cattle feed is sufficient to maintain elk through the winter. Thorne et al.[87] found that approximately 17.5 g of good quality feed per kilogram of body weight was required to maintain body weight of confined pregnant elk in Wyoming. They estimated that the requirement would probably be approximately 22.75 g/kg for free-ranging elk. They used a concentrate of one third each of oats, wheat bran, and corn to increase the quality of the regular native hay ration.

High-quality alfalfa hay is the most common artificial elk food used during the winter. It is high in nutritional requirements and elk readily accept it. However, some feeding problems have occurred when elk in poor body condition were fed this diet, as rumen microorganisms could not convert to the new food before the animal starved.

Two citrus products, dried lemon and orange pulp, proved to be both palatable and nutritious when experimentally fed to elk during the winter and spring in Wyoming.[8]

Table 1
RELATIVE VALUES OF FORB SPECIES EATEN BY ROCKY MOUNTAIN ELK

Plant name	Forage value[a] Winter	Spring	Summer	Fall	Ref.
Achillea				1.00 – 2	48, 67
Achillea millefolium	1.20 – 5	1.00 – 2	1.00 – 6	1.00 – 4	9, 15, 19, 21, 25, 29, 49, 52, 54, 68, 81, 85
Aconitum columbianum			2.00 + 1		52
Actaea spicata			1.00 – 1	2.00 + 1	92
Adenocaulon bicolor			2.00 + 1		52
Agastache urticifolia			1.50 + 2		68, 92
Agoseris glauca		1.50 + 4	2.43 * 7		13, 25, 48, 49, 71, 72, 85
Alectoria fremontii	1.00 – 1				20
Allium textile		2.00 + 1			55
Anemone piperi				1.00 – 1	43
Angelica arguta			1.00 – 1		84
Angelica lyallii			1.00 – 1	1.00 – 1	92
Antennaria	1.00 – 3	1.00 – 2	1.00 – 1	1.00 – 2	29, 36, 67, 72
Antennaria parvifolia		1.00 – 1			85
Antennaria rosea	1.00 – 1				29
Aquilegia flavesens		1.00 – 1	1.00 – 1		13
Arenaria			1.00 – 1		16
Arnica		1.50 + 2	2.00 + 1	1.00 – 1	36, 49, 85
Arnica cordifolia			1.75 + 4		9, 25, 48, 71
Arnica latifolia			2.00 + 1	3.00 * 1	92
Arnica sororia		2.00 + 1			55
Artemisia dracunculus	2.00 + 1				75
Artemisia frigida	1.33 – 6	1.00 – 4			21, 31, 35, 36, 48, 49, 75, 85
Artemisia longifolia	1.00 – 2			1.00 – 2	50, 51, 55
Artemisia ludoviciana	1.00 – 2			2.00 + 1	49, 51, 55
Asplenium felix-femina			3.00 * 1	2.00 + 1	92
Aster	2.00 + 2	2.00 + 2	1.80 + 5	1.44 – 9	29, 31, 32, 34–36, 48, 56, 65, 67, 71, 72, 75, 83
Aster canescens			2.00 + 1	3.00 * 1	92
Aster commutatus				2.00 + 1	55
Aster conspicuous			1.00 – 1	2.00 + 1	43
Aster eatoni			2.00 + 1	3.00 * 1	92
Astragalus	1.50 + 2		3.00 * 1	2.00 + 1	49, 75, 92
Astragalus miser			1.50 + 2		49, 71
Balsamorhiza saggittata	1.60 + 5	1.67 – 3	1.33 – 3		4, 9, 13, 17, 29, 43, 69, 75, 84
Boykinia heucheriformis			3.00 * 1	2.00 + 1	92
Caltha leptosepala			2.50 * 2		12, 66
Castilleja			1.00 – 1		68
Castilleja miniata			1.00 – 1		92
Centaurea maculosa	1.00 – 1		2.00 + 1		9
Cerastium arvense		1.00 – 1			85

[a] Each entry consists of three parts. The first number is the computed value ranking. The second part is the value ranking symbol: –, least valuable; +, valuable; *, highly valuable. The third part is the number of references upon which the ranking is based.

Table 1 (continued)
RELATIVE VALUES OF FORB SPECIES EATEN BY ROCKY MOUNTAIN ELK

Plant name	Forage value[a]				Ref.
	Winter	Spring	Summer	Fall	
Calachortus elgans			1.00 – 1		92
Chamaenerion angustifolium	3.00 * 1		3.00 * 1	2.00 + 1	29, 92
Chimaphila umbellata				1.00 – 1	9
Cirsium	1.00 – 2	1.00 – 3		2.00 + 3	19, 31, 32, 36, 67, 75
Cirsium candaensis		1.00 – 1			81
Cirsium foliosum		1.00 – 1	1.00 – 1	3.00 * 1	49
Claytonia asarifolia			3.00 * 1	3.00 * 1	92
Claytonia sibirica			3.00 * 1	3.00 * 1	92
Clintonia uniflora			1.00 – 1	1.00 – 1	92
Commandra sp.				1.00 – 1	36
Commandra pallida			2.00 + 1		56
Commandra umbellata			3.00 * 1		50
Coptis occidentalis	1.00 – 1	2.00 + 1	1.50 + 2	2.00 + 2	43, 92
Corydalis scouleri			2.00 + 1		92
Crepis modocensus			1.00 – 1		84
Delphinium			2.00 + 1		4
Delphinium bicolor		1.00 – 1	2.00 + 1		25, 31, 71
Delphinium scopulorum		2.00 + 1	2.00 + 2		13, 92
Dodecatheon conjugens			3.00 * 1	3.00 * 1	92
Dodecatheon pauciflorum		1.00 – 1			31
Dodocatheon conjugens			3.00 * 1	3.00 * 1	92
Epilobium angustifolium			1.67 + 3		7, 9, 48
Equisetum	1.00 – 1		1.00 – 1	1.00 – 1	34, 36
Erigeron		2.00 + 1	1.50 + 2	2.00 + 1	25, 49
Eriogonum	2.00 + 1			1.50 + 2	29, 35, 36
Eriogonum heracleoides			1.00 – 1	1.00 – 1	52, 68
Eriogonum multiceps	2.00 + 1				51
Eriogonum umbellatum				1.00 – 1	85
Fragarla	1.50 + 2	2.00 + 1	1.00 – 3	1.00 – 2	29, 36, 49, 92
Fragaria vesca		1.00 – 1			9
Fragaria virginiana			1.00 – 2		85, 71
Fraseria			2.00 + 1		4
Gaillardia			1.00 – 1		92
Galium asperrimum			1.00 – 1		52
Galium boreale	1.00 – 1				75
Geranium		2.00 + 1	1.50 + 2	1.00 – 1	4, 36, 48, 49
Geranium richardsonii		3.00 * 1			49
Geranium viscossissimum		1.67 + 3	2.33 * 9	2.00 + 2	25, 48, 49, 71–73, 81, 85, 92
Geum			2.00 + 1		12
Geum triflorum		1.00 – 1	2.00 + 1		25, 31
Geum turbanitum		1.00 – 1			16
Glycyrrhiza lepidota			2.00 + 2	2.50 * 2	50, 55
Hedysarum sulphurescens			2.00 + 1		48
Helianthus maximilliana				2.00 + 1	55
Heracleum lanatum		1.00 – 1	1.00 – 2		13, 92

Table 1 (continued)
RELATIVE VALUES OF FORB SPECIES EATEN BY ROCKY MOUNTAIN ELK

Plant name	Forage value[a] Winter	Spring	Summer	Fall	Ref.
Hieracium			2.00 + 1		4
Hieracium albertinum			1.00 – 1		84
Hieracium albiflorum			1.50 + 2	1.00 – 1	52, 92
Hieracium chapacanum			2.00 + 1		68
Hieracium cynoglossoides		1.00 – 1	1.00 – 1		85
Hieracium scouleri			1.00 – 1		92
Heuchera glabella			2.00 + 1	2.00 + 1	92
Hydrophyllum capitatum			3.00 * 1		92
Iris missouriensis				1.00 – 1	15
Lactuca pulchella		2.00 + 1			55
Lactuca serriola			1.50 + 2		50, 55
Lathyrus	1.00 – 1	2.00 + 2	2.00 + 1		7, 12
Lathyrus laetivirens			2.00 + 1		11
Ledum groenlandicum	2.00 + 1				7
Lespedeza stipulacea				2.00 + 1	65
Liatris punctata	1.00 – 1			3.00 * 1	49, 75
Ligusticum grayi			1.00 – 1		68
Ligusticum scopulorum			2.00 + 1	1.00 – 1	92
Ligusticum tenufolium			2.00 + 1	1.00 – 1	92
Lithospermum ruderale	1.00 – 2				29, 75
Lupinus	1.50 + 4	2.00 + 1	1.00 – 8	2.14 + 7	9, 19, 21, 25, 34–36, 38, 48, 67, 68, 72, 75, 85, 92
Lupinus leucophyllus			3.00 * 1		56
Lupinus ornatus	2.00 + 1				29
Lupinus sericeus	2.67 * 3	2.00 + 2	1.50 + 2		13, 31, 49, 75, 84
Medicago sativa			3.00 * 1	3.00 * 1	3, 56
Melilotus officinalis	2.00 + 2	1.00 – 1	2.67 * 3	3.00 * 2	50, 51, 55, 56
Mertensia cilata			3.00 * 1	3.00 * 1	92
Microseris		1.00 – 1	1.00 – 2		49, 85
Mitella stauropetala			1.00 – 1	2.00 + 1	92
Musineon divaricatum		2.00 + 1			51
Myosotis alpestris			3.00 * 1	3.00 * 1	72
Oenothera flava				2.00 + 1	72
Opuntia		2.00 + 2			38, 90
Oreoxis alpina			3.00 * 1		16
Osmorhiza occidentalis			3.00 * 2		68, 92
Oxytropis			1.00 – 1		36
Oxytropis splendens		1.00 – 1			48
Oxytropis viscida		3.00 * 1			49
Pedicularis cystopteridifolia			2.00 + 1		25
Pedicularis groenlandica			2.00 + 1		92
Pedicularis racemosa			2.00 + 1	1.00 – 1	92
Penstemon	1.00 – 1	1.50 + 2	1.00 – 1		9, 19, 29
Penstemon attenuatus			1.00 – 1		84
Penstemon confertus				1.00 – 1	92
Penstemon pinetorum			3.00 * 1	3.00 * 1	92

Table 1 (continued)
RELATIVE VALUES OF FORB SPECIES EATEN BY ROCKY MOUNTAIN ELK

Plant name	Forage value[a] Winter	Spring	Summer	Fall	Ref.
Penstemon procerus			1.00 – 1		68
Petasites saggittatus		2.00 + 1			7
Phacelia		1.00 – 1			9
Phacelia heterophylla			1.00 – 1	2.00 + 1	92
Phacelia leucophylla			1.00 – 1		84
Phlox	1.00 – 1			1.00 – 1	36, 67
Phlox hoodii	1.00 – 1				72
Polemonium delicatum			2.00 + 1	1.00 – 1	92
Polygonum phytolaccaefolium			2.50 * 2		68, 92
Potentilla		1.33 – 3	1.25 – 8	1.50 + 2	9, 16, 25, 36, 48, 49, 72, 85
Potentilla glandulosa		1.00 – 1	1.33 – 3	2.00 + 1	68, 81, 92
Potentilla gracilis			1.00 – 1		48
Pteridium aquilinum		1.00 – 1			43
Pyrola	3.00 * 1			1.00 – 1	9, 62
Ranunculus			2.00 + 1		49
Ranunculus glaberrimus		1.00 – 1	1.00 – 1		13
Rumex paucifolius		2.00 + 1	2.00 + 1		13
Sanguisorba sitchensis			2.00 + 1		68
Selaginella densa		1.00 – 1			85
Senecio	1.50 + 2	2.00 + 1	1.00 – 3		7, 29, 66, 72, 81
Senecio columbianum			3.00 * 1		92
Senecio triangularis			2.50 * 2	3.00 * 1	68, 92
Smilacina racemosa				2.00 + 1	92
Smilacina stellata				2.00 + 1	92
Solidago	1.00 – 1				29
Solidago elongata	1.00 – 1				9
Sonchus arvensis			2.00 + 1		7
Sphaeraclea rivularis			3.00 * 1	3.00 * 1	92
Stellaria	1.00 – 1				49
Taraxacum		1.00 – 2	3.00 * 3		25, 48, 85
Taraxacum officinale	1.00 – 1	1.75 + 4	2.20 + 5	2.00 + 2	7, 9, 13, 49, 71, 81, 84
Thermopsis montana		1.75 + 4	1.00 – 1		72
Thermopsis pinetorum				1.00 – 1	15
Tragopogon				1.00 – 1	32
Tragopogon dubius	2.00 + 1	2.00 + 1	2.00 + 1		31, 55, 75
Trautvetteria caroliniensis			2.00 + 1		52
Trifolium		1.50 + 2	2.00 + 3		25, 43, 49, 73
Trifolium dasyphyllum			3.00 * 1		16
Trifolium haydeni		2.00 + 1	2.00 + 1		13
Trifolium repens			3.00 * 1		85
Trifolium rydbergi		2.00 + 1	2.00 + 1		13
Typha				1.00 – 1	36
Valeriana sitchensis			3.00 * 1	3.00 * 1	92
Veratrum californicum			1.00 – 1		52
Veratrum eschscholtzii			1.00 – 1		92

Table 1 (continued)
RELATIVE VALUES OF FORB SPECIES EATEN BY ROCKY MOUNTAIN ELK

Plant name	Forage value[a] Winter	Spring	Summer	Fall	Ref.
Veratrum viride	1.00 – 1		2.00 + 1	2.00 + 1	43
Vicia americana		2.00 + 2	2.00 + 1		7, 51, 55
Viola nuttallii		1.00 – 1	1.00 – 1		13
Wyethia			2.50 * 2	2.50 * 2	67, 72, 73
Xanthium strumarium	3.00 * 1				55
Xerophyllum tenax	1.50 + 2		1.67 + 3	2.50 * 2	9, 29, 43, 92
Zizia aptera		3.00 * 1	3.00 * 1		49

Table 2
RELATIVE VALUES OF GRASS SPECIES EATEN BY ROCKY MOUNTAIN ELK

Plant name	Forage value[a] Winter	Spring	Summer	Fall	Ref.
Agropyron			2.00 + 1		71
Agropyron cristatum		2.00 + 1			48
Agropyron dasystachyum		1.00 – 1			48
Agropyron intermedium			1.00 – 1		84
Agropyron pauciflorum	3.00 * 1				29
Agropyron scribneri			2.00 + 1		16
Agropyron smithii	2.33 * 3	2.00 + 3	1.50 + 4	2.67 * 3	50, 51, 55, 81, 84
Agropyron spicatum	2.82 * 11	2.09 + 11	1.67 + 6	2.00 + 3	9, 14, 17, 21, 25, 29, 31, 38 43, 48, 49, 52, 56, 69, 75, 81, 83–85
Agropyron subsecundum			1.00 – 1	3.00 * 1	25, 72
Agropyron trachycaulum			1.00 – 1	2.00 + 1	72
Agropyron trichophorum			1.00 – 1		52
Agrostis exarata			3.00 * 1	2.00 + 1	92
Agrostis idahoensis		2.00 + 1	1.00 – 2		43, 66
Agrostis oregonensis			1.00 – 1		43
Bouteloua gracilis		1.00 – 1	1.00 – 1	1.00 – 1	38
Bromus	1.00 – 1		1.75 + 4	2.00 + 1	25, 49, 65, 71, 85
Bromus carinatus		2.00 + 1	2.00 + 2	2.00 + 2	43, 52, 92
Bromus inermis	3.00 * 1	2.00 + 1	2.00 + 2	2.00 + 2	38, 48, 56
Bromus marginatus	3.00 * 1		2.50 * 2		29, 71, 84
Bromus tectorum	2.00 + 2	1.00 – 1			14, 20, 43
Bromus vulgaris		3.00 * 1	2.00 + 1	1.00 – 1	43
Calamagrostis canadensis	2.50 * 2	1.00 – 1	1.50 + 2	1.00 – 1	29, 43, 49, 84
Calamagrostis rubescens	2.00 + 2	2.00 + 2	2.00 + 2	1.33 – 3	29, 43, 49, 92
Calamovilfa longifolia	2.00 + 1				51
Camassia quamash		2.00 + 1	1.00 – 1		49
Carex	1.60 + 5	1.30 – 3	1.82 + 11	1.80 + 5	3, 25, 29, 36, 38, 43 49, 52, 66, 68, 69, 71, 72, 84
Carex filifolia	3.00 * 1				29
Carex geyeri	2.40 * 5	2.00 + 2	2.00 + 7	2.50 * 2	9, 20, 29, 43, 48, 49, 50, 52, 68, 84, 92
Carex nubicola		2.00 + 1	2.00 + 1		13
Carex raynoldsii		2.00 + 1	2.00 + 1		13
Dactylis glomerata		2.00 + 1	2.00 + 1		43, 52
Danthonia	2.00 + 1	1.00 – 1	2.00 + 1	1.00 – 1	38, 49
Danthonia intermedia	3.00 * 1		2.00 + 1		29, 71
Danthonia parryi	2.00 + 1				49
Danthonia unispicata	2.00 + 1				17
Deschampsia caespitosa			2.00 + 2		43, 66
Deschampsia stropupurea			1.00 – 1		43
Digitaria ischaemum				2.00 + 1	65
Digitaria sanguinalis				2.00 + 1	65

[a] Each entry consists of three parts. The first number is the computed value ranking. The second part is the value ranking symbol: –, least valuable; +, valuable; *, highly valuable. The third part is the number of references upon which the ranking is based.

Table 2 (continued)
RELATIVE VALUES OF GRASS SPECIES EATEN BY ROCKY MOUNTAIN ELK

Plant name	Forage value[a] Winter	Spring	Summer	Fall	Ref.
Distichlis stricta	2.00 + 1		2.00 + 2		55
Elymus canadensis			2.00 + 1		71
Elymus flavescens	3.00 * 1	3.00 * 1			90
Elymus glaucus	1.50 + 2	1.50 + 2	3.00 * 1	2.50 * 2	72, 81, 92
Elymus innovatus				1.00 – 1	72
Festuca		2.00 + 1	3.00 * 1	2.00 + 1	38
Festuca arundinaceae			1.00 – 1		52
Festuca idahoensis	2.40 * 10	2.33 * 9	1.75 + 4	2.50 * 4	9, 13, 14, 17, 21, 25, 29, 31, 33, 48, 49, 69, 71, 72, 75, 81, 83–85
Festuca ovina		2.00 + 1	1.50 + 2		13, 52
Festuca scabrella	2.17 + 6	2.80 * 5	1.00 – 2	2.50 * 2	17, 25, 29, 31, 48, 49, 75, 81, 85
Juncoides parviflorum			3.00 * 1	3.00 * 1	92
Juncus balticus			3.00 * 1		66
Juncus parryi			3.00 * 1		68
Koeleria cristata	1.67 + 6	1.50 + 4	2.00 + 3	2.50 * 3	17, 25, 29, 31, 50, 51, 55, 69, 84, 85
Melica spectabilis			1.00 – 1		71, 72
Melica sublata		1.00 – 1	1.00 – 1		43
Muhlenbergia	1.50 + 2				17, 69
Muhlenbergia causpidata	1.00 – 2		1.00 – 1	2.00 + 1	50, 51, 55
Muhlenbergia fuliformis			2.00 + 1		43
Muhlenbergia montana	1.00 – 1	1.00 – 1	1.00 – 1		38
Oryzopsis hymenoides	2.00 + 1			3.00 * 1	51, 90
Phleum		2.00 + 1	1.50 + 2		13, 85
Phleum alpinum		2.00 + 1	2.00 + 2		13, 66, 71
Phleum pratense	2.00 + 3	2.50 * 2	1.00 – 3	3.00 * 1	29, 49, 52, 81
Poa	1.83 + 6	2.38 * 8	2.00 + 7	2.50 * 4	20, 21, 25, 31, 38, 48, 49, 56, 71, 75, 81, 84, 85
Poa alpina			2.00 + 1		49
Poa canadensis	3.00 * 1				29
Poa compressa	2.00 + 2	2.50 * 2	1.50 + 2	2.00 + 1	29, 43, 52, 55
Poa epilis		2.00 + 1	2.00 + 2		13, 66
Poa pratensis			1.00 – 1		52
Poa secunda	2.00 + 1	2.00 + 3	1.33 – 3	2.00 + 2	9, 13, 50, 52, 55
Sitanion hystrix			1.00 – 2		68, 84
Stipa	2.25 * 4	2.50 * 2	2.00 + 2	2.00 + 2	17, 38, 43, 49, 69
Stipa columbiana	3.00 * 1		1.00 – 1		29, 68
Stipa comata	1.00 – 1	2.00 + 1		3.00 * 1	51, 90
Stipa viridula	1.00 – 1	2.00 + 1			55
Trisetum spicatum			2.00 + 1	2.00 + 1	25, 48
Trisetum wolfii			2.00 + 1		66

Table 3
RELATIVE VALUE OF SHRUB AND TREE SPECIES EATEN BY ROCKY MOUNTAIN ELK

	Forage value[a]				
Plant name	Winter	Spring	Summer	Fall	Ref.
Abies grandis	1.00 – 2				29, 88
Acer glabrum	2.00 + 6	1.00 – 1	3.00 * 2	2.00 + 2	9, 29, 30, 42, 43, 49, 58, 92
Acer spicatum			2.00 + 1		7
Alnus				1.00 – 1	38
Alnus crispa	1.00 – 1			1.00 – 1	9
Alnus tenuifolia	1.00 – 1		1.50 + 2	2.00 + 1	11, 29, 92
Amelanchier alnifolia	2.50 * 10	1.67 + 3	2.25 * 4	2.00 + 5	2, 7, 9, 11, 29, 30, 37, 42, 43, 49, 58, 75, 81, 84
Arctostaphylos uva-ursi	1.00 – 3	1.00 – 1		1.67 + 3	7, 9, 48, 75
Artemisia cana	2.00 + 1				55
Artemisia tridentata	1.56 + 9	1.20 – 5	1.00 – 2	2.00 + 4	11, 12, 20, 30, 32, 33, 36, 38, 42, 55, 68, 72, 81, 83, 90
Artemisia tripartita	3.00 * 1	2.00 + 1			72
Berberis repens	1.40 – 5	1.83 + 7	1.00 – 3	2.00 + 13	2, 3, 9, 12, 19, 32, 34–36, 38, 40, 43, 46, 48, 49, 67, 72, 88
Betula glandulosa	2.00 + 2		3.00 * 1		29, 49
Betula fontinalis	1.00 – 1				29
Ceanothus americanus			3.00 * 1		5
Ceanothus sanguineus	3.00 * 5	2.00 + 1	2.00 + 2	2.00 + 2	29, 43, 58, 88, 91, 92
Ceanothus velutinus	2.38 * 8	2.00 + 1	2.50 * 2	2.20 + 5	9, 19, 20, 29, 30, 43, 49, 56, 58, 91, 92
Cercocarpus montanus	2.67 * 3	2.00 + 1	2.00 + 1	2.00 + 1	38, 42, 54
Cercocarpus ledifolius	2.50 * 2	2.00 + 1		2.40 * 2	2, 19, 20
Chrysothamnus	1.00 – 1	1.00 – 1		1.00 – 1	38
Chrysothamnus nauseosus	2.00 + 3	1.00 – 1			18, 20, 83, 90
Chrysothamnus viscidiflorus	1.40 – 5	3.00 * 1		1.67 + 3	18, 31, 50, 51, 55, 83, 90
Cornus stolonifera	2.13 + 8		3.00 * 1	1.50 + 2	7, 9, 29, 30, 42, 43, 81, 88, 92
Dasiphora fruticosa	1.00 – 1				29
Elaeagnus commutata	2.50 * 2				29, 49
Eurotia	2.00 + 1			1.00 – 1	38
Fraxinus	2.00 + 1				37
Galax aphylla	2.00 + 1		2.00 + 1		5
Garrya wrightii	3.00 * 1				54
Gaultheria procumbens	2.00 + 1		2.00 + 1		5
Holodiscus discolor	1.00 – 2		1.50 + 2	1.00 – 1	52, 58, 91, 92
Juniperus	1.67 + 3	2.00 + 1		1.50 + 2	35, 38, 54
Juniperus communis	1.00 – 3			1.00 – 1	29, 48, 75
Juniperus horizontalis	1.00 – 3				17, 49, 75
Juniperus occidentalis	2.00 + 1				20

[a] Each entry consists of three parts. The first number is the computed value ranking. The second part is the value ranking symbol: –, least valuable; +, valuable; *, highly valuable. The third part is the number of references upon which the ranking is based.

Table 3 (continued)
RELATIVE VALUE OF SHRUB AND TREE SPECIES EATEN BY ROCKY MOUNTAIN ELK

	Forage value[a]				
Plant name	Winter	Spring	Summer	Fall	Ref.
Juniperus scopulorum	1.50 + 2	1.00 – 1		1.00 – 1	29, 55
Ledum groenlandicum	2.00 + 1				7
Linnaea		1.00 – 1		1.00 – 1	48
Linnaea borealis	1.00 – 1				29
Lonicera involucrata	2.00 + 1		2.00 + 1		29, 92
Lonicera utahensis			3.00 * 1	1.00 – 1	92
Menziesia ferruginea			2.00 + 2	1.00 – 1	43, 92
Odostemon	1.00 – 1				20
Odostemon aquifolium	1.00 – 1				29
Pachistima myrsinites	1.50 + 2	2.00 + 1	1.00 – 1	2.20 + 5	3, 9, 20, 29, 43, 46, 54
Philadelphus lewisii	1.00 – 2				29, 58
Physocarpus malvaceus	1.00 – 2		2.00 + 2	1.00 – 1	52, 58, 92
Pinus	1.00 – 3			1.00 – 3	12, 35, 72, 88
Pinus banksiana	1.00 – 1				62
Pinus contorta	1.40 – 5	1.00 – 1		1.33 – 3	9, 29, 32, 36, 49, 75, 81
Pinus flexilis				1.00 – 1	67
Pinus ponderosa	1.00 – 2				20, 29
Populus	1.00 – 1				9
Populus balsamifera	2.00 + 1				7
Populus tremuloides	2.33 * 9	2.25 * 4	1.74 + 4	2.50 * 8	3, 7, 11, 12, 15, 29, 36, 42, 49, 54, 67, 75, 81, 92
Populus trichocarpa	2.50 * 2			1.00 – 1	29, 81
Potentilla			1.00 – 1		15
Potentilla fruitcosa	1.00 – 1				49
Prunus	1.50 + 2			1.00 – 1	5, 9
Prunus virginiana	2.44 * 9	2.00 + 3	2.00 + 3	1.50 + 2	7, 18, 29, 32, 37, 42, 50, 55, 58, 72, 75, 90, 92
Prunus emarginata	1.00 – 1		3.00 * 1	3.00 * 1	58, 92
Prunus pensylvanica	2.00 + 1				7
Pseudotsuga menziesii var. *glauca*	1.33 – 15	1.00 – 4	1.00 – 1	1.33 – 3	9, 11, 12, 19, 20, 23, 29, 32, 35, 36, 43, 48, 55, 58, 75, 88
Purshia tridentata	3.00 * 4			3.00 * 1	3, 18, 20, 42, 90
Pyrus malus	2.00 + 1				62
Quercus coccinea	2.00 + 1				5
Quercus gambellii	3.00 * 3	2.00 + 1	2.00 + 1		11, 37, 38, 42
Quercus rubra	2.00 + 1				62
Rhamnus purshiana	2.00 + 1		1.00 – 2		43, 92
Rhus typhina	1.00 – 1				62
Ribes	1.00 – 1		1.00 – 1		29, 84
Ribes cereum			1.00 – 1		11
Ribes cognatum			3.00 * 1	2.00 +1	92
Ribes montigenum			3.00 * 1		68
Ribes peticlare			1.00 – 1	1.00 – 1	92

Table 3 (continued)
RELATIVE VALUE OF SHRUB AND TREE SPECIES EATEN BY ROCKY MOUNTAIN ELK

Plant name	Forage value[a]				Ref.
	Winter	Spring	Summer	Fall	
Ribes viscosissmum			3.00 * 1	2.00 + 1	92
Rosa	1.50 + 2	1.00 – 1	2.00 + 2	2.00 * 1	29, 43, 75, 92
Rosa acicularis	1.00 – 2		3.00 * 1	3.00 * 1	7, 69
Rosa arkansana			1.00 – 1		50
Rubus				1.00 – 1	36
Rubus idaeus			2.00 + 1		7
Rubus parviflorus			2.00 + 2	2.00 + 1	84, 92
Salix	2.10 + 10	2.00 + 1	1.75 + 8	2.25 * 8	7, 9, 12, 20, 30, 32, 33, 36, 48, 49, 56, 67, 72, 75, 85, 91, 92
Salix bebbiana	3.00 * 1				29
Salix exigua	2.00 + 1				29
Salix geyeriana	2.00 + 1				29
Salix lutea	3.00 * 1				29
Salix melanopsis	2.00 + 1				29
Salix scouleriana	1.75 + 4	1.50 + 2	1.00 – 1	2.00 + 1	29, 43, 81, 88
Salix Subcoerulea	3.00 * 1		2.50 * 2	1.00 – 1	29
Sambucus		2.00 + 1	2.50 * 2	1.00 – 1	4, 43
Sambucus coerulea			2.50 * 2	2.50 * 2	43, 92
Sambucus melanocarpa	3.00 * 1		3.00 * 1	3.00 * 1	29, 92
Sassafras variifolium	2.00 + 1				5
Shepherdia canadensis	2.00 + 1		1.00 – 1	1.50 + 2	36, 38, 56, 75
Sorbus americana			3.00 * 1	3.00 * 1	92
Sorbus occidentalis	2.00 + 1				29
Sorbus scopulina				2.00 + 1	43
Spiraea	1.00 – 1		1.00 – 1	1.00 – 1	29, 48
Spiraea betulifolia	2.00 + 1				75
Spiraea lucida			2.00 + 1	2.00 + 1	92
Spiraea menziesii			2.00 + 1	2.00 + 1	92
Symphoricarpos	1.40 – 5		1.50 + 2	1.83 + 6	20, 34, 35, 42, 48, 55, 56, 67, 75, 91
Symphoricarpos albus	2.00 + 2	1.00 – 1	1.67 + 3	1.67 + 3	7, 9, 29, 43, 52, 84, 92
Symphoricarpos occidentalis	1.00 – 1		1.00 – 1	2.00 + 1	50, 72
Symphoricarpos orbiculatus				2.00 + 1	65
Symphoricarpos utahensis		1.00 – 1	1.00 – 1		11
Taxus brevifolia			3.00 * 1		52
Tetradymia canescens	1.50 + 2			1.50 + 2	72, 90
Thuja occidentalis	3.00 * 1				62
Thuja plicata	2.00 + 2	1.00 – 1		1.00 – 1	43, 88
Tsuga canadensis	1.00 – 1				62
Vaccinium	1.75 + 4		1.00 – 3	1.00 – 4	9, 12, 20, 36, 49, 62, 67, 85
Vaccinium membranaceum	1.00 – 1	1.00 – 1	3.00 * 1	1.50 + 4	29, 43, 48, 56, 92
Vaccinium scoparium	1.00 – 1		1.00 – 2	1.75 + 4	15, 29, 43, 48, 54, 92
Yucca glauca	2.00 + 1	1.00 – 1		1.50 + 2	38, 50, 51

Table 4
RELATIVE VALUES OF FORBS, FERNS, AND LICHENS EATEN BY ROOSEVELT ELK

	Forage value[a]				
Plant name	Winter	Spring	Summer	Fall	Ref.
Angelica lucida		2.00 + 1			6
Arenaria macrophylla	2.00 + 1	2.00 + 1			39
Athyrium cyclosorum	2.00 + 1	2.00 + 1			6, 89
Blechnum spicant	1.00 – 1	1.00 – 1	1.00 – 1	1.00 – 1	86
Epilobium angustifolium			2.00 + 2	3.00 * 2	6, 89
Epilobium watsonii		1.00 – 1	1.00 – 1		86
Equisetum		2.00 + 1			6
Heracleum lanatum		2.00 + 2	2.00 + 1		6
Hypochoeris radiacata	2.67 * 3	2.33 * 3	2.75 * 4	2.33 * 3	39, 41, 45, 86
Leptarix menziesii	2.00 + 1	2.00 + 1		2.00 + 1	76
Ligusticum hultehii			2.00 + 1		6
Lupinus nootkatensis			2.00 + 1		6
Montia			2.00 + 1		6
Montia sibirica	1.00 – 1	2.00 + 2	2.00 + 2	2.00 + 2	76, 86
Oxalis oregana	2.00 + 1	1.50 + 2	1.25 – 4	1.50 + 2	41, 76, 86
Plantago lanceolata		1.00 – 1	2.00 + 1	2.00 + 1	41
Polystichum munitum	1.75 + 4	1.50 + 4	1.25 – 4	1.50 + 4	39, 45, 76, 82, 86
Prunella vulgaris			1.00 – 1	1.00 – 1	41
Pteridium aquilinum	1.33 – 3	1.50 + 2	1.50 + 2	1.00 – 1	39, 45, 86
Ramalina reticulata	2.00 + 1				76
Ranunculus repens			1.00 – 1	1.00 – 1	41
Sanguisorba sitchensis			2.00 + 1		6
Stachys mexicana		1.00 – 1	1.00 – 1		86
Struthiopteris spicant	2.00 + 1	2.00 + 1		2.00 + 2	76, 82
Tiarella trifoliata	2.00 + 1	2.00 + 1		2.00 + 1	76
Usnea barbata	2.00 + 1	2.00 + 1			76
Usnea plicata	2.00 + 1				76
Veratrum eschscholtzii			2.00 + 1		6

[a] Each entry consists of three parts. The first number is the computed value ranking. The second part is the value ranking symbol: –, least valuable; +, valuable; *, highly valuable. The third part is the number of references upon which the ranking is based.

Table 5
RELATIVE VALUES OF GRASSES AND GRASSLIKE PLANTS EATEN BY ROOSEVELT ELK

	Forage value[a]				
Plant name	Winter	Spring	Summer	Fall	Ref.
Agrostis alba	3.00 * 1	3.00 * 1	2.00 + 1	2.00 + 1	41
Anthoxanthum odoratum	2.00 + 1	3.00 * 1	3.00 * 1	3.00 * 1	41
Bromus marginatus			1.00 – 1		41
Calamagrostis canadensis		2.00 + 1			6
Carex		2.00 + 1			6
Carex ablata			3.00 * 1	3.00 * 1	76
Carex leptopoda		2.00 + 1			76
Carex obnupta	2.00 + 1	1.00 – 1	1.00 – 1	1.00 – 1	41
Dactylis glomerata	3.00 * 1	3.00 * 1	3.00 * 1	3.00 * 1	41
Danthonia californica	3.00 * 1	3.00 * 1		2.00 + 1	41
Deschampsia caespitosa	2.00 + 1				6
Elymus			3.00 * 1	1.00 – 1	41
Festuca rubra	1.00 – 1				6
Holcus lanatus			1.00 – 1	1.00 – 1	41
Luzula parviflora	1.00 – 1		1.00 – 1	1.00 – 1	86

[a] Each entry consists of three parts. The first number is the computed value ranking. The second part is the value ranking symbol: –, least valuable; +, valuable; *, highly valuable. The third part is the number of references upon which the ranking is based.

Table 6
RELATIVE VALUES OF SHRUBS AND TREES EATEN BY ROOSEVELT ELK

	Forage value[a]				
Plant name	Winter	Spring	Summer	Fall	Ref.
Acer circinatum	1.75 + 4	2.50 * 4	3.00 * 4	2.00 + 4	39, 45, 76, 82, 86
Acer macrophyllum				2.00 + 1	82
Alnus oregana				2.00 + 1	82
Alnus sinuata			1.00 – 1		76
Baccharis pilularis			1.00 – 1		41
Berberis nervosa	3.00 * 1	2.00 + 1	2.00 + 1	3.00 * 1	86
Gaultheria shallon	2.50 * 4	1.75 + 4	1.67 + 3	1.75 + 4	39, 45, 76, 82, 86
Oplopanax horridus	2.00 + 1	2.00 + 1		2.00 + 1	6, 82, 89
Picea sitchensis	2.00 + 1				76
Pseudotsuga menzesii	2.67 * 3	1.00 – 2	1.00 – 1	2.00 + 1	39, 76, 86
Rhamnus purshiana	2.00 + 1	1.00 – 1	2.00 + 1	1.00 – 1	86
Rosa californica			1.00 – 1	1.00 – 1	41
Rosa nutkana	2.00 + 1				89
Rubus laciniatus		2.00 + 1	2.00 + 1	2.00 + 1	45
Rubus leucodermis			1.00 – 1	1.00 – 1	86
Rubus parviflorus		2.00 + 2	2.00 + 2	2.00 + 1	45, 86
Rubus pedatus	2.00 + 1				6
Rubus procerus	1.00 – 1	3.00 * 1	1.00 – 1		41
Rubus spectabilis	1.20 – 5	2.33 * 6	2.33 * 6	2.00 + 6	6, 39, 41, 45, 76, 82, 86, 89
Rubus ursinatus	3.00 * 3	2.67 * 3	2.67 * 3	3.00 * 2	39, 45, 86
Rubus vitifolius	3.00 * 1	2.00 + 1	1.00 – 1	2.00 + 1	41
Salix	3.00 * 2			1.50 + 2	6, 82, 89
Salix scouleriana	2.00 + 1	2.00 + 1			76
Salix sitchensis	2.00 + 1	2.00 + 1			76
Sambucus racemosa	3.00 * 2	2.00 + 1	2.50 * 2	2.33 * 3	6, 45, 89
Thuja plicata	3.00 * 1				76
Tsuga heterophylla	2.25 + 4	1.33 – 3	1.50 + 2	1.67 + 3	39, 45, 76, 86
Vaccinium membranaccum		2.00 + 1	2.00 + 1	2.00 + 1	76
Vaccinium ovalifolium	1.33 – 3	1.50 + 2	1.50 + 2	1.50 + 2	6, 76, 89
Vaccinium ovatum	1.50 + 2	1.50 + 2		1.50 + 2	39, 82, 86
Vaccinium parviflorum	2.00 + 1	2.00 + 1		2.00 + 2	76, 82
Vaccinium parvifolium	2.50 * 2	2.00 + 2	2.50 * 2	2.50 * 2	45, 86
Viburnum edule	2.00 + 1			2.00 + 1	6, 89

[a] Each entry consists of three parts. The first number is the computed value ranking. The second part is the value ranking symbol: –, least valuable; +, valuable; *, highly valuable. The third part is the number of references upon which the ranking is based.

DIET QUALITY

The Rumen

As in other ruminants, the digestion process in elk functions efficiently if the rumen community is provided with the basic essentials: optimal levels of temperature, moisture, energy source, nitrogen, and essential minerals. In return, the elk assimilates or further digests the by-products of rumen community fermentation, which include a rich supply of volatile fatty acids, protein, and fat-soluble vitamins in the quantity and variety that the host animal requires for its own body needs. Water-soluble vitamins A, D, E, and K are not synthesized by the rumen community, but deficiencies of these vitamins have not been reported for elk. Vitamin A deficiency is probably rare among wild elk, as its precursor, carotene, is abundant in green forage and winter browse.

Among ruminants, the ability to digest cellulose has been correlated with the relative rate of food passage through the rumen. This, in turn, seems to be associated with the rumen-reticulum:body weight ratio and certain cellulolytic protozoans which cannot maintain their populations in rapid food passage systems. Elk rumen capacity has not been documented; however, Prins and Geelen[70] determined that a close relative, the European red deer (*Cervus elephus*), has a rumen-reticulum:body weight ratio of 23%. This is similar to sheep and cattle, with 25 and 26.4%, respectively, and significantly larger than the 10% ratios for whitetail deer[77,79] and mule deer.[78]

Approximately half the rumen organism biomass consists of ciliate protozoans, while the remaining half consists of bacteria. McBee et al.[57] counted from 57.5 to 66.9 billion bacteria per gram of rumen content from four Yellowstone elk. *Bacterioides* spp., notably *B. succinogenes* and *B. ruminicola,* comprised 48% of the bacteria. *Butyrivibrio* spp. constituted the major portion of the remaining 52%.

Among protozoans, ratios of large:small oligotrichs and small oligotrichs:holotrichs may be indicative of rate of passage through the rumen. Because cellulolysis is a relatively slow process, these ratios may reflect relative ability to digest cellulose.[70] McBee et al.[57] found ratios of large:small oligotrichs to vary from 0.4 to 1.3 in four wild Rocky Mountain elk examined. Prins and Geelen[70] observed smaller ratios among smaller deer: 0.2 in red and fallow deer; no large oligotrichs were found in roe deer, only the cosmopolitan *Entodinium* spp. In addition, holotrichs (*Dasytricha* spp. and *Isotricha* spp.) were found in red deer, but not in the smaller fallow and roe deer.

In summary, it appears that elk may have a rumen-reticulum volume, rumen flora and fauna, and digestion capabilities (at least for cellulose) that are more akin to cattle than to other North American deer species. In estimating elk nutritional needs to a first approximation, cattle parameters may be more appropriate to use than deer parameters.

Forage Consumption Rate

Several ad libitum consumption studies have been conducted with elk in winter. As with other ruminants, the daily consumption rate reflects the relative nutritional need and the size of the animal, as well as relative digestibility of the ration. In four feeding trials in Montana and Wyoming,[1,30,44,47] adult Rocky Mountain elk consumed 2.0 to 2.5 lb (dry weight) per hundredweight of meadow hay, but generally decreased their consumption as fiber content increased. Geis[30] reported that cow elk in several trials consumed 2.04 to 2.09 lb per hundredweight of fair quality meadow hay, but would eat 1.56 to 2.08 lb per hundredweight of deciduous browse. When fed pure diets of lodgepole pine or hay, adult elk consumption rates were only 1.5 and 2.4, respectively.[44] In the same study, when hay was mixed with conifer browse, lodgepole pine, or Douglas fir, the consumption rate increased to 2.55 to 2.97. No separate consumption rates have been determined for bulls and cows, but Murie[64] reports that bulls consumed 1.4 times as much as cows, based on dry weight rumen content. As would be expected, the average calf consumption rate is higher than the adult, ranging from 2.3 to 3.1 when fed meadow hay.[1,10,30,47] Boll[9] fed calves seven rations for two winters. Consumption rates varied from 2.7 and 1.7 for hay and bunchgrass, respectively, to 1.28 for a 1:1 mixture of bunchgrass and conifer browse. Adults and calves generally lost weight on all rations fed in the above-cited studies except meadow hay.

Thorne et al.[87] fed a variety of rations to 46 cow elk throughout three winter-spring periods until parturition. Cows fed 1.75 and 1.94 lb of hay and concentrate per 100 lb body weight were able to maintain or slightly increase their weights. The crude protein contents of these diets were 8.4 and 14.5%, respectively. Apparently, like domestic livestock,[60] elk require less total feed when concentrates are included in their diet. Thorne et al.[87] concluded that if cow elk lose greater than 3% of body weight during gestation (not including the weight of the fetus, fetal membranes, and fluid) calves have less than a 90% chance for survival to age 1 month. Cows fed only native hay at rates of from 1.3 to 1.7 lb per hundredweight lost up to 17% of body weight, and calves averaged less than 30 lb at birth, with approximately 50% survival. Average elk calf weight at birth is about 35.5 lb.[64]

REFERENCES

1. **Anderson, C. C.,** The elk of Jackson Hole: a review of Jackson Hole studies, *Wyo. Game Fish. Comm. Bull.,* No. 10, 1958.
2. **Anderson, C. C.,** Reconnaissance of Deer Habitat in Special Problems Areas, Federal Aid Completion Rep. Proj. W-27-R11, WP-4, J-7, and J-8, Wyoming Game and Fisheries Commission, Laramie, 1958.
3. **Anderson, C. C., Crump, W. I., and Baker, T. C.,** Food Habits of Antelope, Elk, Deer and Moose, Federal Aid Completion Rep. Proj. W-27-R-9, WP-1, J-1, Wyoming Game and Fisheries Commission, Laramie, 1956, 8–31.
4. **Baker, T. C., Anderson, C., and Crump, W. I.,** Food habits study of game animals, *Wyo. Wildl.,* 17(11), 24–31, 1953.
5. **Baldwin, W. P. and Patton, C. P.,** A preliminary study of the food habits of elk in Virginia, *Trans. North Am. Wildl. Conf.,* 3, 747–755, 1938.
6. **Batchelor, R. F.,** The Roosevelt Elk in Alaska – Its Ecology and Management, Alaska Department of Fish and Game, Juneau, 1965.
7. **Blood, D. A.,** Range relationships of elk and cattle in Riding Mountain National Park, Manitoba, *Can. Wildl. Serv. Wildl. Manage. Bull.,* Ser. 1, No. 19, 1966.
8. **Blunt, F. M.,** A study of the flavonoid compounds in game animal rations, Federal Aid Proj. W-69-R-2, Job 1, Wyoming Game and Fisheries Commission, Laramie, 1960.
9. **Bohne, J. R.,** Food Habits, Seasonal Distribution and Habitat Utilization of Elk in the South Fork of Fish Creek, Lolo National Forest, Montana, M.S. thesis, University of Montana, Missoula, 1974.
10. **Boll, L. A.,** Elk Nutrition – The Response of Elk Calves to Various Winter Diets Under Controlled Conditions, M.S. thesis, University of Montana, Missoula, 1958.
11. **Boyd, R. J.,** Elk of the White River Plateau, Colorado, Tech. Bull. No. 25, Colorado Division of Game, Fish and Parks, Fort Collins, 1970.
12. **Boyd, R. J.,** Stomach Contents Data from the Rio Grande Elk Herd in Southwestern Colorado, Colorado Division of Game, Fish and Parks, Fort Collins, 1972.
13. **Brazda, A. R.,** Elk migration patterns and some of the factors affecting movements in the Gallatin River drainage, Montana, *J. Wildl. Manage.,* 17, 9–23, 1953.
14. **Buechner, H. K.,** Winter-range utilization by elk and mule deer in southeastern Washington, *J. Range Manage.,* 5, 76–80, 1952.
15. **Burt, T., Jr. and Cates, G. H.,** Pecos elk reproductive and food habits studies, Federal Aid Completion Rep. Proj. W-93-R-1, WP-3, J-6, New Mexico Department of Game and Fisheries, Albuquerque, 1959.
16. **Capp, J. C.,** Competition among Bighorn Sheep, Elk and Deer in Rocky Mountain National Park, Colorado, M.S. thesis, Colorado State University, Fort Collins, 1967.
17. **Casgranda, L. and Janson, R. G.,** Wildlife Investigations (District 4): Big Game Surveys and Investigations, Federal Aid Completion Rep. Proj. W-74-R-2, J-A1, Part 3, Montana Department of Fish and Game, Helena, 1957.

18. **Chadwick, H. W.**, Plant Succession and Big Game Winter Movements and Feeding Habits in a Sand Dune Area in Fremont County, Idaho, M.S. thesis, University of Idaho, Moscow, 1960.
19. **Claar, J. J.**, Correlations of Ungulate Food Habits and Winter Range Conditions in the Idaho Primitive Area, M.S. thesis, Univeristy of Idaho, Moscow, 1973.
20. **Cliff, E. P.**, Relationship between elk and mule deer in the Blue Mountains of Oregon, *Trans. North Am. Wildl. Conf.*, 4, 560–569, 1939.
21. **Constan, K. J.**, Winter foods and range use of three species of ungulates, *J. Wildl. Manage.*, 36, 1968–1976, 1972.
22. **Corsi, R.**, Elk feedgrounds: some facts and figures, *Wyo. Wildl.*, 36(12), 18–19, 1972.
23. **Cowan, I. M.**, Range competition between mule deer, bighorn sheep and elk in Jasper Park, Alberta, *Trans. North Am. Wildl. Conf.*, 12, 223–227, 1947.
24. **De Nio, R. M.**, Elk and deer foods and feeding habits, *Trans. North Am. Wildl. Conf.*, 3, 421–427, 1938.
25. **Eustace, C. D.**, Food Habits, Range Use and Relationships Between Elk and Livestock in the Gravelly Mountains, Montana, M.S. thesis, Montana State University, Bozeman, 1967.
26. **Ferrel, C. M.**, Food habits investigation, miscellaneous food habits studies, Calif. Federal Aid Rep. Proj. W-52-R-6, Job 4, 1962.
27. **Ferrel, C. M.**, Tule elk and Roosevelt elk diets, Calif. Federal Aid Rep. Proj. W-52-R-6, Job 4, 1963.
28. **Ferrel, C. M.** Tule elk food habits, Calif. Federal Aid Rep. Proj. W-52-R-6, Job 4, 1965.
29. **Gaffney, W. D.**, The effects of winter elk browsing, south fork of the Flathead River, Montana, *J. Wildl. Manage.*, 5, 427, 1941.
30. **Geis, A. F.**, The Food Requirements and Relative Digestibility of a Variety of Winter Diets Fed to Elk (*Cervus canadensis nelsoni*) Under Controlled Conditions, M.S. thesis, Montana State University, Missoula, 1954.
31. **Gordon, F. A.**, Range Relationships of Elk and Cattle on Elk Winter Range, Crow Creek, Montana, M.S. thesis, Montana State University, Boxeman, 1968.
32. **Greer, K. R.**, Analysis of 68 samples from the 1956 Gallatin Canyon elk "hunter kill," in Wildlife Investigations – State: Wildlife Investigations Laboratory, Federal Aid Completion Rep. Proj. W-83-R-2, all Jobs, Montana Department of Fish and Game, Helena, 1959, 5–6.
33. **Greer, K. R.**, Analysis of 23 elk rumen samples collected during 1957 from the vicinity of Gardiner, Montana, in Wildlife Investigations – State: Wildlife Investigations Laboratory, Montana Dep. of Fish and Game, Federal Aid Completion Rep. Proj. W-83-R-2, all Jobs, Montana Department of Fish and Game, Helena, 1959, 8.
34. **Greer, K. R.**, Analysis of twenty-one elk rumens collected from the 1956 hunter-kill and Boyd Ranch, Missoula County, in Wildlife Investigations – State: Wildlife Investigations Laboratory, Federal Aid Completion Rep. Proj. W-83-R-3, all Jobs, Montana Department of Fish and Game, Helena, 1960, 14.
35. **Greer, K. R.**, Analysis of twenty-nine elk rumens collected during 1956–57 from the Sun River Game Range – Lewis and Clark County, Montana, in Wildlife Investigations – State: Wildlife Investigations Laboratory, Federal Aid Completion Rep. Proj. W-83-R-3, all Jobs, Montana Department of Fish and Game, Helena, 1960, 16–17.
36. **Greer, K. R., Kirsch, J. B., and Yeagher, H. W.**, Seasonal Food Habits of the Northern Yellowstone Elk (Wapiti) Herds During 1957 and 1962–67 as Determined from 793 Rumen Samples, Federal Aid Final Rep. Proj. W-83-R-12, J-B-1, Montana Department of Fish and Game, Helena, 1970.
37. **Greffenius, R. J.**, Results of the Copper Ridge Basin elk study, *U.S. For. Serv. Rocky Mount. Bull.*, 21, 14–15, 1938.
38. **Hansen, R. M. and Reich, L. D.**, Diet overlap with elk, deer, and cattle in northern Colorado, *J. Range Manage.*, 28(1), 43–47, 1975.
39. **Harper, J. A.**, Ecological study of Roosevelt elk, Game Research Rep. 1, Oregon Game Commission, Salem, 1966.
40. **Harper, J. A.**, Ecology of Roosevelt elk, Oregon State Game Commission, Corvallis, 1971.
41. **Harper, J. A., Harn, J. H., Bentley, W., and Yocum, C. F.**, The status and ecology of the Roosevelt Elk in California, *Wildl. Monogr.*, No. 16, 1967.
42. **Harris, J. T.**, Analysis of Elk Winter Range, South Fork of the White River, Colorado, M.S. thesis, Colorado State Univeristy, Fort Collins, 1958.
43. **Hash, H. S.**, Movements and Food Habits of the Lochsa Elk, M.S. thesis, University of Idaho, Moscow, 1973.
44. **Helwig, L. L.**, The Value of Conifers as a Winter Feed for Elk (*Cervus canadensis nelsoni*), Determined Under Controlled Conditions, M.S. thesis, University of Montana, Missoula, 1957.
45. **Hines, W. W.**, Ecological study of Roosevelt elk, Progress report, Fed. Aid Proj. W-59-R-7, Job 1, Oregon Game Commission, Salem, 1970.

46. **Hoskins, L. W.,** Browse Utilization Studies on the Pocatello Winter Big Game Range, M.S. thesis, University of Idaho, Moscow, 1952.
47. **Hungerford, C. R.,** The Food Consumption and Weight Response of Elk (*Cervus canadenis nelsoni*) Under Winter Conditions, M.S. thesis, University of Montana, Missoula, 1952.
48. **Kirsch, J. B.,** Range Use, Relationships to Logging, and Food Habits of the Elk in the Little Belt Mountains, Montana, M.S. thesis, Montana State University, Bozeman, 1963.
49. **Knight, R. R.,** The Sun River elk herd, *Wildl. Monogr.,* No. 23, 1970.
50. **Knowles, C. J.,** Range Relationships of Mule Deer, Elk and Cattle in a Rest Rotation Grazing System During Summer and Fall, Federal Aid Proj. W-120-R, Job BG-10.01, Montana Department of Fish and Game, Helena, 1975.
51. **Komberec, T. J.,** Range Relationships of Mule Deer, Elk, and Cattle in a Rest Rotation Grazing System During Winter and Spring, Federal Aid Proj. W-120-K, Job 2, Montana Department of Fish and Game, Helena, 1976.
52. **Korfhage, R. C.,** Summer Food Habits of Elk in the Blue Mountains of Northeastern Oregon Based on Fecal Analysis, M.S. thesis, Washington State University, Pullman, 1974.
53. **Kufeld, R. C.,** Foods eaten by the Rocky Mountain elk, *J. Range Manage.,* 26(2), 106–113, 1973.
54. **Lang, E. M.,** Elk of New Mexico, Bull. No. 8, New Mexico Department of Game and Fish, Albuquerque, 1958.
55. **Mackie, R. J.,** Range ecology and relations of mule deer, elk and cattle in the Missouri River Breaks, Montana, *Wildl. Monogr.,* No. 20, 1970.
56. **Martinka, C. J.,** Population ecology of summer resident elk in Jackson Hole, Wyoming, *J. Wildl. Manage.,* 33, 465–481, 1969.
57. **McBee, R. H., Johnson, J. L., and Bryant, M. P.,** Ruminal microorganisms from elk, *J. Wildl. Manage.,* 33(1), 181–186, 1969.
58. **McCulloch, C. Y., Jr.,** Utilization of winter browse on wilderness big game range, *J. Wildl. Manage.,* 19(2), 206–215, 1955.
59. **McCullough, D. R.,** *The Tule Elk – Its History, Behavior, and Ecology,* University of California Press, Berkeley, 1971.
60. **McDonald, P., Edwards, R. A., and Greenhalgh, J. F. D.,** *Animal Nutrition,* Hafner, New York, 1973.
61. **Moen, A. N.,** *Wildlife Ecology: An Analytical Approach,* W. H. Freeman, San Francisco, 1973.
62. **Moran, R. J.,** The Rocky Mountain Elk in Michigan, Res. Dev. Rep. No. 267, Michigan Department of Natural Resources, Wildlife Division, Lansing, 1973.
63. **Morris, M. S. and Schwartz, J. E.,** Mule deer and elk food habits on the National Bison Range, *J. Wildl. Manage.,* 21, 189–193, 1957.
64. **Murie, O. J.,** *The Elk of North America,* Stackpole, Harrisburg, Pa., 1951.
65. **Murphy, D. A.,** A captive elk herd in Missouri, *J. Wildl. Manage.,* 27(3), 411–414, 1963.
66. **Nichols, L., Jr.,** Forage Utilization by Elk and Domestic Sheep in the White River National Forest, M.S. thesis, Colorado State University, Fort Collins, 1957.
67. **Peek, J. M.,** Wildlife Investigations (District 3): Big Game Survey and Investigation – summer of 1956–1960. Gravelly-Snowcrest Rumen Collections, Federal Aid Completion Rep. Proj. W-73-R-8, J-A-1, Montana Department of Fish and Game, Helena, 1963.
68. **Pickford, G. D. and Reid, E. H.,** Competition of elk and domestic livestock for summer range forage, *J. Wildl. Manage.,* 7, 328–332, 1943.
69. **Picton, H. D.,** A comparison of the results of feeding site examinations with the vegetative composition of the sites in the Sun River Game Range, in Wildlife Investigations (District 4): Big Game Surveys and Investigations, Federal Aid Compliances Rep. Proj. W-74-R-5, J-A-1, Montana Department of Fish and Game, Helena, 1960.
70. **Prins, R. A. and Geelen, M. J. H.,** Rumen characteristics of red deer, fallow deer, and roe deer, *J. Wildl. Manage.,* 35(4), 673–680, 1971.
71. **Probasco, G. E.,** Diet Preferences and Utilization Patterns of Elk on the Bighorn Mountains, Wyoming, M.S. thesis, University of Wyoming, Laramie, 1968.
72. **Rouse, R. A.,** Elk Food Habits, Range Use and Movements, Gravelly Mountains, Montana, M.S. thesis, Montana State University, Bozeman, 1957.
73. **Rouse, R. A.,** Wildlife Investigations (District 3): Elk Investigations (Elk-livestock Relationships), Federal Aid Completion Rep. Proj. W-73-R-3, J-A-2, Montana Department of Fish and Game, Helena, 1958.
74. **Rush, W. M.,** Northern Yellowstone Elk Study, Montana Fish and Game Commission, Helena, 1932.
75. **Schallenberger, A.,** Big Game Forage Competition in Sun River Canyon, M.S. thesis, Montana State University, Bozeman, 1965.

76. **Schwartz, J. E. and Mitchell, G. E.,** The Roosevelt elk on the Olympic Peninsula, Washington, *J. Wildl. Manage.*, 9(4), 295–319, 1945.
77. **Short, H. L.,** Rumen fermentations and energy relationships in white-tailed deer, *J. Wildl. Manage.*, 27, 184–195, 1963.
78. **Short, H. L., Medin, D. E., and Anderson, A. E.,** Ruminoreticular characteristics of mule deer, *J. Wildl. Manage.*, 46(2), 196–199, 1965.
79. **Short, H. L., Remenga, E. E., and Boyd, C. E.,** Variations in ruminoreticular contents of white-tailed deer, *J. Wildl. Manage.*, 33(1), 187–191, 1969.
80. **Short, H. L., Segelquist, C. A., Goodrum, P. D., and Boyd, C. E.,** Rumino-reticular characteristics of deer on food of two types, *J. Wildl. Manage.*, 33(2), 280–383, 1969.
81. **Singer, F. J.,** Wildfire and Ungulates in the Glacier National Park Area, Northwestern Montana, M.S. thesis, University of Idaho, Moscow, 1975.
82. **Skinner, M. P.,** Browsing of the Olympic peninsula elk in early winter, *J. Mammal.*, 17(3), 253–256, 1936.
83. **Snyder, R.,** Forage Competition Between Cattle and Elk in the Gird Creek Drainage of Western Montana, M.S. thesis, University of Montana, Missoula, 1969.
84. **Stark, R. H., Jr.,** Elk-cattle Interrelations on the Colockum Creek Watershed, M.S. thesis, Washington State University, Pullman, 1973.
85. **Stevens, D. R.,** Range relationships of elk and livestock, Crow Creek drainage, Montana, *J. Wildl. Manage.*, 30, 349–363, 1966.
86. **Swanson, D. O.,** Roosevelt Elk – Forest Relationships in the Douglas Fir Region of the Southern Oregon Coast Range, Ph.D. thesis, University of Michigan, Ann Arbor, 1970.
87. **Thorne, T. E., Dean, R. E., and Hepworth, W. G.,** Nutrition during gestation in relation to successful reproduction in elk, *J. Wildl. Manage.*, 40(2), 330–335, 1976.
88. **Trout, L. C. and Leege, T. A.,** Are the northern Idaho elk herds doomed? *Idaho Wildl. Rev.*, 24(3), 3–6, 1971.
89. **Troyer, W. A.,** The Roosevelt elk on Afognak Island, Alaska, *J. Wildl. Manage.*, 24(1), 15–21, 1960.
90. **Wing, L. D.,** Big Game and Livestock Browse Utilization and Feeding Habits on a Sandy Range in Southeastern Idaho, M.S. thesis, University of Idaho, Moscow, 1962.
91. **Young, V. A.,** The carrying capacity of big game range, *J. Wildl. Manage.*, 2, 131–134, 1938.
92. **Young, V. A. and Robinette, W. L.,** A Study of the Range Habits of Elk on the Selway Game Preserve, Bull. No. 9, University of Idaho School of Forestry, Moscow, 1939.

DIETS FOR CARIBOU

A. T. Bergerud

INTRODUCTION

Caribou or reindeer (*Rangifer tarandus*) are distributed circumpolar throughout the Holarctic in Canada, Alaska, the U.S.S.R., the Fennoscandian countries, Spitsbergen, and Greenland. Within this range they are found primarily in the tundra and taiga biomes and at times in the boreal forest.

NATURAL DIET

Caribou have catholic tastes, eating a wider variety of plants than other deer species.[1,2] However, they prefer green vascular plants and mushrooms;[3] when these plants cannot be secured they are opportunistic, and relative availability is the prime factor in species utilization.[3] Their food habits, which have been investigated throughout their range, are similar, even at the plant species level, no doubt because of the similarity of the forage resources between regions. Extremely important foods such as reindeer lichens (*Cladonina alpestris* and *C. rangiferina*), dwarf birch (*Betula nana*), sedge (*Carex aquatilis*), cottongrass (*Eriophorum vaginatum*), and horsetail (*Equisetum fluviatile*) are heavily utilized in both Eurasia and North America. Thus, a synthesis is possible, and one can predict the natural diet of caribou in a given locality based on the composition of the flora and knowledge of the winter snow cover.

Changes in the diet composition are best tabulated on a seasonal basis (spring, summer, fall, and winter). Both plant phenology and snow cover are seasonal phenomena that relate to food availability. The spring diet includes that period from the start of new plant growth until the leaves of deciduous shrubs are abundant – usually May until mid June (6 weeks). The summer diet covers the period from mid June until freezing weather in September (2 months). The fall season stretches from mid September until snow cover becomes general about mid November (2 months). The winter season includes the interval from mid November through mid April (5 months). Often in late April or early May, prior to new plant growth, some south slopes are free of snow, and the diet then resembles that in the fall period.

As the spring thaw advances, caribou quickly change their diet to include new green growth.[2] Preferences follow the primary production progression.[3,4] In mountainous regions, caribou can keep abreast of new plant phenology by altitudinal shifts and visiting receding snowfields. Sedges (*Carex* spp., *Scirpus* spp., and *Eriophorum* spp.) and grasses (*Deschampsia* spp., *Calamagrostis* spp., and *Festuca* spp.) are the most important plant groups in the spring diet (Table 1*), frequently comprising 50% of the diet.[3,7] Many growing plants are utilized only when they first appear, e.g., catkins of willow, needles of larch (*Larix laricina*), leaves of alder (*Alnus* spp. and Solomon's seal, *Smilacina trifolia*), and buds of sweet gale (*Myrica gale*) (Table 2).

The most important plant materials in the summer diet are the green leaves of deciduous shrubs, especially willow, birch, and blueberry (*Vaccinium* spp.).[2–5] Caribou are fastidious feeders that strip the leaves from woody stems and pick only new sprouts and finer stem tips of grasses, sedges, herbs, and horsetail (*Equisetum* spp.). In late summer the animals actively search for mushrooms.[2–5,8]

After the first hard frost the deciduous leaves quickly become unavailable and the

*The tables are found at the end of the text.

animals switch to terrestrial lichens and the leaves of evergreen shrubs. The fruticose reindeer lichen, *Cladonia alpestris*, is probably the most preferred lichen, followed by *C. mitis-sylvatica* and *C. rangiferina*, then Ireland moss (*Cetraria islandica*), and then the *Stereocoulon* species.[9,10] Many shrubs of the Ericaceae family are foraged. Graminales, mushrooms, and horsetails are still heavily utilized, if available.

In winter, snow cover dictates food availability (Table 3). Winter greens are especially important if they can be found by digging holes in the snow.[5,8] Important circumpolar species such as *Deschampsia flexuosa, Equisetum fluviatile, Carex aquatilis*, and *Eriophorum vaginatum* are readily sought,[5,7,8] and the animals also look for the fine twigs of birch, willow, and blueberry.[16,38,39]

If sun crusts that can support the weight of a caribou develop on the snow, the animals will shift from open habitats to forest cover to seek arboreal lichens growing on coniferous trees (Table 3).

In winter, as a necessity the animal will eat most kinds of food, including nonpreferred conifer needles[3,4,12] and green bryophytes.[3,7] Even foliose lichens such as *Umbilicaria hyperborea, Cetraria ciliaris,* and *Parmelia physodies* growing on the bark of trees and rocks serve as foods when supplies are short.[3,4] The leaves of evergreen shrubs are also important (Table 3). Lichens undoubtedly are the primary food in most areas because of their widespread distribution and relative abundance, but lichens are certainly not required, and a winter diet of green sedges and grasses would be more nutritious.

PARTIALLY PURIFIED DIET

Captive caribou and domestic reindeer have been fed a variety of diets, including alfalfa hay, timothy hay, sugar beets, collected *Cladonia* lichens, wild hays and cereal grains, barley, oats, chopped corn, and commercial calf starters.[11,34] Two diets recently used in Alaska are Purina calf starter No. 1 (72% digestible) and mixed reindeer feed (45% digestible; includes 67% mixed *Cladonia* lichens, 25% brome grass [*Festuca altaica*], and 8% cured *Carex aquatilis*).[35] Caution must be used in switching animals from a natural diet to roughages or commercial pellets since the rumen mucosa structure as well as the rumen flora and fauna are sensitive to changes. Physiological disorders and death can result from rapid alterations of diets.

Several high energy diets have been formulated by workers in Norway and Canada (Tables 4 and 5). Several of these diets will allow caribou to maintain adult weights in winter, provide sufficient energy for winter growth, and result in high birth weight of calves and increased milk yield of females (Tables 5 and 6).

CHEMICALLY DEFINED DIET

The chemical composition of one natural diet of caribou is shown in Table 7. Table 8 shows the nutrient composition of many natural foods and Table 4 indicates the composition of four purified diets.

QUALITATIVE AND QUANTITATIVE REQUIREMENTS

The annual diet of caribou shows a marked seasonal variation in quality and quantity. The spring fare is the most nutritious (Table 7). Caribou actively seek the new green growth that is high in crude protein (Table 8). They select forage rich in protein among species, among plants within species, and among plant parts within plants.

By such selection the animals are able to maintain a positive energy balance in the spring and summer (Table 9). They accumulate fat reserves most rapidly in August and early September after insect harassment has declined and the animals are able to feed throughout the day. In winter the animals must mobilize these reserves for maintenance, and it is not uncommon for animals to lose 10 to 26% of the autumn weight prior to June.[1,4,43-46] This negative energy balance occurs because lichens, the most abundant food in the winter diet, are low in nitrogen and minerals although adequate in carbohydrates (Table 8). If free-ranging animals can find adequate winter greens (the buds, shoots, and stems of sedges and grasses), weights can theoretically be maintained. However, it has been documented with captive animals that caribou voluntarily reduce food intake in winter as a physiological adaptation to foods shortages; basal metabolism and caloric intake may be reduced 30 to 40% (Table 10, bottom).[47-50]

Information on the quantity of forage utilized per day has been gathered from feeding trials, observations of free-ranging animals, and calculations based on digestibility of forage. The results are similar and range from 0.04 to 0.06 kg dry matter per kilogram of body weight per day[1,4,34,51] or approximately 5 kg of forage per day per animal.[52,53]

Reindeer milk is especially rich (Table 11) and can result in growth rates of 275 to 450 g/day between birth and weaning.[44,60-62] Maximum initial milk intake of reindeer in Alaska was 2.5 l/day[63] at 2 to 3 weeks of age and declined exponentially to about 0.75 l/day at 90 days postpartum. The milk intake equation was $I = 2.18e^{-0.0116A}$, where I equals milk intake and A is age in days. The total yield of one female during 5.5 months of lactation was 160 l.[63] In another study, male calves 0 to 4 weeks old took 1.44 to 1.65 l/day, with a caloric value of 3500 to 4100 kcal/day.[59]

The amount of water ingested per day has not been reported. Caribou appear to drink water much more readily in the winter than in the summer.[2] Snow and slush are preferred over free water. Seasonal changes in water flux and intake may minimize fluctuations in rumen and body temperature and reduce thermal shock from consumption of frozen vegetation.[64]

The salt and mineral requirements of caribou have not been specified in the literature. Lichens are low in sodium, potassium, magnesium, calcium, and phosphorus. Caribou also chew their antlers and drink salt water when available. Natural salt licks are visited in the spring in Alaska,[2] and the Russians feed captive reindeer NaCl in the winter and spring; they recommend 5 g daily.[45]

Table 1
RESUMÉ OF ANNUAL UTILIZATION OF PLANT GROUPS BY CARIBOU

Plant group	Forage season[a] Spring	Summer	Fall	Winter
Terrestrial lichens	T	T	P	T
Arboreal lichens	T	T	T	T
Bryophytes (green mosses)	A	A	A	T
Filicales (ferns)	–	P	–	–
Equisetales (horsetails)	P	P	P	P[b]
Herbs	P	P	P	P[b]
Evergreen broadleaf shrubs	P	A	T	T
Deciduous broadleaf shrubs	P	P	P	P[c]
Conifers	A[d]	A	A	A[e]
Graminales (sedges and grasses)	P	T	T	P
Aquatics	–	P	–	–
Basidiomycetes (mushrooms)	–	P	P	–
Club mosses (*Lycopodium* spp.)	–	–	–	A

[a] P = preferred; T = taken as available; A = accidental ingestion.
[b] If they can be located in the snow.
[c] Especially twigs of *Vaccinium* spp., *Betula* spp., *Salix* spp., and *Sarbus* spp.
[d] Needles of *Larix* spp. preferred.
[e] *Taxus canadensis* taken as available.

Table 2
A PARTIAL LIST OF SOME OF THE PLANT SPECIES REPORTEDLY EATEN BY CARIBOU

Nomenclature that of Original Sources in Most Instances

Terrestrial Lichens

Alectoria nigricans,[3] *A. ochroleuca*[3,13,14,27]

Cetraria cucullata,[2,8,13,17,30] *C. ericetorum,*[30] *C. hepatizon,*[3] *C. islandica,*[2,3,5,8,14,24,30,32,42] *C. nivalis,*[8,13,14,30,32] *C. richardsonii*[9]

Cladonia alpestris,[2,3,5,8-10,12-15,19,27,30,42] *C. amaurocraea,*[2] *C. amouracra,*[8,13] *C. arbuscula,*[13] *C. belliflora,*[17] *C. boryi,*[3] *C. cornuta,*[8,15] *C. crispata,*[3,27] *C. degenerans,*[8] *C. gracilis,*[3,8,15,30] *C. mitis,*[2,3,10,13,15,20,30] *C. rangiferina,*[2,3,5,8,10,12-15,19,27,30,32] *C. sylvatica,*[2,3,5,8,17,24,27] *C. uncialis*[2,3,10,13,19,24,30]

Cornicularia aculeata[30]

Dactylina arctica[3,42]

Nephroma artica,[11,42] *N. artila*[30]

Peltigera aphtosa,[9,19,32] *P. canina*[3]

Sphaerophorus globosus[3,8,13,24,25]

Stereocaulon grande,[32] *S. paschale,*[3,5,8,9,18,30] *S. tomentosum,*[27] *S. vesuvianum*[3]

Thamnolia vermicularis[7,42]

Umbilicaria hyperborea[4]

Arboreal Lichens

Alectoria altaica,[3] *A. giegana,*[14] *A. jubata,*[3,14,16,29,31] *A. ochrolauca,*[30] *A. pubescens,*[30] *A. sarmentosa*[3,16,29,31]

Bryopogon chalybeiformis,[8] *B. implexum*[8]

Cetraria ciliaris,[3] *C. glauca,*[3] *C. hepaticon*[36]

Table 2 (continued)
A PARTIAL LIST OF SOME OF THE PLANT SPECIES REPORTEDLY EATEN BY CARIBOU

Nomenclature that of Original Sources in Most Instances

Arboreal Lichens (continued)

Evernia mesomorpha,[10,13] *E. prunastri*[14]
Parmelia galbina,[3] *P. caperata,*[3] *P. omphalodes,*[30] *P. physodes,*[3,16,19] *P. saxatilis,*[4] *P. sulcata*[3]
Physica airpolia[3]
Ramalina farinocca[4]
Usnea barbata,[18] *U. comosa,*[13] *U. barbatus,*[19] *U. hirta,*[1] *U. gossypoga,*[8] *U. plicata*[4]

Herbs

Equisetum arvense,[2,8,20,27,30] *E. fluviatile,*[3,5,8,32] *E. hiemale,*[6,8] *E. komarovii,*[8] *E. limosum,*[8] *E. palustre,*[2] *E. pratense,*[2] *E. scirpodes,*[2,6,8] *E. sylvaticum,*[8,20] *E. variegatum*[6,8,42]

Osmundaceae
Osmunda cinnamomea,[3] *O. regalis*[3]

Polypodiaceae
Dryopteris spinulosa[3]

Najadceae
Potamogeton gramineus,[3] *P. natans*[3]

Juncaginaceae
Scheuchzeria palustris[3]

Araceae
Arisaema atrorubens[13]

Liliaceae
Clintonia borealis,[12] *Erythronium grandiflorum,*[31] *Maianthemum canadense,*[3,12] *Smilacina trifolia*[16]

Iridaceae
Iris setosa[8]

Orchidaceae
Spiranthes romanzoffiana[13]

Santalaceae
Geocaulon lividum[7]

Polygonaceae
Oxyria digyna,[5] *Polygonum alaskanum,*[11,28] *P. bistorta,*[2,8,20] *P. viviparum,*[5,8,20,30,42] *Rumex acetosa,*[5,8,20,27,30] *R. acetosella,*[30] *R. artica,*[2] *R. haplorhizus,*[20] *R. occidentalis*[11,28]

Portulacaceae
Claytonia acutifolia[8]

Nymphaeaceae
Nymphaea variegetum[3,12]

Ranunculaceae
Aquilegia brevistyla,[13] *Caltha palustris,*[8,12] *Coptis groenlandica,*[3] *Ranunculus acris,*[33] *R. nivalis,*[33] *R. pallasii,*[11,20,28] *R. repens,*[22] *Thalictrum polygamum*[3]

Cruciferae
Draba alpina[42]

Sarraceniaceae
Sarracenia purpurea[3,13]

Droseraceae
Drosera rotundifolia[3]

Crassulaceae
Sedum roseum[2]

Table 2 (continued)
A PARTIAL LIST OF SOME OF THE PLANT SPECIES REPORTEDLY EATEN BY CARIBOU

Nomenclature that of Original Sources in Most Instances

Herbs (continued)

Saxifragaceae
Saxifraga hirculus,[42] *S. tricuspidata,*[2,4] *Tiarella unifoliata*[31]
Rosaceae
Alchemilla glabra,[30,33] *A. vulgaris,*[8] *Comarum palustre,*[8,20] *Fragaria vesca,*[19] *Geum rivale,*[20] *Potentilla palustris,*[5,8] *P. tridentata,*[3] *Rubus chamaemorus,*[3,7,8,11] *Sanguisorba canadensis*[3]
Fabaceae
Astragalus alpinus,[14,28] *A. umbellatus,*[2] *Hedysarum alpinum,*[2,4] *Lathyrus ochroleucus,*[18] *L. palustris,*[18] *L. venosus,*[18] *Lupinus arcticus,*[2,28] *Oxytropis maydelliana,*[14] *O. nigrescens,*[2] *Trifolium lupinaster,*[8] *Vicia cracca*[8]
Onagraceae
Epilobium angustifolium,[11-13,20,22] *E. latifolium,*[2] *E. palustre*[20,27]
Araliaceae
Aralia hispida,[13] *A. nudicaulis,*[12] *A. racemosa*[13]
Umbelliferae
Angelica archangelica,[5] *Coelopleurum gmelin,*[28] *Ligusticum scothicum*[28]
Cornaceae
Cornus canadensis[3,12,15,31]
Ericaceae
Pyrola grandiflora[17]
Primulaceae
Lysimachia terrestris,[13] *Trientalis borealis,*[3] *T. europea*[30]
Gentianaceae
Gentiana glauca,[2,11] *Menyanthes trifoliata,*[3,5,8,13,26] *Swertia perennis*[2]
Polemoniaceae
Polemonium acutiflorum,[8] *P. boreale*[42]
Scrophulariaceae
Pedicularis amoena,[8] *P. groenlandicum,*[2,31] *P. lapponica,*[8] *P. sudetica,*[8,42] *P. verticillata*[8,11,20]
Lentibulariaceae
Utricularia minor[3]
Valerianaceae
Valeriana capitata,[20,42] *V. eapritata*[11,28]
Lobeliaceae
Lobelia dortmanna[3]
Compositae
Antennaria dioica,[8,22] *A. monocephala,*[2] *Artemsia artica,*[2,11,20,28] *A. trilesii,*[11,20,28] *Aster ciliolatus,*[12] *A. umbellatus,*[3] *Chrysanthemum intergifolium,*[42] *Hieracium alpinum,*[30,33] *Lactuca alpina,*[5] *Leontodon autumnalis,*[30,33] *Petasites frigidus,*[2,8] *Solidago lapponica,*[8] *S. rugosa,*[3] *S. virgaurea*[30,33]

Deciduous Shrubs

Salicaceae
Populus balsamifera,[12,17] *P. tremuloides,*[3,12-14,18,19,27] *Salix alaxensis,*[2,17] *S. arctica,*[2,42] *S. arctophila,*[14] *S. caprea,*[5] *S. cordata,*[19] *S. cordifolia,*[14] *S. discolor,*[3,18] *S. glandulifera,*[8] *S. glauca,*[2,5,8,20,27] *S. hastata,*[5,8] *S. herbacea,*[5,8,30,33] *S. humilis,*[3] *S. lanata,*[5,8,42] *S. lapponicum,*[5,27,33] *S. nigricans,*[5] *S. pellita,*[3] *S. ovalifolia,*[42] *S. phlebophylla,*[17] *S. phylicifolia,*[5,8] *S. planifolia,*[14] *S. polaris,*[5,88] *S. pulchra,*[2,14,17,32,42] *S. reticulata,*[8,14,33,42] *S. rotundifolia,*[27,42] *S. sachalinensis*[8]

Table 2 (continued)
A PARTIAL LIST OF SOME OF THE PLANT SPECIES REPORTEDLY EATEN BY CARIBOU

Nomenclature that of Original Sources in Most Instances

Deciduous Shrubs (continued)

Myricaceae

Myrica gale[3,7,13,16]

Betulaceae

Alnus crispa,[3,4,13] *A. hirsuta,*[8] *A. rugosa,*[3,18] *Betula exilis,*[8] *B. glandulosa,*[2-4,13-15,19] *B. michauxii,*[3] *B. middendorfii,*[8] *B. nana,*[2,5,8,24,25,27,30,32] *B. papyrifera,*[3,13,16] *B. pubescens,*[5] *B. pumila,*[3,18,19] *B. rotundifolia,*[17,22,28] *B. fortuosa,*[5] *Corylus americana*[18]

Rosaceae

Amelanchier alnifolia,[18] *A. bartramiana,*[3] *Dryas intergrifolia,*[4,14,42] *D. octopetala,*[2,11,17,28,32] *Potentilla fruticosa,*[3] *Prunus pensylvanica,*[12,13] *P. virginiana,*[18] *Pyrus americana,*[12] *P. arbutifolia,*[3] *P. floribunda,*[3] *Rosa acicularis,*[12] *Rubus idaeus,*[12] *R. parviflorus,*[12] *R. pedatus,*[31] *Sorbus sambucifolia,*[8] *Spiraea latifolia*[3]

Aquifoliaceae

Nemopanthus mucronata[3,16]

Aceraceae

Acer pensylvanicum,[18] *A. spicatum*

Elaeaginaceae

Shepherdia canadensis[12,15]

Cornaceae

Cornus stolonifera[12,13,18,19]

Ericaceae

Arctostaphylos rubra,[4,7] *Menziesia ferruginea,*[29] *Rhododendron canadense,*[3] *Vaccinium angustifolium,*[3,6] *V. canadense,*[18] *V. myritillus,*[5,7,15,30] *V. uliginosum*[2,3,4,7,11,14,15,17,18,22,25,28]

Caprifoliaceae

Diervilla lonicera,[12,13] *Sambucus pubens,*[12] *Viburnum cassinoides,*[3,16] *V. rafinesquiana*[12]

Evergreen Shrubs

Empetraceae

Empetrum hermaphroditum,[30] *E. nigrum*[2-4,7,11,14-18,20,25,27,28]

Celastraceae

Pachistima myrsinites[29]

Ericaceae

Andromeda glaucophylla,[3,13,17,18] *A. polifolia,*[15,17,27,30] *Arctostaphylos alpina,*[2,14,17,25,27,30] *A. uva-ursi,*[2,4,7,15,17,18,30] *Cassiope lycopodiodes,*[17] *C. mertensiana,*[17] *C. tetragona,*[2,15] *Chamaecistus procumbens,*[17] *Chamaedaphne calyculata,*[3,7,13,15,16,18,19] *Diapensia lapponica,*[11,17] *Gaultheria hispidula,*[3,16] *Harrimanella stelleriana,*[17] *Kalmia angustifolia,*[3] *K. polifolia,*[3,7,16] *Ledum decumbens,*[2,4,14,15,17,28] *L. groenlandicum,*[2-4,7,13,15,20,28] *L. palustre,*[11,20,25,27] *Loiseleuria procumbens,*[7,14,15,30] *Oxycoccus microcarpus,*[7] *Phyllodoce empetriformis,*[17] *P. caerulea,*[14,30] *Rhododendron lapponicum,*[4,14] *Vaccinium oxycoccus,*[3,4,7,16,17] *V. vitis-idaea*[2-4,7,11,14,15,17,22,25,28,30,32,42]

Caprifoliaceae

Linnea borealis[3,8,12,15,19,31]

Table 2 (continued)
A PARTIAL LIST OF SOME OF THE PLANT SPECIES REPORTEDLY EATEN BY CARIBOU

Nomenclature that of Original Sources in Most Instances

Grasses

Graminae

Agropyron repens,[8] *Agrostis borealis,*[14] *Anthoxanthum alpina,*[30] *A. odoratum,*[5] *Alopecurus alpinus,*[42] *Arctagrostis latifolia,*[8,27] *Arctophila fulva,*[8] *Calamagrostis canadensis,*[2,14,32] *C. deschampsoides,*[14] *C. langsdorfii,*[8] *C. neglecta,*[5,8] *C. purpurascens,*[14] *Danthonia spicata,*[17] *Deschampsia caespitosa,*[5,8,14,33] *Dupontia fisheri,*[42] *D. flexuosa,*[5,8,22,30,33] *Elymus arenarius,*[42] *Festuca altaica,*[2,32] *F. ovina,*[5,8,22,30] *F. rubra,*[5] *Hierochloe alpina,*[2,14,32] *H. odorata,*[17] *Poa alpina,*[30] *P. artica,*[8] *P. glauca,*[14] *P. pratensis*[8]

Sedges

Cyperaceae

Carex aquatilis,[2,5,8,20,26,32,42] *C. bigelowii*[2,5,30,32,33] *C. brunnescens,*[30] *C. caespitosa,*[8] *C. concolor,*[14] *C. filiformis,*[5] *C. flava,*[5] *C. glareosa,*[8] *C. gracilis,*[8] *C. hyperborea,*[8] *C. inflata,*[8] *C. juncella,*[5] *C. lachenalii,*[5] *C. membranacea,*[2,14,42] *C. podacarpa,*[2] *C. rariflora,*[8,14] *C. rostrata,*[5] *C. rotundata,*[8] *C. rupestris,*[42] *C. scirpoidea,*[42] *C. socromeana,*[8] *C. stans,*[8] *C. vaginata,*[30] *C. wiluica,*[8] *Eriophorum angustifolium,*[2,8,11,22,28,32,33,42] *E. brachyantherum,*[32] *E. callitrix,*[22,28] *E. chamissonis,*[8,13] *E. polystachyum,*[5] *E. russeolum,*[8] *E. scheuchzeri,*[8,42] *E. vaginatum,*[2,5,8,14,20,23] *Scirpus cespitosus*[3,16,26]

Rushes

Juncaceae

Juncus trifidus[30,33]

Conifers

Taxaceae

Taxus canadensis[12]

Pinaceae

Larix laricina,[7,15] *Picea mariana,*[4,16] *Pinus banksiana,*[7,15]

Cupressaceae

Juniperus communis[4]

Table 3
EFFECT OF WINTER SNOW CONDITIONS ON CARIBOU'S CHOICE OF FORAGE GROUPS IN NORTHERN CANADA[6,7]

Season (months)	Snow condition and feeding behavior	Forage groups most utilized	Possible forage plant species
Early winter (December–January)	Snow soft and shallow, <50 cm; animals move and dig feeding holes	Horsetails, sedges, grasses, terrestrial lichens, and shrubs	*Equisetum fluviatile* *Carex aquatilis* *Calamagrostis canadensis* *Cladonia* spp. *Cetraria* spp. *Stereocaulon* spp. *Myrica gale* *Vaccinium vitis-idaea* *Vaccinium uliginosum* *Kalmia polifolia*
Midwinter (February)	Snow soft and deep, >50 cm; little movement, dig feeding holes	Terrestrial lichens and evergreen shrubs	*Cladonia lichens* *Stereocaulon* spp. *Cetraria* spp. *Ledum groenlandicum* *Vaccinium vitis-idaea* *Vaccinium uliginosum*
Late winter (March–April)	Snow deep with sun crust (animals move and walk on crust)	Arboreal lichens and twigs of deciduous shrubs	*Alectoria* spp. *Usnea* spp. *Everniia* spp. *Betula* spp. *Alnus* spp. *Salix* spp.
Late winter–early spring (March–May)	Snow depth diminishing, alternate crust and crustless conditions, then bare spots on south slopes; animals migrating	Terrestrial lichens and evergreen shrubs	*Cladonia* spp. *Stereocaulon* spp. *Cetraria* spp. *Vaccinium vitis-idaea* *Ledum* spp. *Lycopodium* spp. *Vaccinium uliginosum*

Table 4
COMPOSITION (%) OF FOUR HIGH ENERGY FEEDS TESTED WITH REINDEER

Ingredient	69-A[36]	71-A[36]	Ingredient	36-57[37]	36-69c[37]
Barley (ground)	9.1	40.0	Yellow corn (ground)	30.0	31.5
Oats (ground)	14.3	17.0	No. 5 feed wheat (ground)	12.5	26.3
Wheat bran	–	15.0	Wheat bran (ground)	13.8	13.8
Beet pulp	8.8	–	Molasses (cane)	7.5	5.0
Young grass (ground)	19.5	25.0	Beet pulp (ground)	10.0	10.0
Straw (ground)	46.7	–	Dehydrated grass meal	10.0	10.0
Soybean oil	–	3.0	Soybean meal (50%)	8.8	–
Seaweed	0.67	–	Fish meal (70% herring)	5.5	–
NaCl	0.36	–	Steamed bone meal	1.0	–
$Ca_3(PO_4)_2$	0.67	–	Iodized salt	0.1	–
			Dairy (calf) micromix	–	2.5
			Chromic oxide	–	1.0
Nutrients per Percent Dry Matter					
Crude protein	6.3	13.4		–	–
Crude fat	–	7.2		–	–
Nitrogen-free extract	48.7	62.7		–	–
Crude Fiber	23.9	11.0		–	–
Ash	–	5.7		–	–

Table 5
COMPARISON OF INTAKE, FATTENING FEED UNITS, CHEMICAL COMPOSITION, DIGESTIBILITY, AND WEIGHT GAIN FOR SIX DIETS FED TO REINDEER CALVES IN THE WINTER IN NORWAY[36]

Four Calves; Each Diet Fed for 83 Days

	Lichen (mostly *Cladonia alpestris*)	Lichen plus minerals	Lichen plus minerals and barley	Lichen plus minerals and urea barley	Lichens plus minerals and soybean meal	Lichens plus TF-A
Intake (Grams per Day per Animal)						
Supplementary feed	0	0	148.4	136.0	147.3	100.3
Mixed minerals	0	16.2	18.2	16.3	11.5	42.1
Intake FFU[a]						
In lichen	0.698	0.745	0.746	0.842	0.846	0.878
In supplement	0	0	0.148	0.128	0.145	0.050
Total	0.698	0.745	0.894	0.970	0.991	0.928
Intake (Grams per Day per Animal)						
Crude protein (g)	24.5	26.4	42.2	78.1[b]	101.6	74.4[c]
Calcium (g)	0.7	2.6	3.1	2.8	2.6	6.7
Phosphorus (g)	0.4	1.7	2.3	2.2	2.3	4.5
Magnesium (g)	0.2	0.7	0.9	1.0	1.0	1.9
Digested crude protein (g)	-4.6	-7.0	2.4	36.2	58.5	35.1
Digestibility (%)						
Organic	75	67	73	68	69	65
Ether extract	66	65	69	64	59	66
Crude protein	-24	-40	-0	46	58	47
Crude fiber	75	70	70	67	67	65
Nitrogen-free extract	79	71	78	72	73	69
Weight Gain						
Gain (kg)	-4.2	-2.6	-2.1	1.3	1.6	2.2
Gain (g/day)	-50.6	-31.3	-26.3	15.7	19.3	27.9

[a] Fattening feed units estimated on basis of chemical composition of lichen and digestibility of lichens in trails using Kellner-Mollgoord method (a reduction of 1.04 NKF [net calories for fattening] per gram undigestible organic matter).
[b] Percent of protein from urea 46.2.
[c] Percent of protein from urea 48.9.

Table 6
COMPARISON OF THE WEIGHT OF FEMALES, MILK YIELD, BIRTH WEIGHT OF THEIR CALVES, AND SUBSEQUENT GROWTH BASED ON 20 FEMALES FED TWO DIFFERENT DIETS[36]

Norway, February 16–May 6

	Lichen[a] only	Lichen[a]+ TF-69[a]
Daily consumption[b]		
Lichen	4.225 kg	6.326 kg
TF-69	–	6.117 kg
Change in body weight		
Females, Feb. 2–May 5	–3.65 kg	3.50 kg
Milk Yield (g)		
4 hr after birth	20 g	63 g
13 hr after birth	66 g	118 g
Mean weight of calf (kg)		
1 day	3.74 kg	4.37 kg
16 days	5.64 kg	7.43 kg
Daily gain of calves (g)		
16 days old until July 17	280	322
July 17–Sept. 29	281	293

[a] Mostly *Cladonia alpestris.*
[b] See Table 4 for diet composition.

Table 7
PERCENTAGE DISTRIBUTION OF NUTRIENTS IN FOUR SEASONAL SAMPLE DIETS FROM NEWFOUNDLAND[3]

Season and plant species	Protein	Fat	Fiber	Ash	NFE	Ca	P
	Percentage composition on dry matter basis						
Spring – June							
Alnus crispa	30.85	3.35	13.31	4.11	48.38	0.42	0.37
Cladonia mitis	4.11	2.47	28.22	1.33	63.87	0.17	0.05
Kalmia polifolia	11.43	2.17	16.88	2.75	66.78	0.59	0.15
Larix laricina	18.47	5.03	18.59	3.82	58.47	0.24	0.27
Scirpus cespitosus	15.87	0.85	27.25	2.85	55.40	0.12	0.15
Smilacina trifolia	20.20	2.72	14.52	6.61	59.85	0.27	0.48
Average	15.72	2.77	19.80	3.59	58.79	0.30	0.20
Summer – August							
Amelanchier bartraniana	18.27	2.58	22.59	7.20	29.36	1.32	0.20
Betula pumila	12.82	4.20	22.21	3.33	57.44	0.80	0.11
Cladonia mitis	2.28	1.25	38.12	0.70	57.65	0.09	0.02
Rubus chamaemorus	13.02	1.38	13.05	4.80	63.27	0.27	0.38
Sanguisorba canadensis	14.71	2.76	18.15	7.14	57.24	1.17	0.20
Vaccinium angustifolium	9.47	1.57	17.33	2.80	68.83	0.60	0.12
Average	11.76	2.29	21.91	4.33	55.63	0.71	0.16
Fall – October							
Chamaedaphne calyculata	9.02	4.96	26.54	2.64	56.84	0.76	0.09
Cladonia alpestris	2.72	1.35	43.04	1.82	51.07	0.12	0.04
C. rangiferina	2.56	6.87	43.74	1.49	51.34	0.13	0.05
Cornus canadensis	11.13	4.60	12.13	9.71	62.43	0.33	0.13
Kalmia angustifolia	10.03	3.68	13.69	2.51	70.09	0.56	0.11
Vaccinium angustifolium	7.96	3.84	16.20	4.62	67.38	0.98	0.09
Average	7.24	3.21	25.89	3.80	59.86	0.48	0.09
Winter							
Alectoria jubata	5.78	0.56	7.47	1.45	84.74	0.11	0.07
Chamaedaphne calyculata	9.86	6.31	22.67	2.44	58.72	0.65	0.10
Cladonia mitis	3.90	1.58	53.39	3.16	37.97	0.07	0.02
C. rangiferina	3.27	0.81	41.86	2.07	51.99	0.10	0.03
Kalmia angustifolia	9.46	4.13	12.83	2.22	71.36	0.61	0.11
Parmelia spp.	5.69	1.22	13.61	1.49	77.99	0.19	0.06
Scirpus cespitosus	4.82	1.02	35.87	0.67	57.62	0.13	0.01
Average	6.12	2.23	26.81	1.93	62.91	0.27	0.06

Table 8
CHEMICAL ANALYSIS OF SOME PLANTS EATEN BY CARIBOU

Plant group and species	Month collected	Percentage composition on dry matter basis: Crude protein	Crude fat	Crude fiber	NFE	Ash	Ca	P	Ref.
		Terrestrial Lichens							
Cladonia alpestris	June	2.7	2.0	31.4	62.69	1.23	0.11	0.07	38
	July	1.6	2.5	58.2	36.81	0.89	0.11	0.92	39
	Aug.	1.4	2.8	69.0	26.18	0.62	0.08	0.02	39
	Sept.	2.1	1.6	45.0	50.30	0.92	0.09	0.05	38
	Oct.	2.7	1.4	43.0	51.07	1.82	0.12	0.04	36
	Nov.	1.4	1.5	63.8	32.51	0.79	0.13	0.02	39
	Feb.	2.0	2.9	47.3	47.07	0.73	0.11	0.02	39
	Mar.	2.3	1.3	34.0	61.63	0.85	—	—	38
C. rangiferina	May	1.4	0.9	52.8	44.08	0.82	0.11	0.04	39
	June	2.5	0.7	43.8	52.07	0.90	0.10	0.06	38
	July	1.7	1.0	64.8	31.61	0.89	0.14	0.02	39
	Aug.	2.0	0.4	65.7	31.03	0.86	0.13	0.65	39
	Sept.	3.2	0.7	37.7	57.50	1.00	0.10	0.06	38
	Oct.	2.6	0.9	43.7	51.34	1.49	0.13	0.05	40
	Nov.	1.8	0.7	64.0	32.60	0.90	0.11	0.04	39
	Feb.	1.4	0.7	61.0	35.92	0.98	0.12	0.04	39
C. mitis	May	1.4	1.5	48.6	47.85	0.65	0.12	0.04	39
	June	4.1	2.5	38.2	63.87	1.33	0.70	0.05	40
	July	1.6	1.6	59.7	36.28	0.82	0.11	0.03	39
	Aug.	2.3	1.3	31.1	57.65	0.70	0.09	0.02	40
	Aug.	1.4	1.4	58.8	37.47	0.93	0.15	0.02	39
	Nov.	1.6	1.3	56.4	39.74	0.96	0.13	0.02	39
	Feb.	1.5	1.7	50.2	45.80	0.80	0.14	0.06	39
Cetraria nivalis	May	1.4	1.8	11.2	83.93	1.67	0.32	0.02	39
	June	2.6	2.6	7.2	85.68	1.96	0.31	0.06	38
	July	1.5	2.5	14.3	79.97	1.73	0.40	0.03	39
	Aug.	1.4	2.4	8.1	86.99	1.11	0.13	0.04	39
	Sept.	2.6	2.3	7.9	84.98	2.21	0.38	0.06	38
	Nov.	1.3	1.7	11.9	83.55	1.55	0.36	0.04	39
	Feb.	1.4	2.6	10.3	84.18	1.52	0.36	0.05	39
	[illegible]	[illegible]	1.4	7.4	[illegible]	[illegible]	[illegible]	[illegible]	38

Table 8 (continued)
CHEMICAL ANALYSIS OF SOME PLANTS EATEN BY CARIBOU

Plant group and species	Month collected	Percentage composition on dry matter basis							Ref.
		Crude protein	Crude fat	Crude fiber	NFE	Ash	Ca	P	
Terrestrial Lichens (continued)									
Stereocaulon spp.	June	6.6	1.2	28.3	61.23	2.67	0.06	0.10	38
	Sept.	7.9	1.0	24.6	65.08	1.46	0.05	0.13	38
Peltigera cephthosa	June	17.8	2.0	23.6	54.59	1.96	0.10	0.07	38
	Sept.	18.5	2.2	23.7	53.78	1.90	0.10	0.06	38
P. canina	June	18.6	1.5	20.8	56.85	2.27	0.19	1.1	38
	Sept.	21.9	1.1	23.5	51.15	2.37	0.15	0.8	38
Broadleaf Evergreen Shrubs									
Kalmia polifolia	May	11.1	2.0	16.0	68.31	2.59	0.60	0.15	40
	June	11.5	2.3	17.7	65.60	2.88	0.58	0.15	40
	July	11.7	2.4	18.5	64.57	2.85	0.58	0.15	40
	Aug.	9.6	2.8	19.2	65.66	2.70	0.64	0.13	40
K. angustifolia	Oct.	9.7	3.9	13.3	70.73	2.62	0.58	0.11	40
Chamaedaphne calculata	June	10.4	5.1	21.5	60.59	2.46	0.43	0.12	40
	Aug.	8.7	2.8	17.8	65.51	5.17	0.77	0.10	40
	Oct.	9.4	5.6	24.6	57.78	2.54	0.71	0.09	40
Ledum palustre	May	7.8	10.5	26.9	51.28	3.52	0.62	0.13	39
	July	9.6	10.3	29.8	47.31	2.99	0.53	0.12	39
	Aug.	7.6	10.5	23.8	54.72	3.38	0.63	0.11	39
	Nov.	7.8	9.1	23.6	56.47	3.03	0.58	0.13	39
	Feb.	8.0	10.1	26.2	52.69	3.01	0.58	0.11	39
L. groenlandicum	Sept.	10.5	5.0	24.7	57.23	2.68	0.48	0.19	39
	Mar.	8.7	4.6	23.9	60.52	2.24	0.47	0.15	39

Table 8 (continued)
CHEMICAL ANALYSIS OF SOME PLANTS EATEN BY CARIBOU

Plant group and species	Month collected	Crude protein	Crude fat	Crude fiber	NFE	Ash	Ca	P	Ref.
		Percentage composition on dry matter basis							
Broadleaf Evergreen Shrubs (continued)									
Arctostaphylos uva-ursi	June	5.6	4.1	11.7	76.37	2.20	0.50	0.13	38
	Sept.	6.3	5.4	13.8	71.82	2.68	0.61	0.17	38
Vaccinium vitis-idaea	June	6.3	2.8	15.6	72.91	2.17	0.55	0.16	38
	Sept.	7.9	2.8	20.1	66.24	2.25	0.55	0.11	38
Broadleaf Deciduous Shrubs									
Betula glandulosa	July	10.4	7.3	12.7	66.54	3.06	0.44	0.13	39
	Aug.	12.1	7.9	15.8	60.27	3.93	0.61	0.15	39
B. pumila	Aug.	12.8	4.2	22.2	57.44	3.33	0.86	0.11	40
B. nana	July	19.7	7.1	13.4	56.25	3.63	–	–	41
	Aug.	16.2	8.8	13.1	27.83	3.95	–	–	41
	Sept.	8.2	8.1	26.6	54.20	3.50	–	–	41
Salix hastata	Aug.	13.3	3.8	12.2	–	3.74	–	–	41
S. glauca	July	10.9	4.1	12.4	66.44	6.16	1.00	0.34	39
	Aug.	19.9	4.4	15.8	–	6.71	–	–	37
	Aug.	9.7	2.9	13.5	68.56	5.34	0.99	0.18	39
Alnus rugosa	Aug.	24.7	3.4	20.1	45.72	6.10	0.87	0.70	40
A. crispa	June	30.9	3.4	20.1	45.72	6.10	0.87	0.70	40
Amelanchier bartramiana	Aug.	18.3	2.6	22.6	29.36	7.20	1.32	0.20	40

Table 8 (continued)
CHEMICAL ANALYSIS OF SOME PLANTS EATEN BY CARIBOU

Plant group and species	Month collected	Percentage composition on dry matter basis							Ref.
		Crude protein	Crude fat	Crude fiber	NFE	Ash	Ca	P	
		Broadleaf Deciduous Shrubs (continued)							
Vaccinium angustifolium	June	14.0	1.4	15.5	61.87	3.29	0.31	0.28	40
	July	10.5	1.3	18.2	67.40	2.66	0.47	0.13	40
	Aug.	9.5	1.7	16.3	69.94	2.60	0.63	0.11	40
	Sept.	8.3	1.8	17.3	69.58	3.09	0.75	0.10	40
	Oct.	8.0	3.8	16.2	67.38	4.62	0.98	0.09	40
Nemopanthus mucronata	June	25.4	2.0	12.8	55.23	4.52	0.30	0.50	40
	July	14.4	2.0	11.7	68.27	3.69	0.49	0.19	40
	Aug.	11.9	2.2	10.6	70.87	4.43	0.52	0.17	40
	Sept.	9.5	2.4	10.1	74.19	3.88	0.59	0.13	40
		Herbs							
Epilobium angultifolium	June	14.9	3.6	26.9	47.82	6.84	–	–	41
	July	8.7	1.7	27.8	55.8	5.90	–	–	41
Rumex acetosa	June	–	–	27.5	33.1	10.4	–	–	41
	July	7.80	1.30	42.0	44.0	4.90	–	–	41
Menyanthes trifoliata	Aug.	13.4	3.3	12.8	55.5	7.00	–	–	41
Cornus canadensis	Oct.	11.3	4.6	12.1	62.43	9.71	3.33	0.13	40
Sanguisorba canadensis	June	19.7	3.0	12.0	57.78	7.59	1.30	0.26	40
	Aug.	14.7	2.8	18.2	57.24	7.14	1.17	0.25	40
Smilacina trifolata	May	24.8	2.7	13.9	51.14	7.52	0.26	0.59	40
	June	15.6	2.8	15.2	60.83	5.69	0.28	0.37	40
	July	17.5	4	15.6	55.85	7.73	0.46	0.28	40
	Aug.	17.9	8	13.9	56.15	7.19	0.83	0.26	40

Table 8 (continued)
CHEMICAL ANALYSIS OF SOME PLANTS EATEN BY CARIBOU

Plant group and species	Month collected	Percentage composition on dry matter basis: Crude protein	Crude fat	Crude fiber	NFE	Ash	Ca	P	Ref.
		Herbs (continued)							
Rubus chamaemorus	June	17.5	1.4	13.1	63.28	4.80	0.27	0.28	40
	July	15.1	1.4	17.0	61.86	4.65	0.31	0.18	40
	Aug.	12.1	2.3	12.5	68.20	4.90	0.57	0.13	40
	Sept.	8.8	2.8	15.2	69.51	3.68	0.54	0.10	40
		Arboreal Lichens							
Evernia mesomorpha	June	5.1	3.5	10.1	79.44	1.83	0.07	0.06	38
	Sept.	5.4	3.5	11.7	78.53	1.83	0.08	0.06	38
	Mar.	3.9	3.2	12.5	78.93	1.45	0.08	0.05	38
Usnea nirta	Sept.	5.5	4.9	7.1	81.17	1.41	0.19	0.07	38
	Mar.	4.6	5.7	6.9	81.37	1.41	0.23	0.05	38
Alectoria jubata	June	6.3	1.6	6.6	85.04	1.08	0.10	0.08	38
	Sept.	4.5	0.4	5.7	88.10	1.23	0.15	0.10	38
	Oct.	5.8	0.6	7.5	84.74	1.45	0.11	0.07	40
	Mar.	4.0	0.4	5.4	89.19	1.04	0.13	0.09	38
Parmelia spp.	Oct.	5.7	1.2	13.6	77.99	1.49	0.9	0.06	40
		Bryophytes							
Calliergonella schreberi	June	6.5	2.6	32.8	56.08	2.01	0.37	0.07	40
	Aug.	4.7	2.3	29.9	54.45	8.66	0.10	0.10	40
	Oct.	5.0	1.9	35.3	50.29	7.53	0.32	0.12	40
		Conifers							
Larix laricina	May	11.0	4.4	18.7	61.77	4.12	0.28	0.24	40
	June	15.2	5.6	18.3	57.26	3.80	0.25	0.26	40

Table 8 (continued)
CHEMICAL ANALYSIS OF SOME PLANTS EATEN BY CARIBOU

Plant group and species	Month collected	Percentage composition on dry matter basis							Ref.
		Crude protein	Crude fat	Crude fiber	NFE	Ash	Ca	P	
			Graminales						
Scirpus cespitosus	May	11.5	0.8	29.1	56.16	2.51	0.10	0.14	40
	June	13.2	0.7	27.1	56.39	2.67	0.15	0.13	40
	July	11.8	0.7	38.0	55.59	2.90	0.20	0.08	40
	Aug.	9.3	1.0	26.3	60.27	3.14	0.22	0.08	40
	Sept.	6.9	1.7	27.1	61.80	2.47	0.23	0.07	40
	Oct.	4.8	1.0	35.9	57.62	0.67	0.13	0.01	40
Arctagrostis latifolia	May	3.0	0.4	50.2	42.65	3.75	0.21	0.05	39
	July	8.8	8.6	45.4	40.67	4.53	0.23	0.15	39
	Aug.	5.2	1.7	45.8	43.52	3.78	0.24	0.09	39
	Nov.	2.5	0.5	52.6	41.13	3.27	0.25	0.03	39
	Feb.	2.3	0.3	55.1	39.46	2.84	0.22	0.03	39
Erophorum vaginatum	May	3.6	2.2	46.1	51.39	2.71	0.41	0.05	39
	June	13.0	1.0	24.4	58.97	2.63	0.19	0.19	39
	Aug.	10.3	1.3	33.8	51.61	2.97	0.22	0.20	39
	Nov.	3.7	0.9	46.1	52.42	2.88	0.39	0.05	39
	Feb.	3.2	0.8	40.2	52.65	3.15	0.49	0.05	39

Table 9
DRY MATTER INTAKE AND ENERGY AVAILABLE FOR FATTENING OF CARIBOU PREDICTED FROM ESTIMATES OF AVAILABLE LIVE BIOMASS (52 g/m^2) AND RELATIONSHIPS BETWEEN DAILY INTAKE VS. BIOMASS[42]

Prudhoe Bay, Alaska, July 11–20

	Lactating female, body weight 83 kg	Nonlactating female, body weight 83 kg	Male, body weight 115 kg
Dry matter intake (g/day/kg$^{0.75}$)	131	112	104
Gross energy intake (kcal/day/kg$^{0.75}$)	642	550	508
Metabolizable energy intake[a] (kcal/day/kg$^{0.75}$)	289	248	229
Energy requirements (kcal/day/kg$^{0.75}$)			
Maintenance[b]	190	190	190
Milk production	73	–	–
Energy available for fattening (kcal/day/kg$^{0.75}$)	26	58	39
Body weight gain[c] (g/day)	77	173	148

[a] Metabolizable energy intake = dry matter intake × gross energy of food (4.9 kcal/g) × energy digestibility (0.55) × metabolizability of digested energy (0.82).

[b] Energy requirement for maintenance = 2 × fasting metabolic rate (97 kcal/day per kg$^{0.75}$). Energy requirement for lactation = energy output in milk efficiency of milk synthesis (0.74).

[c] Body weight gain = energy available for fattening × efficiency of fattening (0.39)/energy content of new tissue (3.6 kcal/g[fattening weight]).

Table 10
ENERGY REQUIREMENTS OF REINDEER AND CARIBOU

	Female	Male
Full Production		
Mcal/day[a]	7.0	9.5
Digestible protein[a] (g/day)	315	425
Maintenance Requirement		
Summer		
Mcal/day	3.8	5.1
Digestible protein (g/day)	115	150
Winter		
Mcal/day	5.1	6.2
Digestible protein (g/day)	150	190
Digestible Energy (kcal/$W^{0.75}$)[b]		
Summer		
Caribou	427	480
Reindeer	361	390
Winter		
Caribou	235	265
Reindeer	235	265

[a] Reference for Mcal/day and digestible protein is 47.

[b] Reference for K cal/$W^{0.75}$ is 48. The latter authors estimate 5.1–6.5 Mcal/day for 70-kg reindeer and 8 Mcal/day for free-ranging animals.

Table 11
GROSS COMPOSITION OF REINDEER MILK[54-58]

Constituents (%)	Canada[a]	Canada	Alaska[b]	Scotland	Norway
Total solids	34.8	31.8	31.6	33.1	32.8
Ash	1.07	1.15	1.3	–	1.48
Fat	19.1	16.9	15.5	16.9	17.1
Lactose	2.4	2.5	3.1	2.8	2.8
Protein	11.8	9.7	10.1	11.5	10.4

[a] McEwan and Whitehead[59] give available digestive energy (kcal/100 cm^3 milk) as 205–250.

[b] Also, noncasein nitrogen (mg/100 ml) = 8.2; citric acid (g/100 ml) = 0.312; energy kcal/g dry matter) = 6.50.

REFERENCES

1. **Bergerud, A. T.,** Decline of caribou in North America following settlement, *J. Wildl. Manage.,* 38(4), 757–769, 1974.
2. **Skoog, R. O.,** Ecology of the Caribou (*Rangier tarandus granti*) in Alaska, Ph.D. thesis, University of California, Berkeley, 1968.
3. **Bergerud, A. T.,** Food habits of Newfoundland caribou, *J. Wildl. Manage.,* 36(3), 913–923, 1972.
4. **Kelsall, J. P.,** *The Migratory Barren Ground Caribou of Canada,* Department of Indian Affairs and Northern Development, Canadian Wildlife Service, Ottawa, 1968.
5. **Skuncke, F.,** Reindeer ecology and management in Sweden, *Biol. Pap. Univ. Alaska,* No. 8, 1–82, 1969.
6. **Miller, D. R.,** Seasonal changes in the feeding behaviour of barren-ground caribou on the taiga winter range, in *The Behaviour of Ungulates and its Relationship to Management,* Vol. 2, Geist, V. and Walther, R., Eds., International Union for Conservation of Nature and Natural Resources, Morges, Swit., 1974, 744–755.
7. **Miller, D. R.,** Biology of the Kaminuriak population of barren-ground caribou. Part 3: Taiga winter range relationships and diet, *Can. Wildl. Serv. Rep. Ser.,* 36, 1–42, 1976.
8. **Karaev, G. I.,** Reindeer fodder resources, in Reindeer Husbandry, Zhigunov̌, P. S., Ed., U.S. Department of Commerce, Springfield, Va., 1968, 129–176.
9. **Luick, J. R.,** Lichen species preference in reindeer, in Industrial Development in Alaska and its Effects on the Nutritional and Physiological Status of Arctic Animals, Tech. Prog. Rep. 1975-76, Institute of Arctic Biology, University of Alaska, Fairbanks, 1976, 30–33.
10. **DesMeules, P. and Heyland, J.,** Contribution to the study of the food habits of caribou. Part I – lichen preferences, *Nat. Can.,* 96, 317–331, 1969.
11. **Hadwen, S. and Palmer, L. J.,** Reindeer in Alaska, *U.S. Dept. Agric. Bull.,* No. 1089, 1–74, 1922.
12. **Cringan, A. T.,** Some Aspects of the Biology of Caribou and a Study of the Woodland Caribou Range of the Slate Islands, Lake Superior, Ontario, M.A. thesis, University of Toronto, Ontario, 1956.
13. **Simkin, D. W.,** A Preliminary Report of the Woodland Caribou Study in Ontario, Wildl. Rep. No. 59, Ontario Department of Lands and Forests, Maple, Ontario, 1965, 1–75.
14. **Banfield, A. W. F.,** Preliminary investigation of the barren-ground caribou, *Can. Wildl. Serv. Wildl. Manage. Bull.,* Ser. 1. 10B., 1–112, 1954.
15. **Scotter, G. W.,** The winter diet of barren-ground caribou in northern Canada, *Can. Field Nat.,* 71, 33–39, 1967.
16. **Bergerud, A. T. and Nolan, M. J.,** Food habits of hand-reared caribou *Rangifer tarandus* L. in Newfoundland, *Oikos,* 21(2), 348–350, 1970.
17. **Murie, O. J.,** Alaska-Yukon caribou, *North Am. Fauna,* 54, 1–93, 1935.
18. **Shelford, V. E. and Olson, S.,** Sere, climax and influent animals with special reference to the trans-continental coniferous forests in North America, *Ecology,* 16, 374–402, 1935.
19. **Manweiler, J.,** Woodland caribou study in northern Minnesota, *Parks and Recreation,* 22(2), 74–78, 1938.
20. **Flerov̌, K. K.,** Kobargi i aleni, in *Fauna SSSR Mlekopitayushchie,* Vol. 1, No. 2, Akademiya Nauk SSSR, Moscow, 1952.
21. **Bogdanowskaya-Guiheneut, I. D.,** Natural conditions and reindeer pastures on Kolguev Island, *Trans. Inst. Polar Agric.,* 2, 7–161, 1938.
22. **Glinka, D. M.,** The seasons of reindeer pastures and significance of green forage in the winter feeding of reindeer, *Trans. Inst. Polar Agric.,* 4, 31–46, 1939.
23. **Igoshna, K. N.,** The pasture foods and food seasons of the reindeer-husbandry of the Pri-Ural district, *Sov. Olenevodstvo,* 10, 125–195, 1937.
24. **Sdobnikov, V. M.,** The composition of the reindeer forage in autumn, *Trans. Arctic Inst. U.S.S.R.,* 24, 128–136, 1935.
25. **Sdobnikov, V. M.,** Materials to the problem of winter food for reindeer, *Trans. Arctic Inst. U.S.S.R.,* 24, 137–141, 1935.
26. **Spigul, E. M.,** Digestion of some fall pasture forages by reindeer, *Sov. Olenevodstvo,* 11, 31–42, 1937.
27. **Alexandrova, V. D.,** Forage characteristics of the plants in the far north of the U.S.S.R., *Trans. Inst. Polar Agric.,* 11, 1–96, 1940.
28. **Palmer, L. J.,** Progress of reindeer grazing investigations in Alaska, *U.S. Dept. Agric. Bull.,* No. 1423, 1–36, 1926.
29. **Freddy, D. J.,** Status and Management of the Selkirk Caribou Herd, 1973, M.S. thesis, University of Idaho, Moscow, 1974, 132.

30. **Gaare, E. and Skogland, T.,** Villreinene Naeringsvaner. Report from the grazing project of the Norwegian International Biological Program committee, Trondheim, Norway, 1971, 1–25.
31. **Edwards, R. Y. and Ritcey, R. W.,** Foods of caribou in Wells Gray Park, British Columbia, *Can. Field Nat.,* 74(1), 3–7, 1960.
32. **Dearden, B. L., Hansen, R. M., and Pegau, R. E.,** Plant fragment discernibility in caribou rumens, in *Proc. First Int. Reindeer and Caribou Symp.,* Special Rep. No. 1, Luick, J. R., Lent, P. C., Klein, D. R., and White, R. G., Eds., Biological Papers of the University of Alaska, Fairbanks, 1975, 257–277.
33. **Skogland, T.,** Range use and food selectivity by wild reindeer in southern Norway, in *Proc. First Int. Reindeer and Caribou Symp.,* Special Rep. No. 1, Luick, J. R., Lent, P. C., Klein, D. R., and White, R. G., Eds., Biological Papers of the University of Alaska, Fairbanks, 1975, 343–354.
34. **Palmer, L. J.,** Food requirements of some Alaskan game mammals, *J. Mammal.,* 25(1), 49–54, 1942.
35. **Luick, J. R.,** Comparison of digestibility of various reindeer forages using the nylon bag and *in vitro* technique, Nutrition and Metabolism of Reindeer and Caribou in Alaska, Prog. Rep. 1973-74, Institute of Arctic Biology, University of Alaska, Fairbanks, 1974, 46–52.
36. **Jacobsen, E. and Skjenneberg, S.,** Some results from feeding experiments with reindeer, in *Proc. First Int. Reindeer and Caribou Symp.,* Special Rep. No. 1, Luick, J. R., Lent, P. C., Klein, D. R., and White, R. G., Eds., Biological Papers of the University of Alaska, Fairbanks, 1975, 95–107.
37. **McEwan, E. H.,** personal communication.
38. **Scotter, G. W.,** Chemical composition of forage lichens from Northern Saskatchewan as related to use by barren ground caribou, *Can. J. Plant. Sci.,* 45(3), 246–250, 1965.
39. **Scotter, G. W.,** Chemical composition of forage plants from the Reindeer Preserve Northwest Territories, *Arctic,* 25(1), 21–27, 1972.
40. **Bergerud, A. T.,** personal files, 1976.
41. **Courtright, A. M.,** Range Management and the Genus *Rangifer*: A Review of Selected Literature, M.S. thesis, University of Alaska, Fairbanks, 1959, 172.
42. **White, G. R., Thomson, B. R., Skogland, T., Person, S. J., Russell, D. E., Holleman, D. F., and Luick, J. R.,** Ecology of caribou at Prudhoe Bay Alaska, in Ecological Investigations of the Tundra Biome in the Prudhoe Region, Alaska, Special Rep. No. 2, Brown, J. B., Ed., Biological Papers of the University of Alaska, Fairbanks, 1975, 151–201.
43. **Dauphine, T. C.,** Biology of the Kaminuriak population of barren-ground caribou. Part 4. Reproduction, growth and nutrition, *Can. Wildl. Serv. Rep. Ser.,* in press, 1976.
44. **Preobrazhenskii, B. V.,** Management and breeding of reindeer, in *Reindeer Husbandry,* Zhigunov̌, P.S., Ed., U.S. Department of Commerce, Springfield, Va., 1968, 78–128.
45. **Dmitrochenko, A. P.,** Reindeer feeding experiments on lichens, *Sov. Olenevodstvo,* 4, 17–44, 1935.
46. **Druri, S. M.,** Changes with Age and Season in the Live Weight of Reindeer, Publ. No. 31, A.N. Severtsov Institute of Animal Morphology, Leningrad, 1960.
47. **Steen, E.,** Some aspects of the nutrition of semi-domestic reindeer, in *Comparative Nutrition of Wild Animals,* Crawford, M. A., Ed., Academic Press, New York, No. 21, 1968, 117–128.
48. **McEwan, E. H. and Whitehead, P. E.,** Seasonal changes in the energy and nitrogen intake in reindeer and caribou, *Can. J. Zool.,* 48, 905–913, 1970.
49. **Segal, A. N.,** *The Reindeer of Karelia,* Akademia Nauk SSSR, Moscow, 1962.
50. **Kvitkin, Yu. P.,** Seasonal changes in gaseous and energy exchange in reindeer, *Sov. Zootekh.,* 9, 86–88, 1950 (Abstr. No. 777 in *Nutr. Abstr. Rev.,* 22(1), 153, 1950).
51. **Terent'ev, N. D.,** Experimental stall-feeding of reindeer moss and hay, *Sov. Olenevodstvo,* 8, 69–86, 1936.
52. **Hanson, W. C., Whicker, F. W., and Lipscomb, J. F.,** Lichen forage ingestation rates of free-roaming caribou estimated with Fallout Cesium-137, in *Proc. First Int. Reindeer and Caribou Symp.,* Special Rep. No. 1, Luick, J. R., Lent, P. C., Klein, D. R., and White, R. G., Eds., Biological Papers of the University of Alaska, 1975, 71–79.
53. **Igoshina, K. N. and Florovskaya, F.,** Utilization of pastures and pasturages of reindeer in the subarctic Ural Mountains, *Tr. Leningr. Naucho Issled. Inst. Poliarnogo Zemled. Zhivotnovod. Promysl. Khoziaist.,* 8, 5–163, 1939.
54. **Hatcher, V. C., McEwan, E. H., and Baker, B. E.,** Caribou milk. 1. Barren-ground caribou (*Rangifer tarandus groenlandicus*): gross composition, fat and protein constitution, *Can. J. Zool.,* 45, 1101–1106, 1967.

55. **Luick, J. R., Holleman, D. F., and White, R. G.,** Milk chemistry, Studies on the Nutrition and Metabolism of Reindeer-Caribou in Alaska with Special Interest in Nutritional and Environmental Adaption, Prog. Rep. 1970–71, Institute of Arctic Biology, University of Alaska, Fairbanks, 1971, 65–72.
56. **Aschaffenburg, R., Gregory, M. E., Kon, S. K., Rowland, S. J., and Thompson, S. Y.,** The composition of the milk of reindeer, *J. Dairy Res.,* 29, 325–328, 1962.
57. **Werenskiold, F.,** Rensdyrmelk, *Tidsskr. Norske Landbruk,* 2, 372–375, 1895.
58. **Luick, J. R., White, R. G., Gau, A. M., and Jenness, R.,** Compositional changes in the milk secreted by reindeer. I. Gross composition and ash, *J. Dairy Sci.,* 57, 1325–1333, 1974.
59. **McEwan, E. H. and Whitehead, P. E.,** Measurement of milk intake of reindeer and caribou calves using titrated water, *Can. J. Zool.,* 49(4), 443–447, 1971.
60. **McEwan, E. H.,** Growth and development of the barren-ground caribou. II. Postnatal growth rates, *Can. J. Zool.,* 46, 1023–1029, 1968.
61. **Palmer, L. J.,** Raising Reindeer in Alaska, U.S. Department of Agriculture, Misc. Publ. No. 207, 1934, 1–40.
62. **Krebs, C. J. and Cowan, I. M.,** Growth studies of reindeer fawns, *Can. J. Zool.,* 40, 863–869, 1962.
63. **Holleman, D. F., White, R. G., and Luick, J. R.,** Estimates of milk intake and yield, *Studies on the Nutrition and Metabolism of Reindeer-Caribou in Alaska with Special Interest in Nutritional and Environmental Adaptation,* Prog. Rep. 1973–74, Institute of Arctic Biology, University of Alaska, Fairbanks, 1974, 64–67.
64. **Cameron, R. D., White, R. G., and Luick, J. R.,** The accumulation of water in reindeer during winter, *Proc. First Int. Reindeer and Caribou Symp.,* Special Rep. No. 1, Luick, J. R., Lent, P. C., Klein, D. R., and White, R. G., Eds., Biological Papers of the University of Alaska, 1975, 374–378.

DIETS FOR FREELY GRAZING REINDEER*

J. R. Luick

INTRODUCTION

Reindeer† are indigenous to arctic tundra and taiga range lands where they subsist (and indeed thrive) under widely ranging nutritional circumstances and survive ostensibly unremitting environmental stresses. During the 6- to 8-month winter grazing period, reindeer subsist chiefly on lichens, which are algal/fungal symbionts of poor nutritional quality. Additional stresses include predation by arctic carnivores, violent climatic conditions, scarcity of food, and the necessity of locating and excavating food from beneath deep and frequently encrusted snow.

Summer grazing conditions impose an entirely different set of nutritional and environmental stresses and with the exception of the considerably improved nutritional quality of available food, appear to be no less severe than winter grazing conditions. Female reindeer calve during April and May when food is often scarce and their body condition is extremely poor. The metabolic demands for producing an adequate supply of highly nutritious milk, as well as energy demands for almost continual movement of the herd in search of food impose additional stresses during late spring. Predators are always present, and harassment by mosquitos, black flies, and parasitic warble and nostril flies provide a significant deterrent to the attainment of the peak body condition necessary for winter survival.

Thus, the annual movement patterns of reindeer, and, hence, the plant foods they consume are dictated seemingly by a number of environmental and climatic factors. Comprehensive discussions of these factors and consideration of the adaptive significance of the observed migration patterns have been published recently by Kelsall,[1] Skogland,[2] and White et al.[3] In brief, it seems that motivation for herd movement is dictated mainly by the necessity of following the phenotypic progression and primary production of various plant species. This seasonal usage of phytosociological plant communities in Norway and Siberia is well described by Skogland,[2] and Glinka,[4] respectively (Table 1).

Table 2 lists a number of vegetative types and plant species that are consumed selectively by reindeer during migration. Interestingly, these plants are found in most of the reindeer grazing ranges of the circumpolar nations.[1,6–8] Perhaps the considerable differences found in the chemical composition of various food plants (Tables 3 through 5), the dry matter digestibility of individual plants and mixed forages (Table 6) and especially the seasonal changes that occur in the nutritive value of these plants as they progress through the various stages of maturity are of greater interest (Table 7).[5,9]

* Data reported in this chapter that originated in the author's laboratory resulted from research activities supported by U.S. Atomic Energy Commission Contract AT(04-3)-310; Energy Research Development Administration Contract E(45-1)-2229; and the Office of Polar Programs and the International Biological Program of the National Science Foundation (NSF Grant Number GV-29342). By acceptance of this article, the publisher and/or recipient acknowledges the U.S. Government's right to retain a nonexclusive, royalty-free license in and to any copyright covering this paper.

† Unless otherwise noted, the term "reindeer" as used in this chapter refers to both semidomesticated (i.e., herded) and wild reindeer (in North America, "caribou"), classified taxonomically as *Rangifer tarandus.*

Table 1
RANGE USE BY REINDEER IN SOUTHERN NORWAY[2]

Plant community	Annual use (%)	Period of high intensity use
Lichen heaths	60.0	December through April (95%)
Grassy meadows	14.5	Early spring and late fall (50%)
Salices and herb snow beds	22.5	Early and late summer (95%)
Bogs	3.0	Midsummer (45%)

Table 2
FOOD PLANTS COMMONLY SELECTED BY REINDEER AND CARIBOU[a]

Arboreal lichens
- *Alectoria jubata*
- *A. oregana*
- *Evernia mesomorpha*
- *Parmelia physodes*
- *P. saxatilis*
- *Ramalina farinacea*
- *Usnea hirta*
- *U. plicata*

Fruticose lichens
- *Cetraria islandica*
- *C. nivalis*
- *Cladonia alpestris*
- *C. amaurocraea*
- *C. arbuscula*
- *C. gracilis*
- *C. mitis*
- *C. multiformis*
- *C. rangiferina*
- *C. uncialis*
- *Peltigera* spp.
- *Stereocaulon* spp.
- *Thamnolia vermicularis*

Rock lichens
- *Umbilicaria hyperborea*

Herbs and forbs
- *Artemisia arctica*
- *A. richardsonii*
- *Equisitum* spp.
- *Hedysarum* spp.
- *Heraculum lanatum*
- *Lupinus* spp.
- *Oxytropis* spp.
- *Pedicularis* spp.
- *Petasites frigidus*
- *Polygonium bistorta*
- *Saxifraga oppositifolia*

Shrubs
- *Betula glandulosa*
- *B. nana*
- *Dryas integrifolia*
- *D. octapetala*
- *Salix arctica*
- *S. glauca*
- *S. lanata*
- *S. ovalifolia*
- *S. pulchra*
- *S. rotundifolia*
- *S. scirpoidia*

Grasses
- *Arctophila fulva*
- *Calamagrostic* spp.
- *Deschampsia* spp.
- *Dupontia fischeri*
- *Festuca* spp.
- *Hierochloe* spp.
- *Poa* spp.

Sedges
- *Carex aquatilis*
- *Eriophorum angustifolium*

Woody perennials
- *Arctostaphylos alpina*
- *A. rubra*
- *Empetrum nigrum*
- *Ledum decumbens*
- *L. groenlandicum*
- *Rhododendron lapponicum*
- *Vaccinium oxycoccus*
- *V. uliginosum*
- *V. vitis-idaea*

[a] Relative contribution to daily food intake varies seasonally with availability and stage of maturity of the various plant species.

Table 3
CHEMICAL ANALYSES OF SOME PREFERRED REINDEER AND CARIBOU FOOD PLANTS[1,5,9,31]

Item	n[a]	Crude protein	Crude fat	Crude fiber	Ash	N-free extract	Kcal/100 g	Ref.
Fungi	2	34.76	4.76	20.80	8.12	31.55		1
Musci spp.	2	4.81	3.39	28.26	6.88	56.64		1
Arboreal lichens								
Alectoria jubata	1	4.81	2.55	39.7				31
A. sarmentosa	1	3.50	10.94	3.4				31
Fruticose lichens								
Cetraria cucullata	3	3.77	3.82	11.23	2.14	80.02		1
	2	2.3	2.3	5.3	1.2	89.0	468	5
C. islandica	5	4.60	3.38	7.85	1.51	82.66		1
	3	2.6	1.1	6.2	1.2	89.3	446	5
C. nivalis	4	2.47	3.95	5.11	2.15	86.34		1
Cladonia alpestris	3	2.75	1.52	44.49	1.87	49.12		1
	2	2.67	2.79	32.3	0.76			31
C. crispata	1	3.5	1.1	40.8	1.3	53.3	431	5
C. gracilis	2	2.6	0.8	30.7	1.1	64.9	451	5
C. mitis	1	2.54	1.67	28.00	2.18	65.62		1
C. rangiferina	5	2.21	1.58	42.30	1.49	51.39		1
	1	2.03	1.11	38.3	0.83			31
	3	2.8	0.6	28.3	1.4	66.9	449	5
C. sylvatica	2	1.86	1.16	43.97	2.21	50.80		1
	3	2.3	0.9	23.6	1.1	72.1	435	5
C. uncialis	1	2.3	1.6	19.8	0.9	75.4	437	5
Peltigera sp.	1	19.77	1.29	25.33	9.14	44.47		1
Stereocaulon spp.	4	8.02	1.78	24.57	2.55	63.07		1
Sedges and grasses								
Arctagrostis latifolia	1	22.0	2.6	24.9	6.7	43.8		9
Carex aquatilis	3	10.55	3.21	35.44	5.95	44.84		1
	2	10.4	2.9	25.7	3.5	57.6	457	9
C. bigelowii	3	11.3	1.5	28.5	4.3	54.5	475	5
C. nesophila	1	25.1	2.8	22.3	5.1	44.7	–	9
C. spp.	7	19.4	2.9	30.0	6.1	50.9	468	9
Eriophorum angustifolium	2	14.1	1.7	21.1	4.4	58.8	489	5
E. chamissonis	1	15.1	2.5	15.6	4.0	62.8	472	9
Herbs and forbs								
Artemisia arctica	2	20.0	2.2	9.3	10.2	54.3	451	9
Epilobium angustifolium	4	13.37	3.09	27.08	6.60	49.80		1
Equisetum spp.	6	11.36	3.31	18.81	13.55	52.97		1
Heraculum lanatum	2	38.7	2.9	8.7	16.0	338	461	9
Ligusticum hultenii	3	25.0	2.8	9.9	7.9	57.9	478	9
Potentilla palustris	1	18.8	3.1	17.5	5.1	55.5	463	5
Ranunculus occidentalis	2	20.9	3.7	12.5	6.6	61.5	466	9
Shrubs								
Betula glandulosa and *B. nana*	15	15.59	7.16	16.21	3.32	57.70		1
B. nana	3	15.8	2.7	19.6	2.9	58.0	546	5
Dryas octapetala	2	13.2	5.2	19.3	3.3	54.5	–	9
Salix sp.	1	21.9	1.6	–	4.3	–	–	9
S. spp.	15	20.99	3.10	16.66	6.53	52.74		1
S. pulchra	2	23.8	1.9	10.1	4.4	60.0	497	5

[a] n = Number of analyses used to compute mean values.

Table 3 (continued)
CHEMICAL ANALYSES OF SOME PREFERRED REINDEER AND CARIBOU FOOD PLANTS[1,5,9,31]

Item	n[a]	Crude protein	Crude fat	Crude fiber	Ash	N-free extract	Kcal/100 g	Ref.
Woody perennials								
Arctostaphylos alpina	1	8.55	2.54	9.45	5.37	74.09		1
Empetrum nigrum	4	4.60	9.98	23.11	3.33	58.97		1
	1	5.4	10.2	23.8	1.8	58.8	573	5
Ledum decumbens	3	6.52	6.52	27.30	2.17	57.49		1
	1	7.4	6.2	25.9	1.8	58.7	555	5
Vaccinium uliginosum	2	6.75	3.14	31.39	3.07	55.64		1
	3	14.2	2.4	21.1	2.6	59.6	511	5
V. vitis-idaea	2	5.76	3.13	20.20	2.66	68.24		1
	1	6.2	3.6	26.9	2.3	61.0	518	5

Table 4
ELEMENTAL ANALYSIS OF LICHENS COMMONLY SELECTED BY REINDEER AND CARIBOU[32]

Element	Mixed lichens[a]	*Cladonia alpestris*	*Cladonia rangiferina*	*Cetraria* spp.	*Stereocaulon* spp.
Si	7100	1000	70	300	1000
Na	122	54	63	240	67
K	560	733	693	1510	1625
Ca	790	514	783	1080	450
Mg	390	300	100	500	200
Mn	100	100	150	100	400
P	200	200	100	300	200
Fe	840	300	150	200	200
Zn	18	10	3	50	10
Cu	4.8	2	3	4	2
S	50	100	10	50	200
Pb	1	0.6	3	1	0.6
Co	0.06	0.4	0.1	0.1	0.2
Hg		Trace	Trace	Trace	Trace
Cs	0.2	0.2	0.1	0.3	0.2
Sr	6	6	6	4	4
Al	300	300	300	300	300

[a] *Cladonia alpestris* (60%); *C. rangiferina* (20%); *C. arbuscula* (10%); *Cetraria islandica* (5%); miscellaneous species, including *Cladonia gracilis, Cetraria cucullata, C. laevigata,* and *Stereocaulon alpinum* (5%).

Table 5
DETERGENT FIBER ANALYSIS OF PLANT MATERIAL CONSUMED BY REINDEER AND CARIBOU[14]

Values are Expressed as g/100 g Dry Matter

		Cell wall contents		
Sample	Cell contents	Hemi-cellulose	Cellulose	Lignin
	Shrubs			
Dryas integrifolia				
leaves/stem base	64.6	2.3	14.0	19.1
inflorescences	53.8	-2.0	21.1	27.1
Salix arctica	81.5	2.1	11.9	4.5
S. lanata	79.4	2.7	13.2	4.7
S. ovalifolia	68.8	3.9	18.6	8.7
S. pulchra	72.9	9.8	11.4	6.0
	Sedges			
Carex aquatilis	41.8	36.5	18.6	3.2
Eriophorum angustifolium				
Live leaves/10% dead	31.5	43.8	19.1	5.1
Inflorescence + stem	42.5	36.0	16.3	4.7
	Grasses			
Arctophila fulva	32.3	38.8	27.3	1.7
Dupontia fisheri				
Live leaves/10% inflorescences	54.9	25.4	15.8	3.9
Inflorescences + stem	37.9	31.1	25.5	5.5
	Herbs			
Artemisia richardsonii sp.	57.7	13.1	24.2	5.0
Pedicularis sp.	74.1	5.1	14.3	6.5
Saxifraga oppositifolia	73.1	-8.5	20.5	15.4
	Lichens			
Cetraria cucullata	68.4	27.8	-0.6	4.3
Cladonia alpestris (Coal Creek)	17.0	78.3	1.7	2.9
C. alpestris (Nome)	16.7	81.0	-1.3	3.6

Table 6
DRY MATTER DIGESTIBILITY[a] OF INDIVIDUAL AND MIXED SPECIMENS OF PLANTS SELECTIVELY GRAZED BY REINDEER AND CARIBOU[3,14]

		In vitro digestibility technique			
		Source of Inoculum			
Plant	Sample	Caribou	Reindeer[b]	Mean	Reindeer[b]
		Shrubs			
Dryas integrifolia	Leaves/stem base	33	33	33	
	Heads		21		
D. octopetala	Leaves and stems (alpine)				12 ± 2
Salix arctica	Leaves/buds	71	52 ± 2	62	
S. lanata	Live leaves		34 ± 2		
S. ovalifolia	Leaves/buds	35	46 ± 6	41	
S. pulchra	Leaves/buds		54	54	24 ± 0.3
S. reticulata	Leaves/buds		38		
		Sedge and cotton grass			
Carex aquatilis	Live leaves	56	68 ± 4	62	39 ± 2
Eriophorum angustifolium	Live leaves	48	59 ± 3	55	32 ± 3
	Inflorescences + stem		56 ± 3		
E. vaginatum	Mature leaves				28 ± 5
		Grasses			
Arctophila fulva	Live leaves	56	72	64	
Dupontia fisheri	Live leaves		79 ± 2		
	Inflorescences	67	70	68	
		Herbs			
Artemisia richardsoiana	Leaves	62	66 ± 3	64	
Braya spp.	Whole plant		68		
Oxytropis sp.	Live leaves		70		
Parrya nudicalus	Whole plant		67		
Pedicularis sp.	Inflorescence + leaves		64 ± 6		
Saxifraga oppositifolia	Inflorescence + leaves		33 ± 1		
		Lichens			
Alectoria nigricans					45
Cetraria cucullata		74 ± 3	48 ± 1	68	77 ± 9
C. islandica					48 ± 9
Cladonia alpestris		27 ± 14	16 ± 7	21	16 ± 3
C. arbuscula					10
C. rangiferina					40
C. uncialis					44
Lobaria linita					45
Peltigera aphthosa					44
Stereocaulon alpinum					18
Thamnolia vermicularis					59

[a] Percent ± S E.

[b] Rumen inoculum was obtained from tranquilized caribou and reindeer grazing at Prudhoe Bay, Alaska, or reindeer given a mixed diet containing 67% (dry weight) lichens, 8% *Carex aquatilis*, and 25% brome hay. The food was ground in a hammermill and thoroughly mixed.

Table 6 (continued)
DRY MATTER DIGESTIBILITY[a] OF INDIVIDUAL AND MIXED SPECIMENS OF PLANTS SELECTIVELY GRAZED BY REINDEER AND CARIBOU[3,14]

		In vitro digestibility technique			
		Source of Inoculum			
Plant	Sample	Caribou	Reindeer[b]	Mean	Reindeer[b]
Mosses					
Hylocomium splendens	Entire gametophyte				19 ± 6
Sphagnum magellanicum	Entire gametophyte				6
Other	Entire gametophyte				13
Esophageal egesta					
Dryas heath	Reindeer No. 12		49 ± 3.3		
Dupontia brook bank	Reindeer No. 31		50 ± 0.6		
Eriophorum meadow	Reindeer No. 10		62 ± 0.2		
Eriophorum meadow	Reindeer No. 12		49 ± 1.4		
Combined	Reindeer				45 ± 1.0
	Caribou C2	37 ± 2.2			
	Caribou C3	43 ± 1.6			

Table 7
SEASONAL CHANGES IN THE CHEMICAL COMPOSITION AND CALORIC CONTENT OF IMPORTANT REINDEER FORAGE PLANTS NEAR NOME, ALASKA[5]

Species	Month	Protein (%; N × 6.25)	Fat (%) ether extract	Crude fiber (%)	Ash (%)	Nitrogen-free extract	Kcal/100 g
Betula nana	March	5.3	2.6	29.3	1.3	61.5	560
	June	26.0	2.2	13.2	4.6	54.0	542
	August	16.0	3.5	16.3	2.8	61.4	536
Carex bigelowii[a]	June	16.6	1.3	24.5	4.3	53.3	471
	August	12.9	1.8	28.5	4.0	52.8	479
Cetraria cucullata	June	2.4	1.9	3.6	1.2	90.9	474
	August	2.2	2.6	7.0	1.2	87.0	463
C. islandica	March	2.4	0.8	6.6	1.2	89.0	453
	June	2.6	0.3	3.7	1.1	92.3	426
	August	2.6	1.1	8.4	1.2	86.7	458
Cladonia rangiferina	March	2.7	1.0	30.3	1.1	64.9	451
	June	2.9	0.4	27.7	2.3	66.7	466
	August	2.7	0.4	26.9	0.8	69.2	431
C. gracilis	March	2.8	1.0	29.8	1.1	65.3	471
	August	2.4	0.6	31.6	1.0	64.4	431
C. sylvatica[b]	March	2.5	1.4	21.9	1.1	73.1	426
	June	2.2	0.7	26.8	1.2	69.1	452
	August	2.0	0.6	22.2	0.9	74.3	426
Eriophorum angustifolium	June	16.1	1.3	21.5	5.6	55.5	477
	August	12.1	2.0	20.7	3.1	62.1	501
Salix pulchra	June	25.2	1.6	8.6	4.9	59.7	504
	August	22.4	2.1	11.5	3.8	60.2	490
Vaccinium uliginosum	March	6.0	3.1	35.9	1.3	53.7	543
	June	23.4	2.2	11.0	4.0	59.4	523
	August	13.2	2.0	16.5	2.5	65.8	466

[a] Includes a small portion of *Carex aquatilis.*
[b] Includes *Cladonia mitis.*

SPRING DIETS

In Alaska, spring grazing commences at the time of calving, or shortly thereafter, when snow is disappearing rapidly and green shoots and other vegetative parts of grasses, shrubs, and sedges become available. Reindeer feed extensively on dry heaths and wet grassy meadows, selecting *Dupontia fischeri, Dryas integrifolia,* and especially *Eriophorum angustifolium.*[3] Wet boggy areas, the habitat of *Carex aquatilis,* are not visited widely during the warm, wind-free, daylight hours, presumably because they harbor hordes of mosquitos and other stinging insects. Lactating reindeer attain peak milk production soon after calving; the milk contains high concentrations of fat and protein but little lactose. The calves commence grazing 2 to 3 days following birth;[1,10,11] consequently, they gain body weight rapidly, whereas the adult females remain in poor body condition.[12] During early spring, reindeer spend about 50% of the day searching for and consuming plants. Eating rates, estimated at 70 to 90% of grazing rates, amount to about 10 hr/day during late winter, but increase to nearly 19 hr/day during late spring, when plant foods become more abundant and are of increasing nutritional value.[1,2,5,10]

MIDSUMMER DIETS

During midsummer, reindeer gain body weight and improve in condition rapidly while feeding on grasses such as *Carex aquatilis* and *Dupontia fischeri;* sedges, especially *Eriophorum angustifolium;* shrubs, including *Salix* spp.; and glandular birches, *Betula glandulosa* and *B. nana.*[1,3,13] Lichens are also eaten selectively, when available in sufficient quantity. According to Kelsall, grasses and sedges make up 28.1% of the daily food intake of Canadian reindeer, lichens 31.3%, and shrubs 31.2%; herbs and forbs are little grazed.[1] White et al. reported similar findings for reindeer at Prudhoe Bay, Alaska, with green vegetation, grasses, and sedges being consumed in a somewhat higher proportion (40%) to compensate for the relatively low biomass of lichens.[3] Person[14] and White et al.[3] found that the grasses appeared to be the most highly digestible plants, followed in order by herbs and forbs, sedges, and shrubs. Digestibility of lichens was highly variable, whereas the mosses were of little nutritional value. In fact, the mosses, which often comprise more than 50% of the vegetative biomass, were never found in significant amounts in rumen contents or egesta collected from esophageal fistula.[14] Thus, in midsummer, when food plants attain peak nutritional content and quality and reindeer spend about 35% of the day consuming green vegetation, milk production begins to decline, insect harassment is high, and the reindeer attain the peak body condition necessary for winter survival.[3,5,9,10,15,16]

LATE SUMMER AND FALL DIETS

The grazing behavior of reindeer during late summer and fall is characterized by a shift in plant selection from grasses and sedges to green leafy parts of woody perennials such as *Arctostaphylos alpina, A. rubra, Vaccinium uliginosum, Ledum decumbens, Empetrum nigrum,* and *Rhododendron lapponicum.*[1] *Dryas integrifolia,* several salices, and various herbs and forbs (especially *Pedicularis* sp.) are eaten in large quantities.[1,3] Mushrooms are so highly prized that the intense search for them often leads to straying and even stampeding of commercial herds of semidomesticated reindeer.[33] Of similar concern to reindeer herders are the effects of tramped dry and brittle lichen heaths during the midsummer to late fall grazing season. Pegau has determined that 35% of the edible lichen biomass may be destroyed in this way during close herding of reindeer on the heaths.[17] Fortunately, lichens are so hygroscopic that a brief rain shower or period of high humidity is sufficient to convert the brittle lichen heath to a soggy, sponge-like biomass.

Finally, digestibility studies have shown that reindeer make efficient use of the crude fiber in grass and sedge hays.[18,19]

WINTER DIETS

The singular nutritional characteristic of reindeer that distinguishes them from all other large herbivores and that has enabled them to utilize holoarctic range lands is their preference for and ability to survive on lichens during the long (6- to 8-months) winter grazing period.[13,23] Reindeer seem to have an uncanny ability to locate and excavate lichens, even where these food plants are buried under snow cover of up to 60 cm or more.[20] This does not imply that reindeer thrive under these conditions. On the contrary, the very low protein (~2 to 3% DM), ash (~2% DM), and fat (1.5%) content and the high crude fiber (45%) of lichens (Table 3) results in a considerable loss of body substance (but not necessarily body weight) during prolonged winter grazing.[4,12,21] The data listed in Table 8 are intended to emphasize the great differences in nutritive content that exist between lichens and the common livestock feed, alfalfa hay. The above discussion also should not be construed to imply that reindeer eat only lichens during winter; reindeer also consume variable amounts of frozen grasses, sedges, the leaves of woody plants, and mushrooms which are found in craters that are dug in search of lichens.[4,22,23]

Although competition for grazing on winter lichen ranges is essentially nonexistent, the reindeer's high dependence upon that food source is not without hazard. Massive "die-offs" have occurred when early spring rains freeze and form an impenetratable crust of ice on top of the snow.[23] Skogland states that reindeer spend about 62% of the day searching for and eating lichens, and various workers have estimated that lichen consumption rates are approximately 2 to 5 kg per day per animal.[1,2,24,25] The

Table 8
NUTRIENT CONTENT OF THE PREFERRED LICHEN, *CLADONIA ALPESTRIS* COMPARED TO ALFALFA HAY[a,b]

Ingredient	*Cladonia alpestris*	Alfalfa hay
Crude protein (%)	2.75	18.2
Crude fat (%)	1.52	2.5
Crude fiber (%)	44.49	28.0
Ash (%)	1.87	10.3
N-free extract (%)	49.12	41.0
Na (mg/kg)	54	190
K (mg/kg)	733	2460
Ca (mg/kg)	514	1350
Mg (mg/kg)	300	340
Mn (mg/kg)	100	46.5
P (mg/kg)	200	300
Fe (mg/kg)	300	50
Zn (mg/kg)	10	35.1
Cu (mg/kg)	2	18.7
Co (mg/kg)	0.4	0.24

[a] Values for proximate analysis and mineral content are taken from Kelsall[1] and Luick,[32] respectively.

[b] Sun-cured, ground alfalfa hay (N.A.S. Ref. No. 1-00-111) as reported in United States-Canadian Tables of Feed Composition, National Academy of Sciences Publication 1684, Washington, D.C. 1969.

consumption of even these relatively small amounts of frozen foodstuffs and solid water (snow) is of considerable energetic cost to reindeer, since the necessity of raising the temperature of ingesta to deep body temperature has been estimated at approximately 25% of the resting metabolic rate.[13,26]

Lastly, rations have been compounded and tested in Norway and Siberia for use either as supplements to winter lichen diets or as complete rations during emergency feeding conditions.[23,27,28] Formulas for these rations are listed in Table 9.

THE LICHEN PARADOX

Recent findings at the Cantwell Reindeer Research Station, Cantwell, Alaska seem to establish a paradoxical situation insofar as they apply to the winter grazing habits of reindeer. In preliminary studies on the food and water preferences of penned reindeer fed widely differing basal rations, lichens were always selected (and consumed *in toto*) over nutritionally superior livestock ration (pelletized Purina Cattle Starter No. 1) and/or three species of long hay, *Poa pretensis* (Kentucky Nugget Bluegrass), and *Festuca rubra* (mixed Arctared red fescue), and *Calamagrostis canadensis* (Bluejoint reedgrass). Snow was preferred over cold, warm, or saline water as a source of water intake.[29]

This strong preference for a nutritionally inadequate and energetically taxing diet, especially when nutritional and climatic stresses are most severe, does not fit easily into the concept of natural selection. However, the recent finding in our laboratory, that the daily water flux of freely grazing reindeer diminishes threefold to fivefold during winter suggests a reasonable explanation for the phenomenon. The survival value of ingesting small amounts of dietary protein and minerals such as lichen and distilled water (snow) may be attributed to the conservation of body water and, therefore, body heat by decreasing the amount of water required to eliminate metabolic nitrogen and body electrolytes.[12] This hypothesis is supported by Cameron's finding that decreases in water flux of pen-fed reindeer during severe cold are elicited chiefly by the level of dietary protein intake.[26]

Evidence contrary to this hypothesis stems from reports that reindeer crave urine and sea water, gnaw old bones and shed antlers, and kill and eat lemmings and mice, as well as bird's eggs and nestlings.[30] It is also known that reindeer maintain and even gain body weight during winter when there is an adequate supply of frozen green vegetation to supplement the lichen biomass.[4,21] Similar results have also been noted with reindeer

Table 9
DIETS FOR SUPPLEMENTING OR REPLACING NORMAL FEEDING REGIMES OF REINDEER DURING PERIODS OF SEVERE UNDERNUTRITION

Supplemental ration[a,28]	Complete ration A[23]	Complete ration B[27]
Urea (50%)	Cereal grains (41%)	Ground barley (40%)
Dicalcium phosphate (30%)	Ground hay (25%)	Ground oats (17%)
Minerals (20%)	Ground straw (15%)	Wheat bran (15%)
	Molasses beet chips (10%)	Soybean oil (3%)
	Wheat bran (5%)	
	Dicalcium phosphate (1.7%)	
	Iodized salt (0.3%)	
	Vitamins (2%)	
	Mineral mixture (trace)	

[a] This ration is mixed with lichens and fed ad libitum to reindeer as a supplement to typical winter diets.

that are fed supplemental and/or complete rations during severe winter weather (Table 9).[23,27,28]

In rebuttal to the above, it seems unlikely that significant quantities of urine, sea water, old bones, lemmings, bird's eggs, etc. would be available or accessible during deep winter; hence, they should not be considered as supplements to winter lichen diets. Additionally, Cameron and Luick[12] confirmed the earlier findings that reindeer can maintain body weight while grazing mixed vegetation during winter, but discovered further that this was accomplished by replacing a substantial loss of body solids with an almost equivalent amount of body water. Thus, changes in body weight are deceptive and must therefore be interpreted with great caution.

REFERENCES

1. **Kelsall, J. P.,** *The Caribou,* Queen's Printer, Ottawa, Ontario, Canada, 1969, 1–340.
2. **Skogland, T.,** Range use and food selectivity by wild reindeer in southern Norway, in Proc. 1st Int. Reindeer and Caribou Symp., Special Rep. No. 1, Luick, J. R., Lent, P. C., Klein, D. R., and White, R. G., Eds., Biological Papers, University of Alaska, Fairbanks, 1975, 342–354.
3. **White, R. G., Thomson, B. R., Skogland, T., Person, S. J., Russell, D. E., Holleman, D. F., and Luick, J. R.,** Ecology of caribou at Prudhoe Bay, Alaska, in Ecological Investigations of the Tundra Biome in the Prudhoe Bay Region, Alaska, Special Rep. No. 2, Brown, J., Ed., Biological Papers, University of Alaska, Fairbanks, 1975, 151–201.
4. **Glinka, D. M.,** The seasons of reindeer pastures and significance of green forage in the winter feeding of reindeer, Trans. Inst. Polar Agric., Animal Husbandry and Fishing and Hunting Industry Series, *Reindeer Industry,* 4, 45–46, 1939.
5. **Pegau, R. E.,** Reindeer Range Appraisal in Alaska, M.Sci. thesis, University of Alaska, Fairbanks, 1969, 1–130.
6. **Andreev, A. N., Ed.,** *Botanical Research in Yakutia,* Yakutian Branch SO-AN Academy of Science, Yakuts, USSR, 1975, 1–153.
7. **Skjenneberg, S., Ed.,** *Reindeer and Reindeer Management,* A.s. Fjell-Nytt, Lesjaskog, Norway, 1965, 1–326.
8. **Kuvaev, V. B.,** Possibilities for rationalized use of the Verkhoyansk deer pastures, *Probl. Severa,* 8, 280–289, 1964.
9. **Klein, D. R.,** Nutritive Quality of Alaskan Range Forage and Associated Growth and Welfare of Reindeer and Caribou, Completion Rep., Contract No. 14-11-001-3748, Bureau of Land Management, U.S. Department of the Interior, 1970, 1–30.
10. **Luick, J. R., Ed.,** Nutrition and Metabolism of Reindeer and Caribou in Alaska, Progress Rep. 1973/1974, Contract No. AT(45-1)-2229, U.S. Atomic Energy Commission, 1974, 1–109.
11. **Luick, J. R., White, R. G., Gau, A. M., and Jenness, R.,** Compositional changes in the milk secreted by grazing reindeer. I. Gross composition and ash, *J. Dairy Sci.,* 57, 1325–1333, 1974.
12. **Cameron, R. D. and Luick, J. R.,** Seasonal changes in total body water, extracellular fluid, and blood volume in grazing reindeer, *Can. J. Zool.,* 50, 107–116, 1972.
13. **White, R. G.,** Some aspects of nutritional adaptations of arctic herbivorous mammals, in *Physiological Adaptation to the Environment,* Vernberg, F. J., Ed., Intext Educational Publications, New York, 1975, 239–268.
14. **Person, S. J.,** Digestibility of Indigenous Plants Utilized by *Rangifer tarandus,* Ph.D. dissertation, Institute of Arctic Biology, University of Alaska, Fairbanks, 1975, 1–97.
15. **White, R. G.,** Factors regulating the food intake of caribou and reindeer on tundra ranges, in Proc. Circumpolar Conf. on Northern Ecology, Ottawa, Ontario, Canada, 1975, 1–5.
16. **McEwan, E. H.,** Growth and development of the barren-ground caribou. II. Postnatal growth rates, *Can. J. Zool.,* 46, 1023–1029, 1968.
17. **Pegau, R. E.,** Effect of reindeer trampling and grazing on lichens, *J. Range Manage.,* 23, 95–97, 1970.
18. **Nordfelt, S., Cagell, W., and Nordkvist, M.,** Digestibility experiments with reindeer, *Särtryck Och Förhandsmeddelande,* 151, 1–14, 1961.
19. **Eriksson, S. and Schmekel, J.,** A comparison between the ability of reindeer and sheep to digest their food, *Kungl. Lantbrukshögskolans Annaler,* 28, 175–180, 1962.
20. **Pruitt, W. O., Jr.,** Behavior of Barren-ground Caribou, Biological Papers, Special Rep. No. 3, University of Alaska, Fairbanks, 1960, 1–43.
21. **Terentiev, N. D.,** Experimental stall-feeding of reindeer on reindeer moss and hay, *Soviet Reindeer Industry II STVO,* 8, 86, 1936.

22. **Auramchik, M. N.,** The winter feeding of reindeer on the Yamal North, Trans. Inst. Polar Agric., Animal Husbandry and Fishing and Hunting Industry Series, *Reindeer Industry,* 4, 65–66, 1939.
23. **Steen, E.,** Some aspects of the nutrition of semi-domestic reindeer, in *Comparative Nutrition of Wild Animals,* Crawford, M. A., Ed., Academic Press, New York, 1968, 117–128.
24. **Holleman, D. F. and Luick, J. R.,** Radiocesium cycling in reindeer/caribou, in Proc. 1st Int. Reindeer and Caribou Symp., Special Rep. No. 1, Luick, J. R., Lent, P. C., Klein, D. R., and White, R. G., Eds., Biol. Papers, University of Alaska, Fairbanks, 1975, 64–70.
25. **Hanson, W. C., Whicker, F. W., and Lipscomb, J. F.,** Lichen forage ingestion rates of free-roaming caribou estimated with fallout cesium-137, in Proc. 1st Int. Reindeer and Caribou Symp., Special Rep. No. 1, Luick, J. R., Lent, P. C., Klein, D. R. and White, R. G., Eds., Biological Papers, University of Alaska, Special Rep. No. 1, 1975, 71–79.
26. **Cameron, R. D.,** Water metabolism by reindeer (*Rangifer tarandus*), Ph.D. dissertation, Institute of Arctic Biology, University of Alaska, Fairbanks, 1972, 1–98.
27. **Jacobsen, E. and Skjenneberg, S.,** Some results from feeding experiments with reindeer, in Proc. 1st Int. Reindeer and Caribou Symp., Special Rep. No. 1, Luick, J. R., Lent, P. C., Klein, D. R. and White, R. G., Eds., Biological Papers, University of Alaska, Fairbanks, 1975, 95–107.
28. **Luick, J. R.,** Reindeer research in northern Siberia, *Reindeer Herders Newsl.,* 1, 1–12, 1976.
29. **Luick, J. R., Ed.,** Industrial Development in Alaska and its Effects on the Nutritional and Physiological Status of Arctic Animals, Progress Rep. 1975/76, Contract No. E(45-1)-2229, U.S. Energy Research Development Administration, 1976, 1–60.
30. **Llano, G. A.,** Utilization of lichens in the arctic and subarctic, *Econ. Bot.,* 10, 367–392, 1956.
31. **Pulliainen, E.,** Nutritive values of some lichens used as food by reindeer in northeastern Lapland, *Ann. Zool. Fenn.,* 8, 285–289, 1971.
32. **Luick, J. R., Ed.,** Nutrition and Metabolism of Reindeer and Caribou in Alaska, Progress Rep. 1970/1971, Contract No. AT(04-3)-310, U.S. Atomic Energy Commission, 1971, 1–100.
33. **Borodzin, E.,** personal communication.

DIETS FOR CAPTIVE REINDEER*

J. R. Luick

INTRODUCTION

In recent years there has been an increasing interest in maintaining reindeer in captivity for nonagricultural purposes. The magic of Christmas with Santa Claus and his reindeer certainly lives in the memories of nearly everyone, including entrepreneurs who establish "Santa Claus Villages" and often train reindeer to pull colorful sleds during the Christmas pageantry. Reindeer† have long been seen in zoological gardens and are being used increasingly at research institutes for experimental purposes.[1] In addition, lucrative oriental markets for reindeer antler have stimulated the establishment of reindeer farms where antlers are harvested annually and processed for use as aphrodisiacs and for medicinal purposes.[2,3]

In the United States, live reindeer are usually purchased in Alaska from Eskimo herders and shipped by air to destinations where plant foods that are typically consumed by freely grazing reindeer are not found in sufficient quantities to provide adequate nourishment. Therefore, reindeer must be adapted quickly to commercial livestock rations; this often leads to fatal nutritional and digestive disorders.[4-6]

In considering the dietary habits of freely grazing reindeer and the wide variety of plants that are consumed selectively on a seasonal basis, as well as the exotic nature and unusual chemical composition of several of these arctic plants, it seems hardly surprising that more than 60% of these reindeer may die within 2 weeks of departure from tundra ranges.[6] Other factors contributing to this high mortality rate include the stresses of capture and handling, hyperthermia, regurgitation and inhalation of rumen contents, injury and disease, and overdoses of immobilizing agents and tranquilizers.[4-6]

In the following discussion, consideration is given to the nutritional aspects of hand rearing of orphan calves, capture and shipment, and the maintenance of reindeer on atypical diets. Since there are few published works on which to build a critical and definitive treatise, much of this information is based on personal experiences of the author and on information solicited from veterinarians and managers of commercial reindeer enterprises. Finally, because considerable differences in opinion were expressed in answer to a questionnaire on diets for captive reindeer, the comments of respondents were not summarized, but rather were edited slightly and are presented in tabular form (see tables).

NURSING CALVES

Reindeer calves are born in early spring, often before the snow has disappeared completely. Suckling commences withing a few hours after birth and is both frequent and

* Data reported in this chapter that originated in the author's laboratory resulted from research activities supported by U.S. Atomic Energy Commission Contract AT(04-3)-310; Energy Research Development Administration Contract E(45-1)-2229; and the Office of Polar Programs and the International Biological Program of the National Science Foundation (NSF Grant Number GV-29342). By acceptance of this article, the publisher and/or recipient acknowledges the U.S. Government's right to retain a nonexclusive, royalty-free license in and to any copyright covering this paper.

† Unless otherwise noted, the term "reindeer" as used in this chapter refers to both semidomesticated (i.e., herded) and wild reindeer (in North America, "caribou") classified taxonomically as *Rangifer tarandus.*

of short duration.[7] This characteristic behavior has evolved presumably because of the necessity for correlating with the activity patterns of the herd which, in a freely grazing situation, means spending from 70 to 90% of the day searching for and consuming food and/or fleeing from insects or larger predators.[8] The calves also begin to nibble pelleted rations and/or graze the early green vegetation; in fact, the time spent grazing (7%) already exceeds the time spent nursing (4%) in the daily activity patterns of 1- to 10-day-old reindeer calves.[9]

Milk production and, therefore, milk intake of the calves apparently peak not later than 15 to 20 days post partum at about 2.5 l/day (Figure 1).[10] Reindeer milk is of unusually high nutrient content and contains more total solids, chiefly protein and fat, than the milk of any other large herbivore studied to date.[11] Using data from Figures 1 and 2, it appears that the energy intake from milk solids alone approaches 4000 kcal/day for a very young reindeer calf of about 5 kg body weight.

The relative contribution of various nutrients in milk and green vegetation changes markedly as the summer grazing season progresses. For example, by 3 to 4 weeks post partum, calves spend a greater fraction of the day grazing (24%) than nursing (<1%).[8] However, the marked decline in milk production shown in Figure 1 should not imply that milk nutrients become an insignificant factor in the daily nutrient intake of the nursing calf. On the contrary, as summer progresses, reindeer milk becomes increasingly concentrated, rising to 35% total solids by 90 days post partum; at this time, milk intake has fallen from 2.5 to 0.75 l/day.[10,11] From these data and the energy content of milk during late lactation, it can be calculated that the calf still receives about 1700 kcal/day

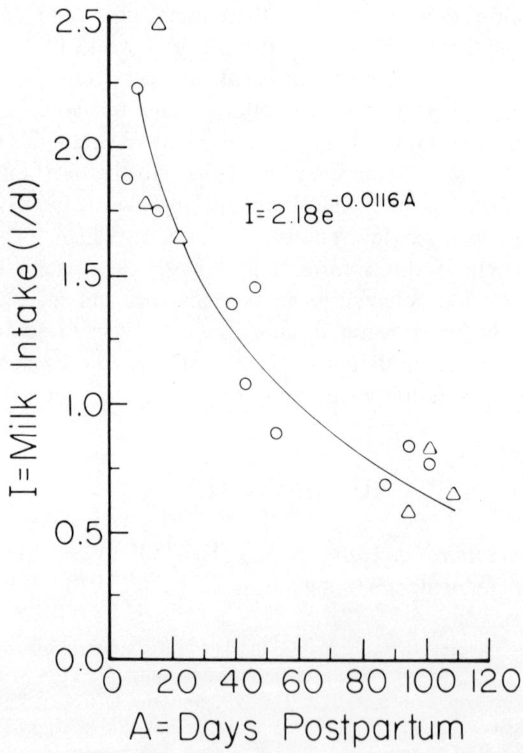

FIGURE 1. Milk production (milk intake of calves) of four lactating female reindeer as determined by the deuterium/tritium oxide (○) and the radiocesium (Δ) methods, respectively.[10,28]

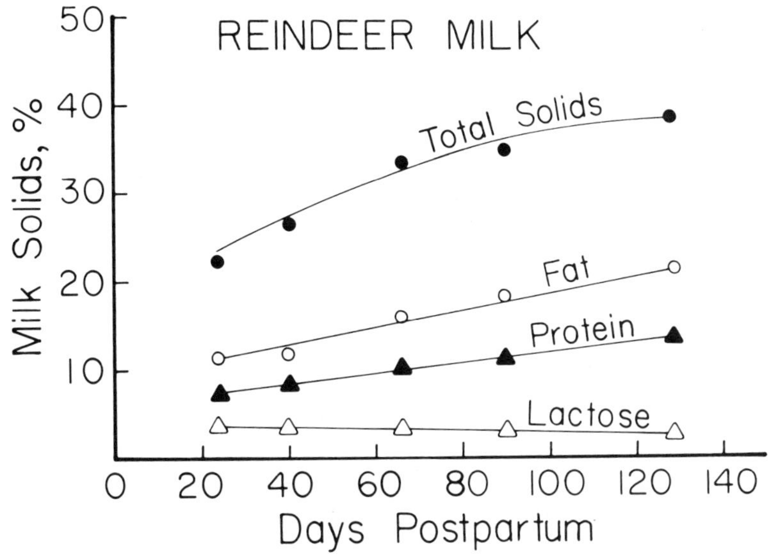

FIGURE 2. Composition changes in major organic constituents of reindeer milk during the lactation cycle.[11]

of milk solids.[12] Thus, the very rapid and early growth of the reindeer calf, averaging about 0.3 to 0.4 kg/day from 0 to 90 days of age, is a consequence of several factors, viz., the high intake of milk solids, the marked improvement in nutrient content of plant foods, and the increasing time spent grazing and/or eating pelleted rations that are fed to the adult animals.[13-16]

ORPHAN CALVES

In contrast to the nursing reindeer calf that grazes within hours of birth, survives the rigors of climatic extremes, and is required to keep up with the movements of the herd, the orphan calf is surprisingly fragile and seems highly susceptible to nutritional and physiological stresses. These stresses can be minimized to some degree through skilled and knowledgeable managerial techniques, i.e., the provision of adequate shade, fresh water, and small paddocks where the calves can graze, exercise, and, most importantly, relax out of sight of humans, dogs, etc. When close confinement is unavoidable, special attention must be given to sanitary conditions, ventilation, and minimizing unnecessary disturbances.

These criteria are of utmost importance during the first 2 weeks of confinement. In two transplants in 1958 and 1959, death losses among wild reindeer calves were 68 and 69%, respectively. All deaths occurred within 2 weeks of capture and, of these, 67% died within 48 hr of capture.[17] All calves showed a loss of alertness, appetite, and strength – a pathologic syndrome typical of excessive physiological stress.

Fortunately, not all shippers have been as unsuccessful, at least during the capture and relocation phases. Thus, reindeer calves captured and flown from tundra ranges to southern Canada experienced low mortality rates. These calves were blindfolded, confined in large burlap sacks (tied around the neck), and force-fed a reconstituted dairy calf-milk replacer during transport.[29]

At least three options can be considered for the feeding of orphan reindeer calves, i.e., imprinting of the orphan calf onto a foster reindeer nurse cow, bottle feeding of reindeer milk, and bottle feeding of reconstituted dry milk solids. However, in practice, only the last option offers a practical solution. The foster cow technique is difficult to achieve, not so much because of a reluctance by the young calf to imprint onto the cow, but rather because of the refusal of the cow to accept another calf (author's personal observation). Ironically, when no overt attempt is made by man to achieve imprinting, most reindeer cows do indeed tolerate nursing by a strange calf (termed "cross-nursing" or "thief-sucking"). Attempts at cross-nursing increase until the calves are 5 weeks old, but are usually unsuccessful in the absence of the cow's own calf.[7]

Bottle feeding of reindeer milk in sufficient quantity to provide adequate nutrition for the calf is most difficult, chiefly because of the difficulties in hand milking untrained reindeer cows.[11] Even with trained cows, the volume of milk obtained by hand stripping, regardless of whether oxytocin is used, is far less than the amount needed (or normally available).[18] In addition, growth of the donor cow's own calf may be compromised if its daily intake diminished significantly.

Table 1 shows two dietary programs that have been used with some success in hand rearing reindeer calves. In each instance, the objective was to simulate natural conditions as closely as possible. Reconstituted cow or sow milk replacer, bottle fed five to seven times daily, provided the basal ration. This was supplemented with a pelleted dairy calf grain ration, green vegetation, and fresh, clean water. The calves also ate hay, straw, and/or excelsior bedding, but even at this early age they characteristically preferred lichens over all other forages.[17]

High mortality rates of bottle-fed reindeer calves is largely due to digestive disorders of bacterial or nutritional origin. The former seems especially lethal to the calves and requires prompt professional treatment. Simple nutritional diarrhea, i.e., scouring, has been controlled by adding small amounts of baby rice cereal* to the reconstituted milk.[1] Jones suggests that this disorder is related to the high fat content of the milk and failure to encourage the consumption of natural forage by limiting the volume of milk offered to the calves.[17]

Table 1
COMMENTS ON DIETS FOR ORPHANED REINDEER CALVES

Diet A

High-energy, medicated, milk replacer,[a] reconstituted as for dairy calves, bottle fed five times daily.
Dairy calf concentrate mix[b] offered ad libitum.
Fresh, clean water and salt block always available.
Hand-cut green vegetation, mostly willow (*Salix* sp.) and fireweed (*Epilobium angustifolium*), fed twice daily.

Diet B

Milk replacer constituted as follows: 26 oz canned evaporated milk, 26 oz reconstituted nonfat dry milk solids, 2 oz (equal parts by volume) cod liver oil and whole egg that have been beaten previously into an emulsion. Bottle fed seven times daily.
Calves ate green vegetation, hay, straw, and excelsior bedding, but refused a dairy calf grain supplement. Lichens were preferred to all other forages.

[a] Purina Calf Starter, Ralston Purina Co., St. Louis, Mo.
[b] Purina Nursing Chow, Ralston Purina Co., St. Louis, Mo.

* Gerber Instant Baby Rice Cereal, Gerber, Freemont, Michigan.

CAPTURE AND TRANSPORT

All shippers seemingly agree that the successful translocation of reindeer depends almost entirely upon two criteria: (1) an adequate supply of lichens ("reindeer moss") must be provided as the principal foodstuff during the early phases of the operation, and (2) excitation must be minimized during the capture, handling, and loading of reindeer.

Lichens are usually hand picked or raked from the tundra range lands and bagged for feeding during the 3- to 4-week interval between capture and successful transition to a livestock ration. Considering that an adult reindeer will eat 2 to 5 kg of lichens (dry weight) per day, a relatively small shipment of ten reindeer should be provided with about 900 kg (1 ton) of bagged lichens.[19,20] Lichen picking is tedious, time consuming, and, therefore, costly. In addition, dried lichens are bulky, will shatter if compressed, and should always be moistened before being offered.[21] Lichens can be pelletized and will still be eaten, but, unfortunately, suitable pellet mills are not generally available in the far North.[22]

The importance of supplying lichens during translocation is related to their high acceptibility by reindeer and to their availability (lichens are the only natural reindeer food that can be harvested in large quantities in the vicinity of the corrals). In addition, the reindeer's ruminal microflora and microfauna are well adapted to fermenting lichens. On the other hand, lichens are a very poor source of essential dietary nutrients.[15] Table 2 lists feeding techniques used during capture and transport, and especially during the important transition from tundra food plants to commercial livestock rations.

TRANSITION RATIONS

The transition from tundra vegetation to a completely atypical dietary regimen is usually accomplished by gradually decreasing the proportion of lichens in the daily feed; the rate should be adjusted to the amount of lichens on hand. Serious digestive and physiological disorders will almost certainly occur if the transition is rapid, and especially if the new ration differs greatly in chemical composition from tundra food plants.[4,5] For example, the transition to a low-protein, high-fiber, pelleted horse ration is more likely to be successful than a change to a high-protein, high-energy dairy cattle concentrate. Reindeer should have access to an abundant supply of fresh water and, if possible, be allowed to feed on grass and/or legume pastures. From a management standpoint, all shippers agree that every effort must be made to minimize handling and other causes of stress during this critical transitional phase.

MAINTENANCE RATIONS

In considering feeding programs for captive reindeer, first it must be acknowledged that there is little likelihood of simulating the nutrient intake patterns of freely grazing reindeer on tundra ranges. In the natural state, reindeer selectively eat perhaps hundreds of different grasses, sedges, herbs, forbs, woody plants, and many species of lichens particularly in winter.[15] Moreover, the chemical and nutrient compositions of food plants vary enormously with season. For example, the crude protein, crude fiber, and calcium content of winter diets, approximately 3, 18, and 2%, respectively, of dry matter shifts to about 20, 10, and 8%, respectively, on lush summer pastures.[23] The ability to tolerate these dietary changes may be viewed as an evolutionary adaptation that has enabled reindeer to occupy a unique ecological niche, and it is now quite clear that they do not reflect a similarly unique, cyclical pattern of dietary requirements.[24,25] Reindeer have long been maintained in captivity on widely differing dietary regimens (Table 3) and climatic circumstances, none of which even closely approximate natural conditions. They

Table 2
COMMENTS ON PROCEDURES TO ENSURE SUCCESSFUL TRANSLOCATION FROM TUNDRA TO FARM

Capture and Transport

1. Capturing reindeer must be done carefully to minimize stress. It should be done in winter, preferably not in summer or fall heat. Blindfold the animals when first transferring from field to crate. For shipping, crate them and bag up enough lichens and mosses to feed enroute. Stress will be too high for them to eat any other domestic feeds during this critical period. Do not use tranquilizers or immobilizing agents because they "mask" respiratory problems, hyperthermia, etc. that develop during handling. Administer long-acting penicillin, 6 to 8 cc per reindeer, to counteract shipping fever. Reindeer will always eat snow in winter, but often will not find a water tub on dry ground in a new biotype such as a zoo. Zoos would do best to box stall new arrivals until such time as they adapt to a new diet rather than to place them immediately into small, cement-floored, paddocks with no retreat.
2. It helps if you have natural reindeer moss (lichens) to mix with grains. Offer free choice and bed down on high-quality alfalfa hay. New deer adapt faster if they are mixed with deer already adapted. Adaptation is not a serious problem. We do not use tranquilizers, immobilizing agents, or shipping fever vaccine but do administer penicillin/dihydrostreptomycin to all reindeer before departure.
3. Immediately after capture, feed lichens and green vegetation and allow the reindeer to "cool off" in an isolated corral for 4 or 5 days, then ship in crates, minimize stress, never ship in warm weather, minimize the use of drugs for restraint and give cattle-type shipping fever vaccine and penicillin/dihydrostreptomycin or long-acting penicillin before departure.
4. When shipping large numbers of hobbled reindeer (not crated), it is usually necessary to administer tranquilizers (xylazine and acepromezine) as well as an immobilizing agent (e.g., etorphine). However, despite close medical attention, death losses can be very high (> 50% within 24 hr of arrival), especially when the reindeer experience excessive stress during capture and handling. Etorphine antagonist diprenorphine (M50-50) should be given upon arrival at destination.

Transition to Maintenance Rations

1. Sprinkle fine-quality leafy alfalfa hay into the lichens and spread on ground. Put new arrivals with reindeer already conditioned to eat commercial foods. Do not put alfalfa hay in a hopper, but spread on ground. It works superbly well. Commence feeding pelletized grain mix and a commercially available deer pellet immediately.
2. Add lichens to the new feeds, then gradually decrease the amount of lichens until the reindeer are completely adjusted to the new feed.
3. Gradually increase the daily ration of livestock-type pelleted feed; no problem.
4. Transition accomplished by adding increasing amounts of a low-protein, low-energy, high-fiber pelleted complete ration to the basal lichen diet.

Table 3
DIETARY PROGRAMS FOR MAINTAINING REINDEER IN CAPTIVITY

1. Best-quality green second-cut alfalfa hay, 10 lb/reindeer/day. Supplemental feed consists of 50% deer pellets and 50% special pelleted grain mixture (60% oats, 20% wheat, 20% barley with vitamin A added). Mineral-protein blocks, browse (willow and white poplar in summer), and fresh water, free choice. Thibenzole® added to water regularly. It has been very successful in every way.
2. Reindeer are fed chiefly a pelleted complete horse ration, "Round-Up"® (Albers Milling Co., Los Angeles, Calif.), herbivore pellets (Albers Milling Co.), Sudan grass hay and occasionally Omolene® (Ralston Purina Co., St. Louis, Mo.). Alfalfa hay has been offered in the past, but reindeer did not eat it.
3. Large amounts of fine grasses and browse in the natural state (pasture), free choice; second- and third-cut alfalfa hay, free choice; pelleted grain supplement, 2 lb/animal/day. Clover pastures are very good.
4. Lichens are the basic ingredient of reindeer diets in most European zoos. Grass and/or alfalfa hays are also fed and supplemented with a concentrated, pelletized feed (Basel Zoological Gardens, Basel, Switzerland).
5. Purina Cattle Starter No. 1 (Ralston Purina, St. Louis, Mo.) is available at all times and provides excellent growth and reproduction. Reindeer are allowed to graze mixed vegetation pastures in summer. Salt blocks and water are also available free choice.
6. Aim for a low-protein diet, supplemented with alfalfa or clover hay. Feed twice daily and provide pastures for grazing.

grow and reproduce on most common livestock feeds, (Tables 4 and 5), are subject to the usual nutritional disorders and diseases of domestic livestock (Table 6), and respond favorably to the medical tactics of the veterinary practitioner (Table 7).

Thus, success in maintaining reindeer in captivity lies chiefly in skilled and knowledgeable management of the herd. As summarized by Abrams, the nutritional aspects of management include the provision of a multicomponent diet with variations of formulation from time to time, gradual changes in dietary programs to prevent digestive upsets, and close attention to food hygiene.[26] Rations should be procured frequently and in relatively small quantities to assure high quality and should not be stored under conditions of high temperature or humidity, nor exposed to air for long periods of time. Similarly, reindeer in temperate climates require shade for protection from thermal stress, trees or rubbing posts to facilitate cleaning and polishing of antlers prior to rut, and an area within the paddock where they can relax with a sense of security and privacy.[27]

ACKNOWLEDGMENTS

The author wishes to thank the following persons who most generously provided much information on practical aspects of raising captive reindeer: D. Branger, Top Hat Ranch, Roscoe, Montana; R. Butcher and G. Carruthers, Santa's Village, Santa Cruz, California; Dr. R. Dieterich, Research Veterinarian, University of Alaska, Fairbanks; Dr. L. Griner, San Diego Zoo, San Diego, California; Dr. I. McEwan, University of British Columbia, Vancouver; Dr. A. Oeming, Alberta Game Farm, Edmonton; Dr. H. Saegesser, Director, City Deer Park "Dählhölzli," Bern, Switzerland; Dr. H. Wackernagel, Basel Zoological Gardens, Basel, Switzerland.

Table 4
COMPOSITION AND PROXIMATE ANALYSES OF SOME SUCCESSFUL REINDEER RATIONS

Basel Zoo[a]
(%)

Oats (15)
Barley (24)
Wheat (14)
Wheat bran (10)
Alfalfa meal, dehydrated (19)
Peanut meal (9)
Soybean meal (11)
Linseed meal (2)
Salt (1)
Dicalcium phosphate (2)
Limestone (1)
Binder (1.5)
Minerals and vitamins (0.8)

Crude protein (18)
Crude fat (2.4)
Crude fiber (7)
Total digestible nutrients (66)

University of British Columbia[b]
(%)

Corn (35)
Bran (15)
Alfalfa meal, dehydrated (11.6)
Soybean meal, 48.5% (8.7)
Herring meal, 72% (6.4)
Iodized salt (1.2)
Rock phosphate, defluorinated (1.2)
Molasses (8.7)
Beet pulp (11.6)
Minerals (0.3)

Crude protein (15 min)
Crude fat (2.5 min)
Crude fiber (6.5 max)
Total digestible nutrients (58–60)

Santa's Village[c]
(%)

Beet pulp (33.4)
Barley, ground (23.9)
Oats, ground (23.9)
Rice bran (14.4)
Salt, trace mineralized (1)
Molasses (1.4)
Mineral mix (2)
Crude protein (10)
Crude fat (5)
Crude fiber (12.6)
Total digestible nutrients (64.6)

Arctic Health Research Laboratory[d]
(%)

Rolled corn and/or barley[e]
Ground oats, barley and/or corn[e]
Cottonseed and/or soybean meal[e]
Wheat, millrun
Wheat, middlings
Beet pulp pellets
Salt
Molasses

Crude protein (16 min)
Crude fat (3 min)
Crude fiber (8 max)

Alberta Game Farm[f]
(%)

Commercial deer pellets (50)
Grain mix (50)
- rolled oats (60)
- rolled wheat (20)
- rolled barley (20)
- added vitamin A

Scandinavian Supplementary Feed[g]
(%)

Cereal grains (41)
Ground hay (25)
Molasses beet chips (10)
Ground straw (15)
Wheat bran (5)
Dicalcium phosphate (11.7)
Iodized salt (0.3)
Vitamins (2)
Trace Minerals

[a] Basel Zoo, Switzerland; fed as pelleted supplement with grass or alfalfa hay.
[b] Vancouver, British Columbia; University of British Columbia deer ration #36-57, Buckerfields Ltd., Vancouver, British Columbia, fed as a pelleted, complete ration.
[c] Santa Cruz, California: "Special Mixture," Gunter Bros., Morgan Hill, Calif., fed as a complete pelleted ration, but supplemented occasionally with alfalfa hay.
[d] Fairbanks, Alaska; "Quality Texture," Fischer Mills Inc., Seattle, Washington, fed as a complete, pelleted ration.
[e] Varying proportions depending upon availability and price.
[f] Edmonton, Alberta; fed as a supplement to choice, alfalfa hay.
[g] Complete ration used to supplement poor range diets, especially during severe winters.

Table 4 (continued)
COMPOSITION AND PROXIMATE ANALYSES OF SOME SUCCESSFUL REINDEER RATIONS

Crude protein (8)
Crude fat (2)
Crude fiber (18)
Nitrogen-free extract (63)
Ash (9)

Norwegian Reindeer Research Institute[j]
(%)

Barley, ground (40)
Young grass, ground (25)
Oats, ground (17)
Wheat bran (15)
Soybean oil (3)

Crude protein (10.9)
Crude fat (2.7)
Crude fiber (9.3)
Nitrogen-free extract (58.6)
Ash (5.9)

Institute of Arctic Biology[h]

Plant protein products
Grain products
Processed grain by-products
Roughage products
Cane molasses
Alfalfa meal
Urea
Calcium carbonate
Defluorinated phosphate
Ammonium sulfate
Iodized salt
Added vitamin A
Minerals

Crude protein (11% min; includes not more than 2–5% equivalent protein from N.P.N.)
Crude fat (1% min)
Crude fiber (21% max)

Top Hat Ranch[i]

Oats
Wheat middlings
Wheat bran
Beet pulp
Barley
Yellow corn
Alfalfa meal, dehydrated
Soybean meal
Cane and beet molasses
Linseed meal
Meat scraps
Added vitamins
Minerals

[h] Fairbanks, Alaska; Purina Cattle Starter (V), #1 (G), fed as a complete ration.
[i] Roscoe, Montana; fed as a supplement to alfalfa hay and grass pastures.
[j] Harstad, Norway; fed as complete ration when winter range feeds are unavailable or inadequate.

Table 5
COMMENTS ON SPECIFIC FEEDS FOR REINDEER

Grass Hays

1. These are useless, unpalatable to the reindeer, and offering no worthwhile nutrition.
2. Over the years the reindeer have been offered various grass hays but only ate small amounts.
3. Very poor unless cut when very immature and fine.
4. We feed common meadow hay.
5. Grass hays seem to be of little nutritional value for reindeer since there is much waste when fed as long hay in bunkers. Recently found that Kentucky nugget bluegrass was almost as good as Cattle Starter No. 1 for a winter ration.
6. Reindeer will eat grass hays, especially brome, wheatgrass, timothy, and bluegrass.

Alfalfa Hay ("Long")

1. For best results we use second-cut, very green and leafy alfalfa hay. It must be spread on the ground, preferably in winter on snow. Reindeer eat the green leaves almost immediately and thrive on them.
2. Reindeer ate only small quantities of alfalfa hay.
3. We use second- and third-cut alfalfa; the reindeer selectively graze the leaves but not the stems.
4. Yes, we feed some alfalfa hay.
5. No experience.
6. We feed either alfalfa or clover hay.
7. Alfalfa hay is "too rich" for reindeer.

Alfalfa Hay (Pelleted or Cubed)

1. This takes more time to become accustomed to, but they will eventually accept it and do well.
2. We do not feed cubed or pelleted alfalfa hay.
3. Very good when adapted to it.
4. No experience.
5. Cubed alfalfa hay has been suspect in laminitis cases.

Cereal Grains (e.g., Crushed Oats)

1. Once conditioned to grains, it is a "must" for them. They consume amazingly high quantities and relish it. Five gal will feed ten reindeer at a serving, twice a day.
2. We feed Omolene® (Ralston Purina Corp., St. Louis, Mo.); three animals receive 8 qt/day.
3. Very good when adapted to it.
4. Are accepted.
5. We never feed as such, but they do make up part of our pelleted ration.
6. Corn is okay; best to crush or roll grains before feeding; cottonseed meal is "too hot."

Salt

1. If you have them in a cobalt-deficient area, use this kind of salt since it has iodine in it. Absolutely necessary at all times.
2. We offer salt blocks, some have been mineralized.
3. When fed free choice, reindeer will eat small amounts.
4. Included in our pelleted ration.
5. Offered free choice as a mineralized salt block.

Mineral Mixtures

1. Use a good protein-mineral mixture in the form of a block that can be chewed.
2. Very poor luck with getting reindeer to eat.
3. Included in pellets.
4. Not fed separately.

Table 5 (continued)
COMMENTS ON SPECIFIC FEEDS FOR REINDEER

Water

1. We regulate the water supply in summer so that we can use a low level vermifuge treatment like Thibenzole® to keep parasite infestations to a minimum. This is impossible to do when reindeer have access to any residual ground water.
2. We use automatic water dispensing devices.
3. Free choice.
4. Accessible at all times.
5. Important to have an ample supply of good fresh water available at all times.

Complete Rations such as Purina or Albers Feeds

1. Once they become accustomed to such feed, it is perfectly fine. Others insist that beet pulp is the answer, but we have not used it.
2. We have fed horse pellets Omolene (Ralston Purina, St. Louis, Mo.), 8 qt/day for three animals; herbivore pellets 8 qt/day for three animals; also "Round-Up"® (Albers Milling Co., Los Angeles, Calif.).
3. Some pellets are very good, especially if they contain sugar-beet pulp pellets.
4. Use of pelleted, concentrated rations to supplement hay rations is standard technique.
5. We feed pelleted rations almost exclusively and prefer a high-roughage, low-protein ration similar to horse feeds, i.e., "Round-Up" (Albers Milling Co., Los Angeles, Calif.) or to early feedlot rations, i.e., Purina Cattle Starter No. 1 (Ralston Purina Co., St. Louis, Mo.).
6. Works fine.
7. We feed complete pelleted rations but take great care to mix well; drying must not occur at a temperature that is too high; watch for mildew, off flavors, etc.

Medicated Feeds

1. Not necessary unless treating for bacterial diarrhea or other disease requiring antibiotic therapy.
2. We do not feed.
3. We do not feed.
4. We do not feed.
5. Have fed medicated feeds unintentionally with no obvious problems or advantages.

Toxic Feeds

1. Never use musty, dusty, moldy alfalfa hay or heavy amounts of barley. Grain feeding should not be overdone. Care should be taken until the animals are accustomed to it. It can cause bloat and foundering.
2. No evidence.
3. No problems when good husbandry practices are used.
4. Never feed "hot rations" to stressed reindeer directly "off the range."

Feeds Leading to Nutritional or Digestive Problems

1. None if the diet is based chiefly on superior quality, green, leafy, tender, second-cut alfalfa. Reindeer approach this as they would lichens.
2. No evidence.
3. Some coarse roughages will plug the digestive tract at the stomach outlet.
4. Reindeer, as well as other ruminants, must be adapted slowly and carefully to new dietary programs that differ significantly from typical arctic range land diets in chemical composition, nutrient content, palatability, etc.

Table 6
COMMENTS ON NUTRITIONAL DISORDERS AND OTHER DISEASES OF REINDEER

Digestive Upsets

1. Occurs from poor hay and excessive grain; reindeer become "semibloated." Drench with mineral oil.
2. Yes, catarrhal gastroenteritis; ulceration and peritonitis.
3. Yes, diarrhea from excessive alfalfa and too little browse.
4. No cases.
5. Several cases occurred when recently shipped reindeer were fed a dairy cattle complete ration (high protein) and/or a feedlot beef cattle finishing ration (high energy) with no adaptation to the feed. Syndrome included bloat and central nervous system involvement, i.e., goose stepping and/or staggering. Laboratory studies showed no pathogens or pathogenic processes.

Bloat

1. Common cause of death.
2. No cases.
3. No cases.
4. No cases.
5. No classical cases, but bloated reindeer are seen when stressed reindeer are fed high-energy, high-protein rations without prior adaptation.

Impaction of Gastrointestinal Tract

1. No cases.
2. It does occur.
3. It can occur when reindeer eat excessive amounts of coarse browse.
4. No cases.
5. At least two cases of abomasal impaction; one blockage of entry into omasam.

Peritonitis and/or Rumenitis

1. No cases.
2. No cases.
3. No cases.
4. Two cases.
5. Several cases of two distinct types: low grade from infestations of *Setaria yehi* (abdominal worm); rumenitis type, usually from enteric type of organisms, e.g., *Escherichia coli,* etc.

Cardiovascular Disease

1. No evidence.
2. No evidence.
3. One case (myocardial aortic calcification).
4. Three cases (cardiac insufficiency, pericarditis, and defective septum).
5. One case (coronary thrombosis with necrosis).

Capture-confinement Stress

1. Can be severe and cause high mortality; most severe in warm weather; isolate and allow to relax.
2. Stress syndrome associated with capture and transport is called "Capture Myopathy" and is similar to White Muscle Disease. Microscopic examination definitely shows muscle involvement.

Table 6 (continued)
COMMENTS ON NUTRITIONAL DISORDERS AND OTHER DISEASES OF REINDEER

Internal or External Parasites

1. Common moose tick, *Dermacentor albipictus,* infests reindeer in wooded areas in their paddocks. If infestation is severe, spray thoroughly with sheep tick eradicator.
2. No cases or not diagnosed.
3. No cases or not diagnosed.
4. No cases or not diagnosed.
5. Yes, including stomach and intestinal nematodes, adult intestinal tapeworms, larval tapeworms in liver and muscle, lung worms, setaria, nasal bots, and warbles.

Face or Jaw Abscesses

1. Occasionally, not typical actinomycosis rather jaw abscesses that headed and were drained; not common.
2. No cases.
3. No cases.
4. Two cases of lump jaw, hard bone; did not respond to treatment; died.
5. Generally resulting from trauma and wearing halters; respond well to treatment.

Brucellosis

1. No cases.
2. No cases.
3. No cases.
4. No cases.
5. One case, sacrificed.

Abortion

1. No cases.
2. No cases.
3. No cases.
4. One case.
5. One case.

Pneumonia

1. Yes, especially during hot summers. Successfully treated with penicillin and streptomycin.
2. No cases.
3. Unspecified number.
4. Six cases diagnosed in 25 years.
5. Several cases.

Johne's Disease (Paratuberculosis)

1. No cases.
2. No cases.
3. No cases.
4. No cases.
5. No cases.
6. Several cases; poor response to treatment.

Table 6 (continued)
COMMENTS ON NUTRITIONAL DISORDERS AND OTHER DISEASES OF REINDEER

Foot and Leg Problems

1. No foot rot but cases of joint abscesses and/or laminitis are treated with antibiotics.
2. Few cases, not a major problem.
3. Few cases, not a major problem.
4. Few cases, not a major problem.
5. Few cases, not a major problem.

Other

1. Navel infections; treat with iodine ointment, not liquid solution; infection quickly involves brain and lack of coordination of extremities.
2. Nitrate poisoning on early spring pastures previously heavily fertilized; several deaths.

Comments on the Most Serious Health Problem

1. Parasite infestation is constant, requires frequent treatment; scouring due to parasites, bacterial infection, and coccidiosis. No other diseases are considered a serious problem.
2. High neonatal deaths of nonspecific causes. Low rate of reproduction.
3. No single disease problem is evident but combined occurrence of the diseases encountered is serious.
4. No specific problem.
5. Parasites are a continuing problem; malnutrition and infections from various stresses associated with use in research studies.

Comments on the Principal Causes of Death in Reindeer Herds

1. Bloat from poor alfalfa hay and old age; highest mortality in neonatal calves.
2. Pneumonia, especially during cold rainy and sleeting weather in spring. Two died from stomach impaction from eating very coarse browse. One death from hair ball. One died from foot rot that did not respond to treatment.
3. Various, no single cause is especially noteworthy. About 50% death loss among young calves; lameness generally precedes death in older animals.
4. Newly arrived animals die chiefly from a capture-stress syndrome. Herd reindeer die from a great variety of causes. No specific disease problem.

Table 7
COMMENTS ON PREVENTATIVE MEDICINE

Vaccinations

1. We vaccinate only for the various clostridia using the commercial named vaccine, Coopvax.®
2. We do not vaccinate.
3. We vaccinate against leptospirosis.
4. We give hyperimmune-gamma globulin to newborn calves (7 cc "Bovigam," Berna schweiz, Serum v. Impfinstitut, Bern C.H.).
5. We do not vaccinate.

Antibiotics

1. Use tetracycline in grain if needed. This is a soluble powder that they normally will not hesitate to eat with grain.
2. Penicillin/dihydrostreptomycin when body temperature indicates (e.g., for pneumonia); good results obtained.
3. Penicillin/dihydrostreptomycin, kanomycin, and tetracycline are used as needed.

Worm Medicines

1. We use either oral drench procedures with Thibenzole® (in drinking water) or, if tapeworms are the problem, we catch the animals and orally drench with Teniatol®. Check feces for evidence of parasites on a regular basis.
2. We do not administer.
3. Yes, Thibenzole.
4. "Telmin" twice a year; 20 mg/kg body weight original Janssen product, Cilag-chemie AG, Schaffhausen C.H.
5. We prefer injectable Ripercol® for control of both intestinal and lung worms.

Control of Warble Flies, Mosquitos, etc.

1. Only animals shipped from the Arctic have a warble fly problem. During the first year, we remove the larvae by hand; after that nothing develops here.
2. No problem at our location.
3. No problem after leaving Alaska.
4. All reindeer are injected with Warbex® in late October for control of nasal and warble bots.

REFERENCES

1. **Dieterich, R. A. and Luick, J. R.,** Reindeer in biomedical research, *Lab. Anim. Sci.*, 21, 817–824, 1971.
2. **Anon.,** Articles of Tibetan Medicine, V/O "Medexport," Moscow, G-200, USSR, 1–35, undated.
3. **Anon.,** Reindeer herds could provide basis for new industry, *Alaska Ind.*, 4, 45–47, 1972.
4. **Luick, J. R., Ed.,** Nutrition and Metabolism of Reindeer-caribou in Alaska with Special Interest in Nutritional and Environmental Adaptation, Prog. Rep. 1967/68, Contract No. AT(04-3)-310, PA-5, U.S. Atomic Energy Commission, 1968, 1–38.
5. **Luick, J. R., Ed.,** Nutrition and Metabolism of Reindeer-caribou in Alaska with Special Interest in Nutritional and Environmental Adaptation, Prog. Rep. 1968/69, Contract No. AT(04-3)-310, PA-5, U.S. Atomic Energy Commission, 1969, 1–32.
6. **Luick, J. R., Ed.,** Industrial Development in Alaska and its Effects on the Nutritional and Physiological Status of Arctic Animals, Prog. Rep. 1975/76, Contract No. E(45-1)-2229, U.S. Energy Research Development Administration, 1976, 1–60.
7. **Espmark, Y.,** Mother-young relationship and ontogeny of behaviour in reindeer (*Rangifer tarandus* L.), *Z. Tierpsychol.*, 29, 42–81, 1971.
8. **White, R. G., Thomson, B. R., Skogland, T., Person, S. J., Russell, D. E., Holleman, D. F., and Luick, J. R.,** Ecology of caribou at Prudhoe Bay, Alaska, in Ecological Investigations of the Tundra Biome in the Prudhoe Bay Region, Alaska, Special Rep. No. 2, Brown, J., Ed., Biological Papers, University of Alaska, Fairbanks, 1975, 151–201.

9. **Thompson, B.,** Wild reindeer activity, Hardangervidda, 1971, report from the grazing project of the Norwegian IBP Committee, Statens viltundersokelser, Direktoratet for Jakt, Viltstell og Ferskvannsfiske, Trondheim, Norway 1973, 1–76.
10. **Holleman, D. F., White, R. G., and Luick, J. R.,** in Nutrition and Metabolism of Reindeer and Caribou in Alaska, Prog. Rep. 1973/74, Luick, J. R., Ed., Contract No. AT(45-1)-2229, U.S. Atomic Energy Commission, 1974, 64–67.
11. **Luick, J. R., White, R. G., Gau, A. M., and Jenness, R.,** Compositional changes in the milk secreted by grazing reindeer. I. Gross composition and ash, *J. Dairy Sci.,* 57, 1325–1333, 1974.
12. **Luick, J. R., Ed.,** Nutrition and Metabolism of Reindeer and Caribou in Alaska, Prog. Rep. 1970/71, Contract No. AT(04-3)-310, U.S. Atomic Energy Commission, 1971, 1–100.
13. **Krebs, C. J. and Cowan, I. M.,** Growth studies of reindeer fawns, *Can. J. Zool.,* 40, 863–869, 1962.
14. **Dobrotvorsky, I. M., Shubskaya, E. I., Ivanov-Drury, S. M., Ponomarev, A. S., and Vinogradov, M. P.,** Growth and development of the reindeer fawns, Trans. Inst. Polar Agric. Anim. Husb. Fish. Hunt. Ind., in *Reindeer Ind.,* Vol. 3 (Russian with English summaries); cited in Krebs, C. J. and Cowan, I. M., Growth studies of reindeer fawns, *Can. J. Zool.,* 40, 863–869, 1962.
15. **Luick, J. R.,** Diets for Freely Grazing Reindeer, found elsewhere in this volume.
16. **McEwan, E. H.,** Growth and development of the barren-ground caribou. II. Postnatal growth rates, *Can. J. Zool.,* 46, 1023–1029, 1968.
17. **Jones, R. D., Jr.,** Raising caribou for an Aleutian introduction, *J. Wildl. Manage.,* 30, 453–460, 1966.
18. **Varo, M. and Varo, H.,** The milk production of reindeer cows and the shape of milk in the growth of reindeer calves, *J. Sci. Agric. Soc. Finl.,* 43, 1–10, 1971.
19. **Holleman, D. F. and Luick, J. R.,** Radiocesium cycling in reindeer/caribou, in Proc. 1st Int. Reindeer and Caribou Symp., Special Rep. No. 1, Luick, J. R., Lent, P. C., Klein, D. R., and White, R. G., Eds., Biological Papers, University of Alaska, Fairbanks, 1975, 64–70.
20. **Hanson, W. C., Whicker, F. W., and Lipscomb, J. R.,** Lichen forage ingestion rates of free-roaming caribou estimated with fallout cesium-137, in Proc. 1st Int. Reindeer and Caribou Symp., Special Rep. No. 1, Luick, J. R., Lent, P. C., Klein, D. R., and White, R. G., Eds., Biological Papers, University of Alaska, Fairbanks, 1975, 71–79.
21. **Terentiev, N. D.,** Experimental stall-feeding of reindeer on reindeer moss and hay, *Soviet Reindeer Industry II STVO,* 8, 86, 1936.
22. **Luick, J. R., Ed.,** Nutrition and Metabolism of Reindeer and Caribou in Alaska, Prog. Rep. 1969/70, Contract No. AT(04-3)-310, PA-5, U.S. Atomic Energy Commission, 1970, 1–87.
23. **Steen, E.,** Some aspects of the nutrition of semi-domestic reindeer, in *Comparative Nutrition of Wild Animals,* Vol. 21, Crawford, M. A., Ed., Academic Press, New York, 1968, 117–128.
24. **Bergerud, A. T. and Noland, M. J.,** Food habits of hand-reared caribou *Rangifer tarandus* L. in Newfoundland, *Oikos,* 21, 348–350, 1970.
25. **Bergerud, A. T.,** Food habits of Newfoundland caribou, *J. Wildl. Manage.,* 36, 913–923, 1972.
26. **Abrams, J. T.,** Fundamental approach to the nutrition of the captive wild herbivore, in *Comparative Nutrition of Wild Animals,* Vol. 21, Crawford, M. A., Ed., Academic Press, New York, 1968, 41–62.
27. **Wackernagel, H.,** The feeding and care of herbivora in the Basel Zoological Gardens, *Vitamins,* 3, 109–115, undated.
28. **Holleman, D. F., White, R. G., and Luick, J. R.,** New isotope methods for estimating milk intake and yield, *J. Dairy Sci.,* 58, 1814–1821, 1975.
29. **Oeming, A.,** personal communication.

PRONGHORN DIETS, SYNTHETIC AND NATURAL

C. C. Schwartz and J. G. Nagy

Synthetic Diets

We have successfully hand-reared pronghorn fawns using a formula of one part evaporated milk and four parts homogenized milk. Detailed feeding regimes and methods have been reported by Schwartz et al.[1]

We have maintained adult pronghorn on a diet of mixed concentrate (Table 1) and second cutting leafy alfalfa hay offered ad libitum.

Natural Diets

According to Humphrey,[3] the pronghorn habitat is very diverse; it includes the deserts of Baja California, Mexico, and the Great Plains, plus parts of Wyoming and Montana. Consequently, pronghorn have diverse food habits throughout their range which have been well documented and reviewed. Excellent reviews of the literature are presented by Yoakum,[4-7] Severson et al.,[8] and Sundstrom et al.[9] In general, most investigators agree that browse and forbs make up the major percentage of the diet on an annual basis, while grasses are utilized in lesser amounts (Table 2). Pronghorn appear to select a diet that contains a wide variety of plant species. Buechner[34] found 228 plant species utilized in the Trans-Pecos region of Texas, and Hoover et al.[12] found 89 plants eaten by pronghorn in Colorado.

Of all plants utilized, big sagebrush (*Artemisia tridentata*) and silver sagebrush (*A. cana*) are the most important when available. In areas where sagebrush does not occur, other shrub species are utilized: snowberry (*Symphoricarpos* sp.) in Montana[20,23] and snakeweed (*Gutierrezia* sp.) in Colorado[12] and Wyoming.[8,32]

Forbs also comprise a large portion of the annual diet. Species composition and utilization by pronghorn vary from range to range and by season. Greatest usage appears to be in the summer and fall. Forbs appear to be of critical importance for antelope survival on the more southerly ranges.[9]

Grasses are not consumed in large quantities on an annual basis (Table 2). The highest usage of grass by pronghorn is during early spring or at the time of the year of rapid growth and production. Species utilization varies with area and availability.

Table 1
COMPOSITION ANALYSIS OF PELLETED BASAL RATION[a]

Ingredient	Percent (wt)
Barley, pulverized	10.0
Corn, pulverized	30.0
Milo, pulverized	5.0
Oats, pulverized	7.5
Wheat middlings	6.5
Beet pulp, shredded	2.5
Brewer's grain	35.0
Dicalcium phosphate	1.0
Molasses, cane	2.5
Vitamins A, D, E, premix	Trace (0.002)
Trace mineral package (Mg, Zn, I, Fe, Co, Cu, Ca)	Trace (0.005)

Analysis (dry matter basis)	Percent	
Dry matter	90.59	
Crude protein	19.34	= 100% (Crude protein through Acid-detergent fiber)
Ether extract	2.83	
Ash	7.86	
Cell-wall constituents	52.81	
Acid-detergent fiber	13.58	
Lignin	3.58	
Apparent dry matter digestibility	70.67	
Gross energy (kcal/g)	4.60	
Apparent digestible energy (kcal/g)	3.45	
Apparent metabolizable energy (kcal/g)	3.05	

[a] Data were presented by Carpenter and Baker.[2]

Table 2
PERCENTAGE OF VARIOUS TYPES OF FORAGE CONSUMED BY PRONGHORN THROUGHOUT THE CURRENT RANGE

State	Method	Period sampled	Sample size	Forage type: Browse	Forbs	Grasses	Other	Ref.
Alberta, Canada	Rumen samples	Annual	112	35	51	13	–	10
		Fall, winter	50	60	17	23	–	
Colorado	–	–	–	84	10	6	–	11
	Rumen samples	Fall	30	72	22	1	5 (cacti)	12
		Winter	74	54	26	5	15 (cacti)	
		Spring	15	25	51	20	4 (cacti)	
		Summer	73	22	66	1	10 (cacti)	
		Annual	192	43	42	4	11 (cacti)	
	Tame animals	Winter	20	47	23	30	–	13
		Spring	20	30	19	49	–	
		Summer	12	15	8	77	–	
	Rumen samples	October	25	73	24	1	–	14
	Tame animals	Annual	152	2–5	43–54	44–52	–	15
	Rumen samples	Annual	262	47	20	7	19 (winter wheat)	16
Idaho	Rumen samples	Winter	6	85	10	5	–	17
Kansas	Feeding minutes	Annual	–	18	20	22	40 (cacti)	18
Montana	Rumen samples	Winter	18	78	19	1	–	19
	Rumen samples	Fall	47	86	7	7	–	20
	Rumen samples	Annual	–	63	34	3	–	21
	Rumen samples	Annual	–	64	20	16	–	22
	Rumen samples	Sept.–Dec.	–	87	6	7	–	23
Nevada	Rumen samples	Summer, fall	11	49	49	2	–	24
New Mexico	Rumen samples	Annual	111	35	57	3	–	25
Oregon	Rumen samples	Annual	26	90	6	2	–	26
	Rumen samples	Annual	189	69	21	7	–	27
South Dakota	Rumen samples	Annual	60	66	23	11	–	28
Texas	Rumen samples	Annual	45	35	58	7	–	29
Utah	Feeding site observation	Aug.–Sept.	31	58	42	–	–	30
Wyoming		Summer, fall	30	86	10	4	–	31
	Rumen samples	Annual	235	57–90	2–74	1–32	–	32
	Rumen samples	Annual	24	81	15	4	–	33

REFERENCES

1. **Schwartz, C. C., Nagy, J. G., and Kerr, S. M.,** *J. Wildl. Manage.*, 40(3), in press, 1976.
2. **Carpenter, L. H. and Baker, D. L.,** Middle Park Cooperative Deer Study-Deer Habitat Evaluation, P-R Rep. W-38-R-29, Colorado Game and Fish Dept., Denver, 1976, 241–263.
3. **Humphrey, R. R.,** *Range Ecology,* Ronald Press, New York, 1962.
4. **Yoakum, J.,** Literature of the American Pronghorn Antelope, unnumbered bulletin, U.S. Dept. of the Interior, Washington, D.C., 1967.
5. **Yoakum, J.,** Literature of the American Pronghorn Antelope, Addendum No. 1, U.S. Dept. of the Interior, Washington, D.C., 1968.
6. **Yoakum, J.,** Literature of the American Pronghorn Antelope, Addendum No. 2, U.S. Dept. of the Interior, Washington, D.C., 1970.
7. **Yoakum, J.,** Literature of the American Pronghorn Antelope, Addendum No. 3, U.S. Dept. of the Interior, Washington, D.C., 1974.
8. **Severson, K. E., May, M., and Hepworth, W. G.,** *Food Preferences, Carrying Capabilities and Forage Competition Between Antelope and Domestic Sheep in Wyoming's Red Desert,* Sci. Monogr. 10, Agricultural Experiment Station, University of Wyoming, Laramie, 1968.
9. **Sundstrom, C., Hepworth, W. G., and Diem, K. L.,** Abundance, Distribution and Food Habits of the Pronghorn, Bull. No. 12, Wyoming Fish and Game Commission, Cheyenne, 1973.
10. **Mitchell, G. J. and Smoliak, S.,** *J. Wildl. Manage.*, 35(2), 238–250, 1971.
11. **Einarsen, A. S.,** *The Pronghorn Antelope and Its Management,* Wildlife Management Institute, Washington, D.C., 1948.
12. **Hoover, R. L., Till, C. E., and Ogilvie, S.,** The Antelope of Colorado, Tech. Bull. No. 4, Colorado Fish and Game Dept., Denver, 1959.
13. **Hoover, J. P.,** Food Habits of Pronghorn Antelope on Pawnee National Grasslands, 1970, M.S. thesis, Colorado State University, Fort Collins, 1971.
14. **Scarvie, O. and Arney, J. A.,** *Food Habits of Pronghorn Antelope, Antilocapra americana, in October in Northern Colorado,* Colorado State University Press, Fort Collins, 1957.
15. **Schwartz, C. C. and Nagy, J. G.,** *J. Wildl. Manage.*, 40(3), in press, 1976.
16. **Tileston, J. V. and Yeager, L. E.,** A Resume of Colorado Big Game Research Projects, 1939–1957, Tech. Bull. No. 9, Colorado Fish and Game Dept., Denver, 1962.
17. **Fichter, E. and Nelson, A. E.,** Study of Pronghorn Population, P-R Proj. Rep. W-85-R-136), Idaho Dept. of Fish and Game, Boise, 1962.
18. **Hlavachick, B. D.,** *Antelope States Workshop Proc.*, 2, 60–65, 1966.
19. **Bayless, S. R.,** *J. Wildl. Manage.*, 33(3), 538–551, 1969.
20. **Buck, P. D.,** The Biology of the Antelope (*Antilocapra americana*) in Montana, M.S. thesis, Montana State University, Bozeman, 1947.
21. **Cole, G. F.,** The Pronghorn Antelope – Its Range Use and Food Habits in Central Montana with Special Reference to Alfalfa, Tech. Bull. 516, Agricultural Experiment Station, Montana State University, Bozeman, 1956.
22. **Cole, G. F. and Wilkins, B. T.,** The Pronghorn Antelope, Its Range Use and Food Habits in Central Montana with Special Reference to Wheat, Tech. Bull. 2, Montana Fish and Game Dept., Bozeman, 1958.
23. **Covey, F.,** *J. Wildl. Manage.*, 10(4), 367, 1946.
24. **Tsukamoto, G. and Diebert, W.,** *Trans. Interstate Antelope Conf.*, 19, 7–13, 1969.
25. **Russell, P.,** Antelope of New Mexico, Bull. 12, New Mexico Dept. of Game and Fish, Santa Fe, 1964.
26. **Mason, E.,** *J. Wildl. Manage.*, 16(3), 387–389, 1952.
27. **Yoakum, J.,** *Trans. Interstate Antelope Conf.*, 9, 47–59, 1958.
28. **Terwilliger, C.,** Food Habits of Antelope, Game Management Prob., Colorado State University, Fort Collins, 1946.
29. **Hailey, T.,** Trans-Pecos Game Management Survey, P-R Proj. Rep. W-57-R-14, Texas Parks and Wildlife Dept., Austin, 1967.
30. **Beale, D. M. and Smith, A. D.,** *J. Wildl. Manage.*, 34(3), 570–582, 1970.
31. **Baker, T.,** *Wyoming Wildl.*, 17(11), 24–26, 1953.
32. **Creek, D.,** A Summary of Pronghorn Antelope Investigations, Wyoming Game and Fish Commission, Cheyenne, 1967.
33. **Taylor, E.,** Methods of Forage Sampling to Determine Food Habits, Forage Requirements and Carrying Capacity for Pronghorn Antelope, P-R Proj. Rep. FW-3-R-18, Wyoming Game and Fish Commission, Cheyenne, 1972.
34. **Buechner, H. K.,** *Am. Midl. Nat.*, 43(2), 257–354, 1950.

DIETS (NATURAL AND SYNTHETIC): BEEF CATTLE

J. K. Matsushima

INTRODUCTION

Formulation of diets for beef cattle is complicated, especially when meeting the specified nutrient requirements is desired. The following conditions or situations exemplify the difficulties which are inherent in diet formulation for beef cattle:

1. Variation in nutrient composition of feeds
2. Different substitutes such as by-products being used to replace normal ingredients
3. Variability in nutritional status or fleshing of the animal when it is acquired for feeding or other productive purpose
4. Variation in environmental conditions such as weather, shelter, etc., from one area or farm to another
5. Difference in age and functional purpose (such as breeding vs. finishing) of the animal, necessitating different nutritional regimes

Since there are a multitude of diets that can be fed to beef cattle, the following guidelines or suggestions are listed for consideration:

1. The daily feed consumption varies from 2 to 3% of the animal's weight. For example, a 600-lb animal will consume around 12 to 18 lb of feed (dry feed equivalent) daily.
2. As animals get fatter and as the diet is enriched in caloric content, the daily feed consumption decreases per unit body weight.
3. The protein content of the diet decreases from 13 to 8.5%, on dry matter basis, as the animal increases in age and body weight. However, the protein level is dependent upon the desired production level. For example, a young calf weighing 500 lb being fed for a maximum gain should be fed a higher protein diet than an animal of similar weight being fed for maintenance or minimum gain.
4. When animals are fed an abundunt quantity of legume roughage such as alfalfa hay a protein supplement is not necessary in most cases.
5. When protein supplementation is required, a nonprotein nitrogen source such as urea may be partially or completely substituted for a natural protein supplement. A general guideline in the use of urea is as follows: (a) Young cattle, weighing between 400 to 500 lb, on a high roughage diet with a minimum of 3 lb of grain in the daily diet, should not be fed more than 45 g (0.1 lb) urea. If no grain is being fed, a natural protein supplement is recommended. (b) For young cattle on high energy diets, the quantity of urea should not exceed 90 g (0.2 lb) per head daily. Higher quantities of urea will depress feed intake. (c) For cattle under range or dry pasture conditions where the feed supply is mainly poor quality roughage and particularly when intake of supplement cannot be carefully controlled, a natural protein supplement is recommended over a protein supplement which contains urea. However, if overconsumption of urea can be controlled, protein supplementation should be limited so that the daily intake of urea does not exceed 90 g (0.2 lb).

The National Research Council (NRC) recommendations are most commonly used as guidelines for minimum and maximum nutrient requirements for beef cattle. The tables are very complete and quite lengthy. In order to illustrate the use of the NRC tables for formulating diets, a few representative requirements have been selected.*

EXAMPLE DIETS

Eight example diets are shown in Tables 2 through 9. These diets may not include the feeds that are typical for a farm or ranch in a given area. Feeds that are commonly available in a given area could be substituted for those feeds shown in a given diet.

Since animals do not consume the same amount of feed from one day to the next, the example diets are shown on a percentage basis as well as the quantity of each ingredient to be fed on a daily basis. The example diet presented in Table 2 shows that feeding 4.22 kg of meadow hay, 1.95 kg of ground ear corn, and 0.21 kg soybean meal will furnish 5.8 kg of dry matter. This is the quantity of dry matter which the NRC recommends. In the event that animals cannot consume this quantity of dry matter, the diet would be reduced so that each of the three ingredients would be decreased in the same proportion, on dry matter basis, as noted in Column 4 (percentage of mixture) of Table 2. However, if one prefers to decrease the three ingredients on a natural basis (sometimes referred to as wet basis), the procedure on a percentage basis would involve dividing the total weight of the diet, 6.38 kg, into each of the weights of the ingredients multiplied by 100 to give the following: 66.1% meadow hay, 30.6% ground ear corn, and 3.3% soybean meal.

NET ENERGY CALCULATIONS

After the protein, mineral, and vitamin requirements are adequately met, the increase in animal gains is usually dependent upon the quantity of energy supplied in the ration. The energy requirements for maintenance are met first, and any additional energy which is left over from the diet will be used for production or animal gain. In certain situations when excess protein is included in the diet, the surplus protein could be utilized for energy. However, this would be an expensive way of supplying energy and therefore should be discouraged.

In calculating the net energy needs for an animal, the following steps are taken (example diet from Table 2):

1. Calculate the kilocalories per gram of diet:

	NEm		NEg	
0.68 kg meadow hay	(1.0)**	0.680 kcal	(0.11)**	0.075 kcal
0.29 kg ground ear corn	(2.23)	0.647 kcal	(1.39)	0.403 kcal
0.03 kg soybean meal	(2.06)	0.062 kcal	(1.37)	0.041 kcal
1.00 kg Diet		1.389 kcal		0.519 kcal

2. From Table 1, the net energy requirement for maintenance (NEm) for this particular group is noted as 4.1 Mcal.

3. Calculate the quantity of this diet that is needed for maintenance:

$$\frac{4.1}{1.389} = 2.95 \text{ kg diet needed for maintenance}$$

4. Calculate the quantity of megacalories supplied by each ingredient in the diet for maintenance, NEm:

* For more complete requirements, see Reference 3.

** Figures taken from Table 10.

$\frac{0.68}{1.389}$ = 0.49 × 4.1 (Mcal required for maintenance) = 2.01 Mcal from 0.68 kg of meadow hay

$\frac{0.647}{1.389}$ = 0.466 × 4.1 = 1.91 Mcal from 0.29 kg ground ear corn

$\frac{0.062}{1.389}$ = 0.045 × 4.1 = 0.18 Mcal from 0.03 kg soybean meal

5. Calculate the quantity of megacalories supplied by each ingredient in the diet for gain or production, NEg:

$\frac{0.075}{0.519}$ = 0.145 × 1.49 (Mcal required for 0.5 kg daily gain) = 0.22 Mcal from 0.68 kg meadow hay

$\frac{0.403}{0.519}$ = 0.776 × 1.49 = 1.16 Mcal from 0.29 kg ground ear corn

$\frac{0.062}{0.519}$ = 0.079 × 1.49 = 0.12 Mcal from 0.03 kg soybean meal

SYNTHETIC DIETS FOR BEEF CATTLE

For optimum or maximum beef production it is not customary to use synthetic diets. However, in situations when certain ingredients are scarce or financially prohibitive a synthetic substitute may be used. The purpose of including these synthetic substitutes would be to supply something of physical nature to stimulate the ruminal wall. Materials such as oyster shells (hen size), polyethylene tabs (commercially sold as Ruff-tabs®), and vermiculite have been used with minimum success. These synthetic or uncommon roughage substitutes are used to replace part or all of the roughage in the diets. When used in high concentrate rations the levels or rates vary from 5 to 10%, by weight, in the diet.

For certain basic research, some purified diets have been fed to cattle with a certain degree of success. Table 11 illustrates a purified diet that was prepared for sheep but that can be fed to beef cattle.

Table 1
NUTRIENT REQUIREMENTS OF BEEF CATTLE (DAILY NUTRIENTS PER ANIMAL)[a]

Body weight[b] (kg)	Average daily gain (kg)	Daily dry matter per animal (kg)	Total protein (kg)	Energy: TDN[c] (kg)	Energy: NEm[d] (Mcal)	Energy: NEg[e] (Mcal)	Ca (g)	P (g)	Vitamin A (IU, thousands)
Growing-finishing Steer Calves and Yearlings									
200	0.5	5.8	0.57	3.4	4.10	1.49	14	13	12
300	1.1	7.6	0.82	5.6	5.5	4.78	25	22	16
400	1.3	8.6	0.90	7.3	6.89	7.16	25	22	19
500	1.2	9.6	0.96	8.2	8.14	7.73	21	21	23
Growing-finishing Heifer Calves and Yearlings									
200	0.5	6.0	0.58	3.5	4.10	1.66	14	13	13
300	1.1	7.5	0.78	6.0	5.55	5.55	23	20	16
350	1.2[f]	8.1	0.81	6.9	6.23	6.91	21	20	18
450	1.0[f]	8.5	0.83	7.4	7.52	6.71	19	19	20
Cows Nursing Calves – Average Milking Ability – First 3 to 4 Months Post Partum									
450		9.3	0.86	5.0	10.5		26	26	23
500		9.8	0.90	5.3	11.1		27	27	24
550		10.5	0.97	5.6	11.9		28	28	26
Cows Nursing Calves – Superior Milking Ability – First 3 to 4 Months Post Partum									
450		11.3	1.23	6.4	13.6		45	42	36
500		11.8	1.29	6.7	14.2		46	43	38
550		12.4	1.35	7.0	14.9		46	44	41
Dry Pregnant Mature Cows – Last 2 to 3 Months of Pregnancy									
450	0.4	8.1	0.48	4.2	9.1		15	15	23
500	0.4	8.6	0.51	4.5	9.7		15	15	24
550	0.4	9.1	0.54	4.8	10.3		16	16	26

[a] Values in this table were taken from Reference 3.
[b] Average weight for a feeding period.
[c] Total digestible nutrient.
[d] Net energy for maintenance.
[e] Net energy for gain.
[f] Most heifers of the weight indicated will not sustain the energy intake necessary to maintain this rate of gain for an extended period.

Table 1 (continued)
NUTRIENT REQUIREMENTS OF BEEF CATTLE (DAILY NUTRIENTS PER ANIMAL)[a]

Body weight[b] (kg)	Average daily gain (kg)	Daily dry matter per animal (kg)	Total protein (kg)	Energy TDN[c] (kg)	Energy NEm[d] (Mcal)	Energy NEg[e] (Mcal)	Ca (g)	P (g)	Vitamin A (IU, thousands)
Dry Pregnant Mature Cows – Middle Third of Pregnancy									
400		6.1	0.36	3.3	6.89		11	11	17
450		6.7	0.39	3.6	7.52		12	12	19
500		7.2	0.42	3.9	8.14		13	13	20
Pregnant Yearling Heifers – Last 3 to 4 Months of Pregnancy									
325	0.4[g]	6.6	0.58	3.5	5.89	0.62	15	15	19
375	0.6[g]	9.3	0.81	4.9	6.56	1.68	19	19	26
425	0.6[g]	10.1	0.88	5.3	7.21	1.84	19	19	28
Bulls, Growth and Maintenance (Moderate Activity)									
400	0.9	11.0	1.03	7.0	6.9	4.1	23	23	43
500	0.7	12.2	1.08	7.5	8.5	3.7	22	22	48
700	0.3	12.9	1.09	7.7	11.0	2.0	23	23	50
900	0.0	11.4	0.99	6.3	13.3	0.0	21	21	44

[g] Approximately 0.4 ± 0.1 kg of weight gain per day over the last third of pregnancy is accounted for by the products of conception.

Table 2
EXAMPLE DIET FOR 200-kg YOUNG GROWING STEER CALVES TO GAIN 0.5 kg/DAY

Assuming that a High Roughage Diet is to be Fed in Dry Lot

	Daily ration			Daily nutrients supplied[a] (percentage figures are on dry matter basis)							
				Total protein		TDN[b]					
Ingredient	Quantity per day, natural basis (kg)	Dry matter equivalent (kg)	Percentage of mixture	kg	%	kg	%	NEm[c] (kcal)	NEg[d] (kcal)	Calcium (g)	Phosphorus (g)
Meadow hay	4.22	3.95	68.0	0.33	5.65	2.37	40.9	2.01	0.22	22.9	6.3
Ground ear corn	1.95	1.66	29.0	0.15	2.66	1.41	24.3	1.91	1.16	0.7	4.3
Soybean meal	0.21	0.19	3.0	0.09	1.54	0.16	2.8	0.18	0.12	0.6	1.2
Total[e]	6.38	5.80	100.0	0.57	9.85	3.94	68.0	4.10	1.50	24.2	**11.8**
National Research Council recommendation[f]		5.8		0.57		3.4		4.1	1.49	14.0	13.0

[a] The kilograms of protein and TDN and the grams of calcium and phosphorus shown under "daily nutrients supplied" were obtained by multiplying the figure shown in the "dry matter equivalent" column by the nutrient composition value for that feed as noted in Table 10. The net energy requirements for maintenance and gain (where needed) were computed according to the procedures discussed under the section in the text entitled "Net Energy Calculations."

[b] Total digestible nutrient.

[c] Net energy for maintenance.

[d] Net energy for gain.

[e] The values in this line that are in boldface type indicate a deficiency, as compared to the NRC recommendation, and therefore a supplement is needed to correct the deficiency.

[f] The NRC recommendations for various weights of animals and specific functions are listed in Table 1.

Table 3
EXAMPLE RATION FOR FINISHING YEARLING HEIFERS WEIGHING 350 kg TO GAIN 1.2 kg DAILY

Assuming that the Diet is Formulated at 75% Concentrate on Dry Matter Basis

	Daily ration			Daily nutrients supplied[a] (percentage figures are on dry matter basis)							
				Total protein		TDN[b]					
Ingredient	Quantity per day, natural basis (kg)	Dry matter equivalent (kg)	Percentage of mixture	kg	%	kg	%	NEm[c] (kcal)	NEg[d] (kcal)	Calcium (g)	Phosphorus (g)
Cracked corn	6.85	6.08	75.0	0.614	7.58	5.52	68.18	5.10	5.67	1.22	17.60
Corn silage	6.19	1.72	21.2	0.137	1.69	1.21	14.99	0.98	1.07	4.82	4.13
Alfalfa hay	0.33	0.31	3.8	0.059	0.73	0.19	2.31	0.15	0.15	5.15	0.78
Total[e]	13.37	8.11	100.0	0.81	10.0	6.92	85.48	6.23	6.89	**11:20**	22.50
National Research Council recommendation[f]		8.1		0.81		6.9		6.23	6.91	21.0	20.0

[a] The kilograms of protein and TDN and the grams of calcium and phosphorus shown under "daily nutrients supplied" were obtained by multiplying the figure shown in the "dry matter equivalent" column by the nutrient composition value for that feed as noted in Table 10. The net energy requirements for maintenance and gain (where needed) were computed according to the procedures discussed under the section in the text entitled "Net Energy Calculations."

[b] Total digestible nutrient.

[c] Net energy for maintenance.

[d] Net energy for gain.

[e] The values in this line that are in boldface type indicate a deficiency, as compared to the NRC recommendation, and therefore a supplement is needed to correct the deficiency.

[f] The NRC recommendations for various weights of animals and specific functions are listed in Table 1.

Table 4
EXAMPLE DIET FOR 500-kg COWS NURSING CALVES WITH AVERAGE MILKING ABILITY

Assuming that a High Roughage Diet is Fed in Dry Lot

Daily ration				Daily nutrients supplied[a] (percentage figures are on dry matter basis)							
Ingredient	Quantity per day, natural basis (kg)	Dry matter equivalent (kg)	Percentage of mixture	Total protein kg	Total protein %	TDN[b] kg	TDN[b] %	NEm[c] (kcal)	NEg[d] (kcal)	Calcium (g)	Phosphorus (g)
Meadow hay	9.63	9.0	90.0	0.75	7.5	5.4	54.0	9.0	–	52.2	14.4
Ground ear corn	0.98	0.84	8.4	0.08	0.8	0.7	7.0	1.87	–	0.3	2.2
Cottonseed meal	0.18	0.16	1.6	0.07	0.7	0.1	1.0	0.27	–	0.3	1.9
Total[e]	10.79	10.0	100.0	0.9	9.0	6.2	62.0	11.14	–	52.8	**18.5**
National Research Council recommendation[f]		9.8		0.9			5.3		11.1	27.0	27.0

[a] The kilograms of protein and TDN and the grams of calcium and phosphorus shown under "daily nutrients supplied" were obtained by multiplying the figure shown in the "dry matter equivalent" column by the nutrient composition value for that feed as noted in Table 10. The net energy requirements for maintenance and gain (where needed) were computed according to the procedures discussed under the section in the text entitled "Net Energy Calculations."

[b] Total digestible nutrient.

[c] Net energy for maintenance.

[d] Net energy for gain.

[e] The values in this line that are in boldface type indicate a deficiency, as compared to the NRC recommendation, and therefore a supplement is needed to correct the deficiency.

[f] The NRC recommendations for various weights of animals and specific functions are listed in Table 1.

Table 5
EXAMPLE DIET FOR 500-kg COWS NURSING CALVES WITH SUPERIOR MILKING ABILITY

Assuming that a High Roughage Diet is Fed in Dry Lot

	Daily ration			Daily nutrients supplied[a] (percentage figures are on dry matter basis)							
				Total protein		TDN[b]					
Ingredient	Quantity per day, natural basis (kg)	Dry matter equivalent (kg)	Percentage of mixture	kg	%	kg	%	NEm[c] (kcal)	NEg[d] (kcal)	Calcium (g)	Phosphorus (g)
Meadow hay	10.10	9.44	80.0	0.78	6.6	5.7	48.0	9.44	–	54.8	15.1
Ground ear corn	1.85	1.58	13.4	0.15	1.2	1.3	11.4	3.52	–	0.6	4.1
Cottonseed meal	0.86	0.78	6.6	0.35	3.0	0.6	5.0	1.32	–	1.4	9.0
Total[e]	12.81	11.8	100.0	1.28	10.8	7.6	64.4	14.28	–	56.8	**28.2**
National Research Council recommendation[f]		11.8		1.29		6.7		14.2		46.0	43.0

[a] The kilograms of protein and TDN and the grams of calcium and phosphorus shown under "daily nutrients supplied" were obtained by multiplying the figure shown in the "dry matter equivalent" column by the nutrient composition value for that feed as noted in Table 10. The net energy requirements for maintenance and gain (where needed) were computed according to the procedures discussed under the section in the text entitled "Net Energy Calculations."

[b] Total digestible nutrient.

[c] Net energy for maintenance.

[d] Net energy for gain.

[e] The values in this line that are in boldface type indicate a deficiency, as compared to the NRC recommendation, and therefore a supplement is needed to correct the deficiency.

[f] The NRC recommendations for various weights of animals and specific functions are listed in Table 1.

Table 6
EXAMPLE DIET FOR DRY PREGNANT MATURE COWS WEIGHING 500 kg DURING LAST 2 TO 3 MONTHS OF PREGNANCY

Assuming that a High Roughage Diet is Fed in Dry Lot

	Daily ration			Daily nutrients supplied[a] (percentage figures are on dry matter basis)								
Ingredient	Quantity per day, natural basis (kg)	Dry matter equivalent (kg)	Percentage of mixture	Total protein kg	Total protein %	TDN[b] kg	TDN[b] %	NEm[c] (kcal)	NEg[d] (kcal)	Calcium (g)	Phosphorus (g)
Meadow hay	8.24	7.7	89.5	0.64	7.43	4.62	53.7	7.7	–	44.6	12.3
Ground ear corn	1.05	0.9	10.5	0.08	0.98	0.76	8.93	2.0	–	0.4	2.3
Total[e]	9.29	8.6	100.0	0.72	8.41	5.38	62.63	9.7	–	45.0	**14.6**
National Research Council recommendation[f]		8.6		0.51		4.5		9.7		15.0	15.0

[a] The kilograms of protein and TDN and the grams of calcium and phosphorus shown under "daily nutrients supplied" were obtained by multiplying the figure shown in the "dry matter equivalent" column by the nutrient composition value for that feed as noted in Table 10. The net energy requirements for maintenance and gain (where needed) were computed according to the procedures discussed under the section in the text entitled "Net Energy Calculations."

[b] Total digestible nutrient.

[c] Net energy for maintenance.

[d] Net energy for gain.

[e] The values in this line that are in boldface type indicate a deficiency, as compared to the NRC recommendation, and therefore a supplement is needed to correct the deficiency.

[f] The NRC recommendations for various weights of animals and specific functions are listed in Table 1.

Table 7
EXAMPLE DIET FOR DRY PREGNANT MATURE COWS WEIGHING 450 kg DURING THE MIDDLE THIRD OF PREGNANCY

Assuming that a High Roughage Diet is Fed in Dry Lot

	Daily ration			Daily nutrients supplied[a] (percentage figures are on dry matter basis)								
				Total protein		TDN[b]						
Ingredient	Quantity per day, natural basis (kg)	Dry matter equivalent (kg)	Percentage of mixture	kg	%	kg	%	NEm[c] (kcal)	NEg[d] (kcal)	Calcium (g)	Phosphorus (g)
Meadow hay	6.10	5.70	85.1	0.47	7.06	3.42	51.06	5.70	–	33.06	9.12
Ground ear corn	1.17	1.00	14.9	0.09	1.39	0.85	12.67	2.23	–	0.40	2.60
Total[e]	7.27	6.70	100.0	0.56	8.45	4.27	63.73	7.93	–	33.46	**11.72**
National Research Council recommendation[f]		6.7		0.39		3.6		7.52		12.0	12.0

[a] The kilograms of protein and TDN and the grams of calcium and phosphorus shown under "daily nutrients supplied" were obtained by multiplying the figure shown in the "dry matter equivalent" column by the nutrient composition value for that feed as noted in Table 10. The net energy requirements for maintenance and gain (where needed) were computed according to the procedures discussed under the section in the text entitled "Net Energy Calculations."
[b] Total digestible nutrient.
[c] Net energy for maintenance.
[d] Net energy for gain.
[e] The values in this line that are in boldface type indicate a deficiency, as compared to the NRC recommendation, and therefore a supplement is needed to correct the deficiency.
[f] The NRC recommendations for various weights of animals and specific functions are listed in Table 1.

Table 8
EXAMPLE DIET FOR PREGNANT YEARLING HEIFERS WEIGHING 375 kg DURING THE LAST 3 TO 4 MONTHS OF PREGNANCY

Assuming that a High Roughage Diet is Fed in Dry Lot

	Daily ration			Daily nutrients supplied[a] (percentage figures are on dry matter basis)								
Ingredient	Quantity per day, natural basis (kg)	Dry matter equivalent (kg)	Percentage of mixture	Total protein kg	Total protein %	TDN[b] kg	TDN[b] %	NEm[c] (kcal)	NEg[d] (kcal)	Calcium (g)	Phosphorus (g)
Corn silage	19.4	5.4	72.0	0.43	5.76	3.82	50.9	4.94	2.20	15.1	13.0
Alfalfa hay	2.3	2.1	28.0	0.40	5.32	1.28	17.1	1.62	0.60	34.9	5.2
Total[e]	21.7	7.5	100.0	0.83	11.08	5.10	68.0	6.56	2.80	50.0	**18.2**
National Research Council recommendation[f]		9.3		0.81		4.9		6.56	1.68	19	19

[a] The kilograms of protein and TDN and the grams of calcium and phosphorus shown under "daily nutrients supplied" were obtained by multiplying the figure shown in the "dry matter equivalent" column by the nutrient composition value for that feed as noted in Table 10. The net energy requirements for maintenance and gain (where needed) were computed according to the procedures discussed under the section in the text entitled "Net Energy Calculations."
[b] Total digestible nutrient.
[c] Net energy for maintenance.
[d] Net energy for gain.
[e] The values in this line that are in boldface type indicate a deficiency, as compared to the NRC recommendation, and therefore a supplement is needed to correct the deficiency.
[f] The NRC recommendations for various weights of animals and specific functions are listed in Table 1.

Table 9
EXAMPLE DIET FOR 500 kg BULLS, MODERATE ACTIVITY, 0.4 kg DAILY GAIN

Assuming that a High Roughage Diet is Fed in Dry Lot

Daily ration				Daily nutrients supplied[a] (percentage figures are on dry matter basis)							
Ingredient	Quantity per day, natural basis (kg)	Dry matter equivalent (kg)	Percentage of mixture	Total protein		TDN[b]		NEm[c] (kcal)	NEg[d] (kcal)	Calcium (g)	Phosphorus (g)
				kg	%	kg	%				
Meadow hay	10.44	9.76	80.0	0.81	6.64	5.86	48.0	5.46	0.47	56.6	15.6
Ground ear corn	2.86	2.44	20.0	0.23	1.89	2.07	17.0	3.04	15.0	1.0	6.4
Total[e]	13.30	12.0	100.0	**1.04**	8.53	7.93	65.0	8.50	15.47	57.6	**22.0**
National Research Council recommendation[f]		12.2		1.08		7.5		8.5	1.97	22.0	22.0

[a] The kilograms of protein and TDN and the grams of calcium and phosphorus shown under "daily nutrients supplied" were obtained by multiplying the figure shown in the "dry matter equivalent" column by the nutrient composition value for that feed as noted in Table 10. The net energy requirements for maintenance and gain (where needed) were computed according to the procedures discussed under the section in the text entitled "Net Energy Calculations."

[b] Total digestible nutrient.

[c] Net energy for maintenance.

[d] Net energy for gain.

[e] The values in this line that are in boldface type indicate a deficiency, as compared to the NRC recommendation, and therefore a supplement is needed to correct the deficiency.

[f] The NRC recommendations for various weights of animals and specific functions are listed in Table 1.

Table 10
FEED COMPOSITION TABLE[a]

Reference no.	Name of feed	Dry matter (%)	Protein (%)	TDN[b] (%)	NEm[c] (Mcal/kg)	NEg[d] (Mcal/kg)	Calcium (%)	Phosphorus (%)
1-00-023	Alfalfa hay	92.7	19.0	60.8	1.31	0.69	1.66	0.25
4-00-549	Barley, grain	89.0	13.0	80.8	2.13	1.40	0.08	0.45
4-00-669	Beat pulp, dried	90.6	9.6	71.9	1.60	1.03	0.75	0.10
5-01-168	Cattle, milk, fresh	12.6	27.8	128.3	4.59	2.01	0.93	0.75
5-01-170	Cattle, skim milk	9.3	35.1	91.0	2.32	1.50	1.37	1.05
1-01-415	Clover, red, hay	78.0	14.7	58.5	1.26	0.62	1.45	0.23
3-02-820	Corn silage	27.8	8.0	70.7	1.56	0.99	0.28	0.24
4-02-849	Corn, ears, ground	85.4	9.3	85.0	2.23	1.39	0.04	0.26
4-02-931	Corn grain, Number 2	88.7	10.1	90.9	2.28	1.48	0.02	0.29
5-01-621	Cottonseed meal	91.1	46.0	75.2	1.69	1.11	0.18	1.16
5-02-045	Flax-linseed meal	91.1	39.4	80.7	1.90	1.27	0.43	0.95
1-03-181	Native meadow hay	93.5	8.3	60.0	1.00	0.11	0.58	0.16
1-03-187	Native prairie hay	91.4	4.9	53.7	1.15	0.44	0.38	0.12
4-03-309	Oats, grain	88.9	13.2	74.5	1.73	1.14	0.10	0.37
4-04-047	Rye, grain	88.2	12.8	80.9	2.04	1.36	0.08	0.39
4-04-383	Sorghum (milo), grain	88.5	10.0	80.6	1.96	1.31	0.03	0.32
3-04-468	Sorghum silage	28.2	6.8	57.0	1.25	0.61	0.33	0.20
1-04-372	Sorghum fodder, dry	90.2	6.9	57.3	(1.25)	(0.61)	(0.33)	(0.20)
5-04-600	Soybean meal, exp.	90.7	46.7	84.9	2.06	1.37	0.31	0.65
5-04-604	Soybean meal, solv.	89.1	52.4	82.4	1.93	1.29	0.33	0.73
5-04-610	Soybean seeds	90.9	41.7	91.4	2.41	1.53	0.27	0.63
4-04-696	Sugarcane, molasses	77.0	5.9	96.0	2.27	1.48	1.05	0.11
4-04-695	Sugarcane, dried molasses	93.5	10.7	85.3	(2.27)	(1.48)	1.23	0.15
1-04-885	Timothy hay, mature	88.3	7.7	58.7	1.24	0.59	0.20	0.17

[a] The values shown are on dry matter basis. Values in parentheses were taken from similar feeds that gave values for these nutrients.
[b] Total digestible nutrient.
[c] Net energy for maintenance.
[d] Net energy for gain.

From National Research Council, *Atlas of Nutritional Data on United States and Canadian Feeds,* National Academy of Sciences, Washington, D.C., 1971. With permission.

Table 11
EXAMPLE OF PURIFIED DIETS
(GRAMS/KILOGRAM DRY MATTER)

Diet	Cellulose	Starch	Sucrose	Urea	Minerals	Nitrogen
High cellulose-low starch	514	124	247	75	40	35.0
Low cellulose-high starch	315	471	117	56	40	26.1

REFERENCES

1. **Beever, D. E., Harrison, D. G., Thomson, D. J., Cammell, S. B., and Osbourn, D. F.,** A method for the estimation of dietary and microbial protein in duodenal digesta of ruminants, *Br. J. Nutr.*, 32, 99–112, 1974.
2. National Research Council, *Atlas of Nutritional Data on United States and Canadian Feeds,* National Academy of Sciences, Washington, D.C., 1971.
3. National Research Council, *Nutrient Requirements of Beef Cattle,* 5th rev. ed., National Academy of Sciences, Washington, D.C., 1976.

DIETS FOR ZEBU CATTLE

G. S. Sidhu and J. S. Ichhponani

INTRODUCTION

Zebu cattle, grouped under the species *Bos indicus,* are humped animals common to tropical and subtropical regions. In the United States they are also known as the Brahman group of cattle. Zebu cattle are wholly domestic beasts, no wild ancestors having been found since the start of historic time. It has been variously estimated that cattle of this type appeared in domesticated form some time between 4000 and 2100 B.C. The zebu is characterized by a hump of fleshy tissue over the withers, a large dewlap, large drooping ears, and a voice that sounds more like a grunt than a low.[1]

Zebu cattle are classified[2] as thoracic (chest) humped or cervicothoracic (neck) humped cattle according to the location of their hump. The former are usually referred to as "zebu," although not all are zebu proper; some, such as the long-horned Fulani cattle of West Africa, are humpless. At present, cervicothoracic humped cattle occur only in, or are derived from, contact areas of thoracic humped zebu and humpless cattle. Examples of such cattle are "Sanga" cattle of East and South Africa, the types derived from Indian zebu and humpless Tibetan cattle of Bhutan, the types derived from crosses between South China zebu and the humpless Mongolian or Manchurian cattle, and the new crossbreeds between zebu and European cattle.

DISTRIBUTION OF ZEBU CATTLE

Available evidence indicates that humped zebus, descended from the Asiatic wild cattle, first evolved in the Indian subcontinent, and from there spread to other areas after the Aryan invasion. Indian zebu cattle were introduced into Southeast Asia, including South China and Taiwan; after the 4th century, Arab and Indian traders introduced zebu from the southern parts of Arabia, Iran, Baluchistan, and India to the east coast of Africa. From here the zebu spread through the savanna belts to the west coast and then south to the Zambesi River. During their spread, zebu cattle absorbed/swamped the earlier humpless and cervicothoracic humped cattle of these regions, incorporating various new characteristics while remaining truly zebu. Today, because of their ability to adapt to semitropical and tropical environments, the zebus are seen in almost every part of the world, excluding cold temperate regions. Some of the modern semitropical breeds contain zebu blood to a varying degree, the outstanding examples being Santa Gertudis, Beefmaster, American Brahman, Jamaica Hope, AMZ (Australian Miling Zebus), and Braford. More than half of the world cattle population of over 1000 million consists of zebu-type cattle or their crossbreeds.

NUTRIENT REQUIREMENTS OF ZEBU CATTLE

Nutrient requirements of cattle have been dealt with elsewhere in this handbook, so no attempt will be made to review them here. However, some salient differences in the nutrient requirements of zebu cattle as compared with European cattle (*Bos taurus*) will be pointed out.

Recommendations regarding nutrient, mineral, and vitamin requirements for different livestock functions such as growth, milk production, and maintenance have been made by the U.S. National Research Council[3] and Morrison.[20] These recommendations are based on data obtained on cattle in Canada, the United States, and Europe. Unfortunately, there is a dearth of information on nutrient requirements of zebu cattle living in tropical and subtropical conditions. Generally, nutritionists have tended to follow the NRC[3] recommendations for zebu cattle, but Krishna and Ranjhan[4] made a recent attempt to compile information about the comparative nutrient requirements of zebu and European cattle living in warm climates. This information is reproduced with some modifications in Tables 1 through 4. Data available on the Indian buffalo are also included for the sake of comparison. From the data in these tables, Krishna and Ranjhan[4] have made recommendations on nutrient requirements of cattle in India, and these are summarized in Table 5. This information is applicable to cattle reared in tropical and subtropical regions of the world.

It is apparent from Table 5 that, consistent with their evolutionary background, zebu cattle have low maintenance requirements. This tendency seems to persist even in nutrient requirements for milk production when the level of production is low. However,

Table 1
SUMMARY OF ENERGY REQUIREMENTS OF CATTLE AND BUFFALO FOR MAINTENANCE

Bullock/ cow/ buffalo	Breed/species	Dry/ lactating	Requirement (kcal/$W^{0.75}$) Digestible energy	Metabolizable energy	Ref.
Cattle	Zebu	Dry	–	61	5
Bullock	Zebu	–	–	97.6	6
Buffalo	Murrah	Dry	–	126.8 (maintenance requirement)	7
Cow	Hariana-zebu	Milch	–	137.8 (all roughage ration)	8
Cow	Tharparkar/ Sahiwal-zebu	Dry	146	120	9
Cow	Sahiwal-zebu	Milch	196 (winter)	160 (winter)	9
			165 (summer)	129 (summer)	
Cow (crossbred)	Brown Swiss X Sahiwal	Dry	–	100 (–ve balance)	10
				103 (+ve balance)	
			Early Lactation[a]		
Cow (crossbred)	Brown Swiss X Sahiwal	Milch	–	128.0 (–ve balance)	
			–	131.7 (+ve balance)	
			–	130.7 (all cows)	
			Midlactation[b]		
			–	131.6 (–ve balance)	10
			–	137.4 (+ve balance)	
			–	134.3 (all cows)	
Cow	Holstein-Friesian	Milch	–	150.2 (–ve balance)	11
				175.1 (all cows)	
Cow (crossbred)	Holstein-Friesian X Hariana	Milch	–	144.4 (+ve balance)	11
				124.5 (–ve balance)	
				133.0 (all cows)	
Cow	European breed	Milch	149	122	12
Cow	European breed	Milch	154	127	3
Steer	Brahman-zebu	300 kg body weight	–	86.4	13
Steer	Africander	266 kg body weight	–	102.5	13

[a] First 100 days post partum.
[b] Period between 100 and 200 days post partum.

Table 2
SUMMARY OF ENERGY REQUIREMENT OF DAIRY CATTLE FOR MILK PRODUCTION

Animal	Breed/species	Requirement (kcal per kg 4% fat-corrected milk)	Ref.
Buffalo	Murrah	1171 ME	7
Cow	Tharparkar/Sahiwal	Winter: 1527 DE,[a] 1230 ME[b]	9
		Summer: 1362 DE,[a] 1063 ME[b]	
Cow (crossbred)	Brown Swiss × Sahiwal	1165 ME (early lactation)	10
		1146 ME (midlactation)	
		1152 ME (all cows)	
Cow (crossbred)	Holstein-Friesian × Hariana	1165 ME (+ve balance)	11
		910 ME (–ve balance)	
		1039 ME (all cows)	
Cow	Holstein-Friesian	864 ME (–ve balance)	11
		1039 ME (all cows)	
Cow	Temperate breed	1188 ME	12
Cow	Temperate breed	1130 ME	3

[a] Digestible energy.
[b] Metabolizable energy.

Table 3
SUMMARY OF CRUDE PROTEIN REQUIREMENT OF DAIRY ANIMALS FOR MAINTENANCE

Animal	Breed	Dry/lactating	Requirement ($g/W^{0.75}$)	Ref.
Bullock (zebu)	–	–	1.84	14
Buffalo	–	Dry	2.38	15
Buffalo	–	Dry	1.70[a]	16
Cow (zebu)	–	Dry	1.19[b]	16
Cow (zebu)	Tharparkar-Sahiwal	Dry	1.97	9
Cow (zebu)	Tharparkar-Sahiwal	Milch	3.83 (winter), 2.60 (summer)	9
Bullock (crossbred)	Brown Swiss × Sahiwal	–	2.10	10
Cow (crossbred)	Brown Swiss × Sahiwal	Milch	2.25 (early lactation)	10
			2.23 (mid-lactation)	10
			2.31 (all cows)	10
Cow (European)	Holstein-Friesian	Milch	4.21	11
Cow (crossbred)	–	Milch	3.73	11
Dairy cattle	European breeds	–	2.84	12
Dairy cattle	European breeds	–	2.84	3

[a] Minimum protein requirement calculated from endogenous urinary nitrogen (EUN).
[b] Minimum protein requirement calculated from nitrogen balance studies.

data available are scanty and inconclusive; more data based on systematic research work are required. Nutrient requirements for maintenance and productive functions increase when the animal is subject to heat stress over 27°C.[17] But zebu cattle are better able to cope with heat stress than are European cattle.

For all practical purposes, mineral and vitamin requirements of zebu cattle may be assumed to be not significantly different from those of European cattle. Marked

Table 4
SUMMARY OF CRUDE PROTEIN REQUIREMENTS OF DAIRY ANIMALS FOR MILK PRODUCTION

Animal	Breed	Requirement	Ref.
Cow	Tharparkar-Sahiwal	130 g digestible nitrogen per 100 g milk nitrogen (winter)	9
		Early Lactation[a]	
Cow (crossbred)	Brown Swiss × Sahiwal	151.1 g digestible nitrogen per 100 g milk nitrogen	10
		Midlactation[b]	
		150.0 g digestible nitrogen per 100 g milk nitrogen	10
		All Lactating Cows	
		149.7 g digestible nitrogen per 100 g milk nitrogen	10
Cow	Holstein-Friesian × Hariana	156.4 g digestible nitrogen per 100 g milk nitrogen	10
Dairy cattle	Temperate breeds	132 g digestible nitrogen per 100 g milk nitrogen	12
Dairy cattle	Temperate breeds	Approx. 147 g digestible nitrogen per 100 g milk nitrogen	3

[a] First 100 days post partum.
[b] Period between 100 and 200 days post partum.

Table 5
RECOMMENDED NUTRIENT REQUIREMENTS OF ZEBU AND CROSSBRED CATTLE UNDER TROPICAL AND SUBTROPICAL CONDITIONS[4]

Nutrient	Function	Animal	Requirement per day
Energy	Maintenance	Zebu cow	120 kcal ME[a] per $kg^{0.75}$
Energy	Maintenance	Crossbred cow	127 kcal ME per $kg^{0.75}$
Energy	Maintenance	Zebu steer/bullock	92 kcal ME per $kg^{0.75}$
Energy	Milk production	Zebu cow	1060 kcal ME per kg 4% fat-corrected milk
Energy	Milk production	Crossbreds	1130 kcal ME per kg 4% fat-corrected milk
Digestible crude protein	Maintenance	Zebu cow	1.97 g per kg $W^{0.75}$
Digestible crude protein	Maintenance	Crossbred cow	2.84 g per kg $W^{0.75}$
Digestible crude protein	Milk production	Zebu cow	130 g digestible nitrogen per 100 g milk nitrogen
Digestible crude protein	Milk production	Crossbred cow[b]	147 g digestible nitrogen per 100 g milk nitrogen

[a] Metabolizable energy.
[b] Recommended for crossbreds giving over 20 kg milk per day.

differences exist between zebu and European cattle in water metabolism and requirements. Water consumption is influenced by both daily dry matter intake and atomspheric temperature. Under heat stress conditions the water turnover rate of zebu cattle is much lower than that of European cattle. In zebu cattle weighing from 150 to 750 kg and kept at 40°C environmental temperature, the turnover rate was 350 ml/$kg^{0.82}$/24 hr, as compared to 600 ml/$kg^{0.82}$/24 hr for European cattle.[18] Data on water consumption per kilogram of dry matter intake by zebu and European cattle at different temperatures are given in Table 6.

Table 6
EFFECT OF TEMPERATURE ON WATER INTAKE[18]

Water Intake, Liters per Kilogram of Dry Matter Consumed

Temperature (°C)	4.4	10.0	15.6	21.1	26.7	32.2	38.7
Zebu (*Bos indicus*)	2.87	2.98	3.24	3.48	4.06	5.75	11.58
European (*B. taurus*)	3.08	3.33	3.83	4.49	5.18	7.33	18.9

ENVIRONMENTAL-HUMAN-CATTLE INTERACTIONS AND NATURE OF FEED MATERIALS AVAILABLE

During the evolutionary process, the ability of zebu cattle to survive and reproduce in tropical and subtropical environments has been the dominant production trait. This has been dictated by rigorous climatic conditions brought about by high temperature; alternate dry and wet periods; alternate scarcity and abundance of feed; great variations in feed quality; severe short winters; and prevalence of insect pests, parasites, and pathogens. Further problems have been created by the demographic changes in human population in these areas. A rapid rise in human population has forced the diverting of most of the available land to the production of food and fiber for man. Only land of marginal productivity is put under pasture for grazing or for growing of cultivated fodder crops. As a result, the productive performance of animals in these areas has suffered.

In large parts of Asia and Africa, draft animal power is essential for planting and harvesting crops and for various other farm operations. In most cases, crop planting and harvesting must be completed within the time limits imposed by weather and climate. Draft animals may be used only a few months of the year, but they must be fed all year round and may consume a substantial portion of feed. This constitutes an extra feed requirement for draft animals in addition to that required to feed other productive animals.

Under the prevailing conditions, the average cattle owner does not have sufficient resources to maintain or breed cattle for different functions such as the production of milk or meat or for use as draft animals.[19] He needs an all-purpose animal which will produce milk, provide bullocks for work, and provide meat when killed. Despite these limitations, some specialization of zebu cattle along functional lines has occurred, but in no way close to that to which European cattle have been subjected. However, the differences in some cases are as great; for example, the Sahiwal and Red Sindi breeds of the Indian subcontinent are dairy breeds as specialized as Jersey or Friesian. Similarly, the Gir, Boran, and Brahman breeds of India, East Africa, and the United States, respectively, represent beef-producing zebus, while the Dhanni and Nagore of India are bred for fast work. The Krishna Valley and Amrit Mahal of South India are bred for heavy draft.

Animal feeding practices in tropical and subtropical areas are very distinct from those followed in developed countries of temperate regions. Since grains are required primarily for consumption by humans, the major feed supply comes through crop by-products, including both grains and other plant parts, animal and industrial wastes, and a few other unusual by-products of forest origin. The feed supply sources, including both concentrates and roughages, can be grouped into the following classes:

I. Field crop seeds and their by-products
   A. Crop grains
   B. Flour milling by-products
   C. Rice sheller mill by-products
   D. Pulse grain by-products
   E. Brewery and distillery by-products

- F. Starch industry by-products
- G. Oil extraction plant by-products
- H. Gum industry by-products
- I. Coffee wastes
- J. Seed husks and hulls
- K. Corn cobs

II. Field crop plant residues and by-products
III. Cultivated fodders
IV. Natural grasses and improved pastures
V. Seeds and fruits of forest trees
VI. Tree leaves and shrubs
VII. Aquatic weeds
VIII. Miscellaneous organic wastes
- A. Poultry wastes
- B. Farm stock wastes
- C. Organic wastes as substitutes for production of protein-rich feed
- D. Cellulosic wastes
- E. Vegetable- and fruit-processing wastes

IX. Silk industry wastes
X. Fish industry wastes

Field Crop Seeds and Their By-products

Some field crops can be directly fed to livestock; in addition, if they are processed to produce human foods they yield valuable by-products. The chemical composition and nutritive value of this class of feedstuff are available from appropriate tables.

Crop Grains

Field crops that can be directly used as animal feeds may be classified into four major groups: grain crops such as rice, maize, sorghum, wheat, triticale, oat, barley, millets; root crops such as sweet potato and cassava, grain legumes such as pulses; and oilseeds. Man directly competes with livestock in the consumption of these feeds, so their availability for livestock feeding depends upon whether the supply is in surplus to man's needs and on the relative prices of these feeds and those of finished animal products. In general, it will only be more economical to feed these to nonruminant livestock such as pigs and poultry, whose efficiency in converting them into animal products is superior to that of ruminant livestock. The feeding value of crop grains is well established. Grains like maize, wheat, and rice are classified as high energy feeds, while oat and barley are classified as medium energy feeds. After threshing, oat and barley grains remain encased in their flower glumes, and because of this contain higher percentages of crude fiber.

Flour Milling By-products

Wheat bran is a major by-product of flour milling. The quality of protein and the high phosphorus content give bran a unique place in livestock feeding. About 40% of the wheat germ is in the bran, which accounts for its high quality protein. In ruminant rations it provides supplementary phosphorus to correct a common shortage in the forage. The bran cellulose and hemicellulose carbohydrates are an acceptable source of energy for these animals. Other wheat milling by-products are shorts, middlings, and feed flour, most of which are hardly available for livestock feeding in less developed countries. Bran is usually fed in combination with grains and oilseed cakes.

Rice Sheller Mill By-products

The world's paddy crop production is currently estimated at 516 million t. Rice hulls are removed from a paddy at the first milling stage or in households by pounding with pestle and mortar. Rice bran is recovered at the second milling stage or during home pounding of rice, and constitutes 9 to 10% of paddy by weight. The quality of bran is influenced by the milling or pounding process and the type of paddy used. Rice bran contains 12 to 14% each of polyunsaturated oil and protein and is rich in B vitamins. It is subject to deterioration due to lipid autoxidation on storage in hot and humid conditions. Solvent extraction of oil produces deoiled rice bran. When mixed with other concentrates and molasses it constitutes a good cattle feed.

Pulse Chuni or Husks

Chuni, which consists of broken grains and hulls, is obtained during manufacture of pulses. For example, kidney bean *chuni* contains 20.5% crude protein and 10.9% crude fiber, and the digestible crude protein (DCP) and total digestible nutrients (TDN) are 16.3 and 66.9%, respectively. The high DCP indicates that *chuni* can be used as a protein supplement in the ration of young and milch cattle. Pulses are widely eaten in Asia and Africa, and considerable quantities of *chuni* from different pulses are available for cattle feeding.

Brewery and Distillery By-products

Brewery by-products are brewer's dried grains, malt sprouts, and spent hops. Distillery by-products are distillers' dried grains and solubles, stillage, and yeast. The nutritive value of these by-products for livestock feed was reported by Morrison.[20] Although available in large quantities in some tropical and subtropical areas, their feeding potential is limited.

Starch Industry By-products

The major starch industry by-products are maize gluten feed, maize gluten meal, cornsteep fluid, and tapioca root by-products (Figure 1).

Maize gluten feed – This consists of maize gluten meal and corn bran, with or without corn solubles. It may also include a part of corn germ meal. Usually the manufacturers include just enough corn gluten meal to safely meet a guarantee of 25% protein or some lower percentage. Maize gluten feed of the usual grades supplies about 70% as much digestible protein as linseed cake, and is nearly equal to linseed cake in content of TDN. Maize gluten feed is not as palatable as corn, oats, or wheat bran. The amount of phosphorus in maize gluten feed depends on whether or not the maize solubles have been included in it; the average phosphorus content is 0.80% but will be considerably lower if

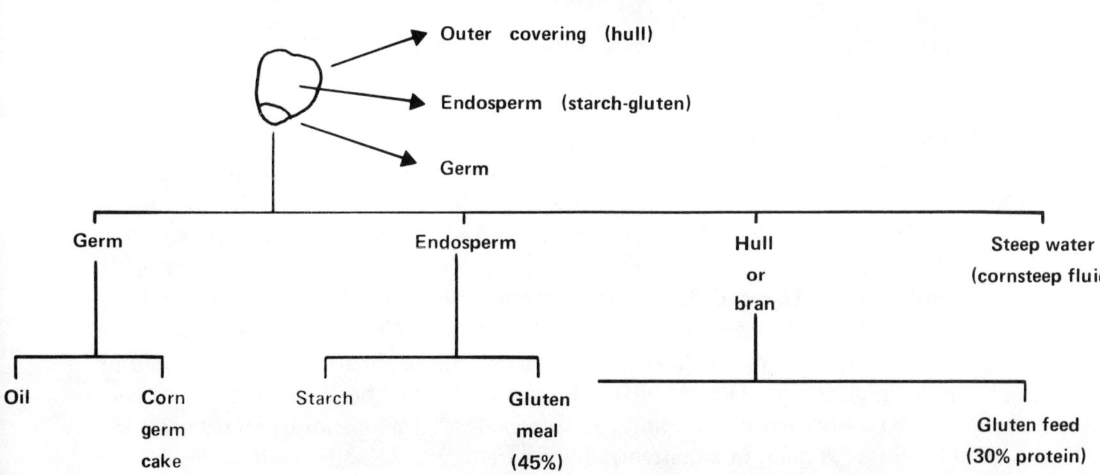

FIGURE 1. Maize by-products.

the corn solubles are not present. Because of the lime used in neutralizing maize solubles, gluten feed usually contains more calcium than corn or other grains.

Maize gluten meal – This usually contains corn gluten separated in the wet-milling process of starch manufacture with practically none of the hull fragments. It may or may not include corn solubles and may occasionally contain some germ oil meal. Gluten meal contains about 45% protein and supplies nearly as much digestible protein as soybean meal. Gluten meal contains a higher amount of TDN than does soybean meal, averaging 79.7%, but is much lower in calcium and phosphorus.

Maize germ oil cake – The corn germ removed in the wet-milling process is dried and the oil extracted. The residual fraction is called germ oil cake or meal, and contains a slightly smaller amount of protein than does corn gluten feed.

Cornsteep fluid – This is another by-product of the starch industry and contains about 40% crude protein and 12% soluble sugars on a dry matter basis. The protein fraction is mostly (about 90%) in the form of nonprotein nitrogen. Amino acid analysis of the hydrolysate of the original cornsteep liquor indicates that 84% of its nonprotein nitrogen is present in the form of free amino acids and peptides.[21] The four major amino acids in the liquor are alanine, leucine, proline, and γ-aminobutyric acid. Thus, cornsteep fluid can provide amino acids as well as nonprotein nitrogen for ruminants. Ligget and Koffler[22] demonstrated the importance of cornsteep fluid in the nutrition of microorganisms when they found that most organisms capable of growing well on simple media containing leaf meal and peptone will grow well on a medium containing cornsteep fluid. Singh and Makkar[23] reported a higher microbial count and trichloroacetic precipitable nitrogen in the rumen of cows fed a cornsteep fluid ration as compared to cows fed a control ration containing urea. They further emphasized that, based on ammonical blood nitrogen, the ration containing cornsteep fluid was utilized more efficiently than the control ration.

Tapioca by-products – Tapioca (cassava) is a widely used English name for dried root products of the cassava plant *Manihot utilissima* or *M. esculenta* of the family Euphorbiaceae (bitter type of cassava). *M. palmata,* synonymous with *M. dulcis* and *Jatropha dulcis,* is the sweet type of cassava. Some of the new commercial varieties contain as much as 38% starch and give an average yield of 40 t/ha. The fresh tubers are highly perishable and deteriorate within 2 or 3 days of harvest if not dried or properly stored. The following are the main dry commercial feed products of the tapioca plant:

Tapioca chips: 4- or 5-cm irregular slices of roots

Tapioca meal: fine powder from the manufacturing of chips and ground cassava roots

Tapioca waste: residual pulp separated from the starch

It is estimated that the annual world production of tapioca roots is about 100 million t, and Brazil (30%), Indonesia (20%), and Zaire (10%) are the main producing countries. Thailand produces only about 3% of the total world supply. The toxicity of tapioca, which is due to the presence of cyanogenetic glycosides, has always been the main factor retarding exploitation and real potential of tapioca as a feed. Free hydrogen cyanide is liberated from the glucoside by the action of the enzyme linamarase, which is naturally present in the plant. The cyanide content in fresh tubers varies between 0.01 to 0.04%, the bitter varieties containing more than the sweet ones. Chemical composition of the root product varies considerably according to variety, age of the plant, and the processing technology employed.[24] The calorific value of roots and the digestibility of tapioca starch are relatively high when compared to cereals.[25] Tapioca root products are very good substitutes for grain in a concentrate mixture for dairy cattle, particularly in areas where grains are in short supply for the human population. The release of energy from tapioca meal and the release of nitrogen from urea have a very close timing, which is very important for maximum utilization of nonprotein nitrogenous compounds in cattle diets; thus, a significant saving in protein foodstuffs for monogastric livestock is effected. Tapioca waste is a by-product from the tapioca starch industry. In contrast to maize starch industry by-products, tapioca waste is very poor in protein content, containing about 2% DCP and 64% TDN on a dry matter basis. Tapioca chips and waste can replace part of the maize in a concentrate mixture, but protein content must be adjusted by including protein-rich feeds in the diet. Even in green lots, a simple feed concentrate, e.g., with a composition of 85% tapioca root meal, 6% molasses, 8% urea, and 1% mineral supplement, would match up well with tropical grasses and supply all deficient nutrients required for optimum performance and utilization of the nutritive potential of green forages. In dry lots, the combination of two feed ingredients such as tapioca pellets and grass meal pellets may totally substitute for cereals. When properly processed, fresh tapioca roots may also serve as a basic energy source for intensive cattle feeding. In tropical regions, cattle development is usually hampered by inadequate forage, the quality of which is changing too rapidly to meet physiological requirements and economic performance of ruminants. Under such conditions, tapioca roots, either in fresh form or preferably as a dry product, may play an important role. Tapioca root, rich in digestible energy and poor in protein and fiber, matches up well with animal wastes that are low in energy and high in other feed constituents like protein, fiber, and ash. Tapioca starch waste replacing maize in a concentrated mixture was found to have no adverse effect on body weight, milk yield, butter fat, and total nonfat solids in zebu cattle. Obviously, the monetary benefit that could accrue as a result of incorporation of tapioca starch waste in the ration of cows is quite appreciable.

Oilseed By-products

Oilseeds are primarily processed for their edible oil constituent, and this traditional treatment alone, whether by pressing, by solvent extraction, or by a combination of these techniques, results in a residue rich in protein (30 to 60%). In tropical areas, the major oilseed cakes are groundnut or peanut, cotton, mustard, coconut, and sesame. Other oilseed crops grown in lesser quantities in these areas include sunflower, soybean, and niger. The majority of the oilseeds in tropical and subtropical areas are pressed by an expeller process for extraction of the oil, and the residual material is available as a cake containing about 7 to 9% oil. These cakes usually form the major protein source in a concentrate mixture.

Gum Industry By-products (Guar Meal)

Guar meal, a by-product of the gum industry, contains 40 to 45% crude protein, and can therefore serve as a good source of protein for both monogastric and ruminant animals. However, it contains natural toxicants, particularly a trypsin inhibitor, and other heat-resistant toxic substances, and is not recommended for use in rations for monogastrics. Toasting the meal reduces some of the toxicity for growing and for laying birds. Kwatra and Sidhu[26] reported considerable improvement in the nutritive value of guar meal when it was treated with hot water and hydrochloric acid. Guar meal has great potential for ruminant feeding as it could replace oilseed cakes in a concentrate ration. Guar crop is well adapted for growth on dry and poor land, and large quantities of it are available in India and other countries.

The use of guar meal as a sole concentrate in cattle ration results in chronic diarrhea, but guar meal as part of a balanced ration does not cause any digestive disorder.[27] The effect of guar meal on animal growth and digestibility of rations with poor quality fodders has been studied by various workers.[28-30] It has been reported[31] that replacing groundnut cake by luar meal in a concentrate mixture had no adverse effect on growth, feed efficiency, and digestibility of protein when fed to 220-day-old crossbred calves.

Coffee Processing Plant Wastes

Coffee processing wastes are coffee pulp and hulls, which represent 29 to 12% of whole coffee cherry, respectively. Coffee pulp has considerable potential as a feed for cattle. The chemical composition of coffee pulp is well documented, although it is not used extensively for various reasons. The most important problem is handling and direct use of the material as obtained from coffee processing plants because of its high moisture content. Furthermore, the relative abundance of other more stable, more easily obtainable feeds probably has contributed in part to the lack of interest in the use of this resource. Wet pulp has good nutritional qualities, but sun-dried pulp loses much of its high quality as a feed.

Dried coffee pulp contains about 10% crude protein, 44% nitrogen-free extract, and 8% ash. A cellular content of 63% indicates that the material has a relatively high level of nutrients, while the levels of lignocellulose, hemicellulose, cellulose, and lignin show the product to be superior to various other types of feeds.[32] Of the protein, 3% is found in the form of lignocellulose.

The results of dietary studies carried out with lactating cows also showed that coffee pulp levels below 20% did not affect milk production. In view of these results, various cattle growers in Central America are already using small quantities of dehydrated pulp as a dairy cattle feed.

Coffee silage could be an adequate solution to both the high moisture content problem and the seasonal production of pulp, since even after 4 months the resulting silage has very good organoleptic characteristics, particularly if ensiled with molasses. However, after exposure to air the coffee silage becomes dark in color with a concomitant decrease in palatability, which reduces feed intake by the animal. The dark coloration is probably due to oxidation of polyphenols present in the coffee pulp. This problem is being investigated since its solution could represent a significant advance in the utilization of coffee pulp. Ensiling and dehydrating do not substantially modify the factors present in coffee pulp that adversely affect the animals. Coffee pulp contains relatively high levels of caffeine which is in lignified form and probably not readily available.

It is interesting to note that the level of lysine present in coffee pulp protein is as high as that found in soybean meal protein. Like soy, coffee pulp protein is deficient in sulfur-containing amino acids. At present, the biological availability of these amino acids is not known. This is an aspect which requires some investigation because of the relatively high tannin content in coffee pulp and the known fact that tannins may combine with protein to make its component amino acids biologically unavailable to the animal.

Seed Hulls or Husks

This class of feed is derived from hulls or husks of paddy, groundnut, oat, barley, gram, coffee, cottonseed, and maize. The hulls are not palatable to cattle and can form only a small portion of rations fed to unproductive animals. Under acute scarcity or famine conditions cattle may be able to consume these feeds if mixed with cereal straw. Hungry cattle may consume between 1 and 2 kg of these hulls. When these feeds form a part of the ration the requirement of energy-rich feeds will be more in order to make up for the poor energy content. These husks are highly lignified.

Groundnut shell is as unpalatable to livestock as paddy husk. In nutritive value, maize husk is superior to paddy husk, groundnut shell, and coffee husk. Maize husk can replace 50% of the dry roughages moiety in adult cattle ration. Feeding these husks in combination with molasses and 1% urea improved palatability and utilization of energy.

Comparison of the chemical composition of coffee hulls with corn cobs and cottonseed hulls shows that protein content is similar among these by-products, while crude fiber is significantly higher in coffee hulls. The nitrogen-free extract content is very low, and lignin content is around 24% in coffee hulls. This limits the utilization of coffee hulls.

Corn Cobs

A large quantity of corn cobs are available in tropical and subtropical and other corn growing areas. Corn cobs contain 32% crude fiber and supply 45.9% of total digestible nutrients for ruminants, which is slightly more than in oat straw or cottonseed hulls. Cobs contain 2.3% crude protein, having no digestibility. They are also very low in minerals, having about 0.04% phosphorus. Like oat hulls, corn cobs are a low-grade roughage.

Crop Plant By-products

The major residual by-products of crop plants are cereal straws, pulses and oilseed plant straws, sugar cane plant by-products, tapioca leaves, sorghum, maize and millet stalks, sweet potato, and haulms. In tropical and subtropical areas the bulk of feed is constituted by these agricultural by-products because good quality forages are in short supply. The potential of these feedstuffs as sources of nutrients, particularly energy, is not fully realized because of high lignification of the cellulose fraction. Furthermore, these roughages are fed to animals with little or no consideration for balancing the diet under practical farm conditions. Attempts are being made to improve the nutritive value of these low quality roughages by chemical and fungal treatments.

Cereal Straws

Cereal straws (wheat, paddy, raye, oat, barley) are generally deficient in digestible protein, poor in energy and minerals, and have a low palatability. The nutritive value of these straws is known to be improved by alkali treatment.[33-35] The soaking of straws derived from wheat, oat, and barley improved the feed intake significantly in zebu cattle and buffaloes. The metabolizable energy (ME) content and production rates of total volatile fatty acids were higher in soaked straws than in dry straws. Oat straw is readily eaten by all classes of cattle, and in some areas is a regular part of the ration of low-producing and dry cattle.

The value of rice straw as a roughage feed for ruminants has never been fully exploited, although traditionally it has been used for feeding buffaloes and zebu cattle in many areas. Its potential has not been realized for two main reasons: (1) only a fraction of the total rice straw produced is fed to animals, the remainder being used for the manufacture of paper or ploughed into the soil or burnt; (2) the rice straw used for feeding is offered with little or no other feed supplement (energy, protein, and minerals) and this results in very poor utilization.

Rice straw is nutritionally comparable to barley and wheat straws and in some cases may surpass these. Its nutritive value is influenced greatly by the stage of harvesting. The protein content is higher if it is harvested when it still retains some of its green color, when it compares favorably with poor-to-average hay made from tropical pasture species. Straws are consumed in small amounts when they constitute the sole feed. There is ample scope for feeding rice straw in tropical areas, and its potential is enhanced by the availability of cane molasses as a readily available source of energy. Starch and protein-containing feed ingredients are also needed; the latter may be fed in part as true protein and in part as nonprotein nitrogen.

Pulse Straws

Most common pulses are *Urad* (*Phaseolus mungo*), *Moong* (*P. radiatus*), *moth* (*P. aconitifolius*), cowpeas (*Vigna catiang*), *masoor* (*Lens esculenta*), *Arhar* (*Cajanus indicus*), and gram (*Cicer arietinum*). After harvesting the crops, the seeds of pulses are thrashed out and the residual product, including the husk of pods, leaves, and tender stems, is classed as pulse straw. The straws of *Urad, Moong,* and *moth* are more palatable than others. Straws of *Arhar* and *masoor,* although of comparable nutritive value, are not as palatable as the other three straws. The energy values of these straws compare with that of cereal straws but they are a fairly good source of digestible protein.

A palatable roughage containing 3% digestible protein and 40% TDN can meet the maintenance requirements of adult cattle and buffaloes. Since the pulse straws contain 40 to 100% more DCP over and above the maintenance requirements, they can meet the production requirements of the animals to a certain extent. However, supplementation of these straws with energy-rich feeds is necessary for maintaining a higher rate of milk production.

Oilseed Plant Straw

Groundnut plant straw, along with leaves, contains a higher amount of DCP than that of nonleguminous hays, and compares well with leguminous hay of cowpeas. The protein content of groundnut straws varies between 12 and 16%. The energy content of groundnut plant straw expressed as TDN is also higher than most of the grass hays; in addition, it is also highly palatable and nutritious. Feeding experiments with milch cows have proven that groundnut straw in conjunction with wheat bran and wheat straw can meet the cows' entire nutritional requirements for maintenance and milk production up to 8 kg/day.

Rape bhoosa or straw is more fibrous and lower in nutrient content than other pulse straws, and should be fed in limited quantities with cereal straws to nonproductive animals.

Sugar Cane Plant By-products

Sugar cane, an important source of income for many countries, is widely grown. The processing of sugar cane yields a number of important by-products for animal feeding:

Molasses – This is a valuable source of energy which is often either underutilized or exported at a low price. It has long been known that inclusion of a small percentage of molasses improves the palatability of the diet, and it is particularly valuable when incorporated with coarse, unpalatable roughages. Lofgreen and Otagaki[37,38] found less net energy in mixed diets containing more than 10 to 15% molasses, although this varied with total feed ration composition. Hatch and Beeson[39] reported that replacing 5% rolled maize with molasses in rations had no apparent effect, but 10 and 15% replacement increased nitrogen retention, energy and dry matter digestibility, and butyric acid production in the rumen. Because of the large quantities of cane molasses produced in tropical areas, it is important to explore the possibility of using higher than conventional levels. Reports[40,41] indicate that molasses could contribute up to 80% of the ME for beef cattle under specific feeding conditions. Results indicated that growing heifers could initially be fed 25% molasses, and the level could later be increased to 33%. In the latter diet, two by-products, rice straw and molasses, accounted for 68% of the total weight of the feed.

Bagasse – Conventional bagasse contains two fractions, an outer portion called the rind and the inner part known as the pith. Millions of tons of bagasse are produced in sugar cane factories and a significant percentage is burned as fuel; thus, a potential source of animal feed is not fully exploited. Bagasse is a low quality feed, principally because of its high percentage of lignocellulose. When mixed with protein and a source of energy such as cane molasses, moderate levels of bagasse can promote satisfactory live weight gains. Levels higher than 30% by weight tend to depress the rate of gain. Reports[42] indicate that bagasse contains 60.60% acid detergent fiber, 45.59% cellulose, 13.22%

lignin, and 1.35% total water-soluble sugars. The soluble carbohydrate content of factory bagasse is insufficient to provide required energy for the proper utilization of urea in ruminants. Treatment of bagasse with sodium hydroxide or steam significantly increases digestibility. Even if chemical treatment does not become a commercial reality, much can be achieved by feeding untreated bagasse at optimum levels in mixed rations. Since bagasse and molasses are usually available in combination, there is a great potential for feeding these in the correct ratio with supplementary protein and nonprotein nitrogen.

Sugar cane leaves – Sugar cane leaves and tops form another major by-product. These leaves are separated from the cane and are burnt. For feeding livestock the leaves should preferably be collected on the day of harvest when they are still green. The feasibility and economics of collecting the leaves vary from one region to another. Where labor costs are low, sugar cane tops and leaves provide a cheap source of feed. The leaves should be chopped to facilitate feeding and to avoid rejection of the fibrous parts. O'Donovan[43] fed chopped sugar cane leaves and tops containing 5.1% crude protein to a herd of commercial dairy cattle and to beef cattle. This feed was found to be sufficient to meet the maintenance requirements of dairy cattle and to provide for the production of about 2 kg of milk per cow – a production level that could scarcely be realized with average quality pangola grass silage. With beef cattle, maintenance requirements were met, and an average daily live weight gain of 0.25 kg was obtained. It was necessary to feed more protein and energy in order to achieve higher rates of gain. There is evidence that satisfactory silage can be made from the chopped leaves, and this is a convenient method of conserving the surplus for subsequent feeding.

Sugar factory residue – Sugar factory residue is a mixture of the first froth of boiled cane juice and fine bagasse or pith, mixed in a ratio of about 5 to 1. It contains 11 to 12% protein, 26 to 35% fiber, and 1.4% calcium and 1% phosphorus on a dry matter basis. Its rather high phosphorus content could be an advantage if it is fed to cattle when herbage phosphorus content is low. The residue mixed with molasses is consumed by the animals without difficulty, and the inclusion of the residue does not significantly depress the rate of live weight gain. Because the residue has a low dry matter content (about 25%), molds grow rapidly on it in hot humid conditions. Economic considerations are likely to rule out drying, so it is necessary to utilize the material in fresh form. Therefore, it must be fed in close proximity to sugar factories. There is little information as yet concerning its real value as a livestock feed. The extent to which it will be fed to livestock will also depend on the availability of alternative sources of better quality feed.

Stalks

Crops like maize, *jowar* (sorghum), and *bajra* (pearl millet) are grown for green fodder as well as for grain production, and in the latter case large quantities of stalks are available as the residue of stemmy plant material. These nonleguminous stalks are too mature and are highly lignified. Voluntary intake of these does not usually exceed more than 1.5% of the liveweight of the cow. Such feedstuffs are primarily intended to feed nonproducing livestock during summer or some other time when their energy requirement is relatively low. These materials are normally cheap and are fed as a means of converting them into something useful. Their efficiency is not a major consideration.

Tapioca Plant By-products

The aerial part of tapioca may offer a new perspective for livestock feeding in tapioca producing areas. The production of tapioca leaves is about 10 to 15 t of dry matter per hectare. This can be used as a source of protein as well as forage for cattle feeding. Although the protein value of tapioca leaves is well established, the usual objection to their use as feed for animals has been the cyanide present in the leaves. In human nutrition this problem was elucidated by Rogers and Milner.[44] Environmental conditions as well as the stage of plant maturity influence the cyanide content. It was found that the cyanide content is at its minimum when the plants are 5 months old. Leaf meal at a level of 0.5 to 0.8% of body weight does not cause any adverse effect on the physiological well being of the animals adjudged in terms of red and white blood cells, hemoglobin, plasma protein, and calcium and phosphorus concentration in blood.[45]

A few studies[46,47] have shown that tapioca leaves are a potential protein source for farm livestock. The aerial part can be processed to prepare plant meal containing about 24% protein or can be separated into leaf fraction containing 30% protein and a stem fraction containing 7 to 10% protein on a dry matter basis. The leaf protein is rich in amino acids like leucine, isoleucine, valine, glycine, lysine, and glutamic acid. Menon[48] reported that tapioca leaves contain about 3% DCP and 14% TDN on a fresh basis. Growth studies revealed that the body weight gains of red *Sindi* calves that were fed a tapioca leaf meal ration compared well with a control group. Comparison between the control ration and a ration in which 50% of groundnut cake was replaced showed that the leaf meal can be incorporated in milch cattle rations at a level of 0.4% of their body weight with appreciable monetary benefits.

Sweet Potato Haulms

Sweet potato, an important crop in many areas of the tropics, is a source of animal and human food. The haulms have not received deserved attention as a feed for ruminants, although in some countries they are fed to buffaloes and zebu cattle. They are available during the dry season when the feed supply is usually critical.

O'Donovan[43] fed sweet potato haulms to dairy cows with good results. They invariably increased milk yields when replacing a diet of sugar cane leaves or silage, principally because of their higher protein content. The haulms were consumed avidly and there were no noticeable ill effects. They may supply a part of the feed for dairy and beef cattle, especially in those countries where labor costs for collection and transport are not too high.

Cultivated Fodders

With the development of specialized dairy farms near large cities to meet the demands for fluid milk and other dairy products of the urban population in developing countries, special emphasis is being placed on stall feeding of milch animals on cultivated green fodders. Specially bred fodder varieties of these crops are sown in heavily fertilized and irrigated soils to produce high yields of green fodder per hectare. The fodder crops are harvested at a stage when they contain optimum levels of digestible nutrients, chaffed and mixed with hays or other grain crop residues, and fed. In this way a substantial saving is effected in the quantities of concentrate required to produce milk. Cheap labor and relatively high prices of grain make stall feeding of cultivated fodders an economically attractive alternative to grazing and concentrate feeding, which is a normal practice in industrialized countries with high labor costs. A major quantity of cultivated green fodders are available during the rainy season and early part of winter. However, in irrigated areas the cropping pattern could be worked out to ensure an ample supply of green fodders throughout the year. During lean periods silage is used to supplement green fodder. According to a report in a relay cropping pattern of fodders worked out at the National Dairy Research Institute, Karnal, India,[49] a yield of 1500 t of fodder could be obtained in a year from a hectare of land. This quantity of fodder is sufficient to feed 10 to 14 cows over a period of 1 year. Suggested cropping rotations and yields expected are applicable to northern India, parts of Pakistan, and middle eastern countries. Table 7 illustrates the potential of intensive fodder production provided the labor costs are not forbidding. The sowing dates and crop rotations could be modified or changed to suit the climatic and soil requirements in other regions of the tropics and subtropics.

Table 7
CROP ROTATIONS, PLAN OF SOWING, AND EXPECTED YIELDS OF FODDER AND NUTRIENTS

Suitable for Irrigated Areas of Northern India and Parts of Pakistan, Egypt, and Other Middle East Countries

Fodder crops[a]	Sowing time	Harvesting time	Fodder yields (t/ha)		Nutrient yields (kg/ha)	
			Green	Dry matter	DCP[b]	TDN[c]
Maize + cowpea	3rd week of April	June–July	30	6.9	481	4,108
Maize + cowpea + teosinte	2nd week of July	August	45	9.5	760	7,150
Berseem + mustard	Last week in September	Six cuttings from December–May	100	15.0	2684	12,580
Total per hectare			175	31.4	3925	23,830
Maize + cowpea	1st week in April	May–June	30	6.9	481	4,100
Sorghum + guar	Mid June	August	45	12.1	930	6,750
Mustard	1st week in September	November–December	25	3.8	458	1,750
Oats + peas	1st week in November	January–February	40	9.5	789	5,475
Total per hectare			140	32.3	2,658	18,075
Sweet Sudan + cowpea	1st week in May	Three cuts in July, August, and September	100	23.5	1,205	11,218
Berseem + oats	End of September	Five cuts from December–May	100	15.0	2,727	13,210
Total per hectare			200	38.5	3,932	24,428
Sweet Sudan grass + cowpea	Mid April	Three cuts in June, August, and September	100	23.5	1,205	11,218
Mustard	Mid September	November–December	25	3.8	458	1,750
Oats + peas	Mid November	Two cuts in February–April	50	12.0	964	7,075
Total per hectare			175	39.3	2,627	20,043

[a] Botanical names of fodders: maize or corn, *Zea mays;* cowpea, *Vigna* spp.; teosinte, *Euchlaena* spp.; berseem, *Trifolium alexandrium;* sorghum, *Sorghum vulgare;* mustard, *Brassica juncea;* guar, *Cyamopsis tetragonoloba;* oats, *Avena sterilis;* peas, *Pipsum sativa;* Sudan grass, *Sorghum sudanese;* pearl millet, *Pennisetum glaucum;* hybrid Napier; cross between *Pennisetum glaucum* and *P. purpureum;* lucerne, *Medicago sativa.*

[b] Digestible crude protein.

[c] Total digestible nutrients.

Table 7 (continued)
CROP ROTATIONS, PLAN OF SOWING, AND EXPECTED YIELDS OF FODDER AND NUTRIENTS

Suitable for Irrigated Areas of Northern India and Parts of Pakistan, Egypt, and Other Middle East Countries

Fodder crops[a]	Sowing time	Harvesting time	Fodder yields (t/ha)		Nutrient yields (kg/ha)	
			Green	Dry matter	DCP[b]	TDN[c]
Teosinte + pearl millet + cowpea	Early May	Three cuts in July, August, and October	100	21.8	1,547	14,550
Berseem + oats	Early October	Five cuts from December –May	100	15.0	2,727	13,210
Total per hectare			200	36.8	4,274	27,760
Maize and cowpea	Last week in February	April	30	6.9	481	4,100
Maize + cowpea	Mid May	July	45	10.3	760	7,150
Maize + cowpea	1st week in August	September–October	30	6.9	481	4,100
Oats + mustard	Mid October	Two cuts in January and February	50	11.0	796	6,650
Total per hectare			155	35.1	2,518	22,000
Hybrid Napier (planted once in 3 years)	2nd week in March	Fodder cut at 6- to 8-week intervals	125	25.0	1,350	17,250
Maize + cowpea (planted between rows of Napier every year)	2nd week in May	Fodder cut at 6- to 8-week intervals	25	5.8	409	4,000
Berseem + mustard (planted between rows of Napier every year)	Mid September	Fodder cut at 6- to 8-week intervals	85	12.7	2,285	10,690
Total per hectare			235	43.5	4,044	31,940

Table 7 (continued)
CROP ROTATIONS, PLAN OF SOWING, AND EXPECTED YIELDS OF FODDER AND NUTRIENTS

Suitable for Irrigated Areas of Northern India and Parts of Pakistan, Egypt, and Other Middle East Countries

Fodder crops[a]	Sowing time	Harvesting time	Fodder yields (t/ha)		Nutrient yields (kg/ha)	
			Green	Dry Matter	DCP[b]	TDN[c]
Hybrid Napier (planted once in 3 years)	Mid March	Fodder cut at 6- to 8-week intervals	125	25	1,350	17,250
Lucerne (alfalfa) (planted once in 3 years)	Mid October	Fodder cut at 6- to 8-week intervals	85	17	2,750	10,200
Total per hectare			210	42	4,100	27,450
Total for 8 ha/year.			1,490	298.6	195,526	
Fodder available per hectare per year.			186	373.0	3,510	24,440
Fodder available per hectare per day.			0.5	0.1	9.6	67.0

Natural Grasses

Natural grasses are found along roadsides, footpaths, and river or canal banks, as field fringe forage in cropping areas, in natural pastures under the crops, and in accessible secondary bush lands and seasonal flooded grasslands. The grazing practices on natural grasses are almost universal in South and Southeast Asia, in humid tropical islands, and in the humid regions of Africa. The supply of forage may be augmented by planting forage trees along roads, footpaths, river and canal banks, and on the field bunds in rice-growing areas.

Natural forage under tree crops such as coconut, rubber, oil palm, and fruit varies widely in species composition, yield, and quality. In Southeast Asia, where such forage is not fully utilized, the dominant grass species is often *Imperata cylindrica*, but where it is well grazed the sward usually consists of a mixture of *Axonopus* spp., *Paspalum* spp., *Mimosa* spp., and *Desmodium* spp. This type of grazing is not particularly productive but it is a palatable and useful feed.

Grasses in accessible secondary bush areas vary even more than those under tree crops in species composition, yield, and quality, though within the same ecological zone approximately the same species are present.

Seasonally flooded grasslands can provide excellent forage for a part of the year. They are very extensive in South America, in some localized areas in Africa such as around Lake Chad, and in certain countries in southern Asia. A very palatable and nutritious grass species found in the seasonally flooded areas of northern South America and Indonesia is *Leersia hexandra.*

Two thirds of the total nutrients to cattle in terms of starch equivalent and protein become available through green grasses.[82] Ruminant livestock can be grazed under tall and hybrid coconuts, though bovines might damage dwarf coconuts. Coconut plantations provide a particularly attractive environment for grazing cattle buffaloes and sheep, and this integration increases total production and return per unit area of land and reduces weeding costs. In the past, the possibility of utilizing forage available in rubber plantations has been almost entirely neglected. The inherent difficulties are that the availability of forage varies with the age of the rubber tree; livestock are required that will not damage the young trees or knock the collection cups off the mature trees; and that no effort has been made to select suitable forage plants for the rubber plantation environment. These difficulties can be overcome, and livestock could be adjusted to plantations in different numbers at different times.

Seeds of Fruit and Forest Trees

In India and many other countries in Asia and Africa, forests have considerable potential as sources of seeds for livestock feeding. There is a growing need to utilize unconventional feed ingredients in ration formulations in order to minimize the use of expensive conventional feed ingredients. The selection of these ingredients and the extent to which they should be incorporated in mixtures depend mainly on the digestible nutrients they contain. Since most of the agro-industrial by-products are not as rich in their energy concentration as grains and oil cakes, these by-products can substitute for only a portion of the conventional feed mixture. Another factor limiting substitution is the low palatability of these by-products.

Mahua (Brassica latifolia) Cake

Mahua cake is a residual by-product left after the extraction of oil from the seed. It is not palatable if fed as such. Palatability was found to improve after keeping it steeped in water for a couple of hours; in actual practice it is suggested that it should be fed in combination with other feed ingredients. Nutritive value of Mahua cake is comparable with that of mango seed kernels. The cake may replace part of either the rice bran or barley in a concentrative mixture.

Mahua Flowers

There seems to be a fairly wide variation in the DCP value of these flowers due to differences in the climatic conditions of their sources. There is some difference in TDN content, too. The flowers can replace a part of wheat bran and gram *chuni* mixture.

Mango Seed Kernels

Estimated annual production of mango in India alone is 7 million t. The waste products of the crop amount to about 1 million t of stones and 0.5 million t of skin. A large proportion of this material is wasted or allowed to rot and cause environmental pollution. Some of it is consumed by scavenging domestic livestock, particularly pigs, but a large proportion of it goes to waste. The kernel can be a useful livestock feed ingredient[50] and can yield industrial starch, but the dehulling of the stones is a laborious, slow job. Fruit factories process large amounts of mangoes, and stones are obtained as a by-product but are not economically utilized. There are two technical problems involved in the effective utilization of mango stones: (1) they must be dried and (2) they must be decorticated. A mango dehuller was developed at Pantnagar.[33] It has been shown that mango seed kernel can be used satisfactorily up to 40% in a concentrate mixture for working bullocks. Its nutritive value is comparable to that of rice bran or barley.

Neem (Melia indica) Seed Cake

Deoiled neem seed cake is an industrial by-product of the soap industry. Neem kernel comprises about 26% of the dry fruit, and the kernel contains about 45% oil. In India, it is estimated that 414,633 t of neem seed is available, which could yield 62,000 t of oil and 352,000 t of residual cake. Neem cake contains 1.36% sulfur, which is higher than other cakes; the ash content is also high. Neem cake also has substantial amounts of residual bitter principles and malodorous constitutents and has been considered unfit for cattle feed. However, in some parts of India, cattle are fed with neem seed cake; although bitter, animals take to it very well when they are adapted to it. The cake's bitterness acts as an appetizer and improves digestion. Neem seed cake is also used as a vermicide.

Watermelon Seed

Watermelon has been cultivated for many centuries in tropical regions. Its seeds are oily and nutritious. It has a fruiting season from May to July, and the creepers which bear many fruits grow well in sandy waterlogged areas of river beds. Watermelon seed contains about 28 to 30% oil and the residual cake contains 25 to 30% protein. It can replace part of the protein-rich feeds in a concentrate mixture.

Cassia tora Seed

This plant grows abundantly during the monsoon season in uncultivated and barren lands and also in the outer hinges of forests. The seeds collected from the pods are not palatable to livestock if used as a sole source of concentrate portion of the ration. Feeding *Cassia tora* as a part of the ration for milking cows has produced encouraging results. Patnaik[51] observed that as the level of its substitution increased, the level of milk production, as well as butterfat content, also increased. Boiled *Cassia tora* seeds can be incorporated at a level of 15% in a concentrate mixture for lactating cows without any detrimental effect on milk yield and composition. However, when the level is increased to 30% the milk yield decreases.

Babul Pods (Acacia arabica)

Babul pod can be used as a component of a concentrate mixture. It contains 5.7% DCP and 62.5% TDN. The pods must be crushed thoroughly before feeding to cattle, or the seeds will pass out undigested in the feces. Crushed pods can replace either wheat bran or rice bran in a concentrate mixture.

Tamarind Seeds (Tamarindus indicus)

Tamarind seeds can be used to replace a part of basal feed, particularly rice bran or wheat bran. In order to make it palatable to cattle, finely ground seeds should be soaked in water an hour before feeding.

Rain Tree Pods (Enterolobium saman)

Finely crushed rain tree pods can conventionally replace a part of the rice bran in a concentrate mixture.

Date Stones

Date stone is a poor source of DCP and falls between concentrate and roughages in TDN. It can replace gram husk in a concentrate mixture.

Tobacco Seed Cake

In feeding experiments it has been possible to gradually replace the entire concentrate component in the ration of bullocks by tobacco seed cake. For dairy cows it should be fed in conjunction with other conventional concentrates in order to make it acceptable.

Aniseed Cake

This cake is a poor source of energy. It is comparable to husks, and the DCP value is similar to that of cereal brans.

Silk Cottonseed Cake

In energy value, silk cottonseed cake is comparable to gram husk and maize husk, though its DCP value is similar to that of cereals. Although the animals receiving silk cottonseed cake maintained positive nitrogen balance, the material was not considered a good source of protein supplement.

Salseed Meal (Shorea robusta)

Sal tree forests cover an estimated 114,000 km^2 in various parts of India. The tree bears fruit almost every year, which falls to the ground in May and June. Annual production potential of salseed kernels is about 5.5 million which may yield 688,000 of fat, with the remaining fraction used as meal or cake. Sal oil is mainly used for making soap. The processing method for salseed before it is used in animal food involves extraction of oil either through an expeller process or a solvent extraction process. The flow sheet for processing salseed is illustrated in Figure 2.

Salseed meal is a good source of carbohydrates and can be used as an energy source, replacing maize in animal feed. Chemically it compares favorably with maize, although the energy value of the meal is about 55% of maize. The meal can be used up to a 20% level in a concentrate mixture for growing and adult cattle. However, it contains fairly large quantities (10 to 12%) of tannins, which have a deleterious effect on growth and productive performance of animals if incorporated in large quantities in a concentrate mixture. Furthermore, the tannins present in the meal interfere with the utilization of nutrients in the total ration.

Panda and Sharma[52] reported a simple technique for removing a major part of tannins by water treatment only. An ammoniation process has been developed recently in the Unilever Laboratories in the United Kingdom, where it has been shown that the detoxification process greatly improved the utilization of nutrients in the ration.

Subramanian[45] reported on a long-range study of feeding value of salseed meal for

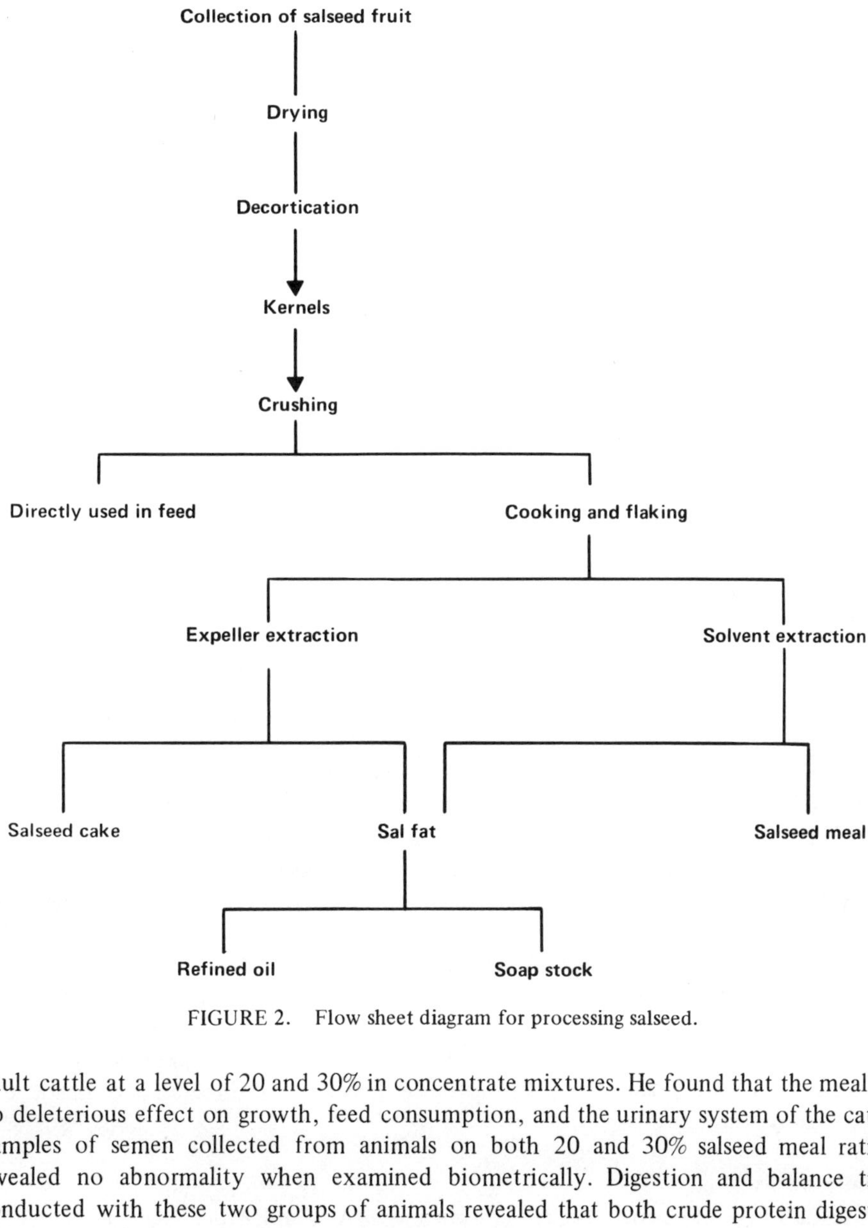

FIGURE 2. Flow sheet diagram for processing salseed.

adult cattle at a level of 20 and 30% in concentrate mixtures. He found that the meal has no deleterious effect on growth, feed consumption, and the urinary system of the cattle. Samples of semen collected from animals on both 20 and 30% salseed meal rations revealed no abnormality when examined biometrically. Digestion and balance trials conducted with these two groups of animals revealed that both crude protein digestion and nitrogen balance were depressed as the percentage of salseed meal increased in the rations. Two animals from the 30% salseed meal group were sacrificed for histopathological studies; no kidney abnormalities attributable to salseed meal toxicity were revealed.

Forest Leaves and Shrubs

An acute shortage of green fodder for ruminants is experienced during different parts of the year in arid, semiarid, dry, and unirrigated areas. The failure of rains adds to the farmers' troubles, not only in providing fodder for livestock but also in growing sufficient food for themselves. A repetition of dry weather or scanty rainfall in the following year results in a near famine condition when cattle turn to alternative sources of livestock feeds for survival. Under such circumstances, forest tree leaves, tender shoots, twigs, and forest grasses make an important contribution to the feed suply of grazing animals in many areas of the less-developed world. In many parts of India, particularly in mountainous regions, the animals are raised to a large extent on tree leaves, fodder, and grasses because of the precipitous nature of the ground and the scarcity of level areas. The cultivation of fodder crops is practically negligible in these areas. The leaves and twigs of certain trees contain a higher level of DCP and TDN and may contribute sufficient quantities of nutrients for growing and lactating cattle. Some nutritional characteristics of tree leaves are

1. The tree leaves contain 40 to 50% more dry matter than does green fodder.
2. Almost all the tree leaves contain two to three times more crude protein than do cultivated fodders and indigenous grasses.
3. Although ether extract appears to be fairly high in tree leaves, in most cases it is not as digestible as that in cultivated fodders.
4. The crude fiber in tree leaves is complex and highly lignified. Although the leaves contain a lower percentage of crude fiber than do grasses and hays, the fiber is less digestible.
5. The high calcium content, a unique feature of tree leaves, is two to three times that of cultivated fodders. The high calcium content is not matched by phosphorus, in most cases resulting in a wide calcium to phosphorus ratio. The calcium content increases and that of phosphorus decreases with progressive maturity.

Therefore, it may be noted that whenever the leaves form a reasonable portion of the cattle ration, a phosphorus-rich feed or mineral mixture in suitable quantities must be added.

During the rainy season, forest grasses are nutritious and contribute a significant portion of nutrient supply to grazing animals. As the winter advances, these grasses become highly fibrous and lignified. In the latter part of winter and in summer, tree leaves could easily replace these fibrous grasses. Tree leaves have a vast potential for supplying nutrients for animal feeding, particularly during the dry season when feed supply is critical.

Large quantities of mulberry leaves stalks as remnants from silkworm feeding are available for livestock feed. These stalks contain 11.4% crude protein, 2.7% ether extract, and 34% crude fiber. The DCP and TDN contents are 7.8 and 48.4%, respectively.

Spent lemon grass (*Cymbopogan flexusux*) is a by-product in the lemon grass industry. It contains 4.37% DCP and 53.3% TDN. The feeding value of this grass is comparable to that of paddy straw.

Aquatic Weeds

Aquatic weeds of various species are available in Asia and Africa, and this source has not been fully exploited. Some weeds are rich in protein while others are rich in minerals. Agar agar is extracted from sargassam seed, and the residue is available for feeding livestock. The chemical analysis of this product shows an ash content of 33%, while the protein content is only 10%. It can be utilized as a mineral supplement.

Organic Wastes

Livestock and poultry generate a vast resource of nutrients in the form of waste products. The use of animal waste as a source of manure for crop production has been the preferred system for recycling nutrients. However, in recent years animal nutritionists have explored the possibility of recycling animal waste as a feed through animals. Animal and plant wastes with a potential as stock feed include:

I. Poultry wastes
   A. Poultry manure
   B. Hatchery wastes
   C. Poultry dressing wastes
   D. Feathers
II. Cattle wastes
   A. Manure
   B. Dead animals
   C. Slaughterhouse wastes
III. Manure as substrates for protein production
IV. Cellulosic wastes
V. Vegetable- and fruit-processing wastes

Animal wastes as such are unfit for livestock feeding due to their high water content. Thus, the development of suitable and economical processing technology so as to remove excessive moisture and destroy harmful pathogens from organic wastes is a challenging opportunity to the engineering technologist. Under this topic the discussion will be oriented towards the nutritional worth of the various types of animal wastes for livestock feeding. In addition, the potential of these wastes as well as cellulosic wastes as substrates for the production of single-cell protein and wastelage is also briefly mentioned.

Poultry Wastes

Poultry Manure

Poultry keeping in tropical and subtropical regions is becoming popular. About 100 hens can produce 1 t of dried poultry manure (DPM) per year. The dried poultry manure is of two types: (1) caged layer droppings and (2) deep litter droppings. The latter is suitable for ruminants, while the former is suitable for poultry feeding. DPM contains a large amount of protein (25 to 30%), with one third as true protein and the remainder as nonprotein nitrogen, mainly uric acid. The uric acid is highly insoluble, which could be used advantageously as a nonprotein nitrogen source for ruminant feeding.

The true protein fraction contains spilled food protein, undigested dietary protein, metabolic fecal protein (digestive juices, cell debris, and bacterial protein), and probably products of bacterial fermentation both in the caeca and in the droppings voided in the poultry houses. The amino acid inventory of the true protein in DPM is broadly equivalent to a cereal protein such as barley.

The composition of DPM is highly variable. The main causes of variability are probably variation in the composition of feed, variable spillage into the manure, differences in the stage of the lay of hens, and feather shedding. Protein nitrogen is less variable than nonprotein nitrogen.

DPM is rich in calcium and phosphorus, both predominantly in highly available forms. Thus, DPM could make a significant contribution in feed formulation, not only as a source of protein and energy to spare cereal grains but also as a source of calcium and phosphorus.

Poultry manure contains about 35% of total feed energy and may be an important source of energy as well as protein for ruminants. Broiler litter was shown to contain 2440 kcal DE/kg and 59.8% total digestible nutrients on a dry matter basis.[53] These values compare well with that of alfalfa hay. The digestibility of laying house litter by cattle was found to be 59.2%.[54] Nitrogen from poultry manure was shown to be efficiently utilized, making possible a saving of vegetable protein feeds. The palatability of the ration was not affected by adding DPM to a concentrate mixture.

Feeding 2.3 to 4.1 kg air-dried caged layer manure had no adverse effect on milk production. A concentrate mixture containing 30% DPM did not affect milk production of dairy cows. Feeding of DPM did not impart undesirable flavors to milk.[55,56]

A study[57] was conducted to evaluate heat-treated poultry manure as a source of nitrogen for ruminants. Poultry manure was subjected to different heat treatments to produce autoclaved (APM) and dried poultry (DPM) manure. The results showed no difference in the digestibility coefficients of dry matter and energy, but the digestibility coefficients of protein were significantly higher for the ration containing soybean meal rather than APM. The rate of gain was almost the same in all the groups. Several experiments have been conducted in Sweden[58] to investigate the feeding value of dried manure for ruminants. The acceptability of DPM was tested, and it was found that most of the animals consumed about 5 to 6 kg/day without any difficulty. In an experiment with beef cattle, one group was fed 55% cereals and 45% DPM and the other 80% cereals and 20% soybean meal. The growth rate was somewhat higher in the first group mentioned, but the differences were not statistically significant. These results indicated that DPM can substantially replace cereals and soybean meal. True protein in DPM could be protected from breakdown in the rumen by treatment with formaldehyde, a treatment that also removes odors and kills pathogens.

Hatchery Waste

Hatchery waste includes unhatched eggs, dead embryo, culled chicks, and eggshell. The waste contains a comparatively high amount of water. If processed in a dehydrating plant, all material could be used in a poultry ration. It contains about 26% protein, 11% ether extract, and 21% calcium on a dry matter basis. Hatchery waste is an acceptable feed ingredient in poultry ration to replace part of the fish meal and maize. The product is also known to contain an unidentified growth factor. The hatchery waste must be boiled at least 30 min prior to drying since some of the eggs may contain the pullorum disease organisms. Rai and Netke[59] reported that heating eggshells at 110°C for 1.5 hr produced eggshell powder free from organisms. Formaldehyde treatment of these wastes could make this product suitable for ruminant feeding.

Poultry Dressing Waste

In dressing and eviscerating plants, blood and poultry offals may accumulate. Offals consist of meat, feed, intestine, lungs, windpipe, and glands. Dressing waste is a high-quality protein source containing 58% protein and 2910 kcal ME/kg. A poultry processing plant could easily maintain another plant to process the poultry by-product waste (this product is also called poultry by-product meal). Where it is readily available it can be used in considerable amounts in rations for both broilers and laying hens. By using formaldehyde to protect the protein from breakdown in the rumen, it can be used as a ruminant feed.

Feather Meal

The feathers are another by-product of poultry dressing plants, and hydrolyzed feather meals have been used in poultry feeds for several years. Approximately 85 to 90% of the protein in feather meal is contributed from keratin. The high cystine content forms the sulfur bridges in feather protein, which are probably a limiting factor in the utilization of the protein. The nutritive value of feather meal largely depends upon the method of processing used. The processing of feathers at 142 to 153°C with 40 to 60 psi is more suitable for feather meal production. Feather meal could be used in poultry and ruminant rations to supply about 20% of the total protein requirement.

Cattle Wastes

Cattle Manure as a Cattle Feed

Cattle manure is a potential source of energy and protein. Studies conducted in the Western countries have shown that the acceptability of dehydrated cattle feces was excellent when fed to growing sheep at 20 to 80% of the mixed ration. Digestibility of ration nitrogen was quite high.

In order to derive maximum benefit from the nutrients present in cattle manure, different types of processing have been tried. Anthony[60] found that rations containing wet cattle manure were readily consumed by fattening steers, and these rations supported gain essentially comparable to cattle fed without manure. Cooking or washing the manure before mixing with a concentrate mixture did not improve its feeding value. However, adding manure appreciably lowered the amount of nonmanure feed required per unit of gain.

Animal Excreta in Wastelage

The ensiling process is another efficient means of using poultry and cattle manure as feed for cattle. The product prepared by ensiling the manure with low-grade roughages is called wastelage.[61] The ensiling process is similar to the conventional silage making. Caswell et al.[62] ensiled high-moisture corn and broiler litter and found that the digestibilities of dry matter, crude protein, crude fiber, and nitrogen-free extract of the ration containing the wastelage were comparable with unsupplemented ration.

A laboratory study[63] was conducted to estimate the nutritive value of ensiled poultry manure. Fresh poultry manure was mixed with grass hay in the ratio of 100:0, 90:10, 80:20, 70:30, and 60:40. The results indicated that a mixture of 60% poultry manure and 40% grass hay was superior to the other combinations for utilization of nutrients. Subsequently, the possibility of putrefaction was avoided by using broiler litter containing fine shavings as bedding material. The analysis showed that the wastelage was a good quality feed. It also observed that during fermentation, uric acid, urea, and creatinin present in poultry manure were converted into high-quality microbial protein in the wastelage. Feeding of this product to calves resulted in a body weight gain of 1 kg/day, which was equivalent to the gain of the control group.

In tropical and subtropical regions, low-grade roughages such as wheat straw, rice straw, maize, and jowar (sorghum) stalks should be used as beddings in poultry and cattle sheds. The beddings impregnated with the urine and feces of cattle and poultry can be converted to a suitable wastelage for ruminants. It is possible to increase the nutritive value of cellulosic feedstuff and reduce the cost of feeding by the ensiling process.

The feeding value of wastelage for breeding ewes and cows was found to be comparable with hay. About 2.3 kg of wastelage, prepared by mixing manure and hay in a ratio of 57:43, replaced 1.78 kg hay per day in a ewe ration. This shows that the wastelage ration conserved 0.79 kg hay per ewe daily, since the wastelage ration included 0.99 kg hay and 1.31 kg manure.[61] Cattle manure can be fermented and ammoniated to produce a high crude-protein-equivalent animal feed.[64]

Dead Carcasses

Dead animals are another potential source of cattle waste to supply meat meal, bone meal, and tallow for livestock feeding. In India alone about 23 million head of cattle and 10 million head of goat and sheep die every year. The dead carcasses are not properly processed and are simply wasted. Each cattle carcass has the potentiality of producing about 40 kg meat meal, 30 kg tallow, and 25 kg bone meal. Thus, the estimated total amounts of meat meal, tallow, and bone meal would be about 920, 690, and 575 million kg, respectively. Similarly, sheep and goats could produce an additional 60 million kg meat meal. The feeding value of these products is well documented, but the economics of carcass utilization varies from one area to another, depending on labor costs.

Manure as Substrates for Protein Production

Nitrogen excreted in poultry and cattle manures is a potential source for conversion to single-cell protein (SCP). Unicellular organisms such as algae, yeasts, fungi, and mixed cultures of microorganisms can utilize animal waste nitrogen for single-cell protein production.[65,66] The efficiency of conversion of nitrogen to protein by some of these microorganisms is as great as or greater than other systems reviewed.

The common housefly pupae and the earthworm have the ability to utilize poultry and cattle manures to produce high-quality protein feedstuffs. The metabolizable energy value of the pupae was found to be 2528 kcal/kg. Chemical analysis showed that the pupae contained 61.4% protein and 9.3% fat, and their amino acid composition was comparable with that of meat meal and fish meal and was better than soybean meal. The feeding value of the housefly pupae is as good as soybean meal for poultry.[67] The nutritional value of earthworms is not available. The costs of these operations have not been determined but the protein yields by housefly larvae and the earthworm were quite low, probably due to their inability to utilize nonprotein nitrogen from the manures.

Cellulosic Wastes

Cellulosic waste materials, e.g., wood, wheat straw, paddy straw, sugar cane bagasse, other low-grade roughages, municipal organic waste, and cereal and vegetable processing wastes, contain a vast store of energy which can be converted into available food energy by means of two methods: (1) chemical or enzymatic hydrolysis to produce sugars, and/or (2) microbial fermentation to produce single-cell protein.

The production of single-cell protein is not a new phenomenon, as torula yeast has been available for several decades. In Britain, yeast has been produced from hydrocarbons (a product of the petroleum industry) and has been shown to be a useful poultry feed ingredient. Annual production of crude petroleum is about 1500 million tons, containing 40 million tons of high-boiling hydrocarbons. This quantity of hydrocarbons can be utilized for the production of SCP.[83] Calculations indicate that 30 to 35 million tons of microbial cell substance could be obtained. The SCP contains about 50 to 65% crude protein. In Finland, a single-cell protein, Pekilo, was produced by growing fungi on wood pulping; it contains about 55 to 60% protein and can economically replace part of either soybean meal or fish meal protein in broiler, layer, or ruminant rations.

Vegetable and Fruit Processing Wastes

A considerable amount of vegetable and fruit processing wastes are available from canneries. The greatest problem in handling these by-products is the higher percentage of moisture (70 to 90%), which poses the economic problem of drying and conserving these materials when they are not utilized in raw form. Part of this material can be dried by hydraulic press and the remainder can be either sun dried or oven dried. In this process it is likely that some of the nutrients would be lost in the liquid separated from solids in the hydraulic press, but drying of the solids would be quicker. In developing countries very little effort has been made in this direction. If these waste products are properly processed and dried they could yield a substantial quantity of material for livestock feeding. This source could spare roughly equivalent tonnage of cereals for human consumption. These waste products can also be used for the production of single-cell protein.

Chemical analysis shows that fruit wastes are mostly rich in soluble carbohydrates. Some of the products like apple peel, apple core, galgal peel, orange peel, mango peel, mango kernel, pea peel, pea core, and pineapple peel contain as much as 70 to 80% nitrogen-free extract and are thus comparable with high-energy feeds. Some of the products like galgal peel, orange peel, and orange waste are also rich in calcium. Thus, it appears that there is considerable scope for commercial utilization of these waste products in a concentrate mixture.

Vegetable wastes compare with protein-rich roughages. Some of the products like cabbage, cauliflower, and radish leaves contain about 25 to 30% protein. Cabbage stem, cauliflower stem with stalks, and tomato pomace are also rich in protein, containing about 20 to 25%; carrot leaves and pea pods contain 15 to 20% protein, and the remaining waste contains 8 to 15%.

Pineapple bran consists of the skin and often the core of the pineapple, and accounts for an estimated 40 to 50% of the total pineapple weight. Because of its rather high crude fiber content, the bran is more suitable for ruminants than for monogastric animals. The net energy values of pineapple bran and pineapple hay are reported to be 118.8 and 85.8 Mcal/100 kg.[68] Fresh pineapple bran usually contains only about 10% dry matter, so it is more convenient to feed it to ruminants in the vicinity of the factory during the canning season. Long-distance transport is both difficult and expensive. However, if pineapple bran is needed elsewhere for feeding, it is possible to employ such techniques as drying (either in the sun or using a conventional dryer) and conserving as silage, in which case the fresh bran is mixed with other ingredients with higher dry matter content.

Citrus pulp is a by-product of citrus canning, and it contains about 25% dry matter. Many of the considerations relating to pineapple bran apply also to citrus pulp. Surplus quantities of pulp may be conserved as silage when mixed with other ingredients such as molasses, urea, and bagasse. A variety of these ingredients may be used, the aim being an end product sufficiently high in dry matter. Citrus pulp can also be sun dried or drum dried, permitting storage for subsequent feeding. However, the latter method is expensive when large quantities of water have to be removed, and may increase the cost beyond its value as a feed. The aim should be to utilize the material without undue processing costs. For this reason the use of fresh or ensiled material seems to offer the best possibilities. Citrus pulp has an important future in livestock feeding, as evidenced by a number of experiments. The dried citrus pulp can comprise up to 40% of the concentrate ration with excellent results.[69] There appears to be little doubt as to the value of citrus pulp in ruminant livestock rations.

Silk Industry Waste

Large quantities of silkworm pupae are available in areas where sericulture is practiced on a large scale for the manufacture of silk. Silkworm pupae are a good source of animal protein. In the manufacture of commercial raw silk, the silkworm cocoons are steamed to stifle pupae and are then sorted out for the reeling operation (preparation of silk strands). The discarded silkworm pupae constitute a good source of animal protein for livestock.

The discarded pupae are collected, transferred to a vessel, and boiled in a lye solution, (1.25%) of a mixture containing sodium carbonate and sodium hydroxide (2:1), for a period of 5 to 10 min. This softens the outer coating and fosters its easy removal in a subsequent operation. The material is then carried to a washing tank, where the outer coat is taken off manually and the excess chemical removed by washing in running water. The residue is collected and pressed by means of a hand press or screw press to remove water. The drying is carried out in a batch operation and takes about 4 to 6 hr per batch. Under favorable climatic conditions, sun drying can also be practiced. The dried worm pupae contain about 25% fat and 55% protein. The flow sheet for processing silkworm pupae is shown in Figure 3.

Deoiled silkworm pupae meal contains 76 to 80% crude protein and is a rich source of lysine and methionine. It can replace 50% of fish meal in livestock rations, particularly in poultry feeds,[70] and can also be used in ruminant rations.

Fish Industry Wastes

Large quantities of marine protein resources, either in the form of fish wastes or fish meal, are available, but a negligible amount, if any, is used for ruminant feeding. Most of

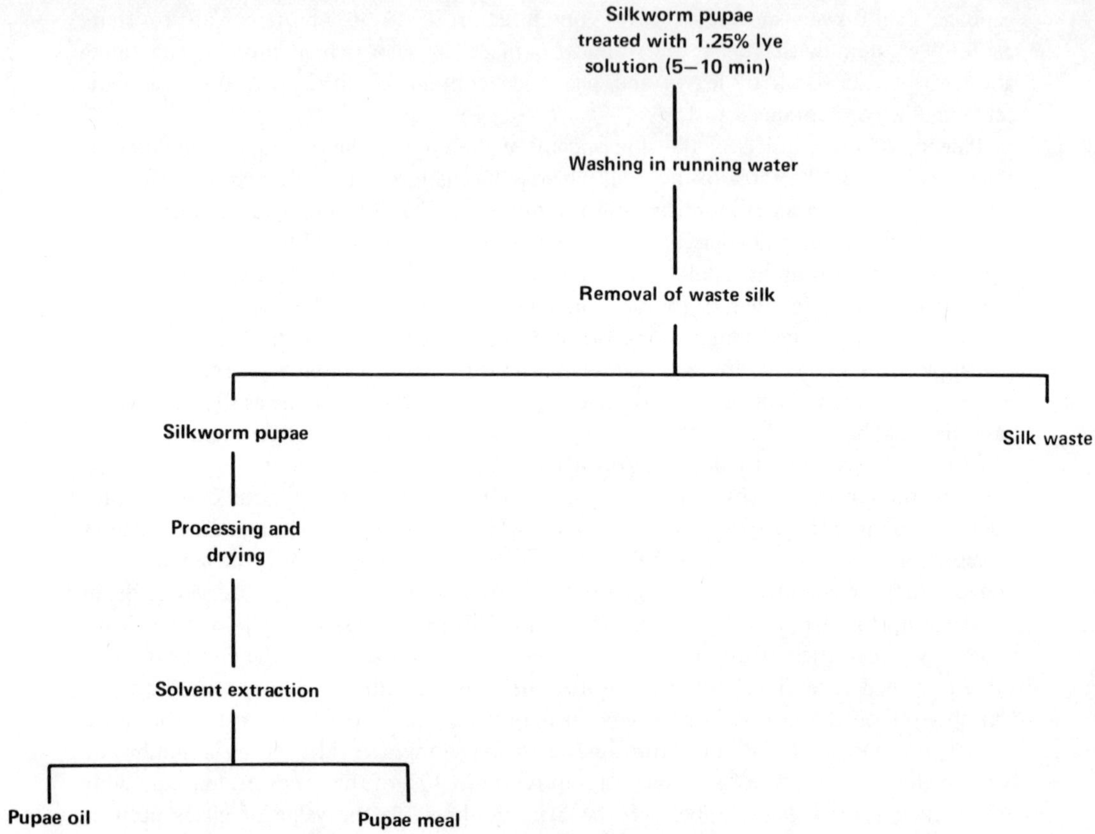

FIGURE 3. Processing of silkworm pupae.

it is utilized in poultry and swine rations because of its high-quality protein. Feeding of low-quality fish meal to ruminants can effectively improve the utilization and dry matter intake of low-grade roughages.

Organic Waste Contaminants and Animal Health

Under practical farm conditions, poultry and cattle inoculate themselves many times daily with all the organisms existing in their fecal matter. This may provide sufficient immunity to the animal, at least for some of the most common organisms present in the fecal matter. Poultry droppings may pose disease problems since DPM may carry *Salmonella pullorum, S. typhimurium, Escherichia coli,* etc., but the problem can be overcome by an appropriate heat treatment. Poultry litter can be sterilized by processing at 150°C for 3 hr or longer. Thus, the proper dehydrating process may further reduce the risk of health hazard. The disease transmission to cattle from DPM is less likely. Available evidence shows that properly processed poultry and cattle manures have no adverse effect on animal health when recycled through poultry or cattle. Cattle manure may possess sufficient estrogenic activity to affect performance in poultry. Chemicals and drugs may accumulate in sufficient quantity in the animal excreta as these additives are allowed in the feed. Residual feed additives and pesticides, if accumulated in larger concentrations in poultry litter, pose a potential health hazard. Information on this aspect is limited, and long-term studies are needed to evaluate the feeding value of manures for livestock and poultry. The dried products should also be examined for microbial and fungal toxins, pathogenic organisms, and parasites. The effect of residual drugs and chemicals on animal health should also be studied.

Published information indicates that animal wastes, particularly poultry manure, cattle manure, and dead carcasses, have the potentiality of supplying protein, energy, and other nutrients for livestock feeding. The contribution of nutrients through animal wastes is sufficiently high and valuable to reduce the pressure on human food and to run the animal industry profitably. In addition, animal wastes can be utilized as substrates for production of single-cell protein suitable for either human food or livestock and poultry feed. The ensiling process is another efficient means of using poultry and cattle manure as feed for cattle. Given the world's present food crisis, the only choice for poultry and dairy industries in developing countries is to raise livestock entirely on agro-industrial by-products and animal wastes (which will be able to make a considerable contribution to the supply of feed). In order to make use of animal wastes as animal feed, it is necessary to develop suitable and economical processing technology. This also depends on the decisions of governments and their concern for providing financial and technical incentives to introduce these new forms of feeds.

PRINCIPLES AND PRACTICE OF PREPARING RATIONS FOR ZEBU CATTLE

A ration is a 24-hr allowance of feed given to meet the nutritional needs of an animal whether it is aimed at maintenance, growth, or production. The farmer is usually interested in a ration that produces efficient growth and production. In order to meet this requirement, the term "balanced ration" was coined to refer to a feed mixture just sufficient to meet the nutritional requirements of a particular class of animal.

The general principles of feeding and computing diets for European breeds of cattle are also applicable to zebu cattle; as a matter of fact, zebu cattle will do well on diets normally fed to European breeds. However, as mentioned previously, climatic, ecological, and socioeconomic factors prevailing in lands where zebu cattle are reared restrict the types and quantities of feedstuffs available and the feeding practices. These factors were kept in mind when formulating diets for zebu cattle.

It is not possible here to indicate even a small proportion of the various computations of diets based on such a wide range of feedstuffs as are available in the different lands and agro-climatic regions where zebu cattle are reared. However, an attempt has been made to enunciate general principles and to illustrate with examples how diets could be computed to meet specific situations. These could be modified to meet other situations. Requisites for computing balanced rations for livestock are

1. Basic considerations
2. Nature and quality of available forages
3. Formulation of meal mixtures from available concentrate feeds
4. Mineral and vitamin supplements to meet common deficiencies
5. Formulation of balanced rations for different classes of zebu cattle

Basic Considerations for Computing Diets or Rations

Proper Balance

It is essential to know nutrient requirements of livestock and composition of feeds available for formulating rations. Energy protein, minerals, vitamins, and nonnutritive feed additives are the main components to be taken into account for feed formulation. Low performance of animals is most usually caused by faulty feed formulations, leading to either insufficient or excessive intake of nutrients.

Appetite of the Animal

Appetite of the animal is measured by voluntary dry matter intake from the ration.

Dry matter consumption varies with the live weight of the animal, its physiological state, and digestibility of the ration. Cattle generally eat about 2.0 to 2.5 kg dry matter per 100 kg live weight. Milch cattle have a slightly higher digestive capacity to consume dry matter per unit-weight.

Supply of Required Nutrients

The quantities of DCP, TDN (energy), and other nutrients required by the animal are calculated from a table of nutrient requirements and are supplied through natural and synthetic feedstuffs in such quantity that the bulk is well within the digestive capacity of the animal.

Forage in the Ration

The major quantum of nutrient requirements (more than two thirds) is normally supplied through the forages, and the rest comes from the concentrates.

Acceptability

Whatever feed is given to the animal, it must be to the animal's liking. For profitable production the animal must be able to eat to the maximum of its appetite, which is only possible if the feed is palatable.

Laxative

Constipation is the cause of most digestive troubles. Therefore, it is necessary to include feeds that are laxative in nature.

Bulk

The stomach of cattle is very capacious. The animals do not feel satisfied unless their rumen are properly filled up, so fibrous and bulky feeds (roughages) should be incorporated in their rations.

Assortment of Feed Ingredients

By combining many feed ingredients in the ration rather than only a few, a better and balanced mixture of nutrients can be furnished.

Quality of the Feed Ingredients

Moldy, musty, and poor-quality feeds in a ration reduce its feeding value and sometimes also adversely affect the health of the animals.

Processing

Some feeds require special processing before feeding in order to render them more digestible and palatable. For example, hard grains should be crushed and coarse fodders should be chaffed before feeding.

Least Cost Ration

The cost of the ration and the labor in feeding should be minimized to the extent that economic efficiency is not affected. Maximum efforts should be made to utilize home-grown grains and fodders. Linear programming for formulating least cost ration may be used.[71]

Nature and Quality of Available Forages

Available forages should be combined to obtain maximum efficiency of production. Some suggested combinations to obtain different levels of DCP are

DCP 10% (10 kg of DM)

Legume (green fodder)	40 kg
Straw/stover	2 kg
Legume hay	11 kg

DCP 8% (10 kg DM)

Legume (green fodder)	32 kg
Straw/stover	4 kg
Oats (green)	40 kg
Nonlegume (green)	20 kg
Legume hay	6 kg
Silage (nonlegume)	15 kg
Legume hay	7 kg
Nonlegume (green)	10 kg
Legume hay	8 kg

DCP 6% (10 kg DM)

Legume (green)	25 kg
Straw/stover	5 kg
Oats (green)	30 kg
Straw/stover	2.5 kg
Nonlegume (green)	30 kg
Legume hay	3 kg
Silage (nonlegume)	25 kg
Legume hay	4 kg

DCP 4% (10 kg DM)

Nonlegume (green)	40 kg
Silage (nonlegume)	35 kg
Legume green	18 kg
Straw/stover	6.5 kg

Note: Recommended forage combinations having a DCP of 8 to 10% can fully meet maintenance requirements and part of the production needs of cattle. Forage combinations having a DCP of 6% will only meet the maintenance requirements of cattle. Combinations having less than a 6% DCP will not meet the maintenance requirements of cattle.

Formulation of Meal or Concentrate Mixtures

The terms *concentrates, grains, grain mixtures,* and *meal mixtures* are used interchangeably to mean grains, seeds, and their by-products. They are low in fiber and rich in TDN. The nutrients available in roughages are usually not sufficient to meet the requirements of high-producing animals, so a concentrate mixture is essential in dairy cattle rations to meet nutrient demand for milk production and in the rations of actively growing cattle.

For formulating a balanced meal mixture for cattle the following points should be considered:

1. DCP in the meal mixture
2. The proportions of main categories of feed ingredients
3. Specifying the feeds to be used in each category and stating the limits of substitution permissible within each category
4. The quality of the roughages available to the animal is the basic factor in determining the percentage of DCP needed in the mixture. When the DCP in roughages is high, the meal mixture may be prepared to contain 12 to 13% DCP. When poor-quality roughages are available, the DCP in the meal mixture should be about 15 to 16%.
5. The feed ingredients of the meal mixture are grouped according to the principal nutrient they furnish. The different groups of feed and their ideal proportions to be added in the concentrate mixture are

I. Basal feeds (energy feeds)
   A. Grains – 30 to 50%
   B. By-products – 30 to 40%
II. Protein supplements – 25 to 35%
III. Mineral supplements
   A. Common salt – 1%
   B. Mineral mixture – 1 to 2%
IV. Vitamin supplements – required only under certain conditions

The proportion of basal feeds and protein supplements can be calculated using Pearson's square; an example is illustrated in Figure 4.

Limits of Certain Ingredients for Use in a Meal Mixture

The commonly available feed ingredients most often used in preparation of cattle meal mixtures with their minimum and maximum limits are shown in Table 8.

Adequacy of Nutrients Other than Protein

The meal mixture should be adequate in total digestible nutrients, i.e., ranging from 70 to 75%. The calcium and phosphorus levels in the concentrate mixture should range between 0.4 and 0.6 and 0.3 to 0.5%, respectively. Besides these, adequate levels of various trace minerals (copper, cobalt, iron, iodine, etc.) in the meal mixture must be ensured. Limits of feed ingredients under normal conditions are indicated in Table 8.

Economics of a Concentrate Mixture

Economics must be kept in mind while selecting feed ingredients for formulation of meal mixtures for cattle. Linear program technique may be profitably used for computation of feed mixtures meeting given nutritional and economic specifications.[71]

The composition of concentrate mixtures for dairy cattle is given in Table 9, while Table 10 lists the composition of concentrate feed formulae as recommended by the Indian Standards Institution[72] for feeding to zebu cattle. Suggested concentrate feed

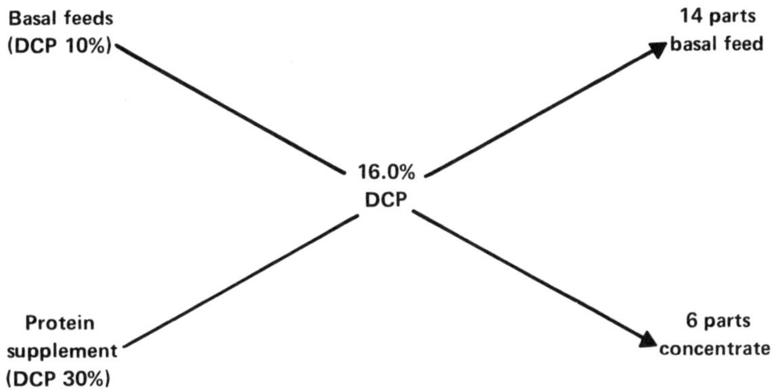

FIGURE 4. Example of calculation of the proportion of basal feeds and protein supplements using Pearson's square. For each 100-kg meal mixture containing 16% digestible crude protein (DCP) the basal feed required is $\frac{14}{20} \times 100 = 70$ kg; the protein concentrate required is $\frac{6}{20} \times 100 = 30$ kg.

Table 8
LIMITS OF CERTAIN INGREDIENTS FOR USE IN A MEAL MIXTURE

Ingredient	Recommended (kg/quintal)		
	Minimum	Maximum	Good level
Basal Feeds			
Grains			
Maize	0	65	35
Barley	0	40	25
Oats	0	40	25
Jowar (sorghum)	0	40	25
Milo	0	40	25
Wheat shorts, etc.	0	40	20
Gram (chickpea)	0	30	20
By-products			
Wheat bran	0	40	25
Rice polish	0	40	25
Molasses	0	15	10
Pulse *chuni*	0	15	10
Protein Supplements			
Groundnut cake	0	30	20
Linseed cake	0	30	20
Cottonseed cake	0	30	20
Rapeseed cake	0	25	20
Soybean cake	0	30	20
Seasame seed cake	0	30	20
Brewer's dried grain	0	20	10
Corn gluten meal, etc.	0	25	15
Mineral Supplements			
Common salt	0.5	1	1
Mineral mixture	1	2.5	1

Table 9
COMPOSITION OF CONCENTRATE MIXTURES FOR DAIRY CATTLE

Ingredient	Concentrate mixture 1[a] (%) 1	Concentrate mixture 1[a] (%) 2	Concentrate mixture 2[b] (%) 1	Concentrate mixture 2[b] (%) 2
Grains				
Maize/wheat (shorts)	30	20	25	15
Barley/oats/pearl millet/sorghum	15	25	13	20
Oilseed cakes				
Groundnut/sesamum/soybean	10	13	20	23
Linseed/rapeseed/cottonseed	13	10	10	10
By-products				
Wheat bran/rice bran	30	30	30	30
Minerals				
Common salt	1	1	1	1
Mineral mixture	1	1	1	1

[a] Digestible crude protein 13%, total digestible nutrients 70%; recommended with high-legume forage.
[b] Digestible crude protein 16%, total digestible nutrients 70%; recommended with high nonlegume forage.

formulae for feeding to zebu cattle under scarcity conditions and incorporating various by-products are given in Table 11.

Mineral Supplements

It has long been known that minerals and vitamins are necessary for the health and well-being of an animal and even for the maintenance of life itself. Unless these are supplied either in natural feed or otherwise, troubles will often appear. These include (1) failure of the animal to grow to full size, (2) failure of the cow to produce as much milk as it should, (3) failure of the cow to breed regularly, (4) failure of the cow to carry her calf for the proper length of time, and (5) development of stiff and enlarged joints. If there is no disease and if the feeding seems to be ample but the animal is not growing or producing according to one's satisfaction, then a lack of minerals or vitamins in ration may be suspected.

Mineral Supplements

The amount of minerals in natural feeds, especially roughages, is less than was formerly thought. This is first due to the fact that the mineral content of the soil tends to decrease because the trace minerals are not normally added in the form of fertilizers. Second, heavy application of fertilizers results in higher forage yields with less mineral content. Also, milk production of cows has been increased through improved breeding techniques, and this increases the mineral requirement. To obtain good performance, it is

Table 10
RECOMMENDED CONCENTRATE FEED FORMULAE FOR ZEBU CATTLE

These formulae have been so computed that each formula is by and large expected to provide 14 to 16% of digestible crude protein and 68 to 74% of total digestible nutrients. Suitable modifications may be made to these formulae according to locality, season, and availability of ingredients. Where the proportion of mineral mixture and common salt to be mixed has not been indicated, in such cases not less than 1% by mass of Mineral Mixture Type 2 from Table 13B and common salt not exceeding 2% by mass should be added.[73]

1.	Cottonseed	20
	Horse gram (*Dolichos biflorus*)	20
	Groundnut oil cake	30
	Wheat bran	30
2.	Maize	20
	Cottonseed	10
	Horse gram (*Kulthi*)	10
	Groundnut oil cake	30
	Wheat bran	30
3.	Wheat bran	50
	Rice polish	10
	Groundnut oil cake	20
	Seasamum (Til) oil cake	20
4.	Cottonseed cake	15
	Groundnut oil cake	20
	Wheat bran	15
	Rice polish	15
	Tur husk (*Cajanus cajau* L.)	15
	Horse gram	20
5.	Sesamum oil cake	15
	Cottonseed	15
	Groundnut oil cake	20
	Rice bran	15
	Moong husk	15
	Wheat bran	20
6.	Cottonseed cake	80
	Wheat bran	10
	Oats	7
	Salt	2
	Mineral mixture	1
7.	Tapioca	40
	Gram	10
	Groundnut oil cake	37
	Wheat bran	13
8.	Tapioca	56
	Groundnut oil cake	44
9.	Wheat bran	27
	Groundnut oil cake	30
	Wheat cuttings (unfit for human consumption)	10
	Gram *chuni*	20
	Gram husk	10
	Salt	2
	Mineral mixture	1
10.	Wheat bran	27
	Groundnut oil cake	30
	Gram *chuni*	20
	Gram husk	10
	Gram	10
	Salt	2
	Mineral mixture	1
11.	Wheat bran	25
	Deoiled rice bran	15
	Gram	12
	Groundnut oil cake	25
	Groundnut oil cake (solvent extracted)	20
	Salt	2
	Mineral mixture	1
12.	Maize cobs (whole)	22
	Wheat bran	30
	Groundnut oil cake	20
	Groundnut oil cake (solvent extracted)	15
	Gram	10
	Salt	2
	Mineral mixture	1
13.	Tapioca	25
	Gram	20
	Groundnut oil cake	25
	Wheat bran	20
	Gram husk	10
14.	Cottonseeds	20
	Guar meal	15
	Groundnut oil cake	20
	Rice polish	20
	Maize bran	20
	Bajra husks	5

Table 10 (continued)
RECOMMENDED CONCENTRATE FEED FORMULAE FOR ZEBU CATTLE

15.	Cottonseeds	25
	Groundnut oil cake	30
	Rice polish	20
	Maize bran	25
16.	Cottonseed oil cake	25
	Groundnut oil cake	25
	Rice polish	20
	Maize bran	30
17.	Cottonseeds	10
	Cottonseed oil cake	30
	Guar meal	30
	Wheat bran	30
18.	Cottonseed oil cake	30
	Guar meal	
	Groundnut oil cake	20
	Wheat bran	30
19.	Cottonseed oil cake	12
	Guar meal	30
	Wheat bran	25
	Maize gluten	8
	Damaged wheat	15
	Wheat sievings	10
20.	Cottonseeds	10
	Cottonseed oil cake	5
	Guar meal	30
	Wheat bran	25
	Maize gluten	5
	Damaged wheat	15
	Wheat sieving	10
21.	Cottonseeds	15
	Guar meal	30
	Wheat bran	25
	Tur *chuni*	10
	Maize gluten	5
	Damaged wheat	5
	Wheat sievings	10
22.	Cottonseed oil cake	17
	Guar meal	25
	Groundnut oil cake	20
	Wheat bran	15
	Tur *chuni*	5
	Urad *chuni* (*Phaseolus mungo*)	18
23.	Cottonseeds	25
	Guar meal	40
	Wheat bran	30
	Maize bran	5
24.	Cottonseeds	10
	Guar meal	33
	Groundnut oil cake	5
	Wheat bran	39
	Maize bran	8
	Tur *chuni*	5
25.	Cottonseed oil cake	5
	Guar meal	30
	Groundnut oil cake	35
	Wheat bran	15
	Maize bran	15
26.	Cottonseed oil cake	20
	Guar meal	30
	Groundnut meal	25
	Wheat bran	10
	Maize bran	15
27.	Cottonseed oil cake	30
	Guar meal	25
	Groundnut oil cake	10
	Wheat bran	30
	Urad *chuni*	5
28.	Cottonseed oil cake	35
	Guar meal	25
	Groundnut oil cake	15
	Wheat bran	25
29.	Guar meal	20
	Groundnut oil cake	30
	Rice polish	25
	Wheat bran	25
30.	Cottonseed oil cake	25
	Groundnut oil cake	25
	Rice polish	20
	Maize bran	30
31.	Groundnut oil cake	25
	Wheat bran	38
	Red gram dust	20
	Gram husk	10
	Molasses	5
	Minerals	2
32.	Groundnut oil cake	35
	Tapioca waste	35
	Wheat bran	30
33.	Groundnut oil cake	40
	Rice bran	20
	Tapioca waste	20
	Molasses	20

Table 10 (continued)
RECOMMENDED CONCENTRATE FEED FORMULAE FOR ZEBU CATTLE

34. Groundnut oil cake	30
Dew gram	20
Gram husk	20
Wheat bran	30
35. Cottonseed oil cake	30
Cottonseed	20
Tamarind seed (powder)	10
Gram husk	10
Rice bran	30
36. Tapioca (chips)	25
Groundnut oil cake	30
Gram *chuni*	25
Rice bran	20
37. Sesamum (til) oil cake	35
Rice bran	20
Gram husk	15
Horse gram	30
38. Cottonseed oil cake	30
Cottonseed	20
Rice bran	20
Gram husk	10
Horse gram	20
39. Jowar	25
Sesamum (til) oil cake	25
Wheat bran	20
Gram husk	10
Horse gram	10
Gram *chuni*	10

	40	41	42	43	44	45	46
Pearl millet/barley/tapioca chips/sorghum/distillery sludge/ragi	25	20	35	25	–	37	–
Wheat bran/rice bran/tapioca meal/*chuni*/oats/paddy	40	30	–	–	37	15	10
Solvent-extracted rice bran	–	–	30	22	–	–	10
Soyabean cake/groundnut cake/sesamum cake/maize gluten meal	32	–	32	–	–	25	–
Guar meal/mustard cake/linseed cake/cottonseed cake	–	47	–	50	30	–	25
Maize gluten feed/coconut cake	–	–	–	–	30	–	52
Groundnut cake/cottonseed cake (undecorticated)	–	–	–	–	–	20	–

Table 11
SUGGESTED FEED FORMULAE FOR FEEDING ZEBU CATTLE UNDER SCARCITY[a]

Mango-seed Kernel

	Parts
A. Groundnut cake	25
Jowar (*Sorghum vulgare*)	25
Rice bran or mango-seed kernel	20
Gram husk	10
Gram	20
B. Groundnut cake	20
Wheat bran	40
Gram	20
Cottonseed	10
Barley or mango-seed kernel	10

Mahua (*Brassica latifolia*) cake

Groundnut cake	25
Rice bran	10
Mahua cake	20
Gram	20

Mahua Flowers

Groundnut cake	30
Wheat bran	27
Gram *chuni* or mahua flowers	20
Gram husk or mahua flowers	10
Gram	10
Salt	2
Mineral mixture	1

Babul (*Acacia arabica*) Pods

Groundnut cake	20
Wheat bran	15
Rice polish	15
Horse gram	20
Cottonseed cake or babul pods	15
Tur, gram husk, or babul pods	15

Tamarind (*Tamarindus indicus*) Seeds

Sesamum (til) cake	35
Horse gram	20
Rice bran or tamarind seeds	20
Tur, gram husk, or tamarind seeds	15

Rain tree (*Enterolobium saman*) Pods

Finely crushed rain tree pods can conveniently replace rice bran in any one of the foregoing feed mixtures.

Sunhamp (*Crotalaria juncea*) Seeds

	Parts
A. Cottonseed oil cake	5
Groundnut oil cake	38
Wheat bran	15
Maize bran	15
Sunhemp seeds (crushed)	27
B. Cottonseed oil cake	10
Groundnut oil cake	35
Wheat bran	15
Maize bran	15
Sunhemp seeds (crushed)	25

Virginia Tobacco-seed Cake

Tobacco-seed cake	25
Cottonseed cake	5
Cottonseed	20
Wheat bran	30
Gram husk	10
Gram	10

Silk Cottonseed Cake

Groundnut cake	20
Gram	30
Silk cottonseed cake	30
Maize	20

Tapioca waste

A. Groundnut cake	30
Tapioca waste	50
Maize/barley	20
B. Groundnut cake	20
Tapioca waste	30
Wheat bran	40
Maize/barley	10

Tomato Pomace

A. Groundnut cake	30
Maize/barley	30
Wheat bran	20
Tomato pomace	20
B. Maize/barley	40
Tomato pomace	30
Groundnut cake	20
Wheat bran	10

[a] Where the proportion of mixeral mixture and salt to be mixed has not been indicated, not less than 1% by mass of Mineral Mixture Type 2, Table 13B, and common salt not exceeding 2% by mass should be added.

Table 11 (continued)
SUGGESTED FEED FORMULAE FOR FEEDING ZEBU CATTLE UNDER SCARCITY[a]

Processed Chicken Excreta from batteries or cages

	Parts
A. Groundnut cake	20
Processed chicken excreta	20
Maize/barley	35
Gram husk	10
Wheat bran	5
Molasses	10
B. Processed chicken excreta	20
Tapioca	40
Rice bran	10
Sesamum cake	30

Aniseed Cake

Groundnut cake	20
Gram	20
Aniseed cake	20
Maize/barley	20
Gram husk	10
Molasses	10

Bijada Cake

Maize or barley	40
Groundnut cake	20
Bijada cake	20
Gram husk	10
Molasses	10

Citrus Pulp and Peel

A. Groundnut cake	30
Citrus by-product meal	30
Wheat/rice bran	20
Gram husk	10
Molasses	10
B. Sesamum cake	20
Groundnut cake	10
Citrus by-product meal	30
Gram *chuni*	20
Wheat/rice bran	20

Tusser Silkworm Pupae

	Parts
A. Silkworm pupae	20
Maize/barley/oats	20
Gram husk	10
Any pulse *chuni*	15
Groundnut cake	10
Wheat/rice bran	15
Molasses	10
B. Silkworm pupae	25
Maize/barley/oats	10
Tapioca	10
Gram husk	10
Any pulse *chuni*	15
Wheat/rice bran	10
Molasses	5

Panewar Seed

A. Barley/oats	20
Groundnut cake	20
Gram	15
Wheat bran	20
Panewar seed (*Cassia tora*)	20
Gram husk	5
B. Wheat bran	15
Panewar seed (*Cassia tora*)	25
Gram husk	5
Cotton seed	10
Maize grit	10
Sesamum cake	25
Tapioca	10

Maize Husk

Groundnut cake	20
Gram	10
Maize husk	20
Wheat bran	30
Barley husk	20

Table 11 (continued)
SUGGESTED FEED FORMULAE FOR FEEDING ZEBU CATTLE UNDER SCARCITY[a]

Concentrate Mixture (Tapioca Leaf Meal)

	Parts
Tapioca leaf meal	30
Maize or barley	30
Groundnut cake	20
Wheat bran	10
Gram	10

Maize and Rice Bran By-products

	Parts
Corn grit	8
Groundnut cake	20
Rice bran	20
Cane molasses	9
Fish meal	1
Wheat bran	34
Gram (*Ceicer arietenum*)	8

Urea and Molasses

	Parts
Corn grit	22
Groundnut cake	15
Rice bran	30
Cane molasses	15
Fish meal	2
Wheat bran	11
Maize grain	3
Urea	2

recommended that a high proportion of good quality forages be included in the rations of zebu cattle so as to supply most of their nutrient requirements for maintenance and production from forages. This practice in some cases necessitates mineral imbalances in the rations which must be corrected by including appropriate mineral supplements.

The minerals that are most likely to be lacking in the ration are sodium, chlorine, calcium, phosphorus, iodine, cobalt, iron, and copper.

Mineral Sources

For preparation of suitable mineral supplements, it seems advisable to indicate the more common supplementary sources of these nutrients, and the information is summarized in Table 12. Two types of mineral mixtures are usually formulated, one containing common salt and the other without salt; requirements for these are given in Table 13.

Formulation of Mineral Supplements

The principal ingredients contributing to the weight of the mineral mixture are the calcium/phosphorus carrier(s) and common salt. The recommended formulae for preparing mineral supplements are given in Table 14.

Problems of Excess Use of Different Minerals in Supplements

Excesses of calcium, phosphorus, and NaCl each lead to different problems:

Salt – Common salt acts not only as a nutrient but also as a condiment to encourage voluntary consumption of the mineral mixture. As the proportion of the salt in

Table 12
SOURCES OF MINERALS AND THEIR POTENCY[71,72]

Nutrient	Common source	Composition	Remarks
Calcium	Bone meal	26% Ca	Also has 18% protein and 11% phosphorus
	Bone meal (steamed)	29% Ca	Has 12% protein and 14% phosphorus
	Bone ash	27% Ca	Contains 13% phosphorus
	Tricalcium phosphate	13% Ca	10% P
	Dicalcium phosphate	24% Ca	20% P
	Monocalcium phosphate	16% Ca	12% P
	Ground limestone	24–36% Ca	Balance is carbonate and magnesium
	Calcium carbonate	40% Ca	–
	Oyster shells	38% Ca	–
Phosphorus	Bone meal and calcium phosphate	See above	
	Rock phosphate	14% P	Not advised unless guaranteed to contain less than 1% fluorine
Iodine	Potassium iodide	76% I	
	Sodium iodide	84% I	
	Potassium iodate	59% I	
	Iodized salt	0.02–0.05% I	
Iron	Ferric oxide	35% Fe	
	Ferrous sulfate	20% Fe	
	Reduced iron	80–100% Fe	
Cobalt	Cobalt sulfate	35% Co	
	Cobalt chloride	46% Co	
Copper	Copper sulfate	25.4% Cu	

Table 13
REQUIREMENTS FOR MINERAL MIXTURES SUPPLEMENTING CATTLE FEEDS

A. Type 1 (Containing Salt)

No.	Characteristic	Requirement
1.	Moisture, percent by weight, max	7
2.	Calcium, percent by weight, min	22
3.	Phosphorus, percent by weight, min	9
4.	Salt (chlorine as sodium chloride), percent by weight, min	22
5.	Iron, percent by weight	0.4–0.6
6.	Iodine (as KI), percent by weight	0.02–0.10
7.	Copper, percent by weight	0.06–0.10
8.	Manganese, percent by weight	0.09–0.12
9.	Cobalt, percent by weight	0.01–0.02
10.	Fluorine, percent by weight, max	0.03
11.	Spores of *Bacillus anthracis, Clostridium* spp.	Nil

Note: The values specified for Characteristics 2 through 10 are calculated on a moisture-free basis.

B. Type 2 (Without Salt)

No.	Characteristic	Requirement
1.	Moisture, percent by weight, max	7
2.	Calcium, percent by weight, min	28
3.	Phosphorus, percent by weight, min	12
4.	Iron, percent by weight	0.50–0.75
5.	Iodine (as KI), percent by weight	0.026–0.130
6.	Copper, percent by weight	0.077–0.130
7.	Manganese, percent by weight	0.12–0.15
8.	Cobalt, percent by weight	0.013–0.026
9.	Fluorine, percent by weight, max	0.04
10.	Spores of *Bacillus anthracis, Clostridium* spp.	Nil

Note: The values specified for Characteristics 2 through 9 are calculated on a moisture-free basis.

the mixture increases, the intake of the other minerals decreases. Calcium and phosphorus may become the limiting factor for growth, production, and maintenance. Thus, the use of excess salt in the mineral supplement should be discouraged.

Calcium – Excessive calcium in the mineral mixture usually affects the quantities of phosphorus and perhaps salt. It can arise mostly because limestone is cheaper than a phosphorus carrier. In general, no supplementary calcium (without phosphorus) is needed in herbivorous rations.

Phosphorus – Excessive use of phosphorus becomes a problem when the carrier contains fluorine, as many carriers do, especially those in the lower price groups.

Requirements for Mineral Supplements

Mineral requirements of zebu cattle can be met by the addition of 1% each of common salt and a mineral supplement in the concentrate mixture when dry matter intake from concentrate is one third of the total. When animals are kept on a daily, high-forage ration, 20 to 30 g each of common salt and mineral supplements or 40 to 60 g of combined supplement should be added for adult cattle.

Table 14
FORMULAE FOR MINERAL SUPPLEMENTS FOR CATTLE

Ingredient	No. 1[71] (using high P carriers)	No. 2[71] (using low P carriers)	No. 3[81] (using chalk)
Monocalcium phosphate	25 kg	–	–
Dicalcium phosphate	20 kg	–	12
Bone meal			
Raw	–	18 kg	–
Steamed	–	20 kg	45 kg
Ash	–	19 kg	–
Rock phosphate	–	20 kg	–
Ground chalk	–	–	10 kg
Salt	50 kg	33 kg	30 kg
Copper sulfate	225 g	150 g	0.1 kg
Potassium iodide	2.8 g	2 g	0.25 kg
Cobalt chloride	8 g	5 g	0.01 kg
Manganese sulfate	210 g	140 g	0.14 kg
Yellow oxide of iron	–	–	0.5 kg
Sodium carbonate	–	–	0.75 kg
Starch	–	–	0.75 kg
Sodium thiosulfate	–	–	0.75 kg

Note: If vitamin supplementation is considered necessary, 200 g of a mineral-stabilized preparation containing at least 300,000 IU of vitamin A and 50,000 IU of vitamin D_3 per gram should be incorporated per 100 kg of mineral mixture.

Vitamin Supplements

Vitamins, like minerals, are essential for growth, production, and maintenance. There are at least 15 or 16 vitamins whose existence has been proven and their functions studied. Only three of these, vitamins A, D (or their precursors), and E, need be furnished under certain conditions to mature cattle in the feed supplement; cattle seem to be able to synthesize others.

Vitamin A

The best way to supply needed vitamin A is to furnish to the animal an adequate supply of green fodder throughout the year. Reports indicate that a mature cow towards the end of her gestation would need 85 to 100 mg of carotene daily. This would be supplied with ½ kg of freshly cut lucerne, 2½ kg of good quality lucerne hay, or 10 to 15 kg of poor quality forage. Supplementary vitamin A may be needed when only stored forage or dry forage and grain are fed, or for higher production. Under high quality roughage feeding programs, vitamin A supplementation in the ration is not recommended. Requirements for vitamin A are higher in young calves, so a suitable supplement must be included in their ration.

Vitamin D

Cattle exposed to sun rays usually will not lack vitamin D since the ultraviolet rays acting upon pro-vitamins in their bodies produce sufficient amounts of the vitamin for their needs. High quality field-cured hay is the best forage for supplying vitamin D. Young, growing calves should be given vitamin D through some suitable vitamin supplement since their requirements are higher than those of adult cattle.

Vitamin E

Requirements of adult cattle for vitamin E are very low and are met with normal feeds. Supplementation may be required under scarcity conditions.

Rations for Different Categories of Zebu Cattle

This section discusses feeding schedules for different categories of zebu cattle under the following categories:

Rations for calves
Rations for growing stock
Rations for milch animals
Rations for dry cows
Rations for bulls
Rations for working bullocks

Rations for Calves

The feeding schedule recommended for young dairy calves from birth to 3 months of age is as follows:

Age 0 to 3 days – A newly born calf is actually a monogastric animal; its rumen starts developing after 3 to 4 weeks of life. Thus, its need for high quality protein (animal protein) and vitamins is obvious. Colostrum is very rich in nutrients and is laxative in nature. The most important contribution of colostrum lies in the presence of antibodies which when ingested protect the calf from diseases in the early stages of its life. The quantity of colostrum a calf should get will vary according to its birth weight, which mainly depends on the breed. Calves from crossbred cows are usually heavier than those from zebu cows. A general rule is to feed the calf daily about 1 kg of colostrum for each 10 kg of its body weight, but the total quantity should not exceed 3 kg. Colostrum should be fed in equal proportions two or three times daily. If no colostrum is available due to death of the dam or otherwise, the calf should be fed whole milk supplemented with vitamin A and about 30 ml of castor oil daily. Castor oil will act as a laxative.

Age 3 to 30 days – Calves should normally be weaned 3 to 4 days after birth. A simple method of feeding the calf is from a pail, and milk should be fed as fresh as possible. As a rule of thumb, milk should be offered at one tenth of the body weight of the calf, and the quantity of whole milk to be fed should be adjusted at weekly intervals. The daily allowance should be divided into at least two feedings. If there is any digestive disturbance, underfeeding is better than overfeeding. When calves are 2 to 3 weeks old, green fodder should be offered along with whole milk. Forage intake will help in the early development of the rumen. Legume fodders are best for calves because they are a rich source of nutrients, palatable, and slightly laxative in nature. If greens are not available, high quality hay may be offered to calves.

Age 1 to 3 months – After 1 month of age, calves will start consuming roughages, so green fodders or high quality hay should be offered ad libitum till the age of 3 months.

Milk replacers – Milk replacer, a constituted feed, is obviously cheaper than milk and can be substituted for it in the feeding of young calves. At an early age a calf cannot consume a great bulk of feed, and thus requires milk replacer, which can be fed on a milk-dry matter equivalent basis. A milk replacer should resemble more or less the broad chemical composition of milk in terms of protein quantity and quality and amino acid, volatile fatty acid, mineral, and vitamin contents. A milk replacer should have a biological value equivalent to milk, and the ingredients used for constituting a milk replacer should be low in crude fiber. The replacement of milk by a milk replacer should be gradual to avoid any drop in growth rate. In addition, a milk replacer should include butyric acid,[74,75] citric acid,[76] and some antibiotics like Aurofac® or TM-5® as additives to stimulate growth and maintain vitality. The polysaccharides are not ordinarily digested, so sufficient time is required to allow the calves to adapt to the added carbohydrates;[78] hence, the change from milk to milk replacer should be very gradual. It is necessary to give a minimum of one liter of milk per day with the milk replacer to impart a milky taste. Milk replacer should be diluted with clean water to the extent of 1:8.[79] The ingredients of a milk replacer are listed in Table 15, and the rate of feeding is given in Table 16. With the milk replacer given in Table 15 it is possible to raise calves that weigh 25 kg at birth to 51 kg at 13 weeks and 95 kg at 25 weeks, which compares favorably with controls given milk at the rate of 1/10th of body weight up to 30 kg body weight and thereafter 1/20th of body weight up to 60 kg body weight. The respective weights in the control group were 53 kg at 13 weeks and 96 kg at 25 weeks.[77] With this schedule calves can be raised with as little as 114 l of milk plus milk replacer, as compared to controls given 299.2 l of milk.

To reduce calf-rearing costs calf starter mixtures are also used. Feeding of a high quality calf starter, as described in Table 17, should start when calves are 1 month old. Milk consumption should be reduced by one liter for every 200 g of the starter consumed; in addition, green legume forages or good quality legume hays should be offered ad libitum.

Care and Feed of Young Stock over 6 Months of Age

After the dairy heifers and male calves have passed the young stage (after 6 months of age) and before they have reached a productive stage, the tendency has been to devote as

Table 15
INGREDIENTS IN MILK REPLACER

Ingredient	Kilograms
Wheat	10.0
Fish meal	12.0
Linseed meal	40.0
Whole milk powder	13.0
Coconut oil	8.0
Linseed oil	4.0
Citric acid	1.5
Molasses	8.0
Mineral mixture[a]	3.0
Butyric acid	0.3
Antibiotic mixture	0.3
Rovimix® A, B_2, D_3	0.015
	99.5

[a] Mineral mixture per 3 kg:

Dicalcium phosphate	1.6500
Sodium chloride	0.9000
Chalk	0.3312
Magnesium carbonate	0.0900
Ferrous sulfate	0.0150
Copper sulfate	0.0021
Manganese dioxide	0.0021
Cobalt chloride	0.0015
Potassium iodide	0.0003
Sodium fluoride	0.0003
Zinc sulfate	0.0075
Total	3.0000

Table 16
RATE OF FEEDING MILK REPLACER TO CALVES

Age (days)	Body weight (kg)	Colostrum	Milk (l)	Milk replacer (g)
0–5		1/10 of body weight	–	–
6–9		–	1/10 of body weight	
10–13			As above	100
14–17			–½[a]	175
18–21			–1	250
22–25			–1½	325
26–29	35		–2	375
30–33			–2½	450
	40		1 (min)	450
	45		1 (min)	500
	50		1 (min)	600
	55		1 (min)	650
	60		1 (min)	700
	65		1 (min)	750

[a] The quantity of milk to be fed on the 13th and subsequent days is obtained by subtracting the value in liters from the quantity of milk calculated on a body weight basis.

Table 17
COMPOSITION OF CALF STARTER MIXTURE

Ingredient	Parts (%)
Maize/wheat (short)	36
Barley/oats (ground)	10
Groundnut cake/sesamum cake/soybean meal	30
Fish meal/meat meal/skim milk powder	7
Wheat bran/rice bran	15
Mineral mixture	2
Vitamins A, B_2, and D supplement (Rovimix® A, B_2, D_3; g/100 kg)	0.015

little care and feed as possible. Malnutrition, especially in stock between the age of 6 months to maturity, affects growth, sexual maturity, and other parameters that may hamper their future usefulness. To compute a ration for young stock the following guidelines are essential:

1. Work out the nutrient requirements of the young stock according to their body weight. The level of nutrition will vary according to the expected daily weight gain of the animal. Adequate care should be taken while formulating rations for crossbred cattle because they grow at a faster rate than zebu cattle.
2. In growing animals the rumen is not fully developed, so the total dry matter to be offered in the ration should be restricted to a maximum of 2% of their body weight.
3. Since growing animals will consume less dry matter a part of their nutrient requirements should be met through concentrate feeds.
4. Animals receiving high energy rations become fat; thus, they develop fatty udders with limited milk secreting tissue and produce poorly. A liberal supply of protein, minerals, and vitamins is beneficial for overcoming this problem.

5. When legume fodders are plentiful, animals over a year of age will be able to consume enough green fodder to meet their protein requirements. Their energy requirements should be met by supplementing the concentrate mixture with a low percentage of digestible crude protein such as cereal grains.

6. A growing animal cannot consume sufficient nonleguminous fodders to fulfill protein or energy requirements. Under such conditions a concentrate mixture with a high percentage of protein is recommended.

7. Heifers within 3 months of calving need additional nutrients for the growth of the fetus. These should be met by feeding a suitable concentrate mixture.

Recommended rations for growing stock having varying body weights and getting optimum body growth are given in Table 18.

The National Dairy Research Institute, Karnal, India[80] has produced feeding schedules for growing zebu stock on cultivated fodders, by-products, and concentrates; some of these illustrations are given in Table 19. By 6 months of age the calves weigh about 100 to 150 kg and can utilize a good deal of forage. More than 70% of the requirements of bulk energy are met by fodder, while the remaining 30% of bulk and energy and about 70% of the protein requirements must be met by a concentrate mixture. If well cared for and well fed, an animal should weigh about 150 to 250 kg at 1 year of age and about 200 to 300 kg at 18 months old. During this period, nutrients for growth must be obtained from fodders and concentrates (as given in Illustrations 1 through 5 in Table 19). For every 50-kg increase in body weight a revised schedule must be prepared; such a schedule is also applicable to fast-growing exotic crossbreds and buffalo calves.

It is evident from Illustrations 1 through 4 that as the body weight increases the quantity of concentrate mixture can be reduced to balance the requirements. Furthermore, certain poor quality roughages such as wheat straw and rice straw can also be incorporated at higher body weights, and the ingredients should meet the dry matter, DCP, and TDN requirements. If good leguminous fodder is not available, there is no

Table 18
RATIONS FOR YOUNG GROWING STOCK

		Ration 1			Ration 2	
Body weight (kg)	Daily gain (g)	Legume (kg)	Straw/ stover (kg)	Concentrate mixture 1[a] (kg)	Nonlegume green (kg)	Concentrate mixture 2[b] (kg)
100	650	7–8	0.5	1.5	6–7	1.5
150	700	15–20	0.5	1.0	10–12	1.5
200	700	15–20	1.0	1.0	12–15	1.5
250	650	20–25	1.5	0.5	20–25	1.0
300	600	15–30	1.5	0.5	25–30	1.0
350	600	30–35	2.0	0.5	30–35	1.0
Pregnant (3 months before calving)		35–40	3.0	2.0	35–40	2.5

Note: During periods when green fodder supply is inadequate, the nutrient requirements of the animals can be met by incorporating a concentrate mixture and dry roughages in their rations. If sufficient green fodders are not available, 5 kg of green fodder can be replaced with 1 kg of dry fodder and 250 g of concentrate mixture.

[a] Digestible crude protein 12 to 13%; total digestible nutrients 70 to 75% (Table 9).
[b] Digestible crude protein 15 to 16%; total digestible nutrients 70 to 75% (Tables 9 and 10).

Table 19
FEEDING SCHEDULES FOR GROWING ZEBU STOCK[80]

Illustration 1

Ingredient	Body weight	Quantity (kg)	DM (kg)	DCP[a] (kg)	TDN[b] (kg)
	100 kg		3.0	0.24	1.5
Sorghum/maize/oats as fodder (crude protein 7%, DM 20%)		11.0	2.2	0.088	1.21
Concentrate mixture (DM 90%) (Illustrations 6–8)		0.88	0.8	0.166	0.58
Total			3.0	0.254	1.79

Illustration 2

Ingredient	Body weight	Quantity (kg)	DM (kg)	DCP (kg)	TDN (kg)
	150 kg		4.5	0.35	2.6
Sorghum/maize/oats fodder (crude protein 7%, DM 20%)		16.6	3.3	0.131	1.815
Concentrate mixture (DM 90%) (Illustrations 6–8)		1.32	1.2	0.234	0.864
Total			4.5	0.366	2.679

Illustration 3

Ingredient	Body weight	Quantity (kg)	DM (kg)	DCP (kg)	TDN (kg)
	300 kg		7.5	0.41	4.10
Wheat straw/rice straw (DM 90%)		2.75	2.5	–	1.00
Sorghum fodder (DM 20%)		17.5	3.5	0.14	1.92
Concentrate mix (DM 90%) (Illustrations 6–8)		1.65	1.5	0.29	1.08
Total			7.5	0.43	4.00

Illustration 4

Ingredient	Body weight	Quantity (kg)	DM (kg)	DCP (kg)	TDN (kg)
	450 kg		11.0	0.48	5.0
Wheat/rice straw (DM 90%)		7.15	6.5	–	2.60
Berseem/lucerne (DM 15%)		22.75	3.5	0.31	1.92
Concentrate mix (DM 90%) (Illustrations 6–8)		1.10	1.0	0.19	0.72
Total			11.0	0.50	5.24

[a] Digestible crude protein.
[b] Total digestible nutrients.

Table 19 (continued)
FEEDING SCHEDULES FOR GROWING ZEBU STOCK[80]

Illustration 5

Ingredient	Body weight	Quantity (kg)	DM (kg)	DCP (kg)	TDN (kg)
	450 kg		11.0	0.48	5.0
Wheat straw (DM 90%)		9.35	8.5	–	3.4
Concentrate mixture (DM 90%)		2.75	2.5	0.49	1.8
Total			11.0	0.49	5.2

Illustration 6

	Parts/ per 100	Crude protein (%)
Maize (crude protein 9%)	32	2.88
Groundnut cake (crude protein 40%)	35	14.00
Wheat bran (crude protein 16%)	20	3.20
Molasses	10	–
Mineral mixture	3	–
Total		20.08

Illustration 7

	Parts/ per 100	Crude protein (%)
Maize (crude protein 9%)	37	3.33
Groundnut cake (crude protein 40%)	25	10.00
Wheat bran (crude protein 16%)	24	3.84
Molasses	10	–
Mineral mixture	3	–
Urea (40% N)	1	2.50[c]
Total		19.67

Illustration 8

	Parts/ per 100	Crude protein (%)
Maize (crude protein 9%)	50	4.50
Groundnut cake (crude protein 40%)	10	4.00
Wheat bran (crude protein 16%)	24	3.84
Molasses	10	–
Mineral mixture	3	–
Urea	3	7.50
Total		19.84

[c] Taken as 6.25 kg protein equivalent of urea nitrogen.

alternative except to increase the concentrate quantity, as explained in Illustration 5, where the total ration will be more expensive.

Use of ammonium compounds is practicable in ruminants from 6 months of age onwards, and their feeding value is well established. Urea is a cheaper source of nonprotein nitrogen which can be incorporated to the extent of 3% in the concentrate mixture or 1% of the total dry matter. The use of such compounds has become obligatory because of the high cost of oilseed meals which are incorporated as a protein source into the rations. A part of the nitrogen source can be economically met with urea in the concentrate ration, as shown in Illustrations 6 through 8 in Table 19.

Rations for Milch Cattle

The following guidelines are recommended for formulating balanced rations for milch cattle:

1. Work out the maintenance requirements of a milch animal according to its body weight.
2. If the animal is in a growing stage, i.e., in her first lactation, then nutrient requirements for growth as well as maintenance should be derived.
3. In addition to maintenance/growth requirements, the lactating animal needs an additional supply of nutrients for milk secretion and to cover the wastage involved in the metabolic process.
4. The lactating cow in an advanced stage of pregnancy requires an extra allowance for the development of the fetus in addition to the maintenance and milk production requirements.
5. It is always economical to meet the major portion of the animal's requirements from green fodders, hay, and silage.

Recommended rations for zebu cows producing 7 kg of milk are given in Table 20.

For cows yielding more than 7 kg milk, additional nutrient requirements for milk production must be met from feeding concentrates. Concentrate mixture 1 is meant to be fed with high-legume fodders, concentrate mixture 2 is intended to be fed with low-legume fodders, and other concentrate mixtures are to be fed with mixed fodders. For each 2.5 kg of milk over and above 7 kg, a lactating cow should be fed 1 kg of additional concentrate mixture in addition to the rations listed in Table 20.

Cows on Good Tropical Pastures

Information on the value of most tropical pastures for milk production is lacking. It can be assumed that a good pasture in tropical and subtropical regions will provide for the maintenance requirements as well as for the production of 4 to 5 kg of milk. Cows producing more than this quantity should be given concentrates according to scale given in Table 21.[81]

Rations for Dry Cows

Most of the nutritional requirements of these animals are met through forages. The scales in Table 22 are recommended for cows weighing 450 to 500 kg.

Rations for Bulls

Sexual maturity in the male is characterized by full development of the testicles and production of viable sperm. Malnutrition can cause sperm production to cease entirely or can lessen the volume of the semen produced. It also affects sperm viability and motility. Overfeeding may fatten the animal and can result in failure of libido.

Table 20
RECOMMENDED RATIONS FOR COWS PRODUCING VARYING AMOUNTS OF MILK

I. Cows producing up to 7 kg milk (fat 4.5%) weighing 450–500 kg[a]
   A. Rations with abundant green fodders and hay
      1. Legume green fodders, 60–65 kg; straws/stover, 3–4 kg; common salt and mineral mixture, 30 g each
      2. Nonlegume green fodder, 35–40 kg; legume hay, 5–7 kg; common salt and mineral mixture, 30 g each
   B. Rations with limited green fodders
      1. Legume green fodders, 25–30 kg; silage, 20–25 kg; concentrate mixture 1, 2 kg
      2. Nonlegume green fodder, 25–30 kg; legume hay, 5–7 kg; concentrate mixture 2, 2 kg
      3. Legume green fodder, 25–30 kg; straw/stover, 7–8 kg; concentrate mixture 1, 3 kg
      4. Nonlegume green fodder, 25–30 kg; straw/stover, 4–5 kg; concentrate mixture 2, 3 kg

[a] All roughage rations under category I.A are capable of meeting nutrient requirements of cows yielding up to 7 kg milk (fat 4.5%), but in early lactation or in hot summer it is recommended that a concentrate mixture up to 1 kg/day (depending on milk yield) be fed to obtain optimum performance from the animal. Even in rations with limited green fodders (Category I.B), extra allowance of a concentrate mixture up to 1 kg/day is recommended to reduce the heat increment in hot summer months. The quantity of roughage dry matter offered may be reduced accordingly.

Table 21
SCALE OF CONCENTRATE FEEDING TO COWS KEPT ON GOOD GRAZING

Daily milk production		Amount of concentrates fed per day	
lb	kg	lb	kg
0–10	0–4.5	0[a]	
10–15	4.5–6.8	2	0.9
15–20	6.8–9.1	4	1.8
20–25	9.1–11.3	6	2.7
25–30	11.3–13.6	8	3.6
30–35	13.6–15.9	10	4.5
35–40	15.9–18.1	12	5.4

[a] When pasture quality is poor, additional supplies of cut fodder, silage, hay, and/or 1.5 kg concentrate mixture is recommended.

Table 22
RATIONS FOR DRY COWS

I. Rations with abundant forages
   A. Legume green fodder, 25–30 kg; straw/stover, 2 kg; concentration mixture 1, 1 kg
   B. Nonlegume green fodder, 25–30 kg; legume hay, 2 kg; concentration mixture 2, 1 kg
   C. Silage, 25–30 kg; legume hay, 3 kg; concentration mixture 2, 1 kg
II. Rations with limited forages
   A. Legume green fodder, 15–20 kg; straw/stover, 3 kg; concentration mixture 1, 1.5 kg
   B. Nonlegume green fodder, 15–20; straw/stover, 2 kg; concentration mixture 2, 2 kg
   C. Silage, 15–20 kg; straw/stover, 2 kg; concentration mixture 2, 2 kg

The following guidelines should be observed while computing rations for bulls:

1. Feed the young bull an abundant well-balanced ration in order to obtain maximum growth.
2. Never let a breeding sire become fat. High-grain rations should not be used in rations for mature bulls in large amounts. Grains such as oats and barley are recommended.
3. Special feeds such as wheat germ oil have proved to be of little value in improving fertility of dairy bulls.

The following rations have been recommended for mature bulls weighing 680 to 700 kg:

I. Mature bulls not in service
   A. Legume green fodder, 35 to 40 kg; straw/stover, 3 kg; concentrate mixture 1 (Table 9), 1.5 kg
   B. Nonlegume green fodder, 35 to 40 kg; straw/stover, 3 kg; concentrate mixture 2 (Tables 9 or 10), 2.0 kg
II. Mature bulls in service
   A. Nonlegume fodder, 35 to 40 kg; straw/stover, 3 kg; concentrate mixture 1 (Table 9), 2.5 kg
   B. Nonlegume green fodder, 35 to 40 kg; concentrate mixture 2 (Tables 9 or 10), 3 kg

Note: If green fodder supply is limited, the quantity of concentrate mixture and the supply of low quality dry roughages, i.e., straw/stover, should be increased.

Rations for Working Bullocks or Oxen

Data on energy and protein requirements of working bullocks are not based on precise measurements because of unpredictable or uncontrollable and variable factors. Consequently, the rations given here are based more on common practice than on precise experimental results. On the basis of work done in India,[82] nutrient requirements of working bullocks are given in Table 23. Table 24 illustrates typical rations required for a bullock weighing 550 kg.

Table 23
NUTRIENT REQUIREMENTS FOR WORKING BULLOCKS PER DAY (MAINTENANCE + WORK)

	Normal work		Heavy work	
Live weight (kg)	DCP[a] (kg)	TDN[b] (kg)	DCP (kg)	TDN (kg)
136	0.15	1.49	0.20	1.90
181	0.21	1.95	0.27	2.50
227	0.27	2.40	0.34	3.10
272	0.32	2.81	0.40	3.60
318	0.37	3.22	0.46	4.21
363	0.42	3.63	0.53	4.80
408	0.46	4.03	0.58	5.30
454	0.51	4.49	0.65	5.88
500	0.56	4.89	0.71	6.31
550	0.61	5.36	0.77	7.00

[a] Digestible crude protein.
[b] Total digestible nutrients.

Table 24
RATIONS FOR WORKING BULLOCKS WEIGHING 550 kg[a]

Feed Ingredients Expressed in Kilograms

	Light work	Heavy work
Winter/Spring		
Green legume fodder	20	24
Straw/stover	8	8
Concentrate mixture	1	1
Summer		
Green fodder legume + nonlegume	20	24
Straw/stover	5.5	5.5
Concentrate mixture	1.3	2.0
Autumn/Winter		
Green fodder legume + nonlegume	15	19
Straw/stover	5	5
Hay	2	2
Concentrate mixture	1	1.7

[a] When animals are not working, the concentrate mixture should be deleted from the ration and the quantity of roughages should be reduced by about 10%.

REFERENCES

1. **Rice, V. A., Andrew, F. N., and Warwick, E. J.,** *Breeding and Improvement of Farm Animals,* McGraw-Hill, New York, 1951.
2. **Epstein, H.,** *World Rev. Anim. Prod.,* 8, 25–29, 1974.
3. National Research Council, *Nutrient Requirements of Dairy Cattle,* National Academy of Sciences, Washington, D.C., 1971.
4. **Krishna, G. and Ranjhan, S. K.,** Energy and Protein Requirements of Indian Dairy Animals (A Review), Indian Veterinary Research Institute, Izatnagar, U. P., India, 1975.
5. **Warth, F. J.,** *Animal Nutrition Research in India,* Macmillan, London, 1926; as cited by Sen, K. C., 1953.
6. **Mullick, D. N. and Kehar, N. D.,** *Proceedings 39th Session Indian Science Congress, Calcutta,* Part III, Science Congress, Calcutta, 1952, 266.
7. **Shrivastava, J. P.,** Studies on the Utilization of Dietary Energy for Maintenance and Milk Production, in Buffaloes, Ph.D. thesis, Indian Veterinary Research Institute, Izatnagar Agra University, Agra, 1970.
8. **Katiyar, R. C.,** Studies on the Utilization of Dietary Energy for Maintenance and Milk Production in Hariana Cows, Ph.D. thesis, Indian Veterinary Research Institute, Izatnagar Agra University, Agra, 1971.
9. **Krishna, G.,** Studies on Energy and Protein Requirements for Milk Production in Indian Dairy Animals, Ph.D. thesis, National Dairy Research Institute, Agra University, India, 1971.
10. **Patle, B. R.,** Studies on the Plane of Nutrition on Utilization of Feed Nutrients for Milk Production in Cross Bred Cows, Ph.D. thesis, National Dairy Research Institute, Karnal, Panjab University, Chandigash, India, 1973.
11. **Ranjhan, S. K., Krishnamopan, D. V. G., and Singh, R.,** *Indian J. Anim. Sci.,* in press.
12. **Sen, K. C. and Ray, S. N.,** Nutritive Values of Indian Cattle Feeds and Feeding of Animals, Bull. 25, Indian Council of Agricultural Research, New Delhi, India, 1964.
13. **Verco, J. E.,** *Br. J. Nutr.,* 24, 599, 1970.
14. **Kehar, N. D., Mukherjee, R., and Sen, K. C.,** *Indian J. Vet. Sci.,* 13, 257, 1944.
15. **Gupta, B. S., Bhargava, V. N., Raina, N. N., and Singh, S. N.,** *Indian J. Vet. Sci. Anim. Husb.,* 36, 90–95, 1967.
16. **Kehar, N. D.,** Setting Up Feeding Standards for Livestock, Division of Agronomy progress report, Indian Agricultural Research Institute, New Delhi, 1969–70.
17. **McDowell, R. E.,** *Improvement in Livestock Production in Warm Climates,* W. H. Freeman, San Francisco, 1972, 110.
18. **Winchester, C. F. and Morris, M. J.,** *J. Anim. Sci.,* 15, 728, 1956.
19. **Williamson, G. and Payne, W. J. A.,** *An Introduction to Animal Husbandry in the Tropics,* Longman, Green & Co., London, 1959, 137–246.
20. **Morrison, F. B.,** *Feeds and Feeding,* 22nd ed., Morrison Publishing, Clinton, Iowa, 1959, 1052.
21. **Christianson, D. D., Cavins, J. F., and Walls, J. S.,** *J. Agric. Food Chem.,* 13, 277, 1965.
22. **Ligget, R. W. and Koffler, H.,** *Bacteriol. Rev.,* 12, 297, 1948.
23. **Singh, B. and Makkar, G. S.,** *Indian J. Anim. Sci.,* in press.
24. **Johnson, R. M. and Raymond, W. D.,** *Trop. Sci.,* 7, 109, 1965.
25. **Yoshida, M., Hoshii, H., Kosaka, K., and Morimoto, H.,** *Jpn. Poult. Sci.,* 3, 29, 1966.
26. **Kwatra, B. L. and Sidhu, G. S.,** *J. Nutr. Diet.,* 5, 57, 1969.
27. **Sadagopan, V. R. and Talpatra, S. K.,** *Indian Vet. J.,* 45, 241, 1968.
28. **Srivastava, S. P. and Singh, G. S.,** *Indian J. Dairy Sci.,* 13, 136, 1960.
29. **Conard, B. E. and Neol, E. M.,** Texas Agric. Exp. Stn. Mimeo. Ser. 676, 1963.
30. **Patel, B. M., Dhani, B. M., and Shukla, P. C.,** *Indian Vet. J.,* 47, 66, 1970.
31. **Sagar, V. and Pradhan, K.,** *Indian J. Dairy Sci.,* 28, 89, 1975.
32. **Bressani, R., Cabezus, M. T., Jarquin, R., and Murillo, B.,** in *Animal Feeds of Tropical and Sub-tropical Origin,* Tropical Products Institute Conference, Tropical Products Institute, London, 1974.
33. **Jackson, M. G.,** Annual Progress Report of All India Co-ordinated Research Project on Waste Materials, Pantnagar, India, 1972.
34. **Walker, H. G., Jr., Kohler, G. O., Graham, R. P., Hart, M. R., and Garrett, W. N.,** Upgrading Cereal Straws for Ruminant Feeds, in Proc. Montana Nutrition Conference, 1975.
35. **Hart, M. R., Graham, R. P., Hanni, P. F., Rockwell, W. C., Walker, M. G., Jr., Kohler, G. O., Waiss, A. C., Jr., and Garrett, W. N.,** *Feedstuffs,* 9, 22–26, 1975.
36. **Chaturvedi, K. L., Singh, V. B., and Ranjhan, S. K.,** Indian Veterinary Research Institute, Izatnagar, unpublished observations, 1972.
37. **Lofgreen, G. P. and Otagaki, K. K.,** *J. Dairy Sci.,* 43, 220, 1960.

38. **Lofgreen, G. P. and Otagaki, K. K.,** *J. Anim. Sci.,* 19, 392, 1960.
39. **Hatch, C. F. and Beeson, W. H.,** *J. Anim. Sci.,* 35, 854, 1972.
40. **Preston, T. R., Willis, M. B., and Elias, A.,** *Rev. Cubana Cienc. Agric.,* 1, 33, 1967.
41. **Elias, A., Preston, T. R., Willis, M. B., and Sutherland, T. M.,** *Rev. Cubana Cienc. Agric.,* 2, 55, 1968.
42. **Kaushal, T. R., Kochar, A. S., and Chopra, A. K.,** *Indian J. Anim. Sci.,* 42, 399, 1972.
43. **O'Donovan, P. B.,** *Livestock Production in Taiwan on Marginal Land,* Food and Agriculture Organization of the United Nations, Rome, 1970.
44. **Rogers, D. J. and Milner, M.,** *Econ. Bot.,* 17, 211, 1963.
45. **Subramanian, C. R. A.,** in *All India Co-ordinated Research Project on Utilization of Agricultural By-products and Industrial Waste Materials, Evolving Economic Rations for Livestock,* Kerala Agricultural University, Mannultry, Trichur, India, 1975.
46. **Ross, E. and Enriquez, E. Q.,** *Poult. Sci.,* 48, 846, 1969.
47. **Eggun, E. O.,** *Br. J. Nutr.,* 24, 761, 1970.
48. **Menon, K. C.,** in *Workshop Proceedings on Agricultural By-products and Industrial Waste Materials, Evolving Economic Rations for Livestock,* Bhuvaneshwar, India, 1969.
49. National Dairy Research Institute, *Dairy Handbook,* Karnal, India, 60–65.
50. **Kehar, N. D. and Chandra, R.,** *Curr. Sci.,* 15, 48, 1946.
51. **Patnaik, N.,** in Proc. Anim. Nutr. Res. Workers Conf., Indian Council of Agricultural Research, Trichur, India, 1966.
52. **Panda, B. and Sharma, N.,** *Poult. Guide,* 11, 19, 1974.
53. **Bhattacharya, A. N. and Fontenot, J. P.,** *J. Anim. Sci.,* 25, 367, 1966.
54. **Brugman, H. H., Dickey, H. C., Plummet, B. E., Goater, J., Heitman, R. N., and Taka, M. R. T.,** *J. Anim. Sci.,* 27, 1132, 1968.
55. **Bucholtz, H. F., Henderson, H. E., Thomas, J. W., and Zindel, H. C.,** *Am. Soc. Agric. Eng. Publ. Proc.,* 271, 408, 1971.
56. **Bull, L. S. and Reid, J. T.,** *Am. Soc. Agric. Eng. Publ. Proc.,* 271, 297, 1971.
57. **El-Sabban, F. F., Bratzler, J. W., Long, T. A., Freer, D. E. H., and Gentry, R. F.,** *J. Anim. Sci.,* 31, 197, 1970.
58. **Claesson, O. and Ahlstrom, B.,** in *Brief Communication, 19th Int. Dairy Sci. Congr.,* Vol. 1E, Thompson Press, India, Ltd., New Delhi, 1974, 88.
59. **Rai, P. S. and Netke, S. P.,** *Indian J. Poult. Sci.,* 8, 7, 1973.
60. **Anthony, W. B.,** *J. Anim. Sci.,* 30, 274, 1970.
61. **Bandel, L. S. and Anthony, W. B.,** *J. Anim. Sci.,* 28, 152, 1969.
62. **Caswell, L. F., Fontenot, J. P., and Webb, K. E., Jr.,** *J. Anim. Sci.,* 39, 138, 1974.
63. **Saylor, W. W. and Long, T. A.,** *J. Anim. Sci.,* 39, 139, 1974.
64. **Moore, J. D. and Anthony, W. B.,** *J. Anim. Sci.,* 30, 324, 1970.
65. **Krauss, R. W.,** *Am. J. Bot.,* 49, 425, 1962.
66. **Singh, Y. K. and Anthony, W. B.,** *J. Anim. Sci.,* 27, 1136, 1968.
67. **Teotia, J. S. and Miller, B. F.,** *Br. Poult. Sci.,* 15, 177, 1974.
68. **Otogaki, K. K., Lofgreen, G. P., Cobb, E., and Dull, G. G.,** *J. Dairy Sci.,* 44, 491, 1961.
69. **Chapman, H. L., Jr., Ammerman, C. B., Baker, F. S., Jr., Hentges, J. F., Hayes, B. W., and Cunha, T. J.,** University of Florida, Gainesville, Bulletin 751, 1973.
70. **Ichhponani, J. S. and Malik, N. S.,** *Br. Poult. Sci.,* 12, 231, 1971.
71. **Crampton, E. W. and Harris, L. F.,** *Applied Animal Nutrition,* W. H. Freeman, San Francisco, 1969.
72. Indian Standards Institution, Specification for Mineral Mixtures for Supplementing Cattle Feeds, New Delhi, 1973.
73. Indian Standards Institution, Compounded Feeds for Cattle, New Delhi, 1975.
74. **Hopper, J. H., Gardner, K. R., and Johnson, B. C.,** *J. Dairy Sci.,* 37, 431, 1954.
75. **Sander, E. G., Warmer, R. G., Harrison, H. N., and Loosli, J. K.,** *J. Dairy Sci.,* 42, 1600, 1959.
76. **Packett, L. V. and Fordam, J. R.,** *J. Anim. Sci.,* 24, 488, 1965.
77. **Arora, S. P., Bajpai, L. D., and Dave, B. K.,** *Ind. J. Anim. Sci.,* 43, 462, 1973.
78. **Bell, J. M.,** *Can. J. Anim. Sci.,* 38, 103, 1958.
79. **Pettyjohn, J. S., Everett, J. P., Jr., and Monchrie, R. D.,** *J. Dairy Sci.,* 46, 710, 1963.
80. National Dairy Research Institute, Calf Raising, Publ. No. 135, Karnal, India, 1975.
81. **Sen, S. K.,** *Handbook of Animal Husbandry,* Indian Council of Agricultural Research, New Delhi, 1962, 210–254.
82. **Enebo, L.,** *Wenner Green Cent. Int. Symp.,* 14, 93–103, 1970.

DIETS (NATURAL AND SYNTHETIC): SHEEP AND GOATS

W. C. Weir

INTRODUCTION

The nutrient requirements for sheep of various weights and life stages were revised in 1975[1] by the Subcommittee on Sheep Nutrition of the Committee on Animal Nutrition, National Research Council, U.S.A. The requirements published by the Agricultural Research Council in England[2] were considered in this revision. A Subcommittee on Goat Nutrition is currently preparing requirements for goats.[3] The publication by Devendra and Burns[4] devotes a chapter to nutrition and feeding of goats, and estimates the requirements as far as known.

The digestive tract of goats and sheep is similar, but their grazing habits differ when a variety of nutrient sources are available to both species. Goats have mobile upper lips and very prehensile tongues which enable them to graze on very short grass and browse on foliage not normally accessible to sheep. When available and accessible, goats will take more browse and less grass than sheep.

When goats and sheep are fed the same diets, the digestibility of the nutrients is generally similar.[5,6] Devendra and Burns[4] suggest that goats may have an increased ability to digest fiber. Gallagher and Shelton[7] compared Angora goats to Rambouillet sheep fed the same diet. The sheep digested a higher percentage of the organic matter and were more efficient in converting feed energy to body gain, while the goats were more efficient at converting feed to animal fiber. For practical or experimental purposes, the same diets may be used for sheep and goats.

DIETS FOR RUMINATING SHEEP AND GOATS

Natural Diets

Naturally roughage-consuming animals, both sheep and goats, will thrive on diets of preserved roughage (hay or silage) grown on fertile soils, harvested before excessively lignified, and stored to prevent destruction of the nutrients. Often, only a supplement of salt is needed for such a diet. If the soil is deficient, if excessive losses of nutrients occur in the preparation and processing of the roughage, or if the quantity of roughage is limiting, deficiencies of energy or any of the nutrients may occur as outlined in *Nutrient Requirements for Sheep.*[1]

With increased nutrient demands of pregnancy, lactation, growth, and/or fattening, energy often becomes the limiting nutrient. This indicates the need for the replacement of low energy roughage (approximately 2.0 Mcal/kg metabolizable energy [ME]) with higher energy concentrates (approximately 3.0 Mcal/kg ME) such as grains or by-product feeds. The need for additional protein in the concentrate will depend on the protein content of the roughage. Instructions for formulating diets for sheep with natural feedstuffs is given on Page 25 of the NRC publication.[1]

Semipurified Diets

In order to isolate individual nutrients and their deficiencies as well as to eliminate variability, semipurified diets have been developed. Diets Ruminant A[8] and A A6[9] were developed for the Animal Breeding Research Organization for long-term genetic studies, while the Bird diet[10] was developed for sulfur studies. These diets based on straw are amenable to long-term studies and include the essential vitamins and minerals to prevent confounding effects due to dietary deficiencies. The straw for Ruminant A and A A6 is chopped to give straw particles with a mean length of about 1 cm. In Diet A A6, the nonhomogenous mixture is prepared as a cube 2 cm in diameter and 2 to 3 cm in length. In the Bird diet the straw was hammermilled through a 5-cm screen.

The Hagsten[11] and Morris[12] diets are higher concentrate diets used for short periods to develop sodium deficiencies. For longer-term studies, care would need to be taken to avoid vitamin and/or mineral deficiencies with such rations.

Weston[13] investigated the minimum roughage requirement for concentrate diets based on wheat. If the wheat was fed whole, 2% of the diet as chopped wheat straw was sufficient, but if the wheat was ground before feeding, it depressed voluntary feed consumption and intake of digestible organic matter, and lowered body weight gains. The importance of the physical form of the diet was demonstrated by Orskov and Frazer.[14] If barley-based supplements were ground and pelleted, the intake of the dried grass as well as the digestibility of the acid detergent fiber fraction was depressed. These authors attribute the adverse effect to excessively acid pH in the rumen following the feeding of ground cereals. Orskov[15] reports increased problems with rumenitis in lambs fed ground barley. Greenhalgh and Reid[16] have reported the effect of pelleting vs. chopping of dried rye grass on feed intake, dry matter digestibility, and development of the digestive tract, as well as body weight changes on both a long- and short-term basis.

Purified Diets

Bunn and Matrone[17] reviewed the development of purified diets for ruminants in 1972. These authors emphasize the importance of sodium and potassium buffering, the essentiality of sulfur in the rumen, and the importance of roughage for saliva stimulation as well as for the development and health of the rumen wall. Among their recommendations was a level of 25 to 30% cellulose and potassium at 0.6% of the dry matter. Jones et al.[18] developed a diet to study the role of both protein and cellulose on the voluntary feed intake and digestibility by sheep. Table 2 shows the constant and variable ingredients as an example of this approach.

The ultimate purified diet for ruminants uses urea or other nonprotein nitrogen as the only source of nitrogen. Two such diets used for sheep are shown in Table 3, along with a comparison diet used by Ludwick et al.[20] to study adaptation of sheep to a urea-based diet. Nimrick et al.[21] used their basal diet to study the response to abomasally infused supplemental amino acids. The limiting amino acids were found to be, in order, (1) methionine, (2) lysine, and (3) threonine.

OTHER DIETS

Diets for Productive Purposes

Diet formulae are not given for each of the productive functions of sheep, i.e., growth, fattening, reproduction, and lactation. The tables of feed requirements in the NRC publication[1] clearly outline these needs, and most sheep production bulletins likewise offer practical guidelines for formulating such diets with feedstuffs that are most available and economical. Robinson and Orskov[22] have published an integrated approach to improving the efficiency of sheep meat production which reports on the system developed at the Rowett Research Institute.

Diets for The Newborn Lamb

Diets to nourish the newborn lamb have been studied extensively. Colostrum is essential to the newborn. Treacher and Penning[23] recommend a minimum of 8 g colostrum per kilogram of lamb birth weight within 2 hr of birth. If the dam does not have colostrum, it can be obtained from another ewe, or a store of frozen ewe or cow colostrum can be utilized in emergencies. After receiving colostrum, the lamb can be reared on milk replacers. Examples of several milk replacers are shown in Table 4.

Historically, orphan lambs have been fed cow's milk, with variable results. Ewe's milk is higher in dry matter (18.5 vs. 12.5%), richer in fat (38 vs. 30% of the dry matter), and lower in lactose (25 vs. 37%), but higher in protein (30 vs. 26%) when compared to cow's milk.[23] The recognition that lambs could be fed ad libitum if the milk replacer was fed cold has stimulated trials that approach commercial scale. Glimp[24] reported improved survival and better gains and efficiency when the fat was increased from 20 to 30% and the lactose level lowered from 42 to 27%. Using the diet shown in Table 4, the death loss in one trial was reduced to 2%. Molénat and Theriez,[25] using the formula shown in Table 4, found that the lambs gained fastest when the milk substitute was reconstituted to 16% dry matter. Penning and Treacher[26] investigated the amount of feed per day and the necessary number of days in the liquid feeding period for most efficient use of feed for gain. They recommended 248 g dry matter daily for 29 days.

Gorrill et al.[27] successfully replaced some of the milk protein with soy protein. Using full-fat soy flour which had been heat treated to destroy the trypsin inhibitor, 30 and 60% of the milk protein was replaced as shown in Table 4 for the 60% replacement formula. Drori and Jordan[28] also compared the regular milk replacer shown in Table 4 with a soy formula in which 10% of the protein came from soy flour. Forty-six of 48 lambs survived on the soy diet but grew somewhat slower than those on regular milk replacer.

To prevent the commonly observed postweaning growth check in lambs weaned at 21 days, Drori and Jordan[28] prepared a liquid supplement, shown in Table 5. Twelve hundred grams of the air-dry supplement was suspended in 2500 ml of H_2O and 80 ml of 1.85% solution of formaldehyde. Each lamb was given 720 ml of this liquid supplement daily from weaning off the mild replacer until 7 to 8 weeks of age. The lambs that had been on the milk replacer with soy protein made better use of the liquid supplement than the lambs formerly on milk proteins only. The authors suggested that the digestive enzymes had become somewhat adapted to the soy protein. Walker[29] had shown that the newborn lamb had enzymes only for glucose, galactose, and lactose. Probably lactase is limiting, too, based on the results of Glimp[24] discussed above. Walker and Stokes[30] investigated the digestibility of various fats for lambs with the following results:

Coconut oil	99%
Peanut oil, cottonseed oil, soy oil, corn oil, safflower oil	97–98%
Lard	90%
Tallow	78%
Rapeseed oil	62%

Solid Diets for Nursing and Weaning Lambs

Dry diets made available to the lamb but not to the dam are commonly referred to as "creep diets." With the increased use of milk replacers and early weaning, these diets are an important source of nutrients for the growing, fattening lamb. Robinson and Orskov[22] emphasize the importance of introducing creep feed at an early age. Table 6 gives two suggested creep diets. Four commonly used in the United States are given in Table 15 of the NRC publication.[1] In their integrated approach to meat production, Robinson and Orskov[22] recommend fattening lambs for slaughter on whole grains and a supplement preferably combined in a small pellet. Table 7 gives two such supplements.

Range and Pasture Supplements

No diets for use in combination with range or pasture forage are presented. Suggested supplements are shown in Table 14 of the NRC publication.[1]

Table 1
SEMIPURIFIED DIETS FOR SHEEP WITH A FUNCTIONAL RUMEN

Ingredient	A[8]	A A6[9]	Bird[10]	Hagsten[11]	Morris[12]
	% dry matter basis				
Chopped barley straw	40	30			
Wheat straw–milled			94.6		
Rice straw–milled					30.0
Ground yellow corn				46.1	
Sorghum grain					65.5
Extracted peanut meal	7	7			
Cottonseed meal				13.2	
Soybean oil meal					3.0
Barley	12.5	22.25			
Wheatfeed	23.25	23.25			
Ground corn cobs				40.0	
Molasses	10.0	10.0			
Urea	1.0	1.0	2.0		
Tallow	1.25	1.25			
Salt (NaCl)	1.5	1.5			
Sodium sulfate			0.40		
Dicalcium phosphate	1.5	1.5			
Limestone		0.25		0.7	
Gypsum					1.5
Sodium bicarbonate	2.0	2.0			
Vitamins					
A	a	a	b		c
D	a	a	b		c
E	a	a	b		
Trace minerals	a	a			c
Iodide	a	a			c
Composition					
Crude protein	13.5	13.5	8.3–9.1	10.49	9.87
Crude fat	2.5	3.0			
Crude fiber	19.0	16.1			
Minerals (%)			3.0		
NaCl			(35.0)		
$CaCO_3$			(10.0)		
$Ca_3(PO_4)_2$			(32.2)		
KCl			(14.3)		
$MgCO_3$			(8.0)		
Trace minerals			(0.5)		
Fe_2O_3			[32.0]		
$CuCl_2 \cdot 2H_2O$			[16.0]		
$MnCl_2 \cdot 4H_2O$			[24.0]		
ZnO			[23.0]		
$Na_2MoO_4 \cdot 2H_2O$			[3.0]		
H_2SeO_3			[1.0]		
$CoCl_2 \cdot 6H_2O$			[0.8]		
KI			[0.2]		

[a] Amounts not specified.

[b] Oral supplement of 100,000 IU vitamin A, 5000 IU vitamin D_3, 40 IU vitamin E given at 3-week intervals.

[c] Retinyl acetate, 2200 IU/kg; cholecalciferol, 220 IU/kg; copper sulfate, 5 mg Cu/kg; cobalt chloride, 0.1 mg Co/kg; calcium pentaiodate, 0.8 mg I/kg.

Table 2
DIET TO STUDY VARIABLE DIETARY PROTEIN AND CELLULOSE LEVELS[18]

	%
Constant ingredients (43.8)	
Sucrose	15
Sodium bicarbonate	6
Potassium bicarbonate	4
Corn oil	4
Dehydrated alfalfa meal	5
Calcium phosphate	1.8
Vitamins	5[a]
Minerals	3[a]
Variable ingredients (56.2)	
Cellulose (Solka-floc BW 20)	7, 14, 21, or 28%
Protein supplement to give 5, 10, 15, 20% dietary crude protein (casein, soy protein and urea)	
Cornstarch to give 100%	

[a] Matrone et al.[19]

Table 3
PURIFIED DIETS USED TO STUDY UREA AS THE ONLY SOURCE OF NITROGEN

	Percentage composition on dry matter basis		
		Ludwick[20]	
Ingredient	Nimrick[21]	Urea	Soy
Cornstarch	57.4	20.9	17.5
Cellulose (Solka-floc BW-40)	25.0	44.5	44.5
Glucose		20.9	17.5
Cane molasses	5.0		
Isolated soy protein			9.4
Urea	4.3	2.6	
Corn oil	3.0	3.9	3.9
Ethoxyquin	0.011		
Antibiotic mixture	0.1[a]		
Minerals (%)	5.0	7.2	7.2
$CaHPO_4 \cdot 2H_2O$	(31.4)		
$CaHPO_4$		(48.94)	(48.94)
K_2SO_4	(26.7)		
K_2CO_3		(31.55)	(31.55)
NaCl	(25.0)	(7.43)	(7.43)
Na_2SO_4	(10.4)		
MgO	(4.08)		
$MgSO_4$		(10.75)	(10.75)
$FeSO_4 \cdot 7H_2O$	(1.79)		
$FeSO_4$		(0.91)	(0.91)
ZnO	(0.186)		
$ZnSO_4$		(0.07)	(0.07)
MnO	(0.18)		
$MnSO_4$		(0.10)	(0.10)
$CuSO_4 \cdot 5H_2O$	(0.141)		
$CuCO_3$		(0.02)	(0.02)
KI	(0.033)	(0.03)	(0.03)

Table 3 (continued)
PURIFIED DIETS USED TO STUDY UREA AS THE ONLY SOURCE OF NITROGEN

	Percentage composition on dry matter basis		
		Ludwick[20]	
Ingredient	Nimrick[21]	Urea	Soy
$CoSO_4 \cdot 7H_2O$	(0.0029)		
$CoCl_2$		(0.0003)	(0.0003)
Na_2SeO_4	(0.0098)		
$Na_2B_4O_7$		(0.19)	(0.19)
MoO_3		(0.0008)	(0.0008)
Vitamins	0.211		
Vitamin A (IU)	(3932)[b]	2700[c]	2700[c]
Vitamin D_2 (IU)	(393.2)[b]	350[c]	350[c]
Vitamin E (IU)	(156.7)[b]		
Choline chloride (mg)	(489.7)[b]		
Niacin (mg)	(19.66)[b]		
D-Ca-pantothenate (mg)	(7.209)[b]		
Thiamine HCl (mg)	(1.425)[b]		
Riboflavin (mg)	(1.311)[b]		
Vitamin B_{12} (μg)	(20.97)[b]		
α-Tocopherol succinate (mg)		50[c]	50[c]

[a] Aureo-10®, American Cyanamid Co., Princeton, N.J., to furnish 22 mg aureomycin per kilogram of diet.
[b] To furnish per kilogram of diet.
[c] Per sheep per day.

Table 4
DIETS FOR EARLY-WEANED LAMB

	Composition of milk replacer diets – dry matter basis					
					Drori[28]	
Ingredient	Glimp[24]	Molénat[25]	Penning[26]	Gorrill[27]	Regular	Soy
Skim milk powder	38.5	71.5	66	32	40	40
Whey					10 Delactosed	5 Whole
Whey, delactosed and demineralized	27.75					
Sweet cheddar whey				15		
Sodium caseinate					10	
Milk protein	2.5					
Lactic ferment		2.5				
Glucose					14.7	9.7
Fat[a]	30.0					
Tallow		16.0		24		
Coconut oil		9.0				
Lecithin		1.0		0.96	2	2
Butter fat			30			
Lard					23	23
Full-fat soy flour				28		
Soy flour (50% protein)						20
Premix	1.25[b]		?[c]			
Dicalcium phosphate				1.0		
Aurofac-10				1.0		
DL-Methionine				0.2	0.3	0.3

[a] 80% homogenized choice white grease + 20% soy lecithin.
[b] 55,000 IU vitamin A/kg, 11,000 IU vitamin D/kg, 33 IU vitamin E/kg, and 22 g neomycin sulfate in addition to B vitamins and trace minerals.
[c] 4% not reported – presumably minerals and vitamins.

Table 4 (continued)
DIETS FOR EARLY-WEANED LAMB

Ingredient	Composition of milk replacer diets – dry matter basis					
					Drori[28]	
	Glimp[24]	Molénat[25]	Penning[26]	Gorrill[27]	Regular	Soy
Vitamins		Per 100 kg		Per kg diet	Per kg dry diet	
A	b	1,200,000 IU		11,000 IU	4,400 IU	4,400 IU
D_3	b	600,000 IU		2,200 IU	1,100 IU	1,100 IU
E	b	5,000 IU		2 IU	66 IU	
C		100 g				
Thiamine		200 mg				
Riboflavin		400 mg				
Pyridoxin		200 mg				
B_{12}		5 mg		8.8 μg		
Biotin		75 mg				
Calcium pantothenate		200 mg				
Choline		180 g				
Furazolidone		5 g				
Ferrous sulfate					660 mg	660 mg
Cuprous oxide					6.6 mg	6.6 mg
Manganese sulfate					66 mg	66 mg
Zinc sulfate					110 mg	110 mg
Cobalt carbonate					0.22 mg	0.22 mg
Chlortetracycline					66 mg	66 mg
Composition, DM basis %						
Protein	24	26		27.25		
Casein	12.7					
Fat	30	27		30		
Lactose	27					
Ash	5.7					
Calcium	1.0					
Phosphorus	0.8					

Table 5
LIQUID SUPPLEMENT INGREDIENTS FOR WEANING OF SMALL LAMBS[28]

Ingredient	% of mix (dry matter basis)
Soybean meal (dehulled)	35.1
Fish meal	25.1
Glucose	36.0
Dicalcium phosphate	2.5
Sodium chloride	1.0
DL-Methionine	0.3

Table 6
CREEP DIETS FOR NURSING AND WEANLING LAMBS

	Dry matter basis	
Ingredient (%)	Robinson[22]	Molénat[25]
Rolled corn	30	
Whole barley	30	
Soybean meal	10	
Rolled oats or barley	30	
Cereal grain		40
Good quality hay	Free choice	
Dehydrated alfalfa		48
Peanut oil meal		10
Minerals and vitamins		2
Straw		Free choice

Table 7
PROTEIN, VITAMIN, AND MINERAL SUPPLEMENT FOR FEEDING TO LAMBS WITH WHOLE CEREAL GRAINS[22]

Percent of total diet	10%	15%
Ingredients (%, dry matter basis)		
Fish meal	80	
Soybean meal		85
Limestone	15	10
Molasses	4	4
Trace mineral/vitamin mix	1	1
to supply per kilogram total feed		
$ZnSO_4 \cdot 7H_2O$	150 mg	
$MnSO_4 \cdot H_2O$	80 mg	
MgO	200 mg	
$CoSO_4 \cdot 7H_2O$	1 mg	
KIO_3	1 mg	
Vitamin A	5000 IU	
Vitamin D	1000 IU	
Vitamin E	20 IU	

REFERENCES

1. **Pope, A. L., Butcher, J. E., Dinusson, W. E., Garrigus, U. S., Hogue, D. E., and Weir, W. C.,** *Nutrient Requirements of Sheep,* 5th rev. ed., National Academy of Sciences, Washington, D. C., 1975.
2. Agricultural Research Council, *The Nutrient Requirements of Farm Livestock, 2, Ruminants,* ARC, London, 1965, 264.
3. **Cunha, T. J.,** personal communication, 1976.
4. **Devendra, C. and Burns, M.,** Goat Production in the Tropics, Tech. Comm. No. 19, Commonwealth Bureau of Breeding Genetics, 1970, pp. xii and 184.
5. **Jones, G. M., Larsen, R. E., Javed, A. H., Donefer, E., and Gaudreau, J. M.,** *J. Anim. Sci.,* 34, 830–838, 1972.
6. **Geoffroy, F.,** *Ann. Zootech.,* 23, 63–73, 1974.
7. **Gallagher, J. R. and Shelton, M.,** *J. Anim. Sci.,* 34, 319–321, 1972.
8. **Wainman, F. W., Blaxter, K. L., and Pullar, J. D.,** *J. Agric. Sci.,* 74, 311–14, 1970.
9. **Wainman, F. W., Smith, J. S., and Dewey, P. J. S.,** *J. Agric. Sci.,* 84, 109–111, 1975.
10. **Bird, P. R.,** *Aust. J. Agric. Res.,* 25, 631–42, 1974.
11. **Hagsten, I., Perry, T. W., and Outhouse, J. B.,** *J. Anim. Sci.,* 40, 329–344, 1975.
12. **Morris, J. G. and Peterson, R. G.,** *J. Nutr.,* 105, 595–598, 1975.
13. **Weston, R. H.,** *Aust. J. Agric. Res.,* 25, 349–362, 1974.
14. **Orskov, E. R. and Fraser, C.,** *Br. J. Nutr.,* 34, 493–500, 1975.
15. **Orskov, E. R.,** *World Rev. Anim. Prod.,* 9, 58–64, 1973.
16. **Greenhalgh, J. F. D. and Reid, G. W.,** *Anim. Prod.,* 19, 77–86, 1974.
17. **Bunn, C. R. and Matrone, G.,** *Nutr. Abstr. Rev.,* 42, 435–444, 1972.
18. **Jones, G. M., Cecyre, A., and Gaudreau, J. M.,** *Can. J. Anim. Sci.,* 53, 445–454, 1973.
19. **Matrone, G., Bunn, C. R., and McNeil, J. J.,** *J. Nutr.,* 84, 215–219, 1964.
20. **Ludwick, R. L., Fontenot, J. P., and Tucher, R. E.,** *J. Anim. Sci.,* 33, 1298–1305, 1971.
21. **Nimrick, K., Hatfield, E. E., Kaminski, J., and Owens, F. N.,** *J. Nutr.,* 100, 1293–1300, 1970.
22. **Robinson, J. J. and Orskov, E. R.,** *World Rev. Anim. Prod.,* 11, 63–67, 1975.
23. **Treacher, R. R. and Penning, P. D.,** *A.D.A.S. Agric. Dev. Advis. Serv. Rev.,* 12, 163–174, 1974.
24. **Glimp, H. A.,** *J. Anim. Sci.,* 1085–1088, 1972.
25. **Molénat, G. and Theriez, M.,** *Ann. Zootech.,* 23, 491–502, 1974.
26. **Penning, P. D. and Treacher, T. T.,** *Anim. Prod.,* 20, 111–121, 1975.
27. **Gorrill, A. D. L., Jones, J. D., and Nicholson, J. W. G.,** *Can. J. Anim. Sci.,* 54, 337–346, 1974.
28. **Drori, D. and Jordan, R. M.,** *J. Anim. Sci.,* 39, 97–100, 1974.
29. **Walker, D. M.,** *J. Agric. Sci.,* 53, 374, 1959.
30. **Walker, D. M. and Stokes, G. B.,** *Br. J. Nutr.,* 24, 425–433, 1970.

DIETS (NATURAL AND SYNTHETIC): GAZELLE

L. H. Blankenship

INTRODUCTION

Gazelle, as used here, are those animals classified in the genus *Gazella.* Various species may be found in eastern and northern Africa, southwest Asia (including parts of southwest U.S.S.R.), India, Tibet, Sinkiang, and Mongolia. Most live in open dry, sometimes stoney country (plains or open bush), some frequent thick bush or mountainous areas, and others live in true deserts. Habitats range from sea level to an elevation of over 5000 m.

Some systematists are undecided about the classification of *Gazella* into various species and subspecies. As many as 25 species have been described, although some of these are now considered subspecies. Haltenorth[22] recognized 12 species and 49 subspecies of gazelle. Walker[46] indicates that this is a valid number of species for *Gazella.* Although there may be some reason for a greater division of the genus, I have chosen to use the classification presented by Haltenorth. The following shows the species considered and a general distribution of each:

Gazella (Nanger) dama Pallas – Sahara and Sahelian zone (generally, Senegal, Lake Chad district, the Sudan, north to Morocco, and Rio de Oro)[12]

Gazella granti Brooke – Eastern Africa (Kenya, Tanzania, Uganda, Somalia, and Ethiopia) and southern Sudan

Gazella soemmeringi Cretzschmar – Somalia, Ethiopia and eastern Sudan[11]

Gazella (Gazella) dorcas L. – Across northern Africa to border of Sahelian zone, Israel, Syria, and Arabia

Gazella gazella Pallas – Northern Africa, Arabia, Israel, and Syria to India and Nepal[12,20] (*G. cuvieri* of some authors included here as *G. g. cuvieri*)

Gazella leptoceros F. Cuvier – Algeria, Libya, Egypt, Sudan, and Arabia[12]

Gazella pelzelni Kohl – Northern Somalia[11]

Gazella rufifrons Gray – Sahelian zone of Africa from Senegal and Gambia to Ethiopia[12]

Gazella spekei Blyth – Somalia[11]

Gazella thomsoni Gunther – Tanzania, Kenya, Ethiopia, and southern Sudan[11]

Gazella (Trachelocele) subgutturosa Guldenstaedt – Arabia and Iraq east and northeast through Iran, Baluchistan, Afghanistan, southwest U.S.S.R., Chinese Turkestan into northern Tibet, Sinkiang, and Mongolia[12]

Feeding habits of *Gazella* spp. depend primarily on their location, season and availability of food, climate (which affects food availability), herd size, and individual needs of the animal; physiological adaptations of some species may also be important.

Diets of gazelle may vary from one part of an animal's range to another. Mountain gazelle (*Gazella g. gazella*), for example, are primarily browsers when living in the desert, feeding on trees and shrubs, but when they live near agricultural areas they may concentrate on crops such as wheat and barley.[28] In one area of Israel, where mountain gazelle are primarily grazers, Mendelssohn[28] found them browsing extensively on *Prosopis farcata* and *Zizyphus lotus* in summer. Some gazelle, such as *Gazella thomsoni,* are primarily grazers,[13,39] some such as *G. subgutturosa,*[43] are basically browsers, but most are mixed feeders taking advantage of edible plants that vary depending on the climate, season, or other conditions.

Probably more intensive studies have been made on food habits of *G. granti* and *G. thomsoni* than on any other gazelle species.[4-6,16,32-34,37-39] However, Baharav[2] and Mendelssohn[28] have contributed much information on the natural diets of *G. gazella* and *G. dorcas.* Information on foods used by these four species was based largely on analyses of rumen and fecal samples, bite counts, and general observations. For the remaining species, general observations have provided the main source of data on food habits. In many cases only a few foods or at most very general descriptions have been recorded.

NATURAL DIETS OF GAZELLES

Members of the genus *Gazella* may have specific feeding habits and yet be capable of varying their habits and diets based upon availability of food. Hofmann[24] has explained in great detail why such capabilities occur in this group of ruminants. Accordingly, there are three main types of feeders (with subdivisions):[24,25]

1. Bulk and roughage eaters (grazers), including roughage grazers, fresh grass grazers that are dependent upon water, and dry region grazers
2. Selectors of juicy, concentrated herbage, including tree and shrub foliage eaters and fruit and dicotyledon (tree, shrub, or forb) foliage selectors
3. Intermediate feeders, including those preferring grasses and those preferring forbs and shrub and tree foliage

East African gazelles are classified as intermediate feeders because of their feeding habits and the morphological features of their stomach. Whereas the diet of many species is rather consistent, the intermediate feeders characteristically have a variable diet.[25] The variable diet becomes necessary for survival due to differences caused by changes in habitat, seasonal weather factors, competition with other species, or drastic changes in vegetation caused by such disturbances as fire and farming. All gazelles seem to fit into this category of intermediate feeders, apparently varying their diet based on availability of food.

Gazella thomsoni Gunther

This gazelle is primarily an animal of the plains and, consequently, is largely a grazer.[5,6,32] However, data in Tables 1 through 5 indicate that Thomson's gazelle will eat a variety of plants. The predominance of grasses is not only due to preference but also to some bias in the analyses. Grasses are easier to identify than most dicotyledons; consequently, grasses are listed by species and dicots are listed often as a group.[5,37,38] Also, the percentage compositions listed often may not be accurate since some plants are more digestible than others, leaving little identifiable material in rumen or fecal samples.[38]

As indicated previously, the food variety is partly due to place, time, and conditions (Tables 1 to 5). For example, very little use was made of *Microchloa kunthii* by *G. thomsoni* in three different areas of Kenya (Tables 1 and 2),[37,38] while in another area a 35% use (Table 4) was indicated.[33] This plant represented only 3.9% of the herb layer in the latter area, which also illustrates a high selectivity of certain foods by this gazelle. Perhaps in such an instance the plant was more palatable due to soil or moisture conditions. *Cynodon dactylon,* although not always overly abundant, usually has a high use rate (Table 6).[5]

Table 1
PERCENT GRASS SPECIES COMPOSITION BY FECAL ANALYSIS OF PLANTS UTILIZED BY *GAZELLA THOMSONI* UNDER DIFFERENT CONDITIONS FROM FIVE SITES IN KENYA[37]

	Location				
Plant species	**Mara 1[a]**	**Mara 2[b]**	**Mara 3[c]**	**Kajiado District[d]**	**Nairobi Park[e]**
Aristida adoensis	13.3	0.5	3.4		
Aristida spp.				0.2	0.2
Bothriochloa insculpta	0.3	0.6	0.9	0.2	0.5
Brachiaria eruciformis				0.1	
Cynodon dactylon	22.1	22.0	11.1		66.4
Cynodon spp.				14.7	
Digitaria milanjiana		0.5			
D. scalarum					0.2
Dinebra retroflexa					0.1
Echinochloa colonum				0.1	
Enneapogon elegans					0.1
Eragrostis spp.	2.4	1.5	0.5	5.1	3.0
Eustachys paspaloides		0.1	0.1		
Harpachne schimperi				4.3	16.0
Ischaemum afrum				0.2	0.2
Microchloa kunthii				0.9	
Panicum coloratum					1.0
P. infestum					0.8
P. massaiense		1.0	1.0		
P. repens		0.6			
Panicum spp.				0.4	
Pennisetum mezianum	3.0	2.2	2.7	18.8	0.7
Rhynchelytrum repens					0.5
Setaria setulosa	3.2				0.2
Sporobolus discosporus				0.1	
S. pellucidus					0.2
S. pyramidalis				0.2	
Themeda triandra	55.5	70.0	79.3	51.0	7.0

[a] Mara 1: 8/13/63; pasture short, green after recent burn.
[b] Mara 2: 8/12/63; pasture partly short and green after burn; partly mature and brown.
[c] Mara 3: 8/12/63; pasture partly short and green after burn; partly mature and mostly brown.
[d] Kajiado District: 7/15/63; pasture mature, brown.
[e] Nairobi National Park: 8/1/63; pasture mature, brown.

Table 2
PERCENT COMPOSITION OF GRASS SPECIES IN DIETS OF *GAZELLA THOMSONI* AT THREE DIFFERENT SEASONS IN NAIROBI NATIONAL PARK, KENYA[38]

Plant species	Season: Following start of rainy season	At end of rainy season	At end of long dry season
Aristida adoensis		0.1	
Aristida spp.[a]	1.9		1.3
Bothriochloa insculpta	1.9	2.3	4.3
Cenchrus ciliaris	0.1		
Chloris gayana			0.4
C. pycnothrix			0.1
Cymbopogon pospischilii			3.9
Cynodon dactylon	7.8	4.1	16.9
Digitaria macroblephara		14.4	0.5
D. scalarum	0.1	0.3	
Dinebra retroflexa		1.4	
Eragrostis spp.[b]	0.2		0.1
Eustachys paspaloides	1.1	1.9	
Harpachne schimperi	0.5		1.0
Heteropogon contortus	1.5		2.5
Hyparrhenia lintonii	0.8		2.0
Ischaemum afrum	16.7	23.0	0.1
Lintonia nutans		2.0	0.3
Microchloa kunthii	0.1		
Panicum coloratum	0.2	0.6	8.1
Pennisetum mezianum	2.0	1.0	
Rhynchelytrum repens			0.1
Setaria phleoides		0.4	0.1
Sporobolus spp.[c]	1.1	0.1	0.4
Themeda triandra	62.8	47.8	48.4

[a] *Aristida adoensis, A. keniensis* Henr.
[b] *Eragrostis braunii* Schweinf., *E. cilianensis* (All.) Lutati, *E. exasparata* Peter, *E. hispida* K. Schum., *E. ?lehmanniana* Nees., *E. superba, E. tenuifolia* (A. Rich.) Steud.
[c] *Sporobolus discosporus, S. festivus* A. Rich., *S. pellucidus, S. phyllotrichus* Hochst., *S. pyramidalis, S. stapfianus* Gandoger.

Table 3
PERCENT PLANT SPECIES COMPOSITION OF RUMEN CONTENTS FROM *GAZELLA THOMSONI* BY SEASON AT KEKOPEY RANCH, GILGIL, KENYA[32]

	1970		1971					
Plant species	**Aug. 18**	**Nov. 12**	**Jan. 27**	**May 12**	**Aug. 6**	**Dec. 11**	**Feb. 19, 1972**	$\bar{x}$
				Grasses				
Aristida spp.	0.3	0.3			0.1	1.6	0.6	0.4
Chloris pycnothrix	0.1	0.9		0.8	0.8	0.6	0.8	0.6
Cymbopogon pospischilii				0.1				<0.1
Cynodon nlemfuensis	53.3	40.7	45.2	46.6	51.9	44.0	43.5	46.5
Digitaria macroblephara							<0.1	<0.1
D. scalarum	14.4	8.2	16.1	20.6	10.8	18.8	13.0	14.6
Enneapogon schimperanus								<0.1
Eragrostis spp.	2.5	11.8	8.2	3.0	3.5	11.5	14.0	7.8
Harpachne schimperi	1.8	5.8	2.4	0.9	1.9	9.2	12.8	5.0
Microchloa kunthii	0.1	3.4		0.6	1.8	1.5	1.2	1.2
Panicum maximum				0.1				<0.1
Pennisetum spp.	0.8	0.6	2.7		0.6	1.0	0.6	0.9
Setaria pallidifusca	0.3		0.2		2.1		0.1	0.4
Sporobolus confinis	1.5	3.2	5.2	3.9	0.6	0.7	2.2	2.5
S. pyramidalis							0.2	<0.1
Themeda triandra	1.1	4.2	9.1	6.5	1.1	3.7	5.8	4.5
Tragus berteronianus							0.1	<0.1
Unidentified	2.9	5.1	3.8	3.6	4.1	4.1	3.6	3.9
Total grass	79.1	84.2	93.0	86.9	79.3	96.7	98.6	88.3
				Other Monocotyledons				
Commelina spp.	0.8		0.5	0.3	0.3			0.3
Cyperus spp.	1.8	6.7	0.6	11.8	10.1	2.9	0.7	4.9
Eleocharis marginulata					1.6			0.2
Total other monocots	2.6	6.7	1.1	12.1	12.0	2.9	0.7	5.4
				Dicotyledons				
Erlangea cordifolia	11.7	5.7	2.0		0.3			2.8
Euphorbia inaequilatera					2.1		0.2	0.3
Hibiscus spp.		0.6	0.1					0.1
Medicago laciniata				0.1	0.6			0.1
Oldenlandia scopulorum	1.5							0.2
Solanum incanum (fruit)	1.0	0.7	0.3		0.2	0.1	0.3	0.4
Unidentified	4.1	2.1	3.5	0.9	5.5	0.3	0.2	2.4
Total dicots	18.3	9.1	5.9	1.0	8.7	0.4	0.7	6.3

Table 4
PERCENT COMPOSITION OF PLANT SPECIES IN RUMEN SAMPLES OF *GAZELLA THOMSONI* FROM KAJIADO DISTRICT, KENYA[33,34]

Plant species	January 1974 (ten samples)	January 1975 (ten samples)
Grasses		
Bothriochloa insculpta	1.7	1.1
Brachiaria eruciformis	0.8	
Cynodon spp.	4.7	14.3
Digitaria macroblephara	1.5	3.6
D. scalarum	1.1	0.7
Eragrostis spp.	1.6	1.2
Eustachys paspaloides	3.5	6.0
Harpachne schimperi	1.7	3.3
Ischaemum afrum	0.1	
Microchloa kunthii	35.0	4.3
Pennisetum mezianum	3.1	2.6
P. stramineum	1.1	1.7
Sporobolus spp.	0.4	1.1
Themeda triandra	14.2	35.0
Minor species	0.9	
Unidentified	4.7	4.1
Subtotal	76.1	79.0
Other Monocotyledons		
Commelina spp.	0.4	0.1
Cyperus spp.	1.5	1.4
Subtotal	1.9	1.5
Dicotyledons		
Acacia spp. (leaves)	2.9	0.9
Acacia spp. (seed pods)	0.6	0.1
Barleria eranthemoides	0.1	6.9
Euphorbia glochidiata		0.2
E. inaequilatera	6.8	
Indigofera spp.	3.2	1.6
Legumes		0.5
Portulaca spp.	1.1	0.3
Minor species	0.9	0.6
Unidentified	6.4	1.4
Unidentifiable		7.0
Subtotal	22.0	19.5

Table 5
SUMMARY OF PERCENT PLANT SPECIES COMPOSITION IN RUMEN SAMPLES FROM 19 *GAZELLA THOMSONI* COLLECTED ON AKIRA RANCH, THE RIFT VALLEY, KENYA, 1970–72[16]

Plant species	Percent
Aristida adscensionis	1.16
Chloris gayana	2.05
Cynodon dactylon	20.37
Digitaria spp.	3.84
Eragrostis spp. / *Harpachne schimperi*	10.74
Hyparrhenia spp.	0.26
Panicum spp.	0.05
Sporobolus spp.	1.05
Themeda triandra	51.58
Sedges	0.11
Acacia spp.	0.47
Fruits	0.16
Other dicotyledons	5.16
Unidentified	3.00

Table 6
FOODS EATEN BY *GAZELLA THOMSONI* COMPARED WITH RELATIVE DENSITIES OF PLANT SPECIES IN HERB LAYER

Plant species	Percent available	Percent eaten
Cynodon nlemfuensis	29	47
Digitaria scalarum	34	15
Eragrostis spp. / *Harpachne schimperi*	6	13
Sporobolus spp.	5	2
Themeda triandra	1	4
Other grasses	6	8
Total grasses	81	89
Commelina spp. / *Cyperus* spp.	14	5
Dicotyledons	5	6

Specific plant parts may also be selected. Earlier work in Tanzania indicated that the grass in diets of collected Thomson's gazelle contained 4.69% leaf, 61.19% sheath, and 34.07% stem.[21] Other work in Kenya showed differing percentages, revealing that availability may be more significant than selectivity.

	Kekopey[5] (%)	**Kajiado**[33] (%)	**Akira**[15] (%)
Leaf	66.0	68.9	60.5
Sheath	25.0	21.0	14.2
Stem	9.0	10.1	25.3

The nutritive value of some of the grasses taken by Thomson's gazelle indicates that this species may obtain only a minimal amount of nutrients from such plants, especially in dry seasons (Table 7). Again, selectivity of species and their parts may allow the gazelle to achieve maintenance, if not to obtain optimum nutrient needs. A mean crude protein percent of 10.6 was calculated for a large sample of known gazelle plant foods taken from an area where gazelle were collected. However, a sample of rumen contents taken from gazelle at the same time indicated a crude protein content of 12.8%, which may represent a selectivity for more nutritious grass parts or species, or a selective browsing of dicotyledons. In a nearby area, the mean crude protein level for four different seasons never exceeded 6.09% for grass leaves, while leaves of two browse plants had a mean crude protein level of about 13%.[16] Digestibility levels of food plants eaten by *G. thomsoni* on Kekopey was 78%, compared to 67% for impala (*Aepyceros melampus*) and 58% for range cattle,[5] another indication of the gazelle's ability to select foods that are most suitable for its physiological needs or digestibility efficiency.

When grass is lush and green, Thomson's gazelle may or may not drink water. When grazing conditions become dry, they require more free water but have the ability to select plants with a high moisture content. Minimum water requirements of these gazelle under experimental conditions was shown by Taylor:[41]

	l/100 kg/day	
	22°C	**22–40°C**
Metabolic	0.72	0.47
Preformed	0.23	0.14
Free	1.25	2.13
Total	2.20	2.74

Table 7
NUTRITIVE ANALYSIS OF LEAVES FROM SOME FOOD PLANTS UTILIZED BY *GAZELLA THOMSONI* ON KEKOPEY RANCH, KENYA, 1971

	May 13 (wet season)					August 5 (dry season)				
Grass	CP[a]	CF[b]	Fat	NFE[c]	Ash	CP	CF	Fat	NFE	Ash
Open Grassland										
Cynodon nlemfuensis	15.1	26.3	2.1	38.0	18.6	8.9	29.1	2.6	47.0	12.4
Digitaria scalarum	14.0	26.1	1.9	40.9	17.0	8.1	31.5	4.2	42.9	13.3
Eragrostis spp.	11.8	31.4	1.5	42.2	13.2	7.9	34.0	2.9	45.2	10.0
Sporobolus confinis	12.1	27.4	0.9	43.9	15.6	7.7	34.0	2.4	41.8	14.1
Themeda triandra	12.8	26.6	1.3	41.2	18.1	6.3	34.3	3.2	42.8	13.4
Mean	13.2	27.6	1.5	41.2	16.5	7.8	32.6	3.1	43.9	12.6
Bushland										
Cymbopogon pospischilii	11.4	30.5	1.9	40.9	15.5	7.0	36.8	2.6	43.1	10.5
Digitaria scalarum	10.9	30.0	2.0	42.4	16.5	9.6	35.2	5.4	36.2	13.6
Eragrostis spp.						5.7	39.9	1.6	43.2	9.6
Hyparrhenia papillipes	12.0	27.1	1.3	37.8	21.8	5.9	37.3	2.0	40.7	14.1
Panicum maximum	10.1	31.3	1.2	36.6	20.5	5.4	43.8	1.5	33.8	15.5
Themeda triandra	10.0	33.6	1.3	39.9	15.0	5.5	38.0	2.4	43.5	10.6
Mean	10.9	30.5	1.5	39.5	17.9	6.5	38.5	2.6	40.1	12.3

[a] Crude protein.
[b] Crude fiber.
[c] Nitrogen-free extract.

Data on food plants and the relative occurrence of each in the diet of Thomson's gazelle in East Africa have been compiled from the quantitative work of many researchers (Table 8).[5,16,32-34,37-39] From this compilation and those in Tables 1 through 6, many of the major natural foods of *G. thomsoni* may be discerned. In addition, Brooks[6] lists some qualitative data, all of which may not be included in the preceding tables; this information is given in the lists below.

Trees eaten:

1. *Acacia campylacantha* (Hochst. ex A. Rich.) Brenan, now designated as *A. polyacantha* Willd. subsp. *campylacantha* – leaves, when low enough
2. *A. drepanolobium* Harms – leaves and tender, growing twig tips
3. *A. seyal* Del. – fruits and leaves
4. *A. spirocarpa* (Hochst. ex A. Rich.), now designated as *A. tortilis* (Forsk.) Hayne subsp. *spirocarpa* – leaves and seed pods; fallen leaves are also eaten.
5. *A. subalata* Vatke, now designated as *A. nilotica* (L.) Del.
6. *Balanites aegyptiaca* (L.) Del. – leaves and pulp of fruit
7. *Borassus aethiopum* Mart. – pulp of fruit
8. *Boscia calonuera* Gilg., probably *B. angustifolia* A. Rich. – leaves and fruit (off ground?)
9. *Commiphora schimperi* (Berg) Engl. – leaves and fresh twigs
10. *Kigelia aethiopica* Dene., probably *K. aethiopum* (Fenzl) Dandy – flowers and fruit from ground
11. *Ormocarpum trichocarpum* (Taub.) Harms – leaves and twigs

Shrubs and herbs eaten:

1. *Cordia gharaf* (Forsk.) Ehrenb. ex Aschers. – leaves
2. *C. ovalis* R. Br. – leaves
3. *Courbonia camporum* Gilg. et Bened., synonym for *Maerua edulis* (Gilg. and Bened.) DeWolf – fruits, flowers, leaves, and tender tips of stems
4. *Harrisonia abyssinica* Oliv. – leaves
5. *Indigofera hochstetteri* Bak.
6. *I. basiflora* Gillett – fresh tips of branches
7. *Indigofera* sp. – tender twig tips during wet season
8. *Justicia elliotii* S. Moore – cropped heavily
9. *Melhania ovata* (Cav.) Spreng. – leaves and fresh twig growth
10. *Solanum panduriforme* E. May ex Dunal (Sodom apple), synonym for *S. incanum* L., – young leaves and berries

Grasses eaten:

1. *Dactyloctenium aegyptium*
2. *Digitaria* spp.
3. *Eustachys paspaloides* – moderate grazing
4. *Pennisetum mezianum* – browse on tufts
5. *Sporobolus marginatus*
6. *Themeda triandra* – fresh growth
7. Limited grazing: *Chloris* spp., *Eragrostis* spp., *Heteropogon contortus, Perotis patens* Gand., *Pogonarthria squarrosa* (Roem. and Schult.) Pilg., *Rhynchelytrum repens, Setaria trinervia,* and *Sporobolus* spp.

Table 8
PLANTS KNOWN TO HAVE BEEN EATEN BY *GAZELLA THOMSONI* IN VARIOUS PLACES AND DURING DIFFERENT SEASONS IN EAST AFRICA[4,5,16,32–34,37–39]

Plant species	Composition of diet (%)						
	Unknown	Trace–1.0	1.1–5.0	5.1–10.0	10.1–20.0	20.1–30.0	30.1+
	Number of times occurring						
	Grasses						
Aristida adoensis Hochst.		2	1		1		
A. adscensionis L.			1				
Aristida spp.		6	3				
Bothriochloa insculpta (A. Rich.) A. Camus		6	4				
Brachiaria eruciformis Griesb.		2					
Cenchrus ciliaris L.		2					
Chloris gayana Kunth.		1	1				
C. pycnothrix Trin.		7					
Cymbopogon pospischilii (K. Schum.) C. E. Hubb.		1	1				
Cynodon dactylon (L.) Pers.			1	1	4	2	1
C. nlemfuensis Vanderyst							7
Cynodon spp.			1		1		

Table 8 (continued)
PLANTS KNOWN TO HAVE BEEN EATEN BY *GAZELLA THOMSONI* IN VARIOUS PLACES AND DURING DIFFERENT SEASONS IN EAST AFRICA[4,5,16,32–34,37–39]

Plant species	Composition of diet (%) Unknown	Trace–1.0	1.1–5.0	5.1–10.0	10.1–20.0	20.1–30.0	30.1+
	Number of times occurring						
Grasses (continued)							
Digitaria macroblephara (Hock.) Stapf		2	2		1		
D. milanjiana (Rendle) Stapf		1					
D. scalarum (Schweinf.) Chiov.		4	1	1	5	1	
Digitaria spp.			1	1			
Dinebra retroflexa (Vahl) Panzer		1	1				
Echinochloa colonum (L.) Link		1					
Enneapogon elegans (Nees) Stapf		1					
E. schimperanus (A. Rich.) Renv.		1					
Eragrostis spp.		3	9	3	3		
Eustachys paspaloides (Vahl) Lanza and Mattei		2	3	1			
Harpachne schimperi A. Rich.		3	6	3	2		
Heteropogon contortus (L.) Roem. and Schult.			2				
Hyparrhenia lintonii Stapf		1	1				
Hyparrhenia spp.		1					
Ischaemum afrum (J. F. Gmel.) Dandy		4			1	1	
Lintonia nutans Stapf		1	1				
Microchloa kunthii Desv.		4	5				1
Panicum coloratum L.		3		1			
P. infestum Anderss.		1					
P. massaiense Mez.		2					
P. maximum Jacq.		1					
P. repens L.		1					
Panicum spp.		2			1		
Pennisetum mezianum Leeke		2	6		2		
P. stramineum Peter			2				
Pennisetum spp.		4	1				
Rhynchelytrum repens (Willd.) C. E. Hubb.		2					
Setaria pallidifusca (Schumach.) Stapf and C. E. Hubb.		1	1				
S. phleoides Stapf		2					
S. setulosa Stapf		1	1				
Sporobolus confinis (Steud.) Chiov.		2	4	1			
S. discosporus Nees		1					

Table 8 (continued)
PLANTS KNOWN TO HAVE BEEN EATEN BY *GAZELLA THOMSONI* IN VARIOUS PLACES AND DURING DIFFERENT SEASONS IN EAST AFRICA[4,5,16,32–34,37–39]

Plant species	Composition of diet (%)						
	Unknown	Trace–1.0	1.1–5.0	5.1–10.0	10.1–20.0	20.1–30.0	30.1+
	Number of times occurring						
		Grasses (continued)					
S. pellucidus Hochst.		1					
S. pyramidalis Beauv.		2					
Sporobolus spp.		3	3				
Themeda triandra Forsk.			5	4	1		9
Tragus berteronianus Schult.		1					
Triticum aestivum L.	X						
Zea mays L.	X						
		Other Monocotyledons					
Commelina spp.		9					
Cyperus spp.		4	5	2	3		
Eleocharis marginulata Steud.		1					
Kyllinga sp.	4						
		Dicotyledons					
Acacia spp.		2	1				
Barleria eranthemoides R. Br. ex C. B. Cl.		1		1			
Courbonia spp.		1					
Enicostema axillare (Lam.) A. Rayn.	2						
Erlangea cordifolia S. Moore		1	2	2	2		
Euphorbia glochidiata Pax		1					
E. inaequilatera Sond.		1	1	1			
Hibiscus spp.		4					
Indigofera spp.			2				
Justicia sp.			1				
Legumes		1	1				
Medicago laciniata (L.) Mill.		2					
Oldenlandia scopulorum Bullock			2				
Portulaca spp.		2					
Rhynchosia sp.		1					
Sida schimperiana Hochst. ex A. Rich.			1				
Solanum incanum L.		9					
Solanum sp.			1				

Gazella granti Brooke

Primarily an animal of the open plains, Grant's gazelle will often frequent open and even relatively dense bush. Because of the wide range in habitat, this species has a choice of a variety of food plants. *G. granti* is a mixed feeder, feeding on forbs and shrubs as well as grasses. It is highly probable that this gazelle feeds on a much greater variety of dicotyledons (especially forbs) than is indicated in Tables 9 through 13. Again, the scarcity of dicots is an indication of the difficulty of identification compared to grasses, as explained previously for *G. thomsoni.* Dicots are a major source of food toward the end of rainy seasons and on into dry seasons, whereas grass makes up a considerable portion of the diet early in the rainy seasons.[38]

Certain grasses, such as *Cynodon dactylon* and *Themeda triandra,* are utilized consistently by Grant's gazelle, whereas others appear to be used seasonally or by area (Tables 9 through 12). *Acacia* spp. are also browsed frequently. Other dicots of importance are usually not known since they are grouped as "unidentified" or as dicotyledons. The compilation in Table 13 presents most of the plants known to be utilized and their relative frequency of occurrence.

Table 9
PERCENT GRASS SPECIES COMPOSITION BY FECAL ANALYSIS OF PLANTS UTILIZED BY *GAZELLA GRANTI* UNDER DIFFERENT CONDITIONS FROM FOUR SITES IN KENYA[37]

	Location			
Plant species	Mara 1[a]	Mara 3[b]	Kajiado District[c]	Nairobi Park[d]
Aristida adoensis	4.8	3.5		
Aristida spp.			0.3	0.2
Bothriochloa insculpta	1.4	2.2	0.4	0.8
Brachiaria eruciformis			0.5	
Cynodon dactylon	55.5	34.0		59.4
Cynodon spp.			30.4	
Dinebra retroflexa			0.2	0.3
Eragrostis spp.	0.9	1.7	1.0	1.5
Eustachys paspaloides	0.3			
Harpachne schimperi			0.4	3.3
Ischaemum afrum			0.4	4.4
Lintonia nutans				0.1
Microchloa kunthii			0.9	
Panicum coloratum				0.7
P. infestum				0.6
P. massaiense	0.2	0.5		
P. poaeoides Stapf				0.1
Panicum spp.			1.8	
Pennisetum mezianum	3.3	3.0	13.7	1.0
Setaria setulosa	3.7			0.3
Themeda triandra	22.3	53.7	42.9	19.0

[a] Mara 1: 8/13/63: pasture short and green after recent burn.
[b] Mara 3: 8/12/63: pasture partly short and green after burn; partly mature and mostly brown.
[c] Kajiado District: 7/15/63; pasture mature, brown.
[d] Nairobi National Park: 8/1/63; pasture mature, brown.

Table 10
PERCENT COMPOSITION OF GRASS SPECIES IN DIETS OF *GAZELLA GRANTI* AT THREE DIFFERENT SEASONS IN NAIROBI NATIONAL PARK, KENYA[38]

	Season		
Plant species	Following start of rainy season	At end of rainy season	At end of long dry season
Aristida spp.[a]	0.8		4.7
Bothriochloa insculpta	1.8	0.8	7.1
Cenchrus ciliaris	0.4	0.4	
Cymbopogon pospischilii	0.1		2.8
Cynodon dactylon	29.6	19.4	11.8
Digitaria macroblephara	1.0	11.8	5.0
Digitaria scalarum	0.8		
Dinebra retroflexa		17.9	
Eragrostis spp.[b]	2.0		1.6
Eustachys paspaloides	0.9	4.3	
Harpachne schimperi			0.5
Heteropogon contortus	0.5		2.0
Hyparrhenia lintonii	1.4		
Ischaemum afrum	8.6	9.8	11.1
Lintonia nutans		2.0	0.3
Microchloa kunthii	0.1		0.5
Panicum coloratum	1.1	0.8	12.9
Pennisetum mezianum	5.5	7.7	0.8
P. stramineum	0.1		
Rhynchelytrum repens			0.5
Setaria pallidifusca		0.4	
S. phleoides	0.5	0.4	1.0
Sporobolus spp.[c]	1.6	1.6	1.2
Themeda triandra	42.5	22.3	32.9

[a] *Aristida adoensis* and *A. keniensis.*
[b] *Eragrostis braunii, E. cilianensis, E. exasparata, E. hispida, E. ?lehmanniana, E. superba,* and *E. tenuifolia.*
[c] *Sporobolus discosporus, S. festivus, S. pellucidus, S. phyllotrichus, S. pyramidalis,* and *S. stapfianus.*

Table 11
PERCENT COMPOSITION OF PLANT SPECIES IN RUMEN SAMPLES OF *GAZELLA GRANTI* FROM KAJIADO DISTRICT, KENYA[33,34]

Plant species	January 1974 (ten samples)	January 1975 (ten samples)
Grasses		
Bothriochloa insculpta		0.3
Brachiaria eruciformis	2.6	
Brachiaria spp.		1.0
Chloris roxburghiana	0.2	
Cynodon spp.	1.8	9.3
Digitaria macroblephara	7.8	1.8
D. scalarum	0.6	
Eragrostis spp.	2.2	0.2
Eustachys paspaloides	4.6	0.5
Harpachne schimperi		0.4
Ischaemum afrum		0.1
Microchloa kunthii	0.6	0.1
Pennisetum mezianum	19.2	1.5
P. stramineum	8.8	1.7
Sporobolus spp.	1.0	
Themeda triandra	5.8	3.4
Minor species	0.6	
Unidentified	7.0	3.3
Subtotal	62.8	23.6
Other Monocotyledons		
Commelina spp.	6.6	2.4
Cyperus spp.	14.2	6.2
Unidentified	1.0	
Subtotal	21.8	8.6
Dicotyledons		
Acacia spp. (leaves)	0.6	3.7
Acacia spp. (seed pods)	0.2	21.0
Barleria eranthemoides	1.8	8.6
Euphorbia glochidiata		3.2
E. inaequilatera	0.6	
Indigofera spp.		2.5
Legumes		1.6
Portulaca spp.		1.0
Minor species	0.8	2.8
Unidentified	11.4	6.3
Unidentifiable		17.1
Subtotal	15.4	67.8

Table 12
SUMMARY OF PLANT SPECIES COMPOSITION IN RUMEN SAMPLES FROM 22 *GAZELLA GRANTI* COLLECTED ON AKIRA RANCH, THE RIFT VALLEY, KENYA, 1970–72[16]

Plant species	Percent
Aristida adscensionis	0.62
Cenchrus ciliaris	0.05
Chloris gayana	1.14
C. pycnothrix	0.05
Cynodon dactylon	23.27
Digitaria spp.	4.32
Eragrostis spp. *Harpachne schimperi*	4.27
Hyparrhenia spp.	0.18
Sporobolus spp.	0.32
Themeda triandra	14.73
Sedges	0.45
Acacia spp.	14.64
Fruits	1.91
Other dicotyledons	31.32
Unidentified	2.64

Table 13
PLANTS KNOWN TO HAVE BEEN EATEN BY *GAZELLA GRANTI* IN VARIOUS PLACES AND SEASONS IN EAST AFRICA[4,16,26,32–34,37–39]

Plant species	Composition of diet (%)						
	Unknown	Trace–1.0	1.1–5.0	5.1–10.0	10.1–20.0	20.1–30.0	30.1+
	Number of times occurring						
			Grasses				
Aristida adoensis			2				
A. adscensionis	6	1					
Aristida spp.		3	1				
Bothriochloa insculpta	3	4	3	1			
Brachiaria eruciformis		1	1				
Brachiaria spp.		1					
Cenchrus ciliaris	8	3					
Chloris gayana			1				
C. pycnothrix		1					
C. roxburghiana Schult.		1					
Cymbopogon pospischilii		1	1				
Cynodon dactylon	15			1	2	2	3
Cynodon spp.			1	1			1
Dactylenium aegyptium (L.) Beauv.	2						
Digitaria macroblephara		1	2	1	1		
D. milanjiana	8						
D. scalarum		2					
Digitaria spp.			1				
Dinebra retroflexa		2			1		
Enneapogon cenchroides (Roem. and Schult.) C. E. Hubb.	1						
Eragrostis rotifer Rendle	3						
E. superba Peyr.	3						
Eragrostis spp.		3	6				
Eustachys paspaloides		3	2				
Harpachne schimperi		3	2	1			
Heteropogon contortus	1	1	1				
Hyparrhenia lintonii			1	1			
Hyparrhenia spp.		1					
Ischaemum afrum		2	1	2	1		
Lintonia nutans		2	1				
Microchloa kunthii		5		1			
Panicum coloratum		2	1		1		
P. infestum		1					
P. massaiense		2					
P. maximum	2						
P. poaeoides		1					
Panicum spp.			1				
Pennisetum mezianum		2	3	2	2		
P. stramineum		1	1	1			
Rhynchelytrum repens		1					
Setaria pallidifusca		1					
S. phleoides		3					
S. setulosa		1	1				
Sporobolus ioclades (Trin.) Nees	14						
Sporobolus spp.		2	3				
Themeda triandra		1	1	1	2	2	4
Tragus berteronianus	2						

Table 13 (continued)
PLANTS KNOWN TO HAVE BEEN EATEN BY *GAZELLA GRANTI* IN VARIOUS PLACES AND SEASONS IN EAST AFRICA[4,16,26,32–34,37–39]

Plant species	Composition of diet (%)						
	Unknown	Trace–1.0	1.1–5.0	5.1–10.0	10.1–20.0	20.1–30.0	30.1+
	Number of times occurring						
			Other Monocotyledons				
Commelina spp.			1	1			
Cyperus spp.		1		1	1		
Kyllinga sp.	2						
			Dicotyledons				
Acacia drepanolobium [Harms ex] Sjostedt	1	2					
A. mellifera (Vahl.) Benth.	2						
A. stuhlmannii Taub.						1	
Acacia spp.					1		
Aster sp.			1				
Balanites aegyptiaca			1				
Barleria eranthemoides			1	1			
Crotolaria spp.	2						
Ecbolium revolutum (Lindau) C. B. Cl.	1						
Enicostema axillare	6						
Euphorbia glochidiata			1				
E. inaequilatera		1					
Indigofera spp.	6		1		1		
Legumes			2				
Portulaca spp.		1					
Sida ovata Forsk.	2						
S. schimperiana					1		
Solanum incanum L.	1						
Solanum sp.	7			1			

The use of various parts of grasses by Grant's gazelle is illustrated in the following table:

	Percent use	
	Athi Kapiti Plains[38]	Akira Ranch[15]
Leaves	64.0	47.3
Sheath	25.2	9.0
Stem	10.8	43.7

The Athi Kapiti Plains collection was made in January 1974, a dry season period, whereas the Akira Ranch (Kenya) collection included both a dry and wet season. Stems would have been more palatable in a wet season.

Like *G. thomsoni,* this gazelle is capable of selecting the more nutritious and digestable foods. The maximum mean crude protein contents of grass and shrubs examined during a rainy season on Akira Ranch were as follows:[16]

	Grass	Shrubs
Leaves	6.09	13.53
Stems	3.71	4.98

During the same study rumen samples from Grant's gazelle showed a mean crude protein analysis of 12.48%, which is probably near the minimal maintenance requirement for this species for the perpetuation of the individual and the population.

Grant's gazelles can survive without water for extended periods, obtaining much of their needed moisture by selective grazing of plants which have the highest moisture content and by fully utilizing available water. Even the driest grasses found in East Africa contain 3 to 5% water by weight, and animals such as this species could obtain most of the water required.[26] Minimum water requirements in liters per 100 kg/day at various temperatures are shown in the following table:[41]

	22°C	22–40°C
Metabolic	0.62	0.62
Preformed	0.21	0.21
Free	1.25	3.03
Total	2.08	3.86

Only the drinking water requirements change under higher heat stresses. Since this species spends much time in the hot sun, under natural conditions the need for free water decreases primarily because of a higher body temperature and lower evaporation rate.[42] Periods of inactivity and infrequent urination also may conserve water. When free water is not available, utilization of preformed and oxidation water seems to satisfy Grant's gazelle's need for at least 4% of its body weight each day.[42] In the Northern Frontier District of Kenya, much of this water may come from a shrub, *Disperma* sp., and other miscellaneous plants.[40] During a drought in September 1965, an analysis of this plant showed it to contain 27.7% ash, 3.14 kcal/g, and 13% crude protein.[40]

Gazella dorcas **L.**

The dorcas gazelle is primarily found in the semidesert and desert areas of northern Africa, extending into Israel, Syria, Arabia, and India. Vegetation variety may be great, but the amount is frequently limited; consequently, the animals will feed on almost any desert vegetation except halophytes. They especially like *Acacia* spp. *G. dorcas* is highly mobile, so it will move to more nutritious pastures as the need arises.

Throughout its range, the dorcas gazelle utilizes *Acacia* species extensively, including leaves, twigs and seed pods.[7,18,28] In Israel's Arava Valley, *A. tortilis* and *A. spirocarpa* were the main foods eaten by the gazelle,[28] while in the Sudan *A. tortilis* was one of the main plants used.[7,18] Throughout much of the range, leaves from these species have adequate nutrients to support life and proper condition through dry seasons, including necessary moisture.[7] A chemical analysis of *Acacia* leaves from the Sudan shows a low-protein content, but evidently such food is sufficient to maintain *G. dorcas* (Table 14).[18] Desert-dwelling gazelles probably require less energy than species from other habitats, but, as with other gazelles, the dorcas also efficiently utilizes its food.

Other plants eaten by *G. dorcas* under natural conditions include a variety from several countries (Table 15). Most of the plants listed are shrubs and forbs.

Mendelssohn[28] found that dorcas gazelle also utilize short-lived annuals following rains, but emphasized that the species do not use halophytes. In India, the gazelle ate some grass,[3] but O'Reagan[48] indicated that in Iran dorcas gazelle apparently are not grass eaters. However, he did find them eating forbs in spring and woody plants in winter. Locusts and larvae were eaten in northern Africa.[11]

Table 14
ANALYSIS OF *ACACIA TORTILIS* FROM FOUR LOCALITIES IN THE SUDAN, EXPRESSED AS PERCENTAGES OF THE WET WEIGHT[18]

Each Figure is the Mean of Ten Samples from One Locality Collected During Winter and Summer

Locality	Moisture	Protein	Carbohydrate	Fiber	Ash	Calories/100 g
Khartoum	49.0	6.0	30.0	8.6	6.7	148
Kosti	49.3	5.2	29.8	9.4	6.2	144
Suakin	48.4	5.7	30.2	8.8	6.4	147
Karima	47.9	5.4	30.8	9.2	7.2	147
Mean value	48.7	5.5	30.2	9.0	6.6	147

Table 15
PLANTS EATEN BY *GAZELLA DORCAS* IN SEVERAL COUNTRIES WITHIN ITS DISTRIBUTION

Plant	Country	Ref.
Acacia tortilis	Sudan, Israel	7,18,28
Zizyphus sp.	Sudan, India	3,18
Aizoon spp.	Northern Africa	11
Acacia spirocarpa	Israel	28
Acacia spp.	Israel	28
Ochradenus baccatus Delile	Israel	28
Panicum turgidum Forsk.	Israel	28
Ephedra spp.	Iran	48
Pteropyrum spp.	Iran	48
Seidlitzia spp.	Iran	48
Acacia leucophloea (Roxb.) Willd.	India[a]	3
Bauhinia sp.	India	3
Diospyros sp.	India	3
Emblica sp.	India	3
Soymida sp.	India	3
Terminalia sp.	India	3

[a] Reference is to *Gazella g. bennetti* but many classifications now tend to place Indian gazelle as *G. dorcas bennetti.*[20]

Although a desert animal, dorcas gazelle could only tolerate dehydration for 3 to 4 days under experimental conditions.[18] Animals maintained on dry sorghum (millet-*Panicum miliaceum* L.) had an average daily water consumption of 4.1 l/100 kg body weight in winter, and 5.6 l/100 kg body weight in summer. From caloric requirements of dorcas fed on sorghum or *Acacia* leaves, the moisture gained as preformed water from these plants was estimated to be 4.7 and 4.4 l/100 kg body weight in winter and summer, respectively; thus, there would be a surplus of 0.6 l water per 100 kg body weight in winter and 1.2 l in summer. Carlisle and Ghobrial[7] found that gazelle must eat 1083 g of *Acacia* leaves per day in November to provide 1592 cal and 206.4 ml metabolic water; this is 143 ml more than the daily requirement. In May, the animal must eat 980 g to provide 1440 cal and 663 ml metabolic water, which shows a deficit of 177 ml.

Gazella gazella **Pallas**

The mountain gazelle (Idmi) is widespread in distribution.[12,20] It lives in habitats ranging from plains and valleys to mountains, and since it is an intermediate-type feeder it consumes a variety of plant species.[28,29] However, this gazelle can survive on only one or a few plant species such as *Acacia* spp. in a natural habitat or a monoculture agricultural field of wheat, potatoes, or cotton. In some natural areas where moisture is more consistent or after seasonal rains, annuals are used extensively, and if soil conditions are right a relatively rich vegetation of small shrubs may exist and be used intensively by gazelle.[28]

As an intermediate feeder, *G. g. gazella* is a browser primarily in the desert, and even in those areas such as Ramot Yissakhar, Israel, where grazing is dominant, several shrubs are browsed extensively during summer months.[2,28] In desert areas where grazing is available, gazelle use grasses and forbs (especially the succulent forbs) as a food and water source.[23] The variety of plants eaten is illustrated in Table 16. Barley also was recorded as a food for *G. g. cuvieri* in Algeria.[30]

Table 16
PLANTS UTILIZED AS FOOD BY *GAZELLA G. GAZELLA* IN ISRAEL[2,28,29]

Plant	Comment
	Agricultural
Capsicum annum L.	Green pepper – leaves and fruit
Cicer arietinum L.	Chick peas – eaten in winter
Citrullus vulgaris Schrad.	Watermelon – leaves and unripe fruit, summer
Citrus spp.	Citrus plants – vegetative parts
Ficus carica L.	Fig – leaves
Gossypium hirsutum L.	Cotton – young plants
Hordeum vulgare L.	Barley – heads and leaves
Medicago sativa L.	Alfalfa
Medicago sp.	Clover
Olea europaea L.	Olive trees – leaves, twigs
Prunus spp.	Peaches – leaves and fruit
Solanum melongena L.	Eggplants – leaves
S. tuberosum L.	Potato – leaves
Triticum aestivum[a]	Wheat – young plants, heads
Vitus spp.	Grapes – leaves and fruits
	Nonagricultural
Acacia spirocarpa	Desert – leaves, twigs, and seed pods
A. tortilis	Desert – leaves, twigs, and seed pods
Allhagi maurorum medik	Summer and fall browse
Asphodelus microcarpus	Monocotyledon – summer and fall
Cynodon dactylon	Grass – warm season; dig up rhizomes with hooves
Nerium oleander L.	Oleander – leaves; poisonous
Ochradenus baccatus	Eaten most of year
Prosopis farcata (Banks and Sol.) Macbride	Summer browse
Urginea maritima (L.) Baker	Mainly taken when leaves are partly wilted
Zizyphus lotus (L.) Desf.	Summer browse
Z. spina-christi (L.) Desf.	Dry-period browse

[a] Also recorded as a food of *G. gazella* by Grau.[19]

Plant species receiving 10% or more use by season in the Upper Galilee area of Israel[64] are listed in the following table:

Spring	Summer	Winter
Calycotoma villosa	*Calycotoma villosa*	*C. siliqua*
Prosopis farcata	*Capparis spinosa* L.	*P. lentiscus*
Zizyphus lotus	*Ceratonia siliqua*	
	Pistacia lentiscus L.	
	Rhamnus palaestina	
	Styrax officialis	
	Z. lotus	

Water is used when and where it is available, but in deserts the gazelle may have to go for extended periods without free water.[28] In such cases water deficits are prevented by use of metabolic and preformed water, inactive periods by the animal, and the slick, shiny coat which reduces evaporation. Gazelles that live in mountains, like *G. g. cuvieri*, probably have fewer water stresses than those that inhabit the desert.

Other Gazelles

Feeding habit studies often require more of a scientist's time and finances than do studies of many other subjects. Also, there has been either a lack of interest or a difficulty in obtaining support for life history studies, including feeding habits, of many wild animals, especially the less accessible or popular ones. Consequently, very little diet information is available for the remaining gazelle species.

Gazella subgutturosa **Guldenstaedt** – O'Reagan[48] lists the following plants as occurring in the diet of the goitered gazelle in Iran:

Artemisia sp. – taken significantly when green
Astragalus sp. – taken significantly when green
Halocnemum sp. – and other chenopods eaten in winter
Triticum aestivum – green heads
Pteropyrum sp. – summer
Ephedra sp. – autumn
Seidlitzia sp. – autumn
Forbs – during spring
Rhanterium eppaposum – This dwarf shrub is an important food plant which also holds considerable moisture in droughts.[43]

Gazella dama **Pallas** – Members of this species browse or graze on *Acacia* spp., various other desert shrubs, and rough desert grasses.[11] They will utilize free water but can survive in long periods of drought, especially if vegetation has sufficient water content.

Gazella soemmeringi **Cretzschmar** – Primarily from very arid zones of Ethiopia and Somalia, this species feeds mostly on grass and shrub leaves, including *Acacia* spp. Water is taken when readily available but the animal can exist for extended periods without access to surface water.

Gazella pelzelni **Kohl** – Pelzeln's gazelle live in very desolate country where vegetation is relatively sparse. Shoots and leaves of *Acacia* spp. as well as other accessible desert bushes are utilized as food.[11] Some grasses and forbs that appear following rains are also eaten when available. Because of its distribution in very arid country, this gazelle can survive without free water for extended periods.

Gazella leptoceros **F. Cuvier** – This species feeds on desert plants in northeast Africa.[11] In Arabia its main food is coarse grass, but it takes tender shoots and young plants when available.[27] Loder's gazelle seldom or never drinks water.

***Gazella rufifrons* Gray** – The red-fronted gazelle of the Sahelian zone and northern Sudan feeds on both grasses and shrubs. Some of the major woody plants eaten are *Acacia* spp., *Balanites* sp., *Boscia* sp., *Calotropis* sp., *Euphorbia* sp., *Leptadenia* sp., *Salvadora* sp., and *Zizyphus* sp.[11]

***Gazella tilonura* Heuglin** – Heuglin's gazelle, of Ethiopia and southern Sudan, is sometimes considered a subspecies of *G. rufifrons* and has similar feeding habits of utilizing grass and forbs when available but depending extensively on woody plants.

***Gazella spekei* Blyth** – Speke's gazelle, found on the interior plateaus of Somalia, must exist on a sparse vegetation for much of the year. *Acacia* spp. and other small shrubs are available for browse. Coarse grass is scattered over the area much of the time but usually annual herbs are available only following rainfalls.

SYNTHETIC DIETS OF GAZELLE

Synthetic foods, as used here, are those specially prepared for animals in captivity. These foods may be naturally grown (e.g., alfalfa), manufactured as concentrates, synthesized as vitamins, or extracted as minerals.

There are two major philosophies regarding zoo or captive animal diets. One thought is that diets as similar as possible to natural foods should be used. The other is that an assimilated food that matches the requirements of an animal for healthy growth and reproduction is sufficient.

The main item in gazelle diets of most zoological gardens today is a concentrate, usually supplemented with minerals, vitamins, and hay. The latter is necessary not so much for nutrients but as a means of occupying an animal's attention and energy, for preventing restlessness and obesity, and satisfying the animal's behavioral feeding drives.[10,35,47]

Commercial standard diets are becoming more common throughout the zoo world. The ease and efficiency of feeding operations, economics, and a decrease in human error caused by continually changing fodder elements are a few of the advantages of such a diet. However, care must be taken in the composition and amounts of the diet fed to certain species. Small ruminants and delicate feeders should have protein levels of 13 to 20%.[45] Gazelle are included among selective browsers, and require high quality leafy legume hay; also, high protein (15 to 19%) dairy conditioners or pellets are a must and should be offered twice a day.[47] However, such food is devoured quite rapidly, leaving the animals with excess time; consequently, the need arises for hay supplement.[10]

Many different diets are used in zoological gardens throughout the world. Differences exist because of personal preferences, food availability, costs, and zoo species composition. Since there is a need to economize with diet ingredients, labor, and supervision, standard diets that will suffice for many species are prepared.[35,44,45] With species composition varying among zoos, alterations often are necessary, either in the composition or amount of a diet being fed. The more modest eaters require smaller amounts of concentrates than those needing a high energy ration.[44]

The various zoo diets presented here illustrate again the great variety of foods gazelle are capable of utilizing with satisfactory results.

Gazella dama

Dama gazelle are located in many zoological gardens. Diets from four zoos are presented since several other species of gazelle are kept at these and fed the same diet.

In the San Antonio Zoological Gardens, Texas, *G. dama* are fed a pelletized concentrate, alfalfa hay, and hydroponic grass (northern white oats, *Avena sativa* L.).[49] The concentrate, prepared by Allied Mills (Table 17), is formed into 3/16-in. pellets and fed at a rate of 1 handful for small animals and 2 handfuls for large animals (this has

Table 17
COMPOSITION OF WAYNE RANGE BREEDER CUBES[a] USED AT SAN ANTONIO ZOOLOGICAL GARDENS[b] IN THE DIETS OF *GAZELLA DAMA, G. GRANTI, G. DORCAS, G. THOMSONI,* AND *G. SUBGUTTUROSA*

Item	Analysis: Guaranteed	Analysis: Actual (%)
Crude protein	20.0% (min)	17.70
Crude fat	2.0% (min)	2.86
Crude fiber	9.0% (max)	6.98
Ash	–	9.02
Nitrogen-free extract	–	52.18
Moisture	–	11.26
Vitamin A	Not less than 30,000 USP units/lb	–
Vitamin D_2	Not less than 4,000 USP units/lb	–

Ingredients

Grain products, soybean meal, cottonseed meal, wheat middlings, corn gluten feed, brewer's dried grains, rice bran, dehydrated alfalfa meal preserved with ethoxyquin, cane molasses, vitamin A palmitate, irradiated dried yeast (source of vitamin D_2), ground limestone, dicalcium phosphate, salt, copper oxide, cobalt carbonate, manganous oxide, iron carbonate, potassium oxide, magnesium oxide, and zinc oxide. Additives: 0.02 ppm selenium and 50 IU vitamin E

[a] Allied Mills, Inc., Chicago, Ill.
[b] 3903 N. St. Mary's Street, San Antonio, Tex.

proven to be convenient measurement for zoo keepers). Alfalfa hay is given free choice. A small amount of hydroponic grass is fed daily to the gazelles. Coastal bermuda hay (*Cynodon dactylon* [L.] Pers.) is placed in pens for bedding and some may be eaten. Analyses of both alfalfa and coastal bermuda hay in early 1974 showed the following composition:*

Item	Coastal bermuda	Alfalfa
Protein	6.63%	18.80%
Moisture	11.26%	11.96%
Fat	0.98%	1.56%
Ash	4.74%	7.28%
Crude fiber	29.08%	14.68%
Nitrogen-free extract	47.31%	45.72%
Vitamin E (mg α-tocopherol/100 g)	1.85	2.55
Vitamin A (IUA/lb)	3,100	101,000
Selenium (ppm)	0.21	0.22

The basic diet used for Dama gazelle at the San Diego Wild Animal Park is a 5/8-in. pelleted hay formula (Table 18). Dr. Lester Nelson[51] indicated that the feeding rate is about 3 lb of pellets per 100 lb of body weight for all gazelle species. The 1/4-in. grain pellets (Table 18) of higher caloric value are added to gazelle feeders during colder months, as well as to feeders designed for young calves.

* From the office of San Antonio Zoological Gardens.

Oat hay is offered free choice and is eaten occasionally by gazelle. Some hybrid bermuda grass is being irrigated for grazing.

Diets for gazelle at Busch Gardens in Tampa, Florida, consist of a grain mix (D & H Chow, shown in Table 19, oats, and a zoo supplement, shown in Table 20), hay mix (alfalfa, one half or less, and timothy *Phleum pratense* L., one half or more), and Rib-Ad® (Table 21). The zoo supplement serves as a special replacement for grass. Hay is provided free choice while Rib-Ad is combined with the grain mix. D & H Chow is fed at a rate of 1 lb/100 lb body weight.

At the National Zoological Park, Washington, D.C., Dama gazelle receive a diet consisting of alfalfa hay, horse checkers (a pelleted alfalfa meal), and Omolene® (a sweet feed mixture of alfalfa pellets, rolled oats, cracked corn, and molasses).[50] A combination of selenium and vitamin E is supplemented at a rate of 2 ppm to the horse checkers and Omolene for prevention of white muscle disease. Mineralized salt blocks are provided.

Gazella granti

Representative diets for this species are shown in Tables 17 through 21. Rates of feeding and the supplementary feeds used are the same as those described previously for Dama gazelle.

The Crandon Park Zoological Garden, Miami, Florida, uses a diet of alfalfa hay, Omolene, and trace mineral salt blocks for Grant's gazelle.[53] There is no supplementation of vitamins or hormones.

At the Hanover Zoo, Hanover, Germany, where several species of gazelle (including *G. granti*) have been kept in the past, the diet was based upon the Basel Zoo feeding specifications for herbivores (Table 22).[9] The Basel Zoo diet, according to Dr. Wackernagel[58] can be given to all herbivores, although forest animals may require more concentrate than savanna or desert animals.

Gazelles at Hanover also were fed rolled oats, fresh fruits such as apples and bananas, carrots, and good mountain hay.[9] Fresh water was provided since gazelle like to drink a little each day when in captivity.

Often it is necessary to hand rear zoo animals. This can be a difficult task, but hand rearing young Grant's gazelle at Lincoln Park Zoo, Chicago, Illinois, was successful.[36] The following formula and schedule were used by curators at the zoo:

Schedule	Formula
Day 1	Colostrum from domestic cow given to young gazelle
Day 2	50% colostrum, 25% tap water, 25% evaporated milk, by volume
Day 3	66% tap water, 34% evaporated milk, 6 drops of a multi-vitamin preparation
Day 4 until the end of bottle feeding	66% tap water, 34% evaporated milk, 18 drops of multi-vitamin preparation, add 3 teaspoons of Pablum® to 225 ml (8 oz) of this formula
From end of 1st week on	Add 0.6 ml cod-liver oil/225 ml of formula (¼ teaspoon Karo® syrup makes formula more palatable)
At 4 weeks	Alfalfa and D & F Chow (Ralston Purina Co.) offered. When readily accepted, cow's milk provided once daily with the solid food

Table 18
COMPOSITION AND ANALYSIS OF FOOD USED AT THE SAN DIEGO WILD ANIMAL PARK[a] FOR *GAZELLA DAMA DAMA*, *G. GRANTI ROOSEVELTI*, *G. LEPTOCEROS*, *G. SOEMMERINGI ERLANGERI*, AND *G. SUBGUTTUROSA SUBGUTTUROSA*

Grain Pellet (¼ in.)

Guaranteed analysis

Item	Percent
Crude protein	18.0 (min)
Crude fat	2.5 (min)
Crude fiber	9.0 (max)
Calcium	1.1 (min)
Calcium	1.3 (max)
Phosphorus	0.6 (min)
Ash	10.0 (max)
Added mineral	5.0 (max)

Ingredients

Ground milo, wheat mixed feed, soybean meal, oat groats, dehydrated alfalfa meal, ground barley, ground corn, dicalcium phosphate, linseed meal, ground limestone, salt, vitamin A palmitate, D-activated animal sterol (source of vitamin D_3), vitamin B_{12} supplement, menadione sodium bisulfite (source of vitamin K activity), pyridoxine hydrochloride, biotin, methionine, riboflavin, niacin, calcium pantothenate, choline chloride, folic acid, ethoxyquin (a preservative), manganous oxide, magnesium sulfate, calcium iodate, iron sulfate, copper oxide, zinc oxide

Hay Pellet (5/8 in.)

Guaranteed analysis

Crude protein	12.0 (min)
Crude fat	1.5 (min)
Crude fiber	25.0 (max)
Ash	9.0 (max)
Added mineral	2.0 (max)

Ingredients

Sun-cured alfalfa meal,[b] beet pulp, grains, cane molasses, salt, monosodium phosphate, vitamin A palmitate, D-activated animal sterol (source of vitamin D_3), vitamin E supplement, ethoxyquin (a preservative), calcium sulfate, manganous oxide, zinc oxide, iron carbonate, iron sulfate, copper oxide, potassium iodide, cobalt carbonate, magnesium oxide, sulfur, potassium chloride

[a] Rt. 1, P.O. Box 725E, Escondido, Cal.
[b] Chopped hay.

Table 19
COMPOSITION OF CUSTOM D & H CHOW (COARSE)[a] USED AT BUSCH GARDENS[b] IN THE DIETS OF *GAZELLA DAMA, G. DORCAS, G. GAZELLA ARABICA, G. GRANTI, G. LEPTOCEROS, G. SOEMMERINGI,* AND *G. THOMSONI*

Guaranteed analysis

Item	Percent
Crude protein	12.5 (min)
Crude fat	2.5 (min)
Crude fiber	10.0 (max)

Ingredients

Soybean meal, peanut meal, crimped oats, cracked yellow corn, alfalfa meal, cane molasses, wheat middlings, dried citrus meal, vitamin A supplement, D-activated animal sterol, defluorinated phosphate, calcium carbonate, iodized salt, potassium sulfate, magnesium sulfate, magnesium oxide, iron oxide, manganous oxide, copper oxide, cobalt carbonate, zinc oxide

[a] Ralston Purina Co., St. Louis, Mo.
[b] P.O. Box 9158 Tampa, Fla.

Table 20
LIST OF INGREDIENTS OF ZOO SUPPLEMENT[a] USED AT BUSCH GARDENS[b] IN DIETS OF *GAZELLA DAMA, G. DORCAS, G. GAZELLA ARABICA, G. GRANTI, G. LEPTOCEROS, G. SOEMMERINGI,* AND *G. THOMSONI*

Ground yellow corn
Brewer's dried grains
Cottonseed meal
Cane molasses
D-activated animal sterol
Ethoxyquin (a preservative)
Calcium carbonate
Iron oxide
Copper oxide
Zinc oxide
Soybean meal
Wheat middlings
Ground peanut hulls
Vitamin A supplement
Vitamin E supplement
Defluorinated phosphate
Iodized salt
Manganous oxide
Cobalt carbonate

[a] Purina Ralston Co., St. Louis, Mo.
[b] P.O. Box 9158 Tampa, Fla.

Table 21
COMPOSITION OF RIB-AD® (NO. 2)[a] USED AT BUSCH GARDENS[b] IN DIETS OF *GAZELLA DAMA*, *G. DORCAS*, *G. GAZELLA ARABICA*, *G. GRANTI*, *G. LEPTOCEROS*, *G. SOEMMERINGI*, AND *G. THOMSONI* AND ST. LOUIS ZOOLOGICAL GARDENS[c] FOR *G. SPEKEI*

Guaranteed analysis

Item	Amount
Vitamin A	180,000 USP units/lb
	400 USP units/g
Vitamin B_2 (riboflavin)	20 mg/lb
Vitamin B_{12}	0.0011 mg/g
Vitamin D_2	175,000 USP units/lb
	385 USP units/g
Vitamin D_3	150,000 IC units/lb
	350 IC units/g
Calcium	32.0000% (max), 28.0000% (min)
Phosphorus	5.0000% (min)
Iodine	0.0625% (min)
Cobalt	0.01225%
Copper	0.0177%
Manganese	0.2188%
Iron	0.177%
Potassium	0.0391%
Zinc	0.0441%
Salt	None

Ingredients

Calcium carbonate; monocalcium phosphate; dicalcium phosphate; defluorinated phosphate; special steamed bone meal (domestic); soybean meal; vitamin B_{12} supplement; Riboflavin; irradiated dried yeast; brewer's dried yeast; gelatin, sugar, and starch surrounding vitamin A palmitate in particle size; active dry yeast; D-activated animal sterol; wheat germ meal; trace minerals; iron (ferrous) sulfate, manganese sulfate, potassium iodide, copper sulfate, potassium sulfate, zinc sulfate, cobalt sulfate.

[a] Manufactured by Vitamineral Products Co., Peoria, Ill.; purchased by St. Louis Zoological Gardens from Ade Manufacturing and Chemical Co., Tiskilwa, Ill.
[b] P.O. Box 9158 Tampa, Fla.
[c] Forest Park, St. Louis, Mo.

Table 22
COMPOSITION OF DIET USED FOR HERBIVOROUS ANIMALS, INCLUDING GAZELLE, AT BASEL ZOOLOGICAL GARDEN, SWITZERLAND[44]

Contents		Composition	
Item	Amount (%)	Item	Amount
Ground oats	15.0	Crude protein	18.0% (approximately)
Ground barley	20.0	Crude fat	2.5%
Ground grain sorghum	8.0	Crude fiber	7.0%
Ground wheat	10.0	Calcium	1.2%
Soybean oil meal	10.0	Phosphorus	0.6%
Groundnut oil meal	8.0	Total digestible nutrients	62.9%
Linseed meal	3.0		
Lucerne meal	10.0	Trace elements added per kilogram of feed	
Wheat bran	5.0	Fe	20.0 mg
Potato flakes	2.0	Mn	10.0 mg
Pomace	3.0	Cu	5.0 mg
Licorice powder	1.0	I	4.0 mg
Ground calcium carbonate	1.0	Co	0.4 mg
Bone meal	2.0	Zn	10.0 mg
Salt	0.9		
Magnesium sulfate	0.2	Vitamins added per kilogram of feed	
Premixture	0.9	A	12,000.0 IU
		D_3	2,000.0 IU
		E	60.0 mg
		C	100.0 mg
		B_1	2.5 mg
		B_2	10.0 mg
		B_6	1.25 mg
		Pantothenic acid	12.5 mg
		Nicotinic acid	25.0 mg
		Choline	500.0 mg
		Folic acid	0.075 mg
		B_{12}	25.0 mg

Gazella soemmeringi

Some of the diets fed to this gazelle are shown in Tables 18 through 22. Additional details are described in the section for *Gazella dama,* since some of the zoos kept both species.

Gazella dorcas

In addition to the diets utilized by the San Antonio Zoological Gardens (Table 17), Busch Gardens (Tables 19 through 21) and National Zoological Park (Table 23), for this and other *Gazella* spp., another diet for *G. dorcas* was provided by Keith F. Cline, Zoological Curator, Detroit Zoological Park.[50] The ruminant ration is fed as a 3/16-in. pellet. Gazelle can also graze upon grass in their exhibit during the warm season.

Dorcas gazelle have been kept at the Hanover Zoo and were fed the standard Basel Zoo diet (Table 22) prepared at Hanover.

Table 23
DIET USED AT DETROIT ZOOLOGICAL PARK[a] FOR *GAZELLA DORCAS* AND *G. SUBGUTTUROSA*

Ration Provided in 3/16-in Pellets

Ingredient	Amount	
	Percent	Pound
Corn cob product (Andersons #4 fines)	35.0	700
Corn, shelled (or Andersons corn screenings)	16.0	320
Soybean meal (44% CP)[b]	26.65	533
Dehydrated alfalfa meal (17% CP)	5.0	100
Cane molasses	5.0	100
Wheat (soft, winter)	10.0	200
Soybean oil	1.0	20
Trace mineral salt (0.8% Zn)	0.5	10
Limestone (38% Ca)	0.4	8
Exotic ruminant VTM premix[c]	0.25	5
Calcium propionate	0.2	4
Total	100.0	2000

Calculated analysis %

Crude protein	15.6
Calcium	0.4
Phosphorus	0.3

[a] Zoological Park Department, 8450 W. Ten Mile Road, P.O. Box 39, Royal Oak, Mich.

[b] If 49% CP soybean meal is used, change shelled corn to 18.7% and soybean meal to 23.95%.

[c] Supplies per pound of mixed feed: 1500 IU vitamin A, 100 IU vitamin D, and 20 IU vitamin E; also supplies 0.1 ppm of selenium in mixed feed from sodium selenite.

Ghobrial[17,18] retained *G. dorcas* as experimental animals in pens and fed them a basic diet of dry sorghum (millet, *Panicum miliaceum* L.). These animals consumed 2.2 kg dry food per 100 kg body weight per day, or 2.6% of body weight.[18] An analysis of sorghum millet (expressed as percentage of the wet weight) used is as follows:

Sample	Moisture	Protein	Carbohydrate	Fat	Fiber	Ash	Calories/100 g
1	7.8	18.9	67.4	2.7	2.0	1.2	379
2	9.6	15.4	68.7	2.6	2.0	1.7	380
3	7.3	13.3	73.6	2.3	2.0	1.5	378
4	8.0	14.5	69.9	2.4	2.0	1.3	379
Mean	8.2	15.5	69.9	2.5	2.0	1.4	379

Note: Each sample figure is the mean of ten samples.

At various times the experimental animals ate *Medicago sativa* L., grass roots (normally rejecting other grass parts), and green or dried dates.[17,18]

Gazella gazella

Diets presented in Tables 19 through 22 are applicable also to this species, as is the information for Busch Gardens included in the section on *G. dama.* The additional rations as well as the regular diet fed to *G. granti* at Hanover Zoo were fed also to *G. g. cuvieri* at that zoo.[9]

G. g. arabica, kept at the New York Zoological Park, ate a usual diet of alfalfa hay, feeding pellets (similar to others described), greens, and vegetables.[8] Fine bone meal and cod-liver oil were added to the diet to provide calcium and vitamin D.

A tame mountain gazelle kept by David L. Harrison[23] ate a variety of foods including sausages, peas, potatoes, pineapple, peaches, native bread, cigarettes, and paper. The gazelle also drank tea and coffee and would lick sweat from human bodies, presumably for its salt content.

Gazella leptoceros

The diets illustrated in Tables 18 through 21 as used at the San Diego Wild Animal Park and Busch Gardens were provided to the slender-horned or Loder's gazelle on the same basis as that described for *G. dama.* Loder's gazelle kept at the Hanover Zoo were fed as described for Grant's gazelle.[9]

Gazella pelzelni

This gazelle is not kept in captivity in very many places, but where it is retained it can thrive on most of the gazelle or herbivore diets described previously. The diet mixture for *G. granti* and other gazelle[9] evidently was fed to Pelzeln's gazelle while it was kept in the Hanover Zoo.[57]

Gazella rufifrons

The basic Basel Zoo diet (Table 22) and supplements previously described were used for the red-fronted gazelle while this species was maintained at the Hanover Zoo.[9]

At the National Zoological Gardens of South Africa, Pretoria, this species is fed green alfalfa and alfalfa and grass hay ad libitum plus a pelleted grain mash ration at the rate of 1 oz pellets per 10 lb body weight.[54] The chemical composition of the grain mash as provided by M. Griessel of Epol (Proprietary) Ltd.[59] is as follows:

Protein	15.85%
Methionine	0.25%
Lysine	0.57%
Fiber	5.30%
Calcium	0.99%
Phosphorus	0.69%
Net energy	1240 cal/kg

Note: The ration is reinforced with a vitamin and a mineral premix.

Gazella spekei

Speke's gazelle kept at the St. Louis Zoological Park, St. Louis, Missouri, are fed the following rations per animal:[52]

D & H Chow	½ lb/adult (Table 24)
Alfalfa hay	¼ flake (1 flake – 3 lb)
Rib-Ad®, Dyna-Fos®	1 teaspoon (Tables 21 and 25)
Hydroponics	½ lb
Carrots	½ lb

Table 24
COMPOSITION AND ANALYSIS OF PURINA D & H CHOW (COARSE) (G)[a] USED IN THE DIET OF *GAZELLA SPEKEI* AT ST. LOUIS ZOOLOGICAL PARK[b]

Guaranteed Analysis

Crude protein	12.5% (min)
Crude fat	2.5% (min)
Crude fiber	10.0% (max)
Ash	4.5%
Calcium	1.5%
Phosphorus	0.6%
Moisture	9.0%
Calories (per pound)	1755
Vitamin A	16 IU/g

Ingredients

Grain products, processed grain by-products, plant protein products, forage products, cane molasses, vitamin A supplement, D-activated animal sterol, defluorinated phosphate, calcium carbonate, iodized salt, magnesium oxide, iron oxide, manganous oxide, copper oxide, cobalt carbonate, zinc oxide

[a] Ralston Purina, St. Louis, Mo.
[b] Forest Park, St. Louis, Mo.

Table 25
ANALYSIS OF DYNA-FOS®[a] USED IN DIET OF *GAZELLA SPEKEI* AT ST. LOUIS ZOOLOGICAL PARK[b]

Guaranteed Analysis

Phosphorus	18.50% (min)
Calcium	24.00% (max)
Calcium	20.00% (min)
Fluorine	00.185% (max)

Ingredients: Dicalcium phosphate, monocalcium phosphate

[a] International Mineral Corp., IMC Plaza, Libertyville, Ill.
[b] Forest Park, St. Louis, Mo.

When kept at Hanover Zoo, this species was fed the basic Basel Zoo diet (Table 22) for herbivores plus rolled oats, fresh fruits such as apples and bananas, carrots, and good mountain hay.[9]

At Giardino Zoologico di Napoli,* Speke's gazelle were fed normal grass hay, leaves of all kinds, carrots, oats, mineral salts, vitamins, and sometimes hard bread or biscuits, but no pellets.[60]

Gazella thomsoni

This is a popular species in many zoos. Some of the diets used for the animal are given in Tables 17, 19 through 22, and 26 with similar rations as described for *G. dama* at the San Antonio Zoological Gardens and Busch Gardens.

The diet used for *G. rufifrons* at the National Zoological Gardens of South Africa is also fed to Thomson's gazelle,[54] as is the diet for Grant's gazelle at Crandon Park Zoological Garden.[53]

The following ration is fed to Thomson's gazelle and other antelope at the Milwaukee County Zoological Park, according to the Assistant Animal Superintendent:

Daily diet/100 lb of body weight
¾ lb D & F or Purina Chow
or
1 lb mixed grain
½ lb diced vegetables
3 slices bread

Plus browse if available
Good grade of second or third crop alfalfa at all times

The diet used at Lincoln Park Zoo, Chicago, for hand rearing young Grant's gazelle is also used for young Thomson's gazelle.[36] Another diet and feeding schedule used for young steinbuck, *Raphicerus campestris,* and Thomson's gazelle was reported by Mrs. G. Chalmers.[62] This diet and schedule is as follows:

1st 3 weeks	Full pasteurized milk, glucose, and Radiostroleum (plus cream if available) – 1 oz every 4 hr, increasing gradually to 4 oz every 4 hr
4th week	Feed 5–6 oz three times daily. Feed grass, dried lucerne hay, sweet potato tops, hibiscus and avocado leaves, bananas, and ground nuts.
Diet continued 4–5 months	The animals seldom drink water.

Gazella tilonura

Heuglin's gazelle was fed the basic Basel Zoo diet (Table 22) while in captivity at Hanover Zoological Gardens. Additional rations fed to the gazelle formerly held at Hanover Zoo[9] were listed in the section on *G. granti.* The diet used at the Naples Zoo for Speke's gazelle was also fed to *G. tilonura,* as well as to other species.[60]

Gazella subgutturosa

Tables 17, 18, and 23 list the diets used by the San Antonio Zoological Gardens, San Diego Wild Animal Park, and Detroit Zoological Park for the goitered gazelle. Feeding rates and supplements are the same as those described for *G. dama* at San Antonio and San Diego and for *G. dorcas* at Detroit.

G. subgutturosa kept by the New Mexico Department of Game and Fish for possible

* 80125 Naples-via-Kennedy, Naples, Italy.

stocking purposes are allowed to graze freely in a pasture near Red Rock. During dry seasons, natural food is supplemented daily with a diet of horse ration pellets (Table 27) and alfalfa hay.[61] When pastures are green, the gazelles are fed only every third day.

General Diets

Besides the Basel Zoo diet (Table 22), which is used generally for many species of herbivores including gazelle, Tables 28 and 29 give two other diets formulated for ruminants. The diet in Table 28 was developed by Dr. Herbert L. Ratcliffe[35] at the Philadelphia Zoological Gardens. A slight variation of this diet is recorded elsewhere[1] as follows for Artiodactyla (except swine) and Perissodactyla (10–40 g of feed) – feed per day per kilogram body weight:

Basic diet – Diet H (deteriorates after 2 weeks under refrigeration): 15% ground yellow corn, 10% ground whole wheat, 10% ground rolled oats, 10% soybean meal, 5.2% brewer's yeast, 26% alfalfa leaf meal, 2.5% oystershell flour, 1% iodized salt, 0.3% A-D feeding oil, 10% linseed meal, 10% brewer's grains. Mix dry beet pulp (45% of total dry weight of feed needed) with equal amount of water; let stand 1–2 hr, then add Diet H ingredients (45% of total dry weight) and 10% chopped cabbage and carrots.

Feed supplement: Hay (free choice), green leaves, or fresh-cut grass (5–10 g/kg body weight); equal quantities of apples, carrots, green lettuce, or cabbage (10–50 g/kg body weight)

The other diet (Table 29) was formulated for a variety of species including gazelle and prepared in a 3/16-in. × 1/2-in. pellet. The recommended feeding rate is 1 lb pellets for 100 lb of body weight every 24 hr, with hay fed free choice.

D. M. Ben-Shaul,[63] Hebrew University, Jerusalem, in feeding baby gazelle used powdered buttermilk the first few days, then later substituted whole milk powder. Fat was added in the form of fresh cream, powdered Preams® or egg yolk with lactose used instead of glucose. The formula was mixed in a consistency and amount that was readily acceptable.

Table 26
DIET USED AT NEW YORK ZOOLOGICAL PARK[1] FOR *GAZELLA THOMSONI* MALE, 26 KG BODY WEIGHT, AT FEEDING RATE OF 9 KG FEED PER WEEK OR 46 G FEED PER DAY PER KILOGRAM BODY WEIGHT

Basic Diet: 55% Alfalfa, 45% GLF[a]

Ingredient	Percent
Hominy feed	25.4
Wheat middlings	15.0
Corn gluten feed	12.0
Ground oats	10.0
Wheat bran	10.0
Dehydrated alfalfa meal	5.0
Corn distiller's dried grains	5.0
Linseed oil	5.0
Wheat germ meal	3.0
Dehulled soybean meal	2.0
Salt	1.0
Dicalcium phosphate	0.75
Ground limestone	0.75
Z-4 mineral mix[b]	0.05
Molasses	5.0
Vitamin premix[c]	0.05

[a] Theoretical analysis: 73% TDN, 16% protein, 4% fat, 8% fiber.
[b] Calcium iodate, cobalt carbonate, copper oxide, iron carbonate, manganolis oxide, zinc oxide.
[c] 2,000,000 units vitamin A; 4,000,000 units vitamin D; 2000 units vitamin E.

Table 27
COMPOSITION OF A COMPLETE HORSE FEED[a] USED FOR *GAZELLA SUBGUTTUROSA* ON RED ROCK PASTURE, NEW MEXICO[b]

Guaranteed Analysis

Item	Percent
Crude protein	12.0 (min)
Crude fat	2.0 (min)
Crude fiber	25.0 (max)
Vitamin A	1500 USP units/lb
Vitamin D	500 USP units/lb

Ingredients

Grain products; plant protein products; processed grain by-products; forage products; roughage products; dicalcium phosphate; salt; cane molasses; vitamin A supplement; vitamin D_3 supplement; vitamin E supplement; vitamin B_{12} supplement; choline chloride; choline pantothenate; folic acid; menadione sodium bisulfite (source of vitamin K activity); methionine hydroxy analogue; niacin; riboflavin supplement; thiamine mononitrate; calcium iodate; cobalt carbonate; copper oxide; ferrous sulfate; manganous oxide; zinc oxide; BHA, BHT, and ethoxyquin (preservatives); ascorbic acid

[a] Ranch-Way Feeds, Ranch-Way, Inc., Fort Collins, Colo.
[b] New Mexico Department of Game and Fish, State Capitol, Santa Fe.

Table 28
COMPOSITION OF DIET B FOR HERBIVOROUS MAMMALS, INCLUDING GAZELLES, IN ZOOLOGICAL GARDENS[35]

Mixture No. 2

Item	Percent
Brewer's grain	10.0
Ground yellow corn	15.0
Ground whole wheat	10.0
Ground whole oats	10.0
Soybean meal	10.0
Linseed meal	10.0
Alfalfa leaf meal	26.0
Brewer's yeast	5.2
Oyster shell flour	2.5
Iodized salt	1.0
A-D feeding oil	0.3

Preparation: Use dry beet pulp (plain or molasses) equal to about 45% of the total dry weight of food needed. Wet with an approximately equal weight of water; let stand for 1 to 2 hr, then add 45% of total dry weight of Mixture No. 2 plus an amount of chopped carrots and uncut kale to equal about 10% of combined weights of dry beet pulp and Mixture No. 2. Mix and feed within 4 to 6 hr. (Mixture No. 2 should not be stored more than 2 weeks). Feed at a rate of 1/4 to 2 oz/lb (10 to 40 g/kg) of body weight. Supplement with hay ad libitum; where good quality hay is not available, give alfalfa pellets at rate of 5 to 10 g/kg body weight.

Table 29
COMPOSITION OF HALL-WOOD "GRAZING RUMINANT DIET"[a] PREPARED FOR GAZELLES AND OTHER RUMINANTS

Analysis

Item	Amount
Protein	19.0%
Fat	3.3%
Fiber	8.7%
Carbohydrate	50.0%
Ash	9.0%
Mineral	
Chlorine	0.171%
Sodium	0.192%
Potassium	1.123%
Calcium	0.475%
Phosphorus	0.441%
Sulfur	0.206%
Magnesium	0.182%
Manganese	6.411 μg
Iron	0.012%
Copper	4.702 ppm
Zinc	Trace
Cobalt	Trace
Iodine	0.04 μg
Vitamin content (per pound)	
Vitamin A	12,217.0 IU
Thiamine B_1	2.6 mg
Riboflavin B_2	1.8 mg
Pantothenic acid	7.2 mg
D_3	249.0 IU
E	25.0 μg
Niacin	19.8 mg
C	137.5 mg
Choline	508.0 mg
Folic acid	0.41 mg
B_{12}	0.1375 μg
Amino acid content (as percent of protein)	
Methionine	0.3
Cystine	0.3
Tryptophane	0.3
Arginine	1.4
Histidine	0.6
Isoleucine	1.4
Leucine	2.1
Phenylalanine	1.2
Threonine	0.9
Valine	1.2
Lysine	1.3
Gross energy (Kcal per kg)	2,829
TDN	79

Ingredients

Dehydrated alfalfa meal, corn gluten feed, ground yellow corn, peanut meal, wheat middlings, ground oats, kelp, cane molasses, corn sugar molasses, soybean meal, and the following: vitamin A palmitate, vitamins B_1, B_2, B_3, B_6, and B_{12} supplement, C (stable); D-activated animal sterol (source of vitamin D_3); E; calcium carbonate; cobalt carbonate; copper oxide; iron carbonate; iron oxide; iron sulfate; manganese sulfate; potassium iodide; salt; zinc oxide; zinc sulfate; dicalcium phosphate; propylene glycol.

[a] Hall-Wood Wild Animal Nutrient, 1148 W. Western Avenue, Muskegon, Mich.

REFERENCES

1. **Altman, P. L. and Dittmer, D. S.,** *Metabolism,* Federation of American Societies for Experimental Biology, Bethesda, Md., 1968.
2. **Baharav, D.,** Energy Flow and Productivity in a Mountain Gazelle (*Gazella gazella* Pallas, 1766) Population, Ph.D. dissertation, Tel Aviv University, Tel Aviv, 1975.
3. **Berwick, S.,** The Gir Forest: an endangered ecosystem, *Am. Sci.,* 64, 28–40, 1976.
4. **Blankenship, L. H. and Field, C. R.,** Factors affecting the distribution of wild ungulates on a ranch in Kenya, preliminary report, *Zool. Afr.,* 7(1), 281–302, 1972.
5. **Blankenship, L. H. and Qvortrup, S. A.,** Resource management on a Kenya ranch, *J. S. Afr. Wildl. Manage. Assoc.,* 4(3), 185–190, 1974.
6. **Brooks, A. C.,** A study of the Thomson's gazelle (*Gazella thomsoni* Gunther) in Tanganyika, *Colonial Res. Publ.,* No. 25, 1961.
7. **Carlisle, D. B. and Ghobrial, L. I.,** Food and water requirements of Dorcas gazelle in the Sudan, *Mammalia,* 32(4), 570–576, 1968.
8. **Crandall, L. S.,** *The Management of Wild Mammals in Captivity,* University of Chicago Press, Chicago, 1964.
9. **Dittrich, L.,** Keeping and breeding gazelles at Hanover Zoo, *Int. Zoo Yearb.,* 8, 139–143, 1968.
10. **Dittrich, L.,** Food presentation in relation to behavior in ungulates, *Int. Zoo Yearb.,* 16, 48–54, 1976.
11. **Dorst, J. and Dandelot, P.,** *A Field Guide to the Larger Mammals of Africa,* 1st ed., Collins, London, 1970.
12. **Ellerman, J. R. and Morrison-Scott, T. C. S.,** *Checklist of Palaearctic and Indian Mammals 1758–1946,* Trustees of British Museum (Natural History), London, 1966.
13. **Estes, R. D.,** The comparative behavior of Grant's and Thomson's gazelles, *J. Mammal.,* 48(2), 189–209, 1967.
14. **Field, C. R.,** The nutritional ecology of large herbivores, in Caesar Kleberg Research Program in Wildlife Ecology, annual progress report, Texas A & M University, College Station, 1972, 254–263.
15. **Field, C. R.,** The nutritional ecology of large herbivores, in Caesar Kleberg Research Program in Wildlife Ecology, annual progress report, Texas A & M University, College Station, 1973, 125–131.
16. **Field, C. R. and Blankenship, L. H.,** Nutrition and reproduction of Grant's and Thomson's gazelles, Coke's hartebeest and giraffe in Kenya, *J. Reprod. Fertil. Suppl.,* 19, 287–301, 1973.
17. **Ghobrial, L. I.,** The water relations of the desert antelope *Gazella dorcas dorcas, Physiol. Zool.,* 43(4), 249–256, 1970.
18. **Ghobrial, L. I.,** Water relation and requirement of the Dorcas gazelle in the Sudan, *Mammalia,* 38(1), 88–107, 1974.
19. **Grau, G. A.,** Behavior of Mountain Gazelle in Israel, Ph.D. dissertation, Texas A & M University, College Station, 1974.
20. **Groves, C. P. and Harrison, D. L.,** The taxonomy of the gazelles (genus *Gazella*) of Arabia, *J. Zool.,* 152, 381–387, 1967.
21. **Gwynne, M. D. and Bell, R. H. V.,** Selection of vegetation components by grazing ungulates in the Serengeti National Park, *Nature,* 220(5165), 390–393, 1968.
22. **Haltenorth, Th.,** Klassifikation der Saugetiere: Artiodactyla, in *Handbook of Zoology,* Vol. 8, No. 32, Walter De Gruyter, Berlin, 1963, 1–167.
23. **Harrison, D. L.,** *The Mammals of Arabia,* Vol. 2, Ernest Benn, London, 1968, 349–365.
24. **Hofmann, R. R.,** Comparisons of the rumen and omasum structure in East African game ruminants in relation to their feeding habits; *Symp. Zool. Soc. London,* 21, 179–194, 1968.
25. **Hofmann, R. R. and Stewart, D. R. M.,** Grazer or browser: a classification based on the stomach-structure and feeding habits of East African ruminants, *Extrait de Mammalia,* 36(2), 226–240, 1972.
26. **Lamprey, H. F.,** Ecological separation of the large mammal species in the Tarangire Game Reserve, Tanganyika, *East Afr. Wildl. J.,* 1, 63–92, 1963.
27. **Meinertzhagen, R.,** *Birds of Arabia,* Oliver and Boyd, London, 1954.
28. **Mendelssohn, H.,** On the Biology and Ecology of Gazelles in Israel, final report to Smithsonian Institution, Washington, D.C., 1972 (mimeographed).
29. **Mendelssohn, H.,** The development of the populations of gazelles in Israel and their behavioral adaptations, in *Proc. Int. Conf. Behav. Ungulates,* Calgary, 1971, International Union for Conservation of Nature and Natural Resources, Morges, Switzerland, 1974, 722–740.
30. **Pease, A. E.,** On the antelopes of the Aures and eastern Algerian Sahara, *Proc. Zool. Soc. London,* Part 4, 809–814, 1897.

31. **Prater, S. H.,** *The Book of Indian Animals,* Bombay Natural History Society, Bombay, 1965.
32. **Qvortrup, S. A.,** Comparative Nutritional Study on Thomson's Gazelle, Impala and Cattle on Kekopey Ranch in the Rift Valley, Kenya, Kenya Wildlife Management Project, United Nations Development Program/Food and Agriculture Organization, 1973 (mimeographed report).
33. **Qvortrup, S. A.,** Dry Season Feeding of Wildlife and Livestock on the Athi Kapiti Plains of Kajiado District, Collection 1 report, Kenya Wildlife Management Project, United Nations Development Program/Food and Agriculture Organization, 1974 (mimeographed).
34. **Qvortrup, S. A.,** Wildlife and Livestock Feeding on the Athi Kapiti Plains of Kajiado District, Collection 3 report, Kenya Wildlife Management Project, United Nations Development Program/Food and Agriculture Organization, 1975 (mimeographed).
35. **Ratcliffe, H. L.,** Diets for zoological gardens: aids to conservation and disease control, *Int. Zoo Yearb.,* 6, 4–23, 1966.
36. **Rosenthal, M. A. and Meritt, D. A., Jr.,** Hand-rearing Grant's gazelle at Lincoln Park Zoo, Chicago, *Int. Zoo Yearb.,* 11, 130, 1971.
37. **Stewart, D. R. M. and Stewart, J.,** Food preference data by fecal analysis for African plains ungulates, *Zool. Afr.,* 5(1), 115–129, 1970.
38. **Stewart, D. R. M. and Stewart, J.,** Comparative food preferences of five East African ungulates at different seasons, in *The Scientific Management of Animal and Plant Communities for Conservation,* Duffey, E. and Watt, A. S., Eds., Blackwell Scientific Publications, Oxford, Engl., 1971, 351–366.
39. **Talbot, L. M.,** Food preferences of some East African wild ungulates, *East Afr. Agric. For. J.,* 27(1), 42–43, 1962.
40. **Taylor, C. R.,** Hygroscopic food: a source of water for desert antelopes? *Nature,* 219, 181–182, 1968.
41. **Taylor, C. R.,** The minimum water requirements of some East African bovids, *Symp. Zool. Soc. London,* 21, 195–206, 1968.
42. **Taylor, C. R.,** The desert gazelle: a paradox resolved, *Symp. Zool. Soc. London,* 31, 215–227, 1972.
43. **Vesey-Fitzgerald, D. F.,** Wildlife in Arabia, *Oryx J. Fauna Preserv. Soc.,* 1(5), 232–235, 1952.
44. **Wackernagel, H.,** Feeding wild animals in zoological gardens, *Int. Zoo Yearb.,* 6, 23–37, 1966.
45. **Wackernagel, H.,** Substitution and prefabricated diets for zoo animals, *Symp. Zool. Soc. London,* 21, 1–12, 1968.
46. **Walker, E. P.,** *Mammals of the World,* Vol. 2, 2nd ed., Paradiso, J. L., Ed., Johns Hopkins Press, Baltimore, 1968, 1461.
47. **Wallach, J. D.,** The nutrition and feeding of captive ruminants in zoos, in *Digestive Physiology and Nutrition of Ruminants,* Vol. 3, Church, D. C., Ed., D. C. Church Publishing, Oregon State University, Corvallis, 1972, 292–307.
48. Letter dated May 30, 1976 from Brian O'Reagan, Division of Conservation, Department of Environment, Tehran, Iran.
49. Personal communication, June 10, 1976, Louis DiSabato, Director, and Ernest Roney, Assistant Director, San Antonio Zoological Gardens, San Antonio, Tex.
50. Letter dated June 16, 1976 from Keith F. Cline, Zoological Curator, for Dr. James C. Savoy, Director, Zoological Park Department, Detroit, Mich.
51. Letter dated June 18, 1976 from Dr. Lester S. Nelson, Director, Veterinary Services, San Diego Wild Animal Park, San Diego, Calif.
52. Letter dated June 21, 1976 from Charles H. Hoessle, Deputy Director, St. Louis Zoological Park, St. Louis, Mo.
53. Letter dated June 23, 1976 from Vernon D. Kisling, Jr., Research Curator, Crandon Park Zoological Garden, Miami, Fla.
54. Letter dated July 15, 1976 from Dr. D. J. Brand, Director, National Zoological Gardens of South Africa, Pretoria.
55. Letter dated July 16, 1976 from William A. Xanten, Jr., Curator, Mammals Unit, National Zoological Park, Washington, D.C.
56. Letter dated July 21, 1976 from Thomas E. Hayes, Curator, Large Mammals, Busch Gardens, Tampa, Fla.
57. Letter dated August 2, 1976 from Dr. L. Dittrich, Zoological Garden, Hanover, Germany.
58. Letter dated August 3, 1976 from Dr. H. Wackernagel, Vice Director, Basel Zoological Garden, Basel, Switzerland.
59. Letter dated August 11, 1976 from M. Griessel, 37 Quinn St., Newtown, P. O. Box 3006, Johannesburg, 200, Republic of South Africa.
60. Letter dated August 23, 1976 from Claire Wenner, Giardino Zoologico Di Napoli, Naples, Italy.
61. Letter dated September 29, 1976 from George Merrill and Alton Ford, New Mexico Department of Game and Fish, Sante Fe.

62. Note by Mrs. G. Chalmers, Nairobi, Kenya, as reported in "Notes on hand-rearing various species of mammals," *Int. Zoo Yearb.*, 4, 330–331, 1962.
63. Note by D. M. Ben-Shaul as reported in "Notes on hand-rearing various species of mammals," *Int. Zoo Yearb.*, 4, 300, 1962.
64. **Baharav, D. and Mendelssohn, H.**, Ecology and Biodynamics of Gazelles in Israel, Proj. Rep. No. 2, Tel Aviv University, Tel Aviv, 1976.

DIETS (NATURAL AND SYNTHETIC): ANTELOPES

J. D. Skinner

Strictly defining the term antelope would probably mean confining this discussion to only eight species in southern Africa. It has therefore been decided to include all 32 species in the family Bovidae except the African Buffalo.[1] Apart from their taxonomic classification, they can be categorized according to size, social behavior, habitat, or nutrition. Detailed knowledge of the nutrition of most species has been confined to studies within particular habitats, and plant species eaten frequently vary from area to area.

NATURAL DIETS

Food selection by antelopes can be classified as falling into two main categories. Some antelope species, notably those belonging to the tribe Tragelaphini, are almost exclusively browsers, feeding on the leaves of trees and shrubs. Others, notably those belonging to the subfamily Alcelaphinae, are exclusively grazers, feeding on monocotyledenous plants. There are also a number of species that graze as well as browse. The feeding habits of antelopes are summarized in Table 1.

Apart from the anatomical differences between grazers and browsers, which have been studied by Hofman and Stewart,[2] there are other major differences, related to the nutritional value of their feed. These are primarily the result of fluctuating nutritional levels, which are most evident in monocotyledenous plants. On the other hand, certain of the most favored fodder trees of browsers, for example, *Acacia* spp. and *Combretum* spp., are deciduous, so that, although nutritional levels remain relatively constant for most of the year, for 2 or 3 months their nutritive value is nil. To counteract this, browsers have to resort to evergreen trees. Perhaps because of their lower nutritive value, or perhaps because the animals have to spend much more time gathering food, the body condition of browsers also declines in late winter.

Grasses show a marked seasonal fluctuation in nutritional value,[3] particularly on the Highveld plains. The nutritive value reaches a maximum in early summer, following the rains, and declines to lowest levels in winter. To adapt to this situation, the grazing species breed in late summer or autumn and calve in early summer or in midsummer, so that the calves are usually born after the spring rains and the first flush of green grass.

Two factors are important with regard to grazers' natural diet. The males decline in body condition during the mating season, as they have little time to feed, and from then on until spring this decline continues because of inadequacies in the natural diet. On the other hand, pregnant females improve in body condition, at least during the first part of the winter, before the demands of late fetal growth and lactation have to be met.[4]

The diet in the wild is, therefore, dependent on the physiological states both of the plants and of the animals. Among mixed feeders, such as springbok,[5] young and succulent monocotyledons, forbs, and dicotyledons are selected. Indeed, succulence would appear to be a prime factor in food selection. More recently[6] it has been found that wildebeest select food plants to obtain an ample supply of calcium for lactation in order to ensure optimal growth rate of calves without the dams losing condition. However, other factors influencing selection are largely unknown. Trace elements and minerals are important,[7] as evidenced by geophagia in deficient areas, but whether antelope can select for these has yet to be proved. Wilson,[7] while investigating the nutritional status of sable and roan antelope, found that protein, phosphorus, and selenium are deficient in nature reserves of the Transvaal. Moreover, sodium, calcium,

Table 1
FEEDING HABITS OF THE ANTELOPE OF SOUTHERN AFRICA

(Class Mammalia, Order Artiodactyla, Family Bovidae)

Subfamily	**Bovinae**	**Cephalophinae**	**Alcelaphinae**	**Antelopinae**
Tribe	Tragelaphini			Antelopini
Species	Eland Kudu Nyala Bushbuck Sitatunga	Grey duiker Red duiker Blue duiker	Blue wildebeest Black wildebeest Red hartebeest Lichtenstein's hartebeest Tsessebe Blesbok/bontebok	Springbok
Feeding method	Browsers by preference	Mainly browsers	Grazers	Mixed feeders tending to browse

Subfamily	**Antelopinae**	**Aepycerotinae**	**Peleinae**	**Hippotraginae**	**Reduncinae**
Tribe	Neotraginae				
Species	Oribi Steenbok Klipspringer Cape grysbok Sharpe's grysbok Dik-dik Suni	Impala	Vaal rhebuck	Gemsbok Sable antelope Roan antelope	Waterbuck Puku Red lechwe Southern reedbuck Mountain reedbuck
Feeding method	Mixed feeders tending to browse	Mixed feeder	Grazer	Grazers	Mainly grazers

magnesium, potassium, iron, and copper are linked to the deficiency syndrome shown by these antelope.

One last aspect of nutrition in the wild that must be mentioned here is the symbiotic relationship between the microfauna and microflora in the rumen and the plants eaten by the different species of antelope. Research is still in its infancy, but it is becoming increasingly apparent that each species of antelope has specific species of ruminal microfauna.[8,9] These are possibly related to diet, but the precise reasons why they differ between antelope species has still to be determined. However, the practice of inoculating introduced antelope with the rumen contents of another species already in the area in order to assist the newcomers in adapting to their new environs may be a waste of time.

SYNTHETIC DIETS

Although antelope in the wild are classified as grazers or browsers, it has been found that in captivity a ration of good quality hay and a well-balanced meal ration will be adequate;[10] however, browsers should also be provided with edible leafy branches.

Lucerne (alfalfa) hay should always be available, and should be of good quality when fed to browsers and small antelopes. In addition, green food, such as oats and barley, is always welcomed.

Young[11] found a concentrate ration containing yellow maize meal (70%), wheaten bran (17%), sunflower cake meal (5%), Rumevite* (5%), and a mineral lick (3%) most satisfactory. The latter consisted of bone meal (30 kg), salt (15 kg), manganese sulfate (120 g), copper sulfate (60 g), cobalt sulfate (15 g), and sodium iodide (5 g).

The only research on energy requirements of captive antelope is that of Meissner.[12] He used a ration containing yellow maize meal (20%), lucerne meal (20%), veldgrass (33%), fish meal (15%), bone meal (1%), salt (0.9%), and minerals and vitamins (0.1%) to determine the energy requirements of young growing blesbok. This diet contained 18.1% protein, 9,600 $kJ \cdot kg^{-1}$ metabolizable energy, and 22.5% crude fiber on a dry matter basis. Meissner[12] found that, although DM intake increased with age, the metabolizable energy intake was relatively constant, suggesting a substantial decrease in energy intake per given unit body mass as the animals aged. It would appear from this research that, unlike farm animals, blesbok take in sufficient energy for maintenance, but very little in excess of this requirement. Similar results have been found in a preliminary trial on the blue duiker.[13]

In earlier work on rearing blesbok lambs, Meissner and Hofmeyr[14] found that, when the lambs were weaned at 7 weeks of age, the solid feed intake was exceptionally low and the animals developed digestive problems on a concentrate diet. Two died, and only two survived when milk was reintroduced into the diet. Post-mortem inspection revealed serious food impaction and a large amount of totally undigested matter, with resultant inflammation of the abomasum. Food must have passed directly into the abomasum without prior fermentation in the rumen. In later experiments,[12] blesbok lambs were weaned only after 10 weeks.

Young[10] has pointed out that the hand-rearing of newborn antelope has always been difficult. Colostrum should be provided during the first 24 hr of life. As an alternative, serum from the same species may be administered orally. All utensils must be kept scrupulously clean to avoid digestive upsets. Only fresh milk should be fed, but where this is not available, the following mixture can be made up: skimmed milk meal (8.5%), brewer's yeast (2%), dicalcium phosphate (1%), salt (0.5%), vitamin A (8.8 g/100 kg milk powder), and vitamin D_3 (1.1 g/100 kg milk powder). This milk mixture is prepared by adding 9 l water to 1 kg of the mixture. To prevent diarrhea, one part (by volume) lime

* Trademark, National Chemical Products.

water (clear supernatant of a saturated solution of calcium hydroxide) should be added to every seven parts of milk or milk mixture.

In conclusion, it is apparent that a great deal more research is required on quantifying natural diets and on establishing optimal diets for antelope in captivity. At this stage it would appear that hand-reared young antelope should be fed milk for about 3 months, with due regard to overfeeding and resultant diarrhea. Where possible, some green food should be provided in the form of oats or barley, and to browsers fresh leafy branches should be made available. When the animals are suffering digestive problems, such as colic or diarrhea, the concentrate should be reduced and relatively more hay should be fed.[10] Green food will also stimulate the appetite.

REFERENCES

1. **Dorst, J. and Dandelot, P.,** *A Field Guide to the Larger Mammals of Africa,* Collins, London, 1970.
2. **Hofman, R. R. and Stewart, D. R. M.,** *Mammalia,* 36, 226–240, 1970.
3. **Weinmann, H.,** in *The Grasses and Pastures of South Africa,* Meredith, D., Ed., Central News Agency, Johannesburg, 1955, pp. 571–600.
4. **Skinner, J. D., Van Zyl, J. H. M., and Oates, G. L.,** *J. S. Afr. Wildl. Manage. Assoc.,* 4, 15–23, 1974.
5. **Skinner, J. D., von La Chevallerie, M., and Van Zyl, J. H. M.,** *Anim. Breeding Abstr.,* 39, 215–224, 1971.
6. **Kreulen, D.,** *East Afr. Wildl. J.,* 13, 297–304, 1975.
7. **Wilson, D. E.,** Factors Affecting Road and Sable Antelope Populations on Nature Reserves in the Transvaal, D.Sc. thesis, University of Pretoria, 1975.
8. **Van Hoven, W.,** *J. Protozool.,* 22, 457–462, 1975.
9. **Wilkinson, R. C. and Van Hoven, W.,** *Zool. Afr.,* 1–22, 1976
10. **Young, E.,** in *The Capture and Care of Wild Animals,* Young, E., Ed., Human & Rousseau, Cape Town, 1973, pp. 134–147.
11. **Young, E.,** *Afr. Wildl.,* 22, 217–230, 1968.
12. **Meissner, H. H.,** *S. Afr. J. Wildl. Res.,* 6, 51–54, 1976.
13. **Cowan, R. L., von Ketelhodt, H. F., and Liebenburg, L. H. P.,** *J. Anim. Sci.,* in press.
14. **Meissner, H. H. and Hofmeyr, H. S.,** *S. Afr. J. Anim. Sci.,* 2, 73–74, 1972.

DIETS FOR WILD RUMINANTS

J. H. Topps

INTRODUCTION

It is now realized that a knowledge of diets that will sustain wild ruminants in good condition and health is needed to promote both the greater use of these animals for human food and the conservation of the natural vegetation in many areas of the world. Many attempts have been made in the last three decades to either assess the food preferences of wild ruminants kept naturally or to measure the dietary needs of certain species that have been partly or completely tamed. Both forms of investigation have yielded valuable knowledge of the feeds that are both acceptable and make a useful contribution to a satisfactory diet. For this reason, published results from each source will be considered, with those concerning food preferences receiving first attention.

FOOD PREFERENCES OF WILD RUMINANTS UNDER NATURAL CONDITIONS

The food habits and eating behavior of a large number of species have been studied. The animals investigated range in size from the largest of the wild ruminants, the giraffe (*Giraffa camelopardalis*) to some of the smallest such as the duiker (*Sylvicapra grimmia*) and steenbok (*Raphiceros campestris*). Several techniques are available for such studies but, unfortunately, all of them are liable to considerable error or bias. Visual observation is probably the most commonly employed technique; either the foods taken by the animals are noted, or the locations where animals eat are recorded and inspected later to see which plants are eaten. In estimating the accuracy of this method for mule deer (*Odocoileus hemionus*), Wallmo et al.[1] found that, at best, the major forage species could be identified, but any quantification of the relative amounts of plant species consumed is difficult. The accuracy of any assessment by visual observation is obviously dependent on the distance between observer and animal, but there are other limitations such as the considerable time needed to obtain comprehensive data for the season or year; also, the preferences of nocturnal feeders cannot be estimated in this way. To gain close proximity for accurate observation, certain investigators have used tamed animals to assess food preferences. Field[2] carried out studies with buffalo (*Synerus caffer*), reedbuck (*Redunca redunca*), and kob (*Adenota kob thomasi*) among other animals in Uganda. As well as being more precise, this method enables samples and parts of plants to be taken for chemical analysis. Unfortunately, there are serious doubts about the similarity of food preferences of tame and wild ruminants within the same species. If the tame animals have been reared from a young age on an unnatural diet, any similarity in preference may be purely incidental. Furthermore, the habit of a single or a few animals may be different from those in large groups or herds.

Certain workers have attempted to avoid the pitfalls associated with visual observation by developing and using the indirect methods of stomach or fecal analysis. The analysis of stomach contents requires the killing of representative animals in large enough numbers for the results to indicate seasonal trends and to show large differences between plant species. This destruction is difficult to justify if samples from the stomach are the sole concern, so frequently animals shot by hunters or during a culling operation are examined to give information on food preferences. The results are likely to be useful but should be treated with caution if a wide view of food habits is being sought. Nevertheless, the analysis of stomach contents, i.e., from the reticulo-rumen in ruminants, has been used successfully for the macroscopic identification of dicotyledons.[3,4] With grasses, a microscope is needed for identification. Several workers in East Africa, e.g., Azavedo and Agnew,[5] have been able to determine on an approximate weight basis the relative amounts of browse and grass eaten by ruminants.

The fecal analysis technique was developed to avoid the slaughter of animals to obtain samples. Stewart[6] carried out the original work on the microscopical identification of epidermal fragments of leaf blades of grasses in the feces. It seems that an estimate can be obtained of the proportions of different grasses eaten by ruminants, but the method is very laborious. Probably for this reason it has not become popular, although a few people[7] have claimed it gave useful results. There are some doubts as to the validity of the method since some plant species appear to vary in the structure of their leaf blade between different areas.

This brief review of the methods employed to study food preferences should highlight the fact that the information obtained can only indicate important trends and cannot be interpreted quantitatively to specify diets for wild ruminants. For example, it is now known that food preferences may be influenced by season, district or area, presence of other species, density of stocking, and age of the animal, all of which are important trends but which confound any idea of establishing an optimum diet for any particular species. A more useful feature which has arisen from studies of food preferences is the important distinction between grazing (eating grass) and browsing (eating woody plants). The food habits of some ruminants place them clearly in one of these categories, while others are mixed feeders. Some eat other parts of the vegetation such as roots, bark, fruit, seed, flowers, and sap. Hofmann,[8] in a morphological study of 20 East African wild ruminants, found that food habits appeared to give rise to one of two basic types of stomachs. Animals with a predominantly grazing habit and which are able to consume large quantities of roughage feeds have a large rumen with an uneven distribution of papillae varying greatly in size and shape. The reticulum is relatively small, while the omasum is distinguished by its large size and numerous laminae of several orders. The abomasum is relatively big and spacious. On the other hand, animals with mainly selective food habits which usually eat softer and succulent forage (including fruit and seeds) have a smaller rumen with a regular distribution of densely arranged papillae that are relatively big, flat, and tongue shaped. The reticulum is relatively large, but the omasum is small or extremely small and is flat and kidney shaped. The abomasum is relatively large and lined by a remarkably thick glandular mucosa. These findings indicate that ruminants that are mainly browsers are significantly different in their gut anatomy from the grazers, and this distinction may be an important consideration in recommending a diet for a particular species. Table 1 gives the food preferences of a number of wild ruminants in terms of grazing and browsing. The classification should not be rigidly accepted since certain species have been found to change their food habits with season or area. For example, Lamprey[9] classifies the eland (*Taurotragus oryx*) as a mixed feeder in the Tarangire game reserve in Tanzania since its diet consisted of roughly three-fourths grasses and one-fourth woody plants. On the other hand, both Van Zyl[10] in the Western Transvaal and Posselt[11] in Rhodesia found that eland were predominantly browsers. Despite these changes, Table 1 may be a useful guide regarding the acceptability to these species of roughages that are likely to be available in conditions where a measure of farming has been imposed.

An overall appraisal of the published results on food preferences of wild ruminants have highlighted two important nutritional considerations. Their diet can be extraordinary varied; as many as 60 plant species have been identified as being eaten in one period by one species. This concept may imply that wild ruminants perform better if their diet is varied, which is a frequent occurrence with domesticated livestock. However, more important is the notion that a diet consisting of many components is unlikely to be deficient in any of the nutrients that are essential to the animal. There appears to be no published record of a nutritional deficiency having been seen in wild ruminants. From this, it would be foolish to assume that their requirements for either major or minor nutrients are any less critical than those of their domesticated counterparts. It is more likely that a diet composed of many plant species insures against any particular deficiency.

The second nutritional inference which may be drawn from food preference data is also concerned with the quality of the diet eaten. The results of chemical analysis indicate that the vegetation that is usually selected has a medium-to-high quality, especially with respect to the content of protein. For example, Roth and Osterberg[12] found that the crude protein contents of dry matter of vegetation selected by eland ranged from 8 to 23%. Cattle under similar conditions were consuming grass with a much lower level of protein. This effect is likely to be general, since the leaves and finer parts of the plant are higher in protein than the whole plant; therefore, typical selection by wild ruminants would result in a diet containing adequate or even high levels of protein. This trend may be particularly pronounced if species of the leguminosae are part of the ruminant's diet, as with giraffe when they eat *Acacia* species. This feature of ruminant's diets, together with another well-known characteristic, their great propensity to produce lean meat (see the review by Ledger et al.[13]), strongly suggest that their protein requirements may be relatively high. In the absence of data to the contrary, it may be well for the present to err on the generous side with respect to protein when diets are being formulated for ruminants that normally live wild.

DIETS FOR THE SUCCESSFUL REARING OF WILD RUMINANTS

In order to tame or domesticate wild ruminants for experimental or productive purposes, it is usually necessary to obtain animals at a very early age and artificially rear them on diets based on milk or milk substitutes. A comparison of several attempts recorded in the literature[11,14-16] illustrate the variable degree of success achieved. However, careful scrutiny of the observations made by the authors on the likely cause of death or a failure to thrive indicates that the nature of the diet is less important than the measures to overcome both the animal's extreme nervousness and its inability to cope with large volumes of feed at one time. In some instances, great difficulty has been experienced in inducing the young ruminant to drink. The animals usually die in a few days if this behavior problem is not solved. In a study of three East African wild ruminants, Hutchison[15] found that some topi (*Damaliscus korrigum*) and Uganda kob (*Adenota kob thomasi*) calves died of extreme dehydration within 3 days because of their refusal to drink or eat. To avoid losing animals, force-feeding from a bottle is sometimes essential. The procedure is not easy but it can be managed if the young animal is under control. One finger is pushed into the angle of the mouth and the teat of the bottle is inserted after it. If the animal fails to swallow the gums are massaged with the finger. Milk is squeezed gently into the mouth from the polyethylene feeding bottle with the teat held downwards. The hole in the teat is made sufficiently large for milk to flow steadily when the bottle is squeezed. For the larger species, a 10-cm teat attached to a 0.5-l bottle is convenient, while a 4- to 5-cm teat is more appropriate for the smaller species. The size of the bottle can be increased when needed, and when the animals are sucking well it should not be difficult to train them to drink from a pan or bucket.

To avoid digestive disorders such as bloat and scouring, it is necessary to feed frequently. At the outset the minimum frequency should be five times a day, but feeding even more often is likely to have some advantage. After a week, the frequency may be reduced to four feeds per day; if weaning has not occurred, longer intervals can be used from about 6 weeks onwards.

It is difficult to say when weaning to solid food should occur. The minimum age appears to be 6 weeks, but if difficulties in feeding are experienced or if the animal is not growing well the transition may need to be delayed for several weeks.

No firm recommendation can be made on the composition of diets for the rearing of wild ruminants. If the chemical composition of the milk of these animals is considered (see the review of Smith[28]), the milk of all species for which results are published is richer than cow's milk in both fat and protein. Indeed, some of the recorded levels of these two constituents are remarkably high, but it is possible that the sample analyzed did not represent all the milk produced by that animal. However, after allowing for such discrepancies, there may be a case for fortifying cow's milk with fat and protein or for using milk substitutes in a more concentrated solution if the preparation is to be fed to young ruminants. Certain investigators have used such a diet, but there is no evidence to suggest that the enrichment or greater concentration was essential. Indeed, Hutchison[15] successfully raised two dik-dik (*Rhynchotragus kirkii*), one on cow's milk and the other on cow's milk enriched with egg yolk, and found no difference in health or weight gain between the two animals. For the present, we may assume that cow's milk given in sufficient amounts and at frequent levels is likely to be a suitable diet for the rearing of most wild ruminants. In some instances either pasteurization or treatment of the milk with an antibiotic may be necessary to counteract the ill effects of virulent bacteria.

DIETS FOR GROWING AND ADULT RUMINANTS

It is reasonable to assume that based on research and practical experience with domestic ruminants, the diet of the wild species should include adequate amounts of energy, protein, essential fatty acids, minerals, vitamins A, D, and E, and fiber. The ingredients selected to supply these nutrients are usually limited in number by their availability and cost, but they should be palatable and compatible with each other and with the animal's digestive system. The limited number of studies which have been undertaken to compare the ability of wild ruminants with that of cattle or sheep indicate the presence of a few significant differences between species.[17-20] Arman and Hopcraft[20] found that digestibility of dry matter and crude fiber in five pelleted diets was higher in hartebeest (*Alcelaphus buselaphus cokii*) and duiker than in sheep, while in Thomson's gazelle (*Gazella thomsonii*) and eland they were lower than in sheep. Similarly, red deer (*Cervus elaphus*), in comparison with sheep, have a lower digestibility (four to five percentage units) of dry matter and cellulose from roughage diets, according to the results of Kay and Goodall.[21] However, the effect of a lower digestibility may be partly offset by ruminants such as eland and red deer having a relatively higher food intake. Therefore, we may assume that diets fed to domestic ruminants are generally suitable for wild ruminants. Even very poor diets, such as low quality hay only, have been given to two species, the hartebeest (*Alcelaphus buselaphus cokii*) and impala (*Aepyceros melampus*), for a limited time under experimental conditions without any apparent ill effects.[22] However, to avoid loss of weight and to possibly assure good health, diets with a large proportion of concentrates, e.g., 60 to 70%, are recommended. On the other hand, the components that are high in fiber make up about 20 to 30% of the diet. This level ensures that the normal acitivity of the reticulo-rumen is maintained, which is likely to be more important for species that graze rather than browse. Due attention should be given to the essential minor constitutents to avoid trace element and vitamin deficiencies. Ullrey et al.,[23] in developing a basal diet for white-tailed deer (*Odocoileus virginianus*), found than an earlier formulation caused signs of vitamin E and selenium deficiency to appear in 20% of the animals after they had consumed the diet for 120 days. The trouble was avoided by including 10% of linseed meal in the diet and adding a supplement which supplied 44 IU vitamin E per kilogram.

Diets that have been developed and used successfully for wild ruminants are given in Tables 2 through 6. As suggested earlier, most of them have a relatively high crude protein content when compared with similar diets for domestic ruminants. Otherwise, there is little to distinguish them from good quality diets that are recommended for either cattle or sheep. If because of a lack of published information there is any doubt as to the nature of a diet that may be suitable for a species of wild ruminant, selection of one of these tabulated formulations is likely to form an appropriate basis for the development of such a diet.

Table 1
CLASSIFICATION OF SOME WILD RUMINANTS INTO GRAZERS, BROWSERS, OR MIXED FEEDERS

African buffalo (*Syncerus caffer*)	Grazer
Black-tailed deer (*Dorcelaphus columbianus*)	Grazer
Blesbok (*Damaliscus albifrons*)	Grazer
Caribou (*Rangifer tarandus*)	Mixed
Duiker, common (*Sylvicapra grimmia*)	Browser
Eland (*Taurotragus oryx*)	Browser
Gazelle, Grant's (*Gazella granti*)	Browser
Gazelle, Thomson's (*Gazella thomsonii*)	Mixed
Giraffe (*Giraffa camelopardalis*)	Browser
Hartebeest, Coke's (*Alcelaphus buselaphus cokii*)	Grazer
Impala (*Aepyceros melampus*)	Grazer
Kob, Uganda (*Adenota kob thomasii*)	Grazer
Kudu, Greater (*Tragelaphus strepsiceros*)	Browser
Mule deer (*Odocoileus hemionus*)	Browser
Oryx (*Oryx gazella*)	Browser
Pronghorn antelope (*Antilocapra americana*)	Browser
Red deer (*Cervus elaphus*)	Grazer
Reedbuck (*Redunca* spp.)	Grazer
Reindeer (*Rangifer tarandus tarandus*)	Mixed
Springbok (*Antidorcas marsuplialis*)	Mixed
Topi (*Damaliscus korrigum*)	Grazer
Waterbuck, Defassa (*Kobus defassa*)	Grazer
White-tailed deer (*Odocoileus virginianus*)	Mixed
Wildebeest (*Connochaetes taurinus*)	Grazer

Table 2
DIETS FOR BLACK-TAILED DEER (*DORCELAPHUS COLUMBIANUS*)[24]

Ingredients	Weanlings (6 weeks–5 months) (%)	Adults (%)
Ground yellow corn	10.0	30.0
Ground wheat feed	33.0	12.5
Ground wheat bran	10.0	13.75
Ground oat groats	13.0	–
Ground beet pulp	–	10.0
Dehydrated grass meal	5.0	10.0
Molasses (cane)	–	7.5
Fishmeal (70% CP), herring	10.0	5.5
Soybean meal (50% CP)	5.0	8.75
Spray-dried skim milk	10.0	–
Steamed bone meal	–	1.0
Dicalcium phosphate	0.5	–
Iodized salt	0.75	1.0
Stabilized animal fat	2.5	–
Vitamin premix	0.25	–

Note: A diet modeled on this formulation has been recommended by Kay et al.[17] for red deer (*Cervus elaphus*).

Table 3
PELLETED DIET[a] FOR RUMINANTS IN BASEL ZOO[25]

Ingredient	%
Ground oats	15.0
Ground barley	20.0
Ground grain sorghum	8.0
Ground wheat	10.0
Soybean meal	10.0
Groundnut meal	8.0
Linseed meal	3.0
Lucerne meal	10.0
Wheat bran	5.0
Potato flakes	2.0
Pomace	3.0
Licorice powder	1.0
Ground calcium carbonate	1.0
Bone meal	2.0
Salt	0.9
Magnesium sulfate	0.2
Trace element and vitamin mixture	0.9
Composition	
Crude protein (%)	ca. 18.0
Ether extract (%)	2.5
Crude fiber (%)	7.0
Calcium (%)	1.2
Phosphorus (%)	0.6
Iron (mg/kg)	20.0
Manganese (mg/kg)	10.0
Copper (mg/kg)	5.0
Iodine (mg/kg)	4.0
Cobalt (mg/kg)	0.4
Zinc (mg/kg)	10.0
Vitamin E (mg/kg)	60.0
Vitamin A (IU/kg)	12,000
Vitamin D_3 (IU/kg)	2,000

[a] The pellets are formulated to suit both grazers and browsers. The former receive fewer pellets, and hay forms the major part of their ration. The browsers are given more of the pelleted diet along with lucerne hay.

Table 4
CONCENTRATE DIET FOR REINDEER IN WINTER[26]

Ingredient	%
Cereals	41.0
Wheat bran	5.0
Molasses beet chips	10.0
Hay meal	25.0
Straw meal	15.0
Dicalcium phosphate	1.7
Iodized salt	0.3
Vitamin and trace element mix	2.0
Composition	
Crude protein	8.0
Ether extract	2.0
Crude fiber	18.0
Ash	9.0

Table 5
DIETS FOR ELAND[27]

Amounts Given in Kilograms

	Winter diet			Summer supplement		
Ingredient	Adult	1–2 years	Less than 1 year	Adult	1–2 years	Less than 1 year
Coarse ground barley	2.0	1.5	0.7	2.0	1.5	0.7
Coarse ground oats	1.0	1.0	0.5	1.0	1.0	0.5
Meadow hay	8.0	6.0	3.0	–	–	–
Sugar beets	4.0	2.0	1.0	20.0	10.0	8.0

Table 6
DIET FOR WHITE-TAILED DEER (*ODOCOILEUS VIRGINIANUS*)[23]

Ingredient	%
Ground corn cobs[a]	34.7
Ground yellow maize	29.5
Soybean meal (49% CP)	18.0
Linseed meal	10.0
Dehydrated alfalfa meal (17% CP)	3.0
Cane molasses	3.0
Corn oil	0.3
Trace element mix, high zinc (0.8%)	0.5
Ground limestone	0.5
Anhydrous sodium sulfate	0.25
Vitamin A, D, and E mix	0.25

Composition	
Dry matter (%)	92.9
Crude protein (%)	17.6
Ether extract (%)	2.0
Ash (%)	4.6
Cell wall constituents (%)	40.5
Acid detergent fiber (%)	17.6
Lignin (%)	3.8
Vitamin A (IU/kg)	3,300
Vitamin D_2 (IU/kg)	220
Vitamin E (IU/kg)	44

[a] Oat hulls may be substituted for corn cobs if a diet with a lower digestibility and nutritive value is required.

REFERENCES

1. **Wallmo, O. C., Gill, R. B., Carpenter, L. H., and Reichert, D. W.,** *J. Wildl. Manage.,* 37, 556-562, 1973.
2. **Field, C. R.,** in *Comparative Nutrition of Wild Animals,* Crawford, M.A., Ed., Academic Press, London, 1968, 135-151.
3. **Wilson, V. J.,** *East Afr. Wildl. J.,* 3, 27–37, 1965.
4. **Wilson, V. J. and Child, G.,** *Arnoldia Rhodesia,* 1(35), 1–2, 1965.
5. **Azavedo, J. C. S. and Agnew, A. D. G.,** *East Afr. Wildl. J.,* 6, 145–146, 1968.
6. **Stewart, D. R. M.,** *J. Appl. Ecol.,* 4, 83–111, 1967.
7. **Kiley, M.,** *East Afr. Wildl. J.,* 4, 153–157, 1966.
8. **Hofmann, R. R.,** in *Comparative Nutrition of Wild Animals,* Crawford, M. A., Ed., Academic Press, London, 1968, 179–194.
9. **Lamprey, H. F.,** *East Afr. Wildl. J.,* 1, 63–92, 1963.
10. **Van Zyl, J. H. M.,** *Zool. Afr.,* 1, 55–72, 1965.
11. **Posselt, J.,** *Rhod. J. Agric. Res.,* 1, 81–87, 1963.
12. **Roth, H. H. and Osterberg, R.,** *Rhod. J. Agric. Res.,* 9, 45–51, 1971.
13. **Ledger, H. S., Sacks, R., and Smith, N. S.,** *Wld. Rev. Anim. Prod.,* 3, 13–37, 1967.
14. **Kellas, L. M.,** *Proc. Zool. Soc. London,* 124, 751–783, 1954.
15. **Hutchison, M.,** *J. Zool.,* 161, 437–442, 1970.
16. **Tschiderer, K.,** *Bodenkultur,* 25(1), 75–87, 1974.
17. **Kay, R. N. B., Maloiy, G. M. O., and Goodall, E. D.,** in *Comparative Nutrition of Wild Animals,* Crawford, M. A., Ed., Academic Press, London, 1968, 101–108.
18. **Allo, A. A., Oh, J. H., Longhurst, W. M., and Conolly, G. E.,** *J. Wildl. Manage.,* 37, 202–210, 1973.
19. **Rogerson, A.,** in *Comparative Nutrition of Wild Animals,* Crawford, M. A., Ed., Academic Press, London, 1968, 153–161.
20. **Arman, P. and Hopcraft, D.,** *Br. J. Nutr.,* 33, 255–264, 1975.
21. **Kay, R. N. B. and Goodall, E. D.,** *Proc. Nutr. Soc.,* 35, 98A–99A, 1976.
22. **Maloiy, G. M. O. and Hopcraft, D.,** *Comp. Biochem. Physiol. A,* 38, 525–534, 1971.
23. **Ullrey, D. E., Youatt, W. G., Johnson, H. E., Kay, L. D., Schoepke, B. L., and Magee, W. T.,** *J. Wildl. Manage.,* 35, 57–62, 1971.
24. **Wood, A. J., Nordan, H. C., and Cowan, I. M.,** *J. Wildl. Manage.,* 25, 295–302, 1961.
25. **Wackernagel, H.,** in *Comparative Nutrition of Wild Animals,* Crawford, M. A., Ed., Academic Press, London, 1968, 1–12.
26. **Steen, E.,** in *Comparative Nutrition of Wild Animals,* Crawford, M. A., Ed., Academic Press, London, 1968, 117–128.
27. **Treus, V. and Krauchenko, D.,** in *Comparative Nutrition of Wild Animals,* Crawford, M. A., Ed., Academic Press, London, 1968, 395–411.
28. **Smith, A.,** *Proc. S. Afr. Soc. Anim. Prod.,* 9, 63–72, 1970.

Other Mammals

DIETS (NATURAL AND SYNTHETIC): MONOTREMATA

L. R. Collins

FAMILY TACHYGLOSSIDAE – ECHIDNAS

Genus *Tachyglossus:* Short-nosed Echidna

Tachyglossus aculeatus

Natural diet - primarily ants and termites, but stomach contents and scats have revealed other prey as well. Griffiths[1] reports that Simon (in litt.) noted that short-nosed echidnas living in Queensland also ate earthworms, small beetles, parts of orthoptera, and larvae of the cossid moth, *Xyleutes.*

Synthetic diet* – Short-nosed echidnas at the National Zoological Park, Melbourne Zoo, Artis Zoo, and Cologne Zoo have been maintained on a ground meat mix containing milk, eggs, and various vitamin and mineral additives. The following diets have been fed to short-nosed echidnas at the institution noted:

Institution: Philadelphia Zoo, Philadelphia, Pennsylvania, U.S.A.

Diet:

Quantity	Ingredient
½ p	Cow's milk (whole)
1	Egg (raw)

Feeding methodology: Milk and egg placed in separate pans. Above quantity per animal fed in late afternoon. One specimen lived for over 49 years on above diet. One teaspoon of lime water was added to the milk for the first 39 years or so but was discontinued for the last 10 years and therefore seems not to have been a vital nutritional ingredient.[2]

Institution: St. Louis Zoo, St. Louis, Missouri, U.S.A.

Diet:

Quantity	Ingredient
1/4 cup	Pablum®
1	Egg (raw)
1/3 teaspoon	RiBad® (multivitamin)
1/3 teaspoon	Salt
1/3 teaspoon	Calcium
1 teaspoon	Lime water
1/2 cup	Milk
1/2 cup	Water

Mixture Preparation: Water is added to Pablum until it has the consistency of a batter. Egg is beaten and mixed with other ingredients (except for the water and milk). The water and milk are mixed and added to the mixture of other ingredients until the resultant mixture takes on the consistency of a thin batter.**

Feeding methodology: Above quantity per animal per day.

Institution: Monash University, Clayton, Victoria, Australia

Diet recipient: Tachyglossus aculeatus

Diet:

Mix 1

Quantity	Ingredient
650 g	Pollard
560 g	Meat and bone meal
4 p	Warm water (40°C)

Mix above and let stand for 10 min.

750 g	Cooked meat (lean beef, not horsemeat)

* Unless otherwise specifically noted, all synthetic diets in this chapter were obtained via either personal communication or from questionnaires mailed to the individual institutions listed in the text.

**One specimen has been maintained on the above diet since June 1938 (still alive as of July 9, 1969).

Prepare cooked meat as follows: Meat should be coarse ground. Cook slowly to remove fat and juices, resulting in a dry cooked meat (steam autoclave 40 min at dry-fast exhaust). Allow to cool and then fine grind. Do *not* use horsemeat – echidnas have been lost because horsemeat was used. Heavy fibers caused a solid ball in the animal's stomach. Ground beef should be used. Add cooked meat to above mixture.

Mix 2

350 g	Glucodine
2 ml	Pentavite (multivitamin mixture)
2 p	Water, cold

Mix above ingredients together.

Mixture preparation: Pour Mix 1 into Mix 2. Add 300 g sandy soil to complete mixture. After final mixing, mixture should have the consistency of pancake batter. Completed mixture may be kept under refrigeration but *not* frozen.

Feeding methodology: Mixture is placed in half-pint milk bottles which are tilted horizontally. The above amount is for ten echidnas.

The following is a milk substitute used in place of raw cow's milk as a part of the echidna diet at the Pretoria Zoo, South Africa.

Institution: National Zoological Gardens of South Africa, Pretoria, South Africa
Milk substitute recipient: Tachyglossus aculeatus
Milk substitute:

Quantity	Ingredient
80 lb (36.32 kg)	Skimmed milk powder
8 lb (3.63 kg)	Whey powder
8.5 lb (3.85 kg)	Fine Kaffir corn meal
2 lb (0.9 kg)	Brewer's yeast
1 lb (0.45 kg)	Dicalcium phosphate
0.5 lb (0.23 kg)	Sodium chloride
4 g/100 lb (4 g/45.4 kg)	Vitamin A
1/2 g/100 lb (1/2 g/45.4 kg)	Vitamin D_3

One pound (0.45 kg) of milk powder is used to prepare 1 gal of milk substitute.

The daily diet of Pretoria's echidnas consisted of a mixture of raw cow's milk, precooked cereal, and raw minced meat. Cow's milk is a potential source of pathogenic bacteria and causes diarrhea in some animals whose diet in captivity contains a large amount of milk. Powdered milk, on the other hand, is less expensive, and the necessary minerals and vitamins can be included in the powdered milk. The echidnas took to powdered milk readily and the zoo was successful in maintaining them in good condition.[2]

Genus *Zaglossus:* Long-nosed Echidna

Zaglossus bruijni

Natural diet – Van Deusen and George[3] suspect that the diet of *Zaglossus* in the wild varies basically from that of *Tachyglossus.* They cite different observers who stressed that captive long-nosed echidnas feed on worms ("Worms are abundant in the loose humus layer of the humid montane forests of New Guinea").[3] Ants are supposedly also taken, although various investigators found that different captive specimens of *Zaglossus bruijni* refused ants. Ripley[4] reports that one captive specimen refused ants but preferred papaya and raw eggs.

Synthetic diet – Long-nosed echidnas at the London Zoo have been maintained on raw meat mixed with milk and earthworms. (Van Deusen and George[3] quote Pocock as saying that specimens of *Zaglossus* sp. are fond of this diet.)

Institution: Taronga Zoological Park, Mosman, Sydney, New South Wales, Australia
Diet:

Quantity	Ingredient
1 tablespoon	Meat (minced)
1 tablespoon	Carrot (minced)
2 oz (56.7 g)	Horse blood
1/2	Egg (boiled) plus shell
1 dessert spoon	Lactogen
1 dessert spoon	Milk (condensed)
1/2 teaspoon	Clay
1	Lettuce leaf
1 drop	Formic acid (every other day)
4 oz (113.4 g)	Water

Mixture preparation: All ingredients blended together.
Feeding methodology: Mixture placed in a dish with holes large enough only for the animals to be able to take the mixture with their tongues. Feed is placed in enclosure at 1500 hours, above quantity per animal per day.

FAMILY ORNITHORHYNCHIDAE – PLATYPUSES

Genus *Ornithorhynchus:* Platypus

Ornithorhynchus anatinus

Natural diet – The food of a platypus in the wild varies with its locale and the yearly seasons. Generally, this semiaquatic monotreme's diet includes various small crustaceans, molluscs, insect larvae, worms, and possibly frogs and tadpoles. The food is encountered and seized with the mandibles as the platypus searches along the bottom of its particular aquatic feeding haunt. A great deal of grit, sand, and soil is secured along with the food, and probably plays an important role in the mastication process. Mastication takes place on the surface while the animal floats along, before diving to secure more foodstuffs.

Synthetic diet

Institution: New York Zoological Park, Bronx, New York, U.S.A.
Diet:

Quantity	Ingredient
1 lb (0.45 kg)	Earthworms (*Lumbricus* or *Pheretima*)
2 dozen	Crayfish (live) (*Cambarus*)
2	Leopard frogs (*Rana pipiens*)
2	Eggs (steamed)
1 handful	Cockroaches (*Blatta*) or mealworms (*Tenebrio*)

Feeding methodology: Worms were furnished in a tin also containing earth, sunk to the bottom of the tank. Earth was always provided because, as suggested by Burrell,[5] it appears to serve an essential function in the digestive process. Crayfish and frogs were liberated in the tank, where they were readily captured by the animals. The steamed eggs were scattered over the bottom of the tank. The above quantity was generally fed per animal per day.

Regarding this diet, Crandall states:

Quantities were varied constantly, of course, to suit the animals' variation in condition, vagaries of appetite and individual whims. In addition to the items listed above, the platypus will take tadpoles, beetle grubs, and some insects. However, these are either seasonal or too difficult to obtain in sufficient quantity to be included in the regular diet. As far as our experience here (New York Zoological Park) goes, platypuses will not take fish, but I have been assured by the late J. E. Ward of Sydney (in litt.) that freshly caught animals will not only eat both fish and beef if it is finely scraped but will thrive on them as part of a general diet.[6]

Institution: Taronga Zoological Park, Sydney, Australia

Diet:

Quantity	Ingredient
1 lb (0.45 kg)	Earthworms
1/2 lb (0.23 kg)	Prawns
3	Egg yolks (hard boiled)

Method of preparation: Heads and tails removed from prawns before feeding.
Feeding methodology: Animals fed above quantity per animal daily at 1545 hours.

Institution: Sir Colin Mackenzie Sanctuary, Healesville, Australia
Diet: During the author's visit to Healesville in February 1969, personal communication with the director, Mr. V. C. Mullett, revealed that the Healesville platypuses received a diet consisting mainly of yabbies during the summer and earthworms supplemented with what living food was available during the winter.

Synthetic diet of lactating female platypus - After breeding the platypus in captivity for the first time, Fleay[7] found that the female with one young (estimated to be 5½ weeks of age) ate the following test meal over a 12+ hr period: "It was found that Jill (the lactating female) in this one night consumed a typical meal of the following items: Worms (native species, larger than European earthworms): 400. Grubs (mainly chaffers): 338. Yabbies: 38. The total weight of these items without soil of any kind was 28½ ounces. (808 g) or 1¾ pounds (0.79 kg): Considering Jill's mere 16¼ inch (41.3 cm) length and her body weight of an exact 2 pounds (0.9 kg) when in her fattest condition, this test gives some idea of the amazing appetite of *Ornithorhynchus* – particularly during some phases of the nursing period."

Concerning this prodigious meal, Fleay comments, "Naturally an animal weighing 2 pounds could not hold a meal of another 1¾ pounds, and it follows that the nursing platypus mother must assimilate the food as she swims and builds up her milk supply to full capacity over the extremely long hours (up to 14) of continual foraging. From December 19 onwards with the youngster then approximately 6½ weeks old, the female's appetite declined with her eating fewer earthworms and concentrating mainly on grubs and yabbies."

Fleay attributes this to the probability that a growth spurt in the early stages of the young's development is followed by a period of quiescence during which less nourishment is required than before. He goes on to state that "In any event, it was true that leading up to the stage of 5 to 6 weeks, Jill's baby required a phenomenal amount of nourishment."

REFERENCES

1. **Griffiths, M.,** *Echidnas,* Pergamon Press, London, 1968.
2. **Young, E.,** A suitable milk substitute for the aardvark *Orycteropus afer* and echidna *Tachyglossus aculeatus, Int. Zoo Yearb.,* 6, 68–69, 1966.
3. **Van Deusen, H. M. and George, G. G.,** Results of the Archbold Expeditions, No. 90: Notes on the Echidnas (*Mammalia, Tachyglossidae*) of New Guinea, *Am. Mus. Novit.,* p. 2383, 1969.
4. **Ripley, S. D.,** *Trail of the Money Bird,* 2nd ed., Harper Brothers, New York, 1942.
5. **Burrell, H.,** *The Platypus,* Eagle Press, Waterloo, Engl., 1927.
6. **Crandall, L. S.,** *The Management of Wild Mammals in Captivity,* University of Chicago Press, Chicago, 1964.
7. **Fleay, D.,** The birth of a baby platypus, *Anim. Kingdom,* 47(3), 51–69, 1944.

DIETS (NATURAL AND SYNTHETIC): MARSUPIALIA

L. R. Collins

FAMILY DIDELPHIDAE – OPOSSUMS

Genus *Caluromys:* Woolly Opossum

Caluromys philander (Type Species), C. lanatus, and C. derbianus

Natural diet – Woolly opossums are omnivorous and probably feed on a variety of insects, fruits, seeds, leaves, soft vegetables, small vertebrates, and carrion.

Synthetic diet* – Captive specimens at the National Zoological Park, New York (Bronx) Zoo, El Pinar Zoo (Caracas), and the National Institute of Mental Health have been maintained on a variety of chopped fruits (banana, apple, orange, grapes, and pawpaw), as well as raw meat or canned dog food. (Investigators at the NIMH found that Hill's® HD prescription moist dog food is fairly well assimilated by didelphids and serves as a good dietary source for weight gain by animals.) American "chameleons" (*Anoles*) fed ad lib; two multivitamin drops (in sugar-sweetened orange juice), evaporated milk (1 oz = 28.35 g), and one diced hardboiled egg are fed daily as dietary supplements at the New York (Bronx) Zoo. When available, four mouse pups (or one skinned adult mouse) are fed three times per week to woolly opossums at the National Zoological Park. Specimens at El Pinar Zoo receive insects in addition to a fruit and meat diet. Woolly opossums in the large colony housed at the University of California, Davis, are maintained on dry Purina® Cat Chow. (Animals were acclimated to this diet by first feeding the chow in a moistened form and then gradually feeding it drier and drier until it was fed directly from the bag.) The dry food assists in breaking up and preventing tartar deposits on the teeth, which can result from the feeding of a soft diet. Altmann[1] reports that specimens of *C. lanatus* maintained at the Berlin Zoo ate mice, mealworms, rice, noodle pudding, eggs, and fruit and berry juice.

Genus *Caluromysiops*: Black-shouldered Opossum

Caluromysiops irrupta

Natural diet – The feeding habits of free-living black-shouldered opossums are not known, although they are thought to be similar to those of the woolly opossums (*Caluromys*). Captive specimens have exhibited highly individualistic dietary preferences.

Synthetic diet

Institution: National Zoological Park, Washington, D.C., U.S.A.

Diet:

Quantity	Ingredient
1/2	Banana
6	Grapes (when available)
1/4	Apple
1/4	Orange
6	Mouse pups (hairless)
1	Chick
1	Mouse (adult)

Diet preparation: Fruit is chopped into convenient wedges for ease of handling by animal. Adult mouse is skinned and fed whole without the skin.

Feeding methodology: Above amounts are fed per animal per day, except mouse pups are fed twice a week, chicks twice a week, and adult mice the rest of the week.

* Unless otherwise specifically noted, all synthetic diets in this chapter were obtained via either personal communication or from questionnaires mailed to the individual institutions listed in the text.

The NZP *Caluromysiops* avidly takes 2- or 3-day-old chicks as well as skinned adult mice and mouse pups. Banana is the most readily accepted fruit, with the apple, orange, and grapes often left uneaten. While being cared for by an animal dealer, this specimen reportedly fed on bread soaked in egg, crickets, and mealworms.

Institution: New York (Bronx) Zoo, New York, U.S.A.

Diet:

Quantity	Ingredient
1/3	Banana
2 oz (56.7 g)	Orange
1 oz (28.35 g)	Apple
1 oz (28.35 g)	Horsemeat (chopped)
1/2 oz (14.18 g)	Liver
2 oz (56.7 g)	Diet "M" (see below)
40	Mealworms
20	Crickets

Diet Preparation: Fruit and meat cut small.

Feeding methodology: Above quantity per animal per day, except cricket ration, which is fed three times per week.

Note: Female specimen maintained on above diet from September 10, 1962 to July 28, 1969.

Diet "M":

Quantity	Ingredient
28 lb (12.71 kg)	Chopped horsemeat
28 cups	Purina dog meal
3/4 cup	Ground limestone
3/4 cup	Paltone®

Method of preparation: The dog meal is soaked in enough water to soften it and then all of the ingredients are mixed together. This diet is prepared daily and fed to a number of different animals.

Institution: Oklahoma City Zoo, Oklahoma, U.S.A.

Diet: a.m. – milk; p.m. – banana, orange, apple, raisins, Ratcliff's diet (horsemeat, dog food, carrots) are fed, with raw ground horsemeat sprinkled on top. Probably roaches are eaten at night. Neither crickets nor mice are eaten (Lentz, personal communication).

Institution: Cincinnati Zoo, Cincinnati, Ohio, U.S.A.

Diet: A variety of fruits, egg, sweet potato, ground meat, and rice bread are fed. Abdec® (multivitamin), Neo-Cal®, and Rib-Ad® are supplements to the diet.

Diet preparation: The egg is hard boiled and the sweet potato cooked before feeding. Rib-Ad is sprinkled on the food prior to feeding. The vitamin supplements are added to the meat.

Genus *Monodelphis*: Short-tailed Opossum

Monodelphis brevicaudata (Type Species)

Also included in this group are *M. adusta, M. americana, M. dimidiata, M. domestica, M. fosteri, M. henseli, M. osgoodi, M. scalops, M. sorex*, and *M. unistrata.*

Natural diet – Free-living short-tailed opossums reportedly feed on fruits, seeds, insects, small rodents, and carrion.[2]

Synthetic diet – Specimens maintained at El Pinar Zoo, Caracas, Venezuela, were fed a diet consisting of pawpaw, banana, chopped raw meat, and (occasionally) insects.

Genus *Dromiciops*: Monitos del Monte

Dromiciops australis

Natural diet – In the wild, *Dromiciops* reportedly feeds on insects and other invertebrates.[3]

Synthetic diet – Greer[4] maintained three specimens on a diet consisting of oatmeal, potatoes, powdered milk mixed with canned peach syrup (this mixture was either lapped from a dish by the animals or was fed through an eye dropper), plum, pear, watermelon, applesauce, blackberries, and hamburger. The hamburger was eaten ravenously (Greer, personal communication). Known quantities of ham were fed to one individual weighing 26.5 g. It ate an average of 4.1 g of ham per day, and, by the end of 5 days, the animal had gained 3 g. A meat mix such as that fed to murine opossums at the National Zoological Park (see below) and supplemented with mealworms, crickets, and hairless mouse pups, as well as a variety of fruit, might be fed to these didelphids.

Genus *Glironia:* Bushy-tailed Opossum

Glironia venusta (Type Species) and G. criniger

Natural diet – The diet of free-living bushy-tailed opossums is unknown but is thought to be similar to that of *Marmosa* sp., consisting of insects, seeds, eggs, and fruit.[2]

Synthetic diet – Specimens of *Glironia* sp. apparently have not been maintained in captivity. A probable diet for captive specimens would include mealworms, crickets, various fruits, and a meat mix similar to that fed to *Marmosa robinsoni* at the National Zoological Park (see below).

Genus *Lestodelphys* – Patagonian Opossum

Lestodelphys halli

Natural diet – Patagonian opossums are probably primarily carnivores feeding on birds, mice, and other small vertebrates. Insects and fruit are apparently rare in this opossum's habitat.[5]

Synthetic diet – To the best of the author's knowledge, patagonian opossums have never been maintained in captivity. A captive diet should probably include newly hatched chicks, hairless mouse pups, skinned adult mice, and a meat mix such as the one fed to *Marmosa robinsoni* at the National Zoo (see below). A fruit supplement to this diet might be tried.

Genus *Marmosa*: Murine Opossum

Marmosa murina (Type Species)

Also included in this group are *M. aceramarcae, M. agilis, M. agricolai, M. canescens, M. cinerea, M. constantine, M. domina, M. dryas, M. elegans, M. emiliae, M. formosa, M. fuscata, M. germana, M. grisea, M. impavida, M. incana, M. invicta, M. karimii, M. lepida, M. leucastra, M. mapiriensis, M. marica, M. mexicana, M. microtarsus, M. noctivaga, M. ocellata, M. parvidens, M. phaea, M. pusilla, M. quichua, M. rapposa, M. regina, M. robinsoni, M. rubra, M. scapulata, M. tatei, M. tyleriana, M. unduaviensis, M. velutina,* and *M. yungasensis.*

Natural diet – In the tropical regions of their range, the diet of murine opossums consists chiefly of insects and soft fruit (especially bananas). Small birds and their eggs, mice, and lizards are also probable dietary components. Specimens that inhabit the colder regions (such as the paramo, where insects and fruit are extremely scarce) tend to exhibit a more carnivorous diet, feeding almost exclusively on small vertebrates.

Some observations have been made on the feeding habits of *Marmosa* in captivity. Captive specimens eat a wide variety of foodstuffs that they would not have access to in the wild. Insects seem to be a favorite source of protein for these small opossums, and Enders[6] found that captive *Marmosa* sp. remained in better condition when insects were introduced into their diet.

Synthetic diet – *Marmosa* sp. at New York (Bronx) Zoo are maintained on a diet consisting of chopped fruit (apple, orange, and banana), Diet "M" (see above), chopped horsemeat, chopped liver, mealworms, and crickets. *M. robinsoni* maintained at the University of California, Davis, in 1970 were fed dry Purina Cat Chow with occasional fruit supplements (Barnes, personal communication). *M. robinsoni* housed at the National Zoological Park received the following diet:

Quantity	Ingredient
1 teaspoon	Meat mix – thoroughly blend the following ingredients: 1 cup finely ground horsemeat 1/4 teaspoon Pervinal® 1/2 egg yolk, hard boiled and finely grated 1 tablespoon powdered milk 1 multivitamin drop
1/2 oz (14.17 g)	Carrot
1/2 oz (14.17 g)	Apple
1/2 oz (14.17 g)	Orange
1 slice	Banana
1/2 oz (14.17 g)	Sweet potato
2	Grapes (quartered)
1/4 teaspoon	Custard
1/4 teaspoon	Beef (strained)
6	Crickets
6	Mealworms

Diet preparation: Carrot and sweet potato cooked well. All fruit and vegetables chopped small.
Feeding procedure: Above quantity per animal per day, except mealworms fed twice a week, crickets once a week, strained beef once a week, and custard once a week. Animals are fed in the late afternoon.

Genus *Philander*: Gray "Four-eyed" Opossum
Philander opossum

Natural diet – Gray "four-eyed" opossums feed on small vertebrates, birds, eggs, insects, fresh-water crustaceans, snails, earthworms, fruit, and probably carrion.[2] Enders[6] notes that captive specimens were more carnivorous in their dietary habits than any of the other didelphids observed. He observed *Philander* eating several different kinds of meat including anteater and rodent flesh as well as grasshoppers, eggs, banana, pineapple, papaya, and figs.

Synthetic diet – Gewalt[7] reports that specimens of *Philander* maintained at the Berlin Zoo ate pigeons, eggs, white mice, and mealworms. Barnes (personal communication) reports that at the University of California, Davis, *Philander* was maintained on several varieties of dry dog and cat foods of different commercial brands. Barnes also reports that *Philander* is not a fastidious animal and relishes fresh meat, fruit, and eggs.

Institution: National Institute of Mental Health, Bethesda, Maryland, U.S.A.
Diet:

Quantity	Ingredient
1 tablespoon	Hill's HD prescription dog food. (Investigators found this dog food to be well assimilated by animals and is a good diet component when weight gain is desired.)
1/2	Banana
1 dozen	Grapes
1/2	Orange (quartered)

Genus *Metachirus*: Brown "Four-eyed" Opossum
Metachirus nudicaudatus

Natural diet – In their natural habitat, brown "four-eyed" opossums feed on fruits, insects, birds' eggs, and small vertebrates. Enders[6] observed that captive specimens were not as carnivorous as *Didelphis* sp. or *Philander* opossum, preferring fruit to meat or carrion. He reports that *M. nudicaudatus* thrives on a frugivorous diet.

Synthetic diet – A brown "four-eyed" opossum was being maintained on a diet consisting of various fruits and green corn at the Barranquilla Zoo in Colombia during 1968.

This didelphid would probably do well on a diet such as that fed to woolly opossums (*Caluromys*) at the National Zoological Park (see above), with the frugivorous components of the diet being emphasized.

Genus *Lutreolina*: Lutrine Opossum

Lutreolina crassicaudata

Natural diet – Lutrine opossums reportedly feed on insects and a variety of small vertebrates including fishes.[2]

Synthetic diet – Davis[8] reports that the New York (Bronx) Zoo's trio of *Lutreolina* thrived on a diet of sliced butterfish and meat mixture as well as frogs, earthworms, shrimp, and mice.

Genus *Didelphis:* Large "Common" Opossum

Didelphis marsupialis (Type Species), D. albiventris, and D. azarae

Natural diet – Large "common" opossums are omnivorous and are by no means fastidious in their dietary habits. They feed on a variety of animal and vegetable matter.

Synthetic diet

Institution: National Zoological Park, Washington, D.C., U.S.A.
Diet recipient: Didelphis marsupialis
Diet:

Quantity	Ingredient
1/2 cup	*Marmosa* meat mix (see above)
1/2	Apple
1/2	Banana
1/2	Orange
6	Grapes
Ad lib	Mixed vegetables
1 tablespoon	Custard
1 slice	Bread, whole wheat

In addition, skinned adult mice, young chicks, and rat pups were fed ad lib.

Institution: New York (Bronx) Zoo, New York, U.S.A.
Diet recipients: D. azarae and *D. marsupialis*
Diet: Crandall[9] reports that *D. marsupialis* and *D. azarae* thrive on a diet consisting of a mixture of dry or moist dog food with chopped raw meat, fine bone meal, and cod-liver oil added. In addition, milk, raw egg, and any available raw fruits and vegetables are taken.

Institution: National Institute of Mental Health, Bethesda, Maryland, U.S.A.
Diet recipient: D. marsupialis
Diet: Red Heart® canned dog food mixed with various dry dog foods has been the main diet for *Didelphis.*

Institution: Barranquilla Zoo, Colombia
Diet recipient: D. azarae
Diet: Azara's opossum has been maintained at various occasions on 4–6 oz (113.4–170.1 g) of meat and one or two bananas, plus various other fruits daily.

Genus *Chironectes:* Yapok

Chironectes minimus

Natural diet – Walker[2] reports that yapoks are strictly carnivorous and feed on crayfish, shrimp, and fish.

Synthetic diet

Institution: National Zoological Park, Washington, D.C., U.S.A.

Diet:

Quantity	Ingredient
1/2 cup	Finely ground horsemeat
1	Hardboiled egg yolk (grated)
1/2 teaspoon	Pervinal

Diet preparation: Mix above ingredients well.

Feeding methodology: Feed above amount per day. In addition, feed six mouse pups (hairless) per week, an occasional chick, frog, crayfish, or skinned mouse. (National Zoological Park's yapok will not eat fish.)

Institution: New York (Bronx) Zoo, New York, U.S.A.

Diet: Davis[8] reports that yapoks maintained at the Bronx Zoo in 1966 thrived on slices of butterfish and a meat mix which contained ingredients that aid in oil production for fur maintenance. Crandall[9] notes that a yapok that lived 3 years was fed on chopped raw meat with finely ground bone meal and cod-liver oil added. Fillet smelts (*Osmerus*) were fed as well.

Institution: Barranquilla Zoo, Colombia

Diet: The yapok maintained at the above institution was fed 3–5 oz of chopped fresh fish daily. It also reportedly fed on white rats and other small mammals.

FAMILY DASYURIDAE – DASYURES

Genus *Phascogale:* Tuan

Phascogale tapoatafa (Type Species) and P. calura

Natural diet – Tuans feed on insects and small vertebrates such as rodents, birds, and lizards.[2]

Captive diet

Institution: Fleay's Fauna Sanctuary, Tallebudgera, Queensland, Australia

Diet recipient: P. tapoatafa

Diet: Fleay fed his breeding pair of tauns on a diet of freshly killed mice, young rats, and sparrows, supplemented with bread and milk that had been sweetened with melon jam or honey.

Institution: Monash University, Melbourne, Australia

Diet recipient: Phascogale tapoatafa

Diet: Young mice, insects, dog cubes, millipedes, and snails

Genus *Antechinus:* Antechinus

Antechinus flavipes (Type Species)

This group also includes *A. apicalis, A. bellus, A. godmani, A. macdonnellensis, A. maculatus, A. melanurus, A. minimus, A. naso, A. rosamondae, A. stuartii, A. swainsoni,* and *A. wilhelmina.*

Natural diet – Antechinuses feed on insects and other small invertebrates as well as mice, lizards, and other small vertebrates.

Synthetic diet

Institution: Australian Museum

Diet recipient: A. stuartii

Diet: Marlow[10] reports that his colony of *A. stuartii* was maintained on carcasses of adult laboratory mice which had been killed previously and kept under refrigeration. Mealworm larvae (*Tenebrio molitor*) were frequently fed as a supplement to this diet. Marlow observes that an adult male which was captured in July 1958 lived for 2 years on a diet of dead laboratory mice and water.

Institution: National Zoological Park, Washington, D.C., U.S.A.
Diet recipients: A. flavipes and *A. melanurus*
Diet: The following meat mix was fed to the small medium-sized dasyures in the marsupial collection of the National Zoological Park's Scientific Research Division:

Quantity	Ingredient
2 cups	Horsemeat (lean and finely ground)
2 tablespoons	Powdered milk
2	Egg yolks (hardboiled and finely ground)
1 teaspoon	Pervinal
1/2 teaspoon	Wheat germ oil
8 drops	Vi-daylin M® multivitamin drops
2 pinches	Iodized salt
2 tablespoons	Dry dog food (finely ground)

Mix above ingredients well by hand.

A. flavipes receives one heaping tablespoon of the above mix per day. In addition, adult mice are fed once a week, hairless mouse pups twice a week (two pups per feeding), and mealworms (*Tenebrio* sp.) or crickets twice a week. An adult male *A. flavipes* was maintained on the above diet for 2 years and 10 months.

Institution: La Trobe University, Melbourne, Australia
Diet recipient: A. apicalis
Diet: Woolley[11] reports that the basic diet fed to her colony of dibblers consisted of "a mixture of raw minced meats, egg and fine ground dried puppy food, in the proportion of 450 g sheep heart, 115 g sheep liver, 1 sheep brain, 1 egg, and 100 g puppy food. The animals were fed once a day, in the late afternoon. Each animal was given 15 g mince daily." Larger quantities were consumed by a gravid female during the later stages of lactation. This basic meat diet was supplemented with insects and *Eucalyptus* and *Banksia* blossoms when available. Honey, diluted 1 to 5 with water, was fed ad lib.

Genus *Planigale:* Planigale

Planigale ingrami (Type Species), P. novaeguineae, and P. tenuirostris

Natural diet – Planigales are avid predators, feeding on such small lizards as geckos, skinks, and small mammals such as *Leggadina* (a murid with head and body length of 60 to 80 mm). Insects are taken, as are spiders. Fleay[12] observed that "the general insect-seeking habit of these midget marsupials gives way under opportunity to a scavenging carnivorous trend." Fleay further reports that *P. ingrami* is not above cannibalism. He observed littermates devour one of their own that had sickened and died. Newly captured females may devour their pouch young.

Synthetic diet – Planigales would probably do well on the meat mix fed to the antechinuses at the National Zoological Park (see above). This meat mix should be finely ground and occasionally supplemented with mealworms, crickets, moths, freshly killed baby chicks (with the entrails exposed), and halved adult mice. Heinsohn[13] reports that two captive planigales thrived on a diet consisting of moths, grasshoppers, raw meat, army worm and scarab beetle larvae, and occasionally small lizards.

Following are two diets fed to *P. ingrami* at the institutions noted.

Institution: Fleay's Fauna Sanctuary, Tallebudgera, Queensland, Australia
Diet: Fleay[12] states that captive planigales will eat mealworms, field crickets, paralyzed spiders, the larvae and pupae of mud-dauber wasps, grasshoppers or locusts, chafer-grubs, any small bird carcass, and steamed egg.

Institution: Australian National Museum
Diet: Marlow (personal communication) reports that his female planigale was maintained on meal-worms and small moths.

Genus *Murexia:* New Guinea Murine Dasyure

Murexia longicaudata (Type Species) and M. rothschildi

Natural diet – There is no information regarding the dietary habits of *Murexia* although it is probably primarily insectivorous. Van Deusen (personal communication) trapped *Murexia* sp. using a regular "museum bait" (a mash consisting of peanut butter, rolled oats, raisins, and fat bacon).

Synthetic diet – There are no published accounts of *Murexia* having been maintained as captive specimens. A meat mix such as that fed to the small dasyures at the National Zoological Park (see above) would probably be acceptable to *Murexia* sp. Hairless mouse and rat pups, 2- or 3-day-old chicks, skinned adult mice, crickets, and mealworms might be offered as supplements to the meat-mix diet.

Genus *Sminthopsis:* Dunnart

Sminthopsis murina (Type Species)

This group also includes *S. crassicaudata, S. froggatti, S. granulipes, S. hirtipes, S. leucopus, S. longicaudata, S. macrura, S. nitela, S. psammophila,* and *S. rufigenis.*

Natural diet – Dunnarts feed on insects such as grasshoppers, moths, and certain beetles. Other small arthropods including centipedes and spiders fall prey to these dasyures. Vertebrates such as small mice and lizards are reportedly caught and eaten.

Synthetic diet

Institution: National Zoological Park, Washington, D.C., U.S.A.
Diet recipient: S. macrura
Diet: Each animal receives one heaping teaspoon of meat mix (see above) per day. In addition, mealworms, crickets, and mouse pups are fed when available. Lactating females always have access to finely grated cuttlebone. All feeding occurs in late afternoon. Four generations have been bred using the above diet.

Institution: University of Adelaide, Australia
Diet recipient: S. crassicaudata
Diet: Minced beef heart (with connective tissue removed) plus added calcium diphosphate (Ewer found that if the connective tissue was not removed, it would lodge between the animals' teeth, causing infections); Nestle's lamb brains plus added Vi-daylin M and calcium diphosphate; egg (cooked as an omelette and finely chopped); live mealworms and other insects.
Feeding methodology: Minced beef heart and additives fed three times per week. Lamb brains and additives fed three times per week. Egg omelette fed once a week. Live mealworms and other insects fed when available. Enough food was fed per day so that there was always a small excess left uneaten. Lactating females and young that were being weaned were always given a few mealworms and a dish of milk with a drop of vitamin concentrate added. Lactating females always had access to finely grated cuttlebone as well. Feeding was carried out in late afternoon.

Institution: Zurich Zoo, Switzerland
Diet recipient: S. crassicaudata
Diet:

Quantity	Ingredients
1 tablespoon	Housefly maggots
1 teaspoon	Mealworms
1	Baby mouse

Feeding methodology: Maggots fed first day, mealworms fed second day, and baby mouse fed third day. Cycle is started over on fourth day. Feeding occurs in the evening.

Institution: Institute of Medical and Veterinary Sciences, Adelaide, Australia
Diet recipient: Sminthopsis sp.
Diet: Except for the mouse supplement, *Sminthopsis* sp. are maintained on the same diet that is fed to *Dasyuroides byrnei* at the above named institution (see below).

Institution: Melbourne Zoo, Australia
Diet recipient: S. crassicaudata
Diet: Dead mice, grasshoppers, and mealworms; 6–8 g of one of a mixture of each of the above ingredients is eaten nightly per animal.

Genus *Antechinomys:* Long-legged Delicate Dasyure

Antechinomys laniger (Type Species) and A. spenceri

Natural diet – In their natural habitat, *Antechinomys* sp. feed on insects, spiders, centipedes, and possibly such small vertebrates as lizards. Following are two synthetic diets fed to captive specimens at the institutions noted.

Institution: National Zoological Park, Washington, D.C., U.S.A.
Diet recipients: A. spenceri and *A. laniger*
Diet: One heaping tablespoon of the Scientific Research Division's dasyurid meat mix daily. In addition, mealworms, crickets, and hairless mouse pups are fed as supplements when available.

Institution: Institute of Medical and Veterinary Sciences, Adelaide, Australia
Diet recipient: A. spenceri
Diet: A. spenceri are maintained on the same diet fed to *Dasyuroides byrnei* (see below) except that mice are not fed as a supplement to this diet.

Genus *Neophascogale:* Neophascogale

Neophascogale lorentzii

Natural diet – Nothing has been recorded regarding the dietary habits of neophascogales in the wild. They are probably primarily insectivorous.

Synthetic diet – Two female neophascogales maintained at the Baiyer River Sanctuary (New Guinea) by Graeme G. George during 1967 were fed a diet consisting of strips of beef as well as cockchafer beetles and cicadas when available. The live insects were consumed greedily. A diet such as that fed to *Antechinus flavipes* at the National Zoological Park (see above) should be appropriate for neophascogales.

Genus *Phacolosorex:* Sorecine Dasyure

Phascolosorex dorsalis (Type Species) and P. doriae

Natural diet – There is no recorded information pertaining to the dietary habits of sorecine dasyures in the wild. Van Deusen (personal communication) reports trapping *Phascolosorex sp.* with the "museum bait" mentioned in the earlier discussion of *Murexia.*

Synthetic diet – There are no published accounts of *Phascolosorex* having been maintained in captivity. A diet such as those fed to *Antechinus* should be appropriate for these dasyures.

Genus *Myoictis:* New Guinea Three-striped Dasyure

Myoictis melas

Natural diet – Nothing has been recorded regarding the feeding habits of *Myoictis* in the wild. They are probably primarily insectivorous.

Synthetic diet – There are no records of *Myoictis* having been maintained in captivity. A diet similar to those fed to captive antechinuses should be appropriate for *Myoictis.*

Genus *Dasyuroides:* Kowari

Dasyuroides byrnei

Natural diet – In the wild, kowaris feed on insects, arachnids, and probably small vertebrates such as birds, rodents, and lizards.

Synthetic diet

Institution: National Zoological Park, Washington, D.C., U.S.A.

Diet:

Quantity	Ingredients
1 heaping teaspoon	NZP dasyurid meat mix (see above for ingredients)
1	2- or 3-day-old chick
4–6	Mouse pups (hairless)
10–12	Mealworms
6–8	Crickets
Ad lib	*Anoles*

Feeding methodology: Each animal receives one heaping tablespoon of meat mix daily except the 2 days per week during which the chick and the mouse pups are fed. *Anoles* are fed when available, and *D. byrnei* apparently relish these small lizards, as they are quickly caught and eaten. If mouse pups are unavailable, a freshly killed adult mouse is fed instead. Mealworms and crickets are fed once per week. Multivitamin drops are added to the drinking water.

Institution: Taronga Zoological Park, Sydney, Australia

Diet:

Quantity	Ingredients
1 oz (28.35 g)	Horsemeat (minced)
1	Mouse (live)
Ad lib	Maggots
Ad lib	Cicadas

Feeding methodology: Above amount per animal per day at 2:20 p.m. Maggots and cicadas are fed when available.

Institution: La Trobe University, Melbourne, Australia

Diet:

Quantity	Ingredient
1 lb (0.45 kg)	Heart (beef or sheep)
1/4 lb (113.4 g)	Liver (preferably sheep)
1 set	Brains (sheep)
1	Egg
100 g	Dog chow (dry, finely ground)

Mixture preparation: Beat egg, add minced brains and liver, and mix well. Add minced heart and mix. Add dog chow and mix.

Feeding methodology: Animals are fed 20 g of above mix per day in the afternoon. In addition, chicks are fed three times per week and mealworms daily. Laboratory mice and young rats are fed occasionally, as are crickets and cockroaches. Gravid females receive 3 g of mealworms daily. Pentavite multivitamin drops are added to the drinking water (one drop per 50 ml of water).

Institution: Institute of Medical and Veterinary Science, Adelaide, Australia

Diet:

Quantity	Ingredient
1 teaspoon	Minced horsemeat
1 teaspoon +	Pal® dog food
1 teaspoon +	Minced hardboiled egg plus milk and vitamin concentrate
Ad lib	Mealworms
Ad lib	Mice

Feeding methodology: The minced horsemeat is fed three times per week, as is the dog food. The egg diet is fed once a week. Mealworms are fed as supplements when available, as are mice.

Institution: Arid Zone Research Institute, Alice Springs, Northern Territory, Australia

Diet: One live white mouse is fed per animal per day.

Weaning diet: *Dasyuroides* young are weaned on a diet consisting of a mixture of steak, kidney, liver, lettuce, and rat pellets. Mixture is fed ad lib at first and then fed in amounts to maintain desirable weight per animal.

Institution: University of Missouri, U.S.A.

Diet: Sorenson[14] reports that *Dasyuroides* at this institution were fed Alpo® canned chicken parts mixed with bananas and freshly killed or frozen mice. This diet was supplemented with lettuce, apples, carrots, melons, and grasshoppers. *Dasyuroides* preferred mice to other foods. Animals were fed daily.

Genus *Dasycercus:* Mulgara

Dasycercus cristicauda

Natural diet – In the wild, mulgaras feed on insects, arachnids, and small vertebrates such as rodents and probably small lizards.

Synthetic diet – Specimens housed at the National Zoological Park, La Trobe University, and the Institute of Medical and Veterinary Sciences, Adelaide, received the same diet as that fed to *Dasyuroides byrnei* at those institutions. Following are two diets fed to specimens at the institutions noted:

Institution: Fleay's Fauna Sanctuary, Tallebudgera, Queensland, Australia
Diet: The pair of mulgaras bred at Fleay's Fauna Sanctuary during 1956 were maintained on a diet consisting of insects, mice, and egg custard.

Institution: University of Adelaide, Australia
Diet: Dr. R. F. Ewer (personal communication) maintained several mulgaras on a diet of minced heart (the staple standby) plus a few mealworms (*Tenebrio*) daily. In addition, a live mouse was offered occasionally, as were live insects, centipedes, scorpions, small snakes, and lizards when available. Moths were taken avidly. Sparrows were baited into an outdoor enclosure by sprinkling grain on the substrate. The mulgaras would catch and eat a sparrow now and then when this method was used. The main meal was fed in the late afternoon, but when young were present a little food was always made available in the morning. A wild-trapped female with pouch young raised her litter successfully on the above diet. Another pair of adult animals was maintained for 9 months, during which time they remained active and healthy.

Genus *Dasyurus:* Viverrine Dasyure

Dasyurus viverrinus (Type Species)

This group also includes *D. albopunctatus, D. geoffroyi, D. hallucatus,* and *D. maculatus.*

Natural diet – Viverrine dasyures feed on a variety of invertebrate and vertebrate prey, including insects, crustaceans, amphibians, reptiles, birds, and small mammals. The smaller species are thought to include a large proportion of insect matter in their dietary regime. George (personal communication) reports that captive specimens of *D. albopunctatus* seem to relish some vegetable matter.

Synthetic diet – Specimens of *D. albopunctatus* maintained at the Berlin Zoo ate sparrows, doves, eggs, chopped raw meat, milk, mice, and bananas.[15]

Institution: Baiyer River Sanctuary, Papua, New Guinea
Diet recipient: D. albopunctatus
Diet: Beef strips, cicadas, beetles, rats, and mice. Occasionally, portions of ripe banana and fresh corn are fed, which the animals reportedly seem to like.

Institution: Monash University, Melbourne, Australia
Diet recipient: D. hallucatus
Diet: Sparrows, hairless newborn mice or rats, crickets, large beetles, locusts, and chicken heads and wings containing pin feathers. In addition, dog biscuits and bones are always available for chewing purposes. These last two items are important for the maintenance of tartar-free teeth.

Institution: Monash University, Melbourne, Australia
Diet recipient: D. viverrinus
Diet: Animals in the large breeding colony are fed one of the following small vertebrates three times per week: young rats, sparrows, or small guinea pigs. In addition, fresh chicken heads and pigeon eggs (twice a week) are fed as dietary supplements. Bones and dog biscuits are always available for chewing purposes.

Institution: National Zoological Park, Washington, D.C., U.S.A.
Diet recipients: D. viverrinus, D. maculatus, and *D. albopunctatus*
Diet: The following meat mix is fed to *Dasyurus* sp. when live food is not available. When available, 3-day-old chicks, rat pups, and baby rabbits are fed.

Quantity	Ingredient
1/2 cup	Finely ground lean horsemeat
1/4 teaspoon	Wheat germ oil
1/2 teaspoon	Powdered milk
1/4 teaspoon	Pervinal
1/2	Finely grated hardboiled egg yolk
1 drop	Vi-daylin M multivitamin

Mixture preparation: Before feeding, mix well (by hand) above proportions of ingredients.
Feeding procedure: For *D. viverrinus,* above amount per animal per day; for *D. maculatus* three-fourths cup of above mixture. Bones are always available for chewing purposes.

Institution: Taronga Zoological Park, Sydney, Australia
Diet recipient: D. maculatus
Diet:

Quantity	Ingredient
1/2 lb (0.23 kg)	Chopped horsemeat
1/4 lb (0.11 kg)	Fish
1	Raw egg

Feeding methodology: Horsemeat fed 5 days per week, fish fed 2 days per week. The raw egg is added to the horsemeat twice a week.

Genus *Sarcophilus:* Tasmanian Devil

Sarcophilus harrisii

Natural diet – Tasmanian devils feed on a wide variety of foodstuffs in the wild, including introduced as well as native species of vertebrates, invertebrates, a small amount of plant matter, and carrion.

Synthetic diet

Institution: National Zoological Park, Washington, D.C.
Diet: One cup per day, 5 days per week, of the meat mix fed to *Dasyurus viverrinus* at this institution (see above). In addition, one-half whole chicken 1 day per week, and six newly hatched chicks 1 day per week. Whole rabbits and calf brains are fed occasionally. (Calf brains are eaten ravenously and were the only food consumed for the first week of the animals' molt.) The author found, however, that an excessive feeding of brains caused loose stools. Rats have never been eaten by this pair of animals. *Sarcophilus* should always have access to large bones for chewing purposes.

Institution: Taronga Zoological Park, Sydney, Australia
Diet: 1 1/2 lb (0.68 kg) of chopped horsemeat daily; 1/2 lb (0.23 kg) of fish is fed weekly, in addition to two raw eggs.

Institution: Melbourne Zoo, Australia
Diet: 400 to 500 g of freshly killed rat, rabbit, or bird flesh is fed three times weekly; in addition, the animals occasionally receive lean horse flesh.

Genus *Thylacinus:* Thylacine

Thylacinus cynocephalus

This species is possibly extinct.

Natural diet – Thylacines prey on small to medium-sized mammals and birds and reportedly are selective when feeding on a carcass, preferring blood, blood-filled nasal membranes, liver, and kidneys. One thylacine was found to have eaten an echidna (*Tachyglossus*).

Synthetic diet – Captive specimens were usually maintained on a diet of chopped meat and small game when available. The last captive thylacine died in the Hobart Zoo in 1933.

Genus *Myrmecobius:* Numbat

Myrmecobius fasciatus

Natural diet – *Myrmecobius* preys on termites and ants in its natural habitat. Occasionally other invertebrates are also eaten.

Synthetic diet – At the Taronga (Sydney) Zoo, large pieces of termite mound are placed in the numbat enclosure. When the animals have consumed the insects from the outer surface of the mound, it is then cut into smaller pieces and these are again placed in the cage. This procedure is repeated five or six times a day.

FAMILY NOTORYCTIDAE

Genus *Notoryctes:* Notoryctid
Notoryctes typhlops (Type Species) and *N. caurinus*

Natural diet – Notoryctids feed mainly on insects, including beetle larvae as well as ants and their pupae.[16]

Synthetic diet – Captive specimens have reportedly eaten earthworms,[2] although specimens at the Arid Zone Research Institute, Alice Springs, Northern Territory, Australia refused them (Howe, personal communication). Specimens at this institution are fed beetle larvae. Howe believes that dried ant eggs may be part of the solution for maintaining these marsupials.

FAMILY PERAMELIDAE – BANDICOOTS

Genus *Perameles:* Long-nosed Bandicoot
Perameles nasuta (Type Species)

This group also includes *P. bougainvillei, P. eremiana, P. fasciata,* and *P. gunni.*

Natural diet – Long-nosed bandicoots are omnivorous, feeding on such plant material as blackberries, plant bulbs, and tubers. Various insects and their larvae as well as other arthropods, earthworms, and small slugs are eaten. Heinsohn[17] reports finding the remains of a small frog in the stomach of one specimen of *P. gunni.* Small rodents might possibly be taken in the wild, as *Perameles* sp. readily take mice in captivity.

Synthetic diet

Institution: Taronga Zoological Park, Sydney, Australia
Diet recipient: P. nasuta
Diet:

Quantity	Ingredient
1/2 lb (226.8 g)	Minced horsemeat
1/2 slice	Chopped fruit bread
1 oz (28.4 g)	Chopped sweet potato
3 dozen	Earthworms

Feeding methodology: P. nasuta was fed above amount per animal per day. The earthworms are placed in an earth-filled dish.

Institution: CSIRO, Canberra, Australia
Diet recipient: P. nasuta
Diet: The breeding group of bandicoots studied by Stodart[18] at Canberra was fed a diet consisting of fresh minced meat, milk, eggs, and bread. The foregoing ingredients were mixed together. During the summer when the ground dried out, dried meat crumbs with added vitamins and bran were also supplied.

Institution: Westbury Zoo, Westbury, Tasmania
Diet recipient: P. gunni
Diet:

Quantity	Ingredient
2 oz (56.7 g)	Minced meat
1 tablespoon	Tea leaves
1 cup	Bread and milk
1	Apple

Method of preparation: The tea leaves are boiled and the apple is quartered before being fed.
Feeding methodology: Above ingredients fed in separate containers per animal per day.

Genus *Echymipera:* Spiny Bandicoot

Echymipera kaluba (Type Species), *E. clara,* and *E. rufescens*

Natural diet – Little is known pertaining to the dietary habits of *Echymipera* sp. and other New Guinea bandicoots. Marlow[16] describes the Australian form of *E. rufescens* as being insectivorous. Most species are probably omnivorous. Van Deusen (personal communication) reports that some of the New Guinea species are trapped by using sweet potatoes as bait. *E. clara* reportedly feeds on the fruits of *Ficus* and *Pandanus* spp.

Synthetic diet – Specimens of *E. kaluba* maintained at the Hallstrom Park Bird of Paradise Sanctuary, New Guinea were fed a diet consisting of beef, earthworms, mealworms, ripe banana, and cooked sweet potato (George, personal communication). The following diet is fed to the National Zoological Park's specimens:

Diet recipient: E. rufescens
Diet: One-half cup of the meat mix fed to the National Zoological Park's carvinorous marsupials (see above). Sliced banana, apple, raisins, and other fruit are fed. Earthworms are placed on the substrate. They burrow into the pine bark and hyperhumus but are rooted out and fed upon (with apparent gusto) by *Echymipera.*

Genus *Macrotis:* Bilby

Marcrotis lagotis (Type Species) and *M. leucura*

Natural diet – Bilbies feed on insects and their larvae, as well as small vertebrates and some vegetation. Captive specimens have been known to eat cake or bread, raw or cooked meat, snails, birds, and mice.

Synthetic diet

Institution: The Institute of Medical and Veterinary Science, Adelaide, Australia
Diet recipient: M. lagotis
Diet:

Day	Quantity	Ingredient
Sunday	60 g	Dried dog food
Monday	60 g	Pal processed dog food
	60 g	Rabbit pellets
Tuesday	1	Apple
	1	Carrot
	1 handful	Shredded cabbage
Wednesday	100 g	Fresh minced steak
Thursday	30 g	Mealworms
Friday	36 g	Heinz® mixed cereal mixed with milk, handful of sultanas + vitamin E + multivitamin supplement + 1 teaspoon Calactate
Saturday	Repetition of above six diets begins.	

Feeding methodology: Feed above amount per animal on an alternating schedule as listed.

Institution: Arid Zone Research Institute, Alice Springs, Northern Territory, Australia
Diet recipient: M. lagotis
Diet: Rat pellets, insect larvae, large insects (e.g., grasshoppers), and chopped beef are fed ad lib per animal. In addition, animals receive water containing Pentavite drops.

Genus *Chaeropus:* Pig-footed Bandicoot

Chaeropus ecaudatus

Natural diet – Pig-footed bandicoots are omnivorous. Captive specimens have been known to feed on lettuces, barley grass, bread, certain bulbous roots, and various insects (especially grasshoppers). Krefft noted that captive specimens would not attack mice nor eat meat, but, according to Jones,[19] "Sturt noted that it showed a much greater partiality for flesh, although it did not flourish on this diet."

Synthetic diet – A mixed diet such as that fed to omnivorous bandicoots (i.e., *Perameles* sp. and *Isoodon* sp.) should be offered to pig-footed bandicoots in case any specimens are captured and held in captivity.

Genus *Isoodon:* Short-nosed Bandicoot

Isoodon obesulus (Type Species), I. auratus, and I. macrourus

Natural diet – Short-nosed bandicoots are omnivorous, feeding on insects and their larvae, earthworms, such small vertebrates as skinks (*Lygosoma* sp.), frogs, and some plant material. Insect matter constitutes a great portion of the diet.

Synthetic diet

Institution: Lincoln Park Zoo, Chicago, Illinois, U.S.A.
Diet recipient: I. macrourus
Diet: Raw meat-mink chow mix (see below), banana, apple, escarole, and cooked sweet potato. Occasionally, large insect larvae, mealworms, earthworms, and freshly killed mice are fed, and these items are eaten with much enthusiasm. Coddled egg mixed with raw ground meat is also acceptable.
Raw meat-mink chow mixture:

Quantity	Ingredient
15 lb (6.8 kg)	Ground, lean horsemeat
15 lb (6.8 kg)	Purina Mink Developer (moistened)
10	Hardboiled eggs (without shells)
1/3 cup	Honey
3/4 cup	Molasses
1/4 cup	Peanut oil
1/3 cup	Vi-Penta® liquid multiple vitamin
1/3 cup	VPC® – Dyna fos
1 tablespoon	Cod-liver oil

The above amount of meat mix is made up fresh daily and fed to all carnivorous small mammals at the Lincoln Park Zoo.

Institution: Melbourne Zoo, Melbourne, Victoria, Australia
Diet recipient: I. obesulus
Diet: Two-thirds cup of high-protein dog biscuit moistened with water. Occasionally this diet is supplemented with dead mice and pieces of fruit or an echidna meat mix consisting of meat, milk, eggs, and nutritional additives.

Institution: Queensland Institute of Medical Research, Queensland, Australia
Diet recipient: I. macrourus
Diet: Raw minced steak, Farex, bran, egg, and milk to which salt and cod-liver oil are added. Vitamin E was added to the diet daily. *Note:* Animals maintained on the above diet should be given something hard (e.g., dry dog biscuits) to chew on as the above diet is soft and tends to cause extensive tartar build-up on the molar teeth, resulting in oral pathologies and the eventual death of the animal.

Genus *Rhynchomeles:* Ceram Island Bandicoot

Rhynchomeles prattorum

Natural diet – No records.

Synthetic diet – No records. Diets such as those fed to omnivorous bandicoots (e.g., *Perameles* sp. and *Isoodon* sp.) should be offered to captive specimens.

Genus *Peroryctes:* "Common" New Guinea Bandicoot

Peroryctes raffrayanus (Type Species), P. broadbenti, and P. longicaudata

Natural diet – Little is known. Van Deusen (personal communication) reports trapping *Peroryctes* sp. with the "museum bait" discussed previously.

Synthetic diet – No records. A mixed diet such as that fed to *Perameles* (see above) and other omnivorous bandicoots should be offered to captive specimens.

Genus *Microperoryctes* Murine Bandicoot

Microperoryctes murina

Natural diet – No records.

Synthetic diet – No records. A mixed diet such as that fed to omnivorous bandicoots, e.g., *Perameles* sp. and *Isoodon* sp., should be offered to captive specimens.

FAMILY CAENOLESTIDAE – CAENOLESTIDS

Genus *Caenolestes:* "Common" Caenolestid

Caenolestes fuliginosus (Type Species)

This group also includes *C. caniventer, C. convelatus, C. obscurus,* and *C. tatei.*

Natural diet – In its natural habitat, *Caenolestes* feeds on caterpillars as well as adult lepidopterans, beetles, ants (Dorylidae), and probably other insects. Centipedes and spiders are also taken.

Osgood[20] and Kirsch (personal communication) report that *Caenolestes* shows a preference for "meat baits." Osgood reports taking *Caenolestes* in traps baited with small birds and bacon rind.

Synthetic diet – Kirsch (personal communication) reports having maintained a male *C. convelatus* alive and well on a diet of earthworms and steak for 5 days before sacrificing the animal.

A meat mix such as that used for the small dasyures at the National Zoological Park in addition to crickets, mealworms, and other available insects, mouse pups, earthworms, and newly hatched chicks would most likely prove to be a suitable diet for these marsupials.

Genus *Lestoros:* Peruvian Caenolestid

Lestoros inca

Natural diet – Peruvian caenolestids feed on insects, grubs, and probably other invertebrates. The dietary habits of these caenolestids probably parallel those of *Caenolestes.*

Synthetic diet – No records. Synthetic diets suggested for *Caenolestes* would probably be apropos for *Lestoros.*

Genus *Rhyncholestes:* Chilean Caenolestid

Rhyncholestes raphanurus

Natural diet – No records. Dietary habits of *Rhyncholestes* probably are similar to those of *Caenolestes* and *Lestoros.*

Synthetic diet – No records. Synthetic diets recommended for *Caenolestes* would probably be apropos for *Rhyncholestes.*

FAMILY PHALANGERIDAE – PHALANGERS

Genus *Phalanger:* Cuscus

Phalanger orientalis (Type Species)

This group also includes *P. celebensis, P. gymnotis, P. maculatus, P. ursinus,* and *P. vestitus.*

Natural diet – Cuscuses are omnivorous, feeding on various fruits and berries, leaves, insects, birds and their eggs, small mammals, and possibly small reptiles.

Synthetic diet – George (personal communication) reports that cuscuses housed at the Hallstrom Park Bird of Paradise Sanctuary, New Guinea are fed a variety of native foodstuffs, including various leaves, pawpaw, passion fruit, bread (occasionally, cuscuses tend to become obese in captivity), young shoots of *Schurmansia* sp. (a native New

Guinea shrub), native celery, succulent shoots of two cultivated native vegetables (*Saccharum* sp. and *Setaria* sp.), raw sweet potato tubers, green bananas in the skin, fresh corn (maize), and the terminal flowering bracts of the banana bunch.

Following are diets fed to captive specimens at the institutions listed:

Institution: National Zoological Park, Washington, D.C., U.S.A.
Diet recipients: P. orientalis and *P. gymnotis*
Diet: The following ingredients are fed ad lib: apple, orange, grapes, kale, carrot, sweet potato, monkey chow (Wayne® or Purina), bread (white and whole wheat), and mixed grain. In addition, each animal receives 1 tablespoon of the dasyure meat mix daily (see above), as well as fresh leaves of the Norway maple (*Acei platanoides*) and sugar maple (*A. saccharum*).
Method of preparation: Fruit and vegetables chopped into small pieces convenient for animal's handling.

Institution: Melbourne Zoo, Melbourne, Victoria, Australia
Diet recipient: P. maculatus
Diet: Oak, Morton Bay fig, and rose leaves are fed ad lib. In addition, 1/2 lb of chopped apple, orange, and cucumber as well as one-half cup of phalanger mixture (see diet for *Petaurus breviceps*, below) are fed daily.

Institution: Taronga Zoo, Sydney, Australia
Diet recipient: P. maculatus
Diet:

Quantity	Ingredient
3	Banana
1	Apple
1/2	Orange
1	Carrot
1/4 head	Lettuce
1/2 slice	Fruit bread
Small branch	Gum or fig leaves

Method of preparation: Fruit and vegetables chopped into small pieces.

Institution: Berlin Zoo, Berlin (west), Germany
Diet recipient: P. maculatus
Diet: Oak, apple, and linden leaves were fed as well as leaves of *Robinia pseudacacia.* White bread soaked in milk, orange, apple, cherries, strawberries, banana, and chopped heart were also eaten, as were egg, mealworms, crickets, beetles, and white mice.[21]

Genus *Trichosurus:* Brush-tailed Possum

Trichosurus vulpecula (Type Species) and T. canis

Natural diet – Brush-tailed possums are omnivorous, feeding on flowers, shoots, leaves, fruits, and seeds of certain trees and shrubs as well as insects and probably small birds.

Synthetic diet

Institution: Westbury Zoo, Tasmania
Diet recipient: T. vulpecula
Diet: Rose, gum and dock leaves, grain, bread, apples, bran, vegetables, sugar cubes, and rose buds (when available) are fed ad lib.

Institution: Artis Zoo, Amsterdam, The Netherlands
Diet recipient: T. vulpecula
Diet: Specimens are maintained on the same diet that is fed to *Potorous* at this institution (see the section below on Macropods).

Institution: Melbourne Zoo, Melbourne, Australia
Diet recipient: T. vulpecula
Diet:

Quantity	Ingredient
Ad lib	Leaves of rose, gum, and dock
1	Chopped apple
1	Chopped orange
1	Chopped cucumber

Feeding methodology: 1/2 lb of above diet per animal per day.

Institution: New York (Bronx) Zoo, New York, U.S.A.
Diet recipient: Trichosurus vulpecula
Diet:

Quantity	Ingredient
1	Apple
1	Orange
10	Grapes
1 1/2	Bananas
1/4	Sweet potato
2 leaves	Lettuce
2 leaves	Kale
6 oz (170.1 g)	Diet "M" (see earlier section on *Caluromysiops irrupta)*
1 oz (28.4 g)	Chopped horsemeat

Method of preparation: Fruit and vegetables chopped into small pieces.
Feeding methodology: Above amount per two animals per day.

Institution: Cologne Zoo, West Germany
Diet recipient: T. vulpecula
Diet:

Quantity	Ingredient
33%	Mixed seed (one part of each of the following: wheat, crushed corn, rolled oats)
33%	Fruits (apples, bananas, pears, grapes)
34%	Other (bread, carrots, lettuce, cauliflower, leaves)

Feeding methodology: Approximately 350 g of above diet per animal per day.

Genus *Acrobates:* Feather-tailed Glider

Acrobates pygmaeus

Natural diet – Troughton[22] believed that insects form the bulk of the feather-tailed glider's natural diet, with soft-bodied termites and white ants being favored. The nectar and probably the sweet exudates and blossoms from eucalyptus as well as other native trees are also consumed by this minute phalanger.

Following are three diets that have been fed to captive specimens at the institutions noted:

Institution: Healesville Sanctuary, Australia (during Fleay's directorship)
Diet: White sugar, honeyed bread and milk, fragments of hardboiled egg yolk, small termites, and meal grubs.

Institution: Taronga Zoo, Sydney, Australia

Diet:

Quantity	Ingredient
1/4	Banana
1/16	Apple
1/12	Orange
1/4	Carrot
Small piece	Fruit bread
4	Mealworms
2 dozen	Fly pupae
1/2 teaspoon	Honey
2 teaspoons	Lactogen
2 drops	Abdec (multivitamin drops)
	Moths, cicadas, cockroaches (when available)

Method of preparation: Fruit and vegetables are chopped fine and the mealworms are cut in half. The honey is poured over fruit and vitamin drops are added. The above amount is fed per animal per day.

Institution: Melbourne Zoo, Melbourne, Australia

Diet: One teaspoonful of the phalanger mixture (see *Petaurus breviceps,* below) is fed per animal per day.

Genus *Burramys:* Broom's Possum

Burramys parvus

Natural diet – Calaby, Dimpel, and Cowan[26] report that the feces from two specimens trapped in the wild consisted predominantly of plant material with some insect remains.

Synthetic diet – Feeding trials on three captive animals indicated that they preferred fruits and seeds, although they also ate some insect matter.[26] The first specimen taken captive in 1966 fed on mealworms and other insect larvae, beetles, moths, chopped fresh steak, sunflower seeds, and apples and other fruits.

Genus *Distoechurus:* Pen-tailed Possum

Distoechurus pennatus

Natural diet – *Distoechurus* reportedly feeds on fruits and blossoms in the wild.[2]

Synthetic diet – Graeme George (personal communication) reports that specimens housed at the Baiyer River Sanctuary, Papua, New Guinea were fed a diet consisting of bananas, pawpaw, dilute skim milk/honey, and mealworms. George suggests that these phalangers might be maintained on the following diet fed to *Dactylopsila* sp. by Mrs. Margaret Mackay of Port Moresby: The mixture is made by cooking up a cupful of oats to form a porridge and adding half a cup of milk and quantities of "Bonox" (a beef extract), "Prolac" (a milk-based infant food high in protein), glucose powder, and honey. This is frozen in small containers for use as required.

Genus *Cercartetus:* Pygmy Possums

Cercartetus nanus (Type Species)

This group also includes *C. caudatus, C. concinnus,* and *C. lepidus.*

Natural diet – Pygmy possums are omnivorous, feeding on various insects and their larvae as well as other arthropods such as spiders and scorpions. Small lizards are also eaten. A variety of fruits, nuts, nectar, and leaves are consumed by these small phalangers. Hickman and Hickman[23] report that their Tasmanian specimens of *C. nanus* showed dietary preferences for insect food. Following are diets that have been fed to pygmy possums at the institutions noted:

Institution: Miss S. D. Bocking, private owner (Reference 22, p. 88)
Diet recipient: C. nanus
Diet: Fresh gum leaves, banksias, and all wild flowers in season, berries (such as blackberries and strawberries), grapes, various fruit juices, and insects such as grasshoppers, moths, flying cockroaches, and white earth-grubs were consumed. Tarantulas and black spiders were eaten, as were rolled oats, bird seed, almonds, honey, and sugar water.

Troughton[22] reports that pygmy possums at Taronga Zoo, Sydney were maintained on insects, fruit, nuts, sponge cake, biscuits, rose leaves, and gum blossoms.

Institution: Melbourne, Zoo, Melbourne, Australia
Diet recipient: C. nanus
Diet: Apple and orange are fed ad lib. In addition, each animal receives five to eight mealworms daily as well as 1 teaspoon of the "phalanger mixture" fed to *Petaurus* (see below).

Institution: University of Tasmania[23]
Diet recipients: C. nanus and *C. lepidus*
Diet: Cicadas; the common mantis (*Orthodera ministralis*); termites (*Porotermes adamsoni*); and both adults and larvae of various coleoptera, especially those of the families Elateridae, Scarabaeidae, and Cerambycidae were eaten, as were many Lepidoptera. Scorpions (*Cercophonius squama*) as well as the large spider (*Delena cancerides*) were consumed, if previously killed. Lizards (*Lygosoma metallicum* and *L. trilineatum*) were caught and eaten.
"An artificial food on which the animals thrived consisted of lightly baked biscuits made from a mixture consisting of self-rising flour (350 g), corn flour (70 g), cane sugar (220 g), butter (110 g), salt (1.5 g), milk (20 ml), and one egg."[23]
Each animal consumed a quantity of food daily equal to approximately 7% of its body weight.

Genus *Gymnobelideus:* Leadbeater's Possum

Natural diet – Little information is available. The stomach contents of one specimen revealed that it had been eating beetles and crickets.[24]

Synthetic diet – No information. The diets fed to *Petaurus* sp. (see below) should be apropos for *Gymnobelideus.*

Genus *Petaurus*

Petaurus australis (Type Species), P. breviceps, and P. norfolcensis

Natural diet – In its natural habitat, *P. australis* feeds on blossoms, nectar, and the sweet sap of certain eucalypt trees (e.g., *E. viminalis* and *E. ovata*). These phalangers have also been observed feeding on clumps of flowering mistletoe (*Loranthus*). *P. norfolcensis* and *P. breviceps* exhibit decidedly more omnivorous dietary preferences than does *P. australis.* These two species of lesser gliders feed on insects and their larvae, arachnids, and small vertebrates such as mice and birds, as well as sweet sap, blossoms, and buds of eucalypts and shrubs.

Synthetic diet

Institution: Fleay's Fauna Sanctuary, Queensland, Australia
Diet recipient: P. australis
Diet: Bread and milk with melon jam, gum blossoms, honey, and slabs of sapwood from preferred eucalypt trees.

Institution: Melbourne Zoo, Australia
Diet recipient: P. breviceps
Diet:

Quantity	Ingredient
Ad lib	Apple
Ad lib	Orange
6–10	Mealworms
2 teaspoons	Phalanger mixture (see below)

Phalanger mixture

Quantity	Ingredient
3 dessert spoons	Honey
3 dessert spoons	Glucose
3 dessert spoons	Complan
1 teaspoon	Bonox
1 teaspoon	Horlicks®
6 drops	Pentavite

Method of preparation: Above ingredients are mixed with milk to liquid consistency in blender. This mixture may be refrigerated (not frozen).

Specimens of *P. breviceps* housed at the U.S. National Zoological Park, New York (Bronx) Zoo, and Chester (England) Zoo receive a wide variety of fruits, grain, and vegetables. In addition, those housed at the National Zoological Park are fed (ad lib) peanuts, Fleishmann's® wheat germ toasted with honey and sugar, bread, diced hardboiled eggs, crickets, mealworms, cicadas, hairless mouse pups, and (twice per week) one teaspoon of the dasyurid meat mix. Specimens housed at the New York (Bronx) Zoo receive liver, chopped horsemeat, mealworms, crickets and Diet "M" (2 oz – 56.7 g). Animals at the Chester (England) Zoo receive minced meat, bread, and pelleted monkey food.

Institution: Fleay's Fauna Sanctuary, Queensland, Australia
Diet recipient: P. norfolcensis
Diet: Boiled bread and milk, sweet jam, strips of fatty meat, the larvae of longicorn beetles, and a variety of insects as well as occasional small birds and mice.

Genus *Dactylopsila:* Striped Possum

Dactylopsila trivirgata (Type Species) and D. megalura

Natural diet – Striped possums are omnivorous, feeding on fruits, leaves, and (mainly) insects and their larvae. Small vertebrates (i.e., birds, lizards, and rodents) are probably also eaten, as a captive specimen was known to have killed and eaten mice, leaving only the skins.

Synthetic diet – Troughton[25] reports that a striped possum at Taronga Zoo, Sydney, was maintained on a variety of foods, including condensed milk, biscuits, fruit, lettuce, honey, and insects. Graeme George (personal communication) reports having maintained these possums "with reasonable success" on a diet consisting of fruit (banana and pawpaw), dilute skim milk/honey, raw egg/milk mixtures, and, when available, large wood-boring insect larvae. He also maintains them on the prepared diet fed to *Distoechurus pennatus* (see above) and reports that striped possums have lived in captivity for "considerable periods" on this diet.

Institution: Healesville Fauna Sanctuary (during Fleay's directorship, 1942), Australia
Diet recipient: D. trivirgata
Diet: A variety of fruits, eucalypt blossoms, coconut, jam, milk, bread and honey, insects and their larvae (especially the large "corrugated" boring larvae of longicorn beetles), and mice are fed.

Institution: New York (Bronx) Zoo, New York, U.S.A.
Diet recipient: D. trivirgata
Diet:

Quantity	Ingredient
1/8	Apple
1/5	Banana
2	Grapes
1/8	Orange
1 leaf	Kale
1/2 oz (14.2 g)	Sweet potato
20	Mealworms
20/week	Crickets
1 oz (28.4 g)	Liver
1 oz (28.4 g)	Chopped horsemeat
2 oz	Diet "M"

Feeding methodology: Feed above amount per animal per day.

Genus *Dactylonax:* Long-fingered Possum

Dactylonax palpator

Natural diet – No information is available, but the diet is probably similar to that of *Dactylopsila.* Van Deusen (personal communication) reports having taken a specimen of *Dactylonax* from a tree, the trunk of which was heavily scarified from the incisor bites of this phalanger.

Synthetic diet – Synthetic diets listed for *Dactylopsila* sp. should be apropros for *Dactylonax palpator.*

Genus *Wyulda:* Scaly-tailed Possum

Wyulda squamicaudata

Natural diet – Scaly-tailed possums are probably omnivorous. One was seen feeding on the blossoms of a bloodwood tree. Natives claim they eat leaves and insects but mostly the fruit of a kind of sandalwood tree that grows among the rocks in which they seek shelter. Butler[27] indicates that one specimen possibly ate the body of a trapped granite rat.

Synthetic diet – A young specimen maintained by Fry[28] has thrived on a mixture consisting of equal parts of rolled oats and crushed nuts, with a small amount of Farex (baby food) mixed with enough honey for a very firm consistency. Eucalypt blossoms (*E. wandoo, E. calophylla, E. todtiana, E. macrandra,* and particularly *E. lehmannii*) are eaten, as are blossoms of *Grevillea, Callistemon, Melaleuca, Hypocalymma, Hibbertia, Calothamnus,* and *Feijoia;* roses are also eaten. Fruits (banana, apple, feijoia, orange, cherry, lilly-pilly, loquat, grapes, pomegranate, and glace cherries) as well as nuts (brazil, barcelona, cashew, pistachio, almond, and walnut) are also eaten by Mrs. Fry's animal. Occasionally, large moths are eaten.

Genus *Tarsipes:* Honey Possum

Tarsipes spencerae

Natural diet – In their natural habitat, honey possums feed on nectar, pollen, and soft-bodied insects. One observer believes that they move about in colonies, their movements being directed by the acceptable plants that are in bloom at the time. Preferred native flowers include banksia, bottlebrushes, hakeas, drandras, and those of eucalypt and Ti trees.

Synthetic diet – Sugar, watered honey, sopped bread, flies, and sprays of acceptable wild flowers were fed to an animal maintained by Glauert at the Western Australian Museum.[29] The phalanger mixture listed previously might be acceptable to these minute nectar feeders.

Genus *Phascolarctos:* Koala

Phascolarctos cinereus

Natural diet – Koalas are highly specialized feeders subsisting on the leaves and shoots of a variety of eucalypt trees, including the following species: manna gum (*E. viminalis*), river red gum (*E. rostrata*), swamp gum (*E. ovata*), yellow box (*E. melliodora*), tallow wood (*E. microcorys*), spotted gum (*E. maculata*), messmate (*E. obliqua*), peppermint gum (*E. radiata*), mahogany (*E. botryoides*), forest red gum (*E. tereticornis*), Sydney blue gum (*E. saligna*), and brush box (*Tristania conferta*). Individual animals consume approximately 2½ lb (1.1 kg) of leaves per day. Captive specimens should have access to leaves from a variety of acceptable food trees.

These phalangers reportedly ingest earth to supplement dietary needs for calcium and other minerals.

Synthetic diet – At the Melbourne Zoo, leaves of the manna gum form the staple portion of the koala diet, with swamp gum and peppermint gum being fed occasionally.

Only mature leaves should be fed, as young leaves (which captives may be forced to eat as a matter of necessity) contain large amounts of prussic acid. Upon mastication of the leaves, this acid may be released in quantities lethal to the animal.

Captive specimens have been known to eat bread sopped with milk and honey in addition to their natural diet.

Genus *Pseudocheirus:* Ring-tailed Possum

Pseudocheirus peregrinus (Type Species)

Also included are *P. albertissi, P. archeri, P. cancescens, P. caroli, P. (Pseudochirops) corinnae, P. (Pseudochirops) cupreus, P. (Petropseudes) dahli, P. forbesi, P. herbertensis, P. (Hemibelideus) lemuroides, P. pygmaeus,* and *P. schlegeli.*

Natural diet – Ring-tailed possums feed on a variety of vegetation in the wild, including the leaves of *Eucalyptus* sp. and the leaves, shoots, and blossoms of other native and introduced plants. Although these possums are regarded as one of the strictest herbivores among the Phalangeridae, Crandall[9] reports that a Victorian ring-tailed possum maintained at the New York (Bronx) Zoo in 1953 ate meat mix as part of its daily diet.

Synthetic diet – At the Melbourne Zoo, specimens of *P. peregrinus* are fed eucalypt leaves, dock, and rose leaves, as well as chopped apple and cucumber.

Genus *Schoinobates:* Greater Glider

Schoinobates volans

Natural diet – Greater gliders are specialized in their dietary habits, feeding mainly on the tender leaves, shoots, and blossoms of preferred eucalypts. Fleay[30] observed, "In this respect it resembles the koala for remarkable to relate, this largest of the gliders is the only one that is a leaf eater, and unless able to get a regular variety of eucalypt leaves according to its own fastidious requirements, it simply will not live."

Synthetic diet – Fleay[30] found that captive specimens preferred the leaves of the following eucalypts (in descending order of preference): common peppermint (*E. australiana*), long-leaved box (*E. elaeophora*), swamp gum (*E. ovata*), manna gum (*E. viminalis*), apple box (*E. stuartiana*), mountain gray gum (*E. goniocalyx*), river red gum (*E. rostrata*) and several species of stringy bark. Troughton[22] reports greater gliders feeding on apple tree leaves and blossoms. In addition, captive specimens may be persuaded to acquire a taste for bread and milk spread with honey or jam (melon and lemon are favorites), but this can only serve as a supplement to the staple diet of eucalypt leaves. *Food should be placed near the top of the enclosure, as greater gliders do not like to descend to the lower regions of their quarters.*

Following is a diet fed to *Schoinobates* at Taronga Zoo, Sydney: A variety of gum leaves and blossoms, bottle brush, wattle, and other native foliage are fed. In addition, each animal receives approximately one-half sliced banana, one chopped apple, one-fourth slice of fruit bread, and one-half sliced carrot. The gum leaves and other foliage are placed in a tub of wet sand to retain freshness. Fruit and vegetables are fed in a dish.

FAMILY PHASCOLOMYIDAE – WOMBATS

Genus *Vombatus:* Naked-nosed Wombat

Vombatus ursinus

Natural diet – Naked-nosed wombats are herbivorous, their diet consisting of grasses, bark, roots, and fungi.

Synthetic diet – Specimens housed at the U.S. National Zoological Park, Cologne Zoo, Melbourne Zoo, and the Wellington Zoo are fed a variety of grasses (e.g., lucern hay and grass), grains (crushed corn, sunflower seeds, wheat, and oats), vegetables (sweet potatoes, assorted greens, and carrots), and fruits (bananas, apples, and pears). In addition, those housed at the U.S. National Zoo receive bread and Omolene®, and Melbourne Zoo specimens are fed calf-growers mash.

Genus *Lasiorhinus:* Hairy-nosed Wombat

Lasiorhinus latifrons (Type Species), *L. barnardi,* and *L. gillespiei*

Natural diet – Hairy-nosed wombats apparently have dietary habits similar to the naked-nosed wombats.

Synthetic diet – Crowcroft (personal communication) maintained captive specimens of *L. latifrons* on a diet of commercial race-horse chow. The animals apparently thrived on this diet and bred in captivity. The diets fed to captive naked-nosed wombats would be apropos for *Lasiorhinus* sp. as well. The following diet is fed to a hairy-nosed wombat at the Budapest Zoo:

Diet recipient: L. latifrons

Diet:

Quantity (g)	Ingredient
50	Barley corn
50	Corn
50	Sunflower seeds
50	Cooked potato
100	Carrots
100	Seasonal fruits

FAMILY MACROPODIDAE – MACROPODS

Genus *Hypsiprymnodon:* Musky Potorine

Hypsiprymnodon moschatus

Natural diet – Musky potorines are omnivorous, including tubers and other plant matter as well as insects and worms in their natural diet. Berries of the palm *Ptychosperma alexandrae* are reportedly eaten. Food is apparently obtained in the same manner as that exhibited by *Perameles,* i.e., by foraging amid ground debris and by digging.

Synthetic diet – A variety of chopped vegetables and fruits as well as a meat mix such as that fed to carnivorous marsupials at the National Zoological Park should provide an adequate diet for captive specimens. In addition, hairless mouse pups, skinned adult mice, crickets, and earthworms should be offered as possible dietary supplements.

Genus *Aepyprymnus:* Rufous Potorine

Aepyprymnus rufescens

Natural diet – In their natural habitat, rufous potorines feed mainly on plant roots but also include other herbage in their diet. They are thought to be strictly herbivorous.

Synthetic diet – A diet consisting of various greens (e.g., kale, turnip greens, and carrot tops), chopped vegetables and fruits, mixed grain, bread, fresh and pelleted alfalfa, as well as a variety of dry commercial pelleted feed (such as rat, rabbit, or sheep chow) should be appropriate for captive specimens of *Aepyprymnus* as well as most other macropods. Animals maintained in outdoor paddocks may supplement their diets with grasses and other native plants. Calcium diphosphate, multivitamin drops, and other nutritional supplements should be added to the daily diet.

Genus *Bettongia:* Bettong

Bettongia gaimardi (Type Species)

Also included are *B. lesueur, B. penicillata,* and *B. tropica.*

Natural diet – Although most species of *Bettongia* are mainly herbivorous, feeding on roots, tubers, seeds, and legume pods, Troughton[22] reports that the typical race of *B. lesueur,* residing on the islands of Shark Bay, was observed to "burrow amongst the cliffs of the seashore, and to feed mainly upon marine refuse; even dead sheep were occasionally partly eaten." Ride[24] reports that captive specimens are fond of meat and gnaw on bones.

Synthetic diet – For the herbivorous species, a diet such as that suggested for *Aepyprymnus* should be apropos. This diet should contain a meat supplement for specimens of *B. lesueur.*

Genus *Caloprymnus:* Desert Potorine

Caloprymnus campestris

Natural diet – Desert potorines are thought to be strictly herbivorous, feeding on the foliage and stems of plants. Roots are thought to compose a smaller percentage of the diet for this species than for some of the other potorines.

Synthetic diet – A diet such as that discussed for *Aepyrpymnus* should be apropos for this species.

Genus *Potorous:* Potoroo

Potorous tridactylus (Type Species)

Also included are *P. apicalis* and *P. platyops.*

Natural diet – Potoroos are omnivorous, feeding on a wide variety of food, including insects, snails, grasses, plant roots, sedges, cranberries, and other food material. Guiler[31] notes that the diet includes significant proportions of insect matter during the spring and early summer. Fungi apparently form a large part of the diet throughout the year.

Synthetic diet – Several feeding loci should be provided in an enclosure containing a group of animals, since potoroos exhibit a well-developed social hierarchy and low-ranking animals are liable to suffer if only one feeding locus is used.[32] The following diets are fed to captive potoroos at the institutions noted:

Institution: National Zoological Park, Washington, D.C., U.S.A.
Diet recipient: P. tridactylus
Diet: Variety of chopped fruits and vegetables.

Institution: University of Tasmania
Diet recipient: P. apicalis
Diet: Ground dehydrated meat, wheat biscuit waste, and fruit (mainly apples) composed the diet of a breeding colony.

Institution: Artis Zoo, Amsterdam, The Netherlands
Diet recipient: P. tridactylus
Diet:

Quantity	Ingredient
40% by weight	Fruit (apple, banana, and orange)
30% by weight	Minced meat
30% by weight	Monkey pellets (protein 19.5%, fat 5%, carbohydrates 57.5%, ash 6%, water 12%)

Genus *Lagorchestes:* Spectacled Wallaby

Lagorchestes leporides (Type Species)

Also included are *L. asomatus, L. conspicillatus,* and *L. hirsutus.*

Natural diet – Spectacled wallabies are herbivorous, feeding on various grasses and herbs.

Synthetic diet – A colony of spectacled wallabies at Alice Springs, Northern Territory, Australia was maintained on a diet consisting of lettuce and other greens, wheat grain, and commercial rabbit food. Pentavite (a multivitamin supplement) was added to the drinking water.

Genus *Lagostrophus:* Banded Gray Wallaby

Lagostrophus fasciatus

Natural diet – Banded gray wallabies are herbivorous, reportedly feeding on grasses, herbs, and fruits in the wild.

Synthetic diet – The diet suggested for captive rufous potorines should be apropos for this species.

Genus *Onychogalea:* Nail-tailed Wallaby

Onychogalea unguifer (Type Species) and O. fraenata

Natural diet – Nail-tailed wallabies feed on plant roots, various coarse grasses, and other vegetation.

Synthetic diet – The diet suggested for captive specimens of *Aepyprymnus* should be apropos for *Onychogalea.*

Genus *Petrogale:* Rock Wallaby

Petrogale penicillata (Type Species)

Also included are *P. brachyotis, P. godmani, P. purpureicollis, P. rothschildi,* and *P. xanthopus.*

Natural diet – Rock wallabies are herbivorous, feeding on grasses, roots, and bark (the latter two items supply much needed moisture during droughts).

Synthetic diet – Specimens of *P. penicillata* maintained at the Wellington and Cologne zoos are fed a variety of chopped vegetables, fruits, bread, hay, and commercial dry pelleted food. Specimens at the Cologne Zoo receive rock salt and those at the Wellington Zoo receive dicalcium phosphate in addition to their daily diets.

Genus *Peradorcas:* Little Rock Wallaby

Peradorcas concinna

Natural diet – No information is available but this diet is probably similar to that of *Petrogale.*

Synthetic diet – No records. The diet listed for captive specimens of *Aepyprymnus* should be apropos for little rock wallabies.

Genus *Dendrolagus:* Tree Kangaroo

Dendrolagus ursinus (Type Species)

Also included are *D. bennettianus, D. dorianus, D. lumholtzi,* and *D. matschiei.*

Natural diet – Tree kangaroos are herbivorous, feeding on various fruits, leaves, creepers, and ferns.

Synthetic diet – The following dietary components are compiled from diets fed to specimens of *Dendrolagus* maintained at Taronga (Sydney) Zoological Park, Melbourne Zoological Gardens, Cologne Zoo, and the National Zoological Park, Washington, D.C.: lucerne hay, green lucerne, mixed fruits (apple, pear, banana), mixed vegetables (carrot, sweet potato, and sweet corn on the cob sans husk), mixed grain (wheat, corn) as well as poplar, willow, fig, and fruit tree leaves, and bread. In addition, specimens of *D. ursinus* (only!) at Taronga (Sydney) Zoo receive cucumber. Specimens of *D. matschiei* at the Melbourne Zoo receive calf-growers mash and one hardboiled egg daily. Salt is available to specimens at both the Taronga and Cologne zoos.

One observer noted that a collection of captive tree kangaroos in New Guinea consumed "immense quantities" of fresh carrot tops, sweet potato vines, and native vegetation. Crandall[9] observes, "It is impossible to escape the conclusion that, like other leaf-eating species, the tree kangaroos require greater food bulk than that afforded by the comparatively concentrated diets usually provided." George (personal communication) reports that specimens maintained at the Hallstrom Park Bird of Paradise Sanctuary in New Guinea are maintained on a diet consisting of sweet potato leaves (the basic component); young shoots of *Schurmansia,* a cress-like celery native to New Guinea; the shoots of two cultivated New Guinea vegetables (*Saccharum* and *Setaria*); raw sweet potato tubers; green bananas in the skin; fresh corn (maize); and the terminal flowering bracts of the banana bunch. The diet is supplemented occasionally with a piece of fresh clayish soil which the tree kangaroos reportedly eat with relish. Every 3 months or so they are given access to a cattle salt lick. Animals that are "out of condition" receive bread in addition to the regular diet.

Genus *Dorcopsis:* New Guinea Forest Wallaby

Dorcopsis muelleri (Type Species)

Also included are *D. atrata* and *D. hageni.*

Natural diet – New Guinea forest wallabies are herbivorous, reportedly feeding on roots, leaves, and grasses.

Synthetic diet – A diet such as the one discussed for *Aepyprymnus* should be apropos for captive specimens of *Dorcopsis.* Those maintained at the Hallstrom Park Bird of Paradise Sanctuary in New Guinea receive a diet similar to that fed to *Dendrolagus.*

Genus *Dorcopsulus:* New Guinea Mountain Wallaby

Dorcopsulus macleayi (Type Species) and D. vanheuri

Natural diet – *Dorcopsulus* is herbivorous, feeding on leaves, roots, and various fruits.

Synthetic diet – A diet such as that discussed for *Aepyprymnus* should be apropos for New Guinea mountain wallabies, with succulent vegetables and fruits emphasized.

Genus *Thylogale:* Pademelon

Thylogale thetis (Type Species)

Also included are *T. billardieri, T. bruijni,* and *T. stigmatica.*

Natural diet – Pademelons feed on leaves, grasses, and shoots.

Synthetic diet – Specimens of *T. billardieri* at the Chester (England) Zoo receive a concentrate mixture consisting of flaked maize, crushed oats, bran, soybean meal, fish meal, and linseed flakes. This mixture is made up in bulk and lasts several weeks. In addition, a supplement of bread, banana, apple, and carrot is also fed. The diet discussed for captive specimens of *Aepyprymnus* should be apropos for pademelons.

Genus *Setonix:* Quokka

Setonix brachyurus

Natural diet – Quokkas are herbivorous, feeding on grasses and other plants.

Synthetic diet – A breeding colony at the University of Western Australia was maintained on a diet consisting of green herbage, chaff, hay, and commercial sheep pellets. The diet outlined for captive specimens of *Aepyprymnus* should also be appropriate for quokkas.

Genus *Wallabia:* Scrub Wallaby

Wallabia bicolor (Type Species)

Also included are *W. agilis, W. dorsalis, W. eugenii, W. greyi, W. irma, W. parma, W. parryi,* and *W. rufogrisea.*

Natural diet – Scrub wallabies are herbivorous, browsing and grazing on grasses and other plants.

Synthetic diet – A diet such as the one outlined for *Aepyprymnus* should be apropos for these marsupials.

Nakazato, Nakayama, and Nakagawa[33] report the successful handrearing of *W. agilis* pouch young on the following formula: Esbilac® (Borden Dairy Co., New York), 10 g; warm water, 100 ml; multivitamin, 0.1 ml. Orphaned pouch young were stimulated to urinate and defecate by digitally massaging their cloacae. Diarrhea was controlled through the administration of Torulamin (a rumen function stimulant manufactured by Toshiba Pharmaceutical Co., Ltd., Kawasaki, Japan).

Wilson[34] reports the successful handrearing of an orphaned Dama wallaby (*W. eugenii*) on this formula: pasteurized cow's milk (1/2 cup), Abdec multivitamin drops (two drops), and glucose D (1/2 teaspoon). The animal weighed 1 lb and sat 9 in. tall when it was first acquired. It was fully developed and furred.

Genus *Megaleia:* Red Kangaroo

Megaleia rufa

Natural diet – Red kangaroos are grazers, feeding mainly on green grasses and short herbs.

Synthetic diet – Most zoos maintaining these animals provide them with a variety of the diet components described earlier for captive specimens of *Aepyprymnus.* The Melbourne Zoo raises the protein content of their macropod diets by the addition of soybean meal. They find that some species of macropods, especially *Macropus major fuliginosis,* do better when the protein level is raised. Dicalcium phosphate is added to the red kangaroo diet at the Wellington Zoo. Those at Taronga (Sydney) Zoo have access to salt licks, and some are given powdered milk in liquid form.

Calaby and Poole[35] report the successful handrearing of young kangaroos (which have reached about 6 lb in weight) on a diet consisting of fresh cow's milk or a suspension of powdered milk, 1 cup; prepared infant cereal (Farex, manufactured by Glaxo-Allenburys), 1 teaspoon; glucose (Glucodine, manufactured by Glaxo-Allenburys), 1 teaspoon added during one daily feeding only; and a few drops of multivitamin formula (Pentavite, manufactured by Nicholas), added during one daily feeding only. Feeding is done every 2 hr during the day, depending on consumption, as well as a late night and early morning feed and one during the night if required. Kaomagma® (manufactured by Wyeth Pharmaceuticals) is administered to relieve diarrhea, as most young animals suffer from this disorder following a change in their diet.

Genus *Macropus:* Forester Kangaroo

Macropus giganteus (Type Species)

Also included are *M. antilopinus, M. bernardus, M. fuliginosus,* and *M. robustus.*

Natural diet – Free-living specimens of *M. giganteus* eat grass like the red kangaroos but usually eat fewer herbs than that species. *M. robustus* feeds on spinifex, soft grasses, and herbs.

Synthetic diet – Analyses of diets fed to forester kangaroos at the following zoos indicate that these animals receive essentially the same diet fed to red kangaroos: Budapest, Zurich, Roeding Park (California), New York (Bronx), Cologne, Taronga (Sydney), and Melbourne.

REFERENCES

1. **Altmann, D.,** *Zool. Gart.,* 35(1, 2), 22–29, 1968.
2. **Walker, E.,** *Mammals of the World,* 3rd ed., Johns Hopkins University Press, Baltimore, 1975.
3. **Mann, G.,** *Invest. Zool. Chil.,* 4, 209–213, 1958.
4. **Greer, J.,** *Publ. Mus. Mich. State Univ. Biol. Ser.,* 3(2), 53–151, 1965.
5. **Cabrera, A. and Yepes, J.,** *Mamiferos Sud Americanos,* 2nd ed., Ediar, Buenos Aires, 1960.
6. **Enders, R.,** *Bull. Mus. Comp. Zool. Harv. Coll.,* 78, 4, 1935.
7. **Gewalt, W.,** *Zool. Gart.,* 37(4, 5), 248–253, 1969.
8. **Davis, J., Jr.,** *Anim. Kingdom,* 69(4), 112–117, 1966.
9. **Crandall, L.,** *The Management of Wild Mammals in Captivity,* University of Chicago Press, Chicago, 1964.
10. **Marlow, B.,** *Aust. J. Zool.,* 9(2), 203–218, 1961.
11. **Woolley, P.,** *J. R. Soc. West. Aust.,* 54(4), 99–102, 1971.
12. **Fleay, D.,** *Victorian Nat.,* 82, 195–204, 1965.
13. **Heinsohn, G.,** *Animals,* 13(5), 220–222, 1970.
14. **Sorenson, M.,** *J. Mammal.,* 51(1), 123–130, 1970.
15. **Gewalt, W.,** *Zool. Gart.,* 32(3), 99–115, 1966.
16. **Marlow, B.,** *Marsupials of Australia,* Jacaranda Press, Brisbane, Australia, 1965.
17. **Heinsohn, G.,** *Univ. Calif. Berkeley Publ. Zool.,* 80, 1–96, 1966.
18. **Stodart, E.,** *Aust. J. Zool.,* 14, 611–623, 1966.
19. **Jones, F.,** *The Mammals of South Australia,* Part II, Government Printer, Adelaide, Australia, 1924, 132–270.
20. **Osgood, W.,** *Field Mus. Nat. Hist.,* 14(1), 1–156, 1921.
21. **Gewalt, W.,** *Zool. Gart.,* 28(5), 213–225, 1964.
22. **Troughton, E.,** *Furred Animals of Australia,* Angus and Robertson, Sydney, Australia, 1967.
23. **Hickman, V. and Hickman, J.,** *Proc. Zool. Soc. London,* 135, 365–374, 1960.
24. **Ride, W.,** *A Guide to the Native Mammals of Australia,* Oxford University Press, Melbourne, Australia, 1970.
25. **Troughton, E.,** *Aust. Mus. Mag.,* 7(12), 431–434, 1942.
26. **Calaby, J. H., Dimpel, H., and Cowan, I.,** Division of Wildlife Research Tech. Paper No. 23, Commonwealth Scientific and Industrial Research Organization, Australia, 1971.
27. **Butler, W. H.,** *Nature Sci.,* 4(7), 3–6, 1966.
28. **Fry, E.,** *Int. Zoo Yearb.,* 11, 44–45, 1971.
29. **Glauert, L.,** *Aust. Mus. Mag.,* 12, 284–285, 1958.
30. **Fleay, D.,** *Gliders of the Gum Trees: The Most Beautiful and Enchanting Australian Marsupials,* Melbourne Bread and Cheese Club, Melbourne, Australia, 1947.
31. **Guiler, E.,** *J. Mammal.,* 52(1), 232–234, 1971.
32. **Guiler, E.,** *Int. Zoo Yearb.,* 11, 21–22, 1971.
33. **Nakazato, R., Nakayama, T., and Nakagawa, S.,** *Int. Zoo Yearb.,* 11, 13–16, 1971.
34. **Wilson, P.,** *Int. Zoo Yearb.,* 11, 20, 1971.
35. **Calaby, J. and Poole, W.,** *Int. Zoo Yearb.,* 11, 5–12, 1971.

FOODS OF WILD AND CAPTIVE INSECTIVORA

G. L. Dryden

The Order Insectivora is composed of structurally diverse but poorly understood small mammals living in a wide variety of habitats. Most of those studied are aggressive, opportunistic hunters and clearly do not restrict their diet to insects. Their principal feeding strategy seems simple: Animals small enough to be overpowered or subdivided are consumed. Prey includes most terrestrial and fresh-water invertebrates, but small vertebrates and carrion are also taken. There is growing realization that insectivores deliberately eat seeds and other plant parts, at least during prey shortfall, though this aspect of their biology remains controversial. Many species in captivity also eat chicken eggs, cow milk, cereal-based mixtures, and other items not regularly encountered in natural conditions. The most realistic conclusion, reflected by the data below, is that insectivores, like many other mammals, have food preferences but eat what is available.

Published data on insectivore diets are very uneven. Most information is derived from examination of wild animals' stomach contents. Exhaustive food preference tests (noted in tabular data) have been conducted on some soricids, less extensively on talpids and erinaceids. Observations on other captive animals confirm the catholic tastes of numerous species. Fecal analyses and direct observation of wild animals feeding add to the available information. Relatively few species have been maintained for prolonged periods on synthetic diets, and when this has been done, little evidence suggests that the formulations used are critical.

Many references give only common or general names of foods taken by insectivores while others specify binomials of consumed organisms. Thus, considerable, probable redundancy must be tolerated by the reader in the interest of comprehensive listings of the various food taxa. An attempt has been made here toward broad coverage of insectivores, and, since mammalian systematics are labile, a few forms may be dually listed. The data emphasize American and European species to the relative neglect of African and Asian forms. This reflects both the preponderance of studies done in America and Europe as well as the difficulty in retrieving literature, especially the Asian, which, when available, treats forms from interesting habitats. On balance, information about food is presented for only a fraction of the many described insectivore species. However, the reader should be able to devise for his purposes a reasonably acceptable insectivore diet. Documentation of foods eaten by additional Eurasian species can be found in Ognev's classic work on the mammals of that region.[1] Another author lists food of various forms.[2] Many accounts cited below include as minor, perhaps incidental, constituents of insectivore diets the animal, plant, and mineral particles known to have been eaten. Readers requiring specific identification of these or an indication of their quantity in the respective diets should consult the papers cited for this information, where available. Trap baits, often exotic and not always essential for trap success, are excluded here. The data thus cover foods known to be eaten primarily by postweaned insectivores. In addition to mothers' milk it can be assumed also that other glandular secretions (e.g., saliva) and feces (own and parental) are potential foods of young insectivores, though consumption of these is poorly documented.

Acknowledgments

I thank A. Buchalczyk and Z. Pucek, Mammals Research Institute, Białowieża, Poland for literature and translation from Russian. J. Eisenberg, National Zoological Park, Washington, D.C.; S. Lahti, University of Helsinki; and J. Meester, University of Natal,

Pietermaritzburg, South Africa were also helpful in tracing literature. A. Kaikusalo, Ojajoki, Loppi, Finland kindly provided unpublished observations, as did the following who additionally helped with literature: R. Baxter and M. Kuyper, University of Natal, Pietermaritzburg; E. Gould, Johns Hopkins University, Baltimore; M. Hawes, University of British Columbia, Vancouver; and G. Rathbun, National Museum, Nairobi, Kenya.

Table 1
FOODS OF WILD (W) AND CAPTIVE (C) INSECTIVORES (LISTED ALPHABETICALLY)

Insectivore		Food	Ref.
		Chrysochloridae	
Amblysomus hottentotus			
	(W, C)	Earthworms; Coleoptera larvae	3, 4; 3
	(C)	Crickets, grasshoppers, snails; tenebrionid larvae, moths, winged termites, raw egg, rolled oats mixed into minced ox heart, mineral and vitamin supplements; locusts, cockroaches	3, 4; 3; 4
		Diet for babies: beaten raw egg, milk, and sugar	3
		Homeria sp. bulbs, potatoes	5
Amblysomus julianae			
	(C)	Ground beef, earthworms, cockroaches, meal worm larvae, grasshoppers, other insects	6
Amblysomus obtusirostris			
	(W)	Termites, tenebrionid larvae	7
Calcochloris obtusirostris			
	(W)	Beetles	8
Chrysochloris stuhlmanni			
	(W)	Snails, earthworms, dipteran and tipulid larvae, other insect larvae	9
Chrysospalax trevelyani			
	(W)	Developing insects, pill-millipedes	8
Eremitalpa granti			
	(W)	Termites, tenebrionid larvae, Lepidoptera, *Pseudococcus* sp.; *Typhlosaurus vermis*	10; 11
	(C)	Crickets, adult Dynastidae beetle abdomens, *Tenebrio molitor* larvae, *Palmatogecko rangei*	10
		Erinaceidae	
Atelerix frontalis			
	(C)	Insects, smoked fish, cheese, raw chopped ox heart	12
Echinosorex gymnurus			
	(W)	Earthworms; arthropods; fish, crabs	13–15; 13, 14; 13, 15
		Land mollusks; frogs, insects	13; 15
		Mature oil palm fruits	16
	(C)	Cockroaches, scarabid beetles, Lepidoptera larvae, millipedes, centipedes, spiders, scorpions, fresh-water crabs, fresh fish, white mice, larval caecilians (*Ichthyophis glutinosus*); snakes	13; 16

Table 1 (continued)
FOODS OF WILD (W) AND CAPTIVE (C) INSECTIVORES (LISTED ALPHABETICALLY)

Insectivore		Food	Ref.
		Erinaceidae (continued)	
Erinaceus dauuricus			
	(W)	Small rodents	17
Erinaceus europaeus			
	(W, C)	Adult, larval, and pupal insects, bird eggs	12
	(W)	Frogs and toads, lizards and snakes, birds, small mammals, carrion, snails, worms, wood lice, spiders, centipedes, millipedes, acorns, plums, pears, apples, berries, fungi	12
		Molluska, Annelida, Orthoptera, Dermoptera, Hymenoptera, Hemiptera, Lepidoptera, Diptera, Carabidae, Sylphidae, Elateridae, Tenebrionidae, Cerambycidae, Chrysomelidae, Curculionidae, miscellaneous beetles	18
		Chrysomyia maggots, *Musca* maggots, *Eristalis* maggots, earwigs, mole crickets, Coleoptera, sweet melon seeds, jujubes, miscellaneous vertebrate and vegetable remains	19[a]
	(C)	*Sorex, Neomys, Microtus*	20
		Milk; condensed milk	12, 21–24; 25
		Custard, bread in milk; for babies: milk, pears, earthworms, oatmeal and pastry; Pablum®, egg noodles, vitamin supplements	12; 24; 25
		Many types of meat; horsemeat; fresh herring	12; 21, 25; 22
		Boiled potatoes and rice, pears, plums, hazelnuts, groundnuts, sunflower seeds, hemp, raisins, other dried fruit, oil beetles (Melöe), cantherides beetles, bees, wasps, hornets	12
		Pelleted rat and chicken feed	26
		Bird eggs (in order of "palatability"): *Rissa tridactyla, Somateria mollissima, Alca torda, Sula bassana, Gallus gallus, Vanellus vanellus, Sterna hirundo, Fulica atra, Columbus palumbus, Athene noctua, Corvus frugilegus, Fratercula arctica, Pica pica, Corvus monedula, Falco subbuteo, Falco tinnunculus, Podiceps ruficollis, Buteo buteo, Sterna macrura, Charadrius hiaticula, Turdus merula, Fringilla coelebs, Acrocephalus schoenobaenus, Sylvia communis, Carduelis cannabina*	27
Erinaceus roumanicus			
	(W)	Insects	28
	(C)	Chicken, sparrows, smoked fish, insects, mealworms, cheese, milk	12

[a] Judged to be *E. europaeus* from locality caught.

Table 1 (continued)
FOODS OF WILD (W) AND CAPTIVE (C) INSECTIVORES (LISTED ALPHABETICALLY)

Insectivore		Food	Ref.
		Erinaceidae (continued)	
Hemiechinus auritus			
	(W)	Mammals, amphibians, *Bucephalus helicopris*, dung rollers and other beetles	29
	(C)	Beetles, crickets, locusts, mantis, Neuroptera, Lepidoptera, miscellaneous insect larvae, termites, dead scorpions, *Bufo arenarius, Rana tigrina, Uromastix hardwickii, Varanus monitor* viscera, *Eryx johnii, Ptyas mucosus, Columba livia,* pigeon eggs, *Rhinopoma kinneari,* goat meat (including lung and liver), cow and buffalo milk, other *Hemiechinus,* viscera of *Rattus rattus, Meriones hurrianae, Tatera indica, Funambulus pennanti, Lepus dayanus*	29
		Insects, chopped raw ox heart, smoked fish, cheese	12
Hylomys suillus			
	(W)	Insects	15
Paraechinus micropus			
	(W)	Mammals, amphibians, *Bucephalus helicopris*, dung rollers and other beetles	29
	(C)	Beetles, crickets, locusts, mantis, Neuroptera, Lepidoptera, miscellaneous insect larvae, *Bufo arenarius, Rana tigrina, Uromastix hardwickii,* viscera of *Varanus monitor* and *Columba livia,* pigeon eggs, *Rhinopoma kinneari,* goat meat (including liver and lung), cow and buffalo milk, other *Paraechinus,* viscera of *Rattus rattus, Meriones hurrianae, Tatera indica, Funambulus pennanti, Lepus dayanus*	29
		Raw horsemeat, Pablum®, egg noodles, condensed milk, vitamin supplements	25
		Milk, rice	30
		Macroscelididae	
Elephantulus intufi			
	(C)	Mealworms, locusts	31, 32
		Minced beef, chicken, *Trinervitermes* sp., Pro Nutro® (constituents/100 g): 22 g protein, 9 g fat, 57 g CHO, 1.4 g fiber, 4.5 g minerals (Ca, P, Fe, Cu, I, Zn), 3500 IU vitamin A, 5 IU vitamin E, 1 mg vitamin B_1, 1.5 mg vitamin B_2, 1 mg vitamin B_6, 0.5 mg vitamin B_{12}, 53 mg vitamin C, 1.5 mg pantothenic acid, 0.5 mg naicin	31
		Baby mice, Calliphoridae larvae, fruit, lettuce, hardboiled egg, Vionate® (mineral supplement)	32

Table 1 (continued)
FOODS OF WILD (W) AND CAPTIVE (C) INSECTIVORES (LISTED ALPHABETICALLY)

Insectivore		Food	Ref.
		Macroscelididae (continued)	
Elephantulus myurus			
	(C)	Locusts, mealworms	32, 33
		Same as in *E. intufi,* Reference 32 (above)	
		Rose beetles, other insects, minced meat, bacon, apple, carrot, cereal, other *Elephantulus*	33
Elephantulus rozeti			
	(C)	Same as in *E. intufi,* Reference 32 (above)	
Elephantulus rufescens			
	(W)	Ants, termites, vegetation; adult and larval beetles, cockroaches, spiders, centipedes, egg cases, winged ants, green plants and seeds; insects	34, 35; 34; 35
	(C)	Meat, mealworms, hardboiled eggs	35, 36
		Grasshoppers, termites, oatmeal and milk	35
		Insects, grapes, oranges, bananas, apples, clover, cabbage, lettuce, cottage cheese, nut meats, "bat diet": one part each yolk of hardboiled egg, ripe banana, mealworms, 6 drops each Jeculin® and wheat germ oil, 3 grains Theragran®	36
Petrodromus tetradactylus			
	(W)	Adult and larval beetles, ants, termites, grasshoppers, crickets, centipedes, millipedes, spiders, winged ants, egg cases, green plants, seeds	34
	(C)	Same as in *E. intufi,* Reference 32 (above)	
		Minced hardboiled egg, beef, invertebrates, apple, carrot, milk	34
Petrodromus torday			
	(C)	Hardboiled egg, milk, mixture: ground horsemeat, bone meal, cod-liver oil	37
Rhynchocyon chrysopygus			
	(W, C)	Adult and larval beetles, centipedes, millipedes, termites, cockroaches, crickets, ants, spiders, earthworms, pupal cases, fly larvae, winged ants	34
	(C)	Same as wild (above) plus locusts, geckos, skinks, colubrid snakes, milk	34
		Potamogalidae	
Mesopotamogale ruwenzorii			
	(C)	Tadpoles, small frogs, fish: *Clarias carsoni, Barbus setifer, Barbus pellegrini, Barbus portali,* fresh-water crabs (*Potamonautes biballensis*), Oligochaeta, adult and larval Hydrophilidae and Hemiptera, larval and pupal ephemeropterans	38
Potamogale velox			
	(C)	Crabs, mixture: dried fresh-water shrimp, minced meat, raw eggs	39

Table 1 (continued)
FOODS OF WILD (W) AND CAPTIVE (C) INSECTIVORES (LISTED ALPHABETICALLY)

Insectivore		Food	Ref.
		Solenodontidae	
Solenodon cubanus			
	(W)	Ants, grubs, insects, reptiles, poultry, fruit, vegetables	41[b]
	(C)	Hardboiled egg, raw egg, mealworms, *Anolis,* milk, lettuce, banana, mixture: ground horsemeat, bone meal, cod-liver oil	37
		White mice, centipedes, insect larvae, mixture: crickets, ground meat, egg yolk, canned milk, Pablum, banana, ABDEC® vitamin supplement	42
Solenodon poeyanus			
	(C)	Lizards, crabs, earthworms, ground meat, eggs, milk	43
		Soricidae	
Blarina brevicauda			
	(W)	Annelids: (W, C) earthworms; insects	44, 45: 46; 47, 48
	(W)	Crustacea; isopods; sowbugs	44; 45; 46
		Mollusks; slugs, snails	44; 45, 46
		Arachnids; spiders; spider cocoons	44; 46; 49
		Ants, insect larvae and pupae; Hemiptera, Coleoptera; Coleoptera larvae	45; 45, 46; 46
		Centipedes, millipedes	44–46
		Hymenoptera, Odonata, Cidadellidae, lepidopterous and dipterous larvae, grasshoppers	46
		Balininus proboscideus; Neodiprion sertifer prepupae	49; 50
		Microtus pennsylvanicus, shrews, other mammals	45
		Considerable plant material; oat and wheat seeds; raspberry seeds, beechnuts; *Endogone* (fungus)	44; 51; 49; 52, 53

[b] Quoting the collector.

Table 1 (continued)
FOODS OF WILD (W) AND CAPTIVE (C) INSECTIVORES (LISTED ALPHABETICALLY)

Insectivore		Food	Ref.
		Soricidae (continued)	
Blarina brevicauda (continued)			
	(C)	Mixture: raw hamburger, ground horsemeat, liver, dry dog food	54
		Vegetables; Ritz® crackers, peanuts, whole wheat bread, cheese, mixture: ground beef, walnuts, sunflower seeds, rolled oats	44; 47
		Bone meal; vitamin supplements	55; 48, 55
		Canned dog food; dry dog food, Purina® cat chow	47, 48, 55–57; 58
		Chicken; raw beef; frozen horsemeat	47; 57; 59
		Turkey breeder ration; rat chow, chicken scratch, cracked corn	48; 55; 60
		Rodents; mice; ground *Rattus rattus*	55; 44, 47; 58
		Birds; frogs, salamanders, fresh-water snails	44, 47; 44
		Fish; whole and ground goldfish	44; 58
		Other shrews; *Blarina brevicauda*	47; 58
		Mealworms; adult and larval mealworms; *Neodiprion sertifer* prepupae	57; 47; 50
		Larval *Sthenopsis argenteomaculata* and *Tabanus,* adult and larval *Ctenicera spinosus,* adult *Tenebrio picipes* and *Agonum sinuatis,* eonymph *Pristiphora erichsonii*	61
Crocidura bicolor			
	(C)	Chopped raw beef, mealworms, Coleoptera, Lepidoptera, Mantidae, Ichneumonidae, *Tabanus* sp., scercopid insects, ants, other *C. bicolor*	62
Crocidura flavescens			
	(W)	Moth pupae, dove (*Streptopelia senegalensis*), bread	31
	(C)	Epol® cat pellets (30% protein, 8% fat, 5% fiber, vitamins, and minerals), minced ox heart, *Helix, Trinervitermes*, earthworms, mealworms, *Mus musculus,* other *C. flavescens,* wheat germ oil, minerals, vitamin supplements, Harveld Swiss Muesli®	31
		Mammals, birds, arachnids, mollusks, termites, beetles, grasshoppers, crickets, millipedes, oligochaetes	63
Crocidura hirta			
	(C)	Crickets, mantises, grasshoppers, flying termites, mealworms, beetles, noctuid and geometrid moths, *Dalaca* sp., *Parasa latistriga, Tylopsis continua, Mastomys natalensis, Rattus rattus, Mus musculus, Rhabdomys pumilla, Mystromys albicaudatus, Miniopterus natalensis,* other *C. hirta,* ground beef and liver.	64
Crocidura leucodon			
	(C)	Fly larvae, mixture: chopped meat, boiled egg, fish meal, cereal, milk, vitamin supplement	65

Table 1 (continued)
FOODS OF WILD (W) AND CAPTIVE (C) INSECTIVORES (LISTED ALPHABETICALLY)

Insectivore		Food	Ref.
		Soricidae (continued)	
Crocidura mariquensis			
	(C)	Epol® cat pellets (30% protein, 8% fat, 5% fiber, vitamins, minerals), minced ox heart, wheat germ oil, mineral and vitamin supplements	31
Crocidura oliveri			
	(C)	Ox heart, mealworms, grasshoppers, beetles, earthworms	66
Crocidura russula			
	(C)	Same as in *C. leucodon,* Reference 65 (above)	
		Same as in *C. oliveri,* Reference 66 (above)	
Crocidura siliacea			
	(C)	Same as in *C. mariquensis,* Reference 31 (above)	
Crocidura suaveolens			
	(C)	*Gryllus domesticus, Tenebrio molitor,* calf liver, minerals, vitamins	67
Cryptotis parva			
	(W)	Adult and larval beetles, earthworms, centipedes, snails	68
	(C)	Crickets, grasshoppers; cockroaches	69–71; 68, 70,[c] 72
		Sowbugs; pill-bugs; insects	69; 70, 73; 74
		Wireworms, carabid beetles, cutworms, whiteworms, snails (*Polygyra*), small mammals, apples, mixture: sardines, salmon, peanut butter, cooked vegetables	68
		Wax moths, sphinx moths, ant lions, beetles, ox beetle grubs, frogs (*Hyla*), toads (*Gastrophryne*), lizards (*Anolis, Eumeces,* and *Leiolopisma*), snake (*Tantilla*), *Peromyscus*	70
		Adult and larval mealworms	72
		Smooth caterpillar, earwig, grub, baked beans, mashed white potato	73
		Horsemeat; hamburger; raw ground beef	59, 75; 69; 70, 74
		Ground chicken; liver	75; 76
		Mice; rats	69, 74–76; 74, 76
		Leopard frogs; rabbit; bird	71; 74; 76
		Various whole small fish, canned mackerel, whole and ground goldfish, dry dog food	58
		Canned cat and dog food, hardboiled eggs, liver powder, brewer's yeast, wheat germ oil, baby vitamins	75
		Mixture: raw hamburger, ground horsemeat, liver, dry dog food	54
		Oatmeal flakes	77
Diplomesodon pulchellum			
	(C)	*Mus musculus,* lizards (*Eremias* and *Phrynocephalus*), scarab beetles	78
Feroculus feroculus			
	(W)	Vegetable matter, tiny roots	79

[c] Specific name given as *floridana.*

Table 1 (continued)
FOODS OF WILD (W) AND CAPTIVE (C) INSECTIVORES (LISTED ALPHABETICALLY)

Insectivore		Food	Ref.
		Soricidae (continued)	
Microsorex hoyi			
	(W)	Jack pine seeds	51[d]
	(C)	Same as in *Blarina brevicauda*, Reference 61 (above)	
		Raw fish and meat, earthworms	80
		Grasshoppers, houseflies, craneflies, beetles, *Microtus* liver, *Sorex cinereus*, other *M. hoyi, Peromyscus maniculatus, Clethrionomys gapperi*, buttered biscuit	81
Myosorex varius			
	(C)	Same as in *C. mariquensis*, Reference 31 (above)	
Neomys fodiens			
	(W)	Tadpoles; frogs; *Rana temporaria, Bufo bufo*	82; 83; 84
		Small fishes, especially salmon parr; *Gasterosteus aculeatus*	82; 84
		Aquatic insects and larvae; adult and larval Coleoptera, Scarabidae, Carabidae, Curculionidae, Lepidoptera, Formicidae, Acarina	82; 83
		Lymnephilus, Notidobia, Nepa, Trichoptera larvae, *Molanna, Anabolia, Dytiscus*	84
		Slugs, Lumbricidae, miscellaneous invertebrates, voles, plant remains	83
		Shrew (*Sorex araneus*), mollus s: *Lymnea stagnalis, Galba corvus, Galba truncata, Radix peregra, Bithynia tentaculata, Planorbarius corneus, Planorbarius planorbis, Anodonta* sp., *Viviparus contectus, Anisus* sp., *Pisidum* sp., *Bathyomphalus contortus, Segmentina nitida*	84
	(C)	Fish, frogs, small mammals, worms	82
		Caterpillars, mealworms; earthworms, oats	85; 86
Notiosorex crawfordi			
	(W)	Centipedes, unidentified insects	87
	(C)	Hamburger, *Passer domesticus;* lizard (*Elgaria multicarinata*)	88; 89
		Honeybee larvae, mealworms; cutworms, isopods	90; 91
		Beetles, earwigs, sowbugs, crickets, grasshoppers, garden cockroaches, moths (including *Erebus*)	92
Scutisorex congicus			
	(W)	Caterpillars; earthworms, frogs	93; 94; 94

[d] Only generic name cited.

Table 1 (continued)
FOODS OF WILD (W) AND CAPTIVE (C) INSECTIVORES (LISTED ALPHABETICALLY)

Insectivore		Food	Ref.
		Soricidae (continued)	
Sorex araneus			
	(W)	Arachnida; Lumbricidae	83, 95; 83, 96, 97
		Coleoptera; adult and larval Coleoptera	95; 96
		Hymenoptera; Hemiptera; Myriapoda	83, 95; 83, 97; 83
		Molluska; Gastropoda; slugs	83; 95, 96; 97
		Diptera; adult Diptera; larval and pupal Diptera; larval and pupal Culicidae and Tabanidae	96; 97; 83, 95; 83
		Lepidoptera larvae; Lepidoptera imagoes and caterpillars	97; 83
		Araneidae, Acarina, Heteroptera, Carabidae, Curculionidae, Formicidae, miscellaneous insects and their eggs	96
		Acari, Araneae, Opiliones, Thysanoptera, miscellaneous insect larvae, plant material	97
		Imagoes and larvae of Elateridae, Carabidae, Chrysomelidae, Curculionidae, Cerambicidae, Scarabaeidae, miscellaneous invertebrates, frogs, miscellaneous vertebrates, plant parts	83
		Small bird eggs, worms, discarded bait fish	82
		Other insectivores; seeds	95; 82, 83
	(C)	*Gryllus domesticus, Tenebrio molitor*, calf liver, minerals, vitamins	67
		Insects, worms, small mammals	82
		Earthworms, oats	86
		Lumbricus, Lepidoptera larvae, *Lithobius variegatus, Ophyiulus pilosus, Polydesmus* sp., *Arion hortensis, Trichonicus pusillus, Oxychilus alliarium, Discus rotundatus,* staphylinids, carabid beetles	97
		Lumbricidae, Molluska, Orthoptera, Heteroptera, Coleoptera, Diptera, Hymenoptera, Lepidoptera, vertebrates: *Phoxinus, Rana, Lacerta, Sorex, Apodemus, Microtus, Clethrionomys,* seeds: *Pinus sibirica, Pinus silvestris, Picea obovata,* etc. (all readily taken)	98[e]
Sorex arcticus			
	(W)	Oligochaeta, Araneida, Acarina, Diptera, Myriopoda, Lepidoptera, Orthoptera, Heteroptera, Coleoptera, Hymenoptera, miscellaneous insects, and beetle larvae	98
	(C)	Same as in *Blarina brevicauda,* Reference 61 (above)	
		Meadow voles, earthworms, fly pupae, mealworms, grasshoppers, hamburger	99
Sorex bairdi			
	(W)	Douglas fir seeds	100

[e] Specific identification of 120 types of foods taken by three species of captive shrews; exhaustive analyses of 2500 stomachs from seven species of wild Siberian shrews.

Table 1 (continued)
FOODS OF WILD (W) AND CAPTIVE (C) INSECTIVORES (LISTED ALPHABETICALLY)

Insectivore	Food	Ref.
	Soricidae (continued)	
Sorex caecutiens		
(W, C)	Small mammals	82
(W)	Lepidoptera, Diptera, Hymenoptera, Lumbricidae, Molluska, Myriapoda, miscellaneous invertebrates, plant parts and seeds, larvae and imagoes of Elateridae, Carabidae, Chrysomelidae, Scarabidae	83
(C)	Lumbricidae, Orthoptera, dipteran larvae and imagoes, hymenopteran and coleopteran larvae, *Rana semiplicata, Apodemus speciosus*	101[f]
	Limnaea stagnalis, Araneida, Acarina, Myriopoda, Odonata, *Lestes, Cordulia,* various Diptera and Orthoptera, Heteroptera, many Coleoptera and Lepidoptera, Hymenoptera, *Phoxinus, Sorex, Apodemus, Microtus, Apodemus* (all readily taken)	98
Sorex centralis		
(C)	Same as in *S. caecutiens,* Reference 101 (above)	
Sorex cinereus		
(W, C)	Earthworms, spiders; *Neodiprion sertifer* prepupae	102, 103; 50
(W)	Insects; beetles, insect pupae	44; 104
	Centipedes, worms, sowbugs, arachnids, mollusks, vegetable matter	44
	Gryllidae, Cidadellidae, Rhagionidae, *Phalangida,* Lepidoptera larvae, snail	102
	Oat and wheat seeds	51
(C)	Same as in *Blarina brevicauda,* Reference 61 (above)	
	Microtus, Ambystoma	58, 103
	Perch, snails, houseflies, hellgramites, centipedes, beetles, grasshoppers, crickets, caterpillars, moths, sowbugs, insect larvae, hamburger	103
	Liver; mouse brains	103, 105; 105
	Mus musculus, ground *Rattus rattus*, small birds, goldfish, canned mackerel, and dog food	58
	Mixture: raw hamburger, ground horsemeat, liver, dry dog food, brains	54
Sorex daphaenodon		
(W)	Oligochaeta, Arachnoidea, Myriopoda, Lepidoptera, Orthoptera, Heteroptera, Hymenoptera, Diptera and larvae, Formicidae, Coleoptera: Carabideae, Elateridae, Chrysomelidae, Curculionidae, Scarabaeidae, unidentified beetles, other insects, their larvae and eggs	98
Sorex dispar		
(W)	Centipedes, spiders, insects	104
	Coleoptera, Arachnida	106

[f] Details of foods selected by six species of captive Soviet-Far East shrews.

Table 1 (continued)
FOODS OF WILD (W) AND CAPTIVE (C) INSECTIVORES (LISTED ALPHABETICALLY)

Insectivore		Food	Ref.
		Soricidae (continued)	
Sorex fumeus			
	(W)	Centipedes; spiders, beetles and larvae	44, 107, 108; 107, 108
		Sowbugs, salamanders	44, 108
		Earthworms, snails, birds, mammals	108
	(C)	Isopods, beetles, earthworms, snails, flies, wireworms, *Plethedon,* baby mice	108
		Plants; *Endogone*	44; 53
Sorex gracillimus			
	(C)	Lumbricidae, Orthoptera, Hymenoptera, larval Diptera and Coleoptera	101
Sorex isodon			
	(C)	Insects, worms, small mammals	82
Sorex longirostris			
	(C)	Various small insects	109
Sorex minutissimus			
	(W)	Myriopoda, Hymenoptera, Coleoptera: Carabidae, Elateridae, Tenebrionidae, Cerambycidae, Chrysomelidae, Curculionidae, Ipidae, unidentified insect larvae, Formicidae, mammal remains	98
	(C)	Lumbricidae, Orthoptera, Geophilidae	101
		Formica rufa, Diprion pupae, *Maerothylacia rubi* larvae, *Phalangida, Lampycis noctiluca* and other small beetles, looper larvae and pupae, *Clethrionomys glareolus,* liver, lung and kidney of *Arvicola terrestris*	110
Sorex minutus			
	(W)	Hymenoptera, Diptera	83, 96
		Lepidoptera, Orthoptera, Myriapoda, larvae and imagoes of Elateridae, Carabidae, Chrysomelidae, Curculionidae, and Scarabidae, miscellaneous invertebrates, vertebrates, plants, and seeds	83
		Araneida, Acarina, Homoptera, Heteroptera, Coleoptera: Carabidae, Chalcididae, Formicidae, Muscidae larvae, miscellaneous insects	96
	(C)	Araneida, Acarina, Myriopoda, *Lestes, Cordulia,* many Orthoptera, Coleoptera, and Lepidoptera, various Heteroptera, Diptera, Hymenoptera, and caterpillars, other insects, *Phoxinus, Sorex, Apodemus, Microtus, Clethrionomys* (all readily taken)	98
		Insects, small mammals; earthworms, oats	82; 86
Sorex mirabilis			
	(C)	Lumbricidae, Orthoptera, Diptera and Coleoptera larvae, Geophilidae, Diplopoda, *Clethrionomys rufocanus*	101
Sorex ornatus			
	(C)	Same as in *S. cinereus,* Reference 54 (above)	
		Mouse, brains, dry dog food	111

Table 1 (continued)
FOODS OF WILD (W) AND CAPTIVE (C) INSECTIVORES (LISTED ALPHABETICALLY)

Insectivore	Food	Ref.
	Soricidae (continued)	
Sorex pacificus		
(W)	Douglas fir seeds	100
(C)	Same as in *S. cinereus,* Reference 54 (above)	
Sorex palustris		
(W)	Insects, stonefly nymphs, mayflies, planarians	44
	Larval Tipulidae, Trichoptera, Simuliidae, and Chironomidae, niads of *Plecoptera* and *Ephemerida,* small arachnids, oligochaetes, fishes, other shrews, vegetable matter	112
	Slugs, snails, earthworms, spider, Gryllidae, Cicadellidae, Diptera and larvae, Hymenoptera, Lepidoptera larvae, *Endogone*	102
(C)	Moths, ants, flies; grasshoppers	113;[g] 113, 114
	Insects, snails, *Microtus, Zapus*	115
	Same as in *S. cinereus,* Reference 54 (above)	
	Aquatic insects, isopods, earthworms, fish, lizards (*Sceloporus graciosus*), baby rats, other *S. palustris,* oatmeal, mixtures: (1) equal parts ground beef, beef brains, and liver, (2) 2/3 beef kidney, 1/3 dog food, (3) 1/3 each beef liver, beef brains, chopped beef suet	114
Sorex sinuosus		
(C)	Same as in *S. ornatus,* Reference 111 (above)	
Sorex trowbridgei		
(W)	Neuroptera, Coleoptera, Diptera, Lepidoptera, Heteroptera, Hymenoptera, spiders, centipedes, annelids, planarians, seeds	116
	Douglas fir seeds	100
(C)	Same as in *S. cinereus,* Reference 54 (above)	
Sorex unguiculatus		
(C)	Lumbricidae, various Orthoptera and Lepidoptera, various Diptera, Hymenoptera, and Coleoptera larvae, Geophilidae, *Clethrionomys rufocanus, Rana semiplicata,* seeds: *Pinus koraiensis, Panax genseng, Zea mays*	101
Sorex vagrans		
(W)	Insect eggs, larvae, pupae, and adults, oligochaetes, lice, fleas, snails	117
	Douglas fir seeds	100
(C)	Canned dog food, raw liver, earthworms, slugs, mice	118
	Same as in *S. cinereus,* References 54 and 58 (above)	
	Same as in *S. ornatus,* Reference 111 (above)	

[g] Generic name given as *Neosorex.*

Table 1 (continued)
FOODS OF WILD (W) AND CAPTIVE (C) INSECTIVORES (LISTED ALPHABETICALLY)

Insectivore		Food	Ref.
		Soricidae (continued)	
Sorex vir[h]			
	(W)	Oligochaeta, Arachnoidae, Hymenoptera, Myriopoda, Diptera and larvae, Lepidoptera, Heteroptera, Coleoptera: Carabidae, Elateridae, Tenebrionidae, Carambicidae, Chrysomelidae, Curculionidae, unidentified vertebrates, insects, their eggs and larvae, plant parts	98
Suncus etruscus			
	(W)	*Armadilla*, young *Procellio, Geophilus, Lithobius, Scutigera coleoptrata*, young *Scolopendra, Argiope lobata, A. fasciata, Segestria florentina, Uroctea durandi*, other spiders, *Loboptera decipiens, Phaneroptera, Tettigonia, Ephippiger, Gryllus, Oedipoda, Melitera, Pieris, Gonepteryx cleopatra, G. rhamni, Papilio, Celerio euphorbiae, Herse concolvulii, Rumina decollata, Helix, Testacella*, other slugs and insects, *Tarentola mauritanica, Psammodromus algirus*, other *S. etruscus*	119
	(C)	*Acheta domesticus, Tenebrio molitor* larvae, beef spleen, small amount preserved olives, *Crocidura russula, C. suaveolens*	119
		Mealworms, flies, grasshoppers, crickets	120
		Fourth stage larval *Tenebrio molitor*	121
Suncus hosei			
	(W)	Insects	15
Suncus murinus			
	(W)	Carrion of *Achatina fulica, Bufo marinus*, and *Mus musculus*	58
	(C)	Bread, jelly, peanut butter, boiled rice, potato and egg, insects	122
		Frozen horsemeat	59
		Canned dog food, houseflies, cockroaches, *Rattus mindanensis, R. exulans, Mus musculus*, cheese, small fish, whole and ground goldfish, copra, Ross-Wells® mink feed: poultry by-product meal, fish meal, meat meal, soybean oil meal, dried egg product, corn, beet pulp, brewer's yeast, molasses, lecithin, preserved animal fat, vitamins A, D, E, B_{12}, K, riboflavin, salt, minerals, with 33% protein, 18% fat, 3% fiber	58
		Purina® cat chow; mixture: beef tripe, liver, and muscle, chicken viscera, muscle, and bones, cooked eggs, cereal, with 54% protein, 39% fat, 2% polysaccharide	123; 124

[h] Originally *S. platycranius.*

Table 1 (continued)
FOODS OF WILD (W) AND CAPTIVE (C) INSECTIVORES (LISTED ALPHABETICALLY)

Insectivore		Food	Ref.
		Talpidae	
Condylura cristata			
	(W)	Slugs; earthworms, larval Diptera, Lepidoptera, and Coleoptera, centipedes, ants	107, 125; 107
		Aquatic and terrestrial oligochaetes, leeches, caddis "worms," Dytiscid and midge larvae, miscellaneous adult and larval insects, shrimps, small minnow	125
	(C)	Earthworms, Tipulidae larvae, *Desmognathus ochrophaeus; Mus musculus, Rattus norvegicus*	107; 58
Desmana moschata			
	(W)	Aquatic mollusks, leeches, aquatic insect larvae, fish and their roe, frogs	1[i]
Galemys pyrenacius			
	(C)	Fragile gastropods, large worms, arthropods (including crustaceans), small fish, mammal viscera	126
Neurotrichus gibbsi			
	(W, C)	Earthworms, isopods, insect pupae	127
	(W)	Springtails, cockroaches, other insects, snails, salamanders, peanut butter	127
		Douglas fir seeds	100
	(C)	Mice; rolled oats; oatmeal	127, 128; 127; 128
		Liver, cat food, mealworms	128
		Adult termites, earthworms, *Tenebrio molitor* larvae	129
Parascalops breweri			
	(W)	Earthworms, centipedes, millipedes, Lepidoptera larvae	107; 130
		Spiders, isopods, Thysanura, Acarina, mollusks, ant eggs, adult and larval Coleoptera and Diptera, plants	107
		Snails, slugs, sowbugs, larvae of tipulids, mycetophilids, leptids, and asilids, Coleoptera larvae: Elateridae, Cantharidae, Scarabidae, Curculionidae, Carabidae, Tenebrionidae	130
	(C)	*Mus musculus, Rattus norvegicus*	58
Scalopus aquaticus			
	(C)	Earthworms; white grubs	131–133; 131, 132
		Mice; ground meat; ground beef	131, 134; 132; 134
		Liver, beef, frogs, oats, wheat, corn, muskmelon and squash seeds	131
		May beetles, raw beef and pork liver, canned dog food	132
		Birds, walnuts, crackers, cheese	133
		English sparrow, squirrel brain, milk	134
		Mealworms, horsemeat; skinned rats	135; 136
		Ground beef, earthworms, adult and larval insects, white grubs, corn, wheat, oats, tomatoes, apples, Irish potatoes (approximate order of preference)	137

[i] Cites another's observation of aquatic plants being eaten.

Table 1 (continued)
FOODS OF WILD (W) AND CAPTIVE (C) INSECTIVORES (LISTED ALPHABETICALLY)

Insectivore		Food	Ref.
		Talpidae (continued)	
Scapanus orarius			
	(W)	Earthworms, Diptera, Orthoptera, Hymenoptera, butterflies, moths, centipedes, beetles, gladiolus and tulip bulbs, wheat seeds, carrot, parsley and parsnip roots	138
Scapanus townsendi			
	(W)	Earthworms, centipedes	138, 139
		Diptera, Orthoptera, Hymenoptera, butterflies, moths, beetles, spiders, vetch, oat and corn seeds	138
		Earthworm cocoons, adult, larval, and pupal insects, slugs	139
	(C)	Ground beef, veal, kidney salmon, chicken, pork	139
Talpa europaea			
	(W)	Earthworms and cocoons, Myriapoda	140, 141
		Wireworms, cockchafer larvae, larval *Curculio, Carabus,* and *Staphylinus,* Hymenoptera larvae and imagoes, Diptera, Lepidoptera	140
		Elaterid and Diptera larvae	141, 142
		Scarabaeid, carabid, staphylinid, and other beetle larvae and imagoes, agrotid larvae	141
		Lumbricidae, Coleoptera and larvae, *Carabus, Aphodius,* Tipulidae larvae, lepidopteran larvae and pupae, mammals, seeds, and plant parts	142
		Frog; truffles (*Ascomycetes*)	143; 144
	(C)	Beetles, snails, slugs, annelids, miscellaneous adult and larval insects, birds, mice, voles, shrews	82
		Lumbricus terrestris, maggots, liver; baby mice, mealworms	145; 145, 146
		Most fresh meats, seafood, chicken eggs	147

Table 1 (continued)
FOODS OF WILD (W) AND CAPTIVE (C) INSECTIVORES (LISTED ALPHABETICALLY)

Insectivore	Food	Ref.
	Tenrecidae	
Centetes ecaudatus		
(C)	Condensed milk, raw horsemeat, Pablum®, egg noodles, vitamin supplements; earthworms	25; 148
Echinops telfairi		
(C)	Mealworms + same as in *Centetes ecaudatus,* Reference 25 (above)	
	Crickets, grasshoppers; baby *Mus musculus*	148; 58
	Same mixture as in *Suncus murinus* Reference 124 (above)	
	Carausius morosus, Schistocerca gregarina, Gryllus domesticus, Plusia gamma, Dilina tiliae, Oryctes nasicornis, Calliphora, Vespa, Eristalis, Tenebrio molitor larvae, pupae, and adults, *Limax, Lumbricus rubellus, Pollachius virens, Gadus morhua,* chicken and turkey fat and viscera, mouse muscle and brains, deer meat, pig liver, kidney, and spleen, beef heart, kidney, and fat, canned milk, vitamin supplements	149
Hemicentetes nigriceps		
(W)	Earthworms; coleopteran larvae; spiders	148, 150; 148; 150
(C)	Same as in *Centetes ecaudatus*, Reference 25 (above)	
Hemicentetes semispinosus		
(C)	Same as in *Centetes ecaudatus,* Reference 25 (above)	
	Same as in *H. nigriceps,* Reference 148 (above)	
	Eathworms, ground meat	150
Limnogale mergulus		
(W)	Aquatic insect larvae, fresh-water shrimp, crayfish, frogs	150
Microgale dobsoni		
(C)	Same as in *H. semispinosus*, Reference 150 (above)	
	Grasshoppers; raw ground horsemeat, powdered milk, ground egg yolk, mealworms, crickets, Pervinal® mineral supplement	148; 151
Microgale talazaci		
(C)	Earthworms + same as in *M. dobsoni,* Reference 151 (above)	
	Raw chopped meat, variety of insects including Coleoptera larvae, Orthoptera	150
Microgale thomasi		
(C)	Raw, chopped horsemeat mixed with condensed milk, Pablum®, ABDEC® vitamin supplement	148
Orizorictyes talpoides		
(W)	Earthworms, insects, grasshoppers, vegetable matter	150

Table 1 (continued)
FOODS OF WILD (W) AND CAPTIVE (C) INSECTIVORES (LISTED ALPHABETICALLY)

Insectivore		Food	Ref.
		Tenrecidae (continued)	
Setifer setosus			
	(C)	Same as in *Centetes ecaudatus,* Reference 25 (above)	
		Crickets, grasshoppers, mealworms; raw ground meat, baby mice	148; 150
		Skinned bird	152
Tenrec ecaudatus			
	(C)	Earthworms, grubs, Orthoptera, raw ground meat, mice	150
		Mealworms, fish, day-old chicks, minced meat, raw eggs, cottage cheese, bananas, Gistocal® mineral supplement	153

REFERENCES

1. **Ognev, S. I.,** *Mammals of Eastern Europe and Northern Asia,* Vol. 1, Pt. 1, 1928 (English translation by Program for Scientific Translations, Jerusalem), National Science Foundation, U.S. Dept. of Commerce, Washington, D.C., 1962, 11–295.
2. **Walker, E. P.,** *Mammals of the World,* Vol. 1, Johns Hopkins Press, Baltimore, 1964, 103–179.
3. **Kuyper, M. A.,** Zoology Department, University of Natal, Pietermaritzburg, South Africa, personal communication, 1975.
4. **Bateman, J. A.,** Laboratory studies of the golden mole and mole-rat, *Afr. Wildl.,* 13, 65–71, 1959.
5. **Bateman, J. A.,** Ratty golden mole, *Zool. Soc. S. Afr. News Bull.,* 2, 16–17, 1961.
6. **Meester, J.,** A new golden mole from the Transvaal (Mammalia: Chrysochloridae), *Ann. Transvaal Mus.,* 28, 35–46, 1972.
7. **Pienaar, U. de V.,** The small mammals of the Kruger National Park – a systematic list and zoogeography, *Koedoe,* No. 7, 1–25, 1964.
8. **Shortridge, G. C.,** *The Mammals of South West Africa,* Heinemann, London, 1934, 40.
9. **Jarvis, J. U. M.,** Notes on the golden mole *Chrysochloris stuhlmanni* Matschie, from the Ruwenzori Mountains, Uganda, *East Afr. Wildl. J.,* 12, 163–166, 1974.
10. **Holm, E.,** Contribution to the knowledge of the biology of Namib Desert golden mole (*Eremitalpa granti namibensis* Bauer and Niethammer 1959), *Sci. Pap. Namib Desert Res. Stn.,* No. 41, 37–42, 1969.
11. **Roberts, A.,** *The Mammals of South Africa,* Trustees, Mammals of South Africa Book Fund, Johannesburg, 1951, 119.
12. **Herter, K.,** *Hedgehogs,* Phoenix House, London, 1965, 39–46.
13. **Lim, B. L.,** Notes on the food habits of *Ptilocercus lowii* Gray (Pentail tree-shrew) and *Echinosorex gymnurus* (Raffles) (Moonrat) in Malaya with remarks on "ecological labelling" by parasite patterns, *J. Zool.,* 152, 375–379, 1967.
14. **Davis, D. D.,** Mammals of the lowland rain-forest of North Borneo, *Bull. Natl. Mus. Singapore,* 31, 1–129, 1962.
15. **Harrison, J.,** *An Introduction to the Mammals of Sabah,* Tien Wah Press, Singapore, 1964, 30–37.
16. **Gould, E.,** Laboratory of Comparative Behavior, Johns Hopkins University, Baltimore, personal communication, 1976.
17. **Loukashin, A. S.,** The mammals of North Manchuria, *J. Mammal.,* 18, 327–332, 1937.
18. **Popov, V. A.,** *Mammals of Volga-Kama Country: Insectivora, Chiroptera, Rodentia,* Academy of Science, USSR, Kazan Branch, 1960, 1–468 (in Russian).
19. **Ch'eng-chao, L.,** Notes on the food of Chinese hedgehogs, *J. Mammal.,* 18, 355–357, 1937.
20. **Loring, J. A.,** Notes on the European hedgehog, *J. Mammal.,* 2, 110–111, 1921.

21. **Allison, M. and Deanesly, R.,** The reaction of anoestrous hedgehogs to experimental conditions, *Proc. R. Soc. Lond. Ser. B,* 116, 170–185, 1934.
22. **Kristoffersson, R.,** Hibernation of the hedgehog (*Erinaceus europaeus*). The ATP and o-phosphate levels in blood and various tissues of hibernating and non-hibernating animals, *Ann. Acad. Sci. Fenn. Ser. A4,* 50, 1–45, 1961.
23. **Morrison, P. R.,** Observation on body temperature in a hedgehog, *J. Mammal.,* 38, 254–255, 1957.
24. **Nowak, W.,** Zur Haltung von spätherbstlichen "Igelkindern" (*Erinaceus europaeus*), *Z. Angew. Zool.,* 59, 285–288, 1972.
25. **Gould, E. and Eisenberg, J. F.,** Notes on the biology of the Tenrecidae, *J. Mammal.,* 47, 660–686, 1966.
26. **Clausen, G. and Ersland, A.,** The respiratory properties of the blood of the hibernating hedgehog *Erinaceus europaeus* L., *Respir. Physiol.,* 5, 221–233, 1968.
27. **Cott, H. B.,** The palatibility of the eggs of birds; illustrated by experiments on the food preferences of the hedgehogs, *Proc. Zool. Soc. London,* 121, 1–41, 1951.
28. **Nowak, W.,** Auftreten eins Ostigels (*Erinaceus roumanicus*) in Oberbayern, *Sitzungsber. Ges. Naturforsch. Freunde Berlin,* 15, 79–81, 1975.
29. **Krishna, D. and Prakash, I.,** Hedgehogs of the desert of Rajasthan. Part 2: Food and feeding habits, *J. Bombay Nat. Hist. Soc.,* 53, 362–366, 1956.
30. **Webb-Peploe, C. G.,** Field notes on the mammals of South Tinnevelly, South India, *J. Bombay Nat. Hist. Soc.,* 46, 630–644, 1946.
31. **Baxter, R. M.,** Zoology Department, University of Natal, Pietermaritzburg, South Africa, personal communication, 1975.
32. **Tripp, H. R. H.,** Capture, laboratory care and breeding of elephant-shrews (Macroscelididae), *Lab. Anim.,* 6, 213–224, 1972.
33. **Critch, P. J.,** The rock elephant shrew, *Afr. Wildl.,* 23, 139–144, 1969.
34. **Rathbun, G.,** National Museum, Nairobi, Kenya, personal communication, 1976.
35. **Hoogstraal, H.,** Mission: malaria, *Nat. Hist. Mag.,* 59, 104–111, 140–141, 1950.
36. **Walker, E. P.,** African elephant shrews, *Nature Mag.,* 48, 295–297, 332, 1955.
37. **Crandall, L. S.,** *The Management of Wild Mammals in Captivity,* University of Chicago Press, Chicago, 1964, 45–51.
38. **Rahm, U.,** Beobachtungen an der ersten in Gefangenschaft gehaltenen *Mesopotamogale ruwenzorii* (Mammalia – Insectivora), *Rev. Suisse Zool.,* 68, 73–90, 1961.
39. **Durrell, G. M.,** Giant water shrew, *Zoo Life,* 8, 145–147, 1953.
40. **True, F. W.,** The almiqui, *Science,* 8, 282, 1886.
41. **Allen, J. A.,** Notes on *Solenodon paradoxus* Brandt, *Bull. Am. Mus. Nat. Hist.,* 24, 505–517, 1908.
42. **Eisenberg, J. F. and Gould, E.,** The behavior of *Solenodon paradoxus* in captivity with comments on the behavior of other insectivora, *Zoologica,* 51, 49–58, 1966.
43. **Barbour, T.,** The solenodons of Cuba, *New Engl. Zool. Club Proc.,* 23, 1–8, 1944.
44. **Hamilton, W. J., Jr.,** The food of the soricidae, *J. Mammal.,* 11, 26–39, 1930.
45. **Eadie, W. R.,** The short-tailed shrew and field mouse predation, *J. Mammal.,* 25, 359–364, 1944.
46. **Whitaker, J. O., Jr. and Ferraro, M. G.,** Summer food of 220 short-tailed shrews from Ithaca, New York, *J. Mammal.,* 44, 419, 1963.
47. **Rood, J. P.,** Habits of the short-tailed shrew in captivity, *J. Mammal.,* 39, 499–507, 1958.
48. **Neal, C. M. and Lustick, S. I.,** Energetics and evaporative water loss in the short-tailed shrew *Blarina brevicauda, Physiol. Zool.,* 46, 180–185, 1973.
49. **Hamilton, W. J., Jr.,** The food of small forest mammals in eastern United States, *J. Mammal.,* 22, 250–263, 1941.
50. **Holling, C. S.,** The selection by certain small mammals of dead, parasitized, and healthy prepupae of the European pine sawfly, *Neodiprion sertifer* (Geoff.), *Can. J. Zool.,* 33, 404–419, 1955.
51. **Criddle, S.,** The granivorous habits of shrews, *Can. Field Nat.,* 87, 69–70, 1973.
52. **Diehl, W. W.,** Endogone as animal food, *Science,* 90, 442, 1939.
53. **Whitaker, J. O., Jr.,** *Endogone, Hymenogaster,* and *Melanogaster* as small mammal foods, *Am. Midl. Nat.,* 67, 152–156, 1962.
54. **Pearson, O. P.,** Keeping shrews in captivity, *J. Mammal.,* 31, 351–352, 1950.
55. **Blus, L. J.,** Reproduction and survival of short-tailed shrews (*Blarina brevicauda*) in capativity, *Lab. Anim. Sci.,* 21, 884–891, 1971.
56. **Mann, P. M.** Keeping short-tailed shrews in captivity, *J. Mammal.,* 37, 274–275, 1956.

57. **Doremus, H. M.,** Heart rate, temperature and respiration rate of the short-tailed shrew in capativity, *J. Mammal.*, 46, 424–425, 1965.
58. **Dryden, G. L.,** unpublished observations.
59. **Gould, E.,** Communication in three genera of shrews (Soricidae): Suncus, Blarina & Cryptotis, *Commun. Behav. Biol.*, 3, 11–31, 1969.
60. **Martinsen, D. L.,** Energetics and activity patterns of short-tailed shrews (*Blarina*) on restricted diets, *Ecology*, 50, 505–510, 1969.
61. **Buckner, C. H.,** Metabolism, food capacity and feeding behavior in four species of shrews, *Can. J. Zool.*, 42, 259–279, 1964.
62. **Ansell, W. F. H.,** Capativity behaviour and post-natal development of the shrew *Crocidura bicolor, Proc. Zool. Soc. Lond.*, 142, 123–127, 1964.
63. **Marlow, B. J. G.,** Observations on the Herero musk shrew, *Crocidura flavescens herero* St. Leger, in captivity, *Proc. Zool. Soc. Lond.*, 124, 805–807, 1955.
64. **Meester, J.,** A systematic revision of the shrew genus *Crocidura* in southern Africa, *Transvaal Mus. Mem.*, 13, 16, 1963.
65. **Hellwing, S.,** Husbandry and breeding of white-toothed shrews (Crocidurinae) in the Research Zoo of the Tel-Aviv University, *Int. Zoo Yearb.*, 13, 127–134, 1973.
66. **Grünwald, A. and Möhres, F. P.,** Beobachtungen zur Jugendentwicklung und Karawanenbildung bei Weisszahnspitzmäusen (Soricidae-Crocidurinae), *Z. Saeugetierkd.*, 39, 321–337, 1974.
67. **Vlasák, P.,** Vergleich der postnatalen Entwicklung der Arten *Sorex araneus* L. und *Crocidura suaveolens* (Pall.) mit Bemerkungen zur Methodick der Laborzucht (Insectivora: Soricidae), *Vestn. Cesk. Spol. Zool.*, 3, 222–233, 1973.
68. **Hamilton, W. J., Jr.,** The biology of the short-tailed shrew, *Cryptotis parva, J. Mammal.*, 25, 1–7, 1944.
69. **Davis, W. B. and Joeris, L.,** Notes on the life-history of the little short-tailed shrew, *J. Mammal.*, 26, 136–138, 1945.
70. **Springer, S.,** Observations on *Cryptotis floridana* in captivity, *J. Mammal.*, 18, 237, 1937.
71. **Hatt, R. T.,** Feeding habits of the least shrew, *J. Mammal.*, 19, 247–248, 1938.
72. **Peterson, C. B.,** Cryptotis parva in western New York, *J. Mammal.*, 17, 284–285, 1936.
73. **Chamberlain, E. B.,** Behavior of the least shrew, *J. Mammal.*, 10, 250–251, 1929.
74. **Pfeiffer, C. J. and Gass, G. H.,** Note on the longevity and habits of captive *Cryptotis parva, J. Mammal.*, 44, 427–428, 1963.
75. **Mock, O. B.,** Reproduction in the Least Shrew (*Cryptotis parva*) in Captivity, Ph.D. thesis, University of Missouri, Columbia, 1970.
76. **Conaway, C. H.,** Maintenance, reproduction and growth of the least shrew in captivity, *J. Mammal.*, 39, 507–512, 1958.
77. **Hamilton, W. J., Jr.,** Habits of *Cryptotis parva* in New York, *J. Mammal.*, 15, 154–155, 1934.
78. **Heptner, V. G.,** The Turkestan desert shrew, its biology and adaptive peculiarities, *J. Mammal.*, 20, 139–149, 1939.
79. **Tutein-Nolthenius, A. C.,** Measurements and notes on some little-known Ceylon mammals, *J. Bombay Nat. Hist. Soc.*, 35, 606–614, 1932.
80. **Saunders, P. B.,** *Microsorex hoyi* in captivity, *J. Mammal.*, 10, 78–79, 1929.
81. **Prince, L. A.,** Notes on the habits of the pigmy shrew (*Microsorex hoyi*) in captivity, *Can. Field Nat.*, 54, 97–100, 1940.
82. **Kaikusalo, A.,** Ojajoki, 12700 Loppi, Finland, personal communication, 1976.
83. **Ivanter, E. V.,** *Population Ecology of Small Mammals in the North-western Taiga of the USSR,* Akad. Sci. USSR, Kerelya Branch, Inst. Biol., Nauka, Leningrad, 1–246 (monograph; in Russian, English summary).
84. **Wolk, K.,** The winter food of the European water-shrew, *Acta Theriol.*, 21, 117–129, 1976.
85. **Hawkins, A. E. and Jewell, P. A.,** Food consumption and energy requirements of captive British shrews and the mole, *Proc. Zool. Soc. Lond.*, 138, 137–155, 1962.
86. **Crowcroft, P.,** Keeping British shrews in captivity, *J. Mammal.*, 32, 354–355, 1951.
87. **Fisher, H. I.,** Notes on shrews of the genus *Notiosorex, J. Mammal.*, 22, 263–269, 1941.
88. **Hoffmeister, D. F. and Goodpaster, W. W.,** Life history of the desert shrew *Notiosorex crawfordi, Southwest. Nat.*, 7, 236–252, 1962.
89. **Cunningham, J. D.,** Ecology and occurrence of *Notiosorex* in southern California, *J. Mammal.*, 37, 108–110, 1956.
90. **Dixon, J.,** Notes on the life history of the gray shrew, *J. Mammal.*, 5, 1–6, 1924.
91. **Gander, F. F.,** A gray shrew in a wood rat's nest, *J. Mammal.*, 9, 247–248, 1928.

92. **Huey, L. M.**, Some habits of a gray shrew in captivity, *J. Mammal.*, 17, 143–145, 1936.
93. **Hollister, N.**, Shrews collected by the Congo Expedition of the American Museum, *Bull. Am. Mus. Nat. Hist.*, 35, 663–680, 1916.
94. **Allen, J. A.**, The skeletal characters of *Scutiosorex* Thomas, *Bull. Am. Mus. Nat. Hist.*, 37, 769–784, 1917.
95. **Lavrov, N. P.**, Contribution to the biology of the common shrew (*Sorex araneus* L.), *Zool. Zh.*, 22, 6, 361–365, 1942.
96. **Mezhzherin, V. A.**, On feeding habits of *Sorex araneus* L. and *S. minutus* L., *Zool. Zh.*, 37, 948–953, 1958 (Russian, English summary).
97. **Rudge, M. R.**, The food of the common shrew *Sorex araneus* L. (Insectivora: Soricidae) in Britain, *J. Anim. Ecol.*, 37, 565–581, 1968.
98. **Yudin, B. S.**, Ecology of shrews (genus *Sorex*) of Western Siberia, *Trudy Biol. Inst.*, 8, 33–134, 1962 (English translation by Office of Technical Services, U.S. Dept. of Commerce, Washington, D.C.).
99. **Clough, G. C.**, The biology of the arctic shrew, *Sorex arcticus, Am. Midl. Nat.*, 69, 69–81, 1963.
100. **Moore, A. W.**, Shrews as a check on Douglas fir regeneration, *J. Mammal.*, 23, 37–41, 1942.
101. **Okhotina, M. V.**, Morpho-ecological features making possible joint habitation for shrews of the genus *Sorex* (*Sorex*, Insectivora), *Proc. Inst. Biol. Pedol. Far East Sci. Cent. USSR Acad. Sci.*, Vol. 17, 1974 (3 unnumbered pages; English).
102. **Whitaker, J. O., Jr. and Schmeltz, L. L.**, Food and external parasites of *Sorex palustris* and food of *Sorex cinereus* from St. Louis County, Minnesota, *J. Mammal.*, 54, 283–285, 1973.
103. **Blossom, P. M.**, A pair of long-tailed shrews (*Sorex cinereus cinereus*) in captivity, *J. Mammal.*, 13, 136–143, 1932.
104. **Richmond, N. D. and Grimm, W. C.**, Ecology and distribution of the shrew *Sorex dispar* in Pennsylvania, *Ecology*, 31, 279–282, 1950.
105. **Kilham, L.**, Mother and young of *Sorex cinereus fontinalis* in captivity, *J. Mammal.*, 32, 115, 1951.
106. **Conaway, C. H. and Pfitzer, D. W.**, *Sorex palustris* and *Sorex dispar* from the Great Smoky Mountains National Park, *J. Mammal.*, 33, 106–108, 1952.
107. **Richmond, N. D. and Rosland, H. R.**, Mammal Survey of Northwestern Pennsylvania, P-R Proj. 20-R, Pennsylvania Game Commission, Harrisburg, 149, 38–39.
108. **Hamilton, W. J., Jr.**, The biology of the smoky shrew (*Sorex fumeus fumeus* Miller), *Zoologica*, 25, 473–492, 1940.
109. **Odum, E. P.**, *Sorex longirostris* at Mountain Lake, Virginia, *J. Mammal.*, 25, 196, 1944.
110. **Kaikusalo, A.**, Beobachtungen an gekäfigten Knirpsspitzmausen, *Sorex minutissimus* Zimmerman, 1780, *Z. Saeugetierkd.*, 32, 301–306, 1967.
111. **Rudd, R. L.**, Notes on maintenance and behavior of shrews in captivity, *J. Mammal.*, 34, 118–120, 1953.
112. **Conaway, C. H.**, Life history of the water shrew (*Sorex palustris navigator*), *Am. Midl. Nat.*, 48, 219–248, 1952.
113. **Gordon, K.**, A pregnant water shrew, *J. Mammal.*, 12, 429, 1931.
114. **Sorenson, M. W.**, Some aspects of water shrew behavior, *Am. Midl. Nat.*, 68, 445–462, 1962.
115. **Svihla, A.**, The mountain water shrew, *Murrelet*, 15, 44–45, 1934.
116. **Jameson, E. W.**, Observations on the biology of *Sorex trowbridgei* in the Sierra Nevada, California, *J. Mammal.*, 36, 339–345, 1955.
117. **Clothier, R. R.**, Contribution to the life history of *Sorex vagrans* in Montana, *J. Mammal.*, 36, 214–221, 1955.
118. **Eisenberg, J. F.**, Studies on the behavior of *Sorex vagrans, Am. Midl. Nat.*, 72, 417–425, 1964.
119. **Fons, R.**, Le repertoire comportmental de la Pachyure etrusque, *Suncus etruscus* (Savi, 1822), *Terre Vie*, 1, 131-157, 1974.
120. **Vogel, P.**, Biologische Beobachtungen an Etruskerspitzmäusen (*Suncus etruscus* Savi, 1832), *Z. Zaeugetierkd.*, 35, 173–185, 1970.
121. **Gereats, A.**, Aktivitätsmuster und Nahrungsbedarf bei *Suncus etruscus, Bonn. Zool. Beitr.*, 23, 181–196, 1972.
122. **Peterson, G. D., Jr.**, *Suncus murinus*, a recent introduction to Guam, *J. Mammal.*, 37, 278–279, 1956.
123. **Dryden, G. L.**, Growth and development of *Suncus murinus* in captivity on Guam, *J. Mammal.*, 49, 51–62, 1968.

124. **Dryden, G. L. and Ross, J. M.,** Enhanced growth and development of captive musk shrews, *Suncus murinus* on an improved diet, *Growth,* 35, 311–325, 1971.
125. **Hamilton, W. J., Jr.,** Habits of the star-nosed mole, *Condylura cristata, J. Mammal.,* 12, 345–355, 1931.
126. **Richard, P. B. and Viallard, A. V.,** Le Desman des Pyreness (*Galemys pyrenaicus*): premieres notes sur sa biologie, *Terre Vie,* 3, 225–245, 1969.
127. **Dalquest, W. W., and Orcutt, D. R.,** The biology of the least shrew-mole *Neurotrichus gibbsii minor, Am. Midl. Nat.,* 27, 387–401, 1942.
128. **Hawes, M. L.,** Zoology Department, University of British Columbia, Vancouver, personal communication, 1975.
129. **Reed, C. A.,** Behavior of a shrew-mole in captivity, *J. Mammal.,* 25, 196–198, 1944.
130. **Eadie, W. R.,** A contribution to the biology of Parascalops breweri, *J. Mammal.,* 20, 150–173, 1939.
131. **Arlton, A. V.,** An ecological study of the mole, *J. Mammal.,* 17, 349–371, 1936.
132. **Henning, W. L.,** Method for keeping the eastern mole in captivity, *J. Mammal.,* 33, 392–395, 1952.
133. **Rood, J. P.,** Notes on a captive mole, *J. Mammal.,* 39, 583–584, 1958.
134. **Christian, J. J.,** Behavior of the mole (*Scalopus*) and the shrew (*Blarina*), *J. Mammal.,* 31, 281–287, 1950.
135. **Walker, E. P.,** Observations on the eastern mole in captivity, *J. Mammal.,* 26, 192, 1945.
136. **Leach, B. J., Bauman, T. R., and Turner, C. W.,** Thyroxine secretion rate of Missouri Valley mole, *Salopus aquaticus, Proc. Soc. Exp. Biol. Med.,* 110, 681–682, 1962.
137. **Hisaw, F. L.,** Feeding habits of moles, *J. Mammal.,* 4, 9–20, 1923.
138. **Moore, A. W.,** Food habits of townsend and coast moles, *J. Mammal.,* 14, 36–40, 1933.
139. **Wight, H. M.,** Food habits of Townsend's mole *Scapanus townsendii* (Bachman), *J. Mammal.,* 9, 19–33, 1928.
140. **Skoczeń, S.,** Stomach contents of the mole, *Talpa europaea* Linnaeus 1758 from southern Poland, *Acta Theriol.,* 11, 551–557, 1966.
141. **Haeck, J.,** Colonization of the mole (*Talpa europaea* L.) in the Ijsselmeerpolders, *Neth. J. Zool.,* 19, 145–248, 1969.
142. **Lapiń, I. M.,** *Biology and Parasites of Small Forest Mammals of the Latvian SSR,* Academy of Sciences of the Latvian SSR, Riga, 1963 (Russian).
143. **Mellanby, K.,** *The Mole,* Taplinger, New York, 1973, 58–67.
144. **Oppermann, J.,** Die Nahrung des Maulwurfs (*Talpa europaea* L. 1758) in unterscheidlichen Lebensraumen, *Pedobiologia,* 8, 59–79, 1968.
145. **Mellanby, K.,** Food and activity in the mole *Talpa europaea, Nature,* 215, 1128–1130, 1967.
146. **Lund, R.,** Discussant in Catching and keeping live moles, Rudge, A. J. B., *J. Zool.,* 149, 42–45, 1966.
147. **Quilliam, T. A.,** The mole's sensory apparatus, *J. Zool.,* 149, 76–88, 1966.
148. **Eisenberg, J. F. and Gould, E.,** The maintenance of tenrecoid insectivores in captivity, *Int. Zoo Yearb.,* 7, 194–196, 1967.
149. **Herter, K.,** Untersuchungen an lebenden Borstenigeln (Tenrecinae); 2 Über das Verhalten und die Lebensweise des Igeltanreks *Echinops telfairi* Martin in Gefangenschaft, *Zool. Beitr.,* 8, 125–165, 1963.
150. **Eisenberg, J. F. and Gould, E.,** The tenrecs; a study in mammalian behavior and evolution, *Smithson. Contrib. Zool.,* No. 27, 1–137, 1970.
151. **Eisenberg, J. F. and Maliniak, E.,** The reproduction of the genus *Microgale* in captivity, *Int. Zoo Yearb.,* 14, 108–110, 1974.
152. **Rand, A. L.,** On the habits of some Madagascar mammals., *J. Mammal.,* 16, 89–104, 1935.
153. **Louwman, J. W. W.,** Breeding the tailless tenrec, *Tenrec ecaudatus* at Wassenaar Zoo, *Int. Zoo Yearb.,* 13, 125–126, 1973.

FOODS OF WILD AND CAPTIVE TUPAIIDAE (TREE SHREWS)

G. L. Dryden

The Tupaiidae comprise a relatively homogeneous group of small, arboreal mammals geographically more restricted (to Southeast and insular Asia) than either the Insectivora or the Primates, both of which groups have been thought to include or exclude them.[1] Most information about their dietary habits is derived from observations on captive specimens, but this is substantiated by limited data from wild tree shrews. Like insectivores, they eat insects and small vertebrates. Like small primates, they also take considerable amounts of soft plant parts. Like both, they are omnivorous within limits imposed by their size and masticatory abilities.

Food preferences of a few species have been recorded.[2] Data reported here are from original sources and cover representative species of four of the six currently recognized genera of tupaiids.

Table 1
FOODS OF WILD (W) AND CAPTIVE (C) TUPAIIDS (LISTED ALPHABETICALLY)

Tupaiid	Food	Ref.
Lyonogale tana[a]	(C) Grasshoppers, mice, bananas, grapes, apples	2, 3
	Crickets, canned horsemeat, lettuce, oranges, carrots	2
	Meat, insects, chicken, herring, earthworms, eggs, pears; drinks tea, milk, and porridge	3
Ptilocercus lowii	(W,C) Cockroaches, beetles	4
	(W) Black ants, earwigs, cicadas, leaf insects, gecko (*Gymnodactylus*)	4
	(C) Mice, liver, grasshoppers, locusts, mealworms, bananas, papayas, oranges, mangoes, apples; crickets	4; 5
Tupaia belangeri	(C) Mealworms, crickets, locusts, carrots, apples, oranges, bananas, pears, melon, grapes, raisins, avocado pears, Cérélac (cereal paste), Vionate® (mineral supplement)	6
Tupaia chinensis	(C) Crickets, grasshoppers, cockroaches, mice, canned and dry dog food, melons, grapes, bananas, apples, oranges, lettuce, carrots, peanuts, sunflower seeds	1
Tupaia glis	(C) Bananas, oranges, apples	1, 7–9
	Carrots, lettuce; grapes, canned dog food	1, 7; 1, 8
	Dry dog food; Purina® monkey chow	1; 8
	Melons, peanuts, sunflower seeds, mice, crickets, grasshoppers, cockroaches; pears	1; 7, 9
	Fresh and preserved cherries, boiled beet root (*Beta*), nuts, soaked maize and rice, *Tenebrio molitor* larvae, biscuits, rusks and maizena pudding, dried milk with oatmeal; drinks tea, fruit juice, a sugar-vitamin-mineral solution	7

[a] Presumed synonymous with *Tana tana* and *Tupaia tana* described by references cited.

Table 1 (continued)
FOODS OF WILD (W) AND CAPTIVE (C) TUPAIIDS (LISTED ALPHABETICALLY)

Tupaiid	Food	Ref.
Tupaia glis (continued)	Prunes, papaya, coconut, pineapple, bark, hermit crabs, flying insects, snails, lizards, raw hamburger, chicken and beef liver, cooked egg, bacon	8
	Insects: *Locusta migratorina, Oxa japonica, Euconocephalus thunbergi, Paratenodera angustipennis, Acrida turrita*	9
Tupaia gracilis	(C) Crickets, grasshoppers, grapes, melons, bananas, apples, dry dog food, lettuce, oranges (order of preference)	2
Tupaia longipes	(C) Crickets, grasshoppers, grapes, melons, lettuce, bananas, apples, carrots, oranges, canned horsemeat, mice (order of preference)	2
Tupaia minor	(C) Crickets, grasshoppers, dry dog food, melons, grapes, bananas, apples, lettuce, oranges (order of preference)	2
Tupaia montana	(C) Same as *Tupaia chinensis* above	1
Tupaia palawanensis	(C) Same as *T. chinensis* above	1
Urogale everetti	(W) Plants: *Spondias purpurea, Ipomea batatas, Ficus carpenteriana, Amomum dalican, Musa* sp.	10
	(C) Meat; rhinoceros beetles (*Orycetes*); insects, bird and turtle eggs, miscellaneous fruits and vegetables	10, 11; 10; 11
	Apples, bananas, milk; mice, lizards, larval and adult insects, dried locusts, earthworms, guava, papaya, mango, prunes, oranges, raisins, pineapple, eggplant, avocado, green corn and popcorn, cooked peanuts, sweet potatoes and rice, coconut, egg, oatmeal	10, 12; 10
	Cherries, peanuts, hardboiled egg, mealworms, lizards (*Anolis*), mixture: ground horsemeat, bone meal, cod-liver oil.	12

REFERENCES

1. **Sorenson, M. W.,** Behavior of tree shrews, *Primate Behav.,* 1, 141–193, 1970.
2. **Sorenson, M. W. and Conaway, C. H.,** Observations of tree shrews in captivity, *J. Sabah Soc.,* 2, 77–91, 1964.
3. **Roelvinck, W. and Immendorf, M.,** Breeding the Bornean tupaia, *Int. Zoo Yearb.,* 4, 76, 1962.
4. **Lim, B. L.,** Note on the food habits of *Ptilocercus lowii* Gray (Pentail tree-shrew) and *Echinosorex gymnurus* (Raffles) (Moonrat) in Malaya with remarks on "ecological labelling" by parasite patterns, *J. Zool.,* 152, 375–379, 1967.
5. **Gould, E.,** Laboratory of Comparative Behavior, Johns Hopkins University, Baltimore, personal communication.
6. **Martin, R. D.,** Breeding tree shrews and mouse lemurs, *Int. Zoo Yearb.,* 15, 35–41, 1975.
7. **Gensch, W.,** Breeding tupaias, *Int. Zoo Yearb.,* 4, 75–76, 1962.
8. **Kaufmann, J. H.,** Studies on the behavior of captive tree shrews (*Tupaia glis*), *Folia Primatol.,* 3, 50–74, 1965.
9. **Asakura, S.,** Breeding common tupaias at Tama Zoo, Tokyo, *Int. Zoo Yearb.,* 10, 70, 1970.
10. **Wharton, C. H.,** Notes on the Philippine tree shrew, *Urogale everetti* Thomas, *J. Mammal.,* 31, 352–354, 1950.
11. **Snedigar, R.,** Breeding of the Philippine tree shrew, *Urogale everetti* Thomas, *J. Mammal.,* 30, 194–195, 1949.
12. **Crandall, L. S.,** *The Management of Wild Mammals in Captivity,* University of Chicago Press, Chicago, 1964, 51–53.

DIETS FOR CHIROPTERA

J. J. Rasweiler IV

INSECTIVOROUS BATS

Efforts have been made to maintain a variety of insectivorous bats in captivity generally on readily available insects (such as mealworms, *Tenebrio molitor*) or mash diets containing some insects. Although the exclusive feeding of mealworms has apparently yielded satisfactory results in the maintenance of adult bats in several laboratories, other investigators have noted that the condition of captive animals receiving such a diet eventually deteriorated.[1] In most cases the factors responsible for these differing results have not been clearly identified; however, one must suspect that the culture methods utilized to rear the mealworms may have often played an important role. Even if the tissue composition of mealworms (or other insects) is not affected by the use of a very simple culture medium, their digestive tract contents would be. This might in turn reduce the nutritional value of the insects to any insectivorous animals. The digestive tracts of some wild caught insects would, for example, probably contain green plant material. In contrast to the grain products generally used as mealworm culture media, green leaves are frequently rich in calcium, have relatively high calcium-to-phosphorus ratios, and contain vitamins (such as ascorbic acid, folic acid, β-carotene = provitamin A, and vitamin K) which are either absent from or poorly represented in the former. If mealworms or other insects are to be cultured for eventual use as bat food, it would seem to be highly advisable to rear or at least fatten the insects on media containing all of the nutrients known to be required by mammals.[1] Rich sources of vitamins and minerals have been both added to the medium used to rear mealworms and dusted on the insects immediately before they were offered to the bats maintained in one laboratory.[2,3] Rat pellets have routinely been broken up and incorporated into the mealworm culture medium utilized in another facility.[4] Presumably, a similar effect could be achieved by adding one of the diets formulated for other domesticated species to a wheat bran base. Such mixtures have been recommended as culture media for mealworms.[5] Of the possible choices, a pelleted food designed for New World monkeys might be the most advisable because it should contain both ascorbic acid and vitamin D_3. There is evidence that several bats, including one insectivorous form (*Vesperugo abramus* = *Pipistrellus* sp.), lack the ability to synthesize ascorbic acid,[6] and it is not clear if bats can utilize vitamin D_2.[1] Regardless of the type of supplementation practiced, it should be kept in mind that the levels of some vitamins may gradually decline due to their degradation following addition to the insect culture medium. Comparable efforts to provide for vitamin and mineral requirements should be made when formulating artificial diets for bats (Table 2).

Since little is known about the specific vitamin requirements of bats, a serious attempt should be made to supply all of their possible needs. Unfortunately, most commercial supplements have been designed for humans or for some of the domesticated animals, and one or more of the vitamins (such as ascorbic acid, folic acid, vitamin K, and biotin) which might possibly be required by bats have frequently been omitted. Furthermore, it has not yet been established that bats can always utilize the same forms of some vitamins as these other species. One of the more complete vitamin preparations currently available would appear to be Vitamin Diet Fortification Mixture.*[1]

Many of the insectivorous bats normally capture their prey while in flight and will not

* ICN Nutritional Biochemicals, Cleveland, Ohio.

immediately take to feeding upon food offered in dishes. These species must therefore either be trained to eat from dishes or hand-fed each day. Several procedures have been developed that facilitate the feeding of such bats. Tesh and Arata[7] devised inexpensive restraining devices from disposable plastic syringes so that a number of bats could be simultaneously immobilized for hand-feedings. Pagels and Blem[8] restrained their bats in the same way but offered food to the animals from dishes positioned beneath each bat's head. Nellis[9] trained his bats to feed from dishes by first placing them directly on top of a layer of mealworms about 10 mm deep in a shallow pan. A mesh cover was placed over the bats to prevent their escape. The squirming of the mealworms generally induced the bats to bite, and once a mealworm had been punctured it was usually consumed. Eventually the bats learned to descend screen partitions to seek mealworms from a similar pan on the cage floor. Several investigators have suggested that insectivorous bats should be housed in small cages (such as cylindrical cardboard containers) while learning to feed. Others have made it a practice to retain as captives only those animals that quickly learned to feed themselves.[1] Finally, the use of elevated feeding dishes has been recommended for those species that are hesitant to descend to food offered on the cage floor.[10]

The provision of an adequate diet is but one of a number of factors that must be considered when maintaining insectivorous or other bats in captivity. Cage designs that have proved to be useful in housing these animals, along with the control of other important environmental parameters, have recently been considered elsewhere.[1,3,10] Special attention deserves to be drawn to the observation that some of the most successful efforts at maintaining certain insectivorous bats in captivity have been obtained by keeping the animals at elevated ambient temperatures. A number of temperate latitude species normally occupy very warm diurnal roosts during the summer, while some of those living in the tropics appear to favor such roosting sites throughout the year.[1]

CARNIVOROUS AND PISCIVOROUS BATS

Although smaller vertebrates and insects have frequently been used to maintain piscivorous and carnivorous bats in captivity, such foods are relatively expensive and inconvenient to provide. Some of these bats may accept one of the well-balanced canned diets developed for the domesticated or captive, wild carnivores. If fresh meat or fish must be utilized, efforts should be made to supplement them with additional sources of vitamins and minerals. In the absence of bones and entrails, such foods would probably lack some required nutrients.[1] Other carnivorous mammals have been maintained at the Basel Zoo on raw minced meat (93%) mixed with a powdered supplement (7%).[11,12] The composition of the latter, which has also been mixed with fruit to feed frugivorous bats, is listed under formula diet 11 in Table 6. A similar approach to feeding many of the carnivorous and piscivorous bats might prove feasible.

FRUGIVOROUS, NECTARIVOROUS, POLLENIVOROUS, AND OMNIVOROUS BATS

It has long been widely assumed that most bats belonging to the families Phyllostomatidae and Pteropodidae feed primarily upon fruits and/or nectar, and these animals have generally been offered similar diets when held in captivity. However, an increasing body of evidence indicates that many members of both groups also include some insects, pollen, floral parts, buds, and/or green foliage in their natural diets. These latter items could contribute very significantly to the fulfillment of protein, vitamin, mineral, and possibly fatty acid requirements.[1] In fact, several recent studies have provided important experimental evidence that pure-fruit or -nectar diets are nutritionally inadequate for some of the phyllostomatid and pteropodid bats.[13-15] While a number of species are listed in Table 5 as having been maintained on only fruits or fruit juices, it would seem to be highly advisable to supplement such foods with additional high quality proteins, vitamins, minerals, and fatty acids. Several formula diets that attempt to provide more balanced rations for members of these families are listed in Table 6. The problem of selecting appropriate nutritional supplements has been discussed in greater detail in the section dealing with insectivorous bats and elsewhere.[1]

REFERENCES

1. **Rasweiler, J. J., IV,** in *Biology of Bats,* Vol. 3, Wimsatt, W. A., Ed., Academic Press, New York, in press.
2. **Kleiman, D. G. and Racey, P. A.,** *Lynx Prague,* 10, 65–77, 1969.
3. **Racey, P. A.,** *Lab. Anim.,* 4, 171–183, 1970.
4. **Wimsatt, W. A.,** personal communication.
5. Ward's Culture Leaflet No. 13, Ward's Natural Science Establishment, Rochester, N.Y.
6. **Chatterjee, I. B.,** *Science,* 182, 1271–1272, 1973.
7. **Tesh, R. B. and Arata, A. A.,** *Health Lab. Sci.,* 4, 106–112, 1967.
8. **Pagels, J. F. and Blem, C. R.,** *Comp. Biochem. Physiol. A,* 45, 497–501, 1973.
9. **Nellis, D. W.,** *J. Mammal.,* 50, 370, 1969.
10. **Racey, P. A.,** in *The UFAW Handbook on the Care and Management of Laboratory Animals,* Williams and Wilkins, Baltimore, 1972, 296–308.
11. **Wackernagel, H.,** *Int. Zoo Yearb.,* 6, 23–37, 1966.
12. **Wackernagel, H.,** in *Comparative Nutrition of Wild Animals,* Crawford, M. A., Ed., Academic Press, London, 1968, 1–12.
13. **Buckland-Wright, J. C. and Pye, J. D.,** *Int. Zoo Yearb.,* 13, 271–277, 1973.
14. **Howell, D. J.,** *Comp. Biochem. Physiol. A,* 48, 263–276, 1974.
15. **Green, R., van Tonder, S. V., Oettle, G. J., Cole, G., and Metz, J.,** *Nature,* 254, 148–150, 1975.

Table 1
DIETS FOR INSECTIVOROUS BATS

	Diet[a]	Supplemental notes	Ref.
	Family Emballonuridae		
Taphozous melanopogon, Taphozous philippinensis	Insects	These bats were maintained only for a few days.	1
	Family Megadermatidae		
Megaderma spasma	Live insects accepted reluctantly	Survived only for a few days in captivity	1
	Family Molossidae		
Tadarida brasiliensis	Meal worms	It was found that *T. brasiliensis* could be trained to consume mealworms offered in dishes or pans.	2
	Mealworms and a vitamin supplement (Stuart® Liquid Formula) at rate of one drop per bat every second day	One animal was kept for over 4 years.	4
	Mealworms and an unidentified vitamin supplement	The breeding success of a colony maintained on this diet was poor.	5
	Mash for insectivorous bats (Krutzsch and Sulkin formula – see Table 2)	*T. brasiliensis* have been maintained for months on this diet.	6, 7

[a] Sources of products mentioned in table:

ABDEC vitamins: Parke, Davis and Co., Detroit, Michigan
Panvitan A: Apparently prepared and marketed in Japan; address of manufacturer unavailable
Poly-Vi-Sol: Mead Johnson Laboratories, Evansville, Indiana
Protovit: Hoffman-LaRoche and Co., Ltd., 4002 Basel, Switzerland
Stress: Phillips Yeast Products, Park Royal, London, N.W. 10, England
Stuart Liquid Formula: Stuart Company, Pasadena, California
Theragran: E. R. Squibb and Sons, New York, New York
Vionate: E. R. Squibb and Sons, New York, New York

Table 1 (continued)
DIETS FOR INSECTIVOROUS BATS

	Diet[a]	Supplemental notes	Ref.
	Family Mormoopidae		
Pteronotus fuliginosus, *Pteronotus macleayii*	Mealworms and a vitamin supplement (Poly-Vi-Sol®) added to the drinking water	These bats have been kept in the laboratory for periods of 4–6 weeks.	8
Pteronotus parnellii (formerly *Chilonycteris parnellii* and *Chilonycteris rubiginosa*)	Mealworms and cockroaches	One bat survived for 6 months in captivity on this diet.	9
	Mealworms and a vitamin supplement (Poly-Vi-Sol) added to the drinking water	*P. parnellii* has been maintained for 4–6 months on this diet.	8
Pteronotus personatus (formerly *Chilonycteris personata* and *Chilonycteris psilotis*)	Fruitflies and mealworms	*P. personatus* would readily learn to accept mealworms offered by hand and, in time, would also take them from dishes.	10, 11
	Family Noctilionidae		
Noctilio albiventris (formerly *Noctilio labialis*)	Mealworms and water containing a vitamin supplement	The bats were successfully trained to feed from dishes.	12
	Fish	*N. albiventris* reportedly thrived on a fish diet in one laboratory.[14]	13, 14
		N. albiventris has generally been considered to be solely insectivorous. Some specimens have been captured, however, which were feeding heavily on fruit, while the stomachs of others held fish parts or insects and pollen.	15
	Family Nycteridae		
Nycteris grandis	Live insects accepted reluctantly	Tolerated captivity for several days	1

Table 1 (continued)
DIETS FOR INSECTIVOROUS BATS

	Diet[a]	Supplemental notes	Ref.
	Family Phyllostomatidae		
Macrotus waterhousii	Mealworms and other (?) insects	Several *M. waterhousii* survived for 1 year in captivity.	10
	Mealworms and crickets	The bats were hand-fed for about 3 days at the initiation of the study. Thereafter, the insects and water were left in each cage for them. When the study was terminated prematurely after 6 months due to the destruction of the animal quarters by a storm, the dead female *M. waterhousii* were found to be in advanced pregnancy.	16
	Family Rhinolophidae		
Hipposideros caffer, *Hipposideros cyclops*	Hand-fed mealworms	These bats would not take mealworms from dishes.	17
Hipposideros cineraceus	Mealworms, cockroaches, termites		18
Rhinolophus ferrum-equinum	Mealworms and beetles	The bats were particularly attracted by buzzing beetles and would drop to the cage floor to catch them. They were much less interested in quiet, crawling beetles.	19, 20
	Moths and hand-offered mealworms	The bats were much more likely to catch and consume rapidly moving moths than those at rest. They readily took to feeding on hand-offered mealworms. A pair were maintained for well over a year on this diet.	21, 22
	Mealworms and water offered in dishes on an elevated platform	The bats were trained to fly to the platform in order to obtain their food and water.	23
		Several investigators have noted that *R. ferrum-equinum* is resistant to learning to feed from dishes. It is not clear whether this feeding problem is attributable only to a reluctance of the bats to descend to the dishes or to a lack of interest in the type of insects (usually mealworms) being offered. There is evidence that these bats will drop to the ground in the wild to catch prey, including flightless beetles. Possibly, captive *R. ferrum-equinum* might be induced to feed	24

Table 1 (continued)
DIETS FOR INSECTIVOROUS BATS

	Diet[a]	Supplemental notes	Ref.
	Family Rhinolophidae (continued)		
Rhinolophus hipposideros	Moths		25
	Fruit flies, houseflies, and blowflies		26
	Flies and other soft-skinned insects; will also consume horseflesh		27
Rhinolophus subrufus	Live cockroaches, grasshoppers, and other large insects		1
	Family Rhinopomatidae		
Rhinopoma hardwickei	Mealworms and a vitamin supplement every 2 or 3 days		28
	Family Vespertilionidae		
Antrozous pallidus	Mealworms and a vitamin supplement (Stuart Liquid Formula) at rate of one drop per bat every second day		4, 29
	Mash for insectivorous bats (Walker formula – see Table 2)		29
	Mealworms and crickets; a liquid multivitamin supplement was added to the drinking water		11
	Unidentified brand of canned dog food		30
		In the wild, *A. pallidus* regularly alights on the ground to capture flightless prey (insects and scorpions).	29, 31–33

Table 1 (continued)
DIETS FOR INSECTIVOROUS BATS

	Diet[a]	Supplemental notes	Ref.
	Family Vespertilionidae (continued)		
Barbastella barbastellus	Handfed mealworms; houseflies were picked off of the ceiling by the bats		34
	Innards of mealworms; flies and soft-skinned insects	It was recommended that dried yeast flakes be added to the culture medium used to rear the mealworms and that a vitamin supplement be provided for the bats.	27
	Hand-fed mealworms	It was noted that the mouth of *B. barbastellus* is very small, and that these bats had difficulties in dealing with the chitinous exoskeletons of the mealworms. The bats never mastered feeding on mealworms offered in dishes.	35
Eptesicus fuscus	Mash for insectivorous bats (Gates formula – see Table 2)		36
	Mealworms and chopped-up honey bees	Mealworms were preferred.	37
	Larvae, pupae, and adults of green-bottle fly		38
	Mealworms and a vitamin supplement (Stuart Liquid Formula) at the rate of one drop per bat every second day		4
	Mash for insectivorous bats (Walker formula – see Table 2)	Crickets or flies could be substituted for the mealworms in the mash.	39
	Mash for insectivorous bats (Krutzsch and Sulkin formula – see Table 2)		6, 7
	Mash for insectivorous bats (Davis and Luckens formula – see Table 2)		40
	Mealworms and mash for insectivorous bats (Gould formula – see Table 2)		41

Table 1 (continued)
DIETS FOR INSECTIVOROUS BATS

	Diet[a]	Supplemental notes	Ref.
	Family Vespertilionidae (continued)		
Eptesicus fuscus (continued)	Larvae, pupae, and adults of *Tenebrio molitor* to supplement diet consisting primarily of wild caught insects; sometimes supplemented with mash for insectivorous bats (Davis and Luckens formula – see Table 2); ABDEC® vitamins were added to the drinking water.		42
	Mash for insectivorous bats (Pagels and Blem formula – see Table 2)		43
Eptesicus nilssoni	Mealworms, mealmoths, and caterpillars	*E. nilssoni* learned to eat from dishes after 10–15 days in captivity	44
Eptesicus serotinus	Mealworms	*E. serotinus* was kept often and for long periods. This species easily learned to eat mealworms from dishes.	35
	Mealworms, blowfly larvae and pupae, and occasional locusts; a powdered vitamin-mineral supplement (Vionate) was sprinkled over the insects before they were offered to the bats; when pregnant or lactating bats were being maintained, a compound containing additional calcium and phosphorus (Stress) was also dusted on the insects	Young bats derived from inseminations or conceptions in the wild were born and reared in captivity. Copulations were observed in captivity. Bats have survived for up to 3 years on this diet.	45–47
Eptesicus tenuipinnis	Hand-fed mealworms		17
Euderma maculatum	Cottage cheese and flies		48
	Mealworms		49
Lasionycteris noctivagans	Banana, bits of raw meat, and insects		32
	Mealworms		50
	Raw hamburger and occasional drops of a multivitamin preparation	Individuals were kept for more than a year.	51

Table 1 (continued)
DIETS FOR INSECTIVOROUS BATS

	Diet[a]	Supplemental notes	Ref.
	Family Vespertilionidae (continued)		
Lasiurus borealis	Mash for insectivorous bats (Gates formula – see Table 2)		36
	Mealworms and vitamin supplement (Stuart Liquid Formula) at the rate of one drop per bat every second day	*L. borealis* did not learn to feed themselves in less than a month and sometimes required much longer periods.	4
	Mealworms and a liquid multivitamin supplement added to the drinking water		12
	Bits of meat, insects, or cottage cheese have been accepted by captives maintained for a few days	*L. borealis* was found to adjust poorly to captivity. When first caged, they thrashed about so violently that injuries were frequently incurred. Captives would not accept mealworms or artificial food.	32
Lasiurus cinereus	Mealworms and a vitamin supplement (Stuart Liquid Formula) at the rate of one drop per bat every second day	*L. cinereus* did not learn to feed themselves in less than a month and sometimes required much longer periods.	4
	Mealworms, moths, and beetles	A single bat was kept for 6 weeks.	52
	Larvae, pupae, and adults of *Tenebrio molitor*		53
	Mealworms	*L. cinereus* were reported to be extremely reluctant to eat or drink by themselves. It was necessary to hand-feed and water the bats two times per day in order to maintain or increase their body weights.	54
Lasiurus seminolus	Mash for insectivorous bats (Gates formula – see Table 2)		36
	Cooked liver, boiled egg, and crickets	A single bat was kept in captivity for a short period.	55
Minopterus medius, *Miniopterus schreibersii*	Mealworms, cockroaches, and termites		18
Myotis austroriparius	Mash for insectivorous bats (Walker formula – see Table 2)	Many *M. austroriparius* would not learn to feed themselves.	39
Myotis bechsteini	Moths		22
Myotis californicus	Termites		38
	Mealworms and a vitamin supplement (Stuart Liquid Formula) at the rate		4

Table 1 (continued)
DIETS FOR INSECTIVOROUS BATS

	Diet[a]	Supplemental notes	Ref.
Family Vespertilionidae (continued)			
Myotis daubentoni	Mealworms and moths		56
	Mealworms		35
		M. daubentoni occasionally consumes small fish in addition to insects in the wild, suggesting that a fish-based artificial diet (see Table 3) might be acceptable to this species.	57, 58
Myotis keenii	Mealworms	One bat was kept for 3 months.	50
Myotis leibii, *Myotis lucifugus*	Mash for insectivorous bats (Gates formula – see Table 2)		36
Myotis lucifugus	Mealworms		59
		Large numbers of *M. lucifugus* have reportedly been very successfully kept for extended periods at elevated temperatures (33.0–33.5°C) on a diet of only mealworms. Under these conditions the bats learned to feed themselves after only one or two hand-feedings.	60–64
	Mealworms and a vitamin supplement (Stuart Liquid Formula) at a rate of one drop per bat every second day		4
	Mealworms and a liquid multivitamin supplement added to the drinking water		12
	Mealworms and chopped-up honey bees	Mealworms were preferred.	37
	Larvae, pupae, and adults of *Tenebrio molitor;* ABDEC vitamins were added to the drinking water.	The bats were hand-fed throughout the study.	42
	Mash for insectivorous bats (Walker formula – see Table 2)	Many of the bats would not learn to feed themselves.	39
	Mash for insectivorous bats (Krutzsch and Sulkin formula – see Table 2)		6, 7
	Mash for insectivorous bats (Gould formula – see Table 2)	This diet was found to be conducive to the growth of young bats in captivity.	41
	Mash for insectivorous bats (Bell formula – see Table 2)	This diet contains no insects, but permitted the rearing of bats from birth to maturity in the laboratory.	65

Table 1 (continued)
DIETS FOR INSECTIVOROUS BATS

	Diet[a]	Supplemental notes	Ref.
	Family Vespertilionidae (continued)		
Myotis macrotarsus	Insects, fish, and crayfish		1
Myotis myotis	Mealworms	It was noted that *M. myotis* would take mealworms on their own accord on their first night in captivity.	66
	Mealworms, other cultured insects (flesh flies, wax moths, walking-stick insects, and larvae of the biscuit weevil) and wild caught insects	It was recommended that dried yeast flakes be added to the culture medium used to rear the mealworms and that a vitamin supplement be provided for the bats as well.	27
	Mealworms and a multivitamin supplement added to the drinking water (Protovit)	The bats were maintained under semi-natural conditions in an attic with temperature and lighting conditions comparable to those existent in their natural summer roosts. These typically become very warm during the day. The adult *M. myotis* learned to feed from dishes. Many young either born in captivity or captured as infants with their mothers were successfully reared.	67
Myotis mystacinus	Mealworms, moths, and spiders	A single bat was kept for 5 weeks.	68
	Moths and small beetles		22
Myotis nattereri	Mealworms, flies, moths, woodlice, and spiders; flying insects left overnight in the cage were caught and consumed	A single bat was kept for 1 month.	69
	Mealworms	One animal was kept for almost 2 months. It would not feed out of a dish.	70
Myotis sodalis	Mash for insectivorous bats (Gates formula – see Table 2)		36
	Larvae, pupae, and adults of the green-bottle fly		38
	Mealworms and a vitamin supplement (Stuart Liquid Formula) at the rate of one drop per bat every second day		4

Table 1 (continued)
DIETS FOR INSECTIVOROUS BATS

	Diet[a]	Supplemental notes	Ref.
	Family Vespertilionidae (continued)		
Myotis thysanodes	Larvae, pupae, and adults of the green-bottle fly		38
	Mealworms and a vitamin supplement (Stuart Liquid Formula) at the rate of one drop per bat every second day		4
Myotis velifer	Chopped beef heart	The meat was placed on the floor of the cage and was readily consumed by the bats.	71
Myotis yumanensis	Larvae, pupae, and adults of the green-bottle fly		38
	Mealworms and a vitamin supplement (Stuart Liquid Formula) at the rate of one drop per bat every second day		4
Nyctalus lasiopterus	(1) Pupae of house fly; or (2) nearly equal volumes of bovine liver and chicken; or (3) a ground mixture of housefly pupae, bovine liver, chicken, and a small amount of a multivitamin supplement (Panvitan)	The bats were hand-fed only on the first 1 or 2 days after capture. Fifteen young were born to nine females in captivity. Six of the young which were fed diets 1 or 3 were reared to adult size. The others were offered meat alone (diet 2) and died.	72
Nyctalus leisleri	Mealworms and other insects		73, 74
Nyctalus noctula	Mealworms, raw lean beef, moths, beetles, and other insects; will accept crickets and cockroaches	It was noted that *N. noctula* would pick up insects while running around on a flat surface.	75
	Mealworms and moths (larvae, pupae, and adults)	An adult female and her infant were kept for several weeks.	76
	Mealworms	A pair of adults were kept in captivity for an extended period. Three young were born, but two died soon after birth. The third survived for a year. When the latter seemed to be growing poorly, it was hand-fed milk and mealworm innards. Eventually, meat and softboiled eggs were added, but mealworms were preferred.	77

Table 1 (continued)
DIETS FOR INSECTIVOROUS BATS

	Diet[a]	Supplemental notes	Ref.
	Family Vespertilionidae (continued)		
Nyctalus noctula (continued)	Mealworms and small pieces of veal liver		78
	Mealworms; flies and cockroaches were refused	Adult bats were kept for a number of months. Seven females bore a total of ten young in captivity. Five of these were stillborn. Two orphans were reared on goat's milk and crushed mealworms.	79
	Mealworms, blowfly larvae and pupae, and occasional locusts; a powdered vitamin-mineral supplement (Vionate) was sprinkled over the insects before they were offered to the bats. When pregnant or lactating bats were being maintained, a compound containing additional calcium and phosphorus (Stress) was dusted on the insects as well.	Young bats were conceived, born, and reared in captivity. The F_1 generation animals were fertile. Animals survived up to 4 years in captivity.	45–47, 80
Nycticeius humeralis	Fine mixture of eggs, cheese, and mealworms; small pieces of vegetables, fruit, and meat were also readily accepted.		81
	Mealworms offered by hand; water containing a vitamin supplement (Theragran®) was administered from a dropper.	Twenty-eight young were born to 14 females in captivity. Growth curves suggestive of reasonably healthy development were recorded through an age of 100 days. Only two adults ever became accustomed to picking up and eating mealworms from a flat surface.	82

Table 1 (continued)
DIETS FOR INSECTIVOROUS BATS

	Diet[a]	Supplemental notes	Ref.
	Family Vespertilionidae (continued)		
Pipistrellus pipistrellus	Mealworms and moths		56
	Mealworms offered at first by hand; the bats later learned to feed themselves.		76
	Mealworms, blowfly larvae and pupae; a powdered vitamin-mineral supplement (Vionate) was sprinkled over the insects before they were offered to the bats. When pregnant or lactating bats were being maintained, a compound containing additional calcium and phosphorus (Stress) was dusted on the insects as well.	Bats were born and reared in captivity, and they proved to be fertile. Animals have survived for up to 2 years on this regime.	45–47, 80, 83
Pipistrellus subflavus	Mash for insectivorous bats (Gates formula – see Table 2)		36
	Mealworms		59
	Mealworms and hamburger	A single bat was kept for over a month.	50
	Mash for insectivorous bats (Walker formula – see Table 2)	Many of the bats would not learn to feed themselves.	39
Plecotus auritus	Mealworms and moths		56
	Moths, flies, spiders, earwigs, small beetles, caterpillars and pupae	It was noted that *P. auritus* would readily learn to pick up food from the cage floor.	84
	Mealworms taken from a dish and flies		85
	Flies and mealworms	A single bat survived on this diet for 428 days.	86
	Large moths, caterpillars, chrysalids, mealworms, grasshoppers, earwigs, and spiders		25
	Mealworms		35, 87

Table 1 (continued)
DIETS FOR INSECTIVOROUS BATS

	Diet[a]	Supplemental notes	Ref.
	Family Vespertilionidae (continued)		
Plecotus auritus (continued)	Mealworms, blowfly larvae and pupae; a powdered vitamin-mineral supplement (Vionate) was sprinkled over the insects before they were offered to the bats. When pregnant or lactating bats were being maintained, a compound containing additional calcium and phosphorus (Stress) was dusted on the insects as well.	Young bats were born in captivity but only one was successfully reared. Maximum longevity recorded in captivity was 1.5 years.	45–47
Plectous townsendii (formerly *Corynorhinus rafinesquii*)	Mealworms and occasional supplement of percomorph oil (a type of fish oil)	It was noted that in general *P. townsendii* did not take well to captivity. It was difficult to induce the animals to feed themselves, and pregnant bats even when eating well aborted or resorbed their conceptuses. Some individuals did learn to feed out of a dish.	88
	Mealworms and a vitamin supplement (Stuart Liquid Formula) at the rate of one drop per bat every second day		4
Plecotus rafinesquii (formerly *Corynorhinus macrotis*)	Mash for insectivorous bats (Gates formula – see Table 2)		36
Tylonycteris pachypus *Tylonycteris robustula*	Mealworms, cockroaches, and termites		18
Vespertilio murinus	Mealworms	This species would feed upon mealworms offered in dishes.	89

Table 1 (continued)
DIETS FOR INSECTIVOROUS BATS

REFERENCES

1. **Novick, A.,** *J. Exp. Zool.,* 137, 443–461, 1958.
2. **Constantine, D. G.,** *J. Mammal.,* 33, 395–397, 1952.
3. **Stephens, R. J.,** *Am. J. Anat.,* 11, 259–286, 1962.
4. **Orr, R. T.,** *J. Mammal.,* 39, 339–344, 1958.
5. **Baer, G. M. and Holguin, G. M.,** *Am. Midl. Natl.,* 85, 515–517, 1971.
6. **Krutzsch, P. H. and Sulkin, S. E.,** *J. Mammal.,* 39, 262–265, 1958.
7. **Sulkin, S. E. and Allen, R.,** *Virus Infections in Bats,* S. Karger, Basel, 1974.
8. **Henson, O. W., Jr.,** personal communication, 1976.
9. **Novick, A. and Vaisnys, J. R.,** *Biol. Bull.,* 127, 478–488, 1964.
10. **Novick, A.,** *Ergeb. Biol.,* 26, 21–26, 1963.
11. **Novick, A.,** *Biol. Bull.,* 128, 297–314, 1965.
12. **Nellis, D. W.,** *J. Mammal.,* 50, 370, 1969.
13. **Bloedel, P.,** *J. Mammal.,* 36, 390–399, 1955.
14. **Suthers, R. A. and Fattu, J. M.,** *Anim. Behav.,* 21, 61–66, 1973.
15. **Howell, D. J. and Burch, D.,** *Rev. Biol. Trop.,* 21, 281–294, 1974.
16. **Burns, J. M., Baker, R. J., and Bleier, W. J.,** *Gen. Comp. Endocrinol.,* 18, 54–58, 1972.
17. **Eisentraut, M.,** *Biol. Zentralbl.,* 199–209, 1940.
18. **Marshall, A. G. and Lim, B. L.,** *Malay. Nat. J.,* 21, 165–170, 1968.
19. **Coward, T. A.,** *Proc. Zool. Soc. London,* pp. 312–324, 1907.
20. **Coward, T. A.,** *Mem. Proc. Manchester Lit. Philos. Soc.,* 52(11), 1–12, 1908.
21. **Whitaker, A.,** *Naturalist,* pp. 71–77, 1909.
22. **Blackmore, M.,** in *The Handbook of British Mammals,* Southern, H. N., Ed., Blackwell Scientific Publications, Oxford, 1964, 224–249.
23. **DeCoursey, G. and DeCoursey, P. J.,** *Biol. Bull.,* 126, 14–27, 1964.
24. **Rasweiler, J. J., IV,** in *Biology of Bats,* Vol. 3, Wimsatt, W. A., Ed., Academic Press, New York, in press.
25. **Poulton, E. B.,** *Proc. Zool. Soc. London,* pp. 277–303, 1929.
26. **Engländer, H.,** *Zool. Gart.,* 10, 221–224, 1939.
27. **Möhres, F. P.,** *Zool. Gart.,* 18, 217–227, 1951.
28. **Vogel, V. B.,** *Z. Vgl. Physiol.,* 64, 324–345, 1969.
29. **Orr, R. T.,** *Proc. Calif. Acad. Sci.,* Ser. 4, Vol. 28, 165–246, 1954.
30. **Beck, A.,** cited by Barbour, R. W. and Davis, W. H., *Bats of America,* University Press of Kentucky, Lexington, 1969, 191.
31. **Ross, A.,** *Proc. West. Found. Vertebr. Zool.,* 1, 205–264, 1967.
32. **Barbour, R. W. and Davis, W. H.,** *Bats of America,* University Press of Kentucky, Lexington, 1969.
33. **Leen, N. and Novick, A.,** *The World of Bats,* Holt, Rinehart, and Winston, New York, 1969.
34. **Oldham, C.,** *Zoologist,* Ser. 4, Vol. 12, 391–392, 1908.
35. **Eisentraut, M.,** *Aus dem Leben der Fledermäuse und Flughunde,* Veb Gustav Fischer Verlag, Jena, E. Germany, 1957.
36. **Gates, W. H.,** *J. Mammal.,* 17, 268–273, 1936.
37. **Wimsatt, W. A.,** *Anat. Rec.,* 88, 193–204, 1944.
38. **Ramage, M. C.,** *J. Mammal.,* 28, 60–62, 1947.
39. **Mohos, S. C.,** *Anat. Rec.,* 139, 369–378, 1961.
40. **Davis, W. H. and Luckens, M. M.,** *Lab. Anim. Care,* 16, 224–227, 1966.
41. **Gould, E.,** *Commun. Behav. Biol.,* Part A, Vol. 5, 263–313, 1971.
42. **Coutts, R. A., Fenton, M. B., and Glen, E.,** *J. Mammal.,* 54, 985–990, 1973.
43. **Pagels, J. F. and Blem, C. R.,** *Comp. Biochem. Physiol. A,* 45, 497–501, 1973.
44. **Schlott, M.,** *Zool. Gart.,* 14, 35–48, 1942.
45. **Kleiman, D. G.,** *J. Zool.,* 157, 187–211, 1969.
46. **Racey, P. A.,** *Lab. Anim.,* 4, 171–183, 1970.
47. **Racey, P. A. and Kleiman, D. G.,** *Int. Zoo Yearb.,* 10, 65–70, 1970.
48. **Handley, C. O., Jr.,** *Proc. U.S. Natl. Mus.,* 110, 95–246, 1959.
49. **Constantine, D. G.,** *J. Mammal.,* 42, 404–405, 1961.
50. **Smith, P. W. and Parmallee, P. W.,** *Trans. Kans. Acad. Sci.,* 57, 200–205, 1954.
51. **Smith, Mrs. P. W.,** cited in Barbour, R. W. and Davis, W. H., *Bats of America,* University Press of Kentucky, Lexington, 1969, 108.
52. **Jackson, H. H. T.,** *The Mammals of Wisconsin,* University of Wisconsin Press, Madison, 1961.
53. **Brisbin, I. L.,** *J. Mammal.,* 47, 719–720, 1966.

Table 1 (continued)
DIETS FOR INSECTIVOROUS BATS

54. **Bogan, M. A.,** *J. Mammal.,* 53, 611–614, 1972.
55. **Constantine, D. G.,** *J. Mammal.,* 39, 64–70, 1958.
56. **Oldham, C.,** *Zoologist,* Ser. 4, Vol. 3, 471–474, 1899.
57. **Brosset, A. and Deboutteville, C. D.,** *Mammalia,* 30, 247–251, 1966.
58. **Brosset, A.,** *La Biologie des Chiroptères,* Masson et Cie, Paris, 1966.
59. **Griffin, D. R. and Welsh, J. H.,** *J. Mammal.,* 18, 337–342, 1937.
60. **Stones, R. C.,** *J. Mammal.,* 46, 681–682, 1965.
61. **Stones, R. C. and Wiebers, J. E.,** in *Mammalian Hibernation III,* Fisher, K. C., Dawe, A. R., Lyman, C. P., Schönbaum, E., and South, F. E., Eds., Oliver and Boyd, Edinburgh, 1967, 97–109.
62. **Bruce, D. S. and Wiebers, J. E.,** *Physiol. Zool.,* 39, 237–243, 1966.
63. **Bruce, D. S. and Wiebers, J. E.,** *Experientia,* 26, 625–627, 1970.
64. **Bruce, D. S. and Wiebers, J. E.,** *J. Mammal.,* 51, 823–824, 1970.
65. **Bell, J. F.,** personal communication.
66. **Eisentraut, M.,** *Biol. Zentralbl.,* 57, 59–74, 1937.
67. **Weigold, H.,** *J. Comp. Physiol.,* 85, 169–212, 1973.
68. **Oldham, C.,** *Zoologist,* Ser. 4, Vol. 3, 49–53, 1899.
69. **Coward, T. A.,** *Zoologist,* Ser. 4, Vol. 9, 51–56, 1905.
70. **Duncker, G.,** *Zool. Gart.,* 4, 17–27, 1931.
71. **Stager, K. E.,** *J. Mammal.,* 20, 225–228, 1939.
72. **Maeda, K.,** *Mammalia,* 36, 269–378, 1972.
73. **Whitaker, A.,** *Naturalist,* pp. 384–388, 1907.
74. **Whitaker, A.,** *Wild Life* (London), 3, 79–90, 1914.
75. **Oldham, C.,** *Zoologist,* Ser. 4, Vol. 5, 51–59, 1901.
76. **Whitaker, A.,** *Naturalist,* pp. 325–330, 1905.
77. **Mohr, E.,** *Zool. Gart.,* 5, 106–120, 1932.
78. **Skreb, N. and Gjulic, B.,** *Mammalia,* 19, 335–343, 1955.
79. **Dittrich, L.,** *Z. Säugetierkd.,* 23, 99–107, 1958.
80. **Kleiman, D. G. and Racey, P. A.,** *Lynx Prague,* 10, 65–77, 1969.
81. **Walker, E. P.,** *Mammals of the World,* Vol. 1, Johns Hopkins Press, Baltimore, 1964.
82. **Jones, C.,** *J. Mammal.,* 48, 1–19, 1967.
83. **Racey, P. A.,** personal communication.
84. **Whitaker, A.,** *Wild Life* (London), 6, 73–79, 1915.
85. **Mohr, E.,** *Z. Säugetierkd.,* 2, 87–92, 1927.
86. **Kummerlöwe, H.,** *Zool. Gart.,* 2, 106–112, 1929.
87. **Eisentraut, M.,** *Zentralbl. Kleintier. Pelztier.,* 13, 1–184, 1937.
88. **Pearson, O. P., Koford, M. R., and Pearson, A. K.,** *J. Mammal.,* 33, 273–320, 1952.
89. **Mathis, J.,** *Z. Morphol. Ökol. Tiere,* 13, 706–721, 1929.

Table 2
MASH DIETS FOR INSECTIVOROUS BATS

Diet	Composition[a]	Supplemental notes
Bell formula[1]	Lean beef (2400 g) Homogenized baby mice (800 g) Homogenized oat meal sufficient to make mixture dry and crumbly ABDEC® vitamins (20 ml)	This diet was prepared in large quantities and then frozen in smaller portions for future use. During their first few days in captivity, the bats (*Myotis lucifugus*) were hand-fed the diet from the end of a sharpened tongue depressor. More of the food was left overnight in their cages. When large numbers of bats were being kept they were fed from stainless steel troughs designed to reduce soiling and subsequent loss of fur from their ventral body surfaces. *M. lucifugus* were raised from birth to maturity on this diet.
Davis and Luckens formula[2]	One part ripe banana One part dry curd cottage cheese or cream cheese One part canned dog food One part mixed insects (obtained from light traps) Several drops of multivitamin preparation	The ingredients were run through a meat grinder and then mixed together by hand. It was noted that the final mash had to be of a firm consistency; if too thin, the bats would eat poorly and develop diarrhea. Insects collected from light traps can be stored in a freezer for future use.[3] It should be cautioned that insectivorous bats are known to be selective with respect to both the species of insects and the specific insect parts that they will consume. One of the factors underlying this selectivity may be the presence of toxic or distasteful substances in certain insects. If light trap collections are utilized that include too many insects containing such substances, they might either sicken or be rejected by the bats.[4]

[a] Sources of products mentioned in table:

ABDEC vitamins: Parke, Davis and Co., Detroit, Michigan
Jeculin: The Upjohn Company, Kalamazoo, Michigan
Theragran: E. R. Squibb and Sons, New York, New York

Table 2 (continued)
MASH DIETS FOR INSECTIVOROUS BATS

Diet	Composition[a]	Supplemental notes
Gates formula[5]	Mixture of bread or crackers and hard boiled egg, moistened with a small quantity of cream cheese, buttermilk, or milk. Varied by the addition of finely chopped meat, vegetables, banana, yeast, dry malted milk, Ovaltine®, or finely chopped nuts.	The addition of chopped-up honey bees to the diet was also tried. Their presence seemed to improve the consistency of the feces and the palatability of the diet. It was recommended that the diet be blended, so that it had the form of a moist, crumbly mash.
Gould formula[6]	One part hardboiled egg yolk One part banana One part cream cheese One part mealworms Drop of liver extract with iron (Jeculin®)	The mixture was ground up in an electric blender.
Krutzsch and Sulkin formula[7]	1 lb large curd cottage cheese One large banana 1 oz cod liver oil About 150 mealworms	The components of this diet were homogenized in an electric blender. In subsequent studies the cod liver oil was deleted from the diet, and a liver extract with iron (Jeculin) and a multivitamin preparation (Theragran®) were added. Furthermore, rather dry cottage cheese was substituted and agar-agar in nutrient broth was added to the mixture in a final concentration of 1% to reduce the consistency to a semi-solid state.[8,9]
Pagels and Blem formula[10]	Mealworms Whole boiled eggs Cream cheese Bananas Small amounts of finely sifted commercial rat chow	The sifted rat chow was blended in with the other components to improve the consistency of the entire mixture.

Table 2 (continued)
MASH DIETS FOR INSECTIVOROUS BATS

Diet	Composition[a]	Supplemental notes
Walker formula[11]	One yolk of hardboiled egg Equal amount by weight of rather dry cottage cheese Equal amount by weight of banana Equal amount by weight of mealworms Six drops of liver extract with iron (Jeculin) Six drops of wheat germ oil 3 g of multivitamin preparation (Theragran)	This mixture was ground to a paste before being served to the bats. Equal parts of beef and beef liver have been substituted for the cottage cheese in a variant of this formula used to maintain bats in another laboratory.[12]

Table 2 (continued)
MASH DIETS FOR INSECTIVOROUS BATS

REFERENCES

1. **Bell, J. F.,** cited in **Rasweiler, J. J., IV,** in *Biology of Bats,* Vol. 3, Wimsatt, W. A., Ed., Academic Press, New York, in press.
2. **Davis, W. H. and Luckens, M. M.,** *Lab. Anim. Care,* 16, 224–227, 1966.
3. **Barbour, R. W. and Davis, W. H.,** *Bats of America,* University Press of Kentucky, Lexington, 1969.
4. **Rasweiler, J. J., IV,** in *Biology of Bats,* Vol. 3, Wimsatt, W. A., Ed., Academic Press, New York, in press.
5. **Gates, W. H.,** *J. Mammal.,* 17, 268–273, 1936.
6. **Gould, E.,** *Commun. Behav. Biol.,* Part A, Vol. 5, 263–313, 1971.
7. **Krutzsch, P. H. and Sulkin, S. E.,** *J. Mammal.,* 39, 262–265, 1958.
8. **Sulkin, S. E., Allen, R., and Sims, R.,** *Am. J. Trop. Med. Hyg.,* 12, 800–814, 1963.
9. **Sulkin, S. E. and Allen, R.,** *Virus Infections in Bats,* S. Karger, Basel, 1974.
10. **Pagels, J. F. and Blem, C. R.,** *Comp. Biochem. Physiol. A,* 45, 497–501, 1973.
11. **Walker, E. P.,** *Nature Mag.,* 47, 125–128, 1954.
12. **Mohos, S. C.,** *Anat. Rec.,* 139, 369–378, 1961.

Table 3
DIETS FOR CARNIVOROUS AND PISCIVOROUS BATS

	Diet[a]	Supplemental notes	Ref.
	Family Megadermatidae		
Macroderma gigas	Live mice, smaller bats, dead birds, geckos, snakes, large insects, and fresh meat	Live prey were preferred to dead food. When raw meat (especially liver) was given, the bats soon became distressed because their semifluid feces fouled their fur.	1
Megaderma lyra	Mice, rats, gerbils, smaller bats, birds, lizards	Although the consumption of a dead mouse by some captive *M. lyra* was observed, this species had no interest in dead animals offered in a subsequent study.[2,4] These bats also feed upon frogs, fish, and insects in the wild.	2–6
	Small live bats, geckos, insects		7
	Mealworms, crickets, mice, and a vitamin supplement (Protovit)	It was noted that a diet of insects alone would not keep *M. lyra* in a healthy condition.	8
	Family Noctilionidae		
Noctilio albiventris (formerly) *Noctilio labialis*)	Captives have been maintained on fish.		See Table 1
Noctilio leporinus	Fish	Several *N. leporinus* were kept for over 6 months.	9
	Mixture of three large mackerel (size range: 350–454 g), 114 g of Zu/Preem® Feline Diet, and 1.2 ml of a vitamin supplement (ViPenta®)	A large colony has been successfully maintained on this diet for an extended period at the New York Zoological Park.	10
		In the wild, *N. leporinus* feeds on both fish and insects.	11–14

[a] Sources of products mentioned in table:

Protovit: Hoffman-LaRoche and Co., Ltd., 4002 Basel, Switzerland.
Stuart Liquid Formula: Stuart Company, Pasadena, California
Vi-Penta: Roche Laboratories, Nutley, New Jersey
Zu/Preem Feline Canned Diet: Riviana Foods, Inc., Hills Division, Topeka, Kansas.

Table 3 (continued)
DIETS FOR CARNIVOROUS AND PISCIVOROUS BATS

	Diet[a]	Supplemental notes	Ref.
	Family Phyllostomatidae		
Phyllostomus hastatus	Raw meats, small mammals, and birds		11, 15
		This species is omnivorous in the wild and had been successfully treated as an omnivore in captivity.	See Table 5
Vampyrum spectrum	Raw meats, small mammals, and birds	Fruit was always rejected. A pair was maintained for 5 years on this diet and successfully reared a young.	14, 16, 17
	Family Vespertilionidae		
Myotis daubentoni	Will occasionally consume fish in the wild		See Table 1
Myotis macrotarsus	Will accept fish as well as small crayfish and insects in captivity	One animal was maintained for about 2 months on insects alone.	7
Myotis vivesi	Frozen whitebait (*Spirinchus starksi,* a small smelt-like fish) cut into slices about 1/4 in. in thickness, mealworms on weekends, and a vitamin supplement (Stuart® Liquid Formula) at rate of one drop per bat every second day	Force feeding was necessary for the first few days, but the bats were then fed from dishes. One animal lived for over 9 years in captivity on this diet.	8, 9
	Frozen shrimp	Some *M. vivesi* were maintained for up to 6 months in captivity.	18
		In the wild, *M. vivesi* consumes fish, small crustaceans, and insects.	9, 19, 20

Table 3 (continued)
DIETS FOR CARNIVOROUS AND PISCIVOROUS BATS

REFERENCES

1. **Douglas, A. M.,** *West. Aust. Nat.,* 10, 125–138, 1967.
2. **Phillips, W. W. A.,** *Manual of the Mammals of Ceylon,* Colombo Museum, Ceylon, 1935.
3. **Gudger, E. W.,** *J. Bombay Nat. Hist. Soc.,* 43, 635–640, 1943.
4. **Prakash, I.,** *J. Mammal.,* 40, 545–547, 1959.
5. **Ramaswamy, K. R.,** *Proc. Natl. Inst. Sci. India Part B,* 27, 287–307, 1961.
6. **Brosset, A.,** *J. Bombay Nat. Hist. Soc.,* 59, 583–624, 1962.
7. **Novick, A.,** *J. Exp. Zool.,* 137, 443–461, 1958.
8. **Möhres, F. P. and Neuweiler, G.,** *Z. Vgl. Physiol.,* 53, 195–227, 1966.
9. **Bloedel, P.,** *J. Mammal.,* 36, 390–399, 1955.
10. **Doherty, J. G.,** New York Zoological Society, Bronx Zoo, personal communication.
11. **Goodwin, G. G. and Greenhall, A. M.,** *Bull. Am. Mus. Nat. Hist.,* 122, 187–302, 1961.
12. **Hooper, E. T. and Brown, J. H.,** *J. Mammal.,* 49, 310–312, 1968.
13. **Fleming, T. H., Hooper, E. T., and Wilson, D. E.,** *Ecology,* 53, 555–569, 1972.
14. **Howell, D. J. and Burch, D.,** *Rev. Biol. Trop.,* 21, 281–294, 1974.
15. **Dunn, L. H.,** *J. Mammal.,* 14, 188–199, 1933.
16. **Ditmars, R. L.,** *Bull. N.Y. Zool. Soc.,* 39, 162–163, 1936.
17. **Greenhall, A. M.,** *J. Mammal.,* 49, 337–340, 1968.
18. **Carpenter, R.,** *Comp. Biochem. Physiol.,* 24, 951–964, 1968.
19. **Burt, W. H.,** *J. Mammal.,* 13, 363–365, 1932.
20. **Reeder, W. G. and Norris, K. S.,** *J. Mammal.,* 35, 81–87, 1954.

Table 4
DIETS FOR SANGUIVOROUS BATS

	Diet	Supplemental notes	Ref.
	Family Phyllostomatidae[a]		
Desmodus rotundus	Defibrinated beef blood	A large captive colony was maintained on this diet for a number of years.[2]	1, 2
	Beef blood defibrinated by stirring and stored at 4°C until offered to the bats in bowls late in the afternoon	Both frozen blood and "instant blood" (shell-frozen and lyophilized, whole defibrinated blood) reconstituted as needed with water also proved to be acceptable to *D. rotundus.* A large captive colony has been maintained for many years on the diet listed to the left.	3, 4
	4500 ml of beef blood plus 500 ml of 3.8% trisodium citrate; the citrated blood was strained through muslin into plastic bottles and frozen at −20°C until the day of use.	Blood treated in this fashion could be stored for up to 6 weeks without losing its palatability. *D. rotundus* has been maintained for extended periods and successfully bred in captivity on this diet.	5, 6
Diaemus youngi	Chicken blood was readily accepted; the blood of goats, donkeys and guinea pigs was also taken; defibrinated cattle blood was refused even when mixed with chicken blood.	Wild *Diaemus* were found by the use of precipitin tests to have fed upon mammalian and/or avian blood. A significant number of bats had been taking cattle blood.	7, 8
	Chicken blood	The bats were given access to live chickens in order to obtain the blood.	9

[a] This classification of *D. rotundus* and *D. youngi* is based upon Forman et al.[10] and Koopman and Jones.[11]

Table 4 (continued)
DIETS FOR SANGUIVOROUS BATS

REFERENCES

1. **Ditmars, R. L. and Greenhall, A. M.,** *Zoologica,* 19, 53–75, 1935.
2. **Trapido, H.,** *J. Mammal.,* 27, 217–219, 1946.
3. **Wimsatt, W. A. and Guerriere, A.,** *J. Mammal.,* 42, 449–455, 1961.
4. **Wimsatt, W. A., Guerriere, A., and Horst, R.,** *J. Mammal.,* 54, 251–254, 1973.
5. **Dickson, J. M. and Green, D. G.,** *Lab. Anim.,* 4, 37–44, 1970.
6. **Quintero, F. and Rasweiler, J. J., IV,** *J. Reprod. Fertil.,* 41, 265–273, 1974.
7. **Goodwin, G. G. and Greenhall, A. M.,** *Bull. Am. Mus. Nat. Hist.,* 122, 187–302, 1961.
8. **Greenhall, A. M.,** *Bijdr. Dierkd.,* 40, 36–39, 1970.
9. **Hawkey, C.,** personal communication.
10. **Forman, G. L., Baker, R. J., and Gerber, J. D.,** *Syst. Zool.,* 17, 417–425, 1970.
11. **Koopman, K. F. and Jones, J. K., Jr.,** in *About Bats,* Slaughter, B. H., and Walton, D. W. Eds., Southern Methodist University Press, Dallas, 1970, 22–28.

Table 5
DIETS FOR FRUGIVOROUS, NECTARIVOROUS, POLLENIVOROUS, AND OMNIVOROUS BATS

	Diet[a]	Supplemental notes	Ref.
	Family Phyllostomatidae		
Anoura caudifera	Formula diet 2 (see Table 6)		1
Anoura geoffroyi	Soft parts of pawpaw	*A. geoffroyi* was kept for 2 months.	2
Anoura sp.	Formula diet 4 (see Table 6)		3
Artibeus jamaicensis	Primarily bananas, cantaloupe, and honeydew melon; occasionally other types of melons, mangos, papayas, and a multivitamin supplement	Six *A. jamaicensis* were maintained in captivity for 13 months. Three infants were conceived and born in captivity. Two bats survived for over 3 years.	4, 5
	Bananas, melons, watermelon, mangos, and papayas		6
	Formula diet 6 (see Table 6)	The *A. jamaicensis* also had access to a liquid diet (formula diet 3, Table 6) prepared for nectarivorous bats maintained in the same cage.	7
	Formula diet 8 (see Table 6)	Twenty-five of 27 *A. jamaicensis* and *Artibeus lituratus* kept on this diet survived for 10 months.	8
	Formula diet 7 (see Table 6)		3
Artibeus lituratus	Zapotes, bananas with mash for insectivorous bats, and papayas		9
	Bananas, melons, watermelon, mangos, and papayas		6
	Formula diet 5 (see Table 6)	A group of 81 females were introduced into captivity and maintained in relatively small cages. During the first 10 days of confinement, 14 bats died (7 of these were females which aborted). Another 9 animals died during the subsequent 10 months.	10

[a] Sources of products mentioned in table:

Complan: Glaxo Laboratories Ltd., Greenford, Middlesex, England
Hepovite: Evans Pharmaceuticals, United Kingdom
Pervinal: U.S. Vitamin and Pharmaceutical Corp., Tuckahoe, New York

Table 5 (continued)
DIETS FOR FRUGIVOROUS, NECTARIVOROUS, POLLENIVOROUS, AND OMNIVOROUS BATS

	Diet[a]	Supplemental notes	Ref.
	Family Phyllostomatidae (continued)		
Artibeus lituratus (continued)	Formula diet 6 (see Table 6)	The *A. lituratus* also had access to a liquid diet (formula diet 3, Table 6) prepared for nectarivorous bats maintained in the same cage.	7
	Formula diet 8 (see Table 6)	Twenty-five of 27 *A. lituratus* and *A. jamaicensis* kept on this diet survived for 10 months.	8
	Formula diet 7 (see Table 6)		3
Carollia perspicillata	Bananas, cantaloupes, melons, and mangos		11
	Bananas		12
	Formula diet 1 (see Table 6) supplemented with bananas	The captive colony increased in size from 19 to 44 animals over a period of 14 months. Ten bats (two adults and eight juveniles) died during this interval; two of these deaths were accidental.	13
	Formula diet 6 (see Table 6)	The *C. perspicillata* also had access to a liquid diet (formula diet 3, Table 6) prepared for nectarivorous bats maintained in the same cage.	7
	Formula diet 7 (see Table 6)	*C. perspicillata* have bred and reared young while maintained on this diet.	3
Carollia sp. (apparently *C. perspicillata* and *Carollia brevicauda*)	Formula diet 2 (see Table 6)	Large numbers of *Carollia* were maintained for extended periods on this diet. About 7% died during their first 100 days in captivity. Twenty-one infants caught with their mothers and three others born in captivity were reared to adulthood. Many fertile laboratory matings were recorded.	1, 14

Table 5 (continued)
DIETS FOR FRUGIVOROUS, NECTARIVOROUS, POLLENIVOROUS, AND OMNIVOROUS BATS

	Diet[a]	Supplemental notes	Ref.
	Family Phyllostomatidae (continued)		
Choeronycteris mexicana	Formula diet 4 (see Table 6)		3
Glossophaga soricina	Sugar or honey water, fruit juices and pulp, and insects caught by the bats in an outdoor flight cage (in El Salvador)		15
	Bananas, cantaloupes, melons, and mangos	The *G. soricina* survived for several months.	11
	Soft, overripe bananas and sugar syrup		16
	Ripe papayas and a vitamin-peptone-sugar solution		9
	Bananas, melon, zapotes, sugar syrup, and a vitamin supplement added to the drinking water		6
	Suspension of mashed banana or mango in a strong sugar solution		17
	Canned fruit nectars, mangos, bananas, honeydew melon, cantaloupe, and watermelon		18
	Formula diet 1 (see Table 6)		19–21
	Formula diet 2 (see Table 6)	Large numbers of *G. soricina* were maintained in the laboratory for extended periods on this diet. When the loss of juveniles and suckling young (both of which fared poorly when introduced into captivity from the wild) was neglected, the mortality rate for adult bats during their first 400 days of confinement was less than 11%. Fourteen of 30 pregnancies ended in abortions, but 16 infants were born and successfully raised in captivity. Eight of the nine laboratory-reared females mated with laboratory-reared males and subsequently delivered F_2 generation young.	1, 21, 22
	Formula diet 3 (see Table 6)	The *G. soricina* also had access to a solid diet (formula diet 6, Table 6) prepared for frugivorous bats maintained in the same cage.	7
	Formula diets 4 and 7 (see Table 6) and mealworms		3

Table 5 (continued)
DIETS FOR FRUGIVOROUS, NECTARIVOROUS, POLLENIVOROUS, AND OMNIVOROUS BATS

	Diet[a]	Supplemental notes	Ref.
	Family Phyllostomatidae (continued)		
Hylonycteris underwoodi	Formula diet 4 (see Table 6)		3
Leptonycteris nivalis	Melons, papayas, guavas, cactus fruit, and fruit juices		5
	Bananas, melons, figs, papayas, mangos, zapotes, sugar syrup, and water		6
Leptonycteris sanborni	Hummingbird food and bananas		23
Leptonycteris sp.	Formula diet 4 (see Table 6)		3
Lonchophylla sp.	Bananas, cantaloupes, melons, and mangos		11
Phyllostomus discolor	Bananas, papayas, zapotes, and mash for insectivorous bats		9
	Formula diet 5 (see Table 6)	A group of 41 *P. discolor* were introduced into captivity and maintained for 4–18 months with only one mortality. Ten young were born and reared. Six laboratory-reared young became pregnant in captivity.	10
	Formula diet 6 (see Table 6)	The *P. discolor* also had access to a liquid diet (formula diet 3, Table 6) prepared for *Glossophaga* maintained in the same cage.	7
	Formula diet 7 (see Table 6) and mealworms	*P. discolor* have bred and reared young while maintained on this diet.	3
Phyllostomus hastatus	Fruit, meat or smaller vertebrates, and insects		2, 11, 24, 25
	Bananas, papayas, zapotes, and mash for insectivorous bats		9
	Formula diet 5 (see Table 6)	Nine *P. hastatus* were maintained on this diet in apparent good health for 8 months. Five females gave birth and successfully reared their young.	22
	Formula diet 7 (see Table 6) and mealworms	*P. hastatus* have bred and reared young while maintained on this diet.	3
	Bananas, melons, ground beef, and a vitamin-mineral supplement (Pervinal®)		26
Stenoderma rufum	Mangos, various fruit nectars, and bananas	Three *S. rufum* were maintained for 3 weeks.	27

Table 5 (continued)
DIETS FOR FRUGIVOROUS, NECTARIVOROUS, POLLENIVOROUS, AND OMNIVOROUS BATS

	Diet[a]	Supplemental notes	Ref.
	Family Phyllostomatidae (continued)		
Sturnira lilium	Formula diet 5 (see Table 6)	A small group of *S. lilium* (n=11) suffered a mortality rate of over 50%, during the period of almost 6 months that it was maintained on this diet.	10
	Formula diet 6 (see Table 6)	The *S. lilium* also had access to a liquid diet (formula diet 3, Table 6) prepared for nectarivorous bats maintained in the same cage.	7
	Formula diet 7 (see Table 6)		3
Vampyrops helleri	Formula diet 1 (see Table 6)	A single bat was maintained on this diet in apparent good health for 18 months.	19
Vampyrops lineatus	Formula diet 6 (see Table 6)	The *V. lineatus* also had access to a liquid diet (formula diet 3, Table 6) prepared for nectarivorous bats maintained in the same cage.	7
	Family Pteropodidae		
Cynopterus brachyotis *Cynopterus horsfieldi*	Soft fruits, condensed milk, and a vitamin supplement (Hepovite)		28
Dobsonia minor	Papayas, bananas, and a nutrient broth of prepared baby food and sucrose		29
Eidolon helvum	Bananas		30
	Bananas and preserved figs		31
	Bananas and oat flakes		32
	Bananas, plantains, avocados, and papayas		33
Eonycteris spelaea	(1) Soft fruits, condensed milk, and a vitamin supplement (Hepovite) or (2) a mixture of Complan (6 g), sugar (6 g), and water (100 ml)		28, 34
Epomops franqueti	Soft fruits		35, 36
	Bananas, plantains, avocados, and papayas		33
Lissonycteris angolensis	Bananas, citrus fruit, pineapple, and melon		37
Micropteropus pusillus	Soft fruits and honey		35
	Bananas, plantains, avocados, and papayas		33

Table 5 (continued)
DIETS FOR FRUGIVOROUS, NECTARIVOROUS, POLLENIVOROUS, AND OMNIVOROUS BATS

	Diet[a]	Supplemental notes	Ref.
	Family Pteropodidae (continued)		
Myonycteris torquata	Soft fruits supplemented occasionally with honey and butter	A pair survived for 3 years in captivity, with the female becoming pregnant four times.	38
Nyctimene albiventer	Papayas, bananas, and a nutrient broth of prepared baby food and sucrose	*N. albiventer* fared poorly on this diet, and their condition declined after a week.	29
Nyctimene major	Papayas, bananas, and nutrient broth of prepared baby food and sucrose	*N. major* was reported to have done well in captivity on this diet.	29
Paranyctimene raptor	Papayas, bananas, and a nutrient broth of prepared baby food and sucrose	*P. raptor* fared poorly on this diet, and their condition declined after a week.	29
Pteropus giganteus	Dates, figs, assorted fresh fruits (apples, pears, cherries, and grapes) varied according to the season, and white bread		39
	Ripe fruits and a "bread" composed of fruits mixed with sugar, milk, and rolled oats		40
	Bananas and figs each night; supplemented twice a week with a mash consisting of bananas, rolled oats, a source of calcium, and a vitamin mix		41
	Formula diet 9 (see Table 6)		7
Pteropus gouldii	Bananas, grapes, melons, high-protein baby cereal, and Pervinal		26
Pteropus hypomelanus	Soft fruits, condensed milk, and a vitamin supplement (Hepovite)		28
Pteropus poliocephalus	Ripe fruits and a "bread" composed of fruits mixed with sugar, milk, and rolled oats		40
	Cultivated fruits such as bananas and pawpaws		42
Pteropus vampyrus	Soft fruits, condensed milk, and a vitamin supplement (Hepovite)		28
Pteropus sp.	Formula diet 7 (see Table 6)		3
	Formula diets 10 and 11 (see Table 6)		43

Table 5 (continued)
DIETS FOR FRUGIVOROUS, NECTARIVOROUS, POLLENIVOROUS, AND OMNIVOROUS BATS

	Diet[a]	Supplemental notes	Ref.
	Family Pteropodidae (continued)		
Rousettus aegyptiacus	Bananas, apples, and rolled oats	*R. aegyptiacus* were kept for several years, and some young were reared.	44, 45
	Bananas	A single bat was kept for 9 months.	46
	Bananas and oat flakes		32
	Bananas, pawpaws, and occasionally other soft, sweet fruits		47
	Bananas supplemented with a rich source of calcium and vitamin D (Complan)		48
	Bananas, pawpaws, and mangos		49
	Bananas, papayas, pears, and oranges	After being maintained for extended periods on this diet, the bats became vitamin B_{12} deficient.	50
Rousettus amplexicaudatus	Soft fruits, condensed milk, and a vitamin supplement (Hepovite)		28
Rousettus leachii	Dates, figs, assorted fresh fruits (apples, pears, cherries, and grapes) varied according to the season, and white bread	A pair of *R. leachii* were successfully bred while maintained on this diet, although the infant apparently exhibited some postnatal growth retardation.	39

Table 5 (continued)
DIETS FOR FRUGIVOROUS, NECTARIVOROUS, POLLENIVOROUS, AND OMNIVOROUS BATS

REFERENCES

1. **Rasweiler, J. J., IV and de Bonilla, H.,** *Lab. Anim. Sci.,* 22, 658–663, 1972.
2. **Goodwin, G. G. and Greenhall, A. M.,** *Bull. Am. Mus. Nat. Hist.,* 122, 187–302, 1961.
3. **Howell, D. J.,** personal communication.
4. **Novick, A.,** *J. Mammal.,* 41, 508–509, 1960.
5. **Novick, A.,** *J. Mammal.,* 44, 44–56, 1963.
6. **Villa, R. B.,** *Los Murcielagos de Mexico,* Instituto de Biologia, Universidad Nacional Antonoma de Mexico, Mexico, 1966.
7. **Doherty, J.,** New York Zoological Society, Bronx Zoo, personal communication.
8. **Farney, J. P.,** personal communication.
9. **McNab, B. K. and Morrison, P.,** *J. Mammal.,* 44, 21–23, 1963.
10. **Rasweiler, J. J., IV and Ishiyama, V.,** *Lab. Anim. Sci.,* 23, 56–61, 1973.
11. **Griffin, D. R. and Novick, A.,** *J. Exp. Zool.,* 130, 251–300, 1955.
12. **Arata, A. A. and Jones, C.,** *Lozania Acta Zoologica Colomb.,* 14, 1–10, 1967.
13. **Kleiman, D. G.,** personal communication.
14. **de Bonilla, H. and Rasweiler, J. J., IV,** *Anat. Rec.,* 179, 385–404, 1974.
15. **Felten, H.,** *Senckenbergiana Biol.,* 37, 179–212, 1956.
16. **Griffin, D. R.,** *Listening in the Dark,* Yale University Press, New Haven, Conn., 1958.
17. **Pye, J. D.,** in *The UFAW Handbook on the Care and Management of Laboratory Animals,* Lane-Petter, W., Worden, A. N., Hill, B. F., Paterson, J. S., and Vevers, H. G., Eds., Williams and Wilkins, Baltimore, 1967, 491–501.
18. **Novick, A.,** personal communication.
19. **Rasweiler, J. J., IV,** *J. Mammal.,* 54, 391–404, 1973.
20. **Rasweiler, J. J., IV,** *J. Reprod. Fertil.,* 31, 249–262, 1972.
21. **Rasweiler, J. J., IV,** *Am. J. Anat.,* 139, 1–36, 1974.
22. **Rasweiler, J. J., IV,** in *Biology of Bats,* Vol. 3, Wimsatt, W. A., Ed., Academic Press, New York, in press.
23. **Carpenter, R. and Graham, J. B.,** *Comp. Biochem. Physiol.,* 22, 709–722, 1967.
24. **Dunn, L. H.,** *J. Mammal.,* 14, 188–199, 1933.
25. **Crandall, L. S.,** *The Management of Wild Mammals in Captivity,* University of Chicago Press, Chicago, 1964.
26. **Thomas, S. P.,** *J. Exp. Biol.,* 63, 273–293, 1975.
27. **Genoways, H. H. and Baker, R. J.,** *Mamm. Species,* No. 18, pp. 1–4, 1972.
28. **Marshall, A. G. and Lim, L. B.,** *Malay. Nat. J.,* 21, 165–170, 1968.
29. **Bartholomew, G. A., Dawson, W. R., and Lasiewski, R. C.,** *Z. Vgl. Physiol.,* 70, 196–209, 1970.
30. **Church, J. C. T. and Noronha, R. F. X.,** *East Afr. Med. J.,* 42, 348–355, 1965.
31. **Kulzer, E.,** *Z. Säugetierkd.,* 34, 129–148, 1969.
32. **Erkert, S.,** *Z. Vgl. Physiol.,* 67, 243–272, 1970.
33. **Jones, C.,** *J. Zool.,* 167, 353–370, 1972.
34. **Ching, L. M. and Marshall, A. G.,** *Malay. Nat. J.,* 21, 171–180, 1968.
35. **Blackwell, K.,** *Int. Zoo Yearb.,* 6, 247, 1966.
36. **Blackwell, K.,** *Int. Zoo Yearb.,* 7, 79–80, 1967.
37. **Novick, A.,** *J. Exp. Zool.,* 137, 443–461, 1958.
38. **Brosset, A.,** *Biol. Gabonica,* 2, 47–86, 1966.
39. **Wunderlich, L.,** *Zool. Gart.,* 32, 78–81, 1891.
40. **Möhres, F. P. and Kulzer, E.,** *Z. Vgl. Physiol.,* 38, 1–29, 1956.
41. **Neuweiler, G.,** *Z. Vgl. Physiol.,* 46, 13–56, 1962.
42. **Nelson, J. E.,** *Anim. Behav.,* 13, 554–557, 1965.
43. **Wackernagel, H.,** *Int. Zoo Yearb.,* 6, 23–37, 1966.
44. **Kulzer, E.,** *Z. Morphol. Ökol. Tiere,* 47, 374–402, 1958.
45. **Kulzer, E.,** *Nat. Volk,* 91, 219–228, 1961.
46. **Griffin, D. R., Novick, A., and Kornfield, M.,** *Biol. Bull.,* 115, 107–113, 1958.
47. **Pennycuick, C. J.,** *J. Exp. Biol.,* 55, 833–845.
48. **Buckland-Wright, J. C. and Pye, J. D.,** *Int. Zoo Yearb.,* 13, 271–277, 1973.
49. **Mutere, F. A.,** *Period. Biol.,* 75, 159–162, 1973.
50. **Green, R., van Tonder, S. V., Oettle, G. J., Cole, G., and Metz, J.,** *Nature,* 254, 148–150, 1975.

Table 6
FORMULA DIETS FOR FRUGIVOROUS, NECTARIVOROUS, POLLENIVOROUS, AND OMNIVOROUS BATS

Diet No.	Composition[a]			Supplemental notes	Ref.
1	Heart's Delight® Peach Nectar		700.00 ml	The cereal and wheat germ were ground to a powder before being mixed with the other components. The dry components were mixed, preweighed, and then refrigerated until the day of use. Final mixing was done in an electric blender.	1
	Beech-Nut® Hi Protein Cereal		29.52 g		
	Wheat germ		4.92 g		
	Whole milk powder		12.81 g		
	Calcium caseinate		20.39 g		
	Sugar		68.00 g		
	Gevral® Protein		3.99 g	It has been suggested that this formula might be further simplified by (1) eliminating the Gevral Protein, a protein-vitamin-mineral supplement, and by elevating the other sources of these nutrients in the diet, (2) eliminating the isopentyl acetate which was utilized to mask the disagreeable odor of the calcium caseinate supplied by one manufacturer, and (3) substituting a commercially prepared, calcium-rich mineral supplement for that listed.	2
	Mineral supplement		1.68 g		
	$NaH_2PO_4 \cdot H_2O$	50.33%			
	$CaHPO_4 \cdot 2H_2O$	47.61%			
	$MnSO_4 \cdot H_2O$	1.81%			
	$ZnCl_2$	0.15%			
	$CuSO_4 \cdot 5H_2O$	0.083%			
	KI	0.014%			

[a] Sources of products mentioned in table:

Alpo Chopped Beef Dinner: Allen Products Company, Inc., Allentown, Pennsylvania
Beech-Nut Hi Protein Cereal: Beech-Nut Life Savers, Inc., Canajoharie, New York
Beef extract (Difco Culture Medium): Difco Laboratories, Detroit, Michigan
Cerpi Tri-Cereal: Comestibles La Rosa, S. A., Santa Rosa de Cabal, Risaralda, Colombia
Ethoxyquin (6-ethoxy-1,2-dihydro-2,2,4-trimethylquinoline): Pfaltz and Bauer, Inc., Flushing, New York
Frutera Colombiana Guava Concentrate: Hartog Trading Corp., New York, New York
Gerber Vegetables and Beef: Gerber Products Co., Fremont, Michigan
Gevral Protein: Lederle Laboratories Division, American Cyanamid Company, Pearl River, New York
Heart's Delight Peach Nectar: California Canners and Growers, San Francisco, California
Instant Protein: Alpine Marine Protein Industries, Inc., New Bedford, Massachusetts
Isopentyl acetate: Matheson Coleman and Bell, Norwood, Ohio
Marine Protein Concentrate: Alpine Marine Protein Industries, Inc., New Bedford, Massachusetts
Mellin's Food: Consolidated Royal Chemical Company, Chicago, Illinois
Pervinal: U.S. Vitamin and Pharmaceutical Corp., Tuckahoe, New York
Polyoxyethylene sorbitan mono-oleate (Tween® 80): Sigma Chemical Company, St. Louis, Missouri
Poly-Vite Drops: Wolin's, Farmingdale, New York
Super Hydramin Powder: Nion Corporation, Los Angeles, California
Vi-Daylin vitamins: Ross Laboratories, Columbus, Ohio
Vi-Penta: Roche Laboratories, Nutley, New Jersey
Vitamin Diet Fortification Mixture: ICN Nutritional Biochemicals, Cleveland, Ohio

Table 6 (continued)
FORMULA DIETS FOR FRUGIVOROUS, NECTARIVOROUS, POLLENIVOROUS, AND OMNIVOROUS BATS

Diet No.	Composition[a]			Supplemental notes	Ref.
1 (continued)	Vitamin Diet Fortification Mixture		0.69 g		
	Corn oil mixture		9.00 ml		
	Corn oil	100.0 ml			
	Polyoxyethylene sorbitan monooleate	2.97 ml			
	Isopentyl acetate	0.22 ml			
	Water		300.00 ml		
2	Frutera® Colombiana Guava Concentrate		194.00 g	This diet was prepared in the same fashion as formula diet 1 (see above).	3
	Cerpi Tri-Cereal (a rice, barley, and corn mixture for infants)		25.38 g		
	Wheat germ		11.93 g	Modifications similar to those suggested for diet 1 might possibly be made in order to simplify preparation of diet 2.	2
	Whole milk powder		15.90 g		
	Calcium caseinate		34.78 g		
	Sugar		183.59 g		
	Gevral Protein		4.97 g		
	Mineral supplement (see formula diet 1 for composition)		2.15 g		
	Vitamin Diet Fortification Mixture		0.86 g		
	Corn oil mixture		11.89 ml		
	Corn oil	100.00 ml			
	Polyoxyethylene sorbitan mono-oleate	2.97 ml			
	Water		806.00 ml		
3	Condensed milk		3/4 cup		4
	Honey		454 g		
	Marine Protein Concentrate		6 teaspoons		
	Beef extract		46 cm (supplied in tubes)		
	Mellin's Food		12 teaspoons		
	Super Hydramine Powder		6 teaspoons		
	Vi-Penta® vitamins		1.8 ml		
	Water		189.2 ml		

Table 6 (continued)
FORMULA DIETS FOR FRUGIVOROUS, NECTARIVOROUS, POLLENIVOROUS, AND OMNIVOROUS BATS

Diet No.	Composition[a]		Supplemental notes	Ref.
4	Eggs	500 g	The mixture is blended before and after adding the water. It can be frozen until the day of use. It should be thoroughly shaken before being served.	5
	Condensed milk (nonsweetened)	510 g		
	Honey or strawberry jam	1370 g		
	Strained baby food (Gerber® Vegetables and Beef)	256 g		
	Instant Protein	130 g		
	Beef extract	85 g		
	Amino acid capsules (lactalbumin hydrolysate)	6 g		
	Poly-Vite Drops	20 g		
	Super Hydramin Powder	140 g		
	Brewer's yeast	135 g		
	Bone meal	15 g		
	Water	5700 g		
5	Sliced bananas	700.00 g	The wheat germ was ground to a powder. The dry components were then mixed, preweighed, and refrigerated until the day of use. Final mixing was done gently by hand so that the banana slices remained reasonably intact.	6
	Wheat germ	9.87 g		
	Whole milk powder	12.90 g		
	Calcium caseinate	24.00 g		
	Sugar	24.48 g		
	Gevral Protein	4.04 g		
	Mineral supplement (see formula diet 1 for composition)	1.71 g	It has been suggested that this formula might be further simplified by (1) eliminating the Gevral Protein, a protein-vitamin-mineral supplement, and by elevating the other sources of these nutrients in the diet, and (2) substituting a commercially prepared calcium-rich mineral supplement for that presently listed. It has also been pointed out that the calcium-to-phosphorus ratio in the diet as presently formulated may fall below the generally desired range of 1:1 to 2:1.	2
	Vitamin Diet Fortification Mixture	0.70 g		
	Corn oil	8.68 ml		

Table 6 (continued)
FORMULA DIETS FOR FRUGIVOROUS, NECTARIVOROUS, POLLENIVOROUS, AND OMNIVOROUS BATS

Diet No.	Composition[a]		Supplemental notes	Ref.
6	Bananas	22		4
	Apple	1		
	Raisins	1 handful		
	Zu/Preem® Feline Canned Diet	604 g		
	Mellin's Food	4 tablespoons		
	Super Hydramin Powder	4 tablespoons		
7	Chopped bananas	10	The diet was mixed gently by hand. Condensed milk and bone meal were added to the food of bats that were obviously pregnant.	5
	Melon pieces (cantaloupe or honeydew)	3–4 cups		
	Cottage cheese	227 g		
	Hamburger or canned dog food (Alpo® Chopped Beef Dinner)	453 g		
	Beef extract	1/4 tube (supplied in tubes)		
	Pervinal®	1/2 cup		
	Eggs	2		
8	Bananas	2–4		7
	Pears, apples, oranges	Small amounts occasionally		
	Skim milk powder	1/4 cup		
	Vi-Daylin® Vitamins	40–50 ml		
9	Apples	214 g		4
	Oranges	214 g		
	Bananas	428 g		
	Grapes	72 g		
	Zu/Preem Feline Canned Diet	72 g		

Table 6 (continued)
FORMULA DIETS FOR FRUGIVOROUS, NECTARIVOROUS, POLLENIVOROUS, AND OMNIVOROUS BATS

Diet No.	Composition[a]		Supplemental notes	Ref.
10	Ground plate maize	120 g	Trace elements added per kilogram of diet (mg):	8
	Ground wheat	100 g	Fe 50.0; I 1.7	
	Barley flakes	100 g	Cu 25.0; Zn 48.0	
	Rolled oats	100 g	Mn 48.0; Co 1.2	
	Groundnut oil meal	100 g	Vitamins added per kilogram of diet:	
	Soya bean meal	80 g	A 60,000 IU; D_3 12,000 IU	
	Wheat germ oil	50 g	B_1 8.0 mg; E 40.0 mg	
	Dried yeast	100 g	B_2 50.0 mg; K 2.0 mg	
	Skimmed milk powder	100 g	B_6 6.0 mg; Folic acid 0.8 mg	
	Soya bean oil	50 g	B_{12} 20.0 μg; Nicotinic acid 40.0 mg	
	Sugar	50 g	C 500.0 μg; Pantothenic acid 20.0 mg	
	Bone meal	20 g	Nine parts of the concentrate listed at left and one part of minced, cooked meat were mixed with meat broth or water. This mixture was then offered to the bats, along with fruit.	
	Salt	10 g		
	Premixture (trace elements and vitamins)	20 g		
11	Fruit dusted with the following supplement		Trace elements added per kilogram of supplement (mg):	8
	Skimmed milk powder	42%	Fe 250.0; Zn 400.0	
	Dried brewer's yeast	42%	Cu 125.0; I 8.6	
	Steamed bone meal	8%	Mn 240.0; Co 6.25	
	Ground calcium carbonate	3%	Vitamins added per kilogram of supplement:	
	Salt	2%	A 100,000 IU; E 1,000 IU	
	Premixture (trace elements and vitamins)	3%	B_2 40.0 mg; K 10.0 mg	
			B_6 50.0 mg; *p*-Aminobenzoic acid 150.0 mg	
			B_{12} 400.0 μg; Biotin 3.0 mg	
			C 1000.0 mg; Folic acid 20.0 mg	
			D_3 7,500 IU; Inositol 3,500.0 mg	
			Nicotinic acid 100.0 mg	
			Pantothenic acid 350.0 mg	
			Antioxidant (Ethoxyquin) added per kilogram of supplement: 150.0 mg	

Table 6 (continued)
FORMULA DIETS FOR FRUGIVOROUS, NECTARIVOROUS, POLLENIVOROUS, AND OMNIVOROUS BATS

REFERENCES

1. **Rasweiler, J. J., IV,** *J. Mammal.,* 54, 391–404, 1973.
2. **Rasweiler, J. J., IV,** in *Biology of Bats,* Vol. 3, Wimsatt, W. A., Ed., Academic Press, New York, in press.
3. **Rasweiler, J. J., IV and de Bonilla, H.,** *Lab. Anim. Sci.,* 22, 658–663, 1972.
4. **Doherty, J.,** New York Zoological Society, Bronx Zoo, personal communication.
5. **Howell, D. J.,** personal communication.
6. **Rasweiler, J. J., IV and Ishiyama, V.,** *Lab. Anim. Sci.,* 23, 56–61, 1973.
7. **Farney, J. P.,** personal communication.
8. **Wackernagel, H.,** *Int. Zoo Yearb.,* 6, 23–37, 1966.

ECOLOGY AND DIETARY HABITS OF THE MALAYSIAN FLYING LEMUR, *CYNOCEPHALUS VARIEGATUS* (Audebert)

Lim Boo Liat

INTRODUCTION

The flying lemur, or colugo (*Cynocephalus*), is a unique animal which is taxonomically classified in its own order, Dermoptera. This order contains two species in a single genus: *Cynocephalus variegatus*, which is common and widely distributed throughout Malaysia, and *Cynocephalus volans*, confined to the Philippines.

The flying lemur is easily confused with the flying squirrel from which it differs in tooth type; the flying lemur's teeth are rodent form, with comb-shaped lower incisors. The color pattern of *C. variegatus* varies. Males have a bright, rich brownish shade, while the females are duller and greyish. In both sexes the back has a generally mottled appearance with soft white or cream-colored spots and lateral fine, black, reticular stripes; the underside of the gliding membrane is pale brownish orange and unspotted. The mottled coloration, resembling the bark of tree trunks, acts as camouflage and may protect the animal against predators.

The habitat of *C. variegatus* ranges from primary forest to cultivated plantations in the open, from lowland to hill forests as high as 2000 m. In the forest, *C. variegatus* lives in holes of tall tree trunks ranging from 30 to 50 m high. In coconut, rubber, cocoa, and other fruit plantations, it hangs upside down on the branches of these trees and palms. It rests spread-eagled against tree trunks or coconut palm branches.

Activity of *C. variegatus* starts immediately after dusk. Specimens were observed to come out of their tree holes at about 18.00 to 19.00 hours. One animal did not make an immediate glide directly from the entrance of the hole, but crawled upwards a few meters above the entrance towards a branch. With outstretched limbs it then glided toward a tree 5 to 15 m away. In the air it floated like a kite and glided at an angle of approximately 60 degrees. The moment it contacted its objective it raised its head slightly and, with all four limbs still outstretched, clung to the trunk of the tree. It then slowly crawled up to a branch and hung upside down (Figure 1) for about 25 to 40 minutes before it started moving to another branch. It continued to glide from tree to tree until it reached a chosen tree where there were young leaves and shoots on which it could feed. Once it had found its feeding area, it remained there for hours, and more often it stayed there hanging upside down until early morning. Before the break of dawn, the animal glided back to its nest. There are instances in which some animals remained in the same tree throughout the day, which accounted for the capture of certain of them in the forest during daylight hours.

FOOD HABITS IN NATURE

A total of 12 specimens were shot during the period from 1969 to 1972 (Table 1). The stomachs with their contents were removed from the carcasses. The contents only were weighed, then placed in a large petri dish, sorted according to coloration, and classified as green matter (green leaves), whitish and yellowish matter (flesh of fruit and floral origin), and insect parts (wings, sclerites, and other remains, probably ants, moths, and beetles). Each sorted part of the contents was weighed, and each weight group of the food components was calculated as percentage of intake.

From examination of the stomach contents (Table 1), it was apparent that the food preference of this animal in nature is green leaves and young shoots; the intake of these items was as high as 80%, the rest being fruits and other floral parts. Insects formed an insignificant part of the food components. A plausible explanation is that while the animal was feeding on the fruit or floral parts it also inadvertently consumed the insects attached to them. It is unlikely that the animal would deliberately hunt for insects.

Information derived from food contents led to further observations of these animals in the field. Five animals were observed during the period of January through April 1973 between 19.00 to 23.00 hours daily in a forest reserve near Kuala Lumpur, Malaysia. These animals were observed to have visited at least 23 trees, and leaves of ten species of fruits and five species of flowers were found to be eaten (Table 2). Because the number of animals observed was too small, it is possible that the observations made could be fairly biased.

FOOD IN CAPTIVITY

The flying lemur has not been known to live for long periods in captivity because it is difficult to substitute its highly restricted natural diet.[1] However, because of the further knowledge gained from field observations, specimens in captivity have been able to thrive fairly well from 15 to 60 weeks. Locally grown fruits such as banana, papaya, orange, and mango are easy to substitute, but green leaves are more difficult. Vegetables such as fresh lettuce and kangkung leaves (*Ipomoea* spp.) are fairly good substitutes as greens, but the leaves of the wild passion fruit (*Passiflora foetida*) and the flowers of coconut palm are preferred. These vegetables are easily purchased in the local markets, and *P. foetida* is obtainable in scrubs, edges of forests, and also in plantations where they grow wild.

In addition to these substitutes, the animals were found to eat grasshoppers that were offered occasionally, i.e., at least once a week. They have been found to like honey, and specimens in captivity took a teaspoonful of pure honey each day. Although the substituted foods mentioned above are still not ideal, for the time being they have been found quite adequate.

REPRODUCTION

Pregnant females were examined in February, April, May, August, September, and December, with only a single embryo found in each individual. The gestation period of *C. variegatus* is not known. In *C. volans* it has been known to be about 60 days, as cited by Walker.[2] A captured pregnant female gave birth to a single young after 7 days in captivity. The young was born with a fully haired back, and it clung with its claws and milk teeth to the belly fur and nipples of the mother. The young started to take soft food after 42 days and was weaned after 3 months. Although it was able to fend for itself, it still clung to the mother until it was about 4½ months old when it eventually became independent. During the period of observation, the young was fed with milk powder, fruits, and fresh vegetables until it was 8 months old, prior to its release.

PARASITES

The flying lemur was found to be infected with helminths, of which the commonest was the oxyurid worm, along with the filarid worm; it also harboured a malaria parasite, *Plasmodium sandoshami*.[3] The presence of these nematodes and a blood protozoan indicates the medical importance of *C. variegatus* in Malaysia.

FIGURE 1. A specimen of *Cynocephalus variegatus* hanging from a tree branch.

Table 1
EXAMINATION OF STOMACH CONTENTS OF *CYNOCEPHALUS VARIEGATUS*

		Stomach contents				
Period	Number of animals examined	Green matter in grams (%)	Whitish and yellowish matter in grams (%)	Insect parts in grams (%)	Total wet weight of contents (g)	Remarks
			1969			
January	1	225 (84.9)	38 (14.3)	2 (0.8)	265	Full
February	1	201 (88.2)	26 (11.4)	1 (0.4)	228	Full
March	1	198 (84.6)	34 (14.5)	2 (0.9)	234	Full
April	1	158 (77.1)	44 (21.5)	3 (1.4)	205	Full
May	1	62 (72.9)	21 (24.7)	2 (2.4)	85	1/3 full
			1970			
August	1	187 (86.6)	28 (12.9)	1 (0.5)	216	Full
September	1	198 (83.2)	38.5 (16.2)	1.5 (0.6)	238	Full
			1971			
June	1	172 (80.0)	40.3 (18.7)	2.7 (1.3)	215	Full
July	1	202 (83.8)	37.8 (15.7)	1.2 (0.5)	241	Full
			1972			
October	1	72 (74.2)	23.7 (24.4)	1.3 (1.4)	97	1/3 full
November	1	199 (82.3)	41.6 (17.2)	1.4 (0.5)	242	Full
December	1	186 (78.8)	48.2 (20.4)	1.8 (0.8)	236	Full

Table 2
LEAVES AND FRUITS OBSERVED TO BE CONSUMED BY *CYNOCEPHALUS VARIEGATUS* IN NATURE

Species	Family	Local name
	Young Leaves	
Sapium buccatum	Euphorbiaceae	Ludai
Parkia speciosa	Mimosaceae	Petai
Planchonella maingayi	Sabotaceae	Minyak burok
Santiria laevigata	Burseraceae	Kendondong kerantai lichin.
Evodio glabra	Rutaceae	Pauh pauh
Koompassia malaccensis	Caesalpiniceae	Kempas
Dialum indum	Legominosae	Keranji
Antiaris toxicaria	Urticaceae	Ipoh
Ranadia densiflora	Rubiaceae	Kaum kopi
Durio sp.	Malvaceae	Durian
	Fruits	
Artocarpsu gomezianus	Urticaceae	Shiny Tampang
Santiria laevigata	Beseraceae	Kendondong kerantai lichin.
Annona sp.	Annonaceae	Mempisang
Artocarpus dadah	Urticaceae	Green Tampang
Randia densiflora	Rubiaceae	Kaum kopi

REFERENCES

1. **Lim Boo Liat,** Observations on the food habits and ecological habitat of the Malaysian Flying Lemur, *Int. Zoo Yearb.*, 7, 196–197, 1967.
2. **Walker, E. P.,** *Genera of Recent Mammals of the World,* 2nd ed., Johns Hopkins University Press, Baltimore, 1968.
3. **Dunn, F. L., Eyles, O. E., and Yap, L. F.,** *Plasmodium sandoshami* sp. nov., a new species of malaria parasite from the Malaysian Flying Lemur, *Ann. Trop. Med. Parasitol.*, 57, 75–81, 1963.

EDENTATE NUTRITION

D. A. Meritt, Jr.

INTRODUCTION

The Class Mammalia has been well studied, with most families thoroughly investigated over time. Although there are some families that have escaped detailed study and query, it is a rare occurrence for an entire mammalian order to receive only cursory study. The Order Edentata is a prime example of such a group of mammals which have not received the investigation and study that is warranted or desirable.

Within the Order Edentata are a diverse and unique grouping of mammals. Members are either toothless or have teeth that are rootless and peglike. Examples show further diversification in having either large numbers of teeth, incisor-like canines, or greatly reduced numbers of teeth. Each animal type has adapted to food and food items that foster life functions and provide elements essential for survival.

Edentates are restricted to the New World, ranging from the extreme southern U.S.A., through Central America, and into the northern half of South America. Some examples may be found as far south as Chile and Argentina.[1] Their habitats include, but are not restricted to, arid and semiarid regions, open woodland, savannah, mixed deciduous forest, and tropical rain forest.[2]

There are three diverse animal groups within this order: anteaters, armadillos, and sloths. Several nutritional approaches have been utilized by these mammals. Anteaters feed primarily on insects, sloths on vegetation, and armadillos on both of these food types as well as fruits and carrion. The anteaters appear to be the least flexible of the three in their choice of food items, sloths somewhat less so, and armadillos the most omnivorous of these mammals.

Within the last decade, techniques have been developed that allow the successful maintenance of these mammals in a captive situation. Some of these efforts have outlined husbandry, nutrition, developmental biology, reproduction, and behavior.[3-8]

ANTEATERS

Since anteaters in nature require large amounts of termites, termite larvae, ants, and other insects, the offering of a palatable, appealing, high protein diet is essential to survival and longevity in a captive state.[6]

The smallest, most delicate, and most secretive of the three anteater genera is the silky or pygmy anteater, *Cyclopes didactylus.* This arboreal anteater is rarely seen in captivity but has variously accepted such food items as raw ground beef, eggs, *Tenebrio* larvae, orange juice, and canned milk.[9] The majority of specimens survive only a few weeks at best, generally as a result of severe debilitation upon arrival. Long-lived captive *Cyclopes* have fed on two liquid diets.* The first is a blend of milk, water, vitamins, protein powder, and Purina® Mink Developer Chow. The second is simply a mixture of water and strained hardboiled egg yolk.[10] Several specimens have exceeded 1 year on these diets, while one example lived 2 years and 5 months, establishing a new captive longevity record which surpassed all other known records by nearly 1 year. It is clear that nutrition alone is not the complete answer to the successful acclimation of this genus to a captive situation. Careful attention to the physical and psychological needs of *Cyclopes* is essential to its captive survival.

* The "Product Information" section following the text describes the various diet ingredients and mixtures mentioned in the text.

The lesser anteater, *Tamandua tetradactyla,* is a medium-sized edentate ranging in weight from 6 to 12 lb as an adult. Equally at home on the ground or in the trees, this mammal feeds on termites, larvae, and ants, as well as insects. Fair numbers have been maintained in captivity but without substantial longevity.[11] Captive examples have been fed and accepted a variety of liquid and semisolid diets with varying degrees of success. The basic diet developed at our facility is a thick, soupy mixture made up of Purina Mink Developer Chow, ground horsemeat, Vidaylin® multivitamins, Paltone® powder with vitamin K added, Gevral-Tone® powder, canned milk, and water. This mixture is fed once or twice daily, depending upon age and weight of the tamandua. It is always offered during the anteater's active period and is placed in the same location daily within the cage or exhibit.[12] It has been demonstrated in this species that the addition of a source of vitamin K activity is essential to its long-term survival in captivity. Without vitamin K added to the diet, subcutaneous hemmorhages have been seen clinically, the result of rather superficial injuries. The daily addition of this vitamin has resulted in the complete absence of these clinical signs and no reoccurrence of these hemorrhagic episodes.

Tamanduas maintained on the palatable, high protein diet have substantially improved upon previous captive longevity records. An adult male and two females, received as babies in 1968, continue to thrive at this time. Several other examples have been on this diet in excess of 3 years without sign of nutritional deficiency. With the onset of sexual maturity and after a successful introductory period, one adult pair produced a nearly full-term, stillborn baby late in 1975. Despite the loss of this first offspring, breeding behavior was observed early in 1976 and a viable offspring was produced early in 1976. This species has been bred in captivity on one other occasion, according to available records. The success of the breeding efforts will ultimately be determined by both behavioral indicators and adequate nutrition in this species.

The largest member of the Myrmecophagidae is the giant anteater, *Myrmecophaga tridactyla.* This essentially terrestrial edentate may weigh in excess of 140 lb and measure more than six ft from tip of nose to end of tail. Food in nature is known to include termites, ants, insects, carrion, and ripe, soft fruit. Captive diets have attempted to duplicate protein levels near those noted in nature. They have generally been offered to captive examples in the form of a thick gruel or semiliquid with varying degrees of success. Our approach has been atypical, the mixture being offered as a semisolid, about the consistency of uncooked meat loaf. The basic ingredients include Purina Mink Developer Chow, ground lean horsemeat, vitamins, minerals, and ground hardboiled eggs. This mixture is readily taken by adult animals, despite the absence of teeth to aid in the breakdown of the solid food. The anteaters use their elongated noses to roll the food into small bits which are then taken into the mouth by the long, thin, moist tongue. The food is swallowed in this form.

One decided advantage of the semisolid diet vs. the liquid diets is the difference in the frequency and consistency of the anteater's feces. Using the former, the stools are firmer and usually passed once a day. No cases of acute diarrhea or rectal prolapse have been observed in anteaters maintained on this mixture.

The need for a source of vitamin K activity in this species has not been clearly demonstrated, but the diet is prophylactically supplemented. One difficulty that has been encountered with this anteater is a drying of the skin, especially during the colder months of the year and during concomitant low relative humidity periods. An increase in the level of humidity and the topical application of baby oil have proven to be an effective treatment for this short-term, seasonal problem in a northern climate.

SLOTHS

The sloths as a family present some unique and intriguing problems in a captive situation. Being entirely arboreal, both the two-toed sloths, *Choloepus didactylus* and *Choloepus hoffmanni,* and three-toed sloths, *Bradypus* sp., require special consideration when confined. Each needs suitable perches, which allow for normal body positions and free access to feeding stations without undue competition. Additionally, both types need a carefully controlled environment, with special attention given to temperature and relative humidity requirements.

Sloths in nature apparently feed exclusively on vegetation. A variety of leaves, buds, shoots, flowers, and fruit are consumed daily. Of the two sloth types, *Choloepus* sp. is the most versatile feeder in captivity, adjusting quite readily to fruits and vegetables as well as meat products. Acceptable food items include cooked sweet potato, white potato, banana, apple, orange, green beans, celery, escarole, lettuce, and grapes. Additionally, ground horsemeat, canned dog food, canned veterinary diets, and/or a meat mixture as offered to our carnivorous small mammals are consumed by *Choloepus* in captivity. These food items are offered on the basis of 70% vegetable-fruit and 30% meat product daily. This diet is supplemented daily with a powdered vitamin-mineral product. Water is available ad lib.

All fruits and vegetables are diced prior to feeding. This form has proven to be the most useful in a captive situation. Sloths can consume their food directly from the feeding pan, without picking it up in their claws. A comparison of diced vs. long, thin, or chunked pieces shows a substantial increase in the amount consumed daily using the diced form. Spillage and waste are reduced considerably using diced food.

Feces and urine are normally expelled at irregular intervals in sloths. The average interval between eliminations ranges from 3 to 5 days. Weight loss at this time may be considerable.[13] Feces are normally pelleted, forming a mass of compressed pellets, elongated and cylindrical in shape.

Three-toed sloths, *Bradypus* sp., although similar in body form and inhabiting similar niches as *Choloepus* sp., are dissimilar animals. In nature these edentates feed on leaves, flowers, buds, shoots, and fruit, preferring young growing leaves. It has been stated that *Bradypus* feeds almost exclusively on *Cecropia* sp. leaves and fruits.[14] This has been refuted in recent studies in the Republic of Panama and the Canal Zone by Montgomery and Sunquist.[15] In fact, several dozen plant species were shown to be included in the sloth's diet during the course of these studies. It should be pointed out that three-toed sloths held under semicaptive conditions in Panama showed a preference for *Cecropia* when this leaf was offered, along with spinach, mango fruit, and leaves of *Ficus* sp.

In a captive situation, these sloths rarely survive longer than a few months at best. Most of these edentates arrive in a debilitated condition and have been deprived of food and water for some days. Animals showing respiratory infections, open infected wounds, and dehydration are the most frequently seen medical complications. Animals receiving prior acclimation arrive in somewhat better condition and may fare generally better in the long run. It has been suggested that the successful acclimation of *Bradypus* sp. may be related to several factors rather than to diet alone. Certainly, attention should be paid to the temperature and humidity needs as well as to nutritional substitutes that may fulfill the requirements of these arboreal herbivores.

Feeding trials carried out in both a northern-climate facility and under semicaptive conditions in Panama demonstrate that *Bradypus* would accept and eat willingly banana, apple, mango fruit, spinach leaves, green beans, lettuce, and orange. All food was finely diced, except for the leafy greens. Meat in any form was rejected. All food was supplemented daily with a vitamin and mineral protein powder combination sprinkled over it. Water was not taken when offered in shallow earthenware crocks, but if it was hand fed with a bulb syringe the sloths drank freely.

Recently caught three-toed sloths either accepted hand feeding and substitute foods within the first 48 hr or refused to feed at all. Those that did not eat were replaced in nature without loss. Of those retained and maintained under captive conditions in their country of origin, nearly all were alive and doing well almost 1 year later. These fed heavily on *Cecropia* sp. but also accepted mixed fruits and vegetables as above and other fresh native leaves.

ARMADILLOS

There are approximately 20 known species of armadillo, ranging in size from the fairy armadillo, *Chlamyphorous truncatus,* the smallest, to the giant armadillo, *Priodontes giganteus.* Their nutritional needs and natural food intake also encompass a wide range. The fairy is known to feed almost exclusively on ants while the giant is known to favor tuberous vegetables, ripe fruit, and carrion.

In captivity, several armadillo types adapt quite readily to a meat mixture composed of ground, raw horsemeat, Purina Mink Developer Chow, ground hardboiled eggs, vitamins, minerals, and fruit pieces.[16] Species that have flourished on the basic diet include the nine banded, *Dasypus novemcinctus;* seven banded, *Dasypus septemcinctus;* six banded, *Euphractus sexcinctus;* La Plata three banded, *Tolypeutes matacus;* small hairy, *Chaetophractus vellerosus;* large hairy, *Chaetophractus villosus;* Bolivian hairy, *Chaetophractus nationi;* Pichiy, *Zaedyus pichiy caurinus;* and naked tail, *Cabassous centralis.*

Giant armadillos, *P. giganteus,* were initially acclimated to a thick soupy mixture consisting of Rivinia Hill's PD diet blended with water, Gevral protein powder, Tone powder, and Vidaylin lemon-flavored multivitamins. This was done since the animals refused other mixtures, including the standard armadillo meat mixture. Within several months of captivity, two of three adult armadillos were acclimated to the meat mixture diet as described above. Growth and weight gain on both diets was excellent. One developing male *Priodontes* increased from about 70 lb on arrival to 130 lb within a 5-month period; a female increased from 73 to 125 lb in the same time. Interestingly enough, the male was fed only the PD mixture while the female received the standard meat mixture during this interval.

Early in the study, two species of armadillo showed hemmorhagic tendencies if scraped, butted, or clawed at by cage mates. In the seven banded, *Dasypus septemcinctus,* the hemmorhage was severe, while in the La Plata three banded, *Tolypeutes matacus,* a bleeding tendency was observed but was not as pronounced as in *Dasypus.* An analysis of the diet at that time revealed an apparent absence of a source of vitamin K activity. As a stopgap until an acceptable dietary source of this vitamin was obtained, ground, macerating ocean smelt were added to the armadillo mixture. This somewhat offensive-smelling food item was eagerly accepted by all previously listed species of armadillo. With this addition, no further complications were observed. Shortly thereafter, K-Sol, at the rate of one tablespoonful per pound of Paltone, was thoroughly mixed into the vitamin mixture. Sprinkled over the food daily, this proved to be an effective and safe prophylactic treatment. Since this addition, no further hemmorhage or bleeding tendencies have been noted in any of these species. In an effort to overcome potential difficulties with other edentate species, all edentates in the project, including anteaters and sloths, receive this source of vitamin K activity daily in their food. Ground raw fish was discontinued with the addition and acceptance of the K-Sol.

CONCLUSION

Although the basic dietary guidelines for the order Edentata have been established, a constant review of their nutritional needs is being made. Adequate nutrition alone is not the answer to the successful maintenance of these mammals in a captive state. Careful attention to the behavioral needs and husbandry guidelines of this group will result in the etablishment and acclimation of captured wild individuals.

The true test of the effectiveness of any husbandry technique and nutritional approach is not only long-term survival, but the production and rearing of offspring.[17] In captivity, breeding and births have occurred at Lincoln Park Zoological Gardens in the following edentate species since the adoption of the foregoing methodology:

Nine-banded armadillo (*Dasypus novemcinctus*)
Large hairy armadillo (*Chaetophractus villosus*)
La Plata three-banded armadillo (*Tolypeutes matacus*)
Six-banded armadillo (*Euphractus sexcinctus*)
Hoffmann's sloth (*Choloepus hoffmanni*)
Lesser anteater (*Tamandua tetradactyla*)

Additionally, we have the potential to produce F_2 offspring in the large hairy armadillo and the La Plata three-banded armadillo this year (1976).

Feeding schedules, dietary ingredients, and recommended amounts follow in Table 1. Sources and manufacturers of dietary additives are listed in the product information section after the text. All animals are fed once daily unless otherwise specified.

Acknowledgment

The continued encouragement and support of Dr. Lester E. Fisher, Director, Lincoln Park Zoological Gardens, Chicago, is appreciated. The assistance and dedication of the animal keepers in the small mammal house and lion house is gratefully aknowledged.

PRODUCT INFORMATION

Gevral Protein Powder: Protein-mineral supplement manufactured by Lederle Laboratories Division, American Cyanamid Co., Pearl River, New York 10965

K-Sol Powder: Vitamin K additive, manufactured by Vet-A-Mix Inc., Shenandoah, Iowa 51601

Paltone Powder: Vitamin-mineral supplement manufactured by Pitmann-Moore Inc., Washington Crossing, New Jersey 08560

PD (Prescription Diet): Dietary animal food manufactured by Hill's Division, Rivinia Foods Inc., Topeka, Kansas 66601

Purina Mink Developer Chow-20: Animal diet manufactured by Ralston Purina Company, St. Louis, Missouri 63188

Tone Powder: Protein-mineral-vitamin supplement manufactured by Burns Biotec Laboratories Inc., Oakland, California 94621

Vidaylin: Lemon-flavored multiple vitamins manufactured by Ross Laboratories, Columbus, Ohio 43216

Meat Mixture Formula: Meat mix – 1 to 1 ground lean horse meat and Purina Mink Developer Chow-20 by volume; to 13.5 kg of this mix add 1 tablespoonful cod-liver oil, 10 hardboiled eggs, 2/3 cup molasses, 1/3 cup honey, 1/4 cup peanut oil, and 1/3 cup multiple vitamins. Blend thoroughly to mix all ingredients.

Giant Armadillo Mixture: Blend 0.9 kg canned PD with water. Add 2 tablespoonfuls Gevral protein powder, 2 tablespoonfuls Tone powder, and 1 tablespoonful Paltone powder with K-Sol added. Blend thoroughly to mix all ingredients.

Lesser Anteater Mixture: 1 to 1 canned milk and water, about 4 oz each. Blend with 2 cups meat mixture. To this add 1 tablespoonful Gevral protein powder, 1 tablespoonful Tone powder, 1/2 teaspoonful Paltone powder, and 1 teaspoonful Vidaylin. Blend thoroughly to mix all ingredients.

Table 1
ANIMAL TYPE AND AMOUNT OF FOOD OFFERED DAILY, AT LINCOLN PARK ZOOLOGICAL GARDENS, CHICAGO, ILLINOIS

Species	Sex	Food type and volume (cups)	Feeding time
Silky anteater (*Cyclopes didactylus*)	M/F	Liquid mix, 1/4	1600 hours
Lesser anteater (*Tamandua tetradactyla*)	M/F	Liquid mix, 4	1400 hours
Giant anteater (*Myrmecophaga tridactyla*)	M/F	Meat mix, 8–12	1500 hours
Two-toed sloth (*Choloepus didactylus*)	M/F	Fruit and vegetables, 2–4	1400 hours
Hoffmann's sloth (*Choloepus hoffmanni*)	M/F	Fruit and vegetables, 2–4	1400 hours
Three-toed sloth (*Bradypus* sp.)	M/F	Fruit and vegetables, 1–2	1000 hours
Nine-banded armadillo (*Dasypus novemcinctus*)	M/F	Meat mix, 1	1400 hours
Six-banded armadillo (*Euphractus sexcinctus*)	M/F	Meat mix, 1	1400 hours
La Plata three-banded armadillo (*Tolypeutes matacus*)	M/F	Meat mix, 1/2	1400 hours
Small hairy armadillo (*Chaetophractus vellerosus*)	M/F	Meat mix, 1/2	1400 hours
Large hairy armadillo (*Chaetophractus villosus*)	M/F	Meat mix, 1	1400 hours
Bolivian hairy armadillo (*Chaetophractus nationi*)	M/F	Meat mix, 1	1400 hours
Pichiy armadillo (*Zaedyus pichiy*)	M/F	Meat mix, 1/2	1400 hours
Naked-tail armadillo (*Cabassous centralis*)	M/F	Meat mix, 1	1400 hours
Seven-banded armadillo (*Dasypus septemcinctus*)	M/F	Meat mix, 1/2	1400 hours

Note: Each armadillo type gets several small pieces of cooked sweet potato and banana and a small amount of ground hardboiled egg sprinkled over the meat mixture at feeding time daily.

REFERENCES

1. **Cabrera, A.,** Catalogo de los mammiferos de America del Sur. I. Buenos Aires, *Rev. Mus. Argent. Cienc. Nat. Bernardino Rivadavia,* 4, 1–307, 1957.
2. **Walker, E. P.,** *Mammals of the World,* 2nd ed., John Hopkins Press, Baltimore, 1968.
3. **Block, J. A.,** Hand rearing seven-banded armadillos, *Dasypus septemcinctus,* at National Zoo, *Int. Zoo Yearb.,* 14, 210–214, 1974.
4. **McCrane, M. P.,** Birth, behaviour and development of a hand-reared two-toed sloth, *Choloepus didactylus,* at National Zoo, Washington, D.C., *Int. Zoo Yearb.,* 6, 153–163, 1966.
5. **Veselovsky, Z.,** A contribution to the knowledge of the reproduction and growth of the two-toed sloth, *Choloepus didactylus,* at Prague Zoo, *Int. Zoo Yearb.,* 6, 147–153, 1966.
6. **Meritt, D. A.,** Edentate diets currently in use at Lincoln Park Zoo, Chicago, *Int. Zoo Yearb.,* 10, 136–138, 1970.
7. **Meritt, D. A.,** *Yearb. Am. Philos. Soc.,* pp. 383–384, 1973.
8. **Meritt, D. A.,** *Yearb. Am. Philos. Soc.,* pp. 358–359, 1974.
9. **Meritt, D. A.,** The silky anteater, *Cyclopes didactylus,* in captivity, *Int. Zoo Yearb.,* 11, 193–195, 1971.
10. **Meritt, D. A.,** The nutrition of edentates, *Int. Zoo Yearb.,* Vol. 16, in press, 1976.
11. **Crandall, L. S.,** *The Management of Wild Mammals in Captivity,* Chicago University Press, Chicago, 1964.
12. **Meritt, D. A.,** The lesser anteater, *Tamandua tetradactyla,* in captivity, *Int. Zoo Yearb.,* 15, 41–44, 1975.
13. **Meritt, D. A.,** Edentate diets II. Two-toed sloths, *Lab. Anim. Sci.,* 23(4), 543–545, 1973.
14. **Beebe, W.,** The three-toed sloth, *Bradypus cuculiger, Zoologica,* 1933.
15. **Montgomery, G. G. and Sunquist, M. E.,** Impact of sloths on neotropical forest energy flow and nutrient cycling, in *Trends in Tropical Ecology: Ecological Studies IV,* Medina, E. and Golly, F., Eds., New York, Springer Verlag, 1975, 69–98.
16. **Meritt, D. A.,** Edentate diets I. Armadillos, *Lab. Anim. Sci.,* 23(4), 540–542, 1973.
17. **Meritt, D. A.,** The development of the La Plata three banded armadillo, *Tolypeutes matacus,* at Lincoln Park Zoo, Chicago, *Int. Zoo Yearb.,* 11, 195–196, 1971.

NATURAL AND SYNTHETIC DIETS FOR PHOLIDOTA

R. R. Tenaza and T. A. Schulz

INTRODUCTION

The seven extant species in the monogeneric order Pholidota are commonly referred to as pangolins or scaly anteaters (*Manis* spp.). Four species occur in sub-Saharan Africa and three in south and southeastern Asia including the Indian subcontinent, Ceylon, southern China, Burma, Thailand, Indochina, Malaya, Sumatra, Java, and Borneo.[1-4] Pangolin habitats vary from dense tropical rain forest to arid, sparsely wooded savannah. The white-bellied tree pangolin (*M. tricuspis*) is diurnal, whereas the remaining six species are nocturnal. Relative amounts of time spent in trees and on the ground vary according to species.

Alimentary specializations accompanying the pangolin's specialized, insectivorous diets include (1) weak, toothless jaws; (2) an elongated (16 to 40 cm) tongue; (3) an elongated xiphisternum for attachment of the complicated tongue musculature; (4) greatly enlarged salivary glands supplying the tongue with a sticky mucous coating to which insect prey adhere; and (5) a thick, muscular stomach lined in the pyloric region with horny projections, apparently functioning to triturate insects swallowed intact by pangolins. Grinding action of the stomach is enhanced by pebbles and sand ingested by the pangolin, as in the gizzard of seed-eating birds.[5-8]

NATURAL DIETS

The principal foods of wild pangolins are termites and formicid ants.[7,9-14] However, grubs, worms, cockroaches, beetles, larvae of bees and wasps, and vegetable matter may also be eaten.[1,6] Arboreal, terrestrial, and subterranean nests of ants and termites are broken open or excavated with the pangolin's stout foreclaws. The pangolin then extracts the insects with its long, sticky tongue.[10-14] Cape pangolins (*M. temminckii*) have also been observed capturing ants and termites beneath dung and pieces of wood on the ground.[14] Although pangolins consume all developmental stages of ants and termites, they may prefer preadult stages.[13,14] When choices are available, pangolins also appear to be rather selective with regard to the species of ants or termites they will eat, consuming some species while ignoring others.[11,13,14]

ARTIFICIAL DIETS

Success maintaining pangolins in captivity has been extremely limited, presumably due to the inadequacy of most artificial diets.[6,7,15-18] Most captive individuals have died within a few weeks after entering captivity. However, some have survived longer, and the longevity record is held by a giant pangolin (*M. gigantea*) which lived for 4 years and grew to adulthood at the Fort Worth Zoo.[15]

To our knowledge, successful breeding of captive pangolins has occurred only at the Bloemfontein Zoo in South Africa, where two pairs of Cape pangolins mated and reared young before dying of accidental poisoning.[19] In other reports of captive pangolins, births occurred from 2 to 40 days after the animals were obtained.[20-23] This indicates that the females were probably pregnant before entering captivity, as gestation in pangolins appears to be approximately 139 days.[19] This is significant, because the ability of a captive pangolin to breed and successfully rear young may be the best indication that it is receiving an adequate diet.

Diets which have been fed to captive pangolins in zoos are described in Table 1. Except for the diluted milk diet fed to a baby pangolin (Diet 4), these diets have in common that each contains animal protein from meat, insects, milk, eggs, or some combinations thereof and roughage in the form of cereal or carrots. Although liquid vitamin supplements are routinely added to several of the diets, one author reports that they may cause a diet to be rejected (see Diet 2, Table 1).

Diet 8 (Table 1) is the only diet on which captive pangolins are known to have completed fertile matings. The unique feature of this diet is that it includes a daily ration of live termites. However, whether the provision of fresh insects is directly responsible for the successful matings is unknown. Female pangolins that were pregnant when obtained have successfully borne and suckled young on diets without fresh insects (Diets 6 and 7, Table 1).

WEANING

Pangolins born in captivity begin showing an interest in solid food when approximately 35 days old and will begin eating live insects at that age.[19,22] When live insects are not provided, weaning may be delayed for 3 to 5 months or more.[21,22] It has been suggested that the stimulus of live food is required to initiate feeding.[22]

PARASITES

Nutrition and health of captive pangolins may be negatively influenced by heavy parasite loads, principally of ticks and cestodes.[15,28] These should be diagnosed and treated as soon as possible after pangolins enter captivity. Precautions must be exercised in treating for ticks or other ectoparasites to avoid accidental poisoning of the pangolins with pesticides.[19]

Table 1
DIETS FED TO FIVE SPECIES OF CAPTIVE PANGOLINS

Species and location	Diet ingredients		Procedure	Remarks	Ref.
1. Giant pangolin (*M. gigantea*), Antwerp Zoo, Belgium	Corn flakes Warm boiled water Chopped meat Wheat germ oil Formic acid[a]	25 g 100 g 150 g ½ tsp 1 drop	Blended together in electric mixer and fed warm (25–26°C) twice daily, morning and afternon	Diet is reported to have been "used successfully"[24]	15, 24
2a. Giant pangolin (*M. gigantea*), Fort Worth Zoo, U.S. and Antwerp Zoo, Belgium	Finely ground raw beef heart Precooked cereal (Pablum®) Reconstituted evaporated milk	2 cups 2 cups 4 cups	Mixed together; raw egg or wheat germ sometimes added; fed in late afternoon	One giant pangolin survived 4 years on the diet at the Fort Worth Zoo, and two lived 2 years on it at Antwerp	15
2b. Indian pangolin (*M. crassicaudata*), Bronx Zoo, U.S.	Warm water Ant "eggs"	1 cup ½ cup	Diet was rejected if cod liver oil or vitamins were added	Three Indian pangolins survived on the diet for 2 months, 4 months, and ca. 30 months, respectively	
3. Indian pangolin (*M. crassicaudata*), Oklahoma City Zoo, U.S.	Dry commercial dog food Ground horse meat Evaporated milk Raw eggs ADBEC vitamins	400 g 200 g ½ can 2 5 drops	Mixed together	A female maintained on this diet gave birth to an infant 40 days after entering the zoo; infant died at 3 days of age, but the female still appeared healthy after 6 months in the zoo	23
4. Indian pangolin (*M. crassicaudata*) baby, India	Cow's milk Water	2 parts 1 part	Fed 3 times daily from feeding bottle with rubber nipple	Baby doubled its weight in 4 months on this diet	25
5a. Indian pangolin (*M. crassicaudata*), Prague Zoo, Czechoslovakia[b] 5b. Chinese pangolin (*M. pentadactyla*), Prague Zoo, Czechoslovakia[b]	Raw and cooked chopped meat Minced boiled egg Carrots Curd Ant pupae Salt		Mixing procedure not described; diet occasionally supplemented with moistended, honey-sweetened oatmeal	None	26

[a] To impart ant odor to food.
[b] Quantities of diet ingredients not given.

Table 1 (continued)
DIETS FED TO FIVE SPECIES OF CAPTIVE PANGOLINS

Species and location	Diet ingredients		Procedure	Remarks	Ref.
6. Chinese pangolin (*M. pentadactyla*), Ueno Zoo, Japan	Ground horse meat Milk Raw egg yolk Precooked cereals Powdered milk (Esbilac®)[c] Calcium Multivitamins	150 g 180 cc 1 5 g 5 g 1 g 0.2 cc	Ingredients mixed together and warmed in a pan	A female maintained on this diet gave birth to an infant 16 days after entering the zoo; the infant was weaned and transferred to this diet at 90 days of age; mother and infant were both living and appeared healthy 6 months after the birth	21
7. White-bellied tree pangolin (*M. tricuspis*), Paignton Zoo, Great Britian and Ife University Zoo, Nigeria	Stimulite®[d] Ground beef	2 parts 1 part	Beef and Stimulite® are well mixed to "a moist but powdery consistency;" vitamins may be added	One half-grown individual survived for 2 months on the diet and another for 13 months; an adult female, pregnant when obtained, successfully suckled her infant for 7 months on this diet	18, 22, 27
8. Cape pangolin (*M. temminckii*); Bloemfontein Zoo, South Africa	Minced meat Boiled cereal ("mealie meal") Raw eggs Liquid vitamins (Embelex®)[e] Fresh cow's milk Live termites	8 oz 8 oz 2 1 tsp 2 pints 2 lb	Ingredients, except termites, blended together in mixer for 15 min; termites fed live.	Two pairs of Cape pangolins maintained on this diet bred in captivity and successfully reared their young; the first pair died after 28 months on the diet, apparently of accidental poisoning; the second pair died after 38 months on the diet, also apparently of accidental poisoning	19
9. Cape pangolin (*M. temminkii*), Basle Zoo, Switzerland[b]	Chopped mealworms Boiled cockroaches Chopped horse heart Fresh eggs Cooked oatmeal		Not described	None	26

[c] Esbilac® is a dried milk product of the Borden Dairy Co., New York.
[d] Stimulite® is a product of P. H. Hastings, Portsmouth, England. It contains dried insects and other ingredients and is intended for insectivorous birds.
[e] Embelex® is a tonic, the product of Maybaker (SA) Ltd., Port Elizabeth, South Africa. It contains caffeine citrate BPC, 0.21 g; potassium glycerophosphate solution BPC, 1.84 g; sodium glycerophosphate solution BPC, 0.92 g; aneurine hydrochloride BP, 0.018 g; nicotinic acid BP, 0.088 g; riboflavine-5-phosphate sodium, 0.024 g; methyl hydroxybenzoate BPC, 0.2 g; flavoring and sweetening q.s., and distilled water to 100 cc.[19]

REFERENCES

1. **Allen, G. M.,** *The Mammals of China and Mongolia,* Vol. 11, Part 1, American Museum of Natural History, New York, 1938, 517–519.
2. **Anderson, S. and Jones, J. K., Jr.,** *Recent Mammals of the World: A Synopsis of Families,* Ronald Press, New York, 1967, 188–191.
3. **Ellerman, J. R. and Morrison-Scott, T. C. S.,** *Checklist of Palaearctic and Indian Mammals 1758–1946,* British Museum of Natural History, 1951, 214.
4. **Walker, E. P.,** *Mammals of the World,* Vol. 3 (A Classified Bibliography), Johns Hopkins Press, Baltimore, 1964, 117–118.
5. **Hatt, R. T.,** Pangolins, *Nat. Hist.,* 23, 725–732, 1934.
6. **Hutton, H. F.,** Notes on the Indian Pangolin (*Manis crassicaudata*), *J. Bombay Nat. Hist. Soc.,* 48, 805–806, 1949.
7. **Aiyappan, A.,** Notes on the Pangolin (*Manis crassicaudata*), *J. Bombay Nat. Hist. Soc.,* 43, 254–257, 1942.
8. **Krause, W. J. and Leeson, C. R.,** The stomach of the pangolin with emphasis on the pyloric teeth, *Acta Anat.,* 88, 1–10, 1974.
9. **Harrison, J.,** *An Introduction to Mammals of Singapore and Malaya,* Malayan Nature Society, Singapore, 1966, 60–62.
10. **Pagès, E.,** Notes sur les Pangolins du Gabon, *Biol. Gabonica,* 1, 209–238, 1965.
11. **Pagès, E.,** Sur l'ecologie et les adaptations de l'orycterope et des pangolins sympatriques du Gabon, *Biol. Gabonica,* 6, 27–92, 1970.
12. **Pages, E.,** Étude éco-éthologique de *Manis tricuspis* par radio-tracking, *Mammalia,* 39, 613–641, 1975.
13. **Prater, S. H.,** *The Book of Indian Animals,* 3rd ed., Bombay Natural History Society, Bombay, 1965, 301–303.
14. **Sweeney, R. C. H.,** Some notes on the feeding habits of the ground pangolin, *Smutsia temminckii* (Smuts), *Ann. Mag. Nat. Hist.,* Series 12, 9, 893–896, 1956.
15. **Crandall, L. S.,** *The Management of Wild Mammals in Captivity,* University of Chicago Press, Chicago, 1964, 195–199.
16. **Hopwood, S. F.,** Some notes on the pangolin (*Manis pentadactyla*) in Burma, *J. Bombay Nat. Hist. Soc.,* 33, 439–440, 1928.
17. **Loveridge, A.,** Pangolins as pets, *Nat. Hist.,* 52, 152–155, 1943.
18. **Menzies, J. I.,** Feeding pangolins (*Manis* spp.) in captivity, *Int. Zoo Yearb.,* 4, 126–128, 1962.
19. **van Ee, C. A.,** A note on breeding the Cape Pangolin *Manis temmincki, Int. Zoo Yearb.,* 6, 163–164, 1966.
20. **Acharjyo, L. N. and Misra, R.,** Birth of an Indian pangolin (*Manis crassicaudata*) in captivity, *J. Bombay Nat. Hist. Soc.,* 69, 174–175, 1971.
21. **Masui, M.,** Birth of a Chinese pangolin *Manis pentadactyla* at Ueno Zoo, Tokyo, *Int. Zoo Yearb.,* 7, 114–116, 1967.
22. **Menzies, J. I.,** A preliminary note on the birth and development of a small-scaled tree pangolin *Manis tricuspis* at Ife University Zoo, *Int. Zoo Yearb.,* 7, 114, 1967.
23. **Ogilvie, P. W. and Bridgwater, D. D.,** Notes on the breeding of an Indian pangolin *Manis crassicaudata* at Oklahoma Zoo, *Int. Zoo Yearb.,* 7, 116–118, 1967.
24. **Anon.,** Antwerp pangolin diet, *Int. Zoo Yearb.,* 4, 128, 1962.
25. **Chevenix-Trench, C. G.,** Notes on a young Indian pangolin or scaly anteater *Manis crassicaudata, J. Bombay Nat. Hist. Soc.,* 24, 590, 192?.
26. **Grzimek, B.,** *Grzimek's Animal Life Encyclopedia,* Vol. 2, Van Nostrand Reinhold, New York, 1972, 182–188.
27. **Menzies, J. I.,** A note on the nutrition of the tree pangolin *Manis tricuspis* in captivity, *Int. Zoo Yearb.,* 6, 71, 1966.
28. **Rahm, U.,** Notes on pangolins of the Ivory Coast, *J. Mammal.,* 37, 531–537, 1956.

DIETS FOR LAGOMORPHS

S. T. Chou

I. RABBITS

The domestic rabbit is primarily herbivorous and will eat most types of greens, grains, and hay. Therefore, in rabbit husbandry, the rations provided consist almost entirely of ingredients from plant sources. Though some rabbit raisers still utilize home-grown feeds and mix their own rations, there is an increasing tendency to utilize commercially prepared rabbit feeds in the form of pellets.

Rabbits habitually practice coprophagy, sometimes referred to as pseudorumination. It refers to consumption of soft fecal pellets directly from the anus. Coprophagy is usually practiced by domestic rabbits as young as 3 weeks of age at night, and by wild rabbits while they are in the burrow, during the daytime. It is generally assumed that coprophagy, in conjunction with fermentation in the large intestine, supplies necessary amounts of most B vitamins, probably provides insurance against an essential-amino-acid deficiency, and may permit further digestion of fiber and other nutrients by a second passage through the digestive tract.

Nutrient requirements[1] and examples of natural and synthetic diets[2,3] are given in Tables 1 to 6.

II. PIKAS

Pikas (*Ochotona refescens refescens*[4] and *Ochotona princeps*[5]) were found to be unusually good cage animals for small-lagomorph experiments. The adult's (*O. refescens refescens*) body weight lies between those of the hamster and the rat; its reproduction rate is as high as that of domesticated rabbits; its sexual maturity is earlier than that of the rabbit; and its diet is easy to prepare.[4]

For the domesticated pikas, during a 2-year study period, the basic diet used was rabbit pellets. Diet 1, containing 14% crude protein and 3.5% fat, was given to weanlings. Diet 2, containing 16% crude protein and 4% fat, was given to adult pikas (including pregnant and suckling females) as well as to the young until weaning. Water was available *ad libitum.*[4]

Like other lagomorphs, the fecal material of pikas occurs in two forms: the hard pellet, which is rarely reingested, and an amorphous mass (soft pellet), which has a high energy value and protein content and is often reingested.

Dietary composition in the natural habitat of pikas (*O. princeps*) in southern Colorado was studied, and stomach contents were analyzed, as shown in Tables 7 and 8.

III. HARES

The food of mountain hares (*Lepus timidus scoticus*)[7,8] and Irish hares (*Lepus timidus hibernicus*)[9] was studied by means of stomach analysis and by observation of grazing habits in regular visits to study areas in Banffshire, Scotland, and Antrim County, Ireland. The main grazing period was at night for the mountain hare, and the Irish hare was often seen feeding during the day. Reflection, that is the production of soft feces that are swallowed directly from the anus, occurs during the period between the end of morning's grazing and the start of the next evening's grazing. The hares studied grazed on a wide variety of herbage, but the main food was ling heather (*Calluna vulgaris*), sedges, and grasses. Stomach contents of hares were as shown in Tables 9 and 10.

Snowshoe hares (*Lepus americanus washingtonii*)[10] in the Pacific Northwest (U.S.A.)

preferred feeding on the seedlings of Douglas fir (*Pseudotsuga menziesii*) and spotted catsear (*Hypochoeris radicata*) over other local herbage. Relative preference of snowshoe hares for different parts of spotted catsear was studied in western Washington. The hares demonstrated preference for open flowers, followed closely by flower buds. Leaves were the least preferred part of the plant. Levels of glucose and fructose, the principal sugars of catsear, appeared to be responsible for the order of preference.

Another informative food habit study on snowshoe hares in Linn County, Western Oregon, was made by Mozejko.[11] Forty-six live trapped hares were observed, upon 15 different species, including trees, shrubs, forbs, and ferns. Bracken-fern (*Pteridium aquilinum*), false dandelion (*Hypochaeris radicata*), and red huckleberry (*Vaccinium parvifolium*) were most commonly eaten. Herbaceous plants comprised the major part of the hares' diet, especially during warmer seasons of the year. Woody plants, including buds, twigs, and bark of trees and shrubs, were utilized heavily during colder seasons of the year when snow often covered herbaceous foods for varying periods. Grasses, sedges, and rushes were least eaten of all plants.

Table 1
NUTRIENT REQUIREMENTS OF RABBITS IN PERCENTAGES OF THE TOTAL RATION [a]

Body weight, lb	Expected daily gain, lb	Total daily feed[b]	Total digestible nutrients, %	Total protein, %	Digestible protein, %
Normal Growth, Does or Bucks, Average 6.5 lb					
4 to 9	–	5.8	60	15	11
Normal Growth and Fattening, Does or Bucks					
4		6.2	65	16	12
5	Average	6.0	65	16	12
6	0.07	5.7	65	16	12
7		5.4	65	16	12
Maintenance, Does or Bucks					
5	–	4.0	55	12	9
10	–	3.3	55	12	9
15	–	3.0	55	12	9
Pregnant Does					
5	–	5.0	58	15	11
10	–	4.1	58	15	11
15	–	3.7	58	15	11
Lactating Doe and Litter of 7					
5	–	–	70	17	13
10	–	–	70	17	13
15	–	–	70	17	13

[a] All feed or rations are based on air-dry weight.
[b] Percentage of live weight.

From National Research Council, *Nutrient Requirements of Rabbits* (Nutrient Requirements of Domestic Animals, No. 9), National Academy of Sciences – National Research Council, Washington, D.C., 1966. With permission.

Table 2
EXAMPLES OF ADEQUATE DIETS[a]

Ingredients	% of total diet
Normal Growth, 4 to 9 lb Average Weight 6.5 lb	
Alfalfa hay	60
Corn, grain	22
Barley, grain	15
Soybean meal	3
Normal Growth and Fattening 4 lb Initial Weight, 8 lb Final Weight[b]	
Alfalfa hay	40
Wheat bran	5
Barley, grain	32
Oats, grain	18
Soybean meal	5
Maintenance, Does and Bucks Average Weight 10 lb	
Alfalfa hay	70
Oats, grain	20
Wheat, grain	10
Pregnant Does Average Weight 10 lb	
Clover hay	50
Oats, grain	44
Soybean meal	6
Lactating Does (Litter of 7) Average Weight 10 lb	
Alfalfa hay	40
Wheat, grain	25
Sorghum, grain	25
Soybean meal	10

[a] To each of the diets 0.5% salt or iodized salt was added in iodine-deficient regions.

[b] May also be appropriate for pregnant and lactating does.

From National Research Council, *Nutrient Requirements of Rabbits* (Nutrient Requirements of Domestic Animals, No. 9), National Academy of Sciences – National Research Council, Washington, D.C., 1966. With permission.

Table 3
NATURAL DIET FOR RABBITS[a]

Ingredients	Percent in diet
Alfalfa meal, dehydrated	77.2
Rolled oats	19.3
Sucrose	2.0
Vitamin and mineral premix	1.5

Vitamin and Mineral Premix

Nutrient supplied	mg/kg diet[b]
Thiamine hydrochloride	5
Riboflavine	5
Pyridoxine hydrochloride	5
Pteroylglutamic acid	2
DL-Calcium panthothenate	25
Nicotinic acid	40
Cyanocobalamin	0.03
Menaquinone-6	0.3
myo-Inositol	200
Choline chloride	2,000
Retinol	30,000 IU
Ergocalciferol	3,000 IU
α-Tocopheryl acetate	9 IU
Ferric chloride·$6H_2O$	300
Calcium acetate·H_2O	10,000

[a] Rabbits given this diet and water (both *ad libitum*) showed growth response and hemoglobin concentration comparable to rabbits fed commercially available pellets, (Master Rabbit Pellets, Maple Leaf Mills Ltd., Toronto, Ontario, Canada). However, feed conversion was poorer than that of the commercial pellets. Ingredients shown were thoroughly mixed and pelleted.[2]

[b] Unless otherwise indicated.

From Chou, S. T. and Robinson, G. A., *Lab. Anim. Sci.*, 22, 48–55, 1972. With permission.

Table 4
NATURAL DIET FOR RABBITS[a]

Ingredients	Percent in diet
Alfalfa meal, dehydrated	74.45
Rolled oats	16.55
Soybean protein[b]	5.0
Sucrose	2.0
Vitamin and mineral premix[c]	1.5
Sodium chloride	0.5

[a] Rabbits given this diet and water (both *ad libitum*) showed growth response, hemoglobin concentration, and feed conversion comparable to those of rabbits fed commercially available pellets (Master Rabbit Pellets, Maple Leaf Mills Ltd., Toronto, Ontario, Canada). Ingredients shown were mixed and pelleted.[2]

[b] Edible 72% (obtained from ICN Nutritional Biochemicals, Cleveland, Ohio, 44128).

[c] The premix composition was the same as that shown in Table 3.

From Chou, S. T. and Robinson, G. A., *Lab. Anim. Sci.*, 22, 48–55, 1972. With permission.

Table 5
SEMISYNTHETIC DIET FOR RABBITS USED IN MOLYBDENUM OR COPPER TOXICITY STUDIES[a]

Ingredients	Percent in diet
Rolled oats	86.5
Alfalfa meal, dehydrated	10.0
Sucrose	2.0
Vitamin and mineral premix[b]	1.5

[a] The rabbits were housed in galvanized screen-bottom cages. Feed and water were given (*ad libitum*) in stainless steel containers.[3]

[b] The premix composition was the same as that shown in Table 3.

From Pepino, A. M., Robinson, F. A., Chou, S. T., and Liptrap, R. M., *Can. J. Physiol. Pharmacol.*, 47, 635–643, 1969. With permission.

Table 6
SEMISYNTHETIC DIET FOR RABBITS[a]

Ingredients	Percent in diet
Rolled oats	78
Alfalfa meal, dehydrated	9
Soybean protein[b]	5
Corn oil	2
Calcium phosphate$\cdot H_2O$, monobasic	1
Calcium phosphate, tribasic	1
Sucrose	1.608
DL-Methionine	0.3
Sodium chloride	0.5
Potassium iodide	0.002
Vitamin and mineral premix[c]	1.5

[a] Ingredients were thoroughly mixed and pelleted. This was an improved diet compared to that shown in Table 5, for rabbits kept in galvanized screen-bottom cages.[2]

[b] Edible 72% (see Table 4 for source of supply).

[c] The premix composition was the same as that shown in Table 3.

From Chou, S. T. and Robinson, G. A., *Lab. Anim. Sci.*, 22, 48–55, 1972. With permission.

Table 7
MEAN PERCENT VOLUME OF FOOD ITEMS IN THE STOMACHS OF 80 PIKAS FROM EIGHT SITES IN SOUTHWESTERN COLORADO

Food item	Lower Lime Creek	Upper Lime Creek	Potato Lake	N. Fork Mancos River	Tomahawk Mine	Kennebec Pass	Stoney Gulch	Stoney Pass
Seeds	2.5	7.2	35.8	1.1	5.1	9.4	2.0	0.4
Forb and shrub	1.7	0	0	1.1	0	8.8	0.2	0
Grass and grass-like	0.8	7.2	35.8	0	5.1	0.6	1.8	0.4
Leaves and stems	97.5	92.8	64.2	98.9	94.9	90.6	98.0	99.6
Forb and shrub	88.7	88.1	57.0	93.0	79.8	81.6	91.0	78.8
Grass and grass-like	8.8	4.7	7.2	5.9	15.1	9.0	7.0	20.8
Sample size	12	16	7	6	17	12	5	5

From Johnson, D. R., *J. Mammal.*, 48, 311–315, 1967. With permission.

Table 8
PERCENT VOLUME OF FOOD ITEMS IN THE STOMACHS OF 89 PIKAS BY SEASON

Food item	Spring (March–May)	Summer (June–August)	Fall (September–November)
Seeds	8.3	12.5	3.3
Forb and shrub	0.3	4.9	0.7
Grass and grass-like	8.0	7.6	2.6
Leaves and stems	91.7	87.5	96.7
Forb and shrub	86.9	78.5	79.5
Grass and grass-like	4.8	9.0	17.2
Sample size	13	34	42

From Johnson, D. R., *J. Mammal.*, 48, 311–315, 1967. With permission.

Table 9
COMPARISON OF THE DIET OF *L. TIMIDUS* IN SCOTLAND AND IRELAND

Plant	Mountain hare, November–March (21 stomachs)	Mountain hare, April–October (26 stomachs)	Irish hare, December–April (20 stomachs)
Calluna vulgaris	90%	48%	28%
Eriophorum spp.	10%	17%	15%
Grasses (mainly *Deschampsia, Festuca,* and *Agrostis* spp.)	Trace	24%	44%
Dicotyledons	–	6%	2%
Carex spp.	–	5%	6%
Scirpus caespitosus	–	–	4%

From Walker, J. and Fairley, J. S., *J. Mammal.*, 49, 783–785, 1968. With permission.

Table 10
FREQUENCY OF OCCURRENCE AND MEAN PERCENTAGE VOLUME OF PLANT SPECIES FOUND IN 20 IRISH HARE STOMACHS

Plant genus	Number of occurrences	Percent occurrence	Mean percent (volume)	
Picea	1	5	Present	
Potentilla and *Galium*	7	35	1	Dicotyledons, 2%
Unidentified dicotyledons	9	45	1	
Calluna	20	100	28	Heather, 29%
Erica	2	10	1	
Agrostis	20	100	19	
Deschampsia and *Festuca*	20	100	13	
Molinia	18	90	9	Upland grasses, 44%
Nardus	1	5	Present	
Unidentified grasses	14	70	3	
Scirpus	16	80	4	
Eriophorum	20	100	15	Sedges, 25%
Carex	12	60	6	
Juncus	2	10	Present	
Moss	3	15	Present	

From Walker, J. and Fairley, J. S., *J. Mammal.*, 49, 783–785, 1968. With permission.

REFERENCES

1. National Research Council, *Nutrient Requirements of Rabbits* (Nutrient Requirements of Domestic Animals, No. 9), National Academy of Sciences – National Research Council, Washington, D.C., 1966.
2. **Chou, S. T. and Robinson, G. A.,** Formulation of a laboratory rabbit diet, *Lab. Anim. Sci.,* 22, 48–55, 1972.
3. **Pepino, A. M., Robinson, F. A., Chou, S. T., and Liptrap, R. M.,** Comparison of ^{131}I retention in the plasma of rabbits fed toxic quantities of cupric or molybdate ions as supplements to a commercial pelleted diet or to an oats-alfalfa laboratory diet, *Can. J. Physiol. Pharmacol.,* 47, 635–643, 1969.
4. **Puget, A.,** The Afghan pika (*Ochotona refescens refescens*): a new laboratory animal, *Lab. Anim. Sci.,* 23, 248–251, 1973.
5. **Underhill, J. E.,** Notes on pika in captivity, *Can. Field Nat.,* 76, 177–178, 1962.
6. **Johnson, D. R.,** Diet and reproduction of Colorado pikas, *J. Mammal.,* 48, 311–315, 1967.
7. **Hewson, R.,** Food and feeding habits of the mountain hare *Lepus timidus scoticus*, Hilzheimer, *Proc. Zool. Soc. London,* 139, 515–526, 1962.
8. **Miller, G. R.,** Evidence for selective feeding on fertilized plots by red grouse, hares, and rabbits, *J. Wildl. Manage.,* 32, 849–853, 1968.
9. **Walker, J. and Fairley, J. S.,** Winter food of Irish hares in County Antrim, Northern Ireland, *J. Mammal.,* 49, 783–785, 1968.
10. **Radwan, M. A. and Campbell, D. L.,** Snowshoe hare preference for spotted catsear flowers in western Washington, *J. Wildl. Manage.,* 32, 104–108, 1968.
11. **Mozejko, A.,** A Food Habits Study of the Snowshoe Hare (*Lepus americanus*), Baird, in Western Oregon, Ph.D. thesis, Oregon State University, Corvallis, June 1972.

DIETS (NATURAL AND SYNTHETIC): TUBULIDENTATA

E. Young

The order Tubulidentata contains only one family (Orycteropodidae), one recent genus (*Orycteropus*), and a single species (the aardvark or African antbear, *O. afer*).[1-3] This species is endemic to Africa, where its distribution ranges from the Sahara and Sudan in the north to the Cape in the south.[1,3-6]

These bulky insectivores may grow to an overall length of 6 ft. They are shy, solitary, and nocturnal in their habits.[2,6] In the adult aardvark the elongate skull has no incisors or canines; the animal uses its long, protractile tongue, which can be extended for a distance of 30 cm and is covered with sticky saliva to collect and lick up its insect food.[1-5]

The natural diet of *Orycteropus* consists almost exclusively of termites and ants. These include species of *Odontotermes, Microtermes, Pseudacanthotermes, Hodotermes,* and *Pheidole.*[1-8] The aardvark also feeds on scarabaeid larvae,[4] locusts,[4] and other insects. Fungus gardens, growing in termite nests, also seem to be utilized during the dry season.[4] In addition, there are indications that the succulent fruits of the wild cucumber, *Cucumis humifractus,* may be consumed by the aardvark.[1,3,4] This plant serves as an important source of water for wild animals in desert and semidesert regions and is said to germinate in buried aardvark excrements. It often grows around the entrance to aardvark holes.

In zoological collections the artificial feeding of the insect-eating group of mammals generally presents major obstacles. The aardvark can, however, be successfully fed on milk and/or semiliquid soups supplemented with raw or finely chopped hard-boiled eggs,[3,6,7,10] raw chopped or minced meat,[3,6,7,9,10] dog meal,[7] bone meal,[7] precooked cereals,[7,9,10] boiled rice,[3,6] scalded bread,[7] cod-liver oil,[7] honey,[7] and/or extra vitamins.[7,9,10] Artificially reared mealworms are taken by the aardvark.[3,10] Dried ants' "eggs" (pupae) and dried "flies" (aquatic insects) are invariably left behind, and dried locusts also do not appear to fall within the tastes of captive *Orycteropus.*[7]

Cow's milk is a potential source of pathogenic bacteria and causes diarrhea in some animals whose diet in captivity contains a large amount of milk. Milk powder with the following composition may be prepared and used to overcome this problem.[9,10]

Skimmed milk powder	80 lb
Whey powder	8 lb
Fine Kaffir corn meal	8.5 lb
Brewer's yeast	2 lb
Dicalcium phosphate	1 lb
Vitamin A	4 g/100 lb
Vitamin D_3	½ g/100 lb

One pound of this milk powder is used to prepare 1 gal of milk substitute. Supplemented with precooked cereal and raw minced meat, this milk replacer has been used with great success in the nutrition and breeding of aardvark and other captive mammalian insectivores.[9,10]

Breeding of *Orycteropus*, which is exceptional in captive animals, has also been achieved by using the following food mixture:[11]

Dry dog food	1802 g
Pablum (precooked cereal)	114 g
Mellins Food[a]	284 g
Honey	232 g
Vitamin-mineral powder	232 g
Condensed milk	2400 ml
Eggs	22
Water, to make	15 l

[a] Mellin's Food is a milk modifier distributed by Mellin's Food Company of North America, Chicago, Ill.

The above-mentioned milk substitute, which has already been used to successfully hand raise young animals of more than 20 different African mammal species,[10] could be equally well applied in the feeding of baby aardvark. The addition of one part by volume of limewater (the clear supernatant of a saturated solution of calcium hydroxide) to every seven parts of milk substitute is recommended as part of a preventive antidiarrheal treatment. Initial prophylactic antibiotic or sulfonamide therapy may also aid in the successful rearing of baby aardvark.

The Crandon Park Zoo successfully hand-reared a young aardvark on a basic diet of Esbilac,* corn syrup, and water. After the first few feeds strained beef liver was added, and the following formula was then used:

Esbilac	1 tablespoon
Dark corn syrup	1 teaspoon
Strained beef liver	1 teaspoon
Warm water	45 ml

The first day it was fed every 2 hr and consumed 15 to 80 ml of formula at each feeding. Feeding intervals were eventually extended to 6-hr periods. After 2 weeks it began to accept the bottle more readily. Various constituents of a food mixture for adults were gradually added to the above formula, and at about 3 months it started to feed on ants.[11]

A proper diet for captive aardvarks, as for other ant- and termite-eating mammals, has always been a matter of great concern. Although much has already been achieved in the feeding and breeding of captive aardvarks, much more scientific research is required to assist in the successful management and care of captive specimens of these unique mammals.

* Esbilac is a powdered milk substitute for bitch's milk manufactured by the Borden Company, Feed Supplements Division, 250 Madison Avenue, New York.

REFERENCES

1. **Dorst, J. and Dandelot, P.,** *A Field Guide to the Larger Mammals of Africa,* Collins, London, 1970, 150–151.
2. **Regan, C. T.,** *Natural History,* Ward, Lock and Co., Ltd., London, 1936, 729.
3. **Walker, E. P.,** *Mammals of the World,* Vol. 2, Johns Hopkins University Press, Baltimore, 1964, 1317–1318.
4. **Kingdon, J.,** *East African Mammals: An Atlas of Evolution in Africa,* Vol. 1, Academic Press, London, 1971, 378–387.
5. **Roberts, A.,** *The Mammals of South Africa,* Central News Agency, South Africa, 1951, 239–240.
6. **Sanderson, I. T.,** *Living Mammals of the World,* Hamish Hamilton, London, 1955, 220–221.
7. **Crandall, L. S.,** *The Management of Wild Mammals in Captivity,* University of Chicago Press, Chicago, 1964, 447–449.
8. **Flower, W. H. and Lydekker, R.,** *An Introduction to the Study of Mammals Living and Extinct,* Adam and Charles Black, London, 1891, 208–211.
9. **Young, E.,** *Int. Zoo Yearb.,* 6, 68–69, 1966.
10. **Young, E., Ed.,** *The Capture and Care of Wild Animals,* Human and Rousseau, Cape Town, 1973, 140–146.
11. **Lucas, J.,** *Int. Zoo Yearb.,* 9, 97–99, 1969.

NUTRITION OF THE HYRAX

N. Fairall and R. P. Millar

INTRODUCTION

Since nutrition is related to the ecological niche occupied by animals, it is relevant to describe the distribution and habitat of the hyraxes. The three extant hyracoid genera, *Procavia, Heterohyrax,* and *Dendrohyrax,* are distributed throughout Africa from the southern tip at 34° south latitude to the Mediterranean and Syria at about 35° north latitude. Within this area they occupy a wide range of habitats from desert and semidesert, through grass savanna, to montane and tropical forest.

The rock hyrax (*Procavia*) is found not only throughout the present range of the order, but also throughout the widest range of habitats, from desert to tropical forest. The bush hyrax (*Heterohyrax*) inhabits the intermediate zone of grassland, bushveld, and tropical forest, while the tree hyrax (*Dendrohyrax*), being truly arboreal, is found only in a true forest.[1-5]

NATURAL DIET

All hyraxes are predominantly vegetarian and have been described as grazers and browsers by different authors.[3,6,7] However, they cannot be simply classified into either of these categories. Most authors discuss their feeding habits in terms of their choice of plant growth form, rather than in terms of plant groupings based on taxonomic relationships. Our studies indicate that this former approach is correct, since it appears that the hyrax adapts to available vegetation; young, succulent plants of a large number of families are randomly eaten rather than a restricted number of "favored" species.

Procavia has been most widely studied in East Africa[3,6,7] and Israel.[2] In South Africa we have investigated their feeding habits in three vastly differing vegetation types by analyzing stomach contents. The studies all indicate that *Procavia* has an extremely varied diet which consists of more than 50 plant families, including a number of species that are either unpalatable or poisonous to domestic stock.[8] However, one report[7] mentions that the animals showed a definite preference for one plant species, while all the reports clearly indicate a preference for grass, when it is available in a young succulent growth stage.

Heterohyrax and *Procavia* are found together in East Africa, and the main studies of *Heterohyrax* have been in comparison with *Procavia.*[3,6,7] All authors indicate that the main difference in the feeding habits of these two genera is that *Heterohyrax* shows a preference for browse as opposed to grass. The most recent and complete study indicates that, although certain plants are not consumed by both genera, there is a clear overlap in food choice. *Heterohyrax* has a more limited spectrum of foods than *Procavia* but, like the other hyraxes, it also shows a marked preference for young buds, shoots, and flowers. A tendency toward selectivity in the choice of food is also evident in this genus.

Dendrohyrax has been studied only superficially. The forest habitat of this genus lacks the diversity of higher plant forms found in savanna and bush habitats. Since it is possible that a tree hyrax could live in one specific tree,[1] its diet would obviously be limited.

Although all hyraxes have a predominantly vegetable diet, both *Procavia* and *Dendrohyrax* have been reported to feed on invertebrates.[1,2] Semicaptive *Procavia* will also eat foods of animal origin by free choice.

ARTIFICIAL DIETS

Since most authors indicate that captive hyraxes eat a wide variety of foods of both vegetable and animal origin, it is generally thought to be extremely easy to sustain the animals if adequate, well-insulated housing is provided.

One report mentions that *Procavia* have been kept on a diet of clover or alfalfa hay, fresh clover or alfalfa, and succulent vegetables supplemented with barley.[2] The animals did so well on this diet that no other was tried. Another colony of *Procavia* was maintained on a diet of 300 g of good quality green alfalfa per animal per day for a 9-year period.[9] We have subsequently fed the same colony commercial rabbit pellets and green alfalfa and found the combination to be satisfactory.

These animals are often kept as pets in South Africa, and are fed available food from the kitchen. Young animals are reared on unsupplemented cow's milk, but are precocious and begin feeding on vegetation from the first day of birth. Older hyraxes are fed a variety of food which range from kitchen greens to cooked maize porridge and bread.

We have conducted some experimental studies on the nutrition of captive *Procavia.* Animals fed green alfalfa at a rate of 79 g dry weight per day gained weight, while a group fed at a rate of 49 g dry weight showed a drastic loss in weight. The dry matter digestibility of the alfalfa was 70%, and this corresponded to an energy intake of 340 and 138 kcal per day on the respective diets. The animals were kept in small cages and had the minimum amount of exercise; while the temperature was low, they would have been required to expend energy to maintain normal body temperature. On the basis of these data and stomach content analyses of wild animals, it is estimated that a hyrax of 2 kg requires a digestible energy intake of approximately 260 kcal per day for maintenance.

In a subsequent experiment with animals fed diets of green alfalfa and rabbit pellets, it was found than an energy intake of 160 kcal per day had a measurable effect on the gonadotrophic and gonadal hormone production,[10] although it did not cause a drastic fall in weight over the 10 weeks of the study.

We have found that in captivity *Procavia* is inclined to obesity; this is also recorded in a study that was conducted in Israel.[2] High-energy diets should therefore be avoided.

Although no detailed figures are available, the protein content of the ration used in our study was clearly adequate. The rabbit pellets had a minimum of 19% protein, while the alfalfa contained 10% protein on a dry base.

No specific vitamin needs have been identified; however, we have treated one case of hairlessness successfully by multi-vitamin injections, suggesting that a vitamin deficiency may have been a cause.

A suitable diet for captive hyraxes could consist of a small amount of concentrate, combined with any available succulent vegetation. The total diet should provide an energy intake of approximately 260 kcal per day for hyraxes weighing 2 kg. A protein content of 10% on a dry base would be adequate. Any combination of feeds supplying the above needs should meet all mineral requirements; however, a vitamin supplement should be added if it is not included in the concentrate ration.

GENERAL ASPECTS OF NUTRITION

Certain aspects of a general nature have a bearing on nutrition. The tendency for the hyrax to become obese in captivity has been mentioned. Although no adverse effects on the health of obese hyraxes have been noticed, we have found cholesterol plaques in captive specimens.

In nature the hyrax is heavily infested with parasites; both round worms and tape worms are found in the intestines and in the bile ducts of the liver. Captive animals are routinely treated, and we have found bile stones in a large number of animals that had been cleaned of internal parasites. No bile stones have been observed in hyraxes living in their natural habitat.

Hyraxes produce a high concentration of calcium in urine, occurring in the form of insoluble calcium carbonate; this precipitates on rocks, causing substantial accumulations. Calcium metabolism appears to be unusual in hyraxes, in that they evidently have an extremely efficient uptake of this element through the gut. We recently fed barium sulfate ("barium meal") to *Procavia* in an experimental study; it poisoned the animals, although it is a widely used clinical procedure in other mammalian groups. Since metallic barium is a known mammalian poison, this further indicates that hyraxes are readily able to metabolize alkaline earth metals such as calcium and barium. Stomach stones consisting of calcium oxalate are also found occasionally in wild animals; this further emphasizes the unusual utilization of calcium.

The previously mentioned preference for succulent vegetation and the hyrax's highly efficient kidneys, which produce a hyperosmotic urine comparable to the camel,[11] are possibly adaptations that permit the hyrax to colonize in arid environments, even in the absence of free water. However, in captivity they drink regularly, and water should be provided.

Procavia has a poorly developed temperature-regulating mechanism.[11] Body temperature is maintained by the hyraxes crowding together; therefore, housing should be designed to allow for this. Where this is not possible, provision should be made for the animals to obtain more energy from their rations.

One final aspect of food provisions for captive hyraxes concerns the appearance of dental caries in one colony of *Procavia*.[2] This was apparently caused by giving cooked barley to the animals; it should be noted that it is quite unnecessary to cook their food.

REFERENCES

1. **Sanderson, I. T.,** *Living Mammals of the World,* Hamish Hamilton, London, 1955.
2. **Mendelssohn, H.,** *Int. Zoo Yearb.,* 5, 116–125, 1965.
3. **Sale, J. B.,** *East Afr. Wildl. J.,* 3, 1–18, 1965.
4. **Bothma, J. du P.,** *Ann. Transvaal Mus.,* 25, 117–152, 1967.
5. **Millar, R. P.,** *Zool. Afr.,* 6, 243–261, 1971.
6. **Turner, M. I. M. and Watson, R. M.,** *East Afr. Wildl. J.,* 3, 49–60, 1965.
7. **Hoeck, H. N.,** *Oecologia* (Berlin), 22, 15–47, 1975.
8. **Sale, J. B.,** *East Afr. Wildl. J.,* 3, 127, 1965.
9. **Thomson, J. K.,** *J.S. Afr. Vet. Med. Assoc.,* 41, 319–322, 1971.
10. **Millar, R. B. and Fairall, N.,** *J. Reprod. Fertil.,* 47, 339–341, 1976.
11. **Louw, E., Louw, G. N., and Retief, C. P.,** *Zool. Afr.,* 7, 413–418, 1972.

FOODS AND FEEDING OF ELEPHANTS

R. T. Reuther

Elephants belong to the order Proboscidea. It has been traditionally thought that the fossil record suggests that this order and the Hyraxes (order Hyracoidea) and the Sirenians (order Sirenia) evolved from a common ancestral stock related to ungulates. However, the more probable affinity of the order Proboscidea to the Perissodactyla has been debated frequently, and with increasing confidence.[1]

Elephants are found today only in Africa south of the Sahara and in parts of Southeastern Asia. The two living species belong to the family Elephantidae. The African elephant (*Loxodonta africana*) and the Asiatic elephant (*Elephas maximus*) are the largest land mammals, reaching weights of 5900 kg. In the wild, elephants occupy forests and semi-open or dense scrub, and are restricted to areas near water. They feed on a variety of trees, shrubs, grasses, and aquatic plants.

The elephant's alimentary system is adapted to a herbivorous diet. The stomach and intestines are simple, lacking a fermentative chamber. The caecum is large, sacculated, and possesses an iliocaecal sphincter. The liver is large and bilobed, and has no gall bladder.

The esophagus is comparatively short, and otherwise possesses no peculiar features differing from that of the ungulates. Mucous glands are abundant in the wall, and its lumen is relatively narrow. Particles of food that actually reach the stomach rarely exceed a maximum size (in the longer axis) of 5 cm.

The stomach is simple, with its main axis almost vertical. It averages about 100 to 140 cm in length in the adult, and about 40 cm in diameter.

The nutritional physiology of a captive female Asiatic elephant weighing 3672 kg was studied in great detail by Benedict in 1936.[2] He found that with a daily food intake of 110 lb (50 kg) hay there was a slight excess of calories and nitrogen; 48 lb (22 kg) of water vaporized daily from lungs and skin (50% of this being from the skin), while 20% total heat given off was lost in water evaporation. The urinary excretion amounted to 4.5 lb (2 kg) per 24 hr. It is known that an elephant may drink up to 22 gal (100 l) of water at one time, and probably up to 40 or 50 gal in a single day.

Mastication of the edible parts of the food is usually very thorough in healthy elephants. An individual may eat up to 410 kg of forage daily. Sykes[1] stated "The wild African elephant usually consumes twice or three times the weight of food daily used by the Asiatic species," while, according to Sedgwick and Fowler,[3] "Indian and African elephants require 15.97 k calories per kg body weight per 24 hours."

Benedict[2] showed that in the Asiatic elephant the passage of undigested food from the time of ingestion to that of defecation occupied between 21 and 54.5 hr. He also measured the apparent digestibility of the timothy hay eaten, and found it to be 44% digestible. New Hampshire cows, using similar hay, showed an average of 60% apparent digestibility, while horses showed 53% and sheep 61%. Similar experiments on the efficiency of the African elephant have apparently not been performed, but it is probable that comparable figures would result.

DIET

According to Crandall,[4] "Elephants in nature, as well as captive animals, consume great quantities of leafy and grassy foods where hay is not available, perhaps as much as 600–700 lbs daily (Blanford, 1888–1891). For zoo animals, fed largely on dried materials, the actual weight consumed is naturally much less." In the New York Zoological Park, a female African "bush" elephant, standing 8 ft 7 in. at the shoulder, required approximately 300 lb of timothy hay daily, given at intervals. Two or three times weekly, 75 to 80 lb of dried alfalfa was substituted for an equal amount of timothy. In addition to hay, this elephant received 16 qt of mixed ground grains containing minerals and salt. All this was in addition to the tidbits constantly solicited from visitors. Two Asiatic females received between them 250 to 300 lb of timothy daily, with the substitution of approximately 75 lb of alfalfa two or three times weekly. Each received 16 qt of ground feed as well.

At the Philadelphia Zoological Garden the following is fed each elephant (three-fourths grown and adults) each 24-hr period:

1 bucket (12 qt) diet B (see "Diet Formulas Developed by the Penrose Laboratory," below)
¼ bucket (3 qt) alfalfa pellets
1 qt crushed oats

In addition, each adult African and Asiatic female consumes two bales each of clover-timothy mixed hay. Each three-fourths grown (approximately) female African and Asiatic consumes one and one-fourth bales of clover-timothy mixed hay. Weight of each bale is between 50 and 60 lb.

In South Africa, about 14 kg of a simple meal mixture consisting of equal parts of "yellow mealie" meal, lucerne meal, and bran, together with ad lib good quality hay, was found to be adequate for keeping adult elephants in excellent condition for short periods of time only. After mixing the meal with chopped hay the ration was sprayed with molasses and water.[5]

From a survey made in 1965 of 12 different zoos,[6] the following foods are recommended for young elephants from 5 weeks and/or 190 lb up to weights and ages when they will accept hay and/or grains:

Milk (cows; fat content approximately 2%)
Bananas
Rice
Pablum® or rice Pablum or oatmeal
Vitamin and/or mineral supplements, especially, vitamins C and D and calcium and phosphorus

Experience with three immature elephants at the Philadelphia Zoo aided in the evolution of the following suggested diet for very young elephants:

Diet A-1
Pablum or other cereal for human infants: 120 g to 4 oz
Rolled oats: 120 g to 4 oz
Raisins: 120 g to 4 oz
Karo® syrup: 120 g to 4 oz
Bananas (three medium size and ripe): 90 g to 3 oz

Mash bananas, blend with above, then mold the mixture into small cylinders 2 to 3 × 0.5 to 0.75 in. Feed immediately, one cylinder at a time, by hand. Repeat above feeding four times in 12 hr.

One of the small Indian elephants mentioned above lacked teeth at first and held each mass in her mouth while making sucking movements until it disintegrated and was swallowed. She would then accept another. As teeth erupted and muscular control of the trunk developed, hand feeding could be stopped and feeding was much more rapid.

In addition to the above, the following should be bottle fed twice daily (3 qt/feeding): a mixture of evaporated milk, water, 4 oz. of Karo syrup, and 2 tablespoons of a multivitamin-mineral preparation. Water from a hose – a gentle stream into the side of the mouth – should be given four times daily.

With growth, the ingredients of the solid food mixture, except the bananas and raisins, should be increased by 10 to 20% at intervals of 3 to 4 weeks, depending on appetite.

At 6 months, the giving of milk should be stopped, chopped escarole should be added to the evening feeding, and Diet B should be offered.

At 12 months Diet A (see below) should replace the special mixture. For the evening feed, escarole and Diet B should be continued, increasing the quantities as the appetite increases. A female Indian elephant grew from 319 to 940 lb in 14 months on this diet.

Table 1 gives an analysis of the milk of female adult Asiatic elephants. Table 2 presents data on the vitamin and amino acid content of the milk of an Asiatic elephant at the Portland Zoo.

DIET FORMULAS DEVELOPED BY THE PENROSE LABORATORY, PHILADELPHIA ZOOLOGICAL GARDEN

Diet A: For Omnivorous Mammals and Birds

Mixture No. 1

Ground yellow corn	15%	Alfalfa leaf meal	5%
Ground whole wheat	15%	Brewer's yeast	10%
Ground whole barley	10%	Dried skim milk	10%
Ground rolled oats	10%	Oyster shell flour	2%
Peanut meal	10%	Iodized salt	1%
Soybean meal	10%	A-D feeding oil	2%

Mixture No. 1 is prepared in bulk as needed, but never stored longer than 2 weeks.

Take nine parts of Mixture No. 1, one part of ground boiled meat, and sufficient meat broth or water to make a stiff mash. Mix thoroughly and press into shallow pans. Refrigerate 24 to 48 hr to harden, then cut into sections for feeding. This ration supplies about 4 cal/g. Allow $\frac{1}{20}$ to ½ oz/lb (1 to 10 g/kg) of body weight, depending upon weight and opportunity for and tendency to spontaneous activity.

The ration alone seems to be completely adequate nutritionally for apes, monkeys, and baboons, although it has not been tested as the sole diet for periods longer than 1 year. Ordinarily, it is supplemented by oranges and carrots for apes, and carrots, kale, or escarole [chicory] for monkeys, baboons, and many other animals.

Diet A-1

Commercial baby cereal (Pablum®, Beechnut®, etc.)	65%	12 lb	5454 g
Whole milk powder	28%	5 lb	2272 g
Wheat germ meal	4%	12 oz	360 g
Gevral® protein[a]	3%	8 oz	40 g

[a] A vitamin-mineral-protein supplement prepared and sold by Lederle Laboratories, Pearl River, New York.

Mix dry and store in refrigerator. Moisten with water to the consistency of firm dough, as much as a week's supply at a time, and refrigerate, but bring to room temperature before feeding. Supplement with oranges and grapes, ½ to 1 oz of fruit per pound of body weight. Feed the mixture in the morning and the fruit late in the afternoon. Allow about ½ oz/lb (about 30 g/kg) of body weight per day. Overfeeding may cause loose stools, but when fed in moderate amounts this ration will check the diarrhea that often is part of malnutrition. Also feed lean ground meat (Diet C, below) with fruit three times a week.

Diet B: For Herbivorous Mammals

Mixture No. 2

Brewer's grains	10.0%	Linseed meal	10.0%
Ground yellow corn	15.0%	Alfalfa leaf meal	26.0%
Ground whole wheat	10.0%	Brewer's yeast	5.2%
Ground whole oats	10.0%	Oyster shell flour	2.5%
Soybean meal	10.0%	Iodized salt	1.0%
		A-D feeding oil	0.3%

Mixture No. 2 is prepared in bulk as needed, but never stored longer than 2 weeks.

To prepare Diet B take dry beet pulp (plain or molasses) equal to about 45% of the total dry weight of the food needed and wet with an approximately equal weight of water. Let stand 1 to 2 hr then add an equal weight (45% of total) of Mixture No. 2 plus an amount of chopped carrots and uncut kale to equal about 10% of the combined weights of dry beet pulp and Mixture No. 2. Mix and feed within 4 to 6 hr. Summer temperatures cause quick fermentation, so this ration must be used promptly or refrigerated. Allow ¼ to 2 oz/lb (10 to 40 g/kg) of body weight. Supplement with hay ad libitum or, where good quality hay is not available, give alfalfa pellets, 5 to 10 g/kg of body weight. Hay of all qualities provides a means of occupying the animals' attention and energy, and prevents restlessness. It also aids in preventing obesity.

Diet C: For Carnivores Except Felidae

Mineral Mixture C-1

Oyster shell flour	50%
Powdered skim milk	40%
Iodized salt	5%
Gevral protein	5%

Diet C

Raw ground meat	86 or 87%
Mineral mixture C-1	12 or 13%
A-D feeding oil	2%

or

Multivitamin preparation (Gevral protein: see Diet A-1 above). (Two formulae allow the use of A-D feeding oil or a more complex vitamin mixture.)

Mix fresh daily as needed; it will keep at least 48 hr under refrigeration. Allow 0.5 to 2 oz/lb (25 to 75 g/kg) of body weight.

Table 1
MILK ANALYSIS OF FEMALE ASIATIC ADULT ELEPHANTS

Time after calving	Water	Specific gravity	Total solids	Fat	Protein	Carbohydrates: Casein	Carbohydrates: Lactose	Ash	Calcium oxide (mg/100 g)	Phosphate (mg/g)	Chloride (mg/g)	Vitamin C (mg/100 g)	Ref.
	83.82	1.038	16.18	3.89		11.82 carbohydrates		0.47	pH neutral				1
				6.7	3.4		6.4	0.4	Vitamins A_1, B_2 D (as cow milk), vitamin C, four times cow milk				2
2 months		1.038	16.4	5.8	4.45	1.75	5.4	0.67	98.5	309	45	0.4	3
2 months		1.035	17.5	7.5	4.4	2.1	4.9	0.57	178.0	186	42	0.25	
4 months		1.035	19.15	8.5	4.65	1.6	5.25	0.63	95.0	245	52	0.4	
1 year 10 days		1.036	21.17	10.0	4.93	2.5	5.4	0.68	84.6	305	46	0.35	
1 year 5½ months		1.032	22.34	12.0	5.1	2.2	4.3	0.8	93.0	245	48	0.3	
1 year 5½ months		1.023	28.02	18.5	5.4	1.4	3.4	0.67	165.2	216	64	0.4	
1 year 6 months		1.024	28.55	19.0	5.25	2.0	3.6	0.64	164.6	244	54	0.3	
5 days			8.27	0.62	3.0		4.01	0.64	pH 7.8				4
Composite of several averages	84.0			4.0	4.0	7.5 sugar		0.4					5

From Reuther, R. T., *Int. Zoo Yearb.*, 9, 168–178, 1969. With permission.

REFERENCES

1. **Benedict, F. G.,** *Carnegie Inst. Washington Publ.*, No. 474, 1969.
2. **Rome Zoo,** *Zoo Life,* 5(1), 1950.
3. **Simon, K. J.,** *Indian Vet. J.*, 36, 500–503, 1959.
4. **Marks, J.,** unpublished, 1964.
5. **Ratcliffe, H. L.,** *Int. Zoo Yearb.*, 6, 4–37, 1966.

Table 2
VITAMIN AND AMINO ACID CONTENT OF THE MILK OF AN ASIATIC ELEPHANT AT THE PORTLAND ZOO[7]

Vitamin and amino acid content	24 hr before birth	36 hr after birth	5 days
Vitamin A(IU/ml)			0.53
Carotene			0
Vitamin C (mg/100 ml) (skimmed milk)	4.0	11.0	10.4
Vitamin B_1 (μg/100 ml)	8.0	12.0	1.0
Vitamin B_2 (μg/100 ml)	192.0	134.0	84.8
Vitamin B_6 (μg/100 ml)	0.935	0.935	trace
Amino acids (μmol/ml)			
Proline			15.60
Glutamic acid			5.85
NH_3			5.40
Aspartic acid			4.02
Leucine			3.66
Serine			3.10
Valine			2.68
Tyrosine			2.67
Alanine			2.58
Threonine			2.15
Glycine			1.92
Phenylalanine			1.84
Isoleucine			1.79
Lysine			1.64
Arginine			0.90
Histidine			0.64
½ Cystine			0.58
Methionine			0.51

REFERENCES

1. **Sykes, S. K.,** *The Natural History of the African Elephant,* Weidenfeld and Nicolson, London, 1971.
2. **Benedict, F. G.,** *Carnegie Inst. Washington Publ.,* No. 474, 1936.
3. **Sedgwick, C. J. and Fowler, M. E.,** Methods of Feeding Mammals, American Association of Zoo Veterinarians, Annual Proceedings, Hills Division, Rivinia Foods, Topeka, Kans., 1972–73.
4. **Crandall, L. S.,** *The Management of Wild Mammals in Captivity,* University of Chicago Press, Chicago, 1964.
5. **Young, E., Ed.,** *The Capture and Care of Wild Animals,* Ralph Curtis Books, Hollywood, Fla., 1973.
6. **Reuther, R. T.,** *Int. Zoo Yearb.,* 9, 168–178, 1969.
7. **Marks, J.,** unpublished, 1964.
8. **Ratcliffe, H. L.,** *Int. Zoo Yearb.,* 6, 4–37, 1966.
9. **Young, E. and Oelofse, J.,** *Int. Zoo Yearb.,* 9, 179–184, 1969.

DIETS (NATURAL AND SYNTHETIC): SUBHUMAN PRIMATES

G. R. Kerr

DIET IN HABITAT

The order Primates consists of some 250 species which include prosimians such as tree shrews, the various New and Old World "monkeys," and the anthropoid species, which includes chimpanzees, gorillas, and man.[1] Evolution of this order has been associated with many changes in feeding behavior and diet *in habitat*,[2-5] and several recent publications reviewed the dietary peculiarities and nutritional requirements of the major species.[6,7]

The 50 to 60 "essential" nutrients consist of water, amino acids, fatty acids, carbohydrates, vitamins, and minerals that are necessary for life, but which cannot be synthesized by the species in adequate amounts and therefore must be obtained in the diet. Water is available in all natural foods. Energy is provided by the carbohydrate, fat, and protein that are found in varying amounts in all animal and vegetable foods. The essential minerals and trace elements are also widely distributed in foods. Fat-soluble vitamins are derived from green and yellow vegetables and certain organ meats (vitamin A), sunlight and fish liver oils (vitamin D), vegatables oils (vitamin E), and a variety of vegetable and animal sources (vitamin K). Water-soluble vitamins are found in a wide variety of animal and vegetable products. Therefore, any diet composed of a variety of foods from animal and vegetable sources will probably contain an adequate amount of the essential nutrients, if eaten in amounts sufficient to meet energy requirements.

In considering the nutritional needs of subhuman primates, it is important to remember that the ad libitum intake of any diet is largely determined by its energy content. Diets supplying the same amounts of energy may have widely differing contents of each of the essential nutrients. Hence, the nutrient content of a diet consumed ad libitum may have a limited relationship to the needs. Moreover, a number of investigations suggest that inadequate intake of dietary protein or a diet imbalanced in its content of essential amino acids (and perhaps other nutrients) will result in a reduction of "appetite" and reduced efficiency of energy utilization.[8] Therefore, any diet that is limited in the content of one essential nutrient may cause changes in the metabolism and utilization of other essential nutrients. For this reason, nutrition-related data from such experiments and field experiences may be difficult to interpret.

The members of the prosimian suborder include species such as tree shrews, lemurs, galagos, and tarsiers. They are usually found in Southeast Asia and the tropical forests of Africa, and their natural diets appear to consist of fruit, leaves, insects, and small animals.[2-6] Experience from zoos suggests that most of these species are omnivorous. Most of the subhuman primates used in laboratories belong to the suborder Anthropoidea and its three superfamilies: Ceboidea (New World monkeys), Cercopithecoidea (Old World monkeys), and Hominoidea (gibbons, orangutans, chimpanzees, and gorillas). The New World monkeys include marmosets, squirrel monkeys, cebus monkeys, spider monkeys, woolly monkeys, and howler monkeys. The Old World species include langurs, colobus monkeys, mangabeys, baboons, and the many species of macaques. Langurs, colobus monkeys, and howler monkeys have voluminous upper digestive tracts, reflecting adaptation to a primarily vegetarian diet.[5] With these exceptions, most Old and New World monkeys are omnivorous.[2-7] The Hominoidea species prefer a diet of fruits and vegetables, but most have also been noted to eat animal products, and most adapt readily to a more "balanced" diet that includes animal sources of protein. Although observations of diet *in habitat* are of value in identifying the normal behavior of the species, they may provide erroneous conclusions as to its nutrient requirements. The same diets will probably be accepted in captivity, but Bilby[9] and Wackernagel[10] have shown that standardized diets maintain the health of a wide variety of subhuman primates.

One of the reasons why primates have succeeded in becoming the "highest" order relates to their ability to adapt to changing environmental conditions, including those that involve food supply. In the same way that these species have adapted to available diets in areas ranging from snowy mountain sides to city slums, they have also adapted to a wide variety of laboratory diets. Species such as the leaf-eating langur require a primarily vegetarian diet and must be able to synthesize the few "essential" nutrients such as vitamin B_{12} that are not found in vegetable foods.[5] With these exceptions, most of the natural, semipurified, and elemental diets that have been of proven success in feeding subhuman primates under laboratory conditions include all of the 50 to 60 nutrients known to be essential for man. Each diet and dietary approach has unique advantages and disadvantages, and the final choice is usually made on the basis of economics, availability, versatility, etc., or on the goals of a particular experiment.

DIETS FOR NEWLY IMPORTED ANIMALS

Newly imported animals have high incidences of morbidity and mortality, particularly for infectious diseases and gastroenteritus; problems of hydration and refusal to eat a balanced laboratory diet are also frequently encountered. Ready access to a source of water and a diet that includes familiar fruits and vegetables resolve the immediate concerns. Since such diets may be accepted in preference to those that are of higher nutritional quality, it is preferable to introduce a new, more balanced diet as soon as possible, coincident to procedures directed to control infectious diseases and infestations. Moreland[11] has recently reviewed the diets used by several commercial importers of New and Old World monkeys (Tables 1 and 2).* They are well accepted, contain all the essential nutrients, and are capable of maintaining health and nutrition during the critical period of importation and conditioning.

"NATURAL" DIETS

Many zoos and research laboratories have extensive experience with diets blended from a variety of natural foods. They are relatively inexpensive and palatable, but variable in nutrient content depending on the formulation of the diet. For this reason they are often fortified with specific mineral and vitamin supplements. Schmidt's protocol[12] for rhesus monkeys is presented in Table 3; the diet developed for baboons by the Southwest Foundation for Research and Education is presented in Table 4;[13] and Table 5 shows a diet that has been proven successful for chimpanzees.[14] The nutrient content of these diets may show little relationship to the nutrient needs of the species, but they are clearly adequate to maintain health, reproduction, growth, and development.

COMMERCIAL DIETS

Liquid Diets

Many laboratories caring for newborn monkeys feed milk diets prepared for human consumption, but somewhat similar in nutrient composition to monkey breast milk.[6,15] One such diet and its nutrient composition are presented in Table 6. Smaller species tend to have a greater quantity of protein in breast milk, and formula for their infants may have to be modified accordingly. Deinhardt[16] routinely feeds newborn marmosets a

*All tables appear at end of text.

modified milk combined with a therapeutic nutrient supplement (Sustagen® – Mead Johnson Laboratories, Evansville, Indiana) with subsequent supplementary feedings of baby cereal, baby foods, fresh fruit, and monkey chow. These diets are easy to prepare, relatively economical, and can be modified with ease; also, the volumes consumed are easily measured, and the diets generally result in satisfactory growth and development. The major disadvantages are that they tend to produce loose stools and are often poorly accepted by adult animals.

Commercial "Chows"

Several commercial firms have developed chows that are based on recipes similar to those of "natural" diets, which have been proven successful in feeding subhuman primates. They are capable of supporting health, reproduction, and growth of most subhuman primate species and are widely used in zoos and research laboratories. Protein is usually provided by skim milk, corn, soybean meal, or fish flour; carbohydrate is provided by cereal products and added sucrose; and lipid is provided as preserved animal fat and vegetable oil. The nutrients are mixed, pressure cooked, blended, shaped into "chow," covered with vitamin A, vitamin C, and fat, and dried. Table 7 presents the approximate nutrient composition of several such chows,[6] but these must be expected to vary somewhat with each "batch." A particular shipment may contain "20% protein," but the amino acid content of that protein may vary widely depending on whether the protein is derived from vegetable or animal products. For these reasons, the chows are often supplemented with fruits, vitamins, vegetables, and periodic high-quality protein in the form of milk, meat, or eggs. The major disadvantage of such diets is that, because of such variability, they cannot be used for studies in which nutritional status or nutrient metabolism may influence the experimental results.

SEMI-PURIFIED DIETS

Since the nutrient content of many of the above diets may show major variations from batch to batch, laboratories concerned with nutrition-related research have found it necessary to develop diets that contain a known amount of each nutrient. These diets are expensive, since they involve much tedious work to produce satisfactory consistency and taste, but they offer nutritional assurance that is not possible with any other option. Although diets are generally based on an estimate of the nutritional requirements of a particular species and shown by trial and error to support normal health, reproduction, and growth, they also carry the risks that they may not contain an optimal "balance" between nutrients, and may be deficient in some trace nutrient not currently known to be essential for the particular species. Tables 8 and 9 present formulations that were found satisfactory for both New and Old World monkeys,[7,17-20] and Table 10 presents a comparable formula for use in infant New World monkeys.[20]

ELEMENTAL OR CHEMICALLY DEFINED DIETS

These diets are composed of mixtures of individual nutrients and with specific amino acids rather than intact protein. They have been primarily developed for humans with gastrointestinal disorders and are used in subhuman primates only for crises involving acute maldigestion and malabsorption states.

Each of the dietary approaches considered has unique advantages and disadvantages. As reviewed by Harris,[21] the primary advantages of natural foods and chows include low cost and palatability; the disadvantages include lack of consistency in nutrient composition, uncertainty as to digestion and absorption of the nutrients, and the potential adverse effects of antinutrients and toxicants. For most primate facilities that are not performing nutrition-related research, such diets are generally quite satisfactory. The various semipurified diets are designed for nutrition-related research, and concerns over digestion, absorption, and toxicity are largely resolved. On the other hand, the costs of semipurified diets can be prohibitive, and they are usually inferior to natural diets with regard to palatability. The design of the experiment, the goals of the institution, and economic considerations generally determine the diet selected.

Table 1
FEEDING OF NEWLY IMPORTED NEW WORLD MONKEYS[a,11]

Marmosets and tamarins

8:30 a.m. – Milk and cereal formula
(1 gal formula feeds 75–100 marmosets)
12 oz nonfat powdered milk
4 oz high-protein cereal[b]
1 can condensed milk
4 raw eggs

Add above to 1/2 gal warm water, mix. Add water to complete gallon of formula. To 1 gal add 2 teaspoons oxytetracycline.[c] Give for 1 week. If animals are not eating well, continue up to 30 days.

Noon – Full course meal
3 fruits, bananas always being given daily (whole and in the peel), oranges, apples, grapes, melons, etc. (in season); shredded lettuce, 4-min boiled eggs (given twice a week), bread, commercial diet,[d] peanuts, and very small portions of fresh ground horsemeat. Fed on disposable trays.

For long-term holding (3months or more)
Animal protein, such as chameleons, crickets or newborn mice, is added to the diet. Vitamins[e] are added to the milk formula.

Water
Electrolyte water given for 1 week after arrival or until hydrated and taking food readily.

Electrolyte formula
1 gal water
1 tablespoon electrolyte[f]
5 tablespoons sugar

Squirrel, *Cebus*, and spider monkeys

8:30 a.m. – Milk and cereal formula
Same as for marmosets

Noon
Same variety as that given to marmosets.

4:00 p.m. – Milk and cereal formula
Plus replenishing of commercial diet and peanuts, (spider monkeys receive fruits instead of commercial diet and peanuts).

After 2 weeks
Gradually cut down on the amount of commercial diet and peanuts. The milk and cereal formula is given in the morning daily regardless of how long the squirrel monkeys are held.

Electrolytes
Given about 1 week if animals are dehydrated.

Woolly monkeys

Same as for squirrel monkeys with exception of amount of commercial diet and peanuts. Woolly monkeys eat very little of these; they eat a great deal more leafy vegetables.

Douroucouli monkeys

Eat extreme amounts of food for their size. Do not care very much for commercial diet and peanuts, but consume large amounts of fruits, leafy vegetables, and small amounts of lean horsemeat. Milk and cereal formula is offered; however, they take very little.

Electrolyte water
Give for first week after arrival.

[a] Protocol of Tarpon Zoo Inc., Tarpon Springs, Fla.
[b] Pablum®, Mead Johnson Company, Evansville, Ind.
[c] Terramycin, Soluble Animal Formula, Chas. Pfizer & Co., New York, N.Y.
[d] Monkey Chow 25, Ralston Purina, St. Louis, Mo.
[e] Becotin®, Eli Lilly & Company, Indianapolis, Ind.
[f] Polysol Veterinary Powder, Cutter Laboratories, Berkeley, Calif.

Table 2
FEEDING OF NEWLY IMPORTED MONKEYS[a,11]

Macaca Mulatta, Speciosa, Irus, and *Nemistrina*	African monkeys, baboons, and chimpanzees	New World monkeys
Commercial monkey diet[b]	**Commercial monkey diet[b]**	**Commercial monkey diet[e]**
Offer twice daily 1/2 average-size apple to each animal, every other day. For animals showing depressed attitude (such as is often seen in the large males) with no mental aberrations, a daily feeding of banana is routine until commercial ration is accepted	Galagos do well using same methods as for others. Occasionally they are coaxed into eating by soaking biscuits in water. Chimpanzees require greatest care in caging and feeding. Young ones especially must be able to see others nearby. Best for same caretaker to provide continuous daily care. Commercial diet is offered with supplement of orange, apple, and banana daily. Also give one slice of white bread smeared with multivitamin[d] and a large portion of candied sweet potato per day.	Supplemental powdered food.[f] A small amount of apple every other day seems to be a good appetite stimulant.
The first 2 days after arrival		
Quiet and routine feeding-cleaning schedules are maintained with no heroic medical treatment attempted. Of course, sick animals are treated. Most newly arrived animals are voracious eaters the first 2 or 3 days with diminution of appetite following. The most critical date, medically speaking, is approximately 14 days after arrival, at which time close attention is paid to food intake.		
Water	Water	Water
Always available. Animals showing anything less than vigorous attitude are given a small bowl of drinking water containing an electrolyte material.[c]	Always available. Watch to make sure they are drinking from the fonts. If in doubt, use watering pans in cages. For chimpanzees, place watering dishes containing electrolytes within cage during first month.	Always available. Watering devices are either the bottle type or the direct-nipple system.

[a] Protocol of Primate Imports Corp., Port Washington, N.Y.
[b] Rockland Monkey Diet, Teklad, Inc., Monmouth, Ill.
[c] Oralyte-A, Elanco Products Co., Div. of Eli Lilly & Co., Indianapolis, Ind.
[d] Irradol-A, Parke Davis, Inc., Ann Arbor, Mich.
[e] Monkey Chow 25, Ralston Purina Company, St. Louis, Mo.
[f] Sustagen®, Mead Johnson Laboratories, Evansville, Ind.

Table 3
"NATURAL DIET" FOR RHESUS MONKEYS[12]

Basal mixture	Composition[a] (%)
Ground wheat	76.25
Soybean meal	12.50
Dried milk	4.50
Dehydrated alfalfa	2.00
Sucrose	0.75
Dicalcium phosphate	0.75
Calcium carbonate	1.50
Ferrous sulfate	0.10
Trace mineral salt	0.50
Brewer's yeast (type 50B)	1.25
Thiamine HCl	0.0011
Riboflavin	0.0011
Niacin	0.0022
Parvo® (3% folic acid)	0.0136
Vitamin B_6	0.0011
d-Calcium pantothenate	0.0022
Vitamin E	0.0011

[a] The above basal mixture is combined with the following to prepare the final diet:

Ingredient	Amounts (g or ml)
Basal mixture	1700
Brown rice	300
Water (cooking)	340
Salt, iodized	20
Eggs, liquid	150
Ascorbic acid	3
Dry vitamin A and D_2 beadlets	0.12
Vitamin B_6	0.05
Isoniazid	1
Water	700

Table 4
"NATURAL DIET" FOR BABOONS[a,13]

Ingredient	Ration (%)
Corn, yellow ground	35
Standard wheat middlings	20
Meat and bone meal	10
Soybean oil meal	10
Skim milk, dried	10
Alfalfa meal, dehydrated	2
Bone meal, steamed	2
Corn oil	6
Molasses, cane	2
Salt, iodized	1
Vitamin B, premix	1
Vitamin A, stabilized concentrate, sufficient to furnish 10,000 UI/lb ration	1
Irradiated yeast, sufficient to furnish 70 IU vitamin D/lb ration	1
Ascorbic acid, 300 mg/lb or 30 g/100 lb of ration	1
Isonicotinic acid hydrozide, 0.015%	1

[a] This ration provides the following approximate analysis: 20.85% protein (40% animal protein), 9.35% fat, 46.70% carbohydrate

Table 5
"NATURAL DIET" FOR CHIMPANZEES[14]

Ingredient	Percent
Ground wheat	10.00
Dehydrated alfalfa	1.00
Ground corn	41.32
Dried skim milk	8.00
Soybean oil meal (44%)	6.00
Sucrose	5.00
Animal fat	2.50
Brewer's yeast	2.00
Meat and bone	2.00
Wheat bran	15.00
Bran cellulose	3.00
Calcium carbonate	2.00
Vitamin-mineral mix	1.00
DL-Methionine	0.05
L-Lysine	0.025
Sodium chloride	0.1
Dicalcium phosphate	1.0

Table 6

REPRESENTATIVE LIQUID DIET[a] FOR SUBHUMAN PRIMATES[6]

Gross composition (g/l)	Fat, 35.4; lactose, 70.5; protein, 18.2; minerals, 5.2
Mineral composition (mg/l)	Calcium, 770; potassium, 1110; sodium, 351; magnesium, 57; chloride, 641; phosphorus, 510; iron, 12.7
Fatty acid composition (% of total)	Caprylic, 0.7; capric, 1.7; lauric, 20.2; myristic, 8.1; palmytic, 11.0; stearic, 2.4; oleic, 22.9; linoleic, 32.8
Amino acid composition (mg/l)	Arginine, 630; histidine, 400; isoleucine, 1070; leucine, 1800; lysine, 1300; methionine, 440; phenylalanine, 870; threonine, 770; tryptophan, 250; valine, 1150; cystine, 160; alanine, 380; aspartic, 820; glutamic, 3370; proline, 1240; serine, 790; tyrosine, 940; glycine, 60.
Vitamin composition (content/qt):	Vitamin A, 2500 USP units; vitamin D, 400 USP units; thiamine, 0.65 mg; riboflavin, 1.00 mg; vitamin C, 50 mg

[a] Similac®, Ross Laboratories, Columbus, Oh.; prepared at the recommended dilution of 132 g Similac powder/l of reconstituted milk (composition kindly provided by manufacturer).

Table 7
COMPOSITION OF REPRESENTATIVE COMMERCIAL CHOW DIETS FOR SUBHUMAN PRIMATES[6]

	Diet A[a]	Diet B[b]	Diet C[c]	Diet D[d]
Gross composition				
Protein (%)	24.50	15.2	25.5	19.67
Fat (%)	5.30	5.2	5.1	6.38
Fiber (%)	3.20	2.5	2.7	2.33
Ash (%)	6.40	4.8	6.1	6.97
Nitrogen-free extract (%)	50.60	63.3	51.6	56.27
Energy (kcal/lb)	1350	1897.7	1956.7	NS[e]
Amino acids (% of diet)				
Threonine	0.89	0.57	1.1	0.70
Valine	1.18	0.79	1.4	0.94
Cystine	0.41	0.25	0.40	0.18
Methionine	0.46	0.39	0.45	0.52
Isoleucine	1.12	0.80	1.5	1.03
Leucine	2.40	1.30	2.3	1.63
Tyrosine	0.91	NS	NS	0.75
Phenylalanine	1.33	0.74	1.3	0.84
Lysine	1.26	0.73	1.6	1.60
Histidine	0.62	0.41	0.73	0.45
Arginine	1.46	0.77	1.7	0.93
Tryptophan	0.25	0.18	0.32	0.32
Glycine		0.71	1.2	
Minerals				
Calcium (%)	1	0.86	0.96	1.02
Phosphorus (%)	0.76	0.47	0.56	0.58
Potassium (%)	0.92	0.56	1.03	0.83
Magnesium (%)	0.16	0.11	0.12	0.13
Sodium (%)	0.36	0.22	0.35	0.26
Chlorine (%)	0.55	0.27	0.41	0.38
Iron (ppm)	300	144	300	76.0
Zinc (ppm)	40	19.9	21.1	4.7
Manganese (ppm)	90	46.2	40.5	120
Copper (ppm)	18	11.3	13.4	16
Cobalt (PPM)	4	0.53	0.25	1.3
Iodine (ppm)	3.60	1.6	1.6	2.4
Vitamins				
Carotene (ppm)	7.00	NS	NS	3.9
A (IU/g)	22.40	28.0	30.0	7.00

[a] Diet A, Wayne Monkey Diet, Allied Mills, Inc., Libertyville, Ill.
[b] Diet B, Purina Monkey Chow (protein 15.0 min), Ralston Purina Co., St. Louis, Mo.
[c] Diet C, Purina Monkey Chow (protein 25.0 min), Ralston Purina Co., St. Louis, Mo.
[d] Diet D, Rockland Laboratory Primate Diet, Teklad, Inc., Monmouth, Ill.
[e] NS, not stated.

Table 7 (continued)
COMPOSITION OF REPRESENTATIVE COMMERCIAL CHOW DIETS FOR SUBHUMAN PRIMATES[6]

	Diet A[a]	Diet B[b]	Diet C[c]	Diet D[d]
Vitamins (continued)				
D (IU/g)	8.20[f]	6.6[g]	6.6[g]	2.95[h]
Thiamine (ppm)	20	7.7	12.0	11.78
Riboflavin (ppm)	12	9.3	10.0	8.74
Niacin (ppm)	82	98.2	89.1	89
Pantothenic acid (ppm)	40	63.3	67.3	22.21
Choline (ppm)	2415	1200	1800	560
Pyridoxine (ppm)	17.20	12.7	14.1	7.76
Folic acid (ppm)	19	4.0	6.0	2.9
Biotin (ppm)	0.27	0.14	0.34	0.04
B_{12} (mg/lb)	21	5.1	15.0	27.0
Sodium menadione bisulfite (ppm)	18.50	NS	NS	NS
C (ppm)	2000	758	750	1830
E (IU/100 g)				1.9
α-Tocopherol (ppm)	34.0	25.0 IU/lb	25.0 IU/lb	

[f] D_2 = 4.10 IU/lb, D_3 = 4.10 IU/lb.
[g] D_3.
[h] D_2.

Table 8
COMPOSITION OF SEMIPURIFIED DIETS FOR NONHUMAN PRIMATES

	New World monkeys		Old World monkeys	
	Portman[7]	Mann et al.[17]	Rinehart and Greenberg[18]	Waisman et al.[a,19]
Casein	25%	25%	18%	18%
Sucrose	61%	60.6%	73%	73%
Salts (Hawk Oser)	4%	4%	4%	4%
Corn oil	8%	8%	2%	3%
Calcium stearate			1%	
Thiamine (mg/day)	1	1	0.5	1
Riboflavin (mg/day)	1	1	1	1
Nicotinic acid (mg/day)	4.9	4.9	5.0	5.0
Calcium pantothenate (mg/day)	3.0	3.0	3.0	3.0
Ascorbic acid (mg/day)	50	25	25	25
p-Aminobenzoic acid (mg/day)		50–100	100	100
Choline (mg/day)	500	100–200	100	50
Inositol (mg/day)	100	50–100	100	100
Biotin (μg/day)	20	20	10	
Pyridoxine HCl (mg/day)	1	1	1	1
Vitamin A (IU/day)	1250	Cod liver oil, 1–2 mg/day	100,000[b]	Cod liver oil, 2% of diet
Vitamin D (IU/day)	400 (D_3)		10,000	
Vitamin E (mg/day)	10		12	
Vitamin K (mg/day)	4		b	
Folic acid (μg/day)	100	100		
Vitamin B_{12} (μg/day)	2			

[a] Supported growth when supplemented with 3% whole liver or liver extract.
[b] Later modified to contain 1800 IU vitamin A, 180 IU vitamin D_3, and 110 μg folic acid.

Table 9
PURIFIED DIET FED IN AGAR GEL VEHICLE[20]

Ingredient	Grams (%)	Kilocalories
Nonnutritive fiber[a]	20.00	
Lactalbumin[b]	15.88	63.52
Dextrin	25.69	102.76
Sucrose	25.69	102.76
Salt mix[c]	4.00	
Vitamin mix[d]	0.40	1.60
Corn oil	8.00	72.00
Choline chloride	0.24	
Inositol	0.10	
Total	100.00	342.64

Add 100 g of agar-agar to 4000 cc of boiling water. Stir well. Let cool slightly and pour into 5000 g of premixed diet. Mix well for 5 min and pour into pans.

[a] Teklad Mills, Madison, Wisc.
[b] Teklad Mills, Madison, Wisc. (78% protein by Kjeldahl analysis, N × 6.25.
[c] Hegsted IV salt mix, Teklad Mills, Madison, Wisc. (+ 4.36 mg Na_2SeO_3 and 48.995 mg $Cr(C_2H_3O_2)_3 \cdot H_2O$ per kg Hegsted IV salt).
[d] The complete vitamin mix contains (in g/kg): thiamine (0.8), riboflavin (1.6), calcium pantothenate (5.0), pyridoxine (0.8), niacin (8.0), folate (0.2), biotin (0.04), vitamin B_{12} (0.03), menadione (1.0), α-tocopherol acetate (500 IU/g) (20), vitamin A acetate (500,000 IU/g) (5.0), vitamin D_3 (200,000 IU/g) (1.25), ascorbic acid (122.5), dextrin (833.78).

Table 10
COMPOSITION OF INFANT SQUIRREL MONKEY DIET[20]

Ingredient	Nonaqueous diet (g/100 g)
Casein[a]	18.76
Lactose[b]	48.76
Corn oil[c]	25.98
Salt mix[d]	3.49
Vitamin mix[e]	0.58
Choline	0.35
Inositol	0.12
Tween® 80[f]	1.47
Suspending agent[g]	0.49

[a] Casein (vitamin-free, fine ground, Teklad Mills, Madison Wisc. (87.7% protein [N × 6.25] by Kjeldahl analysis).
[b] D-Lactose hydrate, Grand Island Biological Co., Grand Island, N.Y.
[c] Mazola®, Best Foods, Englewood Cliffs, N.J.
[d] Hegsted, D. M. et al., *J. Biol. Chem.*, 138, 459, 1941 (modified by the addition of 4.36 mg sodium selenite and 49.0 mg chromium acetate/kg salt mix).
[e] Ausman, L. M. et al., *J. Nutr.*, 102, 1519, 1972 (each animal was given in addition 50 mg ascorbic acid orally three times per week.
[f] Polyoxyethylene sorbitan monooleate, ICN Pharmaceuticals, Inc., Cleveland, Oh.
[g] Bio-Serv. Inc., Frenchtown, N.J.

REFERENCES

1. **Buettner-Janusch, J.**, An introduction to the primates, in *Evolutionary and Genetic Biology of Primates,* Vol. 1, Buettner-Janusch, J., Ed., Academic Press, New York, 1963, 1–64.
2. **Napier, J. R. and Napier, P. H.**, *Handbook of Primates,* Academic Press, New York, 1967.
3. **Devore, I., Ed.**, *Primate Behavior. Field Studies of Monkeys and Apes,* Holt, Rinehart & Winston, New York, 1965.
4. **Jay, P. C., Ed.**, Primates. *Studies in Adaptation and Variability,* Holt, Rinehart & Winston, New York, 1968.
5. **Hill, W. C. O.**, Primates. *Comparative Anatomy and Taxonomy,* Vol. 1–8, University Press, Edinburgh, 1953–1970.
6. **Kerr, G. R.**, Nutritional requirements of subhuman primates, *Physiol. Rev.,* 52, 415–467, 1972.
7. **Portman, O. W.**, *Nutrient Requirements of Monkeys,* Publ. 990, National Academy of Sciences – National Research Council, Washington, D.C., 1971.
8. **Kerr, G. R., el Lozy, M., and Scheffler, G.**, Malnutrition studies in *Maccaca mulatta.* IV. Energy and protein consumption during growth failure and "catch up" growth, *Am. J. Clin. Nutr.,* 28, 1364–1376, 1975.
9. **Bilby, L. W.**, A pilot scheme to investigate the diets of some of the animals at the London zoo, *Symp. Zool. Soc. London,* 21, 63–75, 1968.
10. **Wackernagel, H.**, Substitution and prefabricated diets for zoo animals, *Symp. Zool. Soc. London,* 21, 1–12, 1968.
11. **Moreland, A. F.**, Nutritional status and problems of feeding newly imported primates, in *Feeding and Nutrition of Nonhuman Primates,* Harris, R. S., Ed., Academic Press, New York, 1970, 29–43.
12. **Schmidt, L. H.**, Appendix B, in *Proc. 2nd Conf. Procurement and Production of Rhesus Monkeys,* National Academy of Sciences – National Research Council, Washington, D.C., 1955.
13. **Hummer, R. L.**, Observations of the feeding of baboons, in *Feeding and Nutrition of Nonhuman Primates,* Harris, R. S., Ed., Academic Press, New York, 1970, 183–203.
14. **Hodson, H. H., Jr., Mesa, V. L., and Van Riper, D. C.**, Protein requirements of the young, growing chimpanzee, *Lab. Anim. Care,* 17, 551–562, 1967.
15. **Ausman, L. M., Hayes, K. C., Lage, A., and Hegsted, D. M.**, Nursery care and growth of Old and New World infant monkeys, *Lab. Anim. Care,* 20, 907–913, 1970.
16. **Deinhardt, F.**, Nutritional requirements of marmosets, in *Feeding and Nutrition of Nonhuman Primates,* Harris, R. S., Ed., Academic Press, New York, 1970, 175–182.
17. **Mann, G. V., Watson, P. L., and Adams, L.**, Primate nutrition. 1. The cebus monkey: normal values, *J. Nutr.,* 47, 213–224, 1952.
18. **Rinehart, J. F. and Greenberg, L. D.**, Vitamin B_6 deficiency in the rhesus monkey with particular reference to the occurrence of atherosclerosis, dental caries, and hepatic cirrhosis, *Am. J. Clin. Nutr.,* 4, 318–328, 1956.
19. **Waisman, H. A., Rasmussen, A. F., Elvehjem, C. A., and Clark, P. F.**, Studies on the nutritional requirements of the rhesus monkey, *J. Nutr.,* 26, 205–218, 1943.
20. **Ausman, L. M.**, Department of Nutrition, Harvard School of Public Health, Boston, personal communication. (These diets are easily modified to permit a wide variety of nutrition related experiments.)
21. **Harris, R. S.**, Natural versus purified diets in research with nonhuman primates, in *Feeding and Nutrition of Nonhuman Primates,* Harris, R. S., Ed., Academic Press, New York, 1970, 251–262.

DIETS (SEMIPURIFIED AND NATURAL): HORSES

R. M. Jordan

INTRODUCTION

The composition and nature of the diet fed horses is affected by many factors, including age, production status (reproduction, work, and desired weight gains), palatability, storage problems, costs, toxic substances (molds, antienzymes, laxatives, etc.), and even the value of the horse to which the diet is to be fed. An inexpensive feedstuff (e.g., corn silage) that may contain a toxic substance 0.1% of the time might be fed to a $100 horse but not a $10,000 horse. Diets or diet components may be satisfactory for horses of one age or stage of production but entirely unsuited for others. A race horse or a weanling could not eat enough hay to meet its energy needs, whereas such a diet could be very adequate for maintenance of a mature horse.

The contents of the concentrate or grain portion of the diet are affected by the quality of the forage and the proportion or ratio of forage to concentrate. A weanling with limited feed intake capacity but high nutrient requirements would be fed a forage:concentrate ratio of about 1:1. Conversely, a pregnant mare may have an exacting vitamin, mineral, or protein requirement but if fed liberal amounts of concentrates would get too fat. Therefore, if the concentrate is to provide the major source of phosphorus or protein, those nutrients must be at high levels in the concentrate since only small amounts are fed daily.

Basic characteristics of feedstuffs must also be considered. Grains are higher in energy (50% more) and phosphorus and lower in calcium, vitamins, and trace minerals than good quality forage. Within the framework of the above remarks the following diets which contain a wide range of diet components are suggested.

Table 1
CREEP AND WEANLING DIETS FOR HORSES[1,2]

	Legume hay 1	Legume hay 2	Grass hay 1	Grass hay 2
Ingredient (kg)				
Corn	40			25
Oats	40		30	50
Sorghum, milo		20		
Barley		45		
Wheat			20	
Soybean meal	19	20	20	20
Dehydrated alfalfa		10	20	
Molasses, beet		4	9	4
Dicalcium phosphate	1	1	1	1
Protein (%)	19.1	20.9	20.4	19.8
DE (Mcal/kg)[a]	3.51	3.46	3.23	3.41

[a]Digestible energy expressed in megacalories per kilogram.

Table 2
DIETS FOR YEARLINGS AND 2-YEAR-OLD HORSES[a;2,3]

	Legume hay 1	Legume hay 2	Grass hay 1	Grass hay 2
Ingredient (kg)				
Corn	30		20	
Oats	50			83
Sorghum		20	20	
Barley		55	20	
Wheat		19		
Wheatbran	9		3	
Soybean meal	5	5	20	15
Dehydrated alfalfa			10	
Molasses	5		5	
NaCl	1	1	1	1
Dicalcium phosphate			1	1
Protein (%)	14.5	15.1	20.3	19.0
DE (Mcal/kg)[b]	3.38	3.67	3.05	3.16

[a] Legumes may be alfalfa, clover, lespedeza, etc.; grass hay may be timothy, brome, bluegrass, reed canary grass, etc. These grain mixtures should be fed in the following ratios with hay: yearlings, 1:1.5 to 2; 2-year-olds, 1:2 to 2.5.
[b] Digestible energy expressed in megacalories per kilogram.

Table 3
DIETS FOR GESTATING MARES[a;7–9]

	Legume hay 1	Legume hay 2	Grass hay 1	Grass hay 2
Ingredient (kg)				
Corn	30		60	
Oats	58		13	13
Sorghum		20		60
Barley		53		
Wheat		20		
Wheatbran	10		10	
Soybean meal			10	5
Dehydrated alfalfa				20
Molasses		5	5	
Dicalcium phosphate			2	2
Monosodium phosphate	2	2		
Protein (%)	12.5	12.7	15.1	15.6
DE (Mcal/kg)[b]	3.25	3.62	3.61	3.17

[a] During the last 90 days feed 1 to 2 kg daily along with respective hay.
[b] Digestible energy expressed in megacalories per kilogram.

Table 4
DIETS FOR LACTATING MARES[a;5,7,9]

	Legume hay 1	Legume hay 2	Grass hay 1	Grass hay 2	Grass pasture 1	Grass pasture 2
Ingredient (kg)						
Corn	60			20	90	
Oats	18	18		48		90
Sorghum, milo		50	60			
Barley		20	20			
Wheat						
Wheatbran	10					
Soybean meal	10	5			8	5
Linseed meal				15		
Cottonseed meal			10			
Dehydrated alfalfa				10		
Molasses		5	8	5		3
Dicalcium phosphate			2	2	2	2
Monosodium phosphate	2	2				
Protein (%)	15.4	14.5	15.2	16.3	13.1	14.84
DE (Mcal/kg)[b]	3.57	3.49	3.50	3.22	3.88	3.08

[a] Feed 2 to 5 kg daily along with respective forage in keeping with body weight changes.
[b] Digestible energy expressed in megacalories per kilogram.

Table 5
DIETS FOR HORSES AT HARD WORK (JUMPING, RACING, ETC.)[a;4,5,9]

	Legume hay 1	Legume hay 2	Grass hay 1	Grass hay 2
Ingredient (kg)				
Corn	30			
Oats	60	80	50	
Sorghum				25
Barley		10	30	50
Wheat				10
Wheatbran		10	9	
Soybean meal	5			5
Dehydrated alfalfa			10	5
Molasses	5			4
Dicalcium phosphate			1	1
Protein (%)	14.0	13.7	14.1	14.6
DE (Mcal/kg)[b]	3.12	3.10	3.10	3.14

[a] Feed 2 to 8 kg daily along with respective hay depending on condition desired and intensity of work.
[b] Digestible energy expressed in megacalories per kilogram.

Table 6
SEMIPURIFIED DIET FOR HORSES

Ingredient	%
Casein	10.0
Cornstarch	46.0
Glucose	5.0
Cellulose[a]	24.0
Corn oil	4.0
Mineral mixture[b]	7.0
Vitamin mixture[c]	4.0

From Stillions, M. C., Teeter, S. M., and Nelson, W. E., *J. Anim. Sci.*, 32, 252–255, 1971. With permission.

[a] Solka®, Floc BW-40, Brown and Co., Chicago, Ill.
[b] Provided the following per kilogram of diet: (in grams) NaCl, 9.1; $K_2HPO_4 \cdot 3H_2O$, 24.5; $CaHPO_4$, 2.5; $MgSO_4$, 14.0; $CaSO_4 \cdot 2H_2O$, 7.0; $CaCo_3$, 5.8; $FeCO_3$ (saccharinated), 0.7; KCl, 6.3; (in milligrams) KI, 0.3; $MnSO_4$, 63.0; $ZnCl_2$, 11.9; $CuSO_4 \cdot 5H_2O$, 18.9; $CoCo_3$, 10.5; $Na_2MoO_4 \cdot 2H_2O$, 0.4.
[c] Provided the following per kilogram of diet: vitamin A, 5500 IU; vitamin D, 1000 IU; vitamin E, 220 IU; (in milligrams) vitamin C, 25.0; menadione, 40.0; thiamine HCl, 10.0; riboflavin, 14.0; niacin, 50.0; calcium pantothenate, 50.0; pyridoxine HCl, 10.0; folic acid, 0.9; *p*-aminobenzoic acid, 50.0; inositol, 20.0; biotin, 0.6; (in grams) choline chloride, 1.8.

Table 7
EXAMPLE OF FORMULATED RATION FOR 6-MONTH-OLD COLT

Expected Mature Weight: 500 kg; Roughage Type: Timothy Hay

	Total ration			
Ingredient	Dry basis (kg)	As fed (kg)	Percentage	Percentage of concentrate
Timothy, hay, suncured, midbloom	2.250	2.545	38.9	–
Oats, grain, grade 1 U.S. min wt 34 lb/bushel, max 2% foreign material	3.080	3.385	51.8	84.7
Soybean, seed solvent extracted ground, max 7% fiber	0.440	0.494	7.5	12.4
Calcium phosphate, dibasic, commercial	0.059	0.061	0.9	1.5
Limestone, ground, min 33% calcium	0.056	0.056	0.9	1.4
Total	5.885	6.541	100.0	100.0

From National Research Council, *Nutrient Requirements of Horses,* National Academy of Sciences, Washington, D.C., 1973, 30. With permission.

Table 8
EXAMPLES SHOWING THE EFFECT OF CLASS OF ANIMAL (AGE, WEIGHT, MATURITY, AND PHYSIOLOGICAL CONDITION) AND KIND OF ROUGHAGE ON FORMULATING RATIONS

Requirements, ingredients, and intake	12-month-old colt; expected mature weight 500 kg			Mare at peak of lactation; weight 500 kg		
Daily requirements						
Digestible energy (Mcal)	16.81			27.62		
Digestible protein (g)	472			829		
Calcium (g)	26.0			47.0		
Phosphorus (g)	17.4			38.6		
Vitamin A (IU, thousands)	11.0			25.0		
	Total ration, dry basis (kg)			Total ration, dry basis (kg)		
	Grass hay	Mixed hay	Legume hay	Grass hay	Mixed hay	Legume hay
Ingredients of ration						
Alfalfa, hay, suncured, midbloom	–	1.62	3.25	–	2.50	5.00
Timothy, hay, suncured midbloom	3.25	1.63	–	5.00	2.50	–
Oats, grain, grade 1 U.S. min wt 34 lb/bushel, max 2% foreign material	–	1.35	3.10	5.28	2.25	–
Corn, Dent Yellow, grain, grade 2 U.S. min wt 54 lb/bushel	2.14	1.35	–	–	2.25	3.95
Soybean, seed, solvent extracted ground, max 7% fiber	0.55	0.12	–	0.46	0.35	0.16
Calcium phosphate, dibasic, commercial	–	0.01	–	0.05	0.06	0.07
Limestone, ground, min 33% calcium	0.04	–	–	0.03	–	–
Vitamin A supplement, 40,000 IU/g	–	–	–	0.0001	–	–
Total	5.98	6.08	6.35	10.82	9.91	9.18
Daily intake						
Digestible energy (Mcal)	16.83	16.82	16.89	27.64	27.69	27.65
Digestible protein (g)	476	475	618	829	830	831
Total protein (g)	767	789	968	1354	1340	1332
Calcium (g)	29.3	32.8	46.7	48.6	61.6	85.0
Phosphorus (g)	17.8	18.6	17.4	39.7	39.4	39.1
Vitamin A (IU, thousands)	15.8	29.7	34.3	25.6	44.1	69.8

From National Research Council, *Nutrient Requirements of Horses,* National Academy of Sciences, Washington, D.C., 1973, 30. With permission.

REFERENCES

1. **Jordan, R. M. and Myers, V.,** Effect of protein levels on the growth of weanling and yearling ponies, *J. Anim. Sci.,* 34, 578–581, 1972.
2. **Nelson, M. L. and Jordan, R. M.,** The effect of kind of forage and grain and pelleting on growth of ponies and their ability to tolerate stress, in *Proc. 4th Equine Nutr. and Phys. Symp.,* Vol. 4, California State Polytechnical University, Pomona, 1975, 98–108.
3. **Potter, G. D. and Huchton, J. D.,** Growth of yearling horses fed different sources of protein with supplemental lysine, in *Proc. 4th Equine Nutr. and Phys. Symp.,* Vol. 4, California State Polytechnical University, Pomona, 1975, 19–20.
4. National Research Council, *Nutrient Requirements of Horses,* National Academy of Sciences, Washington, D.C., 1973.
5. **Baker, J. P.,** Nutrient requirements of the horse, in *Proc. Montana Nutr. Conf.,* Vol. 25, Montana State University, Bozeman, 1974, 53–61.
6. **Jordan, R. M.,** Modern horse nutrition, in *35th Minnesota Nutr. Conf.,* Vol. 35, University of Minnesota, St. Paul, 1974, 61–69.
7. **Jordan, R. M.,** Horse nutrition research at Minnesota, in *36th Minnesota Nutr. Conf.,* Vol. 36, University of Minnesota, St. Paul, 1975, 108–115.
8. **Smith, J. D. and Jordan, R. M.,** Some effects of increased dietary copper on ponies, in *Proc. 4th Equine Nutr. and Phys. Symp.,* Vol. 4, California State Polytechnical University, Pomona, 1975, 49–59.
9. **Ensminger, M. E.,** *Horses and Horsemanship,* 4th ed., Interstate, Danville, Ill., 1969.
10. **Stillions, M. C., Teeter, S. M., and Nelson, W. E.,** Utilization of dietary B_{12} and cobalt by mature horses, *J. Anim. Sci.,* 32, 252–255, 1971.

DIETS FOR MARINE MAMMALS

D. Van Dyke and S. H. Ridgway

It is only relatively recently that the animal husbandry necessary to successfully keep a wide variety of marine mammals in captivity has been achieved. Much of the progress in improving the ability to keep these animals in captivity has been based on trial-and-error procedures. Considerable data have been accumulated relative to successful feeding of captive marine mammals.[1-3] However, experimental data are not available relative to degree of utilization of the various nutrients or requirements for specific nutrients.

Marine mammals include about 130 species, give or take a dozen or so, depending on whose nomenclature and taxonomy are adopted. Their diets and habitats are extremely varied, from polar ice to the warm waters of the equator. We have worked with various pinnipeds, primarily the California sea lion (*Zalophus californianus*), and a few species of small cetaceans, notably the Atlantic bottle-nosed dolphin (*Tursiops truncatus*). The sirenians (manatees and dugongs) are consumers of aquatic plant material. The sea otter eats abalone, sea urchins, and other marine invertebrates. The larger baleen whales primarily consume krill which consists of small shrimp-like organisms up to about 3 cm in length.

The California sea lion also consumes a wide variety of fish and relishes squid (*Loligo* sp.). Our captive gray seals (*Halichoerus grypus*) have supplemented their normal diet of herring, squid, and smelt with an occasional sea gull, jellyfish, or mollusk from the sea floor. The walrus's diet consists almost exclusively of mollusks. The killer whale occasionally eats sea lions, seals, and porpoises, but the great majority of its diet consists of fish. The pilot whale appears to favor squid but frequently consumes various types of fish as well.

In the wild, bottle-nosed dolphins tend to feed on the most abundant food which happens to be available at a given place and time.[4] They consume a large variety of fish, often shrimp and occasionally even mollusks. In some parts of the world squid are consumed, but bottle-nosed dolphins in captivity in the United States usually reject squid when it is offered to them.

Groups of captive bottle-nosed dolphins have been maintained in apparent good health for years on an exclusive diet of a single species of fish, the blue runner (*Caranx crysos*).[5] At the present time it is generally regarded as good husbandry practice to feed several different species of fish. Most facilities feed those species that are economical and commercially available in their particular area.[5]

In this paper we will present a quantitative estimate of the various nutrients consumed by those marine mammals with which we are familiar. Our approach to finding what nutrients the animals consume consisted first of noting the species and quantity of fish fed and calculating the number of calories in the food, based on available analyses. Table 1 was compiled by adapting the classic procedures developed by Atwater[6] to marine mammals, whereby a factor is applied to ingested nutrients to estimate the energy available to the animal after losses in digestion and metabolism. Factors utilized were 9.02 kcal for fish lipids and 4.37 kcal for fish protein. Average analytical values of a given fish species were used in the calculations. It must be recognized, however, that fish can vary considerably in nutrient content depending on season and location of catch, storage temperature, and many other factors.

The animal and feeding records of three facilities were used: the Naval Undersea Center at both San Diego and Hawaii and Sea Life Park of Hawaii. In addition, data reported by Sergeant[1] were used to calculate the kilocalories consumed by certain species

Table 1
ENERGY, IN KILOGRAM CALORIES, CONSUMED PER KILOGRAM OF BODY WEIGHT PER DAY

No. animals	Species	Range (kcal/kg)	Mean (kcal/kg)	Mean body weight (kg)
1[a]	Dall porpoise *Phocoenoides dalli*	–	160	120
8[a]	Common porpoise *Phocoena phocoena*	–	146	40[b]
2[a]	Pacific white-sided dolphin *Lagenorhynchus obliquidens*	–	116	87
5	Long-beaked porpoise *Stenella longirostris*	71–97	80.4	62
	Beluga whale *Delphinapterus leucas*			
1[a]	Adult	–	68.5	300[b]
1[a]	Calf	–	116	165[b]
36	Bottle-nosed dolphin *Tursiops truncatus*	40–98	61.6	129
1	Rough-toothed dolphin *Steno bredanensis*	–	59.5	125
1[a]	Killer whale *Orcinus orca*	–	52.8	1136[b]
3	Gill's bottle-nosed dolphin *Tursiops gilli*	44–53	48.6	175[b]
	False killer whale *Pseudorca crassidens*			
1	Adult	–	40	454[b]
1	Juvenile	–	50	363[b]
1	Pacific pilot whale (juvenile) *Globicephala scammoni*	–	42.4	454[b]
10	California sea lion *Zalophus californianus*	61–176	108	50.5
9	Gray seal *Halichoerus grypus*	61–121	98.7	57

[a] From Sergeant.[1]
[b] Animal weight estimated.

not found at these facilities. Silver smelt (*Osmerus mordax*), herring (*Clupea harengus*), Spanish mackerel (*Trachurus symmetricus*), and Columbia river smelt (*Thaleichthys vetulus*) were the fish most commonly fed.

After calculating as closely as possible the calories consumed by the animals, we wanted to arrive at numerical values for individual nutrients. These values are of potential use to those interested in marine mammal nutrition or comparative nutrition. We devised a composite "reference fish" representing a composite of analytical values of different species of fish fed to marine mammals. While considerable variation occurs in the laboratory analyses of the nutrients in various fish species, the common practice of feeding two to four species of fish over a period of time makes the concept of a fictional "reference fish" useful as a means of estimating the nutrients consumed. It must be emphasized that the "reference fish" as devised is compatible with the fish fed at the three facilities surveyed (typically one lean and one moderately fat species each day) in the study, and for comparative nutritional purposes the listed values of Table 2 are estimated quantities consumed by apparently healthy animals at these three facilities. Keepers and trainers are instructed to weigh food and record the amount eaten in the daily record. In practice some individuals are very thorough in this regard, while most

Table 2
ESTIMATE FROM A "REFERENCE FISH" OF NUTRIENTS CONSUMED PER KILOGRAM OF BODY WEIGHT FOR VARIOUS LEVELS OF CALORIC CONSUMPTION

Nutrient	50 kcal/kg	100 kcal/kg	150 kcal/kg
Protein (g)	7.27	14.55	21.8
Lipids (g)	2.09	4.19	6.28
Ash (g)	0.95	1.90	2.85
Water (g)	29.21	58.42	87.63
Calcium (mg)	274	549	823
Phosphorus (mg)	206	411	617
Potassium (mg)	121	242	362
Sodium (mg)	81	161	241
Zinc (mg)	0.783	1.57	2.35
Iron (mg)	1.41	2.81	4.22
Magnesium (mg)	13.4	26.8	40.2
Copper (mg)	0.103	0.206	0.309
Cobalt (mg)	0.021	0.042	0.063
Manganese (mg)	0.084	0.168	0.251
Biotin (mg)	0.0047	0.0095	0.014
Choline (mg)	55	109	164
Folic acid (mg)	0.012	0.024	0.036
Niacin (mg)	1.58	3.16	4.74
Pantothenic acid (mg)	0.138	0.276	0.414
Riboflavin (mg)	0.047	0.094	1.14
Thiamine (mg)	0.04	0.08	0.12
α-Tocopherol (mg)	0.67	1.34	2.02
Pyridoxine (mg)	0.20	0.40	0.59
B_{12} (mg)	0.0035	0.0069	0.0104
Vitamin A (IU)	99	198	256
Vitamin D (IU)	198	395	593
Ascorbic acid (mg)	1.19	2.37	3.56

facilities which feed fish with average higher lipid content, the quantity of most other values will decrease, assuming the same total caloric quantities of fish are fed.

Table 2 lists nutrients consumed per kilogram of body weight, for different levels of caloric consumption, of the "reference fish." This is, to our knowledge, the first attempt to present comprehensive quantitative values of the various nutrients consumed by captive marine mammals. We invite and expect others to offer closer estimates of both nutrient consumption and animal requirements as more data become available. We appreciate the fact that variation from the quantity of any specific nutrient listed in Table 2 may be considerable. This is particularly true of the listed vitamins. It is common practice for facilities maintaining marine mammals, including the three facilities in this survey, to supplement the diet with vitamins. The listed vitamins do not include these supplements. During fish transport, processing, freezing, storage, and defrosting, vitamin E and perhaps other vitamins may be lost or become unavailable to the animal. In addition, an induced thiamine deficiency due to the presence of thiaminase in some fish commonly found in captive marine mammal's diets has resulted in reports of thiamine deficiency in marine mammals.[7-10] Geraci[11] has recommended the addition of 35 mg of thiamine per kilogram of fish fed to overcome thiamine deficiency in captive harp seals (*Phoca groenlandica*) induced by eating herring (*Clupea harengus*) and smelt (*Osmerus mordax*).

SYNTHETIC MARINE MAMMAL RATIONS

The Naval Undersea Center has been active for some time in the development of synthetic rations for marine mammals. Our first experience was with a ration that was formulated in cooperation with a commercial company and led to the development of a product for pinnipeds (now produced by Reliable Protein Products, 4166 South Main Street, Los Angeles, California 90037). We later participated in ration development trials in cooperation with the Bureau of Commercial Fisheries, U.S. Department of Interior, Seattle, Washington, which resulted in the formulation of a pasteurized whole fish ration.[12] An in-house project has resulted in a ration which we have been testing with California sea lions and Atlantic bottle-nosed dolphins.[13,14] We have had animals of both species on synthetic rations for rather extensive periods of time (3 years in the case of one dolphin). Although superior to those tested previously, the present ration is still considered to be experimental. In addition to being nutritionally adequate, synthetic marine mammal rations must have organoleptic properties adequate to keep the animals accepting the food and in many cases actively working for a food reward. Considerable effort toward development of better synthetic rations will be required before convenient multiple species rations are commercially available. A synthetic ration may offer the best means of determining the nutritional requirements of marine mammals.

REFERENCES

1. **Sergeant, D. E.,** *Fiskeridir. Skr. Havunders.,* 15, 246–258, 1969.
2. **Keys, M. C.,** The nutrition of pinnipeds, in *The Behavior and Physiology of Pinnipeds,* Vol. 1, Harrison, R. J., Hubbard, R. C., Peterson, R. S., Rice, C. C., and Schusterman, R. J., Eds., Appleton-Century-Crofts, New York, 1968, 359.
3. **Ridgway, S. H.,** in *Mammals of the Sea: Biology and Medicine,* Ridgway, S. H., Ed., Charles C Thomas, Springfield, Ill., 1972.
4. **Leatherwood, S.,** *Mar. Fish. Rev.,* 37, 10–16, 1976.
5. **Wood, F. G., Norris, K. S., Dudok Van Heel, W. H., and Ray, C.,** in *Whales, Dolphins and Porpoises,* University of California Press, Berkeley, 1966, 649.
6. **Atwater, W. O. and Bryant, A. P.,** Conn. Storrs Agric. Exp. Stn. 12th Annu. Rep. (1899), 1900, 73–110.
7. **Myers, B. J.,** *Can. Field Nat.,* 69, 151–153, 1955.
8. **Rigdon, R. H. and Drager, G. A.,** *J. Am. Vet. Med. Assoc.,* 127, 453–455, 1955.
9. **White, J. R.,** *J. Am. Vet. Med. Assoc.,* 157, 559–262, 1970.
10. **Geraci, J. R.,** *J. Am. Vet. Med. Assoc.,* 165, 801–803, 1974.
11. **Geraci, J. R.,** *Can. J. Zool.,* 50(2), 179–195, 1972.
12. **Patashnik, M., Seidel, D., and Kudo, G.,** MIPR No. 63198-8-2002, Bureau of Commercial Fisheries Technological Laboratory, Seattle, 1970.
13. **Van Dyke, D.,** NUC Tech. Publ. 317, San Diego, 1972.
14. **Van Dyke, D.,** Proc. Reg. AAZPA Conf., 1973, 52–54.

DIETS FOR GERMFREE MAMMALS

M. H. Beaver, D. G. Govaert, and B. S. Wostmann

Diets for germfree (GF) animals can be divided into four basic groups:

Chemically defined (CD) diets – These diets are composed of pure low molecular weight chemicals. Total definition of dietary intake is possible, since no modification of nutrients by intestinal bacteria can occur. At present, CD diets for mice and rats are available (Table 1). These diets are in soluble or water-dissolved form, and are sterilized by filtration through both Millipore® filters (using 90 mm YY3009000 housing with Sartorius membrane filter SM 11107) and Amicon filters (using three TC2 modules with two PM10 150 mm Diaflow® ultrafiltration membranes). This procedure removes all bacteria, most macromolecular impurities over 5000 daltons, and all impurities over 10,000 daltons.[1] As a result, dietary antigenicity has been reduced to a virtually undetectable minimum.[2]

Semisynthetic diets – These diets consist of starch, relatively pure sources of protein and lipid, vitamins, and minerals. Each nutrient can be changed independently of other components to afford control of intake of protein, carbohydrate and lipid components, and of individual vitamins and minerals. They are used in studies that require a fair degree of nutritional definition. These diets are sterilized with free-flowing steam or, if better definition is needed, by irradiation.[3] Several quite similar formulations for GF rats and mice have been developed.[4] The one used since the early Sixties at the Lobund Laboratory (L-474E_{44}) is given in Table 2.

Practical-type diets – These are used in colony production or for experiments that require a relatively low degree of nutritional definition. They include various commercially available diets for GF rats, mice, rabbits, and guinea pigs, which are given in Table 2. Also included is the practical-type pig diet described recently by O. P. Miniats of the University of Guelph, Ontario, Canada. These diets are sterilized as mentioned above.

Diets based mostly on commercially available products – These are fortified to compensate for losses due to sterilization or for inherent nutritional deficiencies. They include diets for dogs based on solid dog food, and diets for cats, dogs, sheep, goats, calves, and monkeys based on canned artificial or cow's milk formula or canned prescription diets, supplemented by additional feeding of vitamins and/or minerals (Table 3).

Recently, solid diets containing 20 to 30% roughage for GF lambs have been reported. The diets, described in more detail by Soares et al.,[5] are based on skim milk solid, cellulose, and soybean meal, with glucose and corn oil added, and are supplemented with vitamins and minerals. Best performance was obtained after the addition of 4% fish meal. However, at this time, these diets are at best in an experimental stage, and no long-term data are available.

Table 1
CHEMICALLY DEFINED WATER-SOLUBLE DIET L-489E8 FED TO GERMFREE C3H MOUSE COLONY

Amounts in 100 g Total Nutrients, Including Separately Fed Lipids

Amino Acids[a] (NRC grade) (g)		Calcium pantothenate	10.00
		Biotin	0.20
L-Leucine	1.53	Folic acid	0.30
L-Phenylalanine	0.56	Cyanocobalmin	0.12
L-Isoleucine	0.92	*p*-Aminobenzoic acid	60.00
L-Methionine	0.87		
L-Tryptophan	0.28	**Minerals (g)**	
L-Valine	0.73		
L-Asparagine	0.79	Ferrous gluconate	0.04
L-Arginine HCl	0.62	Salt mix 35C[b]	0.063
L-Threonine	0.56	Ca glycerophosphate	4.0
L-Lysine HCl	1.35	$CaCl_2 \cdot 2H_2O$	0.14
L-Histidine $HCl \cdot H_2O$	0.56	Mg glycerophosphate	1.10
Glycine	0.23	$MgCl_2 \cdot 6H_2O$	1.10
L-Proline	1.12	NaCl + KI[c]	0.066
L-Serine	1.02	Potassium acetate	1.42
L-Alanine	0.45		
L-Na glutamate	2.60	**Carbohydrate (g)**	
L-Tyrosine ethyl·HCl	0.47		
Total amino acids	14.66	Glucose	72.0
Water-soluble Vitamins (g)		**Lipids[d] (mg)**	
Choline chloride	230.0	Trilinolein-triolein (1:1)	5000.0
Thiamine HCl	1.00	Vitamin A palmitate	0.0956
Riboflavin	1.50	Vitamin D_3	0.00043
Pyridoxine	1.25	DL-α-Tocopherol	48.9
Niacin	7.50	DL-α-Tocopherol acetate	97.8
i-Inositol	50.00	Vitamin K_1	1.067

[a] The nutrients are added, in the order given, to 156 ml of 70° H_2O purified in the Millipore Milli-Q2,3 system. The solution is cooled to 45° for the addition of the water-soluble vitamins and the minerals. The glucose is dissolved in 118 ml of similar H_2O at 70°. The two solutions are mixed and filtered through an Amicon TC2 thin-channel ultrafilter with PM-10 filter membrane presterilized with commercial bleach. The diet is filtered weekly into the isolator through a presterilized Sartorius filter membrane SM 11107 (pore size 0.2 μm) attached to an inlet pipe at the top of the isolator.

[b] 0.063 g salt mix 35C contains 45 mg $Mn(acetate)_2 \cdot 4H_2O$, 11 mg $ZnSO_4 \cdot H_2O$, 3 mg $Cu(acetate)_2 \cdot H_2O$, 2 mg $Cr(acetate)_3 \cdot H_2O$, 0.5 mg NaF, 0.3 mg $(NH_4)MO_7O_{24} \cdot 4H_2O$, 0.09 mg $Co(acetate)_2 \cdot 4H_2O$, 0.018 mg Na_2SeO_3, 0.3 mg $SnSO_4 \cdot 2H_2O$, 0.3 mg $NiCl_2 \cdot 3H_2O$, and 0.09 mg Na_3VO_4.

[c] 65.48 mg NaCl previously mixed with 0.52 mg KI.

[d] The lipid mixture is filtered weekly through a Sartorius SM 11107 membrane into the isolator and dispensed in a dish once a day (twice a day for pregnant and lactating females) to give a daily intake of 0.25 ml (0.225 g), amounting to 5% of total solids in the diet.

Table 2
SEMISYNTHETIC AND PRACTICAL-TYPE DIETS FOR GERMFREE MAMMALS

Amount per Kilogram

Ingredient	Rat: L-474E-44[6]	Rat and mouse: L-485[7]	Rabbit:[a] L-478[8]	Guinea pig[a] L-445[3]	Guinea pig[a] L-477[3]	Pigs[a,9]
Whole wheat, ground	–	–	200.0 g	–	320.0 g	–
Corn, yellow, ground	–	590.0 g	150.0 g	–	145.5 g	630.0 g
Rolled oats	–	–	180.0 g	80.0 g	220.0 g	–
Purina® Laboratory Chow, ground	–	–	–	80.0 g	–	–
Rice starch	604.0 g	–	–	–	–	–
D-Glucose (anhydrous)	–	–	–	8.0 g	–	–
Dried skim milk	–	–	–	–	–	50.0 g
Fish meal	–	–	–	–	–	50.0 g
Soybean meal (45% crude protein)	–	300.0 g	250.0 g	–	–	200.0 g
Casein (vitamin-free test)	240.0 g	–	–	–	78.0 g	–
Lactalbumin	–	–	50.0 g	–	100.0 g	–
Alfalfa, dehydrated	–	35.0 g	50.0 g	–	20.0 g	–
Corn oil	30.0 g	30.0 g	30.0 g	–	30.0 g	20.0 g
Yeast, active dry	–	–	–	12.0 g	–	–
Liver, dessicated	–	–	20.0 g	–	20.0 g	–
Lycine, feed grade	–	5.0 g	–	–	–	–
Methionine, feed grade	–	5.0 g	–	–	–	–
DL-Methionine	3.0 g	–	–	–	–	–
i-Inositol	0.1 g	–	0.5 g	–	0.5 g	–
BHT (butylated hydrotoluene)	–	0.125 g	–	–	–	–
NaCl	–	10.0 g	–	4.0 g	–	–
NaCl, iodized	5.15 g	–	5.3 g	–	5.3 g	3.0 g
$CaCO_3$	17.0 g	5.0 g	8.0 g	–	11.0 g	–
Na_2HPO_4	10.0 g	–	–	–	–	–
K_2HPO_4	10.0 g	–	–	–	–	–
$CaHPO_4 \cdot 2H_2O$	–	10.0 g	7.1 g	–	4.0 g	27.0 g
$Mg(C_2H_3O_2)_2 \cdot 4H_2O$	–	–	8.0 g	–	9.5 g	–
CH_3COOK	–	–	12.0 g	–	9.0 g	–
Cellophane spangles	50.0 g	–	–	–	–	–

[a] For preweaning diets for rabbits, consult Reference 11. For guinea pigs, L-445 can be considered a preweaning diet. For pigs, Trexler[20] recommends diluted canned evaporated milk preparations with vitamins added.

Table 2 (continued)
SEMISYNTHETIC AND PRACTICAL-TYPE DIETS FOR GERMFREE MAMMALS

Amount per Kilogram

Ingredient	Rat: L-474E-44[6]	Rat and mouse: L-485[7]	Rabbit:[a] L-478[8]	Guinea pig[a] L-445[3]	Guinea pig[a] L-477[3]	Pigs[a,9]
Salt Mixes (Trace Minerals)						
Commercial trace mineral mix	–	0.25 g[b]	–	–	–	–
Vitamin and trace mineral mix	–	–	–	–	–	20 g[c]
Supplement VC	–	–	–	Footnote d	Footnote d	
$MgSO_4$	4,000 mg	Footnote b	1,000 mg	–	4,000 mg	
$MgCO_3$	–		3,000 mg	–	–	
$FeC_6H_5O_7 \cdot 5H_2O$	600 mg		800 mg	–	600 mg	
CuO	25 mg		38 mg	–	25 mg	
$MnCO_3$	200 mg		42 mg	–	42 mg	
ZnO	25 mg		25 mg	–	25 mg	
MoO_3	0.05 mg		0.15 mg	–	0.15 mg	
NaF	10 mg		0.22 mg	–	0.22 mg	
$CoCl_2 \cdot 6H_2O$	0.5 mg		4.0 mg	–	4.0 mg	
KBr	0.1 mg		0.1 mg	–	0.1 mg	
Na_2SeO_3	0.1 mg		0.1 mg	–	0.1 mg	
$Cr(C_2H_3O_2)_3 \cdot H_2O$	12.0 mg		–	–	–	
Vitamin Mixes						
Lipid soluble						
Vitamin A	16,000 IU	26,000 IU	8,000 IU	–	8,000 IU	
Vitamin D_2	–	–	1,000 IU	–	1,000 IU	
Vitamin D_3	1,000 IU	1,000 IU	–	–	–	
Vitamin E (mixed tocopherols)	375 mg	–	380 IU	–	380 IU	
Vitamin E (α-tocopherol acetate)	–	225 mg	100 mg	–	100 mg	
Vitamin K (menadione)	100 mg	90 mg	100 mg	–	100 mg	
Corn oil carrier for fat-soluble vitamins	18 g	–	18 g	–	18 g	

[b] Commercial trace mineral mix to contain the following per kilogram of diet: Mn as manganous oxide, 65 mg; Fe as ferrous carbonate, 20 mg; Cu as copper oxide, 2 mg; Zn as zinc oxide, 15 mg; I as calcium iodide, 13 mg; and Co as cobalt chloride, 0.6 mg. Manufacturer's assay of total minerals in Diet L-485 as follows: Cl, 6.3 g; Na, 5.0 g; K, 6.9 g; Ca, 5.6 g; P, 5.3 g; Mg, 2.0 g; Mn, 86.9 mg; Fe, 77.5 mg; Zn, 30.2 mg; Cu, 7.5 mg; Co, 0.7 mg; I, 1.3 mg.

[c] Commercial vitamin and trace mineral mix, containing per kilogram: vitamin A, 1500 IU; vitamin D, 250 IU; vitamin E, 10 mg; menadione Na-bisulfate, 60 μg; vitamin B_2, 2.5 mg; Ca pantothenate, 11 mg; vitamin B_{12}, 20 μg; niacin, 20 mg; choline HCl, 900 mg. Also, Mn, 44 mg; Cu, 11 mg; Fe, 100 mg; Zn, 50 mg; and Se 0.1 mg.[10]

Table 2 (continued)
SEMISYNTHETIC AND PRACTICAL-TYPE DIETS FOR GERMFREE MAMMALS

Amount per Kilogram

Ingredient	Rat: L-474E-44[6]	Rat and mouse: L-485[7]	Rabbit:[a] L-478[8]	Guinea pig[a] L-445[3]	Guinea pig[a] L-477[3]	Pigs[a,9]
Water soluble						
Ascorbic acid	–	–	–	Footnote d	Footnote d	
Thiamine HCl	60 mg	65 mg	30 mg	Footnote d	30 mg[d]	
Riboflavin	30 mg	30 mg	15 mg	–	15 mg	
Nicotinamide	50 mg	–	25 mg	–	25 mg	
Nicotinic acid	50 mg	65 mg	25 mg	–	25 mg	
Pantothenic acid	–	285 mg	–	–	–	
Calcium pantothenate	300 mg	–	–	150 mg	150 mg	
Choline chloride	2,000 mg	2,000 mg	1,000 mg	–	1,000 mg	
Pyridoxine HCl	20 mg	20 mg	10 mg	–	10 mg	
Pyridoxamine·$2H_2O$	4.0 mg	–	2 mg	–	2 mg	
Biotin	1.0 mg	–	0.5 mg	–	0.5 mg	
Folic acid	10 mg	10 mg	5.0 mg	–	5.0 mg	
p-Aminobenzoic acid	50 mg	50 mg	25 mg	–	25 mg	
B_{12} (0.1% trituration in manitol)	250 mg	2.0 mg	125 mg	–	125 mg	
Cornstarch carrier for B vitamins	–	–	1,080 mg	–	1,080 mg	
Rice starch carrier for B vitamins	2,175 mg	–	–	–	–	
H_2O	350 ml[f]	–	Footnote f	816 ml[e]	Footnote f	

[d] Supplement VC2. Dispense in 1-, 2-, or 5-ml ampules; add 1 ml per animal per day either to estimated daily food intake or water intake. Composition: ascorbic acid, 1 g/kg; thiamine HCl, 0.05 g/kg; distilled H_2O, 100 ml/kg diet.

[e] Prepared as a mash with 816 ml boiling distilled water to feed to newborn. Diminish the water after 30 days. Supplement with pelleted Diet L-477 after 3 or 4 days. After 10 days, limit Diet L-485 to one meal daily and discontinue after 2+ days weaning to Diet L-477 entirely.

[f] 350 ml or 350 to 500 ml water added, depending on sterilization technique and storage facilities.

Table 3
DIETS RECOMMENDED FOR FEEDING OF LARGE GERMFREE MAMMALS

	Preweaning	Postweaning
Dogs	Esbilac®[a]-H_2O 1:2 with vitamins and methionine added[12]	Purina® Dog Meal (autoclaved); necessity of vitamin addition in question
Cats	Varamel®[b], with added vitamins and amino acids[13]	Hill's® prescription diet C/D[c,13]
Monkeys	Varamel-H_2O 2:1 with 5% glucose[14] or SMA®[d] baby formula (diluted)[15]	Hill's prescription diet C/D[c,14] Purina Monkey Chow (autoclaved) softened with milk[15]
Sheep and goats	Reconstituted canned evaporated cow's milk, with added vitamins and minerals[16-18]	
Calves	Reconstituted canned evaporated cow's milk[20]	
Foals	Canned evaporated cow's milk-H_2O 1:2 with 25 g glucose/l added[19]	

[a] Borden, Inc., New York, N.Y., U.S.A.
[b] Baker Milk Co., Cleveland, Oh., U.S.A.
[c] Hill Packing Co., Topeka, Kans., U.S.A.
[d] Wyeth Laboratories, Inc., Philadelphia, Pa., U.S.A.

REFERENCES

1. **Pleasants, J. R., Reddy, B. S., and Wostmann, B. S.,** *J. Nutr.,* 100, 498–508, 1970.
2. **Wostmann, B. S., Pleasants, J. R., and Bealmear, P.,** *Fed. Proc. Fed. Am. Soc. Exp. Biol.,* 30, 1779–1784, 1971.
3. Subcommittee on Standards for Gnotobiotes, ILAR, Wostmann, B. S., Chairman, *Gnotobiotes. Standards and Guidelines for the Breeding, Care and Management of Laboratory Animals,* National Academy of Sciences- National Research Council, Washington, D.C., 1970.
4. **Wostmann, B. S.,** Nutrition and metabolism of the germfree mammal, *World Rev. Nutr. Diet.,* 22, 40–92, 1975.
5. **Soares, J. H., Leffel, E. C., and Larsen, R. K.,** *J. Anim. Sci.,* 31, 733–740, 1970.
6. Lobund diet, unpublished.
7. **Kellogg, T. F. and Wostmann, B. S.,** *Lab. Anim. Care,* 19, 812–814, 1969.
8. **Reddy, B. S., Pleasants, J. R., Zimmerman, D. R., and Wostmann, B. S.,** *J. Nutr.,* 87, 189–196, 1965.
9. **Miniats, O. P.,** personal communication, 1976.
10. **Bayley, H. S. and Carlson, W. E.,** *J. Anim. Sci.,* 30, 394–401, 1970.
11. **Reddy, B. S., Wostmann, B. S., and Pleasants, J. R.,** in *The Germ-free Animal in Research,* Coates, M. E., Ed., Academic Press, London, 1968, 87–111.
12. **Heneghan, J. B., Floyd, E., and Cohn, I.,** *J. Surg. Res.,* 6, 24–31, 1966.
13. **Rohovsky, M. W., Griesemer, R. A., and Wolfe, L. G.,** *Lab. Anim. Care,* 16, 52–59, 1966.
14. **Reyniers, J. A.,** *J. Bacteriol.,* 43, 778–784, 1942.
15. **Kalter, S. S., Eichberg, J., and Heberling, R. L.,** in *Germfree Research: Biological Effects of Gnotobiotic Environments,* Heneghen, J. B., Ed., Academic Press, New York, 1973, 611–618.
16. **Alexander, T. J. and Lysons, R. J.,** *Br. Vet. J.,* 127, 349–357, 1971.
17. **Smith, C. K.,** Ph.D. thesis, University of Notre Dame, Notre Dame, Ind., 1966.
18. **Smith, C. K.,** personal communication, 1970.
19. **Drummond, A. J., Trexler, P. C., Edwards, G. B., Hillidge, C., and Cox, J. E.,** *Vet. Rec.,* 92, 555–557, 1973.
20. **Trexler, P. C.,** *Vet. Rec.,* 88, 15–20, 1971.

Index

INDEX

A

B

D

E

F

H

I

N

O

Q

R

S

T

U

V

CRC PUBLICATIONS OF RELATED INTEREST (continued)

CRC MONOTOPIC REPRINTS:

FLEXIBLE PACKAGING OF FOODS
By **Aaron L. Brody, S.B., M.B.A., Ph.D.,** Arthur D. Little, Inc.
The aim of this book is to describe food products employing flexible packaging, the requirements dictating the flexible packaging being used and the current state of flexible packaging in the food industry.

FREEZE-DRYING OF FOODS
By **C. Judson King, B.E., E.M., Sc.D.,** University of California.
This review concentrations on several papers published in recent years, relates them to the rest of the field, gives an evaluation of their findings, and the conclusions they draw.

LABORATORY TESTS FOR THE ASSESSMENT OF NUTRITIONAL STATUS
By **Howerde E. Sauberlich, Ph.D.,** Fitzsimmons Army Medical Center, Denver.
This book considers the development and application of the various biomedical techniques currently available for use in assessing nutritional status.

SOYBEANS AS A FOOD SOURCE, 2nd Edition
By **W. J. Wolf, B.S., Ph.D.,** and **J. C. Cowan, A.B., Ph.D.,** Northern Marketing and Nutrition Research Division, Peoria, Illinois.
This book summarizes the conversion of soybeans to food. Emphasis is given to the protein content of soybeans because of the current high level of interest.

STORAGE, PROCESSING, AND NUTRITIONAL QUALITY OF FRUITS AND VEGETABLES
By **D. K. Salunkhe, B.S., M.S., Ph.D.,** Utah State University.
This book reviews the nutritional value and quality of fruits and vegetables as influenced by chemical treatments, storage and processing conditions.

THE USE OF FUNGI AS FOOD AND IN FOOD PROCESSING, Parts 1 and 2
By **William D. Gray, A.B., Ph.D.,** Northern Illinois University. Edited by **Thomas E. Furia,** Dynapol, Palo Alto, California.
This two-volume work describes how filamentous fungi have been used in the past and present as a food source in various areas of the world.

CRC CRITICAL REVIEW JOURNALS:

CRC CRITICAL REVIEWS IN FOOD SCIENCE AND NUTRITION
Edited by **Thomas E. Furia,** Dynapol, Palo Alto, California.

Please forward inquiries to CRC Press, Inc., 18901 Cranwood Parkway, Cleveland, Ohio 44128.

CRC FORUMS

The CRC Forums represent a new concept in scientific information. Each Forum brings together recognized authorities to discuss topical, scientific and medical issues, while being tape-recorded. Each Forum is moderated by a designated chairperson. Within a few weeks, edited Casette tapes of the discussion, as well as copies of research papers and appropriate tabular data prepared by the participants, are available for purchase. Each Forum economically provides a total information package that allows scientists and researchers to evaluate and study an issue from a range of perspectives. This is accomplished without the expense of attending costly seminars, and the tapes can be used repeatedly at your listening convenience.

CRC FORUM ON OBESITY

Chairperson, JOHN R. K. ROBSON, M.D., D.P.H., Professor of Nutrition, Medical University of South Carolina.

Participants, GEORGE V. MANN, Sc.D., M.D., Professor of Nutrition, School of Medicine, Vanderbilt University, and DEREK MILLER, Ph.D., Senior Lecturer, Department of Nutrition, Queen Elizabeth College, University of London (England). HENRY A. JORDAN, M.D., Behavioral Weight Control Program, University of Pennsylvania.

Obesity is one of the most prevalent conditions in the Western World. While it undoubtedly occurs when the intake of food energy exceeds energy expenditure many aspects of the condition are not clearly understood.

In an effort to assist physicians in dealing with obesity, a CRC Forum has been held to discuss four of its aspects, namely the genetic, metabolic, and dietary basis of obesity; human behavior and weight gain; obesity in clinical practice, and management of obesity.

These topics are the basis of scientific papers prepared by the participants with expertise in these areas. Subsequent discussions and interaction among the members of the Forum cover important subtopics, including the cultural influences of society on diet, as well as individual attitudes towards food. The discussions on the clinical aspects of obesity cover the difficulties in defining obesity and measuring the extent of the problem that the threats of obesity pose to the health of the community. The discussions on management are concerned with the care of the obese in the Family Practitioners Office as well as the problems of the very obese hospitalized patient.

OUTLINE for CRC FORUM ON OBESITY

Side 1. Introduction and participants statements
Prevalence of obesity

Side 2. Genetics
Thermogenesis, gene, cellular

Side 3. Environment
Early feeding, glandular, cultural, and social effects

Side 4. Implications
Diabetes, cardio-vascular disease, hypertension, wear and tear, psychology, minor irritations, menarche growth

Side 5. Treatment
Diet, weight control, total fast, protein modified fast, quality, slimming groups, bizarre treatments

Side 6. Treatment (cont.)
Drugs – amphetamines, T3, saccharin, bulking agents, thermogenic drugs
Activity – exercise and energy dissipation
Surgery – procedures
Prognosis for treatment

Side 7. Behavior modification
Life modification, material and personnel needs
Prognosis for treatment

Side 8. Management
Who to treat
Criteria for obesity – weight, height, skinfold
Standard for obesity – relative weight, Quetelets Index, emotional indices
Preventive measures
New and established treatment regimes – individualized and holistic approach, diet quality, activity and energy dissipation, drugs, role of paramedical personnel
Future developments – thermogenic drugs
Warnings – hazards of surgery, starvation, bizarre diets, dangers during pregnancy and growth
Summary